AN INTRODUCTION TO GRANULAR FLOW

The flow of granular materials such as sand, snow, coal, and catalyst particles is a common occurrence in natural and industrial settings. They are important since a large fraction of the materials handled and processed in the chemical, metallurgical, pharmaceutical, and food-processing industries are granular in nature. The mechanics of these materials' flows is not well understood. This book describes the theories for granular flow based mainly on continuum models, although alternative discrete models are also discussed briefly. The level is appropriate for advanced undergraduates or beginning graduate students. The goal is to inform the reader about observed phenomena and some available models and their shortcomings and to visit some issues that remain unresolved. There is a selection of problems at the end of the chapters to encourage exploration, and extensive references are given.

K. Kesava Rao is a professor of chemical engineering at the Indian Institute of Science in Bangalore. He is a Fellow of the Indian Academy of Sciences and the author of numerous archival publications. His primary area of research after receiving his Ph.D. from the University of Houston has been the flow of granular materials.

Prabhu R. Nott is a professor of chemical engineering at the Indian Institute of Science in Bangalore. He has held visiting positions at Caltech and the University of California–San Diego after receiving his Ph.D. from Princeton. He is the author of numerous archival publications. His primary research concerns the flow and dynamics of complex fluids, such as dry granular materials and particlulate suspensions.

CAMBRIDGE SERIES IN CHEMICAL ENGINEERING

Series Editor:
Arvind Varma, *Purdue University*

Editorial Board:
Alexis T. Bell, *University of California, Berkeley*
Edward Cussler, *University of Minnesota*
Mark E. Davis, *California Institute of Technology*
L. Gary Leal, *University of California, Santa Barbara*
Massimo Morbidelli, *ETH, Zurich*
Athanassios Z. Panagiotopoulos, *Princeton University*
Stanley I. Sandler, *University of Delaware*
Michael L. Schuler, *Cornell University*

Books in the Series:

E. L. Cussler, *Diffusion: Mass Transfer in Fluid Systems*, Second Edition

Liang-Shih Fan and Chao Zhu, *Principles of Gas-Solid Flows*

Hasan Orbey and Stanley I. Sandler, *Modeling Vapor-Liquid Equilibria: Cubic Equations of State and Their Mixing Rules*

T. Michael Duncan and Jeffrey A. Reimer, *Chemical Engineering Design and Analysis: An Introduction*

John C. Slattery, *Advanced Transport Phenomena*

A. Varma, M. Morbidelli, and H. Wu, *Parametric Sensitivity in Chemical Systems*

M. Morbidelli, A. Gavriilidis, and A. Varma, *Catalyst Design: Optimal Distribution of Catalyst in Pellets, Reactors, and Membrances*

E. L. Cussler and G. D. Moggridge, *Chemical Product Design*

Pao C. Chau, *Process Control: A First Course with MATLAB®*

Richard Noble and Patricia Terry, *Principles of Chemical Separations with Environmental Applications*

F. B. Petlyuk, *Distillation Theory and Its Application to Optimal Design of Separation Units*

G. L. Leal, *Advanced Transport Phenomena: Fluid Mechanics and Convective Transport Processes*

T. Russell, A. Robinson, and N. Wagner, *Mass and Heat Transfer*

K. K. Rao and P. R. Nott, *An Introduction to Granular Flow*

An Introduction to Granular Flow

K. Kesava Rao
Indian Institute of Science

Prabhu R. Nott
Indian Institute of Science

CAMBRIDGE
UNIVERSITY PRESS

CAMBRIDGE
UNIVERSITY PRESS

University Printing House, Cambridge CB2 8BS, United Kingdom

One Liberty Plaza, 20th Floor, New York, NY 10006, USA

477 Williamstown Road, Port Melbourne, VIC 3207, Australia

314-321, 3rd Floor, Plot 3, Splendor Forum, Jasola District Centre, New Delhi - 110025, India

103 Penang Road, #05-06/07, Visioncrest Commercial, Singapore 238467

Cambridge University Press is part of the University of Cambridge.

It furthers the University's mission by disseminating knowledge in the pursuit of education, learning and research at the highest international levels of excellence.

www.cambridge.org
Information on this title: www.cambridge.org/9780521571661

First published 2008

A catalogue record for this publication is available from the British Library

Library of Congress Cataloging in Publication data
Rao, K. Kesava.
An introduction to granular flow / K. Kesava Rao and Prabhu R. Nott.
p. cm. – (Cambridge series in chemical engineering)
Includes bibliographical references and index.
ISBN 978-0-521-57166-1 (hardback)
1. Granular materials – Fluid dynamics. I. Nott, Prabhu R. II. Title. III. Series.
TA418.78.R36 2008
620′.43–dc22 2007050026

ISBN 978-0-521-57166-1 Hardback

To our teachers,

Professors M. S. Ananth and Roy Jackson

Contents

Preface

The flow of granular materials such as sand, snow, coal, and catalyst particles is a common occurrence in natural and industrial settings. Unfortunately, the mechanics of these materials is not well understood. Experiments reveal complex and, at times, unexpected behavior, whereas existing theories are often tentative and do not represent the entire range of observed behavior. Nevertheless, significant advances have been made in the understanding of the mechanics of granular flows, and the time is ripe for an account of experimental observations and theoretical models pertaining to flow in relatively simple geometries.

The importance of understanding granular flows need not be overstated – a large fraction of the materials handled and processed in the chemical, metallurgical, pharmaceutical, and food-processing industries are granular in nature. The flow and transportation of these materials are often critical operations in these processes. In most cases, the design of processes and equipment is based largely on experience and empirical rules. An appreciation of the underlying principles may be helpful in developing better design and operating procedures.

Some of the early investigations of granular flow were motivated by the need to understand the deformation of soils subjected to external loads, such as large structures. The deformation rates in these processes are usually very small. Theoretical models for these slow flows have increased in sophistication and complexity over the years, borrowing concepts from metal plasticity and soil mechanics. A contrasting picture of granular flow has emerged over the last three decades. This is believed to be applicable to rapid flows, where the deformation rates are large. Models for rapid flows have been based mainly on the kinetic theory of dense gases, with suitable modifications to account for the inelasticity of interparticle collisions and particle roughness. These models assume that momentum transfer occurs by collisions of short duration between particles and by free flight of particles between collisions. In contrast, momentum transfer during slow flow occurs via contacts of a longer duration between particles that slide and roll relative to each other. A realistic picture of particle interactions that encompasses these extremes is not available at present, barring some tentative attempts.

The theories described here for granular flow are based mainly on continuum models, although alternative discrete models are also discussed very briefly. Chapter 1 describes the qualitative behavior of granular materials in various situations and formulates balance laws. Chapter 2 discusses constitutive equations for slow, plane flow. Chapters 3 and 4 deal with flow in wedge-shaped hoppers and bunkers, respectively; after summarizing some of the experimental observations, the equations formulated in Chapter 2 are used to construct approximate solutions. Chapter 5 deals with constitutive equations for slow three-dimensional flow; these are applied to flow through axisymmetric hoppers and bunkers in Chapter 6. Using kinetic theory, constitutive equations for rapid granular flow are formulated in Chapter 7 and applied to flow between parallel plates and down inclined chutes in Chapter 8. The kinetic theory is extended to accommodate particle roughness in Chapter 9. Finally, a few tentative attempts to model flow in the intermediate regime, where a range of interparticle contact times is present, are desribed in Chapter 10. For the reader interested in details, extensive references are given. Some problems are included at the end of most of the chapters, to enable the reader to either fill in details that are omitted in the book or extend the material.

Overall, we wish to inform the reader about observed phenomena, some of the available models and their shortcomings, and unresolved issues in granular flow. We describe in detail only a very small number of "animals" in the vast and rapidly expanding granular zoo, but we hope that the material presented here will stimulate readers to learn more about this field. Many interesting phenomena and puzzles lurk in topics such as vibrated beds, segregation in rotating cylinders, landslides and sand dunes, the behavior of cohesive powders, and fluid–particle flows, which either have not been discussed here or are mentioned only in passing.

This book deals with granular flow at a level that should be suitable for senior undergraduate and postgraduate students. A knowledge of undergraduate-level fluid and solid mechanics is desirable but not essential. As the book provides a reasonably self-contained introduction to the subject, it may also be useful to new entrants to the field. It is hoped that a study of the simple problems discussed here may prepare the reader for a better understanding of more realistic and complex situations, which are often encountered in industries. Parts of this book are based on graduate courses taught at Princeton University, Indian Institute of Science, and California Institute of Technology.

We request readers to inform us by email, (kesava@chemeng.iisc.ernet.in or prnott@chemeng.iisc.ernet.in) if they come across errors. We shall also be grateful for suggestions that may improve the book. An errata for the book and supplementary information, such as solutions to problems (for instructors) and additional problems, may be found at the catalogue page of the book (http://www.cambridge.org/us/catalogue/catalogue.asp?isbn=0521571669).

We are grateful to our teacher Prof. R. Jackson for introducing us to this field and to Dr. R. M. Nedderman for prodding one of us to write a book on it. Prof. S. Sundaresan had originally planned to be one of the coauthors. He opted out later, but we are very grateful to him for his comments and suggestions on some of the chapters. We are grateful to professors B. Ananthanarayan, I. Goldhirsch, R. Jackson, C. S. Jog, H. S. Mani, S. Ramaswamy, and D. Sen for helpful discussions. We express our heartfelt thanks to our editors, Ms. Florence Padgett, Mr. Roger Astley, and Mr. Peter Gordon, who showed extraordinary patience in dealing with our repeated requests for extensions. We are grateful to Ms. B. G. Girija, Mr. K. Venugopal, Mr. Gautam Parthasarathy, and Mr. Vishwajeet Mehandia for drawing some of the figures; Mr. P. T. Raghuram, Ms. Shruti Seshadri, Mr. Alok Srivastava, Dr. S. Venugopal, and Prof. Sanjeev Gupta for assistance with photographs; the staff of Aptara for promptly responding to our queries regarding LaTeX, and Mr. Anoop Chaturvedi of Aptara for responding to many queries during our review of the copyedited manuscript. Special thanks are due to various publishers and societies who have permitted us to reproduce figures from their journals, to professors A. Drescher and R. P. Behringer for permitting us to use some of their photographs, and to Prof. D. W. Agar for translating the titles of some articles from French and German to English. Finally, we wish to record our gratitude to and deep appreciation for our families, who tolerated our preoccupation with this work for an extended period of time.

Notation

a_1–a_6	constants of $O(1)$ in the heuristic high-density theory for rapid flows, §7.2 and 8.1.1
a_n, b_n	coefficients in the expansion of $\mathcal{A}$, $\mathcal{B}$ in (7.126)
$\mathbf{b}$	body force per unit mass
$\hat{\mathbf{b}}$	scaled body force per unit mass, defined in (7.79)
b_{*0}	$\equiv b_*(\theta = 0)$, constant occurring in the boundary conditions for the radial stress field (see (6.32) and (6.48))
c	cohesion
c_w	adhesion
c_1, c_2, c_3	principal compressive rates of deformation, i.e., the eigenvalues of $\mathbf{C}$
c_*	$= (c_1 + c_2)/2$
$\mathbf{c}$	translation velocity of particles
$\hat{\mathbf{c}}$	scaled translation velocity, defined in (7.79)
$\mathbf{c}'$	postcollisional translation velocity of a particle
$\mathbf{c}''$	precollisional translation velocity of a particle in an "inverse collision," as defined in §7.3.2
d_p	particle diameter
d_w	diameter of a wall hemisphere (see Fig. 8.2)
$\overline{d}$	$\equiv \frac{1}{2}(d_\mathrm{p} + d_\mathrm{w})$
det	determinant of a tensor or a matrix, defined in (A.37) and (A.41)
e_p	coefficient of restitution for particle collisions, defined in (7.3)
e_w	coefficient of restitution for particle–wall collisions
$\mathbf{e}_i$	basis vector for a Cartesian coordinate system
f, $f^{(1)}$	single-particle probability distribution function of particle position and velocity, sometimes referred to as the singlet distribution
f^0	Maxwell–Boltzmann distribution based on the local density, velocity, and temperature
$\hat{f}$	scaled singlet distribution, defined in (7.79)
f'	singlet distribution for the postcollision velocity of a "direct collision" (see §7.3.2)
f''	singlet distribution for the precollision velocity of an "inverse collision"
$f^{(2)}$	probability distribution function of the position and velocity of a pair of particles
$\dot{f}_{\mathrm{coll}}$	rate of change of $f^{(1)}$ due to collisions
f_d	drag force exerted by the fluid on a single particle
g	acceleration due to gravity; $\equiv n^{(2)}/n^2$, pair distribution function

$\mathbf{g}$	translation velocity of particle 1 relative to particle 2
$\hat{\mathbf{g}}$	$\equiv \mathbf{g}/v_s$, scaled relative translation velocity
$\mathbf{g}'$	postcollisional value of $\mathbf{g}$
$g_0(\theta)$	function defining the radial stress field
$g_0(\nu)$	equilibrium pair distribution function at contact (Chapters 7–10)
$g_w(\nu)$	enhancement in the number density of particles in contact with the wall, defined in §8.1.2
$\mathbf{h}$	relative velocity of the point of contact of particle 1 with respect to that of particle 2, given by (9.1)
k_x, k_y, k_z	wavenumbers for the x, y, and z directions (§8.4)
$\mathbf{k}$	unit vector along the line joining the center of particle 1 to the center of particle 2 (see Fig. 7.3); unit vector along the line joining the center of a wall hemisphere to the center of a colliding particle (see Fig. 8.2)
$\mathbf{k}''$	$\mathbf{k}$ for an "inverse collision" (see Fig. 7.6)
ℓ	$\equiv (1 - e_w^2)\, a_6/a_3$, constant introduced in §8.2.1 and §8.3.3
m	$\equiv a_1^2/(a_3\, a_5 \varphi)$, constant introduced in §8.2.1 and §8.3.3
m_i	mass of particle i
m_p	mass of particle
$\dot{m}$	mass flow rate per unit width in chute flow (§8.3 and 10.2)
$\dot{m}^*$	dimensionless mass flow rate, defined in (8.81)
m_ℓ	loading, that is, mass of material per unit area of the chute base (§8.3.1 and 10.2)
m_ℓ^*	dimensionless loading, defined in (8.81)
n	number density of particles
$n^{(2)}$	probability distribution function of the positions of a pair of particles
$\mathbf{n}$	unit normal to a bounding surface
p	mean stress, defined in (5.6); pressure in rapid granular flow
p_c	mean stress at a critical state
p_f	fluid pressure
q	invariant of the stress tensor, defined in (5.6)
$\mathbf{q}$	pseudothermal energy flux; energy flux (Chapter 1)
$\mathbf{q}^s, \mathbf{q}^c$	streaming and collisional contributions to the pseudothermal energy flux
$\hat{\mathbf{q}}$	$\equiv \mathbf{q}/(\rho_p v_s^3)$, scaled pseudothermal energy flux (Chapter 7)
$\mathbf{q}_{pt}$	pseudothermal energy flux; referred to as $\mathbf{q}$ after (7.11)
$\mathbf{q}_r$	flux of rotational fluctuational kinetic energy, defined in (9.29)
$\mathbf{q}_w$	flux of fluctuational kinetic energy to the wall due to particle–wall collisions
$\mathbf{q}'$	flux of true thermal energy
r	radial coordinate measured from the virtual apex of the hopper (see Fig. 3.13); $\equiv \mathcal{N}/\mathcal{S}$, stress ratio introduced in §8.2.1
r_e	radial coordinate corresponding to the edge of the exit slot
$\mathbf{r}$	position vector of particle
$\hat{\mathbf{r}}$	scaled position vector, defined (7.79)
s	$= (c_1 - c_2)/2$, the magnitude of the maximum shear rate in plane flow; mean free path of particles (Chapters 7 and 8)
s_j	dimensionless arc length measured along the jth characteristic
$\bar{s}_w$	mean spacing between wall hemispheres (see Fig. 8.2)
s_{ij}	deviatoric stress component, $= \sigma_{ij} - p\delta_{ij}$
$\mathbf{s}$	$\equiv \hat{\mathbf{\Omega}}/(2\hat{T})^{1/2}$, rescaled peculiar spin (Chapter 9)
$\mathbf{s}'$	deviatoric stress tensor, $= \boldsymbol{\sigma} - p\,\mathbf{I}$

$\hat{t}$	scaled time, defined in (7.79)
$\mathbf{t}_{(\mathbf{n})}$	stress vector, = force per unit area exerted on a surface with unit inward normal **n**
tr	trace of a tensor, defined by (A.18)
u_w	pore pressure or pressure in excess of the atmospheric pressure which is exerted by the fluid on the particles
v_s	velocity scale, used to scale variables in (7.79)
v^*	dimensionless velocity, defined in (8.71) for plane Couette flow and (8.104) for chute flow
$\mathbf{v}$	velocity of a material point, or mean velocity $\langle \mathbf{c} \rangle$ of a collection of particles
$\mathbf{v}_i$	velocity of particle i
$\hat{\mathbf{v}}$	$\equiv \mathbf{v}/v_s$, scaled mean velocity
$\mathbf{v}_w$	velocity of a wall
$\mathbf{v}_{slip}$	$\equiv \mathbf{v} - \mathbf{v}_w$, velocity slip at a wall
w	root-mean-square fluctuation velocity of particle (Chapters 7 and 8)
$\mathbf{w}$	local averaged velocity of the fluid phase relative to the particle phase
y^*	dimensionless distance, defined in (8.71) for plane Couette flow and (8.104) for chute flow
A	dimensionless cross-sectional area, scaled by W^2
$\mathcal{A}, \mathcal{B}, \mathcal{R}$	functions of ξ in the expression for Φ_K in (7.125) and (9.70)
C	cohesion (Section 2.2); rate of compression in the direction of the normal to a plane, defined in (2.74)
C_{ij}	component of the rate of deformation tensor **C**
$\mathbf{C}$	rate of deformation tensor, defined in the compressive sense by (2.49); $\equiv \mathbf{c} - \mathbf{v}$, peculiar velocity of a particle
$\hat{\mathbf{C}}$	$\equiv \mathbf{C}/v_s$, scaled peculiar velocity of particles
$\mathrm{D}/\mathrm{D}t$	material derivative, defined by (1.23)
D	either the diameter or the width of the exit slot of a hopper or bunker (see Fig. 3.1)
$\mathbf{D}$	$= -\mathbf{C}$, rate of deformation tensor, defined by (1.50)
$\hat{\mathbf{D}}$	$\equiv \mathbf{D}H/v_s$, scaled rate of deformation tensor
$\mathbf{F}_d$	drag force per unit volume of the suspension, exerted by the fluid phase on the particle phase
$\hat{\mathbf{G}}$	$\equiv (\hat{\mathbf{C}}_1 + \hat{\mathbf{C}})/2$, scaled mean peculiar velocity of colliding particles
H	head or height of the granular material above the exit slot of a hopper or bunker (see Fig. 3.2); characteristic macroscopic length scale in the Chapman–Enskog expansion (Chapters 7 and 9); distance between plates in plane Couette flow; thickness of flowing layer in chute flow
H^*	$\equiv H/d_p$, dimensionless Couette gap in plane Couette flow, §8.2.1; dimensionless flow depth in chute flow, §8.3
H^*_{hd}	value of H^* for chute flow predicted by the high-density theory at a given angle of inclination θ
I	moment of inertia of a particle (Chapter 9)
I_1, I_2, I_3	principal invariants of the stress tensor, defined in (5.2)
$\hat{I}$	$\equiv 4I/(m_p d_p^2)$, dimensionless moment of inertia of a particle (Chapter 9)
$\mathbf{I}$	unit (identity) tensor
J_2	second invariant of the deviatoric stress tensor, defined by (5.9)
$\mathcal{J}$	Jacobian for the transformation of one set of variables to another

$\mathbf{J}$	impulse per unit mass on particle 2 during collision with particle 1
K	coefficient of earth pressure at rest, i.e., the ratio of the horizontal normal stress to the vertical normal stress; $\equiv Kn^{-1}$, inverse Knudsen number (Chapters 7–10)
$\mathcal{K}(\nu)$	$\equiv \kappa/(\rho_p d_p \sqrt{T})$, thermal conductivity function (Chapters 8 and 10)
Kn	$\equiv H/s$, Knudsen number, ratio of macroscopic to microscopic length scales
$\dot{M}$	mass flow rate
L	$\equiv a_4(1 - e_p^2)/a_3$, constant introduced in §8.2.1 and §8.3.3
$L_n^{(p)}(x)$	Generalized Laguerre polynomials, or Sonine polynomials, used in §7.3.10
$\mathcal{L}$	linearized Boltzmann operator, defined in (7.113)
$\mathcal{L}r$	linearized Boltzmann operator for rough particles, defined in (9.47)
$\mathcal{L}^*$	linearized operator defined in (9.74)
M	$\equiv a_1^2/(a_2\, a_3)$, constant introduced in §8.2.1 and §8.3.3
$\mathbf{M}$	couple stress, defined in (9.27)
N	normal stress
N_b	normal stress exerted on the bin wall
N_h	normal stress exerted on the hopper wall
N_w	normal stress exerted on the wall of a bin or hopper
$\mathcal{N}$	normal stress on the walls in plane Couette flow (see Fig. 8.4)
P	dimensionless perimeter of the bin section, scaled by W
$\mathcal{P}$	$\equiv p/(\rho_p T)$, pressure function
$\mathbf{Q}$	orthogonal tensor, defined in (5.60)
R	particle radius
Re	Reynolds number
S	rate of shear in the direction of a unit vector $\mathbf{t}$ which is tangential to a surface, defined in (2.74)
S_m	bounding surface of a material volume
$\mathcal{S}$	shear stress on the walls in plane Couette flow (see Fig. 8.4)
$\boldsymbol{\mathcal{S}}_w$	stress transmitted to a wall by the particles adjacent to it, introduced in §8.1
T	shear stress; thermodynamic or grain temperature (Chapters 7–10)
T_r	$\equiv I\langle\Omega^2\rangle/(3m_p)$, rotational temperature (Chapter 9)
T_{tot}	$\equiv T + T_r$, total temperature (Chapter 9)
T_w	wall shear stress
T^*	dimensionless grain temperature defined in (8.71)
$\hat{T}$	$\equiv T/v_s^2$, scaled grain temperature
$\hat{U}$	internal energy per unit mass of the material
$\hat{U}'$	true thermal internal energy per unit mass of the material
V_m	material volume
V_D	dimensionless mass flow rate, defined in (3.5)
W	either the half-width or the radius of the bin section of a bunker, depending on whether the cross section is rectangular or circular
$\mathbf{W}$	vorticity tensor or spin tensor, defined by (5.62)
$\mathbf{X}$	position vector of a material point in the reference configuration
α	$\equiv (\ell - m\, r^2)$, constant introduced in §8.2.1 and §8.3.3
β	filling angle (see Fig. 1.14); roughness coefficient, defined in (9.3)

β_r	angle of repose, the angle formed by the free surface of a heap to the horizontal
γ	orientation of the major principal stress axis (see Fig. 2.11)
γ'	orientation of the major principal stress axis relative to the circumferential direction (see Fig. 3.13)
γ_s	surface tension
$\gamma(\nu)$	$\equiv \Gamma d_p/(\rho_p T^{3/2})$, dissipation function (Chapters 8 and 10)
$\gamma_w(\nu)$	$\equiv \Gamma_w/(\rho_p T^{3/2})$, wall dissipation function (Chapters 8 and 10)
δ	angle of wall friction, defined by (2.4);
	parameter measuring departure from equilibrium, defined in (7.49)
δ_{ij}	Kronecker delta, defined by (A.6)
ϵ	$\equiv 1 - e_p^2$, inelasticity of particle collisions
ε	$\equiv 1 - \|\beta\|$, parameter characterizing particle roughness;
	parameter introduced in (8.75)–(8.78) to elucidate the effect of the walls
ζ	spin viscosity, defined in (9.90);
	magnitude of $\boldsymbol{\zeta}$ (Chapters 8 and 9)
$\boldsymbol{\zeta}$	$\equiv \hat{\mathbf{g}}/(2\hat{T})^{1/2}$, rescaled velocity of particle 1 relative to particle 2 in Chapters 7 and 9;
	$\equiv (\mathbf{c} - \mathbf{v}_w)/(2T)^{1/2}$, scaled velocity of a particle relative to the wall in Chapter 8
$\eta(\nu)$	$\equiv \mu/(\rho_p d_p \sqrt{T})$, viscosity function (Chapters 8 and 10)
η'	dimensionless vertical coordinate, $= y'/W$ (see Fig. 4.17)
η_1	$\equiv (1 + e_p)/2$ (Chapters 8 and 10)
η_2	$\equiv \frac{1}{2}(1 + \beta)\hat{I}/(1 + \hat{I})$ (Chapter 9)
$\eta_w(\nu)$	wall viscosity function, defined in $\boldsymbol{S}_w = \eta_w \rho_p T^{1/2} \mathbf{v}_{slip}$ (Chapters 8 and 10)
θ	circumferential coordinate, (see Fig. 3.13);
	angle of inclination of a chute from the horizontal (§8.3 and 10.2)
θ_0	maximum angle between $\mathbf{k}$ and the inward normal of a wall (see Fig. 8.2)
θ_w	wall angle of the hopper, or of the hopper section (see Figs. 3.13 and 4.1)
$\boldsymbol{\theta}(\psi)$	flux of $\langle\psi\rangle$ due to collisions
ϑ	distribution function of particle spin (Chapter 9)
κ	pseudothermal conductivity
κ_r	conductivity of rotational fluctuational kinetic energy, defined in (9.94)
κ^*	thermal conductivity of a dilute gas of elastic spheres
$\dot{\lambda}$	scalar factor of proportionality, which occurs in the flow rules (2.65) and (5.25)
μ	shear viscosity of a granular material in rapid flow
μ^*	shear viscosity of a dilute gas of elastic spheres
μ_b	bulk viscosity of a granular material in rapid flow
μ_f	shear viscosity of the fluid
μ_r	transport coefficient characterizing the diffusion of intrinsic angular momentum, defined in (9.92)
ν	solids fraction, or volume fraction of solids
ν_d	angle of dilation, defined in (2.83)
ν_{lrp}	solids fraction corresponding to loose random packing
ν_{drp}	solids fraction corresponding to dense random packing
ν_s	the scale for the solids fraction, defined below (7.80)

$\overline{\nu}$	average value of ν across the Couette gap in plane Couette flow
ξ	dimensionless horizontal coordinate (Chapters 2 and 4); dimensionless radial coordinate (Chapter 3); natural strain (Chapter 5); magnitude of $\boldsymbol{\xi}$ (Chapters 7–9);
ξ	$\Omega y/d_p$, scaled distance from lower wall in plane Couette flow §8.2.1; $\Omega(H-y)/d_p$, scaled distance from the free surface in chute flow §8.3.3
ξ_m	$\equiv \Omega H^*/2$, value of ξ at the midplane in plane Couette flow §8.2.1
$\boldsymbol{\xi}$	$\equiv \hat{\mathbf{C}}/(2\hat{T})^{1/2}$, rescaled peculiar velocity of particles (Chapters 7–9)
ρ_p	particle density, i.e., the density of the solid material forming the particles
ρ_f	fluid density
ρ_b	bulk density, $= \rho_f(1-\nu)+\rho_p\,\nu$
ρ	density of the granular material, $= \rho_p\,\nu$
ϱ	$\equiv \mathrm{sgn}(dv_x/dy)$ in §10.2
$\boldsymbol{\sigma}$	stress tensor, defined in the compressive sense
$\boldsymbol{\sigma}^s, \boldsymbol{\sigma}^c$	streaming and collisional contributions to the stress tensor in rapid flow
$\boldsymbol{\sigma}^f$	"frictional" stress, defined in §10.1
$\boldsymbol{\sigma}^k$	"kinetic" stress, defined in §10.1
$\hat{\boldsymbol{\sigma}}$	$\equiv \boldsymbol{\sigma}/(\rho_p v_s^2)$, scaled stress tensor (Chapters 7 and 9)
$\sigma_1, \sigma_2, \sigma_3$	principal stresses
σ_i'	effective principal stress, $= \sigma_i - u_w$
σ	mean stress for plane flow, $= (\sigma_1+\sigma_2)/2$
$\overline{\sigma}$	dimensionless normal stress, $= \sigma/(\rho g r_e)$ (Chapter 3); $\sigma/(\rho g W)$ (Chapter 4)
$\overset{\circ}{\boldsymbol{\sigma}}$	Jaumann derivative of $\boldsymbol{\sigma}$, defined by (5.64)
τ	deviator stress, $= (\sigma_1-\sigma_2)/2$; relaxation time, defined in (7.26)
τ_c	mean free time, i.e., average time between particle collisions
$\overline{\tau}$	dimensionless deviator stress, $= \tau/(\rho g r_e)$ (Chapter 3); $\tau/(\rho g W)$ (Chapter 4)
$\boldsymbol{\tau}$	external torque per unit mass on the particles (Chapter 9)
ϕ	angle of internal friction, defined by the slope $\sin\phi$ of the critical state line (see (2.47))
ϕ_*	angle of internal friction, defined by (2.15)
ϕ_μ	angle of friction between grains, such that the coefficient of friction is $\tan\phi_\mu$
φ	specularity coefficient, defined in §8.1.1
$\chi(\psi)$	volumetric source of $\langle\psi\rangle$ due to collisions
ψ	any particle property
ψ'	postcollisional value of particle property ψ
$\dot{\psi}_{\mathrm{coll}}$	rate of change of $\langle\psi\rangle$ due to collisions
$\dot{\psi}_w$	rate of transmission of ψ to a wall by particle–wall collisions
$\boldsymbol{\omega}$	angular velocity (Chapter 10); particle "spin," i.e., its angular velocity about its own axes (Chapter 9)
$\overline{\boldsymbol{\omega}}$	mean particle spin
$\boldsymbol{\omega}''$	precollisional spin of a particle in an "inverse collision," as defined in (9.15)
$\hat{\boldsymbol{\omega}}$	$\equiv \boldsymbol{\omega}(I/m_p)^{1/2}/v_s$, scaled particle spin
Γ	volumetric dissipation rate of pseudothermal energy due to inelastic particle collisions

$\hat{\Gamma}$ $\equiv \Gamma d_\mathrm{p}/(\rho_\mathrm{p} v_\mathrm{s}^3)$, scaled rate of dissipation of pseudothermal energy

Γ_w rate of dissipation of pseudothermal energy due to particle–wall collisions, per unit area of the wall

Γ_r volumetric dissipation rate of rotational fluctuational kinetic energy, defined in (9.30);
volume, due to inelastic particle collisions

Ω constant, $= (\pi/4) - (\phi/2)$;
$\equiv [\pm(L - Mr^2)]^{1/2}$, introduced in §8.2.1

$\boldsymbol{\Omega}$ $\equiv \boldsymbol{\omega} - \overline{\boldsymbol{\omega}}$, peculiar spin of a particle

$\hat{\boldsymbol{\Omega}}$ $\equiv \boldsymbol{\Omega}(I/m_\mathrm{p})^{1/2}/v_\mathrm{s}$, scaled particle spin

Φ perturbation of the singlet distribution from the local Maxwellian, defined in (7.83)

$\dot{\Phi}$ $= -\boldsymbol{\sigma}^\mathrm{T}\!:\!\nabla\mathbf{v}$, stress power

Subscripts

b bin section of a bunker
c critical state
f   fluid property
h hopper section of a bunker
p   particle property
0   value of variable at $\mathrm{O}(K^0)$, i.e., O(1) (Chapters 7–9)
K value of variable at $\mathrm{O}(K)$ (Chapters 7–9)
ϵ value of variable at $\mathrm{O}(\epsilon)$ (Chapters 7 and 9)
ε value of variable at $\mathrm{O}(\varepsilon)$ (Chapters 7 and 9)

Superscript

T   transpose

1

Introduction

A *granular material* is a collection of solid particles or grains, such that most of the particles are in contact with at least some of their neighboring particles. The terms "granular materials," "bulk solids," "particulate solids," and "powders" are often used interchangeably in the literature. Common examples of granular materials are sand, gravel, food grains, seeds, sugar, coal, and cement. Figure 1.1 shows the typical size ranges for some of these materials.

Granular materials are commonly encountered in nature and in various industries. For example, with reference to the chemical industry, Ennis et al. (1994) note that about 40% of the value added is linked to particle technology. Similarly, Bates (2006) notes that more than 50% of all products sold are either granular in form or involve granular materials in their production. In spite of the importance of granular materials, their mechanics is not well understood at present. Nevertheless, some progress has been made during the past few decades. The goal of this book is to describe some of the experimental observations and models related to the mechanical behavior of *flowing* granular materials. As studies in this area are increasing rapidly, our account is necessarily incomplete. However, it is hoped that the book will provide a useful starting point for the beginning student or researcher.

A material is called a *dry* granular material if the fluid in the interstices or voids between the grains is a gas, which is usually air. On the other hand, if the voids are completely filled with a liquid such as water, the material is called a *saturated* granular material. If there is a liquid in some of the voids, and the rest of the voids are filled with a gas, the material is said to be *partially saturated.* For example, the upper region of a soil in its natural environment is usually partially saturated, whereas the lower region is saturated. In the recent literature, both saturated and partially saturated materials are called *wet* granular materials.

If the particles lose contact with each other, and each particle is surrounded by a fluid, the granular material becomes a *fluid–particle suspension.* At present, a unified theory for the mechanical behavior of granular materials and suspensions is lacking. In this book, attention will be confined to the former, barring a brief discussion of some aspects of the latter in this chapter. The reader is referred to Gidaspow (1994), Fan and Zhu (1998), and Jackson (2000) for a detailed discussion of suspensions.

When granular materials are at rest, or in motion, they exhibit many features which cannot be anticipated on the basis of our experience with fluids such as air and water, and solids such as steel and wood. Like solids, they can sustain shear stresses at rest, as in the case of a heap or "sandpile." In a heap formed by pouring a free-flowing granular material from a funnel onto a flat surface, the inclination of the free surface of the heap to the horizontal cannot be arbitrary but is limited by a maximum value called the *angle of repose.* Like liquids, they flow from vessels under the action of gravity, but the mass flow rate is approximately independent of the height of material above the discharge orifice. This

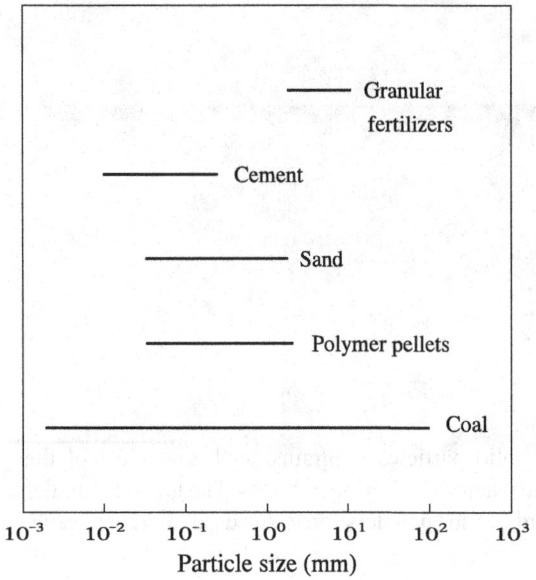

Figure 1.1. Size ranges of various granular materials. (Data taken from Feda, 1982, p. 22, and Brown and Richards, 1970, pp. 4–8.)

feature accounts for the use of hourglasses containing sand as clocks a few centuries ago. Unlike water, granular materials are compressible in the sense that the space between the particles often changes during flow. For example, if the particles are tightly packed, they tend to move apart when they flow. Whereas water and alcohol can be readily mixed to obtain a liquid of uniform composition, attempts to mix particles of two sizes or of two materials often lead to segregation or separation of the constituents.

In some cases, particularly when dealing with coarse materials, the effect of the interstitial fluid can be ignored. In other cases, such as the flow of fine granular materials through a long vertical pipe connected to a storage vessel above it, fluid–particle interaction strongly affects the flow rate of particles.

Overall, granular materials share some features of common fluids and solids but also differ from them in many ways. Various aspects of the behavior of granular materials have been discussed in review articles written by Jackson (1983, 1986), Jaeger and Nagel (1992), Hutter and Rajagopal (1994), Jaeger et al. (1996), de Gennes (1998), Roberts (1998), Sundaresan (2001), Hill and Selvadurai (2005), and Campbell (2006).

The outline of this chapter is as follows. Some examples of granular statics and flow are presented below, followed by a discussion of interparticle and fluid–particle forces. Some of the available models will be described briefly. Here it may be noted that the focus of the book is on continuum models. Finally, balance laws for such models will be formulated.

Some of the material presented in this and subsequent chapters requires a knowledge of vectors and tensors. Readers who are not familiar with these concepts will find a brief discussion in Appendices A and B.

1.1. EXAMPLES OF GRANULAR STATICS AND FLOW

(a) *Heaps*

Consider a conical heap or pile of granular material resting on a horizontal surface. Several workers have measured the normal stress N exerted by the material on the base of the heap as a function of radial distance r measured from the center of the base. Here we discuss the results of Vanel et al. (1999) for heaps of sand. The heap was constructed by pouring sand from a funnel onto a base plate (Fig. 1.2a), with the funnel being moved upward so that its

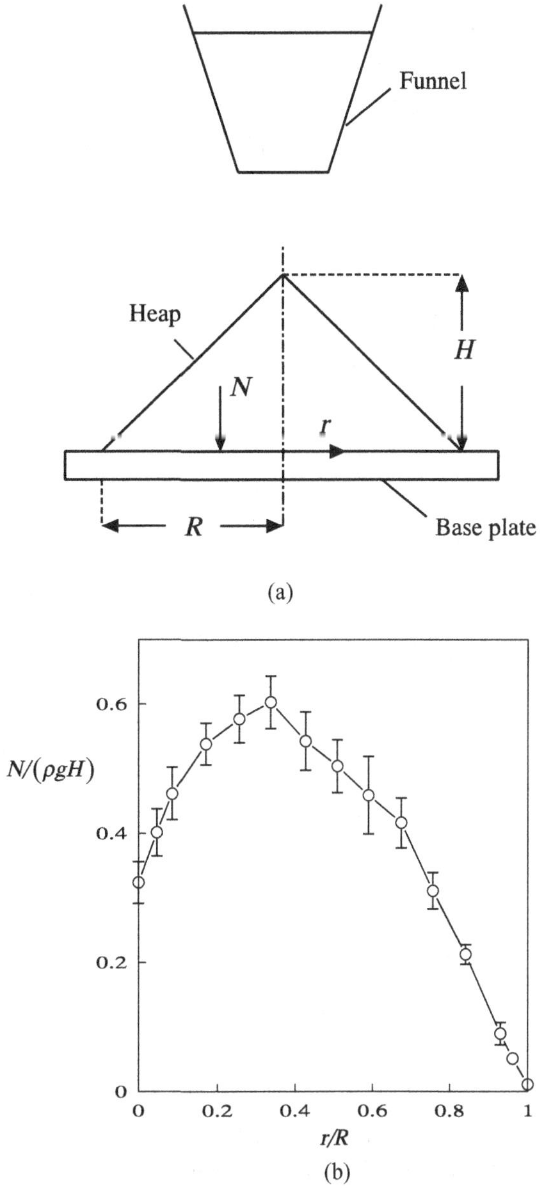

Figure 1.2. (a) A heap formed by pouring material from a funnel onto a circular metal plate. The funnel is raised so that its tip is always slightly above the apex of the heap. (b) Profile of the normal stress N exerted by the heap. The symbols represent the data of Vanel et al. (1999) for sand, with the vertical lines representing the standard deviation of several (typically 10–12) independent experiments. The piecewise linear curve is drawn to guide the eye. Here ρ is the density and g is the acceleration due to gravity. The lengths H, R, and r are measured as shown in (a). (Figure 1.2b has been reproduced from Vanel et al., 1999, with permission from Prof. R. P. Behringer and the American Physical Society. Copyright (1999) by the American Physical Society.)

tip was always slightly above the apex of the heap. The stress profile shows a minimum at the center of the base (Fig. 1.2b), contrary to the intuitive expectation that the normal stress should be a maximum at this location. This is called the *stress dip*, and several models have been proposed to explain such profiles (see, e.g., Cates et al., 1998; Didwania et al., 2000). It should be noted that the stress dip is not a universal feature. For example, when the heap was constructed by pouring material through a sieve that was slowly raised (Fig. 1.3a), the stress dip was not observed (Fig. 1.3b). Similarly, in the experiments of Brockbank et al. (1997), the dip occurred for small glass beads (mean diameter = 0.18 mm, standard deviation = 0.02 mm), but not for large glass beads (mean diameter = 0.56 mm, standard deviation = 0.05 mm).

Given a granular material and a procedure for constructing the heap, it is not yet possible to predict whether a stress dip will occur. Similarly, a simple physical explanation for the dip is not available. An explanation due to Vanel et al. (1999) is given below.

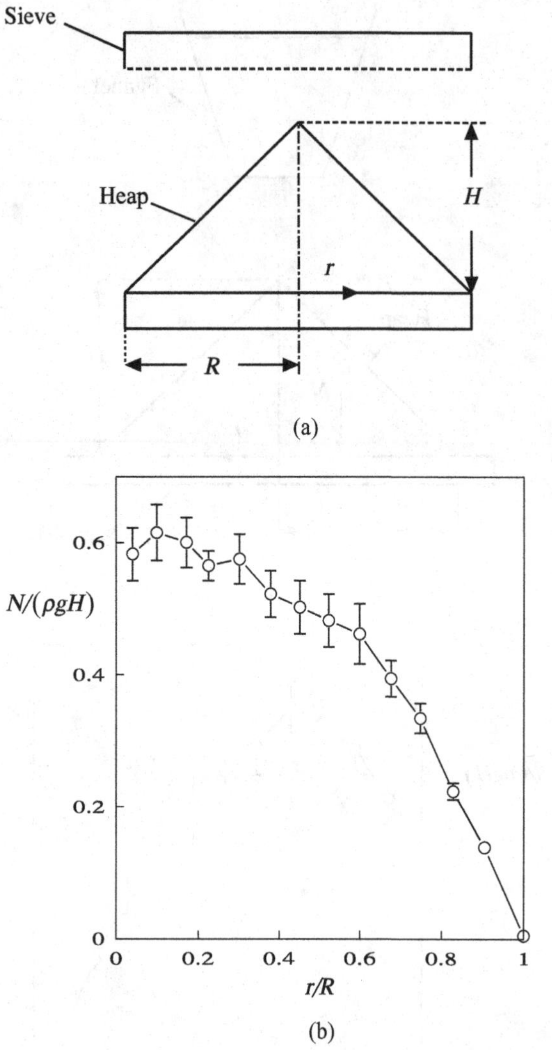

Figure 1.3. (a) A heap formed by pouring material from a sieve onto a circular metal plate. The sieve initially rests on the plate and is then raised gradually. (b) Profile of the normal stress N exerted by the heap. The symbols represent the data of Vanel et al. (1999) for sand, with the vertical lines representing the standard deviation of several (typically 10–12) independent experiments. The piecewise linear curve is drawn to guide the eye. Here ρ is the density and g is the acceleration due to gravity. The lengths H, R, and r are measured as shown in (a). (Figure 1.3b has been reproduced from Vanel et al., 1999, with permission from Prof. R. P. Behringer and the American Physical Society. Copyright (1999) by the American Physical Society.)

Experiments on static beds of granular materials show that the stresses are transmitted along preferred directions, which are called *force chains* or *stress chains* (see, e.g., Wakabayashi, 1957; Liu et al., 1995; Vanel et al., 1999). Figure 1.4 shows a two-dimensional pile of photoelastic disks. When viewed between crossed polarizers, the stress chains are visible as the bright stripes. As the chains are not vertical, Vanel et al. (1999) suggest that they may deflect a part of the weight away from the center of the base, leading to a stress dip. (A similar suggestion was made by Trollope (1968) and Edwards and Oakeshott (1989), who visualized the heap as being composed of arches, each of which supported its own weight.) Unfortunately, Vanel et al. (1999) do not show a picture of the chains for the case where a dip is not observed. Hence it is not known whether the pattern of chains differs significantly from that observed when there is a dip.

The occurrence of a stress dip has no analog in hydrostatics. Further, the measurements of Smid and Novosad (1981) show that unlike fluids, granular materials exert *nonzero* shear stresses on the base of a heap, even when they are at rest. Thus the statics of these materials is qualitatively different from that of fluids.

Figure 1.4. A heap formed by pouring photoelastic disks of two sizes from a funnel onto a plate. The apex of the heap is about 0.3 m above the base, and the diameters of the disks are 7.4 and 9 mm. The heap is viewed between crossed polarizers. (Reproduced from Vanel et al., 1999, with permission from Prof R. P. Behringer and the American Physical Society. Copyright (1999) by the American Physical Society.)

(b) *Silos*

Silos are vessels used for storing granular materials. Depending on their shapes, silos are also called bins, hoppers, or bunkers (Fig. 1.5). It has been found experimentally that the mass flow rate of coarse materials from silos is approximately independent of the head or height of material above the exit slot, provided the head is larger than a few multiples of the size of the exit slot. This is an unexpected result, if we note that the flow rate of water from vessels depends on the head. As discussed in Chapter 3, head independence arises from dry friction between the individual grains and between the grains and the wall of the silo.

Another feature of interest is the occurrence of several flow patterns when materials flow through bunkers. If the walls of the hopper section are sufficiently steep and the material is free flowing, *mass flow* occurs. Here all the material in the bunker is in motion (Fig. 1.6a). On the other hand, if the walls of the hopper section are shallow, *funnel flow* or *core flow* occurs. In the lower part of the bunker, there is a central core of rapidly moving material, surrounded by shoulders of material that is either stagnant or moving very slowly (Fig. 1.6b). The presence of stagnant and flowing zones makes it difficult to model such systems. It is likely that different sets of governing equations may be required for the two zones.

(c) *Chutes*

A chute is used to transport material by gravity flow from one point to another point at a lower level (Fig. 1.7a). Some features of chute flow can be illustrated by considering the experiments of Johnson et al. (1990). The chute is connected to a supply bunker containing glass beads by a chamber having valves at either end. By suitably adjusting the valves,

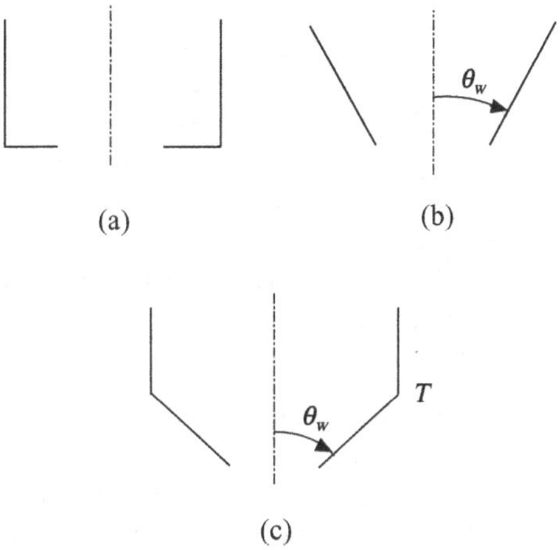

Figure 1.5. Shapes of silos: (a) bin, (b) hopper, and (c) bunker. In (c), T denotes the transition point.

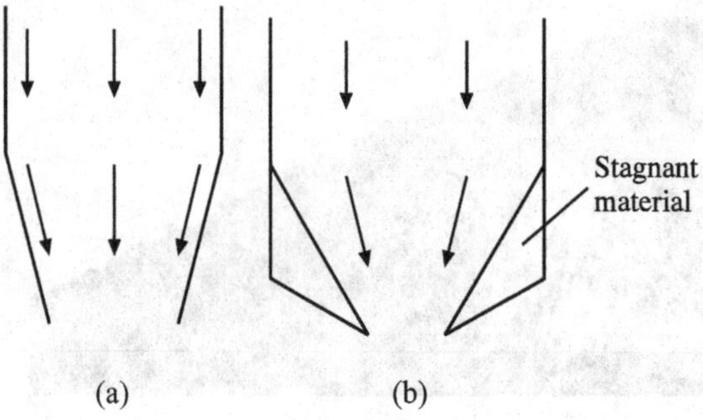

Figure 1.6. Flow patterns in bunkers: (a) mass flow and (b) funnel flow.

it is possible to regulate the mass flow rate of the beads and to permit two types of inlet conditions at the entry of the chute. These are termed (i) the dense entry condition and (ii) the loose entry condition. In case (i), the material enters the chute as a dense bed which moves slowly, whereas in case (ii), it enters as a low-density "cloud" of bouncing particles.

At locations that are not too close to the ends of the chute, it is found that quantities do not vary significantly in the flow direction, i.e., the flow is fully developed. This occurs for a range of inclinations of the chute. For fully developed flows, Fig. 1.7b shows the variation of the dimensionless mass flow rate per unit width of the chute $\dot{m}^*$ with the dimensionless mass holdup m_T^*. The mass holdup is the mass of material contained between two planes normal to the flow direction (see the broken lines in Fig. 1.7a), divided by the corresponding area of the base. At low values of the flow rate, the holdup is found to be independent of the entry condition. As the flow rate increases from point 1, the density of the flowing layer increases. For flow rates higher than point 3, the holdup for the loose entry condition is much less than that for the dense entry condition, and the velocity profiles are also qualitatively different in the two cases. The horizontal arrow in the $\dot{m}^* - m_T^*$ plot in Fig. 1.7b indicates an abrupt transition from the loose entry branch to the dense entry branch as the flow rate is increased. This happens for certain values of the inclination θ. Considering the velocity profiles, it is seen that the slip velocity at the base of the chute can be quite large in some cases, in contrast to the behavior of liquids. Further, for dense entry flows (see the points 4′ and 5′ in Fig. 1.7b), only a part of the material above the base shears, and the rest appears to move like a plug.

Thus chute flows reveal many interesting features, such as multiple steady states, velocity slip at solid boundaries, and the occurrence of many types of density and velocity profiles.

(d) *Vertical channels*

Consider the flow of granular materials through a channel under the action of gravity (Fig. 1.8a). Far from the ends of the channel, the profile of the vertical velocity is found to be approximately independent of the vertical coordinate, and hence a fully developed state is attained, as in the case of fluids. The data of Nedderman and Laohakul (1980) for the flow of glass beads through a channel of rectangular cross section are shown in Fig. 1.8b. Near the center of the channel, the velocity profile is almost flat and the material moves like a plug. In the *shear layer* near the wall of the channel, the velocity varies significantly over a length scale that is of the order of 10 particle diameters. Flow fields containing both plug layers and shear layers occur in many devices such as bunkers, chutes, rotary drums, and shear cells.

(e) *A standpipe connected to a hopper*

A standpipe is a pipe used to convey particles from fluidized beds and hoppers. Figure 1.9a shows a hopper connected to a vertical standpipe and also a bare hopper, both of which are filled with sand. A photograph, taken soon after the sand was allowed to discharge

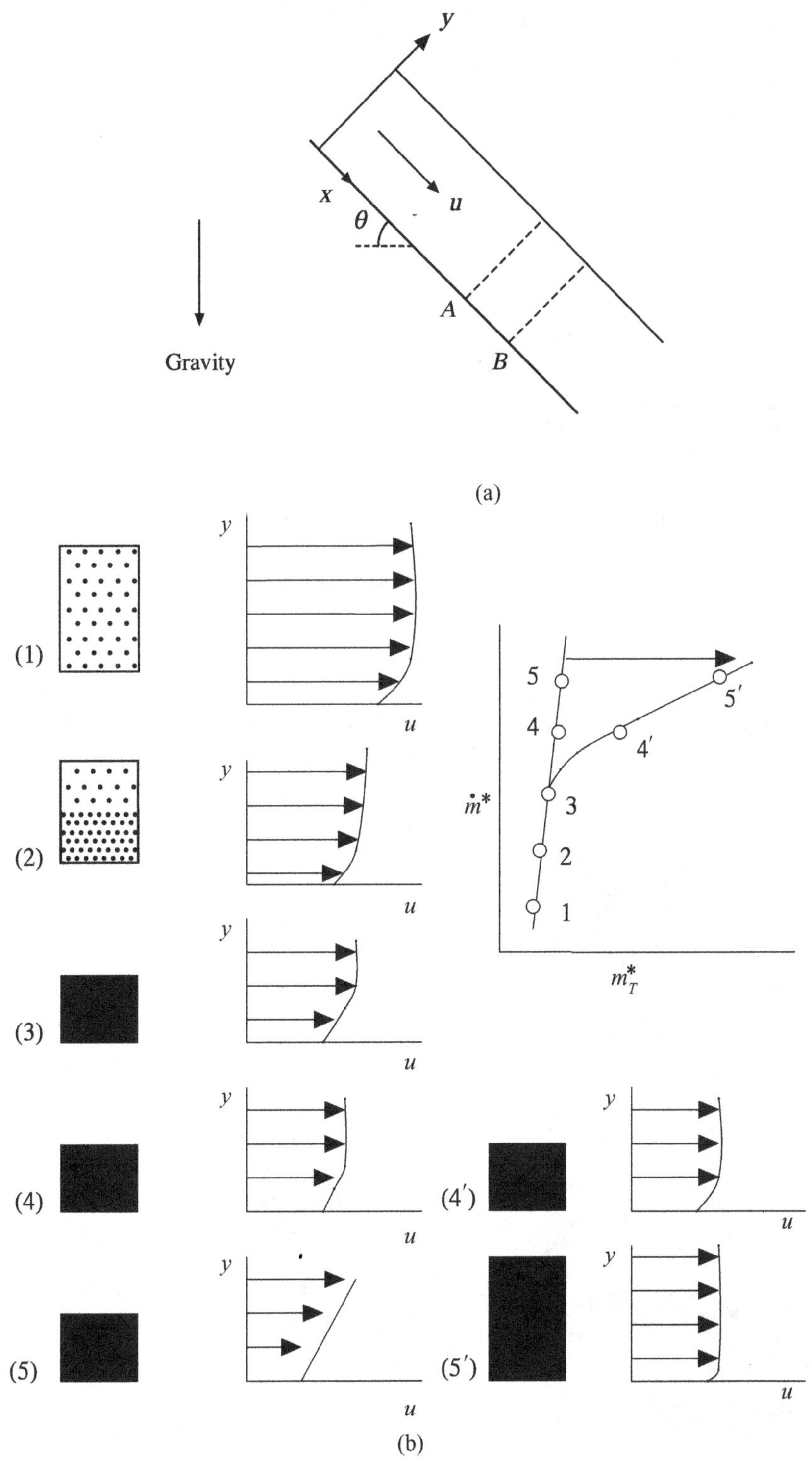

Figure 1.7. (a) Flow down a chute. (b) Variation of the dimensionless mass flow rate per unit width $\dot{m}^*$ with the dimensionless mass holdup m_T^* for the flow of glass beads (Johnson et al., 1990). The flow rate is scaled by $\rho_p\, d_p \sqrt{g\, d_p}$, and the holdup by $\rho_p\, d_p$, where ρ_p is the particle density, d_p is the particle diameter, and g is the acceleration due to gravity. Points 4 and 5 correspond to the loose entry condition and 4′ and 5′ to the dense entry condition. For points 1–3, the holdup is independent of the entry condition. The profiles of the x component of velocity u are also shown. The shaded boxes indicate the distribution of the solids fraction in the y direction. The particle diameter is 1 ± 0.1 mm. (Figure 1.7b has been reprinted from Johnson et al., 1990, with permission of Cambridge University Press.)

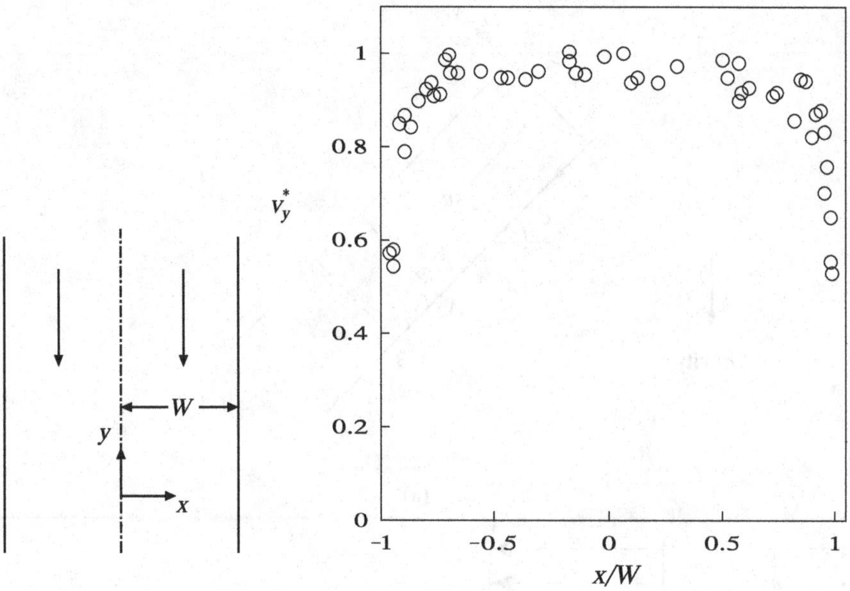

(a) (b)

Figure 1.8. (a) Flow through a vertical channel. (b) Profile of the scaled vertical velocity $v_y^* \equiv v_y(x, y_1)/v_y(0, y_1)$ for glass beads, at some vertical position $y = y_1$ within the fully developed region. Data of Nedderman and Laohakul (1980) for a channel with glass faces and wooden side walls. Parameter values: particle diameter $d_p = 2$ mm, half-width of the channel $W = 60$ mm.

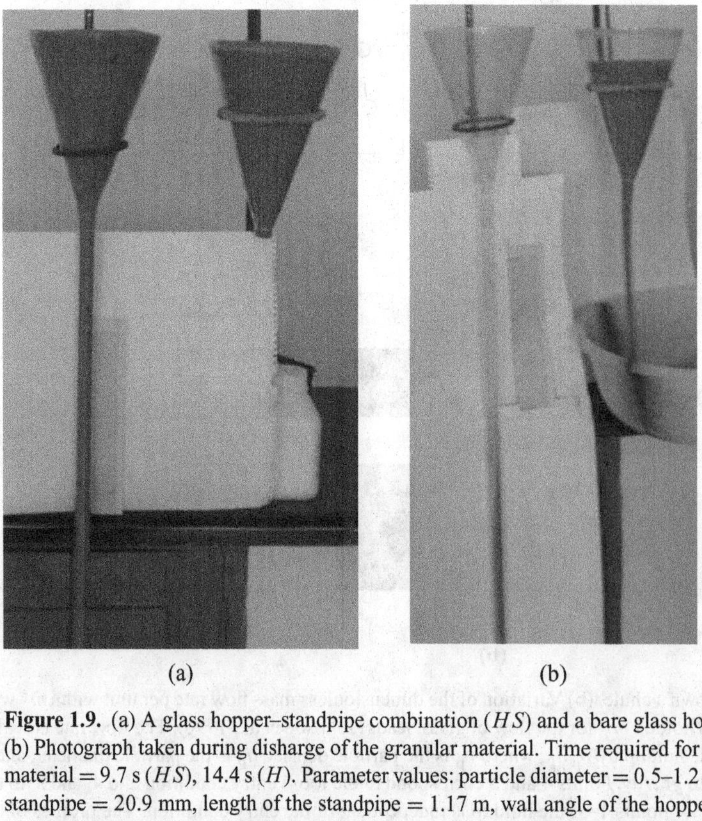

(a) (b)

Figure 1.9. (a) A glass hopper–standpipe combination (HS) and a bare glass hopper (H) filled with sand. (b) Photograph taken during disharge of the granular material. Time required for complete discharge of the material = 9.7 s (HS), 14.4 s (H). Parameter values: particle diameter = 0.5–1.2 mm, inner diameter of the standpipe = 20.9 mm, length of the standpipe = 1.17 m, wall angle of the hopper section of the standpipe $\approx 18°$, diameter of the exit orifice of the hopper = 21.1 mm, wall angle of the hopper $\approx 16°$, initial mass of the granular material = 2.85 kg (HS), 2.55 kg (H). This experiment was set up by Mr. J. Ravi Prakash and Mr. P. T. Raghuram.

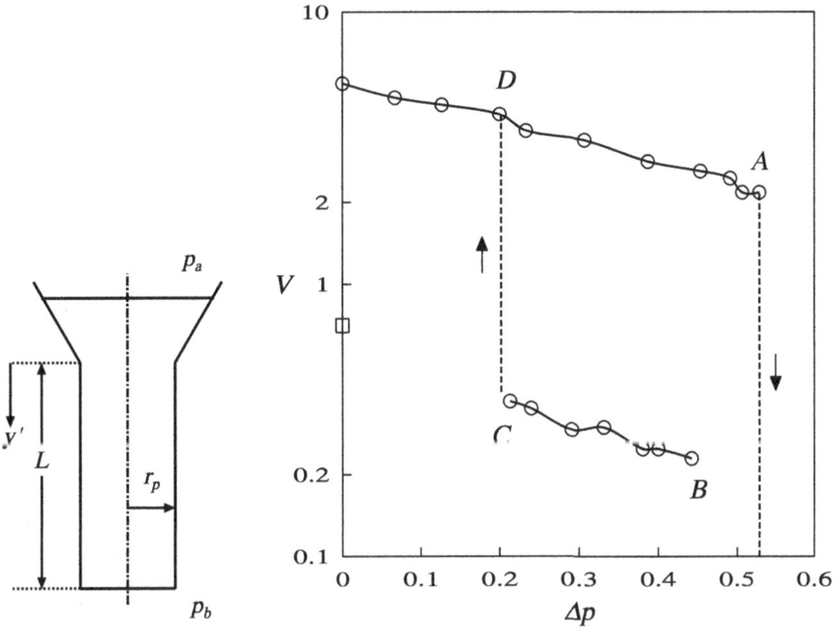

(a) (b)

Figure 1.10. (a) A standpipe connected to a hopper. The pressure of the air is p'_a at the top of the hopper and p'_b at the bottom of the pipe. (b) Variation of the dimensionless mass flow rate V with the dimensionless pressure rise $\Delta p \equiv p_b - p_a$ for the flow of sand: ○, data of Chen et al. (1984); —, curve drawn to guide the eye. Here p_a and p_b are the dimensionless air pressures. The mass flow rate is scaled by $\rho_p \, \nu_0 \, \pi \, r_p^2 \sqrt{g \, r_p}$ and the pressure by $\rho_p \, \nu_0 \, g \, L$, where ρ_p is the particle diameter, g is the acceleration due to gravity, r_p and L are the inner radius and length of the pipe, respectively, and ν_0 is the solids fraction of a moving bed of particles. The square in (b) denotes the dimensionless mass flow rate for a bare hopper. Parameter values: $\rho_p = 2{,}620$ kg m^{-3}, $r_p = 25.4$ mm, $L = 3.27$ m, $\nu_0 = 0.64$, mean diameter of sand = 154 μm.

from both the devices, shows that the hopper section of the hopper–standpipe combination was empty, whereas the bare hopper was nearly full (Fig. 1.9b). It is striking that the average mass flow rate from the former was about 70% larger than that from the latter. The use of a long pipe to increase the flow rate of granular material from a hopper was first reported by Bingham and Wikoff (1931).

Before discussing the reasons for this behavior, let us consider the experiments of Chen et al. (1984) on the flow of sand through a vertical standpipe connected to a hopper (Fig. 1.10a). The flow rate of the particles can be controlled by adjusting the pressure rise $\Delta p' \equiv p'_b - p'_a$, where p'_a and p'_b are the pressure of the air at the top of the hopper and the bottom of the pipe, respectively. Alternatively, the flow rate can be controlled by the orifice size of a slide valve located at the bottom of the pipe. Here we consider the case of a fully open valve, i.e., the orifice diameter is equal to the inner diameter of the pipe.

The circles in Fig. 1.10b show the variation of the dimensionless mass flow rate of particles $V = \dot{m}/(\rho_p \, \pi \, r_p^2 \sqrt{g \, r_p})$ with the dimensionless pressure rise ($\Delta p = (p'_b - p'_a)/(\rho_p \, \nu_0 \, g \, L)$), where $\dot{m}$ is the mass flow rate, ρ_p is the particle density, r_p is the radius of the pipe, g is the acceleration due to gravity, and $\nu_0 = 0.64$ is the solids fraction corresponding to the random close packing of uniform spheres. Let us first discuss the case $\Delta p = 0$. The flow rate obtained using the bare hopper, shown by the square in Fig. 1.10b, is about seven times *lower* than the value obtained using the hopper–standpipe combination.

The reason for the enhanced flow rate in the latter case can be understood by examining the profiles of the pressure of the interstitial air. Fig. 1.11a shows the profile

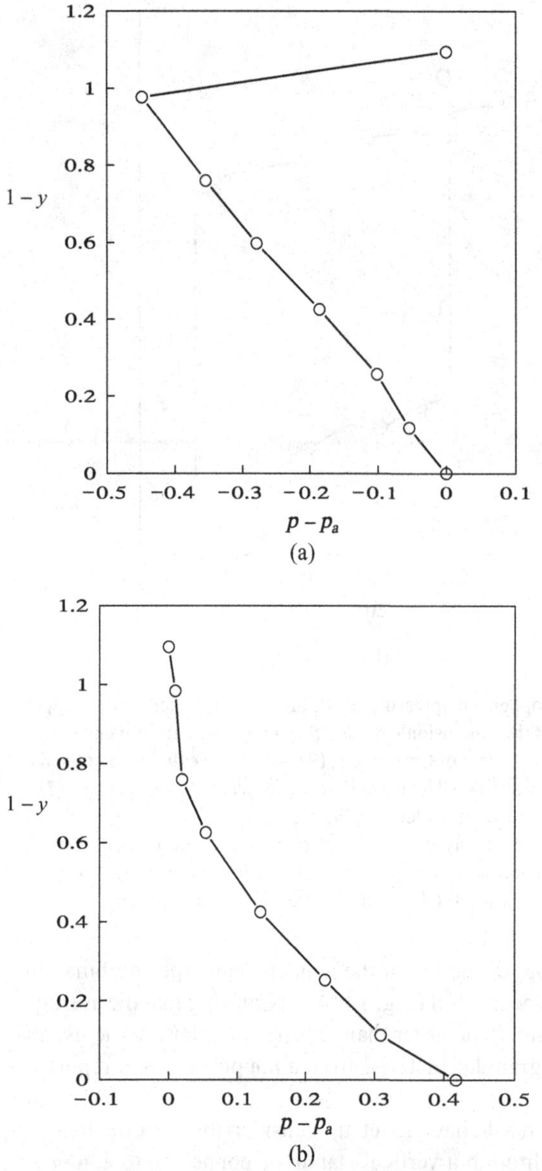

Figure 1.11. Profile of the dimensionless pressure p of the interstitial air for (a) the point on the upper branch of the curve shown in Fig. 1.10b, with $\Delta p \equiv p - p_a = 0$ and (b) the point on the lower branch of the curve with $\Delta p = 0.4$: ○, data of Chen et al. (1984); —, curve drawn to guide the eye. Here p_a is the dimensionless air pressure at the top of the hopper, $y = y'/L$, y' is the vertical coordinate measured as shown in Fig. 1.10a, and L is the length of the standpipe.

obtained by measuring the pressure of the interstitial air at various points along the wall of the pipe during the discharge of sand. The dimensionless pressure at $y = 0$, the junction between the standpipe and the hopper, is less than the atmospheric pressure p_a, whereas it is equal to p_a at the top of the hopper. Hence the pressure gradient exerted by the air on the particles aids gravity within the hopper and opposes it within the standpipe. In contrast, just above the exit slot of a bare hopper, there is an adverse pressure gradient of the interstitial air (Spink and Nedderman, 1978). Hence the material discharges faster from the hopper section of the hopper–standpipe system than from a bare hopper. Now consider the pipe, which is equivalent to a hopper with vertical walls. As discussed in §3.3, the pipe is likely to discharge material at a higher rate than the hopper. If this enhancement is larger than the reduction caused by the adverse pressure gradient of the air in the pipe, the flow rate will be limited by the rate at which material can flow from the hopper section. The net result is that the flow rate is likely to be higher for the hopper–standpipe system than for the bare hopper.

Along the upper branch DA (Fig. 1.10b), there is a packed bed in the hopper and a suspension in the upper part of the pipe. For certain parameter values, the suspension may be bounded from below by a packed bed (Chen et al., 1984). As Δp increases, the flow rate decreases, in accord with our intuition that an adverse pressure gradient should retard the flow of particles. An attempt to increase Δp beyond $(\Delta p)_A$ causes an abrupt decrease in the flow rate. Thus the upper branch terminates at A. Along the lower branch BC, there is a packed bed in both the hopper and the pipe when the valve at the bottom of the pipe is fully open. The flow rate increases as Δp decreases. This branch terminates at C, and an attempt to decrease Δp below $(\Delta p)_C$ causes an abrupt increase in the flow rate. For $(\Delta p)_C \leq \Delta p \leq (\Delta p)_B$, there are two flow rates corresponding to each value of Δp, which bracket the flow rate for the bare hopper. The model of Chen et al. (1984) predicts three flow rates, but one of them lies on an unstable branch of positive slope in the V–Δp plane. Hence only two flow rates are observed experimentally.

At points on the lower branch, the flow rate is *less* than that for the bare hopper. This is because an adverse pressure gradient exists throughout the device (Fig. 1.11b), leading to a reduction in the flow rate.

Overall, the interaction of the particles with the interstitial fluid gives rise to a complex variation of the flow rate with the pressure rise.

(f) *Segregation of a mixture of granular materials in a rotating cylinder*
Consider a horizontal cylinder partly filled with a mixture of fine and coarse sand (Fig. 1.12a). Surprisingly, rotation of the cylinder causes the formation of bands in the axial direction (Fig. 1.12b,c). The dark and light bands contain more coarse sand and fine sand, respectively. This phenomenon of *axial segregation* was first reported by Oyama (1939). Experiments suggest that *radial segregation*, or stratification of the mixture in the radial direction, is a precursor to axial segregation (Donald and Roseman, 1962; Das Gupta et al., 1991; Nakagawa et al., 1997). Donald and Roseman (1962) have suggested the following mechanism for radial segregation. At any cross section of the cylinder, there is a thin shear layer adjacent to the free surface of the bed of granular material. Below this layer, the motion of the material can be approximated by rigid body rotation (Khakhar et al., 1997). In the shear layer, smaller or heavier particles can percolate downward through the gaps between the larger or lighter particles, leading to the formation of a core of the former type of particles. The experiments of Thomas (2000) show that radial segregation does not always occur. In an experiment with particles having a large size ratio, the coarse particles were found to be distributed uniformly across the cross section of the bed.

Assuming that radial segregation occurs, rotation of the cylinder can cause this pattern to become unstable, leading to the formation of bands in the axial direction. Das Gupta et al. (1991) suggest that the inclination of the free surface of the rotating bed must differ for fine and coarse particles if bands are to form. Overall, a simple qualitative explanation for axial segregation is lacking.

The preceding examples provide glimpses of the wide variety of flow behavior that is encountered in practice. Let us now briefly consider various forces acting between the particles.

1.2. INTERPARTICLE FORCES

1.2.1. Electrostatic and van der Waals Forces

Consider two spherical particles A and B, which are separated by a distance S (Fig. 1.13) and are at rest in air. In addition to the external force exerted on A to keep it at rest, the other

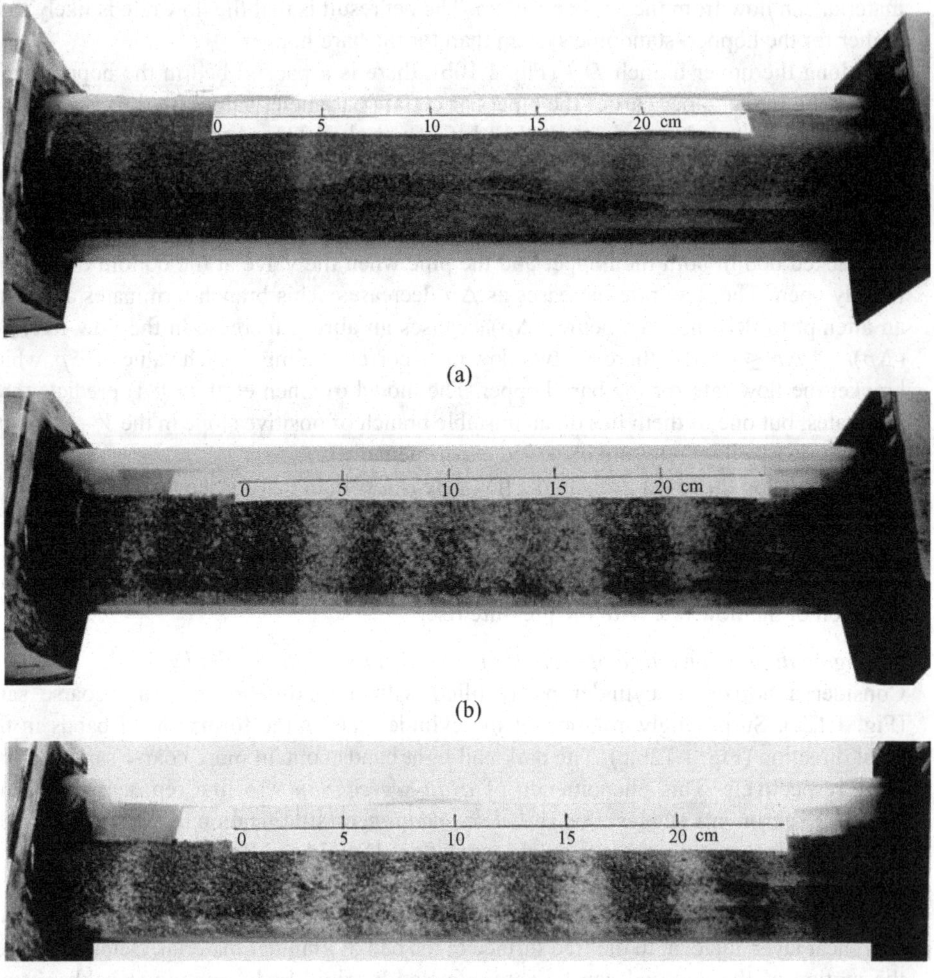

Figure 1.12. Formation of bands in a horizontal glass cylinder of inner diameter 0.075 m and length 0.55 m. The cylinder was partly filled with a mixture of fine sand (particle diameter = 1.0–1.6 mm, weight fraction = 0.58) and coarse sand (particle diameter = 0.2–0.5 mm), such that the total mass of sand was 0.885 kg. The photographs were taken after the cylinder had been rotated about its axis by hand for n revolutions in time t_r and then brought to rest: (a) $n = 89$, $t_r = 74$ s; (b) $n = 93$, $t_r = 44$ s; (c) $n = 52$, $t_r = 33$ s. The angular velocity of the cylinder was approximately in the range 6–13 rad s^{-1} (1–2 revolutions per second). This experiment was set up by Mr. P. T. Raghuram.

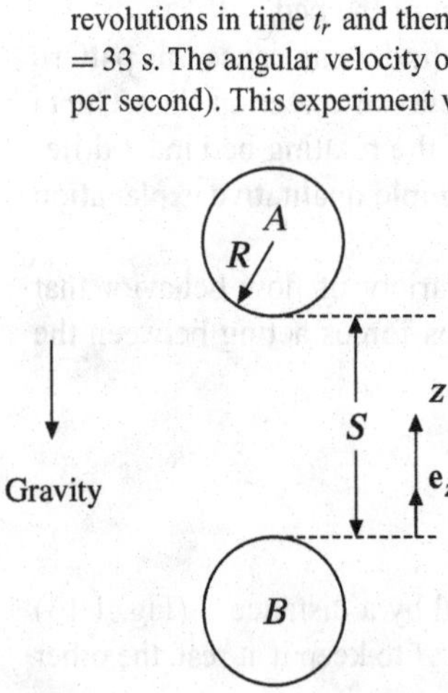

Figure 1.13. Two identical spherical particles of radius R, which are separated by a distance S in air. In addition to gravity, external forces are imposed to prevent the particles from moving. Here $\mathbf{e}_z$ is a unit vector in the z direction.

forces acting on A are its weight $\mathbf{F}_w$, buoyancy $\mathbf{F}_b$, and the interparticle force between A and B. The last involves several types of forces, such as (i) the gravitational force between A and B, which can be neglected for typical grain sizes, (ii) the electrostatic or Coulomb force $\mathbf{F}_c$ between charged particles, and (iii) the van der Waals force $\mathbf{F}_v$ between A and B. Both (ii) and (iii) are examples of *electromagnetic* forces.

It is helpful to compare the magnitudes of these forces. The weight and buoyancy are given by

$$\mathbf{F}_w = -\rho_{\mathrm{p}}\, g\, (4\,\pi\, R^3/3)\, \mathbf{e}_z; \quad \mathbf{F}_b = \rho_{\mathrm{f}}\, g\, (4\,\pi\, R^3/3)\, \mathbf{e}_z \tag{1.1}$$

where ρ_{p} is the particle density, g is the acceleration due to gravity, R is the radius of the particle, ρ_{f} is the density of the fluid, and $\mathbf{e}_z$ is a unit vector in the z direction (Fig. 1.13). The Coulomb force exerted by B on A is given by (Halliday and Resnick, 1969, p. 653; Israelachvili, 1992, p. 33; Griffiths, 2002, p. 77)

$$\mathbf{F}_c = \frac{q_A\, q_B}{4\,\pi\, \epsilon_0\, \epsilon_*\, (S + 2\,R)^2}\, \mathbf{e}_z \tag{1.2}$$

where q_A and q_B are the electric charges on A and B, respectively, ϵ_0 is the permittivity of free space, ϵ_* is the relative permittivity or dielectric constant, and $S + 2\,R$ is the distance between the *centers* of the spheres. In SI units, $\epsilon_0 = 8.85 \times 10^{-12}\ \mathrm{C}^2\ \mathrm{N}^{-1}\ \mathrm{m}^{-2}$ and $1/(4\,\pi\,\epsilon_0) \approx 9 \times 10^9\ \mathrm{N\,m}^2\ \mathrm{C}^{-2}$.

Consider the van der Waals force for the special case of two molecules. This force consists of the orientation, induction, and dispersion contributions, which arise as follows (Israelachvili, 1992, pp. 62, 75, 83). The time-averaged charge distribution around a molecule can be either symmetric or asymmetric. The former corresponds to a *nonpolar* molecule and the latter to a *polar* molecule. For a polar molecule such as carbon monoxide, the charge distribution is equivalent to two unlike point charges equal in magnitude and separated by some distance. This is called a (permanent) *dipole*. Interactions between dipoles of two polar molecules give rise to the *orientation* or *Keesom* contribution. If a polar molecule is near a nonpolar molecule such as methane or argon, the dipole of the former induces a dipole in the latter, and the interaction between these dipoles provides the *induction* or *Debye* contribution. Let us now consider pairs of nonpolar molecules. Even though these molecules do not have dipoles, they have instantaneous dipoles owing to the fluctuating charge distribution. The instantaneous dipole of one molecule induces an instantaneous dipole in the other molecule, and vice versa. The interaction between these dipoles provides the *dispersion* contribution. The induction and dispersion contributions are nonzero even for a pair of polar molecules (Hirschfelder et al., 1964, p. 983).

For macroscopic bodies, the interactions between the molecules in the two bodies must be suitably summed to obtain the total van der Waals force $\mathbf{F}_v$. When both bodies are made of the same material, the van der Waals force is *attractive*. For two identical spheres of radius R separated by a distance S (Fig. 1.13), the force exerted by B on A is given by (Israelachvili, 1992, p. 177)

$$\mathbf{F}_v \approx -\frac{A\,R}{12\,S^2}\, \mathbf{e}_z, \quad S/R \ll 1 \tag{1.3}$$

where A is a constant called the Hamaker constant. For particles in air, the typical range of A is 10^{-20}–10^{-19} J. Equation (1.3) is not valid for $S > 5$ nm, as the force decays more rapidly than assumed in the derivation of (1.3). It is also not valid in the limit $S \to 0$, as it predicts an infinite force. In practice, the overlapping of electron clouds of the atoms on the surfaces of the two particles results in a repulsive force. This ensures that S remains nonzero. For smooth particles in "contact" with each other, S is of the order of atomic dimensions, i.e., a few Å. (1 Å = 0.1 nm.)

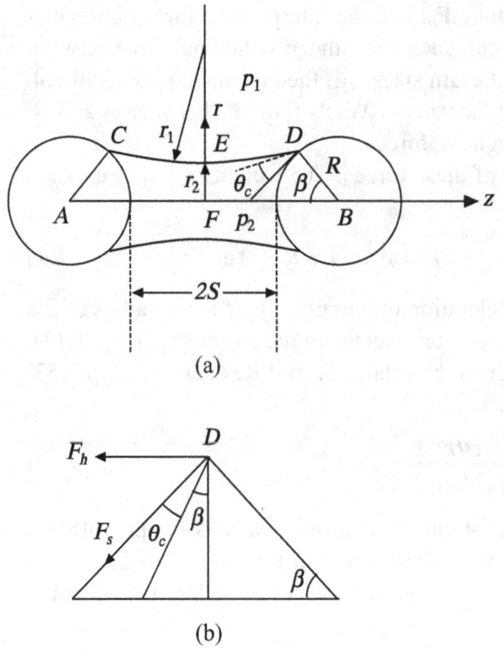

Figure 1.14. (a) A liquid bridge between two spherical particles, each of radius R. Here p_1 is the pressure of the air outside the bridge, p_2 is the pressure of the liquid in the bridge, θ_c is the contact angle between the air–liquid interface and the surface of the particle, and β is the filling angle. (b) The horizontal component F_h of the force due to surface tension F_s, which is exerted at the point D on the particle B.

Let us calculate the magnitudes of the preceding forces for two spherical particles of radius $R = 50\ \mu\text{m}$, with a particle density $\rho_p = 2{,}650\ \text{kg m}^{-3}$, separation $S = 1$ nm, charge density $= 10^{-5}\ \text{C m}^{-2}$, and Hamaker constant $A = 10^{-19}$ J. For air, the dielectric constant $\epsilon_* = 1$. The weight is 1.4×10^{-8} N, which is much less than the Coulomb force (8.9×10^{-8} N) and the van der Waals force (4.2×10^{-7} N). For larger spheres with $R = 500\ \mu\text{m}$, the weight is 1.4×10^{-5} N, which exceeds the Coulomb force (8.9×10^{-6} N) and the van der Waals force (4.2×10^{-6} N). In general, the weight exceeds the other two forces for large particles. Seville et al. (2000) note that if the particles are rough, the effective value of R to be used in (1.3) will depend more on the radius of the asperities on the surfaces of the particles rather than the nominal radius of the particles. In such a case, the van der Waals force is likely to be smaller than that estimated using (1.3).

When the air contains moisture, an adsorbed layer of water molecules forms on the surfaces of the particles at a low humidity. At a high humidity, condensation of water leads to the formation of a *liquid bridge*, provided the particles are not too far apart. The force associated with the liquid bridge is discussed below.

1.2.2. Liquid Bridge or Capillary Forces

A liquid bridge is a layer of liquid connecting two particles (Fig. 1.14a). Considering particle B, the surface tension of the air–liquid interface exerts a horizontal force F_h to the left. Let γ_s denote the surface tension of the liquid, β the filling angle, and θ_c the contact angle between the interface and the surface of the particle. Then F_h is given by the product of the horizontal component of the surface tension $\gamma_s \sin(\beta + \theta_c)$ and the length of the contact line $2\pi R \sin\beta$ (Lian et al., 1993) (see Fig. 1.14b). Thus surface tension causes an *attractive* force between the particles.

In addition, the curvature of the interface causes the pressure p_2 in the liquid to differ from the pressure p_1 of the ambient air (Fig. 1.14a). (The pressure varies within the bridge on account of gravity, but this can be neglected if p_1 is not too low.) The pressure difference $\Delta p \equiv p_1 - p_2$ causes an additional horizontal force to be exerted on the particles. This force is given by the product of Δp and the projected area $\pi\,(R \sin\beta)^2$ associated with the

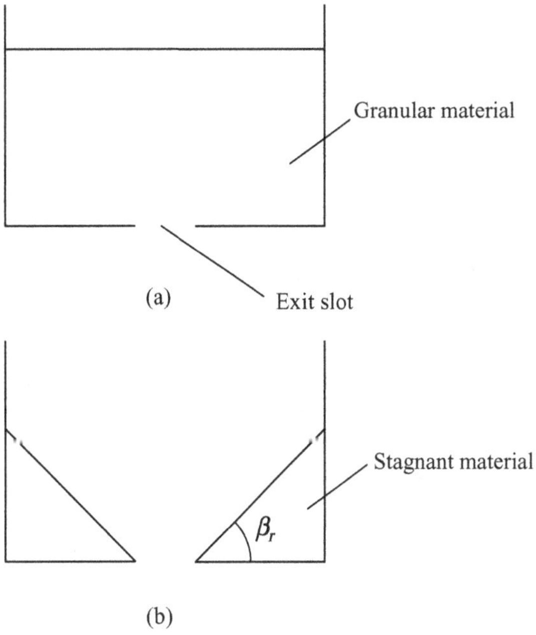

Figure 1.15. (a) Granular material filled in a bin. (b) Wedges of material which remain in the bin after flow stops. The angle of repose β_r is the inclination of the sloping face of the wedge to the horizontal.

three-phase contact line (Lian et al., 1993). For stable liquid bridges, the numerical results of Lian et al. (1993) suggest that Δp is usually >0, leading to an *attractive* force between the particles.

The total force $\mathbf{F}_l$ exerted by the liquid bridge on B is given by

$$\mathbf{F}_l = -2\pi\, R\, \gamma_s\, \sin\beta \left(\sin(\beta + \theta_c) + \frac{R\,\Delta p\, \sin\beta}{2\,\gamma_s}\right) \mathbf{e}_z \tag{1.4}$$

where $\mathbf{e}_z$ is a unit vector in the z direction (Fig. 1.14a).

The pressure difference Δp must be obtained by solving the Young–Laplace equation (1.68) numerically. However, an approximate analytical solution can be obtained by assuming that the meridian curve CD in Fig. 1.14a is a circular arc (see Problem 1.2). This approximation is due to Fisher (1926) (cited in Lian et al., 1993). For the special case of a zero contact angle ($\theta_c = 0$), the approximate solution shows that the maximum value of $F_l \equiv |\mathbf{F}_l|$ is obtained in the limit of particles in contact, with $\beta \to 0$ (Lian et al., 1993). The magnitude of the limiting value is given by $F_{l,\,\max} = 2\pi\, R\, \gamma_s$. For water at 20°C ($\gamma_s = 0.073$ N m^{-1}), a particle radius of 50 μm, and a particle density of 2,650 kg m^{-3}, $F_{l,\,\max} = 2.3 \times 10^{-5}$ N, which is about 1,700 times larger than the weight of the particle. For large particles, the weight exceeds F_l.

The ratio of $F_{l,\,\max}$ to the weight defines a modified *Bond* number (Nase et al., 2001)

$$Bo \equiv F_{l,\,\max}/|\mathbf{F}_w| = (3\,\gamma_s)/(2\,\rho_p\, g\, R^2) \tag{1.5}$$

Even though this number has been introduced for a single liquid bridge between a pair of particles, it may be expected that if $Bo \gg 1$, liquid bridges are likely to strongly influence the mechanical behavior of wet granular materials. This conjecture is supported by the experiments of Nase et al. (2001), which are discussed below.

Suppose that a granular material is filled in a bin and then allowed to discharge slowly through the exit slot (Fig. 1.15a). After the material stops flowing, wedges of stagnant material remain at the bottom of the bin (Fig. 1.15b). The angle β_r is called the *drained angle of repose*. Nase et al. (2001) used glass beads in the size range 0.5–10 mm and varied the surface tension in the range 0.038–0.073 N m^{-1} by adding surfactants to water. For

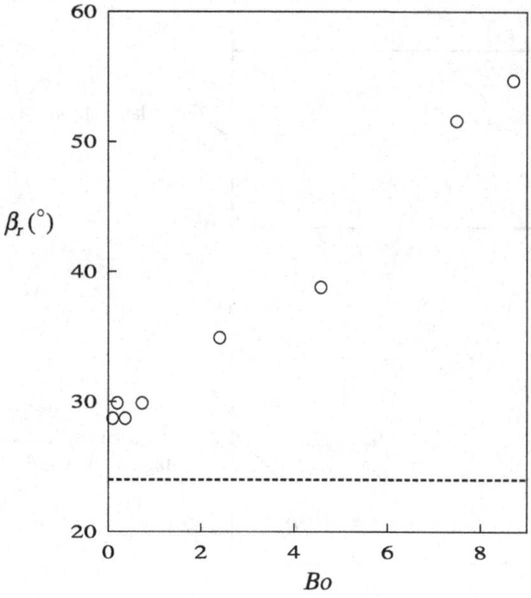

Figure 1.16. Variation of the angle of repose β_r with the Bond number $Bo = (3\,\gamma_s)/(2\,\rho_p\,g\,R^2)$ for the discharge of glass beads through a bin of rectangular cross section. Here γ_s is the surface tension, ρ_p is the particle density, g is the acceleration due to gravity, and R is the particle radius. The wetting liquid is water containing surfactants. The broken line represents the angle of repose of dry beads. (Adapted from Fig. 5 of Nase et al., 2001.)

$Bo < 1$, the angle of repose β_r is slightly higher than that for a dry material (Fig. 1.16). However, it increases sharply with Bo for $Bo > 1$. Even though the volume fraction of the liquid is only about 0.68%, its presence strongly affects the slope of the wedge for $Bo > 1$.

A possible explanation for the increase of β_r with Bo (P. A. Kralchevsky, private communication, 2002) will now be discussed. Consider the forces acting on a particle located at the free surface of the heap. Its weight can be resolved into forces acting normal and tangential to the surface of the heap. An additional normal force acts owing to the liquid bridges connecting it to the particles below. This increases the normal force exerted by the particle on those below, and hence the frictional force opposing its downward motion increases. This in turn permits the heap to have a steeper slope than in the absence of the liquid.

1.2.3. Contact Forces

Consider the case where the spheres are in contact, bearing in mind that "contact" denotes a separation of the order of atomic dimensions between the surfaces of the particles. In general, each particle is subjected to *contact* forces in addition to its weight and buoyancy. These contact forces arise from electromagnetic interactions between the particles (Griffiths, 2002, p. 14). Unfortunately, it is not possible at present to calculate all the contact forces from first principles, and hence they are evaluated using phenomenological models. This applies for example to the normal and tangential forces exerted by granular materials on the walls of a storage vessel. Two types of contact forces, arising from *cohesion* and *friction*, are discussed below.

(i) *Cohesion*

Cohesion is the tendency of particles of the same material to stick together because of the action of interparticle attractive forces, such as the van der Waals force, liquid or solid bridges, Coulomb force between particles with unlike charges, chemical bonds, and sintering or agglomeration of particles. (In the literature on contact mechanics, the term *adhesion* is often used in place of cohesion. Here we follow Brown and Richards (1970, p. 5) and use the term adhesion to denote the tendency of particles to stick to a wall or substrate.) Suppose that two bodies made of the same material are in contact across an interface. The material is said to be *cohesive* if nonzero forces have to be applied in a direction *normal* to

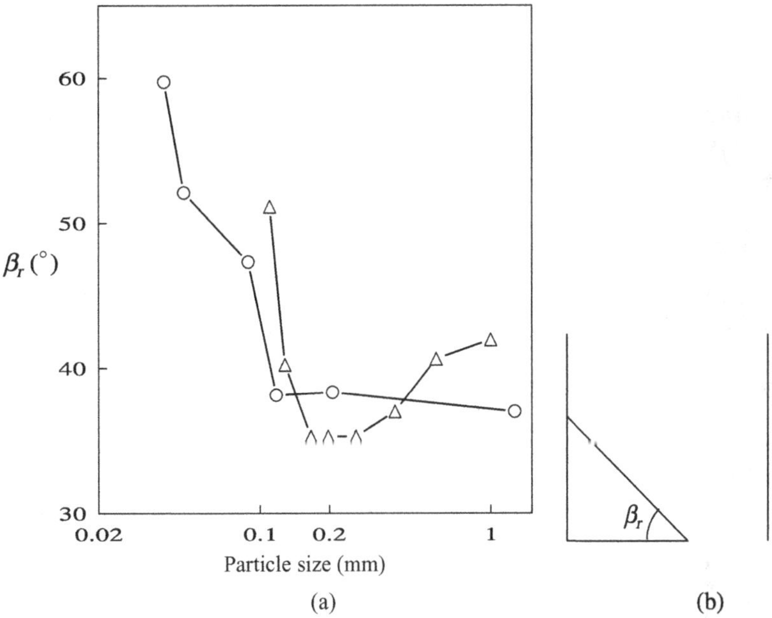

Figure 1.17. (a) Variation of the angle of repose β_r with the particle size for sand (Δ) and coal (○). (b) Bin used to determine β_r. (Adapted from Fig. 2.9 of Brown and Richards, 1970, p. 32.)

the interface to pull the bodies apart. Conversely, a *cohesionless* material is defined as one which cannot sustain tensile normal forces. Materials such as coarse sand and coarse glass beads (with a particle diameter greater than about 100 μm) can be regarded as cohesionless when they are dry, whereas materials such as clay, cement, toner powder for photocopiers (particle diameter $\approx$ 10 μm), and fine limestone powder are cohesive.

The net effect of cohesion is that particles in contact can sustain *tensile* normal forces up to some limiting value. As discussed earlier, cohesion becomes important for fine particles and also for wet granular materials. As a rule of thumb, the particle size below which cohesive effects are important for dry materials can be taken as 100 μm (Brown and Richards, 1970, p. 39). An example is provided by the the variation of the angle of repose β_r with particle size for sand and coal (Fig. 1.17a). (Here β_r is determined by allowing the material to discharge through a bin with an eccentric exit slot, as shown in Fig. 1.17b.) For sizes below 100 μm, β_r increases sharply, indicating the increasing importance of cohesive effects. As shown by the curve for sand, β_r varies with particle size in a complicated manner. At present, it is not possible to predict this variation from first principles. (For smooth spherical particles, Albert et al. (1997) have developed a simple model for β_r, which is discussed in Problem 1.9)

In general, the handling and storage of cohesive materials is fraught with difficulties. For example, cohesive forces impede the flow of fine granular materials from storage vessels. In extreme cases, they lead to the phenomenon of *arching*, i.e., the formation of a stable "arch" of material at the exit region of storage vessels. Figure 1.18 shows an arch formed by filling a funnel with chalk powder. There is no flow until the arch is destroyed by mechanical or other means.

Another problem with cohesive materials is that their mechanical properties tend to be *time dependent*. They often gain strength, i.e., the cohesive force increases, as the contact time between the particles increases. This is called *ageing*, and is caused by various factors such as the escape of entrained air, rupture of particles leading to a larger contact area, external vibration leading to a denser packing, migration of water, and evaporation of water

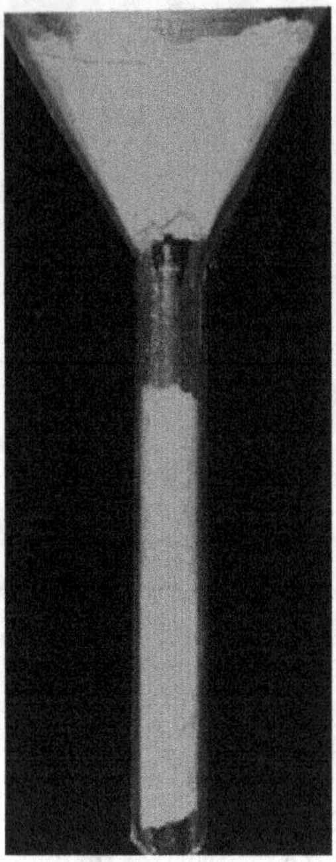

Figure 1.18. Arches in a funnel filled with chalk powder. Inner diameter of the stem = 5 mm.

along with precipitation of dissolved salts, leading to the formation of solid bridges between particles (Jenike, 1964b, p. 24).

An example of ageing is provided by the experiments of Bocquet et al. (1998). A horizontal cylinder partly filled with glass beads is slowly rotated about its axis (Fig. 1.19). The inclination of the free surface of the material to the horizontal increases as the cylinder rotates. Eventually, the free surface becomes steep enough to trigger an avalanche of material. The inclination β_r of the surface at this stage of the experiment is taken as the angle of repose.

The angle β_r was measured after exposing the beads in the stationary cylinder to moist air for a time t and then rotating the cylinder. For fine particles with a diameter <250 μm, $\sin\beta_r$ was found to vary as $\log t$ for t in the range 5–7,000 s. Further, β_r increased with the humidity of the air. For particles with a diameter >500 μm, no ageing was observed except at a high humidity. Bocquet et al. (1998) attribute ageing to the condensation of moisture and formation of liquid bridges in the contact zones between particles.

(ii) *Friction*

Suppose that two bodies are in contact across an interface. *Friction* is the tendency of the interface to resist relative motion of these bodies in any direction that is *tangential* to the interface. The resistance can arise in various ways. Experiments with metals sliding on metals (Bowden and Leben, 1939; Bowden and Tabor, 1939; Bowden et al., 1943) suggest that the *asperities* or projections on the surfaces of the bodies adhere to form junctions. Work must be done to deform and break these junctions, and this is accompanied by *wear* or erosion of material in the interfacial region. Additional work is associated with deformation of the material in a larger region near the interface. This mechanism is called

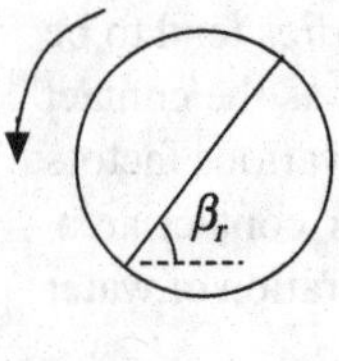

Figure 1.19. A rotating cylinder partly filled with a granular material. The inclination β_r of the free surface when the first avalanche is observed is taken as the angle of repose.

plowing (Bowden et al., 1943), by analogy with a plow being dragged across the soil, forcing the soil to be displaced from the path of the plow. As noted by Briscoe (1982), the plowing component can also be important when one body rolls over the other. Experiments with mica surfaces, which are flat over regions containing several million atoms, suggest that friction can occur even in absence of wear (Krim, 1996, 2002). In this case, the frictional resistance is believed to arise because some of the applied mechanical energy is converted to heat, i.e., the vibration of the atoms near the interface. Because of the inherent roughness of the surfaces of most granular materials, the first two mechanisms, namely, adhesion and plowing, are likely to be more relevant.

1.2.4. Interparticle Forces in Saturated Granular Materials

When the voids between the particles are completely filled with a liquid, there are no liquid bridges. On the other hand, the van der Waals and Coulomb forces are present, but the liquid often weakens their effect relative to dry systems (Israelachvili, 1992, pp. 38, 101). As in the case of dry materials, it will be assumed that saturated materials can also be classified as cohesive or cohesionless. In general, coarse materials can be assumed to be cohesionless.

A material which is cohesionless when it is either dry or wet can become cohesive when it is partially saturated. This can be seen by filling a conical hopper with coarse glass beads whose particle diameter is about 1 mm. If the exit slot of the hopper has a diameter of about 20 mm, the material flows freely when it is dry, and also when the hopper is submerged in water. (Experiments involving underwater hoppers are discussed in Zeininger and Brennen (1985).) On the other hand, suppose that the hopper is filled with dry beads and just enough water is added to fill the voids between the beads. When the hopper is allowed to discharge in air, the beads do not flow freely.

1.3. PACKING CHARACTERISTICS

Before formulating models, it is helpful to introduce some terms related to the packing of particles and to discuss briefly different types of packings.

Consider two beakers resting on a table, one filled with water and the other with a granular material. The density of water is independent of the manner in which the beaker is filled. On the other hand, the volume occupied by the granular material (and hence its density) can be changed within certain limits by various means, such as allowing the material to flow, shaking the beaker, or tapping it. Thus unlike water, a granular material does not have a unique density, but a range of densities. This is illustrated by the experiment shown in Fig. 1.20. The beaker contains a fine powder. When the powder is poured into a hopper and then allowed to discharge into the beaker, the volume occupied by the powder increases as it is now loosely packed.

The *particle density* ρ_p is the density of the solid material forming the particles. Assuming that the particles are nonporous, the *bulk density* ρ_b is defined by

$$\rho_b \equiv \rho_p \, \nu + \rho_f (1 - \nu) \tag{1.6}$$

where ρ_f is the density of the interstitial fluid and ν is the *solids fraction* or the volume of particles per unit total volume. If the interstitial fluid is a gas, $\rho_p/\rho_f \approx 10^3$, and hence $\rho_b \approx \rho_p \, \nu$. A quantity related to ν is the *porosity* or *voids fraction* ϵ, defined by

$$\epsilon \equiv 1 - \nu \tag{1.7}$$

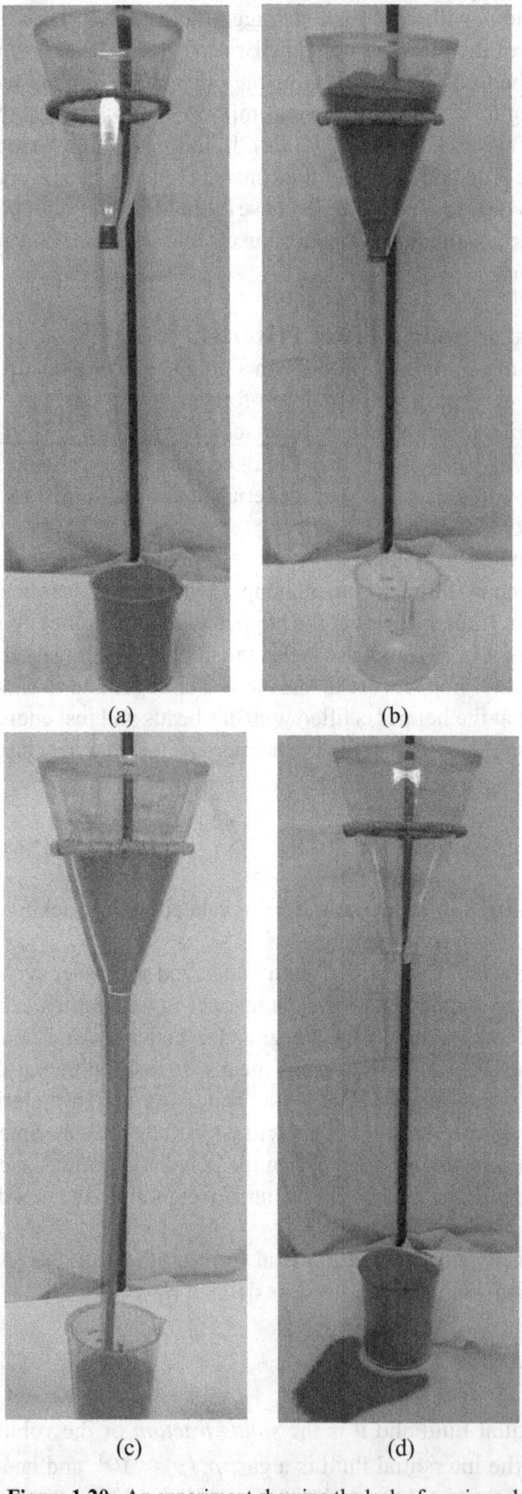

Figure 1.20. An experiment showing the lack of a unique density for a granular material (sand): (a) sand at rest in a beaker, (b) poured from the beaker into a hopper, (c) allowed to discharge from the hopper into the beaker, and (d) the state at the end of the experiment. In (d), sand overflows from the beaker as its density is lower than that in the initial state in (a). Parameter values: particle diameter of sand = 0.14–0.22 mm, orifice diameter of the hopper = 21.1 mm, wall angle of the hopper ≈ 16°.

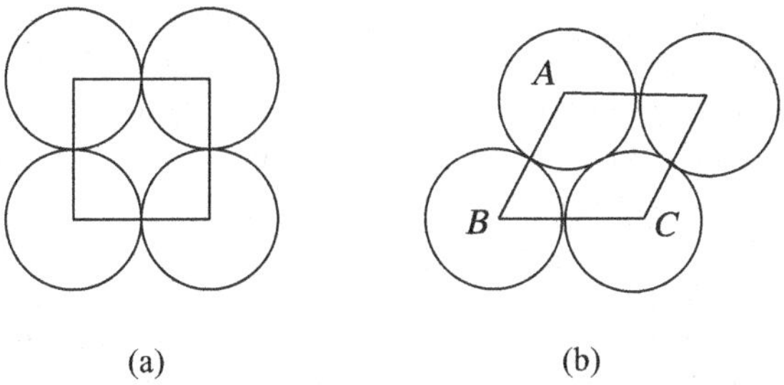

Figure 1.21. Packing of a monolayer of spherical particles to form (a) a square layer and (b) a simple rhombic layer. In (b), $\angle ABC = 60°$.

Thus ϵ is the volume fraction of the interstitial fluid. Finally, the density of the particle phase, i.e., the mass of particles per unit total volume, is defined by

$$\rho \equiv \rho_{\mathrm{p}}\, \nu \tag{1.8}$$

Henceforth, ρ will be called the density of the granular material.

In the theories discussed in this book, the voids fraction $1 - \nu$ is based on the voids *between* the particles, ignoring any voids *within* the particles. Hence for a porous particle, the particle density ρ_{p} represents the effective density of the particle.

With reference to continuum models, the solids fraction ν is the most important feature of packings. In the case of discrete and statistical models, additional features such as the *coordination number* or number of nearest neighbors, the radial distribution of particles around any chosen particle, and the orientation of contact normals are also of importance. Here we confine attention to the values of ν and the coordination number. Two types of packings have been examined extensively in the literature: (i) *regular* packings and (ii) *random* packings. Let us first discuss regular packings, confining attention to spheres of uniform size.

1.3.1. Regular Packings

Two important building blocks of regular packings are *square* layers and *simple rhombic* layers (Fig. 1.21). Starting with a square layer, a packing can be generated by stacking square layers on top of each other. Graton and Fraser (1935) note that this can be done in three ways, leading to three different packings. Similarly, three packings can be obtained by stacking simple rhombic layers on top of each other. Some properties of these packings are listed in Table 1.1. The cubic packing has the lowest value of ν (0.52) and also the lowest

Table 1.1. Properties of Regular Packings of Spheres of Radius R†

(a) *Square layers*			
Name	Cubic	Orthorhombic	Rhombohedral
Spacing of layers	$R\sqrt{4}$	$R\sqrt{3}$	$R\sqrt{2}$
Coordination number	6	8	12
Solids fraction	0.52 ($=\pi/6$)	0.60	0.74 ($=\pi/(3\sqrt{2})$)
(b) *Simple rhombic layers*			
Name	Orthorhombic	Tetragonal–sphenoidal	Rhombohedral
Spacing of layers	$R\sqrt{4}$	$R\sqrt{3}$	$2R\sqrt{2/3}$
Coordination number	8	10	12
Solids fraction	0.60	0.70	0.74 ($=\pi/(3\sqrt{2})$)

Adapted from Graton and Fraser (1935).

† The rhombohedral packing is also referred to as the face-centered cubic packing in the case of square layers, and as hexagonal close packing in the case of simple rhombic layers.

coordination number (6). The rhombohedral packing has the highest value of ν (0.74) and the highest coordination number (12). It may be expected that both regular and random packings of spheres of uniform size will have solids fractions and coordination numbers between these extremes. This is in accord with the available experimental evidence (Scott, 1960; Brown and Richards, 1970, pp. 16–17; Onoda and Liniger, 1990).

1.3.2. Random Packings

Berryman (1983) suggests that a random packing should contain no significant short-range or long-range order. Given a packing, there is no simple way to check whether these conditions are met. Let us consider the issue from a practical viewpoint. Scott (1960) found that when steel ball bearings were poured into containers, the measured values of ν fell between two limits. These limits were called *dense random packing* (DRP) and *loose random packing* (LRP). The DRP limit was obtained by shaking the container gently after filling and corresponded to a solids fraction $\nu_{drp} \approx 0.64$. The LRP limit was obtained as follows. A cylinder filled with granular material was placed with its axis horizontal and slowly rotated about its axis. It was then placed with its axis vertical and the solids fraction was measured. This gave $\nu_{lrp} \approx 0.60$. The rotation causes the material to shear and dilate, and hence the solids fraction is likely to be lower than it is in the experiment used to obtain a dense packing.

Subsequent measurements of Scott and Kilgour (1969) and also the computer simulations of hard-sphere fluids (Berryman, 1983) have corroborated the preceding estimate of ν_{drp}. On the other hand, Pouliquen et al. (1997) have conducted experiments by pouring spheres into a container that was shaken in the horizontal direction. At low flow rates of particles and large amplitudes of the shaking motion, a solids fraction of 0.67 was obtained in some cases. Thus the value of ν_{drp} is not uniquely determined. Indeed, Torquato et al. (2000) argue that it is not possible to define DRP precisely, as "randomness" and "dense packing" are opposing tendencies.

Consider LRP, which is also not a precisely defined state. Onoda and Liniger (1990) suggest that LRP is the loosest random packing which is mechanically stable. They conducted sedimentation experiments with glass spheres in liquids of various densities, and also sheared these suspensions in an annular shear cell. It was found that in the limit of vanishing density difference between the solid material and the fluid, i.e., $\rho_p - \rho_f \to 0$, $\nu_{lrp} \approx 0.56$. This limit corresponds to a vanishing effective body force acting on the particles. It may be expected that $\nu_{lrp} > 0.56$ for $\rho_p - \rho_f > 0$, as the body force would tend to compact the material; this is in accord with their experiments. Scott's experiments, which were cited earlier, used air as the interstitial fluid. This may be one of the factors resposible for the higher vaule of ν_{lrp} ($\approx$0.6) compared to the value reported by Onoda and Liniger (1990).

The ranges of the solids fraction ν obtained for various types of packings are shown in Fig. 1.22. All the lines represent experimental data, except for the line labeled 3, which corresponds to regular packing of uniform spheres. As shown by the line labeled 1, materials composed of irregularly shaped grains have a wider range of ν values than those composed of spherical grains. The lines labeled 4 and 5 represent fine cohesive materials, for which loose arrangements of particles are stabilized by the action of interparticle forces. Hence the values of ν can be much lower than those for cohesionless materials.

1.4. MODELS

Consider how granular materials can be modeled. The available models can be broadly classified into two groups: (i) *discrete* models and (ii) *continuum* models. The former treat the material as a collection of particles, whereas the latter treat it as a continuous medium. For the typical particle size range of interest, say, 0.1–10 mm, the materials appear discrete

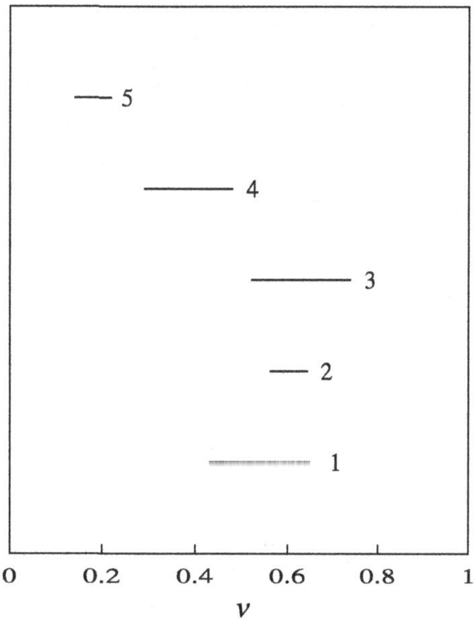

Figure 1.22. Ranges of the solids fraction ν for various packings: 1, random packing of closely graded irregular particles (two grades of glass beads of mean sizes 0.27 mm and 1.1 mm, sand (0.55 mm), tapioca (1.51 mm), and five grades of coal of mean sizes 0.12, 0.24, 0.48, 0.63, and 1.35 mm); 2, random packing of closely graded spherical granules (lead shot (3 mm), oil-coated steel balls (3.15 mm), glass beads (3.2 mm), and polystyrene beads (3.05 mm)); 3, regular packing of uniform spheres; 4, cement (5–15 μm); 5, TiO_2 (0.5 μm). The numbers in brackets represent mean sizes, except in the case of cement. (Adapted from Brown and Richards, 1970, pp. 16–17, 22, and Onoda and Liniger, 1990.)

to the eye. So it may seem evident that discrete models are more realistic than continuum models. This is a reasonable premise. However, imperfect knowledge of contact forces between particles and practical limitations of computing time constrain discrete models to be less realistic than they appear at face value. Despite this drawback, many of the particle-scale attributes such as shape, size distribution, and deformation characteristics can be incorporated more easily into these models than into continuum models. Further, flow in systems with complicated geometries can be readily examined. On the other hand, the simulation of solids handling systems which usually contain a very large number of particles is currently beyond the reach of discrete models, and continuum models have been used in such cases. In contrast to continuum models, it is difficult to construct approximate analytical or semi-analytical solutions to discrete models for flow problems. Thus neither class of models is superior to the other in all respects. This book is largely confined to continuum models.

1.4.1. Discrete Models

The application of discrete models to problems of granular flow appears to have been initiated by Cundall (1971, 1974) (cited in Cundall and Strack, 1979). Here Newton's laws are applied to each particle, and its motion is followed in time. Particles are permitted to overlap, but this is resisted by normal and shear forces which depend on the extent of overlap. Thus the material is idealized as a collection of "soft" particles with a repulsive core. The preceding approach has been termed the *distinct element method* (DEM) by Cundall; it is also called the *discrete element method* in the literature.

Considering spherical particles, the linear and angular momentum balances for the ith particle are given by (Iwashita and Oda, 1998; Zhou et al., 1999)

$$m_{(i)}\frac{\mathrm{d}\mathbf{v}_{(i)}}{\mathrm{d}t} = m_i\,\mathbf{b} + \sum_{j=1}^{k_i}\mathbf{F}_{ij}$$

$$I'_{(i)}\frac{\mathrm{d}\boldsymbol{\omega}_{(i)}}{\mathrm{d}t} = \sum_{j=1}^{k_i}(\mathbf{T}_{ij} + \mathbf{M}_{ij}) \qquad (1.9)$$

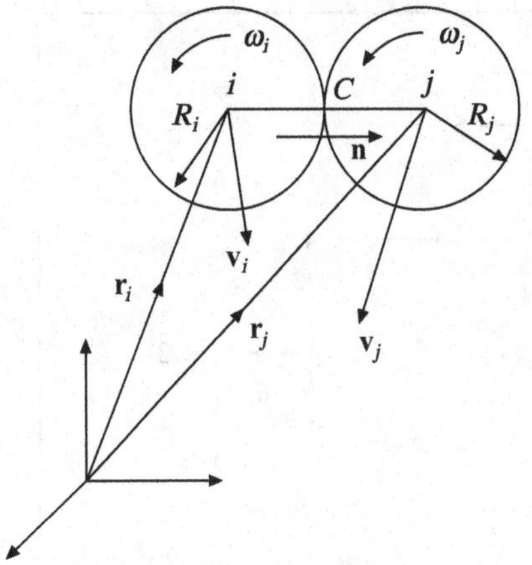

Figure 1.23. Two spherical particles i and j in contact at the point C. Here $\mathbf{v}_i$ is the velocity of the center of mass of particle i, $\boldsymbol{\omega}_i$ is its angular velocity, and $\mathbf{n}$ is a unit vector directed along the line joining the centers of the particles.

where m_i, $\mathbf{v}_i$, I'_i, and $\boldsymbol{\omega}_i$ are the mass, linear velocity of its center of mass, moment of inertia, and angular velocity, respectively, of the particle i, $\mathbf{b}$ is the body force per unit mass, $\mathbf{F}_{ij}$ is the force exerted on particle i by a particle j which is in contact with it, k_i is the number of particles in contact with particle i, $\mathbf{T}_{ij}$ is the torque exerted on particle i due to the tangential component of the contact force between particles i and j, and $\mathbf{M}_{ij}$ is the rolling friction torque exerted by particle j on particle i. The brackets around the indices i and j indicate that the summation convention (see Appendix A) does not apply. The torque $\mathbf{T}_{ij}$ is given by

$$\mathbf{T}_{ij} = (R_{(i)}\mathbf{n}) \times \mathbf{F}_{(i)j}$$

where R_i is the radius of sphere i and $\times$ denotes the cross product of two vectors (see §A.7.1).

The linear velocity is measured with respect to a coordinate system which is at rest relative to the surface of the earth. (Moving coordinate systems can also be used, provided the body force is modified suitably (Problem G.4).) On the other hand, the angular velocity $\boldsymbol{\omega}_i$ is measured with respect to a coordinate system whose origin always coincides with the center of mass of particle i and whose orientation is independent of time. Even though $\boldsymbol{\omega}_i$ is denoted by a curved arrow in Fig. 1.23, it is actually a vector directed along the axis of rotation. Following the usual convention (Resnick and Halliday, 1966, p. 250), the vector is directed out of the plane of the figure for rotation in the anticlockwise direction. This convention corresponds to the right-hand rule, i.e., if the fingers of the right hand are curled in the direction of rotation, the thumb points in the direction of $\boldsymbol{\omega}_i$.

The rolling friction torque $\mathbf{M}_{ij}$ is believed to arise mainly because of hysteresis losses associated with the deformation of the particles during rolling (Tabor, 1955). The need for including this term is shown by the following numerical experiment (Zhou et al., 1999). Suppose that a sphere is dropped onto a flat surface with a zero angular velocity and a nonzero linear velocity parallel to the surface. If the rolling friction torque is omitted in (1.9), calculations show that the sphere moves indefinitely. Hence it is important to include this term in problems such as the formation of a heap. Expressions for $\mathbf{M}_{ij}$ are given in Zhou et al. (1999).

To complete the model, expressions must be specified for the contact forces in the normal and tangential directions. The overlaps between the particles and their relative velocities in these directions are evaluated and then related to the forces through a suitable model, as explained below.

For simplicity, consider two spherical particles i and j in contact (Fig. 1.23). Let $\mathbf{n}$ denote a unit vector along the line joining the centers of the particles, and directed from i

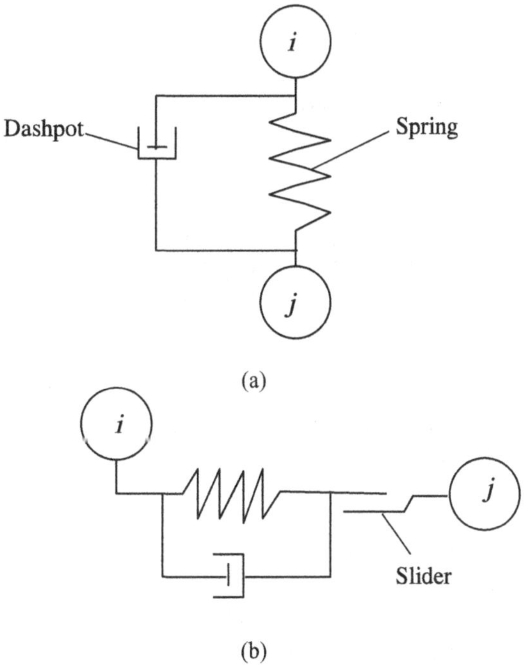

Figure 1.24. Models for the contact forces between two particles in (a) the normal direction and (b) the tangential direction. (Adapted from Fig. 2 of Tsuji et al., 1992.)

to j. The velocity of i relative to j at the contact point C is given by

$$\mathbf{v}_{ij} = \mathbf{v}_i - \mathbf{v}_j + (R_{(i)}\ \boldsymbol{\omega}_{(i)} + R_{(j)}\ \boldsymbol{\omega}_{(j)}) \times \mathbf{n} \tag{1.10}$$

where R_i is the radius of particle i, $\times$ denotes the cross product, and the brackets around the indices i and j indicate that the summation convention does not apply. Thus the relative velocity in the normal direction, i.e., in the direction of **n**, is given by

$$\mathbf{v}_n = (\mathbf{v}_{ij} \cdot \mathbf{n})\,\mathbf{n} \tag{1.11}$$

where $\mathbf{v}_{ij} \cdot \mathbf{n}$ denotes the scalar product or dot product of the vectors $\mathbf{v}_{ij}$ and $\mathbf{n}$. (The dot product is defined by (A.7).) Hence at the point C (Fig. 1.23), the velocity of i relative to j in the tangential direction is given by

$$\mathbf{v}_t = \mathbf{v}_{ij} - \mathbf{v}_n = (\mathbf{n} \times \mathbf{v}_{ij}) \times \mathbf{n} \tag{1.12}$$

where the second equation follows from (A.84). The overlap between the particles in the normal direction is given by

$$\xi_n = R_i + R_j - (\mathbf{r}_j - \mathbf{r}_i) \cdot \mathbf{n} \tag{1.13}$$

where $\mathbf{r}_j$ is the position vector corresponding to particle j (Fig. 1.23).

As the systems of interest usually contain a large number of particles, it would be extremely time consuming to solve for the deformation of the individual particles. Instead, it is assumed that the contact forces can be calculated using spring and dashpot models, wherein the forces exerted by the spring and the dashpot depend on the overlap and the relative velocity, respectively. For example, Tsuji et al. (1992) use an elastic spring and a viscous dashpot in parallel (Fig. 1.24a) to calculate the normal force $\mathbf{F}_n$ exerted on particle i in the direction of **n** by particle j. Thus

$$\mathbf{F}_n = -(k_n\,\xi_n\,\mathbf{n} + C_n\,\mathbf{v}_n) \tag{1.14}$$

where k_n and C_n are the stiffness (or "spring constant") and the damping coefficient, respectively, in the normal direction. For two elastic spheres in contact under the action of normal forces, i.e., forces directed along **n**, the Hertz solution (Hertz, 1882, cited in Johnson, 1987a, p. 90; Landau and Lifshitz, 1986, pp. 26–31; Johnson, 1987a, pp. 90–95; Fan and Zhu, 1998, pp. 59–62) shows that $k_n \,\alpha\, \sqrt{\xi_n}$, and the proportionality constant can be expressed in terms of the radii and elastic properties of the spheres.

To calculate the tangential force, the tangential direction is defined by the unit vector

$$\mathbf{t} \equiv \mathbf{v}_t/|\mathbf{v}_t| \tag{1.15}$$

where $\mathbf{v}_t$ is the relative velocity in the tangential direction, given by (1.12), and $|\mathbf{v}_t|$ denotes the magnitude of $\mathbf{v}_t$. If the current collision begins at time t_0, the overlap in the tangential direction at the current time t is given by

$$\xi_t = \int_{t_0}^{t} |\mathbf{v}_t| \mathrm{d}t \tag{1.16}$$

Finally, using a spring and dashpot in parallel (Fig. 1.24b), the tangential force $\mathbf{F}_t$ exerted on particle i by particle j in the direction of **t** is given by

$$\mathbf{F}_t = -(k_t\, \xi_t\, \mathbf{t} + C_t\, \mathbf{v}_t) \tag{1.17}$$

where k_t, C_t, and ξ_t are the stiffness, the damping coefficient, and the overlap, respectively, in the tangential direction. For two elastic spheres which are pressed into contact by normal forces and are then subjected to tangential forces, the solution due to Cattaneo (1938) (cited in Mindlin, 1949) and Mindlin (1949) (see also Johnson, 1987a, pp. 216–224; Fan and Zhu, 1998, pp. 63–69) shows that $k_t\, \xi_t$, the tangential force arising from the deformation of the spheres, depends on the normal force, the radii and elastic properties of the spheres, and μ, the coefficient of sliding friction between the surfaces of the spheres. Expressions for the stiffness and damping coefficients are given in Tsuji et al. (1992), Brilliantov et al. (1996), Cleary (1998), and Zhou et al. (2002).

Friction between the particles is incorporated as follows (Cundall and Strack, 1979). If the tangential force $|\mathbf{F}_t|$ exceeds $\mu\, |\mathbf{F}_n|$, where $\mathbf{F}_n$ is the normal force and μ is the coefficient of friction, (1.17) is replaced by

$$\mathbf{F}_t = -\mu\, |\mathbf{F}_n|\, \mathbf{t} \tag{1.18}$$

Variants of this model have been used to study shear flow between parallel plates (Walton and Braun, 1986; Thompson and Grest, 1991; Popken and Cleary, 1999), formation of heaps (Zhou et al., 1999, 2002), axisymmetric compression (Thornton, 2000), flow through bins and bunkers (Negi et al., 1992; Langston et al., 1995; Yuu et al., 1995; Rotter et al., 1998), flow down inclined chutes (Drake and Walton, 1995), vibrated beds of particles (Duran, 2000, pp. 190–202), gas–particle flow in a pipe (Tsuji et al., 1992), and flow in devices such as ball mills (Cleary, 1998).

As an example of the application of the DEM, let us briefly discuss the results of Drake and Walton (1995) for flow down an inclined chute (Fig. 1.7a). They conducted experiments on the flow of cellulose acetate spheres in a chute whose transverse dimension was marginally greater than the particle diameter. The flows were photographed using a high-speed camera facing the transparent side wall. The field of view was a rectangular area of size $14\, d_\mathrm{p} \times 18\, d_\mathrm{p}$, where d_p denotes the particle diameter, and the center of this area was about 0.7 m ($\approx 170\ d_\mathrm{p}$) from the exit of the chute. Using the particle positions in the field of view, the profile of the x component of velocity u and the "density" ρ_a (mass of particles per unit area of the base) were determined as a function of the distance y measured normal to the base of the chute (Fig. 1.7a). This problem was also solved numerically using DEM. The measured and predicted velocity profiles agree well except near the upper part of the flowing layer (Fig. 1.25a). On the other hand, there is a larger discrepancy between measured and predicted profiles of the density (Fig. 1.25b). Drake

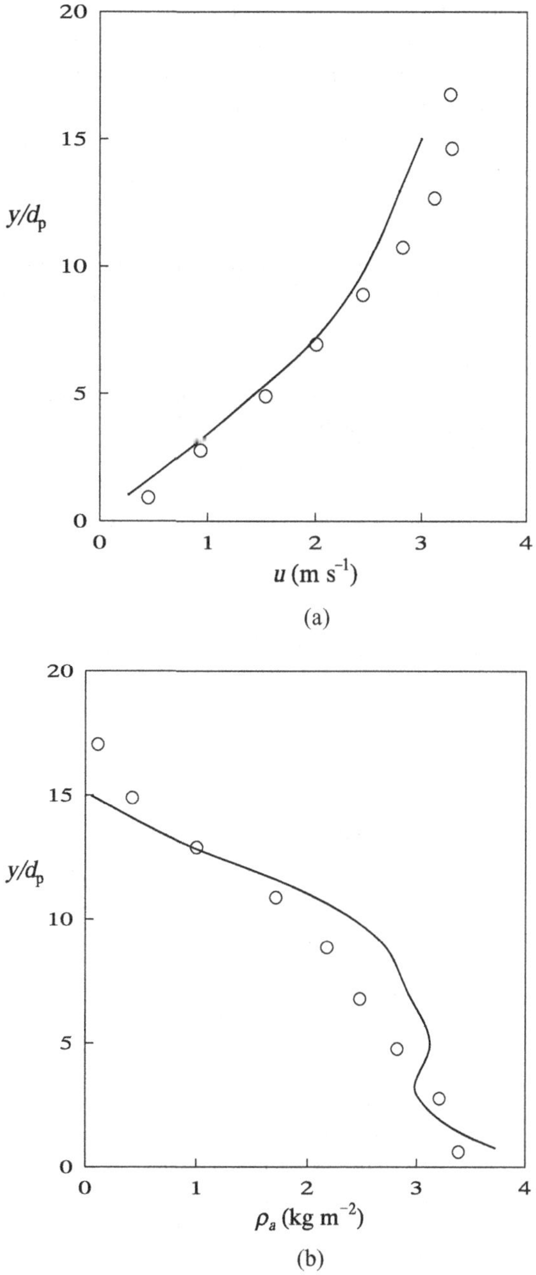

Figure 1.26. Transverse profiles of the (a) flow velocity u, and (b) "density" ρ_a (mass of particles per unit area of the base) for the flow of cellulose acetate spheres down an inclined chute having glass walls and an aluminium base: data (○) and DEM simulations (—) of Drake and Walton (1995). A monolayer of spheres was glued to the base. The y coordinate represents the distance measured normal to the base of the chute (see Fig. 1.7). Parameter values: particle diameter = 6 mm, width of the chute = 6.7 mm, length of the chute = 3.7 m, inclination of the chute to the horizontal = 42.75°. The DEM profiles correspond to a time of 4 s after the start of the simulation.

and Walton (1995) attribute this to the use of a crude model for air drag, which assumes that the air is stationary relative to the chute. In the actual case, the particles entrain air, and hence the latter is also in motion.

The DEM simulations are usually computationally intensive. For the example discussed earlier, the computational time on a Cray YMP machine was 1–10 hours per second of real time, depending on the parameter values used and the mass flux. The total number of particles handled during the time required to attain a steady state ($\approx$5 s) was about 10,000. Unlike the present case, the width of an actual chute is usually much greater than the particle diameter. For example, in the experiments of Johnson et al. (1990), the width was about 63 particle diameters. Hence a much larger number of particles have to be simulated for comparison with typical experimental results. Similarly, with reference to flow through bunkers, the size of the exit slot was chosen as 5–10 particle diameters in most of the studies cited earlier. This was done mainly to reduce the computation time.

An approach similar to DEM, but with the assumption of instantaneous collisions between hard particles, has been developed by Campbell (1982) and Campbell and Brennen (1985). Here the spring–dashpot–slider model (Fig. 1.24) is replaced by assumptions about the relative velocity after impact. This approach is called the *event-driven, rigid particle* method. Application of this method to vibrated beds is described in Herrmann and Luding (1998) and Duran (2000, pp. 76–84, 189–190). Herrmann and Luding (1998) also discuss size segregation, stresses in sandpiles, and flow through hoppers.

Other types of discrete models, called *cellular automata* models, which are based on the movement of particles on a lattice of sites, are described in Baxter and Behringer (1990) and Herrmann and Luding (1998).

With the increasing availability of faster and more powerful computers, the discrete models are likely to become quite popular. At the moment, they are limited by the number of particles that can be handled and by complexities associated with (i) a proper description of contact forces for interparticle and particle–wall collisions, and (ii) the shapes of nonspherical particles.

1.4.2. Continuum Models

Continuum models have been used extensively for both static and flow problems. In these models, the particles are replaced by a continuous medium, and quantities such as velocity and density are assumed to be smooth functions of position and time. As granular materials appear discrete to the eye, it may seem surprising that continuum models can be used. Further, some of the relevant length scales of the apparatus used in laboratory experiments rarely exceed 50–100 particle diameters (Brown and Richards, 1970, pp. 139, 174, 185; Brennen and Pearce, 1978; Johnson et al., 1990). Despite these features, continuum models have been used with varying degrees of success in many situations. The motivation for using these models has been eloquently stated by Truesdell and Muncaster (1980, pp. xvi–xvii):

> However discrete may be nature itself, the mathematics of a very numerous discrete system remains even today beyond anyone's capacity. To analyze the large, we replace it by the infinite, because the properties of the infinite are simpler and easier to manage. The mathematics of large systems is the infinitesimal calculus, the analysis of functions which are defined on infinite sets, and whose values range over infinite sets. We need to differentiate and integrate functions. Otherwise, we are hamstrung if we wish to deal effectively, precisely with more than a few dozen objects able to interact with each other. Thus somehow, we must introduce the continuum.

The preceding discussion does not indicate the conditions under which it is reasonable to use continuum models for granular materials. At present, there is no satisfactory answer. With reference to fluids, several authors (Batchelor, 1967, pp. 4–6; Condon, 1967, pp. 3–8; Whitaker, 1968, p. 1; Fung, 1977, pp. 3–4; Chung, 1988, p. 2) believe that these models are valid if the characteristic length scale of the system is much larger than the mean spacing between the constituent molecules or "particles". (In the context of granular materials, the spacing should be taken as the distance between the centers of the particles, as the gap between the surfaces of the particles can be much less than the particle size in some cases.) On intuitive grounds, it appears that timescales must also be important in flow problems. Indeed, with reference to gases, Chapman and Cowling (1964, pp. 25–26) suggest that (i) an infinitesimal volume element $\mathrm{d}V$ corresponding to the macroscopic or continuum model should contain a large number of molecules, and (ii) an infinitesimal time interval should be large compared to the mean time required for molecules to cross $\mathrm{d}V$, if they are not deflected by collisions. As noted earlier, condition (i) is often violated in typical granular flows. However, if (ii) holds, a small volume element ΔV may effectively contain a large number of particles. For example, consider a medium consisting of particles of diameter 1 mm, and let ΔV be a cube of side 1 mm. Then, at any time, ΔV contains less than two particles. However, if the particles traverse the volume with a mean velocity of 0.2 m s^{-1} (which is comparable to the typical vertical velocities attained in the exit region of laboratory-scale hoppers), about 200 particles will pass through ΔV in 1 s. As long as we confine attention to phenomena which occur on timescales that are much larger than 1 s, ΔV will effectively contain a large number of particles. Perhaps this provides a clue regarding the apparent success of continuum models even in cases where the length scales are not much larger than the particle size.

Once the granular material is idealized as a continuum, it is subject to the balance laws of continuum mechanics. These equations contain too many variables and hence must be supplemented by constitutive equations which describe the behavior of the material. At present, the lack of constitutive equations valid over a wide range of densities and shear rates poses a major hurdle. Further, as noted by Truesdell and Muncaster (1980, p. xvii), it is difficult to incorporate information about the nature of the particles into continuum equations, except in a heuristic manner. Despite these shortcomings, there have been numerous attempts to use continuum models for granular materials. Some of these models will be described later.

1.5. BALANCE LAWS FOR CONTINUUM MODELS

A balance law is postulated for each of the following quantities: (i) mass, (ii) linear momentum, (iii) angular momentum, (iv) energy, and (v) entropy. If there is a discrepancy between theory and experiment, the constitutive equations are modified, but the balance laws are usually left unchanged.

Unless otherwise stated, the granular material will be treated as a single-phase continuum, ignoring the presence of the interstitial fluid. A brief discussion of fluid–particle interactions is given in §1.7. Henceforth, the density of the granular material will be denoted by $\rho = \rho_p \nu$, where ρ_p is the particle density and ν is the solids fraction. It will be assumed that ρ_p is a constant, but ν will be permitted to vary.

A few remarks on the specific balance laws are in order. The mass balance expresses the law of conservation of mass. The linear momentum balance states that the rate of change of linear momentum of a "body" (as defined in §1.5.1) is equal to the sum of the surface or contact forces and body forces acting on it. For materials such as air and water, there are no localized torques. In such cases, it can be shown (Batchelor, 1967, pp. 10–11; Whitaker, 1968, pp. 119–121; Fung, 1977, pp. 37–38, 252–253) (see also Problem 1.8) that the angular momentum balance is identically satisfied by requiring the stress tensor to be *symmetric*. In keeping with the bulk of the literature on granular flow, a symmetric stress tensor will be used in Chapters 1–6. In Chapter 9, it is shown that collisions between rough particles lead to asymmetric stresses. The reader is referred to Kanatani (1979), Lun and Savage (1987), Tejchman and Wu (1993), and Mohan et al. (1999) for a discussion of models based on an asymmetric stress tensor. The entropy balance is rarely used directly, but together with the energy balance, can be used to obtain an expression for the entropy production rate $\dot{S}$ in terms of the gradients of field variables such as velocity and temperature. The *Clausius–Duhem inequality* (Truesdell and Toupin, 1960, pp. 643–644; Malvern, 1969, p. 255; Woods, 1975, pp. 144–148), which represents the second law of thermodynamics for a continuum, states that $\dot{S}$ must be ≥ 0. As shown in books on irreversible thermodynamics and continuum mechanics, this inequality can be used to guess the forms of the constitutive equations and to place constraints on the parameters

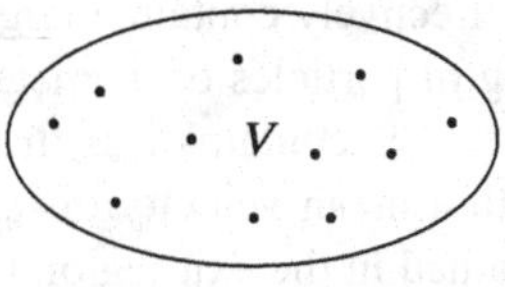

Figure 1.26. A volume V which represents the configuration of a body at a time t. The dots denote material points.

occurring in them. For example, it can be shown that the shear viscosity of a Newtonian fluid must be nonnegative (Woods, 1975, p. 156), as also the Young's modulus of an elastic solid (Malvern, 1969, p. 293).

The balance laws will be formulated in tensor form, and also in terms of components relative to a rectangular Cartesian coordinate system. The equations are given in cylindrical and spherical coordinates in Appendix A.

1.5.1. The Velocity

Before writing the mass balance, the velocity of the medium must be defined. As we are dealing with a continuum, there are no particles. Hence the velocity is defined in terms of the motion of *material points*, which are the continuum analogs of particles. Consider the flow of a granular material through a pipe, and assume that the material can be treated as a continuum. At any time t, consider a volume V (Fig. 1.26). The continuum is assumed to be composed of primitive entities called material points, which are shown by the dots in Fig. 1.26. The material points contained in V constitute a *body*, and the region of physical space occupied by the body is called a *configuration*. The motion of the body consists of the sequence of configurations at various times. For example, the volume V in Fig. 1.26 is the configuration of the body at time t. If t is the current time, the cofiguration is called the *current* configuration. In the absence of phase changes and chemical reactions, each configuration of a body contains the same set of material points. The volume corresponding to each of these configurations is called a *material volume* and is denoted by $V_m(t)$. Thus the shape and size of a material volume can change with time, but it always contains the same set of material points.

To proceed further, the concept of a reference frame must be introduced. A *reference frame* consists of a spatial coordinate system and a clock (R. Jackson, private communication, 2001). For many applications, it is convenient to choose a reference frame that is at rest relative to the surface of the earth. Such a frame is called a *laboratory frame*.

Let us select the configuration of the body at some time t_0 as a *reference* configuration. Consider a Cartesian coordinate system in the laboratory frame. With respect to this coordinate system, the material point P in the reference configuration A (Fig. 1.27) corresponds to the position vector $\mathbf{X}$. The components X_1, X_2, and X_3 of $\mathbf{X}$ are called the *material coordinates* of the material point P. Thus material points can be uniquely identified by the corresponding position vectors in the reference configuration or equivalently, by their material coordinates. At some other time t, the body has the configuration B (Fig. 1.27), and the position vector corresponding to P is $\mathbf{x}$. The *motion* of the body is described by the vector equation

$$\mathbf{x} = \mathbf{x}(\mathbf{X}, t) \tag{1.19}$$

where each vector $\mathbf{X}$ corresponds to a material point.

The *velocity* of a material point is defined by

$$\mathbf{v} \equiv \frac{\mathrm{D}\,\mathbf{x}}{\mathrm{D}\,t} \equiv \left(\frac{\partial \mathbf{x}}{\partial t}\right)_{\mathbf{X}} \tag{1.20}$$

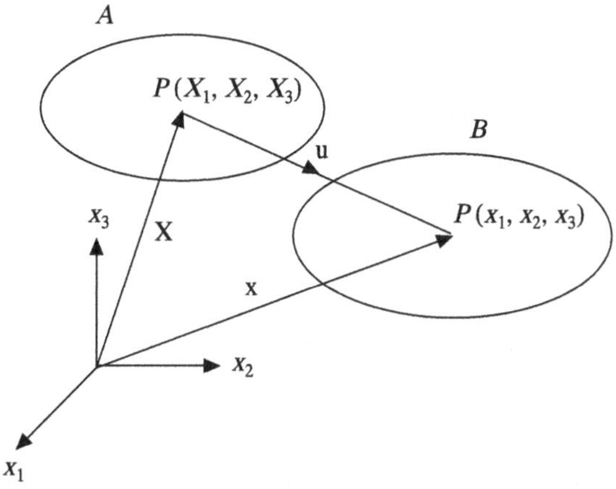

Figure 1.27. The reference configuration A of a body, and the configuration B of the same body at some time t. The material coordinates of P are (X_1, X_2, X_3) and its coordinates at time t are (x_1, x_2, x_3). The vector $\mathbf{u} \equiv \mathbf{x} - \mathbf{X}$ is the displacement vector.

where **X** is held constant while evaluating the derivative. Using index notation (see §A.2), the component form of (1.20) is

$$v_i = \left(\frac{\partial x_i}{\partial t}\right)_{\mathbf{X}} \tag{1.21}$$

The derivative

$$\frac{\mathrm{D}}{\mathrm{D}t} \equiv \left.\frac{\partial}{\partial t}\right|_{\mathbf{X}} \tag{1.22}$$

is called the *material* or *substantial* derivative. It is the time derivative taken while following a material point. An alternative expression for the material derivative is useful in some cases. Changing the independent variables from $(\mathbf{X}, t)$ to $(\mathbf{x}(\mathbf{X}, t), t)$, and choosing a Cartesian coordinate system (Fig. 1.27), (1.22) can be rewritten using index notation as

$$\frac{\mathrm{D}}{\mathrm{D}t} = \left.\frac{\partial}{\partial t}\right|_{\mathbf{x}} + \left.\frac{\partial x_i}{\partial t}\right|_{\mathbf{X}} \frac{\partial}{\partial x_i}$$

or, using (1.21), as

$$\frac{\mathrm{D}}{\mathrm{D}t} = \frac{\partial}{\partial t} + v_i \frac{\partial}{\partial x_i} = \frac{\partial}{\partial t} + \mathbf{v} \cdot \nabla \tag{1.23}$$

where ∇ denotes the gradient operator (see §A.7.8). The partial time derivative on the right-hand side of (1.23) is evaluated keeping **x** constant.

It follows from (1.19) and (1.20) that $\mathbf{v} = \mathbf{v}(\mathbf{X}, t)$. Assuming that (1.19) is invertible, this can also be written as $\mathbf{v} = \mathbf{v}(\mathbf{x}, t)$. The velocity is defined relative to a chosen reference frame, which is the laboratory frame in the present case. The velocity relative to any other frame can be calculated by a suitable transformation of **v**, as discussed in §G.2.2.

1.5.2. Integral and Differential Balances

The balances can be expressed either in *integral* or *differential* form. The form chosen depends on the application of interest to the user. Integral balances provide estimates of overall quantities such as the total mass flow rate from a hopper and the total force exerted on the hopper wall. Differential balances provide detailed information about the density, stress, and velocity fields at every point in the domain of interest, usually at the expense of greater effort compared to the integral balances. The two forms of the mass and momentum balances are discussed below.

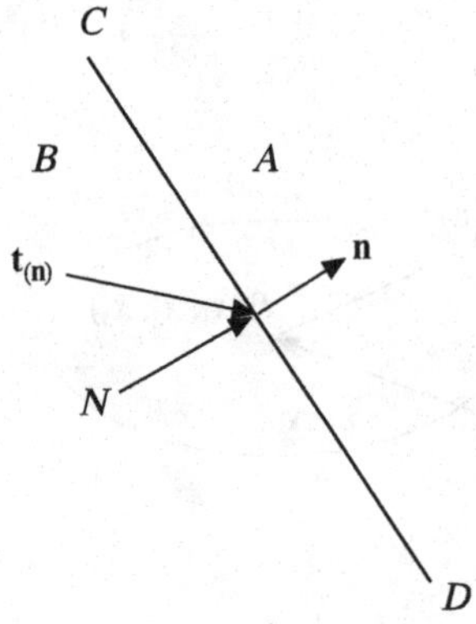

Figure 1.28. The stress vector $\mathbf{t}_{(\mathbf{n})}$ is the contact force exerted across unit area of the plane CD on the material into which the unit normal $\mathbf{n}$ points by the material on the other side. Here N is the normal stress exerted by B on A.

1.5.3. The Mass Balance

Considering a material volume $V_m(t)$, the *integral* mass balance is given by (Fung, 1977, p. 250; Slattery, 1999, p. 18)

$$\frac{\mathrm{d}}{\mathrm{d}t}\int_{V_m(t)} \rho \, \mathrm{d}V = 0 \tag{1.24}$$

where $\rho = \rho_\mathrm{p}\, \nu$ is the density of the granular material.

The *differential* mass balance or *continuity* equation takes the form (Batchelor, 1967, p. 74; Whitaker, 1968, p. 93; Fung, 1977, p. 250; Bird et al., 2002, p. 77)

$$\frac{\partial \rho}{\partial t} + \nabla \cdot (\rho \, \mathbf{v}) = 0 \tag{1.25}$$

It can be obtained from (1.24) by using the transport and divergence theorems (Problem 1.14).

Choosing a Cartesian coordinate system, (1.25) can also be written in Cartesian tensor notation (see §A.5) as

$$\frac{\partial \rho}{\partial t} + \frac{\partial}{\partial x_i}(\rho \, v_i) = 0 \tag{1.26}$$

Here the repeated indices imply summation, as discussed in Appendix A. An alternative form of (1.25) is given by

$$\frac{\mathrm{D}\rho}{\mathrm{D}t} = -\rho \, (\nabla \cdot \mathbf{v}) \tag{1.27}$$

where $\mathrm{D}/\mathrm{D}t$ represents the material derivative defined by (1.23).

1.5.4. The Stress Vector and the Stress Tensor

Before formulating the momentum balance, some quantities must be defined. Consider a plane separating two regions A and B (Fig. 1.28). Let $\mathbf{n}$ be a unit normal to the plane, pointing from region B into region A. The stress vector $\mathbf{t}_{(\mathbf{n})}$ is defined as the contact force exerted across a unit area of the plane on the material into which $\mathbf{n}$ points (region A) by the material on the other side (region B). This convention is the opposite of that commonly used in continuum mechanics (Whitaker, 1968, p. 36; Fung, 1977, p. 64). For cohesionless materials, the present choice is convenient, as it implies that compressive normal stresses are *positive*. It can be shown (Whitaker, 1968, pp. 110–115; Fung, 1977, pp. 70–72) (see also Appendix B) that $\mathbf{t}_{(\mathbf{n})}$ can be expressed in terms of a *stress tensor* $\boldsymbol{\sigma}$ by

$$\mathbf{t}_{(\mathbf{n})} = \mathbf{n} \cdot \boldsymbol{\sigma}; \quad t_{(\mathbf{n})i} = n_j \, \sigma_{ji} \tag{1.28}$$

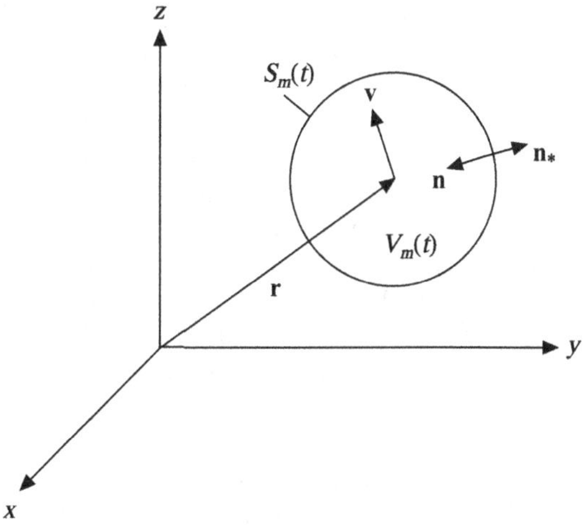

Figure 1.29. A material volume $V_m(t)$ with bounding surface $S_m(t)$, unit inward normal $\mathbf{n}$, and unit outward normal $\mathbf{n}_*$. Here $\mathbf{r}$ is the position vector of a material point relative to a coordinate system which is at rest in the laboratory frame and $\mathbf{v}$ is the velocity of this point.

(Here $\mathbf{n} \cdot \boldsymbol{\sigma}$ denotes the dot product of a vector $\mathbf{n}$ and a tensor $\boldsymbol{\sigma}$; it is defined by (A.12).) With the preceding convention for $\mathbf{t}_{(\mathbf{n})}$, the stress tensor is said to be defined in the *compressive* sense.

1.5.5. The Linear Momentum Balance

In the laboratory frame, the *integral* linear momentum balance is given by (Fung, 1977, p. 251; Slattery, 1999, p. 31; Bird et al., 2002, p. 112)

$$\frac{\mathrm{d}}{\mathrm{d}t} \int_{V_m(t)} \rho\, \mathbf{v}\, \mathrm{d}V = \int_{S_m(t)} \mathbf{t}_{(\mathbf{n})}\, \mathrm{d}S + \int_{V_m(t)} \rho\, \mathbf{b}\, \mathrm{d}V \tag{1.29}$$

where $V_m(t)$ is a material volume with a bounding surface $S_m(t)$ (Fig. 1.29), $\mathbf{t}_{(\mathbf{n})}$ is the stress vector or force per unit area exerted across $S_m(t)$ on the material in $V_m(t)$ by the material outside it, $\mathbf{b}$ is the body force per unit mass, and $\mathbf{n}$ is the unit *inward* normal to $S_m(t)$. Henceforth, we assume that the body force arises solely because of the earth's gravitational field. For most applications involving terrestrial flows, the body force per unit mass $\mathbf{b}$ can then be approximated by $\mathbf{g}$, the acceleration due to gravity. In general, there can also be other types of body forces, such as the force exerted on a charged material by an external electric field.

As discussed in Problem G.4, (1.29) can also be used in reference frames which move relative to the laboratory frame, but the expression for the body force involves several terms.

The *differential* linear momentum balance is given by

$$\rho \frac{\mathrm{D}\mathbf{v}}{\mathrm{D}t} = -\nabla \cdot \boldsymbol{\sigma} + \rho\, \mathbf{b} \tag{1.30}$$

or, in Cartesian tensor notation, by

$$\rho \frac{\mathrm{D}v_i}{\mathrm{D}t} = -\frac{\partial \sigma_{ji}}{\partial x_j} + \rho\, b_i, \quad i = 1, 3 \tag{1.31}$$

where $\mathrm{D}/\mathrm{D}t$ represents the material derivative defined by (1.23). As in the case of the mass balance, (1.30) can be obtained from (1.29) by using the transport and divergence theorems (Problem 1.14).

In some problems, (1.30) may not be adequate to describe motion relative to the laboratory frame, if $\mathbf{b}$ is taken as the acceleration due to gravity $\mathbf{g}$. The body force has to be modified by including additional terms arising from various sources, such as the rotation of the earth about its axis and the motion of the earth about the sun. For example, Draad and Nieuwstadt (1998) measured the profile of the axial velocity of water flowing in a long pipe and found that it was not parabolic, whereas (1.30) predicts a parabolic profile for a Newtonian fluid such as water. Suppose that (1.30), with $\mathbf{b} = \mathbf{g}$, is assumed to be valid in a reference frame which is fixed relative to the axis of the earth. When the velocity $\mathbf{v}$ in (1.30) is expressed in terms of the velocity relative to the laboratory frame, additional terms such as the Coriolis force and the centrifugal force arise because of the rotation of the earth about its axis (see Problem G.4). By including these terms, Draad and Nieuwstadt (1998) were able to match the observed and predicted velocity profiles.

Equation (1.30) is given in most books on fluid and continuum mechanics (see, e.g., Batchelor, 1967, p. 137; Whitaker, 1968, pp. 121–123; Fung, 1977, pp. 250–252; Bird et al., 2002, p. 84), but the frame in which $\mathbf{b} = \mathbf{g}$ is often not stated explicitly.

The first term on the right-hand side of (1.30) differs in sign from the term given in books on continuum mechanics, because of the use of the compressive definition for the stress tensor. The term on the left-hand side of (1.30) is the *inertial term*, i.e., the product of the density and the acceleration of the material, and the first and second terms on the right-hand side are the stress gradient and the body force, respectively.

Using the mass balance (1.25), (1.30) can be rewritten as

$$\frac{\partial(\rho\,\mathbf{v})}{\partial t} + \nabla\cdot(\rho\,\mathbf{v}\,\mathbf{v}) = -\nabla\cdot\boldsymbol{\sigma} + \rho\,\mathbf{b} \tag{1.32}$$

or in Cartesian tensor notation as

$$\frac{\partial}{\partial t}(\rho\,v_i) + \frac{\partial}{\partial x_j}(\rho\,v_j\,v_i) = -\frac{\partial\sigma_{ji}}{\partial x_j} + \rho\,b_i \tag{1.33}$$

1.5.6. The Angular Momentum Balance

As mentioned in §1.5, it is assumed in Chapters 1–6 that there are no localized torques. Then the angular momentum balance is identically satisfied by requiring the stress tensor to be symmetric, i.e.,

$$\boldsymbol{\sigma} = \boldsymbol{\sigma}^{\mathrm{T}} \text{ or } \sigma_{ij} = \sigma_{ji} \tag{1.34}$$

Here the superscript T denotes the transpose of the tensor (see §A.7.4).

The occurrence of localized torques, leading to an asymmetric stress tensor, is discussed in Chapter 9.

1.5.7. The Energy Balance

Neglecting radiative transfer, the energy balance takes the form (Whitaker, 1968, pp. 391–393; Malvern, 1969, pp. 266–230)

$$\rho\,\frac{\mathrm{D}}{\mathrm{D}t}(\hat{U} + \frac{1}{2}\,|\mathbf{v}|^2) = -\nabla\cdot(\boldsymbol{\sigma}\cdot\mathbf{v}) + \rho\,\mathbf{b}\cdot\mathbf{v} - \nabla\cdot\mathbf{q} \tag{1.35}$$

where $\hat{U}$ is the internal energy/unit mass, $|\mathbf{v}|^2/2$ is the kinetic energy/unit mass, and $\mathbf{q}$ is the energy flux vector. The terms $-\nabla\cdot(\boldsymbol{\sigma}\cdot\mathbf{v})$ and $\rho\,\mathbf{b}\cdot\mathbf{v}$ represent the rate of working per unit volume of the material by the stresses and the body force, respectively, and the term $-\nabla\cdot\mathbf{q}$ represents the rate of energy transfer per unit volume of the material by conduction.

Equation (1.35) can also be written in Cartesian tensor notation as

$$\rho\,\frac{\mathrm{D}}{\mathrm{D}t}(\hat{U} + \frac{1}{2}\,v_i\,v_i) = -\frac{\partial}{\partial x_j}(\sigma_{jk}\,v_k) + \rho\,b_i\,v_i - \frac{\partial q_i}{\partial x_i} \tag{1.36}$$

or, using the mass balance, as

$$\frac{\partial}{\partial t}[\rho\,(\hat{U}+\frac{1}{2}\,|\mathbf{v}|^2)]+\nabla\cdot[\rho\,\mathbf{v}\,(\hat{U}+\frac{1}{2}\,|\mathbf{v}|^2)]=-\nabla\cdot(\boldsymbol{\sigma}\cdot\mathbf{v})+\rho\,\mathbf{b}\cdot\mathbf{v} -\nabla\cdot\mathbf{q} \tag{1.37}$$

Another form of the energy balance, which involves only the internal energy, is useful in some cases. This can be obtained as follows. The *mechanical energy balance* is obtained by taking the scalar product of $\mathbf{v}$ with the momentum balance (1.30). The result is

$$\rho\,\frac{\mathrm{D}}{\mathrm{D}t}(\frac{1}{2}\,|\mathbf{v}|^2)=-(\nabla\cdot\boldsymbol{\sigma})\cdot\mathbf{v}+\rho\,\mathbf{b}\cdot\mathbf{v} \tag{1.38}$$

Subtracting (1.38) from (1.37), we obtain the *internal energy balance* (Whitaker, 1968, pp. 395–396)

$$\rho\,\frac{\mathrm{D}\hat{U}}{\mathrm{D}t}=-\nabla\cdot(\boldsymbol{\sigma}\cdot\mathbf{v})+(\nabla\cdot\boldsymbol{\sigma})\cdot\mathbf{v}-\nabla\cdot\mathbf{q} \tag{1.39}$$

The right-hand side of (1.39) can be written more compactly as follows. Using Cartesian tensor notation, we have

$$-\nabla\cdot(\boldsymbol{\sigma}\cdot\mathbf{v})+(\nabla\cdot\boldsymbol{\sigma})\cdot\mathbf{v}=-\frac{\partial}{\partial x_i}(\sigma_{ij}\,v_j)+\frac{\partial\sigma_{ij}}{\partial x_i}\,v_j=-\sigma_{ij}\,\frac{\partial v_j}{\partial x_i} \tag{1.40}$$

Introducing the *velocity gradient tensor*, defined by (see §A.7.9)

$$\nabla\,\mathbf{v}\equiv\frac{\partial v_j}{\partial x_i}\,\mathbf{e}_i\,\mathbf{e}_j \tag{1.41}$$

where $\mathbf{e}_i$ is a basis vector for a Cartesian coordinate system, (1.40) can be rewritten as

$$-\nabla\cdot(\boldsymbol{\sigma}\cdot\mathbf{v})+(\nabla\cdot\boldsymbol{\sigma})\cdot\mathbf{v}=-\boldsymbol{\sigma}^{\mathrm{T}}:(\nabla\,\mathbf{v}) \tag{1.42}$$

Here $\boldsymbol{\sigma}^{\mathrm{T}}:(\nabla\,\mathbf{v})$ is the scalar product of the tensors $\boldsymbol{\sigma}^{\mathrm{T}}$ and $\nabla\,\mathbf{v}$ (see §A.7.3), and the superscript T denotes the transpose of the tensor.

The quantity

$$\dot{\Phi}\equiv-\boldsymbol{\sigma}^{\mathrm{T}}:(\nabla\,\mathbf{v})=-\sigma_{ij}\,\frac{\partial v_j}{\partial x_i} \tag{1.43}$$

is called the *stress power* (Malvern, 1969, p. 228) and can be interpreted as follows. Using (1.40), the mechanical energy balance (1.38) and the internal energy balance (1.39) can be rewritten as

$$\rho\,\frac{\mathrm{D}}{\mathrm{D}t}(\frac{1}{2}\,|\mathbf{v}|^2)=-\nabla\cdot(\boldsymbol{\sigma}\cdot\mathbf{v})-\dot{\Phi}+\rho\,\mathbf{b}\cdot\mathbf{v} \tag{1.44}$$

and

$$\rho\,\frac{\mathrm{D}\hat{U}}{\mathrm{D}t}=\dot{\Phi}-\nabla\cdot\mathbf{q} \tag{1.45}$$

As $\dot{\Phi}$ occurs with opposite signs in (1.44) and (1.45), it represents the rate of interconversion per unit volume of kinetic and internal energies. If $\dot{\Phi}$ is > 0, mechanical energy is converted to internal energy. Further, if $\dot{\Phi}$ dominates the conduction term $-\nabla\cdot\mathbf{q}$ in (1.45), the internal energy of the material increases.

1.6. STATICS

When the material is at rest, the momentum balance reduces to the *force balance* or *equation of equilibrium*

$$-\nabla \cdot \boldsymbol{\sigma} + \rho\, \mathbf{b} = 0 \tag{1.46}$$

which must be solved for the stress tensor $\boldsymbol{\sigma}$.

It is instructive to consider the difference between fluid and granular statics. In a fluid at rest, the stress tensor is given by

$$\boldsymbol{\sigma} = p_f \mathbf{I}; \quad \sigma_{ij} = p_f \delta_{ij} \tag{1.47}$$

where $\mathbf{I}$ is the unit tensor, δ_{ij} is the Kronecker delta defined by (A.6), and p_f is the pressure of the fluid. For an incompressible fluid, the density ρ is a known constant, and hence (1.46) suffices to determine the pressure to within an arbitrary additive constant. If the density ρ_f varies, but the fluid temperature T_f is a constant, the force balance must be supplemented by an *equation of state*

$$\rho_f = \rho_f(p_f, T_f) \tag{1.48}$$

Equations (1.46)–(1.48) then suffice to determine the pressure.

In a granular material at rest, the situation is quite different. As discussed in the example of a heap (§1.1), the material can support nonzero shear stresses at rest. Hence the stress tensor cannot be expressed in the simple form (1.47), as the latter implies that there is no shear stress on *any* plane. Therefore, the force balance must be supplemented by an additional constitutive equation, even in cases where the density does not vary significantly. This is an extremely difficult task, and there is no consensus at present on the form of the constitutive equation. For example, the stress exerted by a heap on the base depends on the manner in which the heap is constructed (see Figs. 1.2 and 1.3). In this respect, problems involving flow are slightly easier to handle than those involving statics, as the flow of the material partly erases the effect of the initial packing. The reader is referred to Rajchenbach (2001) and Geng et al. (2003) for recent attempts to examine the statics of granular materials.

1.7. FLUID–PARTICLE INTERACTION

Consider a single particle that is completely immersed in an unbounded fluid. When there is no relative motion between the particle and the fluid, the only force exerted by the fluid on the particle is *buoyancy*. In this case, buoyancy arises from the variation of the fluid pressure over the surface of the particle. When the particle and the fluid move relative to each other, there are additional forces such as the *drag*, *virtual mass or added mass*, *lift*, and *Basset or history* forces. The drag force arises because of the resistance offered to the flow of the fluid by the particle and is parallel to the direction of the relative velocity $\mathbf{w}_\infty$ between the fluid and the particle. Here $\mathbf{w}_\infty \equiv \mathbf{U}_\infty - \mathbf{V}$, where $\mathbf{U}_\infty$ is the *free-stream* velocity, i.e., the velocity of the fluid far from the particle, and $\mathbf{V}$ is the velocity of the center of mass of the particle. The lift force is perpendicular to the direction of $\mathbf{w}_\infty$ and arises from the variation of the stress over the surface of the particle. The virtual mass force arises when there is a relative acceleration between the fluid and the particles. The reader is referred to Maxey and Riley (1983), Fan and Zhu (1998, pp. 87–101), Joseph and Ocando (2002), and Bagchi and Balachander (2003) for a discussion of these forces. In many applications, the drag force is the dominant fluid–particle force. Some aspects of the drag force are discussed below.

Consider the drag force on a single spherical particle of radius R, which is stationary relative to the laboratory frame and is submerged in an unbounded, incompressible Newtonian fluid flowing steadily past it with a constant free-stream velocity $\mathbf{U}_\infty$. The constitutive equation for an incompressible Newtonian fluid is given by (see, e.g., Slattery, 1999, p. 41; Bird et al., 2002, p. 19)

$$\boldsymbol{\sigma}_\mathrm{f} = p_\mathrm{f}\mathbf{I} - 2\,\mu_\mathrm{f}\mathbf{D} \tag{1.49}$$

where $\boldsymbol{\sigma}_\mathrm{f}$ is the stress tensor for the fluid defined in the compressive sense, p_f is the pressure of the fluid $\mathbf{I}$ is the unit tensor, μ_f is the shear viscosity of the fluid, and $\mathbf{D}$ is the rate of deformation tensor, defined by

$$\mathbf{D} \equiv \frac{1}{2}\,(\nabla\mathbf{u} + (\nabla\mathbf{u})^\mathrm{T}) \tag{1.50}$$

Here $\mathbf{u}$ is the velocity of the fluid and $\nabla\mathbf{u}$ is the velocity gradient tensor, defined by (A.58).

Using (1.49), the momentum balance (1.30) reduces to

$$\rho_\mathrm{f}\mathbf{u}\cdot\nabla\mathbf{u} = -\nabla p_\mathrm{f} + \mu_\mathrm{f}\nabla^2\mathbf{u} + \rho_\mathrm{f}\mathbf{b} \tag{1.51}$$

where ρ_f is the density of the fluid and ∇^2 is the Laplacian operator, defined by (A.52). Let the *Reynolds* number Re be defined by

$$Re \equiv 2\,R\,|\mathbf{U}_\infty|\,\rho_\mathrm{f}/\mu_\mathrm{f} \tag{1.52}$$

In the limit of *creeping flow*, i.e., $Re \to 0$, it can be shown that the inertial term in (1.51) (the term on the left-hand side) can be omitted in comparison with the viscous term (the second term on the right-hand side), except at points that are far from the sphere (see, e.g., Batchelor, 1967, p. 232). In other words, the fluid acceleration in the vicinity of the sphere is negligible in this limit, and the viscous stress gradient is balanced by the pressure gradient and the body force. In creeping flow, the drag force exerted by the fluid on the particle is given by *Stokes' law* (Stokes, 1851; Batchelor, 1967, pp. 230–233; Bird et al., 2002, p. 58–61)

$$\mathbf{f}_d = 6\pi\,R\,\mu_\mathrm{f}\mathbf{U}_\infty; \quad Re \to 0 \tag{1.53}$$

This relation agrees well with the observed drag force for $Re < 0.5$ (Batchelor, 1967, p. 234). As Re increases, $\mathbf{f}_d$ becomes a nonlinear function of $\mathbf{U}_\infty$, and empirical or semiempirical expressions have to be used for large values of Re (Clift et al., 1978, p. 112).

Suppose that the particle translates with a constant velocity $\mathbf{V}$ relative to the laboratory frame, and let the free-stream velocity of the fluid relative to this frame be constant ($=\mathbf{U}_\infty$). By rewriting the mass and momentum balances for the fluid with respect to a moving coordinate system whose origin coincides with the center of the sphere, it follows that the drag force in the limit of creeping flow is given by (1.53), with $\mathbf{U}_\infty$ replaced by the *relative velocity* $\mathbf{U}_\infty - \mathbf{V}$ (Problem 1.10). Thus

$$\mathbf{f}_d = 6\pi\,R\,\mu_\mathrm{f}(\mathbf{U}_\infty - \mathbf{V}); \quad Re \to 0 \tag{1.54}$$

where Re is now based on the relative velocity. The drag force is proportional to the relative velocity in this limit.

Consider the motion of a Newtonian fluid relative to a granular material. One approach consists of using the Navier–Stokes equations in the regions occupied by the fluid and Newton's second law of motion for each of the particles. In addition, constitutive relations for the interparticle contact forces arising either from sustained contact or from rapid collisions must be specified. This is a topic of ongoing research, and there is still considerable uncertainty about the proper expressions for the contact forces. Owing to the complicated and possibly time-dependent geometry of both the fluid–particle interfaces and the contact zones between adjacent particles, it is difficult to implement this rigorous approach for systems containing a large number of particles.

Various approaches have been proposed to circumvent this difficulty. One approach, namely, *local volume averaging* (see, e.g., Anderson and Jackson, 1967; Whitaker, 1973; Hassanizadeh and Gray, 1979a,b; Gray and Hassanizadeh, 1998; Jackson, 2000, pp. 17–26) will now be discussed briefly. The local equations for the fluid and for the particles are averaged over a volume that contains a large number of particles, but whose characteristic length is small compared to the macroscopic length scale of interest.

The mass balances are given by (Gray and Hassanizadeh, 1998)

mass balance for the fluid phase

$$\frac{\partial}{\partial t}(\epsilon\, \rho_{\rm f}) + \nabla \cdot (\epsilon\, \rho_{\rm f}\, \mathbf{u}) = 0 \tag{1.55}$$

mass balance for the particle phase

$$\frac{\partial}{\partial t}(\nu\, \rho_{\rm p}) + \nabla \cdot (\nu\, \rho_{\rm p}\, \mathbf{v}) = 0 \tag{1.56}$$

where ϵ is the porosity or volume fraction of the fluid phase, $\rho_{\rm f}$ and $\mathbf{u}$ are the local average density and velocity of the fluid phase, respectively, $\nu = 1 - \epsilon$ is the solids fraction, and $\rho_{\rm p}$ and $\mathbf{v}$ are the local average density and velocity of the particle phase, respectively. The density $\rho_{\rm f}$ is obtained by averaging the intrinsic density of the fluid over the region occupied by the fluid in the averaging volume.

The linear momentum balances are given by (Gray and Hassanizadeh, 1998; Jackson, 2000, pp. 22–27)

momentum balance for the fluid phase

$$\frac{\partial}{\partial t}(\epsilon\, \rho_{\rm f}\, \mathbf{u}) + \nabla \cdot (\epsilon\, \rho_{\rm f}\, \mathbf{u}\, \mathbf{u}) = -\nabla \cdot \boldsymbol{\sigma}_{\rm f} + \epsilon\, \rho_{\rm f}\, \mathbf{b} - \mathbf{F} \tag{1.57}$$

momentum balance for the particle phase

$$\frac{\partial}{\partial t}(\nu\, \rho_{\rm p}\, \mathbf{v}) + \nabla \cdot (\nu\, \rho_{\rm p}\, \mathbf{v}\, \mathbf{v}) = -\nabla \cdot \boldsymbol{\sigma}_{\rm p} + \nu\, \rho_{\rm p}\, \mathbf{b} + \mathbf{F} \tag{1.58}$$

where $\boldsymbol{\sigma}_{\rm f}$ and $\boldsymbol{\sigma}_{\rm p}$ denote the stress tensors associated with the particle and fluid phases, respectively, $\mathbf{F}$ is the force exerted per unit total volume *by* the fluid *on* the particle phase, and $\boldsymbol{\sigma}_{\rm f}$ and $\boldsymbol{\sigma}_{\rm p}$ are defined in the compressive sense. The stress tensors depend on many quantities, and explicit expressions in terms of local average variables are available only in special cases (see, e.g., Jackson, 1997; Koch and Sangani, 1999; Jackson, 2000, pp. 34–46).

By using the mass balances, the inertial term in the momentum balances can be rewritten in an alternative form. For example, (1.57) can be rewritten as

$$\epsilon\, \rho_{\rm f} \frac{{\rm D}_{\rm f}}{{\rm D}t}(\mathbf{u}) = -\nabla \cdot \boldsymbol{\sigma}_{\rm f} + \epsilon\, \rho_{\rm f}\, \mathbf{b} - \mathbf{F} \tag{1.59}$$

where

$$\frac{{\rm D}_{\rm f}}{{\rm D}t} = \frac{\partial}{\partial t} + \mathbf{u} \cdot \nabla$$

is the material derivative based on the fluid-phase velocity.

The fluid–article interaction force is given by (Jackson, 2000, pp. 29, 48)

$$\mathbf{F} = -\nu\, \nabla \cdot \boldsymbol{\sigma}_{\rm f} + \mathbf{F}_d + \mathbf{F}_o \tag{1.60}$$

where the first term on the right-hand side of (1.60) represents buoyancy, $\mathbf{F}_d$ is the drag force exerted by the fluid on the particle phase, and $\mathbf{F}_o$ represents the other forces such as lift and virtual mass forces. All these forces are defined per unit total volume. As noted by Jackson (2000, p. 27), buoyancy cannot be defined unambiguously for a granular material or a suspension. For example, the buoyancy can also be defined as $-\rho_{\mathrm{f}}\, \nu\, \mathbf{b}$, where $\mathbf{b}$ is the body force per unit mass. If this is used, the expression for $\mathbf{F}$ becomes

$$\mathbf{F} = -\rho_{\mathrm{f}}\, \nu\, \mathbf{b} + \mathbf{F}_d + \mathbf{F}'_o$$

where $\mathbf{F}'_o$ represents the fluid–particle forces other than buoyancy and drag. In general, $\mathbf{F}_o \neq \mathbf{F}'_o$.

An empirical expression for the drag force $\mathbf{F}_d$ will now be discussed. Richardson and Zaki (1954) conducted experiments on the settling of a suspension of spherical particles in liquids. They found that the z component of the local average velocity of the particle phase could be related to the solids fraction ν of the suspension by

$$v_z = -v_t\,(1-\nu)^n \tag{1.61}$$

where the z direction coincides with the upward vertical direction, v_t is the terminal velocity of an isolated particle in an unbounded liquid (i.e., the magnitude of the steady vertical velocity eventually attained by an isolated particle falling under the action of gravity in a fluid), and n is a constant. The value n depends on a Reynolds number

$$Re_t \equiv 2\, R\, v_t\, \rho_{\mathrm{f}}/\mu_{\mathrm{f}} \tag{1.62}$$

as follows

$$\begin{aligned} n &= 4.65;\ Re_t < 0.2 \\ &= 4.35\, Re_t^{-0.03};\ 0.2 < Re_t < 1 \\ &= 4.45\, Re_t^{-0.1};\ 1 < Re_t < 500 \\ &= 2.39;\ 500 < Re_t \end{aligned} \tag{1.63}$$

Even though (1.61) is widely used, it should be noted that the range of solids fractions for which it is valid depends on the size and nature of the particles. For example, glass beads of diameter 1 mm were used in some of the experiments of Richardson and Zaki (1954). The solids fractions were in the range 0.11–0.4 for settling in bromoform and 0.09–0.53 for fluidization using water. Thus the use of (1.61) for high values of ν is suspect. An alternative correlation, based on results obtained using lattice Boltzmann simulations of fluid flow past a bed of spheres, is discussed in Beetstra et al. (2007).

The local average velocity of the particle phase v_z can be related to the drag force as follows. Assuming that the liquid and the particles are incompressible, the net volumetric flux across any plane vanishes, i.e., $(1-\nu)\, u_z + \nu\, v_z = 0$, where u_z is the z component of the local average velocity of the fluid-phase. Hence the z component of the local relative velocity is given by $w_z \equiv u_z - v_z = -v_z/(1-\nu)$. Guided by the form of the expression (1.54) for the drag force for an isolated particle, and the expression obtained for the limiting case of dilute suspension of particles moving slowly relative to the fluid (Jackson, 2000, p. 35), it is usually assumed that $\mathbf{F}_d$ is proportional to the relative velocity $\mathbf{w} = \mathbf{u} - \mathbf{v}$ whenever $|\mathbf{w}|$ is small. The relative velocity is now calculated using the fluid-phase and particle-phase velocities at the *same* location, in contrast to the case of an isolated particle. Using z components of the momentum balances (1.57) and (1.58), it can be shown that (Jackson, 2000, p. 49) (see also Problem 1.16)

$$\mathbf{F}_d = \frac{(\rho_{\mathrm{p}} - \rho_{\mathrm{f}})\, \nu\, g}{v_t\,(1-\nu)^{n-2}}\, \mathbf{w} \tag{1.64}$$

for small values of $|\mathbf{w}|$. On the other hand, for large values of $|\mathbf{w}|$, the drag force is expected to vary quadratically with $\mathbf{w}$, i.e., $\mathbf{F}_d \propto |\mathbf{w}|\,\mathbf{w}$. (This is motivated by the experimental observation that the drag force on an isolated particle is approximately proportional to the square of the free-stream velocity for Reynolds numbers in the range 500–2 $\times$ 10^5 (Bird et al., 2002, p. 187).) In this limit, it can be shown that (Jackson, 2000, p. 50) (see also Problem 1.16)

$$\mathbf{F}_d = \frac{(\rho_\mathrm{p} - \rho_\mathrm{f})\,\nu\, g}{v_t^{\,2}\,(1-\nu)^{2n-3}}\,|\mathbf{w}|\,\mathbf{w} \tag{1.65}$$

Given a value of the relative velocity $\mathbf{w}$, it appears difficult to decide whether (1.64) or (1.65) should be used. It seems reasonable to use (1.64) if the Reynolds number (Re') based on $\mathbf{w}$ and a suitable length scale L is $\ll 1$. Conversely, if $Re' \gg 1$, (1.65) should be used. As noted by R. Jackson (private communication, 2001), L can be chosen as the particle diameter d_p if the suspension is dilute ($\nu \ll 1$). Conversely, for a dense suspension, L may be much less than d_p and should be based on the spacing between the particles.

The z component of the drag force F_{dz} can be compared with the weight of the particles per unit total volume, which is given by $F'_w = \rho_\mathrm{p}\,\nu\, g$. Consider a suspension of particles of radius R =0.5 mm in air. The ratio F_{dz}/F'_w is of order 10^{-3} for a relative velocity w_z =0.01 m s^{-1}, and of order 1 for w_z = 6 m s^{-1} (Problem 1.16). Thus the drag force is negligible compared to the weight for low values of w_z but becomes comparable to, or can exceed, the weight for high values of w_z. Similarly, comparison of particles immersed in air and water shows that for a given value of w_z, the effect of drag is more pronounced in the latter case (Problem 1.16). This arises mainly because the shear viscosity of water is about 40 times larger than that of air, leading to a much lower value for the terminal velocity of an isolated particle in water.

Consider the effect of fluid–particle interactions on the flow of granular materials. Some examples involving air as the interstitial fluid will be discussed. The presence of air retards the discharge of fine particles from storage vessels. Spink and Nedderman (1978) note that the effect of air drag can be significant for particles whose diameter is below 0.4 mm. Conversely, in the case of standpipes (see §1.1, example (e)), air drag causes a large enhancement in the flow rate of particles. The transient behavior is also interesting. When a material containing fine particles is rapidly poured into a storage vessel such as a hopper, it becomes *aerated*, i.e., it contains more air than in the state of packing attained eventually. The entrained air escapes slowly, leading to deaeration of the material. If the hopper outlet is opened when the material is still aerated, it *floods* or discharges much faster than expected (Rathbone et al., 1987).

1.8. SUMMARY

Some examples illustrating the behavior of granular materials at rest and during flow have been discussed. These examples reveal many interesting features such as the stress dip in heaps, several types of flow patterns in silos and chutes, head-independent mass flow rate from silos, enhancement of the mass flow rate in hopper–standpipe systems, and segregation of binary mixtures in rotating cylinders. Some of these features can be explained by existing models, while others await the development of suitable models.

Interparticle forces arise from Coulombic and van der Waals interactions and also due to the presence of liquid bridges. As the particle size increases, these forces decrease relative to the weight of the particle. When the particles are in contact in air, the van der Waals and liquid bridge forces and also other forces arising from chemical bonds and solid bridges contribute to the cohesion. For saturated materials, where the particles are submerged in a liquid, the origin of cohesion is similar, except for the absence of liquid bridges. Cohesion causes the particles to stick together and enables them to sustain tensile normal stresses. Cohesive effects become important as the particle size decreases and for particles in air, as the moisture content of the air increases.

When two bodies are in contact across an interface, relative motion tangential to the interface is resisted by friction. This resistance is believed to originate in the work required to break adhesive junctions between asperities of the two surfaces in the contact zone, plowing of the asperities of one body through the other, and dissipation of energy associated with the vibration of the atoms near the interface.

In addition to the preceding forces, the relative motion between the particles and the interstitial fluid gives rise to various fluid–particle forces. The drag force acting on a suspension of particles is proportional to the relative velocity $\mathbf{w}$ between the particles and the fluid at low values of $|\mathbf{w}|$ and to $|\mathbf{w}|\,\mathbf{w}$ at high values of $|\mathbf{w}|$. At a fixed value of $\mathbf{w}$, the drag force decreases in magnitude relative to the weight of the particle as the particle size increases. Hence in applications such as discharge from silos, coarse particles are less affected by the drag force than fine particles.

In the literature, both discrete and continuum models have been used to describe granular materials. Discrete models can readily incorporate information about the particle size distribution, and possibly the particle shape, and have been used to examine flow in complex geometries. The major drawback is that they are computationally intensive. On the other hand, continuum models can be applied to systems containing a large number of particles but are hampered by the lack of satisfactory constitutive relations.

The solids fraction ν for a random packing of identical spherical particles tends to lie between two limits, which are termed DRP and LRP. Experiments suggest that $\nu_{\text{drp}} \approx 0.64$, but larger values can be obtained in some cases. The experiments of Onoda and Liniger (1990) suggest that $\nu_{\text{lrp}} \approx 0.56$, but the state of LRP is not well defined.

Treating the granular material as a continuum and ignoring the effect of the interstitial fluid, the mass, linear momentum, and energy balances have been formulated. If there are no localized torques, the angular momentum balance is identically satisfied by requiring the stress tensor to be symmetric. Finally, for problems involving relative motion between the particles and the interstitial fluid, local volume-averaged balance equations have been formulated for the two phases.

In the next chapter, experimental observations are used to guide the choice of constitutive equations for quasi-static or slow flows. Under certain conditions, these equations are adequate to examine problems of statics and slow flow.

PROBLEMS

1.1. An estimate of the height of a heap

Calculate the height of a conical heap of granular material that contains the same number of grains as the number of molecules in a hemispherical drop of water. Data: diameter of the drop $= 2$ mm, diameter of the particle $= 1$ mm, angle of repose $\beta_r = 30°$, average solids fraction $= 0.6$.

1.2. Fisher's toroidal approximation for a liquid bridge

(a) Consider an axisymmetric liquid bridge between two spheres of radius R (Fig. 1.14). Let F denote the midpoint of the line joining the centers of the two spheres. Using a cylindrical coordinate system with origin at F, let r_2 denote the radius of the neck FE.

Assuming that the meridian curve CD is a circular arc of radius r_1, show that

$$R_1 \equiv r_1/R = (S_* + 1 - \cos\beta)/\cos(\theta_c + \beta)$$
$$R_2 \equiv r_2/R = \sin\beta - R_1\,[1 - \sin(\theta_c + \beta)] \tag{1.66}$$

where β is the filling angle, θ_c is the contact angle, $2\,S$ is the separation between the spheres, and $S_* \equiv S/R$.

(b) Show that the coordinates $r_* \equiv r/R$ and $Z = z/R$ of a point (r_*, Z) on the surface of the bridge satisfy

$$r_* = R_1 + R_2 - \sqrt{R_1{}^2 - Z^2} \tag{1.67}$$

(c) The *Young–Laplace* equation is given by (Adamson and Gast, 1997, pp. 6–8, 13)

$$2\,H \equiv \frac{R\,\Delta p}{\gamma_s} = \frac{r''_*}{(1+r'^2_*)^{3/2}} - \frac{1}{r_*\sqrt{1+r'^2_*}} \tag{1.68}$$

where H is the dimensionless *mean curvature* of the interface, $\Delta p \equiv p_2 - p_1$, p_2 is the pressure in the liquid phase, p_1 is the pressure in the gas phase outside the bridge, γ_s is the surface tension, and

$$r'_* \equiv \frac{\mathrm{d}r_*}{\mathrm{d}Z}; \quad r''_* \equiv \frac{\mathrm{d}r'_*}{\mathrm{d}Z}$$

Using (1.67) and (1.68), show that

$$H = (2\,r_* - R_1 - R_2)/(2\,r_*\,R_1) \tag{1.69}$$

(d) Calculate the value of H at the neck ($Z = 0$) and at the base (i.e., at the three-phase contact line) for $R = 50$ μm, $S_* = 0$, $\beta = 10°$, and $\theta_c = 0°$. Do these results reveal any inconsistency in Fisher's approximation?

(e) For $\gamma_s = 0.073$ N m^{-1}, and other parameter values as in part (d), calculate the force using the value of H evaluated at the base and compare this with the maximum force $F_{l,\,\max}$. For $S_* = 0$ and $\theta_c = 0$, show that the maximum force is obtained in the limit $\beta \to 0$.

(f) The liquid bridge results in an attractive force between the spheres. What prevents the spheres from coming into contact?

(g) In some books, the Young–Laplace equation is written as

$$\frac{(R\,\Delta p)}{\gamma_s} = -\left[\frac{Z''}{(1+Z'^2)^{3/2}} + \frac{Z'}{r_*\sqrt{1+Z'^2}}\right] \tag{1.70}$$

where

$$Z' \equiv \frac{\mathrm{d}Z}{\mathrm{d}r_*}; \quad Z'' = \frac{\mathrm{d}Z'}{\mathrm{d}r_*}$$

Show that (1.70) is equivalent to (1.68).

Parts (a)–(e) of this problem have been adapted from Lian et al. (1993).

1.3. The variation of the drained angle of repose with the Bond number

Consider the experiments of Nase et al. (2001) on the variation of the angle of repose β_r of a heap with the Bond number *Bo* (Fig. 1.16). A crude model for the effect of *Bo* on β_r can be constructed as suggested by P. A. Kralchevsky (private communication, 2002). The free surface can be regarded as a layer of spherical particles resting on a plane inclined at an angle β_r relative to the horizontal (Fig. 1.30). Consider a particle in this layer, that is connected to particles in the layer below by n liquid bridges. The force F_l exerted by each bridge is inclined at an angle ψ relative to the y direction and can be approximated by its maximum value $F_{l,\,\max}$. Assume that the bridges do not exert a net force on the particle in the x direction and that the friction condition (1.18) is satisfied.

(a) Derive an expression for the variation of β_r with Bo. It may be helpful to introduce an angle β_{r0}, defined by $\tan \beta_{r0} = \mu$, where μ is the coefficient of friction.
(b) Attempt to fit your expression to the data of Nase et al. (2001) (Fig. 1.16).

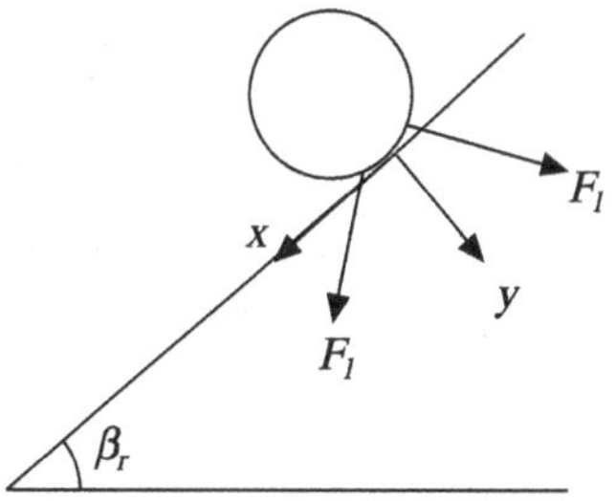

Figure 1.30. A spherical particle resting on an inclined plane and acted on by forces F_l arising from liquid bridges.

1.4. The electrostatic force between two particles
Two steel spheres of diameter 1 mm and density 7,600 kg m^{-3} are shaken in a plastic beaker and the beaker is then placed on a table. The spheres stick to the vertical inner surface of the beaker, with their centers separated by a height $H = 30$ mm (Fig. 1.31). Considering the electrostatic interaction between the spheres, estimate the charge density on the upper sphere.

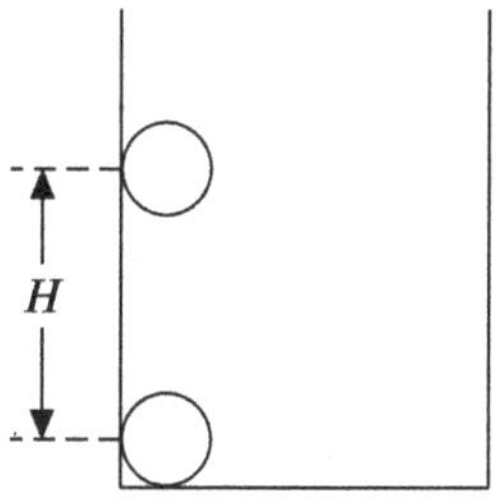

Figure 1.31. Two charged spherical particle at rest in a beaker.

This problem has been adapted from Duran (2000, p. 25).

1.5. The angular momentum balance for a rigid body
A rigid body of mass m_{p} and constant density ρ_{p} is translating and rotating relative to a coordinate system that is stationary in the laboratory frame (Fig. 1.32). The body is subject

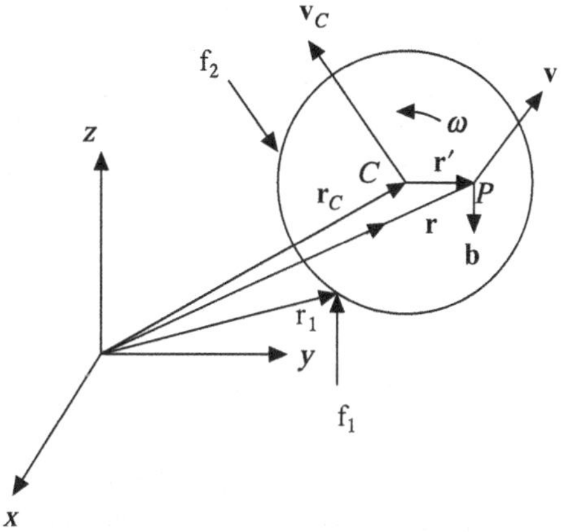

Figure 1.32. A rigid body which translates and rotates relative to a coordinate system that is at rest in the laboratory frame. Here $\mathbf{v}_C$ and $\mathbf{v}$ are the velocity of the center of mass C of the body and the velocity of the point P, respectively, relative to this coordinate system, $\mathbf{b}$ is the body force per unit mass, and $\boldsymbol{\omega}$ is the angular velocity of the body relative to a coordinate system whose origin coincides with C, and whose orientation is independent of time.

to contact forces $\mathbf{f}_i$, $i = 1, k$, which act at the points with position vectors $\mathbf{r}_i$, $i = 1, k$, relative to this coordinate system, and a body force per unit mass $\mathbf{b}$. The angular momentum balance is given by (Malvern, 1969, p. 215; Fung, 1977, p. 252)

$$\frac{\mathrm{d}}{\mathrm{d}t} \int_V \mathbf{r} \times (\rho_\mathrm{p}\, \mathbf{v})\, \mathrm{d}V = \sum_{i=1}^{k} (\mathbf{r} \times \mathbf{f}_i) + \int_V \mathbf{r} \times \rho_\mathrm{p}\, \mathbf{b}\, \mathrm{d}V \tag{1.71}$$

where V is the volume of the body, and $\mathbf{r}$ and $\mathbf{v} = \mathrm{d}\mathbf{r}/\mathrm{d}t$ are the position vector and velocity, respectively, of any point P in the body relative to this coordinate system. The linear momentum balance is given by

$$\frac{\mathrm{d}}{\mathrm{d}t} \int_V \rho_\mathrm{p}\, \mathbf{v}_C\, \mathrm{d}V = \sum_{i=1}^{k} \mathbf{f}_i + \int_V \rho_\mathrm{p}\, \mathbf{b}\, \mathrm{d}V \tag{1.72}$$

where $\mathbf{v}_C$ is the velocity of the center of mass C relative to the fixed coordinate system. The position vector corresponding to C is defined by

$$\mathbf{r}_C \equiv \int_V \rho_\mathrm{p}\, \mathbf{r}\, \mathrm{d}V \Big/ \int_V \rho_\mathrm{p}\, \mathrm{d}V \tag{1.73}$$

(a) Using (1.73), and noting that $\mathbf{r} = \mathbf{r}_C + \mathbf{r}'$, where $\mathbf{r}'$ is the position vector of P relative to C, show that

$$\int_V \rho_\mathrm{p}\, \mathbf{v}'\, \mathrm{d}V = 0 \tag{1.74}$$

Here $\mathbf{v}'$ is the velocity of P relative to C.

(b) Using (1.72), and noting that $\mathbf{v}' = \boldsymbol{\omega} \times \mathbf{r}'$, where $\boldsymbol{\omega}$ is the instantaneous angular velocity vector of the body, show that the angular momentum balance reduces to

$$\frac{\mathrm{d}}{\mathrm{d}t}(\mathbf{I}' \cdot \boldsymbol{\omega}) = \sum_{i=1}^{k} (\mathbf{r}' \times \mathbf{f}_i) \tag{1.75}$$

Here $\mathbf{I}'$ is the *inertia tensor*, defined by (Goldstein, 1980, p. 195)

$$\mathbf{I}' = \int_V \rho_\mathrm{p}\, (r'^2\, \boldsymbol{\delta} - \mathbf{r}'\,\mathbf{r}')\, \mathrm{d}V$$

where $\boldsymbol{\delta}$ is the unit tensor and r' is the magnitude of $\mathbf{r}'$. The identity (A.83) may be helpful in deriving the preceding result.

(c) Consider a sphere of radius R. Show that $\mathbf{I}' = (2\, m_\mathrm{p}\, R^2/5)\, \boldsymbol{\delta}$. Hence show that the angular momentum balance (1.75) reduces to

$$\frac{2\, m_\mathrm{p}\, R^2}{5} \frac{\mathrm{d}\,\boldsymbol{\omega}}{\mathrm{d}t} = \sum_{i=1}^{k} (\mathbf{r}' \times \mathbf{f}_i) \tag{1.76}$$

1.6. The divergence theorem

(a) Consider a rectangular surface $ABCD$ with unit normal $\mathbf{n}_*$ (Fig. 1.33a). The edge AB lies in the x_1–x_2 plane and is of length L, whereas the edge AD is of length M. The *projection* of this surface onto the x_1–x_2 plane is given by the rectangle $ABEF$, which lies in the x_1–x_2 plane and is obtained by drawing lines CE and DF perpendicular

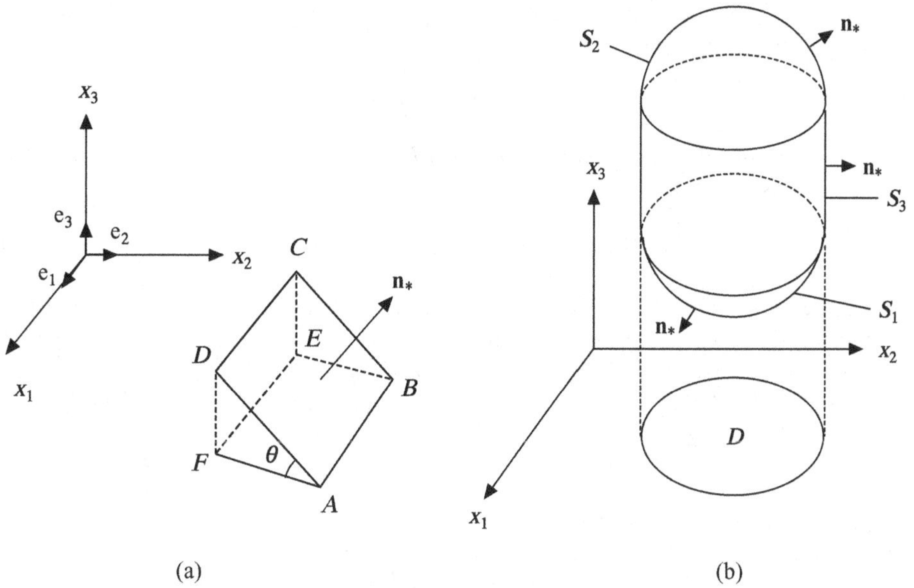

Figure 1.33. (a) The rectangle $ABCD$ has a unit normal $\mathbf{n}_*$ and is inclined at an angle θ to the x_1–x_2 plane. It intersects this plane along AB. The lengths of AB and AD are L and M, respectively. The rectangle $ABEF$ is the projection of $ABCD$ onto the x_1–x_2 plane. (b) A volume V which is bounded by the surface $S = S_1 + S_2 + S_3$. Here $\mathbf{n}_*$ is the unit outward normal to S. The domain D represents the projection of V onto the x_1–x_2 plane.

to this plane. If G and G_{12} denote the areas of $ABCD$ and of its projection $ABEF$, respectively, show that

$$G_{12} = G\,(\mathbf{n}_* \cdot \mathbf{e}_3) = G\,n_{3*}$$

where $\mathbf{e_3}$ is the unit vector along the x_3 axis.

(b) Consider a regular three-dimensional region V which is bounded by a closed surface S and whose projection onto the x_1–x_2 plane is a regular two-dimensional region D (Fig. 1.33b). (A region is said to be *regular* if a straight line passing through any point in its interior does not intersect the boundary at more than two points.) The surface S has a unit outward normal $\mathbf{n}_*$. Assume that S can be divided into three parts S_1, S_2, and S_3, such that

$$x_3 = h_1(x_1, x_2) \text{ on } S_1; \quad x_3 = h_2(x_1, x_2) \geq h_1(x_1, x_2) \text{ on } S_2 \tag{1.77}$$

and

$$n_{3*} \equiv \mathbf{n}_* \cdot \mathbf{e}_3 = 0 \text{ on } S_3 \tag{1.78}$$

Considering a function $F_3(x_1, x_2, x_3)$ that has continuous first derivatives at every point of V, show that

$$\int_V \frac{\partial F_3}{\partial x_3}\,\mathrm{d}V = \int_V \frac{\partial F_3}{\partial x_3}\,\mathrm{d}x_1\mathrm{d}x_2\mathrm{d}x_3 = \int_D [F_3(x_1, x_2, h_2(x_1, x_2))$$

$$-F_3(x_1, x_2, h_1(x_1, x_2))]\,\mathrm{d}x_1\mathrm{d}x_2 \tag{1.79}$$

Using (1.78) and the result obtained in part (a), and noting that $S = S_1 + S_2 + S_3$, show that (1.79) can be rewritten as

$$\int_V \frac{\partial F_3}{\partial x_3} \mathrm{d}V = \int_S F_3\, n_{3*}\, \mathrm{d}S$$

Similarly, considering two other functions $F_1(x_1, x_2, x_3)$ and $F_2(x_1, x_2, x_3)$ that have continuous first derivatives, we have

$$\int_V \frac{\partial F_1}{\partial x_1} \mathrm{d}V = \int_S F_1\, n_{1*}\, \mathrm{d}S; \quad \int_V \frac{\partial F_2}{\partial x_2} \mathrm{d}V = \int_S F_2\, n_{2*}\, \mathrm{d}S \tag{1.80}$$

Let the vector field **F** be defined by $\mathbf{F} = F_i(x_1, x_2, x_3)\mathbf{e}_i$, where $\mathbf{e}_i$, $i = 1, 3$, are the basis vectors along the coordinate axes, and the summation convention applies. Using the earlier results, derive the *divergence theorem* (Kaplan, 1952, pp. 269–271; Piskunov, 1969, pp. 697–699; Fung, 1977, pp. 242–244; Slattery, 1999, pp. 680–682)

$$\int_V \left(\frac{\partial F_1}{\partial x_1} + \frac{\partial F_2}{\partial x_2} + \frac{\partial F_3}{\partial x_3} \right) \mathrm{d}V = \int_S (n_{1*}\, F_1 + n_{2*}\, F_2 + n_{3*}\, F_3)\, \mathrm{d}S$$

or equivalently,

$$\int_V \nabla \cdot \mathbf{F}\, \mathrm{d}V = \int_S \mathbf{n}_* \cdot \mathbf{F}\, \mathrm{d}S \tag{1.81}$$

This problem has been adapted from Kaplan (1952, pp. 269–271) and Piskunov (1969, pp. 697–699). As noted by Kaplan, the theorem also holds for any region that can be divided into a finite number of regions of the type described in part (b). Further, it also holds if **F** is a second-order tensor (Slattery, 1999, p. 682). The divergence theorem is attributed to Gauss and Ostrogradsky.

1.7. The generalized transport theorem and the transport theorem

Consider a volume $V(t)$ occupied by a continuum. The boundary of $V(t)$ moves with a velocity **w** relative to a coordinate system that is stationary in the laboratory frame (Fig. 1.34a). Note that **w** can in general differ from the velocity **v** of the continuum at that

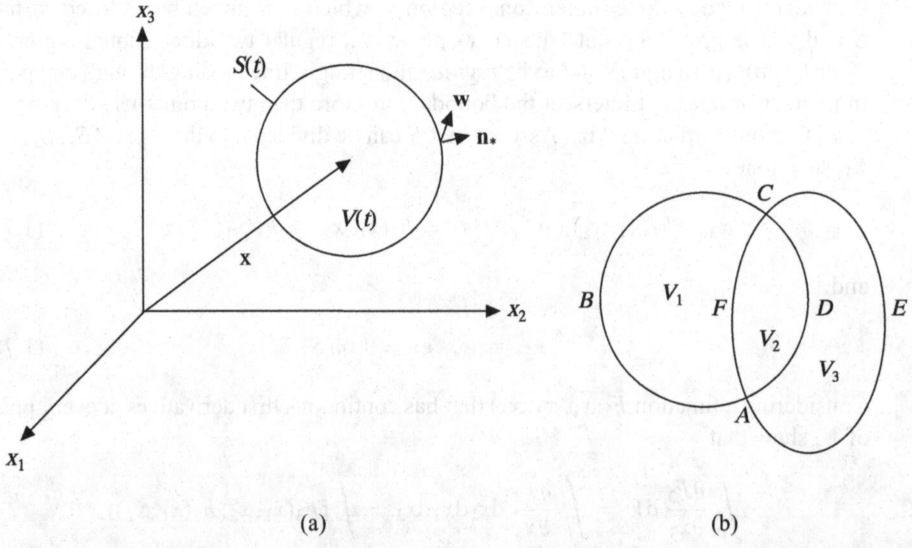

Figure 1.34. (a) A volume $V(t)$, whose boundary $S(t)$ moves with a velocity **w** relative to the laboratory frame. Here $\mathbf{n}_*$ is the unit outward normal to $S(t)$. (b) The volume $V(t)$ at two times t and $t + \Delta t$. Here $V(t) = V_1 + V_2$ and $V(t + \Delta t) = V_2 + V_3$.

point. For example, a part of the boundary can be stationary ($\mathbf{w} = 0$) and permeable to the continuum, and the latter can cross the boundary ($\mathbf{v} \neq 0$). Let ψ denote any quantity defined per unit volume, such as the density ρ, or the linear momentum per unit volume $\rho\mathbf{v}$. Consider the location of the volume at two times t and $t + \Delta t$ (Fig. 1.34b). Let V_1, V_2, and V_3 denote the volumes bounded by the surfaces $ABCFA$, $AFCDA$, and $ADCEA$, respectively, so that $V(t) = V_1 + V_2$, $V(t + \Delta t) = V_2 + V_3$. We wish to derive the *generalized transport theorem*

$$\frac{\mathrm{d}}{\mathrm{d}t}\int_{V(t)} \psi \,\mathrm{d}V = \int_{V(t)} \frac{\partial \psi}{\partial t}\,\mathrm{d}V + \int_{S(t)} (\mathbf{n}_* \cdot \mathbf{w})\,\psi\,\mathrm{d}S \tag{1.82}$$

where $\mathbf{n}_*$ is the unit *outward* normal to the bounding surface $S(t)$ of the volume $V(t)$.

(a) Using the definition of a derivative, show that

$$\frac{\mathrm{d}}{\mathrm{d}t}\int_{V(t)} \psi \,\mathrm{d}V = \int_{V(t)} \frac{\partial \psi}{\partial t}\,\mathrm{d}V + \lim_{\Delta t \to 0}\frac{1}{\Delta t} \times \left(\int_{V_3} \psi(\mathbf{x}, t + \Delta t)\,\mathrm{d}V - \int_{V_1} \psi(\mathbf{x}, t)\,\mathrm{d}V\right) \tag{1.83}$$

where $\mathbf{x}$ is the postion vector corresponding to a point in the volume (Fig. 1.34a).

(b) Consider the second term on the right-hand side of (1.83). Let $\mathrm{d}S$ denote an infinitesimal element of surface area on the surface ADC. Noting that $\mathrm{d}V = \mathrm{d}S\,\mathrm{d}Y$, where $\mathrm{d}Y$ is an infinitesimal length measured along the outward normal to $\mathrm{d}S$ (i.e., the normal pointing into V_3), show that for $\Delta t \ll t$,

$$\int_{V_3} \psi(\mathbf{x}, t + \Delta t)\mathrm{d}V \approx \int_{ADC} \psi(\mathbf{x}, t + \Delta t)\,\Delta Y\,\mathrm{d}S$$

Expressing ΔY in terms of Δt and the velocity $\mathbf{w}$ of the surface ADC, show that

$$\int_{V_3} \psi(\mathbf{x}, t + \Delta t)\mathrm{d}V \approx \int_{ADC} \psi(\mathbf{x}, t + \Delta t)\,(\mathbf{n}_* \cdot \mathbf{w})\,\mathrm{d}S\,\Delta t$$

(c) Using a similar procedure, show that

$$\int_{V_1} \psi(\mathbf{x}, t)\mathrm{d}V \approx -\int_{ABC} \psi(\mathbf{x}, t)\,(\mathbf{n}_* \cdot \mathbf{w})\,\mathrm{d}S\,\Delta t$$

(d) Substituting these results in (1.83), and taking the limit $\Delta t \to 0$, show that it reduces to the generalized transport theorem (1.82). Identifying $V(t)$ with a material volume $V_m(t)$, so that $\mathbf{w} = \mathbf{v}$, the velocity of the continuum at a point on the boundary, (1.82) reduces to the *transport theorem*

$$\frac{\mathrm{d}}{\mathrm{d}t}\int_{V_m(t)} \psi \,\mathrm{d}V = \int_{V_m(t)} \frac{\partial \psi}{\partial t}\,\mathrm{d}V + \int_{S_m(t)} (\mathbf{n}_* \cdot \mathbf{v})\,\psi\,\mathrm{d}S \tag{1.84}$$

where $S_m(t)$ is the bounding surface of $V_m(t)$.

This problem has been adapted, with some modifications, from Shames (1962, pp. 78–81). The generalized transport theorem is also derived in Slattery (1999, p. 21), and the transport theorem is derived in Fung (1977, pp. 248–250) and Slattery (1999, pp. 18–20). The latter uses a different approach.

1.8. Symmetry of the stress tensor for a classical continuum

Consider a fluid which does not have any localized torques. Relative to the laboratory frame, the integral form of the angular momentum balance is given by (Slattery, 1999, p. 31)

$$\frac{\mathrm{d}}{\mathrm{d}t}\int_{V_m(t)} \mathbf{r} \times (\rho\,\mathbf{v})\,\mathrm{d}V = \int_{S_m(t)} \mathbf{r} \times \mathbf{t}_{(\mathbf{n})}\mathrm{d}S + \int_{V_m(t)} \mathbf{r} \times (\rho\,\mathbf{b})\,\mathrm{d}V \tag{1.85}$$

where $V_m(t)$ is a material volume with bounding surface $S_m(t)$ and unit inward normal $\mathbf{n}$ (Fig. 1.29), $\mathbf{r}$ is the position vector corresponding to a material point, $\mathbf{v}$ is the velocity of this point, $\mathbf{t}_{(\mathbf{n})}$ is the stress vector, and $\mathbf{b}$ is the body force per unit mass.

(a) Using the transport theorem (1.84), the divergence theorem (1.81), and the mass balance (1.25), show that

$$\frac{\mathrm{d}}{\mathrm{d}t}\int_{V_m(t)} \rho\,\psi\,\mathrm{d}V = \int_{V_m(t)} \rho\,\frac{\mathrm{D}\psi}{\mathrm{D}t}\,\mathrm{d}V \tag{1.86}$$

where ψ is any scalar, vector, or second-order tensor. Hence (1.85) can be rewritten as

$$\int_{V_m(t)} \rho\,\frac{\mathrm{D}}{\mathrm{D}t}(\mathbf{r}\times\mathbf{v})\,\mathrm{d}V = \int_{S_m(t)} \mathbf{r}\times\mathbf{t}_{(\mathbf{n})}\mathrm{d}S + \int_{V_m(t)} \mathbf{r}\times(\rho\,\mathbf{b})\mathrm{d}V \tag{1.87}$$

where $\mathrm{D}/\mathrm{D}t$ is the material derivative defined by (1.22).

(b) Show that the integrand on the left-hand side of (1.87) can be rewritten as

$$\rho\,\frac{\mathrm{D}}{\mathrm{D}t}(\mathbf{r}\times\mathbf{v}) = \rho\,\mathbf{r}\times\frac{\mathrm{D}\mathbf{v}}{\mathrm{D}t} \tag{1.88}$$

(c) Expressing the stress vector in terms of the stress tensor, and using Cartesian tensor notation, show that

$$\mathbf{r}\times\mathbf{t}_{(\mathbf{n})} = -n_{i*}\,r_k\,\sigma_{ij}\epsilon_{kjl}\,\mathbf{e}_l \tag{1.89}$$

where n_{i*} is the component of the unit *outward* normal to the bounding surface $\mathbf{n}_*$ (see Fig. 1.29), ϵ_{kjl} is the component of the alternating tensor defined by (A.22), and $\mathbf{e}_l$ is a basis vector for the Cartesian coordinate system.

(d) Using the divergence theorem (1.81), the linear momentum balance (1.30), and (1.87)–(1.89), show that the angular momentum balance reduces to

$$\int_V \sigma_{ij}\,\epsilon_{ijl}\,\mathbf{e}_l\,\mathrm{d}V = 0 \tag{1.90}$$

(e) Assuming that the integrand in (1.90) is continuous, show that

$$\sigma_{ij} = \sigma_{ji}$$

Hence the stress tensor is symmetric.

This problem has been adapted from Slattery (1999, p. 35).

1.9. The angle of stability of a heap

(a) Consider three circular disks of diameter D resting on a plane which is inclined at an angle θ to the horizontal (Fig. 1.35a). The configuration is mechanically stable if the vector representing the weight of the upper disk C intersects the line AB joining the centers of the lower two disks at some point between A and B. Determine the *maximal angle of stability* θ_c, i.e., the maximum value of θ that satisfies the stability criterion. The angle θ_c can be regarded as a rough estimate of the angle of repose β_r.

(b) Consider four spheres of diameter D in contact with each other to form a regular tetrahedron, such that three spheres forming the base are in contact with the inclined plane $FGIH$ (Fig. 1.35b). Here A, B, and C are the centers of the spheres forming the base. Show that $\tan\theta_c = 1/(2\sqrt{2})$.

(c) In general, the line AB joining the centers of two of the basal spheres may be inclined at an angle ψ relative to the x axis (Fig. 1.35c). Show that $\tan\theta_c = 1/(2\sqrt{2}\cos\psi)$. Hint: Let K denote the centroid of the triangle ABC. If $\psi = 0$, the line joining K to the midpoint J of AB is perpendicular to FG. For $\psi \neq 0$, the line KL is perpendicular to FG, where $\angle JKL = \psi$.

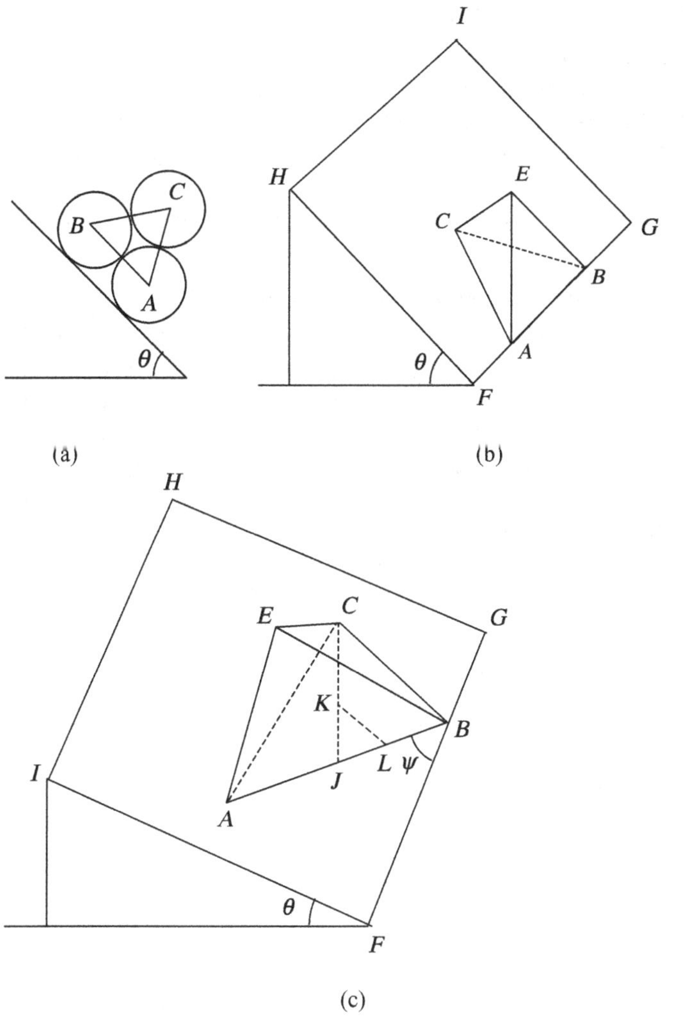

Figure 1.35. Stability of a system of (a) three disks, and (b, c) four spheres resting on an inclined plane. Here A, B, and C denote the centers of the disks in (a), and the centers of the spheres which are in contact with the inclined plane $FGHI$ in (b) and (c). The point E denotes the center of the upper sphere in (b). In (c), AB is inclined at an angle ψ to FG. The point J denotes the midpoint of AB, K is the centroid of triangle ABC, and KL is perpendicular to FG.

(d) Calculate $\overline{\tan\theta_c}$, the average value of $\tan\theta_c$ for $0 \leq \psi \leq \pi/3$. Why has the upper limit on ψ been specified as $\pi/3$? Assuming that $\tan\overline{\theta}_c \approx \overline{\tan\theta_c}$, calculate $\overline{\theta}_c$, the average value of θ_c. Compare this value with that obtained by numerical integration.

This problem has been adapted from Albert et al. (1997).

1.10. The translation of a sphere in a fluid, in the limit of creeping flow

The velocity of the center of a sphere relative to the laboratory frame is $\mathbf{V}$ and the free-stream velocity of the fluid relative to this frame is $\mathbf{U}_\infty$. For the case where both $\mathbf{V}$ and $\mathbf{U}_\infty$ are constants, consider two Cartesian coordinate systems: (i) a system which is at rest relative to the laboratory reference frame and has coordinates (x_1, x_2, x_3) (ii) a system which moves relative to (i) such that its origin coincides with the center of the sphere and has coordinates (x_1', x_2', x_3'). The systems coincide at time $t = 0$, and the coordinates are related by

$$x_i' = x_i - V_i\, t \tag{1.91}$$

Equation (1.91) is called a *Galilean transformation* (Resnick, 1968, p. 5).

(a) The form of the linear momentum balance (1.31) is assumed to remain unchanged when it is expressed in terms of the moving coordinate system. Using (1.91) and assuming that $\sigma_{ij} = \sigma'_{ij}$ and the fluid is incompressible, show that $b'_i = b_i$. Here σ'_{ij} and b'_i denote the components of the stress tensor and the body force per unit mass, respectively, relative to the moving coordinate system.

(b) Hence show that in the limit of creeping flow, the drag force on the sphere is given by (1.54).

1.11. A discrete model for the stress dip at the base of a heap

(a) To simulate the stress dip observed at the base of some sandpiles, several simplified models have been developed, one of which is as follows. Consider a two-dimensional heap formed by a close packed lattice of rigid, circular disks of mass m and with the diameter scaled to unity (Fig. 1.36a). It is assumed that the contact forces between the disks are directed along lines parallel to the coordinate axes, with L and R denoting the forces parallel to the y and x axes, respectively. Applying force balances to the disk A, whose center has the coordinates (x, y), show that $L(x, y) = (y + 1)(W/2)$ and $R(x, y) = (x + 1)(W/2)$. Here $R(x, y)$ and $L(x, y)$ are the forces exerted on disk A by the disks B and C, respectively, $W \equiv m\,g/\cos\theta$, and g is the acceleration due to gravity.

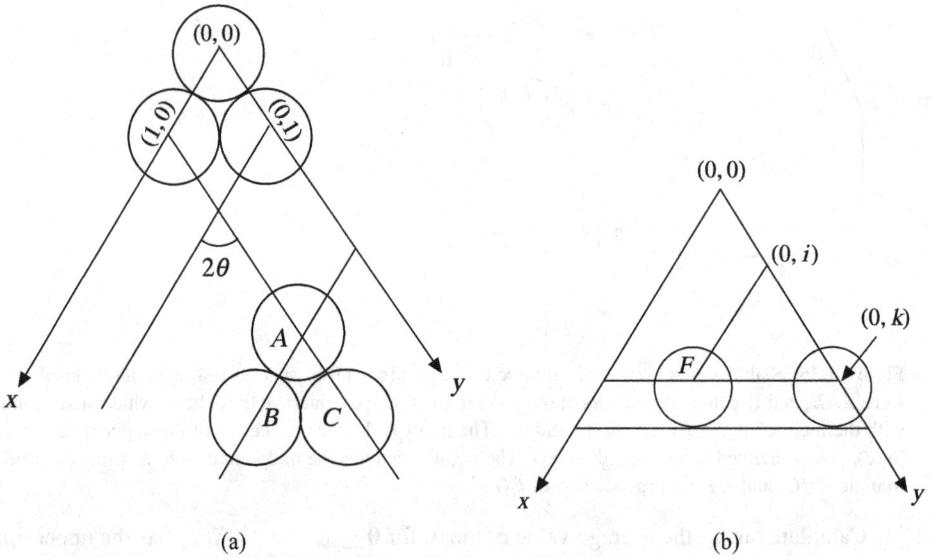

Figure 1.36. (a) A two-dimensional heap formed by circular disks. (b) The y coordinate of the disk F, which is one of the disks forming the base of the heap, is i.

(b) Consider the layer of disks in contact with the ground. For a particle F with y coordinate i (Fig. 1.36b), derive expressions for the normal force N and the horizontal force H exerted on it by the ground.

(c) Sketch N and H as functions of i.

This problem has been adapted from Liffman et al. (1992).

1.12. The dynamics of a spring–mass–dashpot system

(a) A block of mass m slides on a frictionless surface and is connected to a fixed support by a spring and a dashpot acting in parallel (Fig. 1.37). Here k is the spring constant,

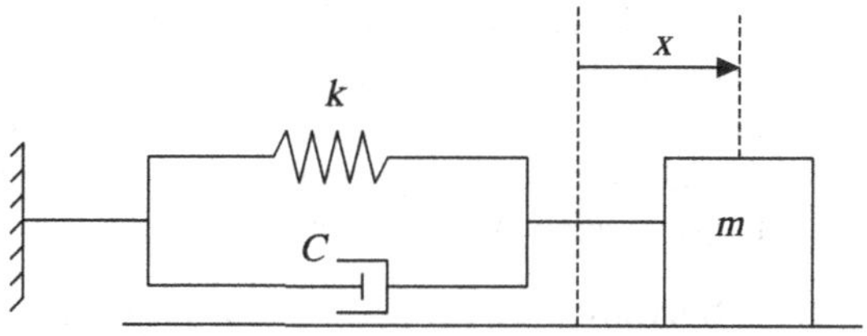

Figure 1.37. A block of mass m is connected to a spring and a dashpot in parallel. The constants k and C denote the spring constant and the damping coefficient, respectively. The block slides on a frictionless surface, and x is the displacement of the center of mass of the block from its equilibrium position.

C is the damping coefficient associated with the dashpot, and x is the displacement of the center of mass of the block from its *equilibrium* or rest position. The forces exerted by the spring and the dashpot are proportional to x and the velocity of the block, respectively. Formulate the equation of motion for the block in terms of $\omega \equiv \sqrt{k/m}$ and $2\,h \equiv C/m$.

(b) Derive the general solution for (i) $h^2 < \omega^2$ and (ii) $h^2 > \omega^2$.

(c) For case (i), sketch the solution for the initial condition $x(0) = x_0$, $\frac{\mathrm{d}x}{\mathrm{d}t}(0) = v_0 > 0$.

(d) For case (ii), identify initial conditions under which $x(t)$ exhibits a maximum, and sketch $x(t)$ for this case.

(e) The case $h^2 = \omega^2$ represents *critical damping*. Derive the general solution for this case.

(f) For long times, show that the approach to the steady state is faster for $h^2 = \omega^2$, than for $h^2 > \omega^2$.

This problem has been adapted from Andronov and Chaikin (1949, pp. 15–16, 26).

1.13. The force exerted on a load cell during the emptying of a hopper

A conical hopper is suspended from a load cell as shown in Fig. 1.38.

(a) The exit orifice is closed with a small plug of cotton, and the hopper is filled with sand by pouring it from a funnel. At the end of this process, the upper surface of the sand is horizontal and level with the top of the hopper. Water is then poured slowly so that it fills all the voids between the particles. Estimate approximately the net force exerted by the system (i.e., sand, water, hopper, and supporting frame) on the load cell in (i) the horizontal direction and (ii) the vertical direction. The mass M_0 of the frame and the empty hopper is 1.5 kg, the particle density of sand is 2,600 kg m^{-3}, and the dimensions of the hopper are $D_0 = 20$ mm, $D_1 = 160$ mm, and $H = 200$ mm.

(b) Suppose that the interstitial fluid is air, and the sand is allowed to discharge by removing the cotton plug. Consider the volume $V(t)$ occupied by the sand in the hopper. At time $t = 0$, when the sand is at rest, $V(t)$ corresponds to the volume $ABCD$ in Fig. 1.38. For simplicity, assume that the upper *free surface* AB of $V(t)$ remains horizontal as the sand

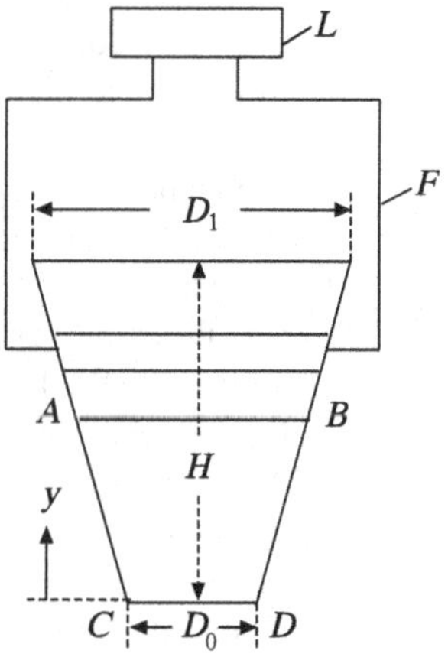

Figure 1.38. A conical hopper suspended from a load cell L by a frame F. For part (b) of the problem, the sand occupies the volume $ABCD$ at time $t = 0$.

flows. The effect of fluid–particle interaction can be ignored. Integrate the differential linear momentum balance over $V(t)$ to derive an expression for net force F_y exerted in the y direction *by* the sand *on* the hopper. Even though it is not possible to evaluate all the integrals explicitly, simplify the expression to the extent possible. In this context, the generalized transport theorem (1.82) may be helpful.

1.14. Derivation of the differential mass and momentum balances from the integral balances

(a) Using the transport theorem (1.84) and the divergence theorem (1.81), show that the integral mass balance (1.24) can be written as

$$\int_{V_m(t)} \left[\frac{\partial \rho}{\partial t} + \nabla \cdot (\mathbf{v}\,\rho)\right] \mathrm{d}V = 0 \tag{1.92}$$

where $V_m(t)$ denotes a material volume with a bounding surface $S_m(t)$.

(b) Under what conditions does (1.92) reduce to the differential mass balance (1.25)?

(c) Using (1.84), and the relation (1.28) between the stress vector and the stress tensor, show that (1.29) reduces to the differential linear momentum balance (1.30).

1.15. A relation between densities in two configurations

Consider a material volume that has a volume V_0 in the reference configuration and a volume V in the current configuration at time t. The density field is given by $\rho_0(\mathbf{X})$ and $\rho(\mathbf{x}, t)$ in the reference and current configurations, respectively, where $\mathbf{X}$ and $\mathbf{x}$ represent the position vectors corresponding to a material point in the reference and current configurations, respectively, and the motion of the material is described by $\mathbf{x} = \mathbf{x}(\mathbf{X}, t)$. Show that

$$\rho(\mathbf{x}(\mathbf{X}, t), t)\, J = \rho_0(\mathbf{X}) \tag{1.93}$$

where J denotes the magnitude of the determinant of the Jacobian matrix $[J_{ij}]$, and

$$J_{ij} \equiv \frac{\partial x_i}{\partial X_j}$$

Note that infinitesimal volume elements in the two configurations are related by $\mathrm{d}V = J\,\mathrm{d}V_0$.

This problem has been adapted from Hunter (1983, p. 30).

1.16. An application of the local volume-averaged equations

(a) Consider the steady sedimentation of a suspension of spherical particles in an incompressible fluid contained in a cylindrical vessel. The diameter of the vessel is much larger than the particle diameter d_p. Using the local volume-averaged momentum balances (1.57) and (1.58), and (1.60) with $\mathbf{F}_o = 0$, derive expressions for the z component of the drag force per unit volume of the suspension F_d. Consider two limiting cases of small and large relative velocities. The z coordinate is directed in the upward vertical direction. It can be assumed that (i) all quantities depend only on the z coordinate, (ii) the local average fluid and particle velocities are directed along the z direction, (iii) $\boldsymbol{\sigma}_\mathrm{f} = p_\mathrm{f}\,\mathbf{I}$, where p_f is the fluid-phase pressure, (iv) the particles are not in contact, i.e., $\boldsymbol{\sigma}_\mathrm{p} = 0$; (v) the local average particle velocity is a constant, and (vi) the Richardson–Zaki equation (1.61) can be used.

(b) Using the results of part (a), calculate the ratio of F_d to the weight of particles per unit volume of the suspension F_w, for suspensions of particles in air and water at 20°C. A suitable drag relation can be used to estimate the terminal velocity of an isolated particle v_t. For example, Turton and Levenspiel (1986) give the following relation for drag coefficient C_D:

$$C_D = \frac{24}{Re_t}(1 + 0.173\, Re_t^{0.657}) + \frac{0.413}{1 + 1.63 \times 10^4\, Re_t^{-1.09}}, \quad Re_t < 2 \times 10^5 \quad (1.94)$$

Here Re_t is the single-particle Reynolds number defined by (1.62). The drag force on an isolated particle is given by $F_d' \equiv C_D\,(\pi\, R^2)\,(\rho_{\rm f}\, v_t^2/2)$, where R is the radius of the particle and $\rho_{\rm f}$ is the density of the fluid. Data: $R = 0.5$ mm, $\rho_{\rm p} = 2{,}600$ kg m^{-3}, $\nu = 0.5$. Values of the relative velocity w_z in m s^{-1}: air: 0.01, 6; water: 5 $\times 10^{-4}$, 0.1.

Part (a) of this problem has been adapted from Jackson (2000, pp. 49–50).

2

Theory for Slow Plane Flow

At present there are no constitutive equations that are valid over the entire range of densities and velocities encountered in the storage and handling of granular materials. Most of the available equations fall into one of two regimes: (i) *slow* flow and (ii) *rapid* flow. In the slow flow regime, the solids fraction ν is high and forces are exerted across interparticle contacts which last for a long time compared to the contact time in the rapid flow regime. The contacts occur during the sliding and rolling of particles relative to each other. In the rapid flow regime, the solids fraction is low, and momentum is transferred mainly by collisions between particles and by free flight of particles between collisions. Consider the flow of a granular material between two parallel plates. If V is the relative velocity of the plates and H is the gap between them, the stresses are found to be approximately independent of the nominal *shear rate* $\dot{\gamma} \equiv V/H$ in the slow flow regime (small $\dot{\gamma}$ and high ν) and to increase strongly with $\dot{\gamma}$ in the rapid flow regime (large $\dot{\gamma}$ and low ν).

In devices such as hoppers and chutes, both the regimes can occur in different spatial regions, and there can also be transition regions where the nature of the flow changes from one regime to the other. Given a device and a set of operating conditions, it is difficult to determine a priori the type of flow regime that is likely to prevail. However, the following criterion can be used as a very rough guideline (Savage and Hutter, 1989).

For steady shear flow, the normal stress arising from interparticle collisions between spherical particles of diameter d_p is of the order of $\rho_p\, d_p^2\, \dot{\gamma}^2$, where ρ_p is the particle density and $\dot{\gamma}$ is the shear rate (see §7.1). Let

$$R_* \equiv (\rho_p\, d_p^2\, \dot{\gamma}^2)/N_T \tag{2.1}$$

denote the ratio of the collisional normal stress to the total normal stress N_T in the direction of the velocity gradient. Then the slow flow regime is characterized by $R_* \ll 1$.

For example, consider the shearing of granular materials in a cylindrical Couette cell (Fig. 2.1). One of the cylinders, say, the inner cylinder, rotates with an angular velocity ω, and the other cylinder is stationary. Assuming that the gap between the cylinders W is small compared to the radius of the inner cylinder R, the nominal shear rate is given by $\dot{\gamma} = R\,\omega/W$. Bocquet et al. (2002) state that in their experiments, the shear stress exerted on the wall of the cell did not vary much with $\dot{\gamma}$. Hence the data can be classified under the regime of slow flow.

The value of R_* can be estimated as follows. In this problem, $N_T = \sigma_{rr}$, the radial normal stress. As it is difficult to estimate the latter, we assume that $N_T \approx N_a \equiv \overline{\sigma}_{zz}(z=0)$, the average value of the axial normal stress at the bottom of the cell. Alternative estimates for the normal stress exerted on the wall are discussed in Problem 4.2. Assuming that (i) the upper surface of the material is stress free, (ii) the vertical shear stress σ_{rz} is negligible,

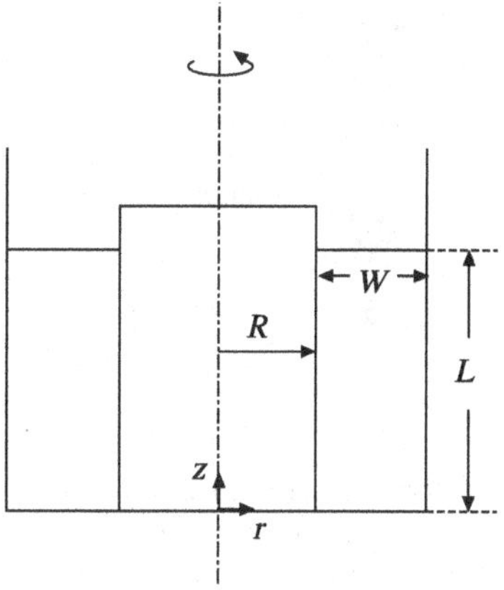

Figure 2.1. A cylindrical Couette cell. The inner cylinder of radius R rotates with a constant angular velocity, and the outer cylinder is stationary. Here L is the height of the granular material.

and (iii) the density of the granular material $\rho = \rho_p \nu$ is a constant, the z component of the momentum balance implies that $N_a \approx \rho_p \nu g L$, where g is the acceleration due to gravity and L is the height of the granular material (Fig. 2.1). As ν is of order 1, and the actual value is not reported in some cases, we replace the preceding estimate by $N_T \approx N_a \approx \rho_p g L$. Even though the dependence of N_a on L is in accord with the model of Tardos et al. (1998), it must be regarded as a very rough estimate.

The experiments of Bocquet et al. (2002) were based on glass beads, with a particle diameter $d_p = 0.55$–0.95 mm and a Couette gap $W = 0.1$ m. For the data shown in Fig. 2 of their paper, the shear rate $\dot{\gamma}$ was in the range 0.13–12 s^{-1}. As the value of L was not reported in their paper, we set $N_T \approx \rho_p g W$. With these parameter values, $R_* = 5 \times 10^{-8}$–5×10^{-4}. Similarly, the data of Mueth et al. (2000) for mustard seeds (Fig. 1 of their paper) correspond to $R_* = 7 \times 10^{-8}$–3×10^{-3}.

The data of Savage and Sayed (1984) and Hanes and Inman (1985) show a strong dependence of the shear stress on the shear rate and hence can be classified under the regime of rapid flow. For these data, the estimated values of R_* are in the range 0.04–0.4 (Savage and Hutter, 1989). The criterion of Savage and Hutter (1989) should be used with caution, as it is difficult to estimate the total normal stress in some cases. Further, all of the material may not be shearing, and this may cause the actual shear rate to differ significantly from the nominal shear rate.

Constitutive equations for slow flow will be discussed in this chapter, deferring consideration of equations for rapid flow to Chapter 7. Attention will be confined to *plane flow*, i.e., a flow for which one component of velocity (measured relative to a Cartesian coordinate system) vanishes throughout the domain.

The outline of this chapter is as follows. Qualitative observations relevant to slow flow are discussed first. The concept of a yield condition is then introduced and is used to construct an approximate solution for the static stress field in a bin. The behavior of samples in shear tests is used to develop a constitutive theory, called the critical state theory. Some alternative approaches are also discussed briefly.

2.1. QUALITATIVE OBSERVATIONS

In his book *Dialogues Concerning Two Sciences*, which was published in 1638, Galileo Galeli made the following remarks regarding granular materials (cited in Galeli, 2003, p. 429):

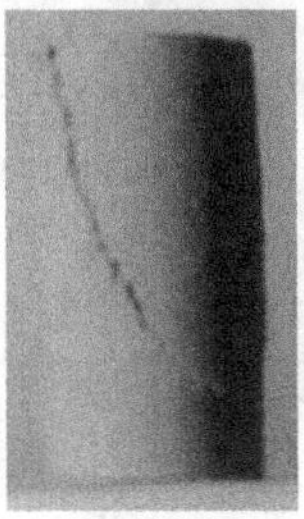

Figure 2.2. Rupture layer in a soil sample subjected to a triaxial compression test. The sample was provided by Dr. A. Vatsala.

When I take a hard substance such as stone or metal and when I reduce it by means of a hammer or fine file to the most minute and impalpable powder, it is clear that its finest particles, although when taken one by one are, on account of their smallness, imperceptible to our sight and touch, are nevertheless finite in size, possess shape, and capability of being counted. It is also true that when once heaped up they remain in a heap; and if an excavation be made within limits the cavity will remain and the surrounding particles will not rush to fill it; if shaken the particles come to rest immediately after the external disturbing agent is removed; the same effects are observed in piles of larger and larger particles, of any shape, even if spherical, as is the case with piles of millet, wheat, lead shot, and every other material.

These observations suggest that inelastic or irrecoverable effects are readily induced in granular materials. Further, the motion of these materials is highly dissipative.

Subsequently, Delanges (1788) (cited in Herbst and Winterkorn, 1965, p. 2) found that the normal force acting on the base of a cylinder filled with materials such as sand, gravel, and millet was not equal to the weight of the material. This deviation from the expected hydrostatic result was attributed to the frictional force exerted on the material by the wall of the cylinder. The effect of wall friction was also demonstrated by Phillips (1910) using a simple experiment. One end of a cylindrical tube was closed with a sheet of thin paper and the tube was then filled with sand. The paper remained intact even when a large normal load was applied to the upper surface of the sand. This shows that stresses are not transmitted very far in columns of granular materials bounded by rough walls.

In his classic paper on soil statics, Coulomb (1776) (cited in Schofield and Wroth, 1968, p. 207) (see also Heyman, 1972) examined the deformation of soils and rocks. His observations have been summarized by Schofield and Wroth (1968, p. 207) as follows:

Coulomb considered that soil was a rigid homogeneous material which could rupture into separate blocks. His experiments on solid rock specimens and on mechanisms with sliding contact suggested that both friction and cohesion must be overcome during slip along the rupture surface.

The occurrence of *rupture surfaces* or *rupture layers* is common in the slow deformation of granular materials. For example, Fig. 2.2 shows a rupture surface formed when a cylindrical sample of clay is subjected to normal stresses σ_r and $\sigma_a > \sigma_r$ in the radial and axial directions, respectively. Similarly, the light bands in Fig. 2.3 represent rupture surfaces spanning the transition region between the vertical and converging sections of a wedge-shaped bunker filled with sand. The flow of sand was stopped after allowing a small amount to discharge, and a radiograph of the bunker was taken using X-rays. In the light regions, deformation and flow of the sand has resulted in a lower density than in the dark regions. As discussed later, the rupture "surface" is actually a three-dimensional

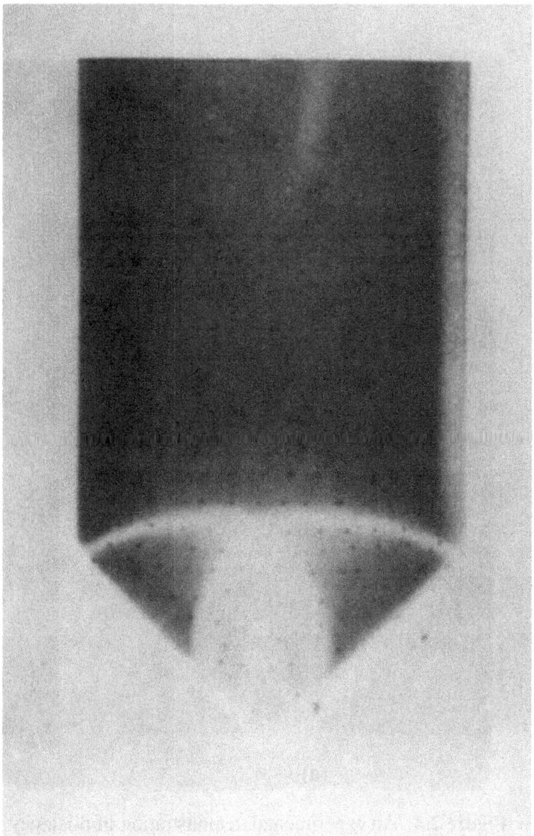

Figure 2.3. Radiograph of a wedge-shaped bunker, showing rupture layers in sand. The radiograph was taken after the upper free surface of the sand had descended by 4 mm and the flow was stopped. The density of the sand is less in the light regions than in the dark regions. The initial value of the solids fraction was 0.66–0.67. (Reproduced from Bransby and Blair-Fish, 1975, with permission from the Institution of Civil Engineers, London.)

region whose thickness in the direction perpendicular to the surface is of the order of a few particle diameters.

The elegant experiments of Reynolds (1885) revealed the phenomenon of *dilatancy*. He states that "the property consists of a definite change in bulk, consequent on a definite change of shape or distortional strain, any disturbance whatever causing a change of volume and generally dilatation." Dilatancy may be illustrated by the following experiment involving plastic bottles (Fig. 2.4a). The bottle on the right is filled with water and the one on the left is filled with sand. Water is then added to fill the gaps between the grains of sand and rise to a certain level in the tube connected to the top of the bottle. When the bottles are squeezed by hand, the water level in the tube connected to the bottle filled with water rises, as expected (Fig. 2.4b). On the other hand, if the sand is densely packed initially, the grains move apart when the bottle is squeezed. Hence water flows down from the tube into the bottle to occupy the extra volume created, and the water level in the tube decreases (Fig. 2.4b). Thus the density of granular materials often changes during deformation, unlike the case of common liquids such as water at moderate pressures.

Another interesting property of granular materials was shown by the experiments of Huber-Burnand (1829) (cited in Herbst and Winterkorn, 1965, p. 3) and Hagen (1852) (cited in Wieghardt, 1975). They found that the rate of flow of materials through orifices was independent of the head of material for large heads. For materials consisting of coarse particles, this has been subsequently confirmed by several investigators. As discussed in Chapter 3, head independence arises because of frictional forces between particles and between particles and the wall of the container. In contrast, the rate of flow of water through

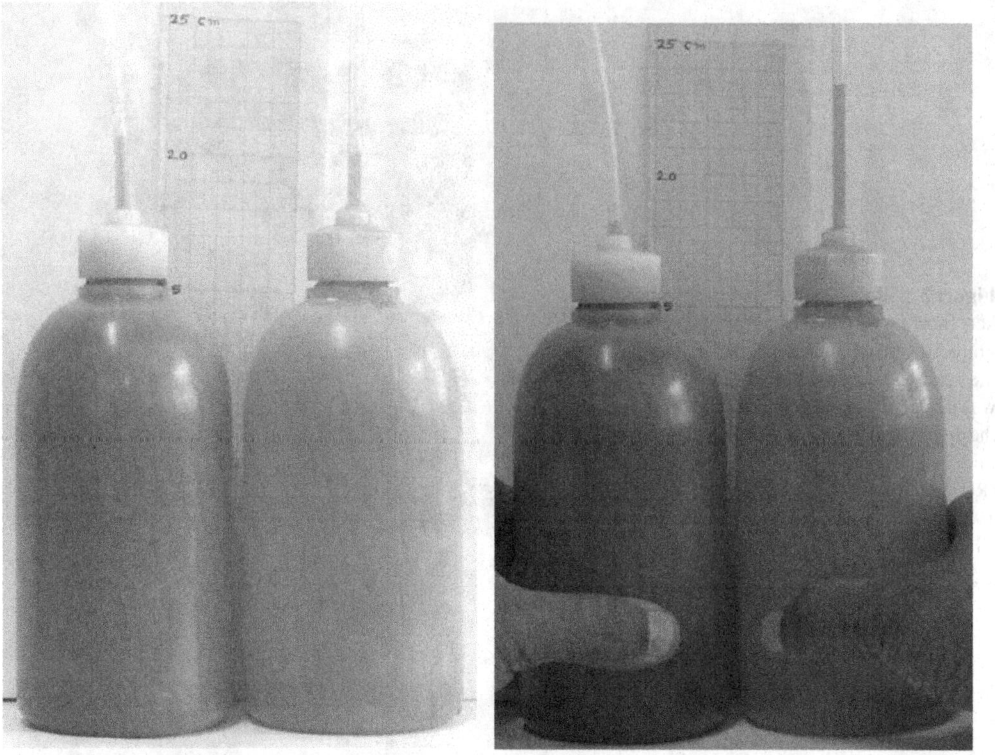

(a) (b)

Figure 2.4. An experimental demonstration of dilatancy using plastic bottles. (a) The bottle on the left is filled with sand and water, and the one on the right is filled with water. (b) When the bottles are squeezed by hand, the water level in the tube decreases relative to its initial state for the bottle filled with sand and water, and increases for the bottle filled with water. This experiment was set up by Mr. P. T. Raghuram, based on the description given in Feilden (1986).

an orifice depends on the head. For fine particles, interaction between the particles and the interstitial fluid gives rise to a head dependence (Gu et al., 1993).

So far we have discussed some qualitative aspects of statics and flow. It is seen that the mechanical behavior of granular materials differs markedly from that of common fluids. The development of constitutive equations which capture some of the above features will now be discussed.

2.2. THE WALL YIELD CONDITION

Consider a block of weight W, which is at rest on a horizontal plane (Fig. 2.5). The applied force H is balanced by the frictional force F_t exerted by the plane on the block. Coulomb's law (Coulomb, 1785) (cited in Dowson, 1979, p. 219) states that the frictional force is related to the normal force F_n exerted on the block by

$$F_t \leq \mu_s F_n + C \tag{2.2}$$

where μ_s and C are constants called the *coefficient of static friction* and the *cohesion* or the *adhesion*, respectively. The latter represents the force of adhesion between the block and the plane. Leonardo da Vinci (cited in MacCurdy, 1956, pp. 570, 576) and Amontons (1699) (cited in Dowson, 1979, pp. 99, 154–156) had proposed (2.2) earlier, but without the cohesion term. MacCurdy's translation of Leonardo's notebooks does not contain any

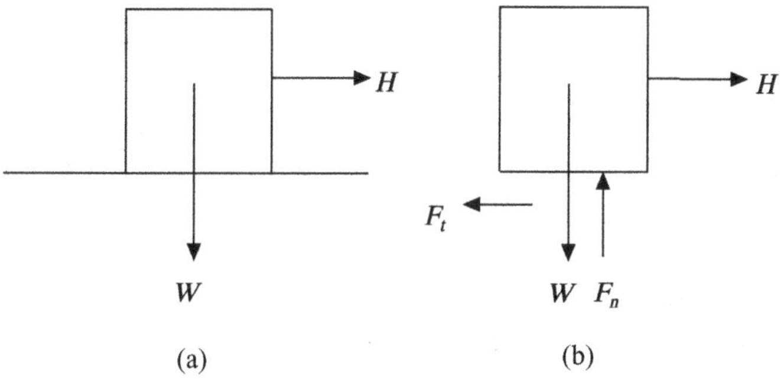

Figure 2.5. (a) A block of weight W at rest on a plane under the action of an applied horizontal force H. (b) Free-body diagram of the block. The plane exerts a normal force F_n and a frictional force F_t on the block.

equations, but the statements regarding friction are equivalent to (2.2). An alternative to Coulomb's law is discussed in Problem 2.5.

If the applied force exceeds $\mu_s F_n + C$, the block begins to slide, and (2.2) must be replaced by

$$F_t = \mu_k F_n + C \tag{2.3}$$

where μ_k is a constant called the *coefficient of kinetic friction*.

Let us extend (2.3) to a granular material moving relative to a solid surface such as the wall of a storage vessel. As it is convenient to work with stresses instead of forces, (2.3) can be replaced by

$$|T_w| = N_w \tan\delta + c_w \tag{2.4}$$

where T_w and N_w are the shear and normal stresses, respectively, exerted on the granular material by the wall, δ is a constant called the *angle of wall friction*, and c_w is called the *adhesion*. Here T_w acts in a direction which opposes the direction of motion of the material relative to the wall. The parameter δ depends on various factors such as the roughness of both the material and the wall, the stress level, and the density of the granular material. The parameter c_w is a measure of adhesion between the granular material and the wall. In the absence of adhesion, a lizard would be unable to move on the ceiling of a room. Equation (2.4) is called the *wall yield condition*. It provides a relation between the stresses acting on the granular material adjacent to a solid surface and is applicable only when the material moves relative to the surface. If the material is at rest relative to the surface, (2.4) must be replaced by

$$|T_w| \leq N_w \tan\delta' + c_w \tag{2.5}$$

where δ' is the static value of the angle of wall friction. As the values of δ' are often not reported for granular materials, we shall set $\delta' = \delta$ when (2.5) is used to examine statics. In most of the applications, we shall neglect the adhesion term and replace (2.4) by

$$|T_w| = N_w \tan\delta \tag{2.6}$$

For granular materials such as glass beads and sand, and wall materials such as aluminum and Lucite, typical values of δ are in the range 8–25° (Brennen and Pearce, 1978; Kaza, 1982; Tüzün and Nedderman, 1985b). For these systems, c_w is negligible when the materials are dry. For coal with a moisture content of 1–3.7%, and in contact with a steel or a Perspex substrate, the value of c_w was found to be about 6 Pa (Brown and Richards, 1970, p. 112).

A simple example illustrating the use of the wall yield condition is discussed in the next section.

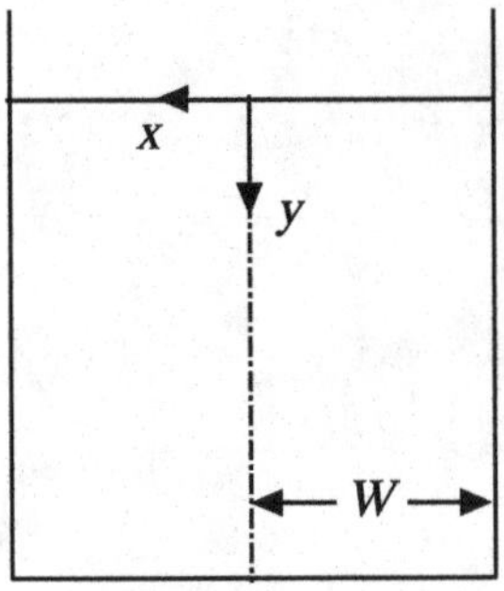

Figure 2.6. Granular material at rest in a bin of rectangular cross section. The bin is of infinite extent in the z direction.

2.3. THE JANSSEN SOLUTION FOR THE STATIC STRESS FIELD IN A BIN

Consider a granular material at rest in a bin of rectangular cross section (Fig. 2.6). For ease of exposition, it is assumed that (i) the bin is of infinite extent in the z direction, (ii) the stresses depend only on the x and y coordinates, and (iii) the density of the granular material is a constant. Subject to certain other assumptions, Janssen (1895) (see Sperl (2006) for an English translation) derived an expression for the variation of the cross-sectional averaged vertical normal stress with the depth y. As shown later, the key result is that the normal stress levels off at large depths, in accord with the experimental observations discussed in §4.2.1.

The y component of the momentum balance (1.31) reduces to

$$\frac{\partial \sigma_{xy}}{\partial x} + \frac{\partial \sigma_{yy}}{\partial y} = \rho g \tag{2.7}$$

where ρ is the density of the granular material and g is the acceleration due to gravity. Introducing the dimensionless variables

$$\xi = x/W; \quad \eta = y/W; \quad \overline{\sigma}_{xy} = \sigma_{xy}/(\rho\, g\, W); \quad \overline{\sigma}_{yy} = \sigma_{yy}/(\rho\, g\, W) \tag{2.8}$$

where W is the half-width of the bin (Fig. 2.6), and integrating (2.7) from $\xi = 0$ to $\xi = 1$, we obtain

$$\frac{\mathrm{d}}{\mathrm{d}\eta}\left(\int_0^1 \overline{\sigma}_{yy} \mathrm{d}\xi\right) = 1 - \overline{\sigma}_{xy}(\xi = 1, \eta) + \overline{\sigma}_{xy}(\xi = 0, \eta) \tag{2.9}$$

Considering symmetric solutions, the shear stress $\overline{\sigma}_{xy}$ vanishes along the centerline $\xi = 0$. In view of the convention adopted for the stress tensor (see §1.5.4), $\overline{T} \equiv \overline{\sigma}_{xy}(\xi = 1, \eta)$ is the dimensionless shear stress exerted in the y direction by the material on the left-hand wall. Assuming that the wall yield condition (2.5) is satisfied, and neglecting adhesion, we have

$$\overline{T} = \overline{N} \tan \delta \tag{2.10}$$

where $\overline{N} \equiv \overline{\sigma}_{xx}(\xi = 1, \eta)$ is the dimensionless normal stress exerted by the material on the wall. Substituting (2.10) in (2.9), we obtain

$$\frac{\mathrm{d}V}{\mathrm{d}\eta} = 1 - \overline{N} \tan \delta \tag{2.11}$$

where $V = \int_0^1 \overline{\sigma}_{yy} \mathrm{d}\xi$ is the cross-sectional averaged vertical normal stress.

Janssen assumed that

$$\overline{N} = K\, \overline{\sigma}_{yy}(\xi = 1, \eta) \approx K\, V \tag{2.12}$$

where K is a constant called the coefficient of earth pressure at rest (Atkinson and Bransby, 1982, p. 134). For a fluid at rest, $K = 1$, as all the normal stresses at a point are equal.

However, for granular materials, $K \neq 1$ in general. Even though there is no sound basis for (2.12), its use permits a simple solution to be obtained. Some information regarding the value of K is given in §2.9.

Using (2.12) and the initial condition $V(0) = V_0$, (2.11) can be integrated to obtain

$$V = (1/M) + [V_0 - (1/M)]\, e^{-M\eta} \tag{2.13}$$

where $M \equiv K\ \tan\delta$. Equations (2.10), (2.12), and (2.13) imply that

$$\begin{aligned} \overline{T} &= \overline{\sigma}_{xy}(\xi = 1, \eta) = M\,V = 1 + (M\,V_0 - 1)\, e^{-M\eta} \\ \overline{N} &= \overline{\sigma}_{xx}(\xi = 1, \eta) = K\,V = (1/\tan\delta) + \\ &\quad K\,[V_0 - (1/M)]\, e^{-M\eta} \end{aligned} \tag{2.14}$$

If $M\,V_0 < 1$, the shear and normal stresses exerted on the wall increase with the depth and level off at large depths. Conversely, if $M V_0 > 1$, the stresses decrease as the depth increases and eventually level off. In the limit $\eta \to \infty$, $V \to 1/M = 1/(K\ \tan\delta)$, $\overline{T} \to 1$, and $\overline{N} \to 1/\tan\delta$. The limiting value of the shear stress can also be obtained directly from the momentum balance (2.7) by assuming that the stress field is "fully developed," i.e., the stress components do not vary in the y direction. In this limit, the shear force exerted on the material by the wall balances the weight of the material.

It should be emphasized that wall roughness is essential for the stress to level off as the depth increases. For a bin with smooth walls ($\delta = 0$), the vertical normal stress V increases linearly with depth, as seen from (2.13) by considering the limit $\delta \to 0$ or equivalently, the limit $M \to 0$. In this case, the normal stress varies hydrostatically with depth, as in a fluid at rest. When the walls are rough ($\delta > 0$), the shear stress exerted by the walls on the material opposes gravity. This causes the normal stress to increase less rapidly with the depth than is required by the hydrostatic variation.

The Janssen solution will be extended to bins of other cross sections and compared with experimental data in §4.2.1.

2.4. THE COULOMB YIELD CONDITION

In order to relax some of the assumptions used in the Janssen solution, it is necessary to introduce the concept of a yield condition. For this purpose, consider a rupture layer separating two regions of a granular material (Fig. 2.2). By analogy with (2.2), Coulomb (1776) (cited in Schofield and Wroth, 1968, p. 207) postulated that the shear stress T_r acting on a rupture layer must satisfy

$$|T_r| \leq N_r\ \tan\phi_* + c \tag{2.15}$$

where N_r is the normal stress acting on this surface, and ϕ_* and c are material constants called the *angle of internal friction* and the *cohesion*, respectively. The normal stress N_r is assigned a positive value when it is compressive.

The parameter ϕ_* is a measure of the frictional resistance to shearing. It depends on the surface roughness, geometry, and packing of the particles, and on the extent of deformation of the particle assembly. Values of ϕ_* are in the range 20–40° for glass beads and sand (Brennen and Pearce, 1978; Nedderman, 1992, p. 25). The parameter c is a measure of the force of cohesion between particles. For Weald clay, the value of c can be estimated using the data of Roscoe et al. (1958) to be about 105 kPa (see Problem 5.11); for limestone powder and kaolinite, the values of c are in the range 1–10 kPa (Feise, 1998). It should be noted that the cohesion depends on the density of the sample.

When (2.15) holds as an inequality, there is no slip between adjacent "blocks" and the value of T_r is indeterminate. On the other hand, when slip occurs, or is about to occur,

(2.15) holds as an equality. In this case, the yield condition is said to be satisfied. Suppose that $|T_r| < c$. Then, if the yield condition is satisfied, $N_r < 0$. Thus cohesive materials, for which $c > 0$, admit tensile normal stresses.

As the velocity of sliding does not occur explicitly in (2.15), it is called a *rate-independent* relation. An interesting example of rate independence is cited by Wieghardt (1975): "When draught animals were replaced by tractors, engineers were surprised to find that the drag on a plow was almost independent of the speed." Data related to rate independence are presented in §2.13.1.

Both (2.15) and the more restricted form

$$|T_r| = N_r \tan\phi_* + c \tag{2.16}$$

are referred to as the *Coulomb failure equation* or equivalently, as the *Coulomb yield condition* (Schofield and Wroth, 1968, p. 207; Nedderman, 1992, p. 23). (The terms "yield condition," "yield criterion," and "yield function" are used interchangeably in the literature.)

2.5. GENERALIZATION OF THE COULOMB YIELD CONDITION

In this section, the bin problem discussed in §2.3 will be used as an example to motivate a generalization of the Coulomb yield condition. Suppose that it is desired to construct a solution that is more accurate than the Janssen solution, in the sense that the approximation (2.12) is not used. The y component of the momentum balance (2.7) must be supplemented by the x component of (1.31), which is given by

$$\frac{\partial \sigma_{xx}}{\partial x} + \frac{\partial \sigma_{yx}}{\partial y} = 0 \tag{2.17}$$

Assuming that the density ρ is a known function of position, the balances involve three stress components and hence cannot be solved unless an additional equation is specified. One possibility is to use an ad hoc assumption of the Janssen kind, i.e., $\sigma_{xx} = K\sigma_{yy}$, where K is a constant. Bouchaud et al. (1995) have used this approach to predict the stress field in a sandpile. Unfortunately, the value of K is not known a priori, and it is not clear how the equation should be modified for other situations.

As an alternative to the Janssen assumption, let us assume that the material satisfies a yield condition. The Coulomb yield condition (2.16) cannot be used directly, as the orientation of the rupture layer is not known. So we formulate the yield condition in terms of the stress components as

$$\tilde{F}(\sigma_{xx}, \sigma_{yy}, \sigma_{xy}, \nu) = 0 \tag{2.18}$$

where ν is the solids fraction. (As mentioned earlier, the stress tensor is assumed to be symmetric.) Even though the Coulomb yield condition does not involve ν, it has been included in (2.18) keeping in mind the applications to be discussed later.

In this book, attention will be confined to *isotropic* materials. An isotropic material is one whose constitutive equation is unaffected by rotation and reflection of the coordinate axes (Fung, 1977, p. 188). In other words, at any point in the material, the mechanical properties are independent of direction. (However, the properties can vary with position.) In constrast, if the properties vary with direction, the material is said to be *anisotropic*. Water and air are isotropic, whereas wood, graphite, and metal crystals are anisotropic.

In the context of granular materials, the following thought experiment illustrates the consequences of isotropic behavior. Consider two identical samples of a granular material confined between horizontal plates and subjected to a constant normal stress N (Fig. 2.7) by applying loads to the upper plates. The lower plates are kept stationary, and the upper plate is moved at constant speed in the x direction for one sample (Fig. 2.7a) and in the y

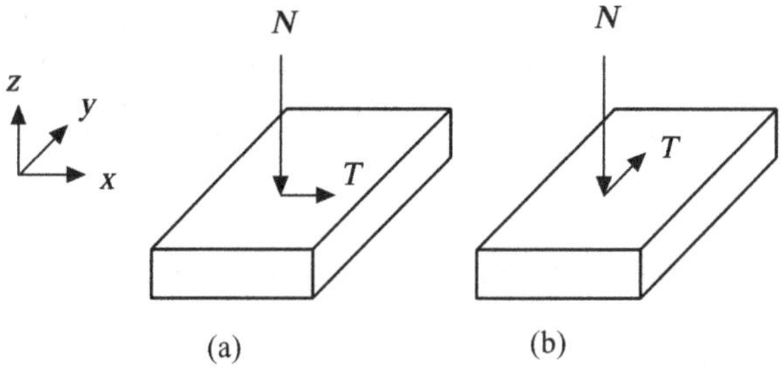

Figure 2.7. For an isotropic material, both the shear tests (a) and (b) will give identical shear stress–displacement curves.

direction for another sample (Fig. 2.7b). If the material is isotropic, the plots of the shear stress T versus the displacement of the upper plate will be identical for both the samples.

There is evidence for the anisotropic behavior of sands, particularly at low stress levels (Feda, 1982, pp. 257–261). The reader is referred to Feda (1982, pp. 122–126, 236–237, 257–261), Desai and Siriwardane (1984, pp. 371, 395), Bauer et al. (2004), and Ovarlez and Clément (2005) for a discussion of anisotropic effects. A material which is initially isotropic can exhibit anisotropy on being sheared. In this book, the assumption of isotropy is mainly used to simplify the analysis. Even with the simplification, the modeling of granular flow is a formidable task.

For isotropic materials, the invariance requirements discussed earlier impose certain constraints on the yield condition (2.18). For example, suppose that the yield condition is given by $\sigma_{xx} + 2\,\sigma_{yy} = 2\,k$, where k is a constant, and a sample of the granular material is subjected to a state of uniaxial compression along the x direction, i.e., all the stress components except σ_{xx} vanish. Then, if the yield condition is satisfied, we obtain $\sigma_{xx} = 2\,k$. Conversely, if the sample is subjected to uniaxial compression along the y direction, the yield condition implies that $\sigma_{yy} = k$. Thus the material response depends on the direction chosen, in violation of the assumption of isotropy. The problem can be overcome by formulating the yield condition in terms of the invariants and principal stresses associated with the two-dimensional stress tensor, as discussed below.

2.5.1. Invariants and Principal Stresses

Here the concept of principal stresses is introduced, and it is shown that the invariants arise naturally in the process of determining the principal stresses. Considering a Cartesian coordinate system (Fig. 2.8), let AB denote the line of intersection of the x–y plane and a plane P which is perpendicular to it. The stress vector $\mathbf{t}_{(\mathbf{n})}$ exerted at a point O on P is given by (see Appendix B)

$$\mathbf{t}_{(\mathbf{n})} = \mathbf{n} \cdot \boldsymbol{\sigma} \tag{2.19}$$

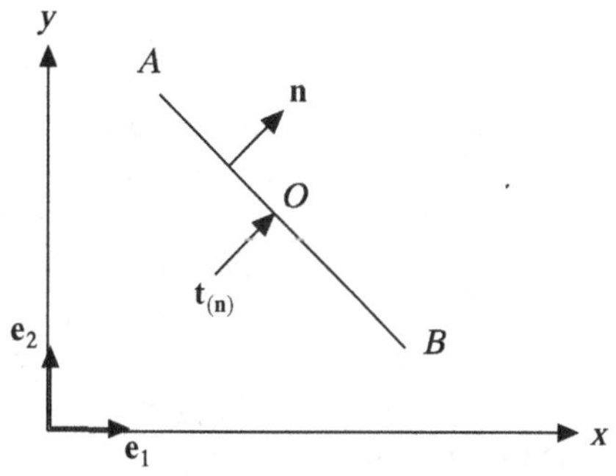

Figure 2.8. The plane AB is oriented such that the traction vector $\mathbf{t}_{(\mathbf{n})}$ is perpendicular to it. Here $\mathbf{n}$ is the unit normal to AB.

where $\mathbf{n}$ is the unit normal to AB and $\boldsymbol{\sigma}$ is the two-dimensional stress tensor. Suppose that it is desired to choose the orientation $\mathbf{n}$ of P so that there are no shear stresses acting on P. In this case, the stress vector $\mathbf{t}_{(\mathbf{n})}$ must have the same direction as $\mathbf{n}$, as shown in Fig. 2.8. Thus

$$\mathbf{t}_{(\mathbf{n})} = a\,\mathbf{n} \tag{2.20}$$

where a is an arbitrary constant. Equations (2.19) and (2.20) imply that $\mathbf{n} \cdot \boldsymbol{\sigma} = a\,\mathbf{n}$, or, as $\boldsymbol{\sigma}$ is assumed to be symmetric

$$\boldsymbol{\sigma} \cdot \mathbf{n} = a\,\mathbf{n} \tag{2.21}$$

It is convenient to rewrite (2.21) as

$$(\boldsymbol{\sigma} - a\mathbf{I}) \cdot \mathbf{n} = 0 \tag{2.22}$$

where $\mathbf{I}$ is the unit tensor.

If the tensor $\boldsymbol{\sigma} - a\mathbf{I}$ is invertible, i.e., if it has an inverse (see §A.7.5), (2.22) implies that

$$(\boldsymbol{\sigma} - a\mathbf{I})^{-1} \cdot ((\boldsymbol{\sigma} - a\mathbf{I}) \cdot \mathbf{n})) = \mathbf{I} \cdot \mathbf{n} = \mathbf{n} = 0$$

Hence there is no direction or nonzero vector $\mathbf{n}$ satisfying (2.21). Therefore, for (2.21) to have a nontrivial solution $\mathbf{n} \neq 0$, $\boldsymbol{\sigma} - a\mathbf{I}$ should not be invertible. It can be shown (Problem A.4) that this is equivalent to requiring

$$\det(\boldsymbol{\sigma} - a\mathbf{I}) = 0 \tag{2.23}$$

where "det" denotes the determinant of the tensor (see §A.7.6). Equation (2.23) can also be written as

$$\det([\boldsymbol{\sigma} - a\mathbf{I}]) = 0 \tag{2.24}$$

where the square brackets denote the matrix representation of the tensor (see §A.4).

Using (A.40) to evaluate the determinant, and identifying the 1 and 2 directions with the x and y axes, respectively (Fig. 2.8), the *characteristic equation* (2.24) can be rewritten as

$$a^2 - \mathrm{tr}(\boldsymbol{\sigma})\,a + \det(\boldsymbol{\sigma}) = 0 \tag{2.25}$$

where $\mathrm{tr}(\boldsymbol{\sigma}) \equiv \sigma_{xx} + \sigma_{yy}$ and $\det(\boldsymbol{\sigma}) = \sigma_{xx}\sigma_{yy} - \sigma_{xy}\,\sigma_{yx} = \sigma_{xx}\sigma_{yy} - \sigma_{xy}{}^2$. Here $\mathrm{tr}(\boldsymbol{\sigma})$ denotes the *trace* of a tensor $\boldsymbol{\sigma}$, defined by (A.18), and $\{\sigma_{ij}\}$ are the components of $\boldsymbol{\sigma}$ relative to the Cartesian coordinate system (Fig. 2.8).

Equation (2.25) has two roots for a, which are called the *eigenvalues* of $\boldsymbol{\sigma}$. The eigenvalues are also called the *principal stresses* and are denoted by σ_1 and σ_2. Corresponding to an eigenvalue σ_k, there is an *eigenvector* $\mathbf{n}_k$, which is the solution of

$$(\boldsymbol{\sigma} - \sigma_{(k)}\mathbf{I}) \cdot \mathbf{n}_{(k)} = 0, \quad k = 1, 2 \tag{2.26}$$

or, in component form,

$$(\sigma_{ij} - \sigma_{(k)})\,n_{j(k)} = 0, \quad i, k = 1, 2 \tag{2.27}$$

where the brackets around the subscript k indicate that the summation convention does not apply to this index, $\mathbf{n}_k = n_{jk}\,\mathbf{e}_j$, and $\mathbf{e}_j$, $j = 1, 2$, are the basis vectors for the coordinate system (Fig. 2.8). In view of (2.23), the two equations (2.27) are not independent, and it suffices to use any one of them to determine the eigenvectors.

In general, any scalar multiple of $\mathbf{n}_k$ is also an eigenvector as it satisfies (2.26). Thus the direction of an eigenvector is determined (to within a multiplicative factor of ± 1), but not its magnitude. In the present case, the eigenvector has been chosen as a unit vector and hence has unit magnitude.

The symmetry of the stress tensor ensures that the principal stresses are real numbers and the eigenvectors corresponding to distinct eigenvalues are orthogonal, i.e., if $\sigma_1 \neq \sigma_2$, then $\mathbf{n}_1 \cdot \mathbf{n}_2 = 0$ (Problem 2.7). Hence if $\sigma_1 \neq \sigma_2$, the directions represented by the eigenvectors $\mathbf{n}_1$ and $\mathbf{n}_2$ can be chosen as Cartesian coordinate axes. These axes are called the *principal stress axes* or the *principal stress directions*. The choice of the principal axes is not unique as they can be chosen parallel to any of the vectors such as $(\mathbf{n}_1, \mathbf{n}_2)$ and $(-\mathbf{n}_1, \mathbf{n}_2)$. If $\sigma_1 = \sigma_2$, every direction in the x–y plane corresponds to an eigenvector, and hence any two orthogonal directions can be chosen as the principal axes (Problem 2.7).

The characteristic equation (2.25) is a polynomial in a, and the coefficients of the polynomial are called the *principal invariants* of the stress tensor (Truesdell and Noll, 1965, p. 23; Slattery, 1999, p. 41). It can be shown that the principal invariants are unaffected by rotation and reflection of the coordinate axes (see, e.g., Pipes, 1963, pp. 45–46; Hunter, 1983, p. 87, and also Problem 2.8). Hence for isotropic materials, the yield condition (2.18) can be rewritten as

$$F_*(I_1, I_2, \nu) = 0 \tag{2.28}$$

where I_1 and I_2 are the principal invariants of the two-dimensional stress tensor.

Using (2.25), the principal invariants can be chosen as

$$I_1 \equiv \mathrm{tr}(\boldsymbol{\sigma}) = \sigma_{xx} + \sigma_{yy}\,; \quad I_2 \equiv \det(\boldsymbol{\sigma}) = \sigma_{xx}\sigma_{yy} - {\sigma_{xy}}^2 \tag{2.29}$$

2.6. THE MOHR–COULOMB YIELD CONDITION

Using the momentum balances (2.7) and (2.17), the relations (2.29), and an explicit form for the yield condition (2.28), problems of plane statics can be solved. A yield condition which has been widely used is the *Mohr–Coulomb yield condition* (see, e.g., Shield, 1953; Spencer, 1964; Collins, 1990)

$$(I_1/2)^2 - I_2 - [(I_1/2) + c\,\cot\phi_*]^2\,\sin^2\phi_* = 0$$

or equivalently,

$$[(\sigma_{xx} - \sigma_{yy})/2]^2 + \sigma_{xy}^2 - [(\sigma_{xx} + \sigma_{yy})/2 + c\,\cot\phi_*]^2\,\sin^2\phi_* = 0 \tag{2.30}$$

where ϕ_* and c are constants called the *angle of internal friction* and the *cohesion*, respectively. The relation between (2.30) and the Coulomb yield condition (2.16) becomes clearer on expressing the normal stress N and the shear stress T acting on a plane in terms of the principal stresses (eigenvalues) σ_1 and σ_2 of the two-dimensional stress tensor. This is done below.

2.7. THE MOHR'S CIRCLE FOR THE TWO-DIMENSIONAL STRESS TENSOR

Let the principal stresses be numbered such that $\sigma_1 \geq \sigma_2$. Then σ_1 and σ_2 are called the *major* and *minor* principal stresses, respectively. Consider a Cartesian coordinate system whose axes coincide with the eigenvectors $\mathbf{n}_1$ and $\mathbf{n}_2$ of $\boldsymbol{\sigma}$ (Fig. 2.9). Thus the x' and y' axes represent the major and minor principal stress axes, respectively. Here we choose one of the two possible directions for the x' axis, and the y' axis is chosen to be inclined at an angle of 90° measured anticlockwise from the x' axis. As the eigenvectors are of unit length, they can be chosen as the basis vectors for this coordinate system.

It will now be shown that the off-diagonal components of $\boldsymbol{\sigma}$ vanish with respect to the principal stress axes. As discussed in §A.4, $\boldsymbol{\sigma}$ can be expressed in terms of its components relative to this coordinate system by $\boldsymbol{\sigma} = \sigma'_{ij}\mathbf{n}_i\mathbf{n}_j$. Hence (2.19) and (2.20) imply that

$$\mathbf{t}_{(\mathbf{n}_1)} = \mathbf{n}_1 \cdot \boldsymbol{\sigma} = \sigma'_{11}\mathbf{n}_1 + \sigma'_{12}\mathbf{n}_2 = \sigma_1\,\mathbf{n}_1$$

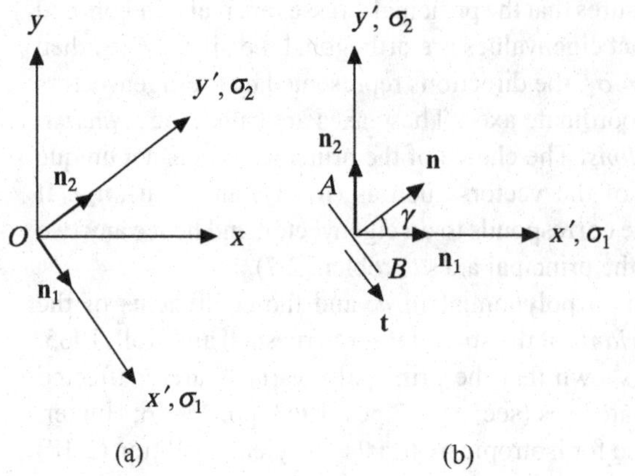

Figure 2.9. (a) The σ_1 or major principal stress axis and the σ_2 or minor principal stress axis. (b) The plane AB on which the shear and normal stresses exerted by the material to the left of AB are to be determined. Here **n** is the unit normal to AB, **t** is a unit vector which is normal to **n**, and $\mathbf{n}_1$ and $\mathbf{n}_2$ are the eigenvectors of the stress tensor $\boldsymbol{\sigma}$. The direction of **t** is chosen so that it is inclined at an angle of 90° measured clockwise from **n**. The angle γ is measured anticlockwise from the σ_1 axis to **n**.

Hence $\sigma'_{11} = \sigma'_{xx} = \sigma_1$ and $\sigma'_{12} = \sigma'_{xy} = 0$. Similarly, we have $\sigma'_{21} = \sigma'_{yx} = 0$ and $\sigma'_{22} = \sigma'_{yy} = \sigma_2$. Thus the stress tensor has the matrix representation (see §A.4)

$$[\boldsymbol{\sigma}] = [\sigma_{ij}] = \begin{pmatrix} \sigma_1 & 0 \\ 0 & \sigma_2 \end{pmatrix} \tag{2.31}$$

with respect to the principal stress axes.

The normal and shear stresses acting on any plane will now be related to the principal stresses. Consider a plane which is perpendicular to the x'–y' plane, such that its unit normal **n** is inclined at an angle γ measured anticlockwise from the σ_1 axis (Fig. 2.9b). Using (2.19), the normal stress N exerted across this plane on the material into which **n** points is given by

$$N \equiv \mathbf{t}_{(\mathbf{n})} \cdot \mathbf{n} = (\mathbf{n} \cdot \boldsymbol{\sigma}) \cdot \mathbf{n}$$

As $\mathbf{n} = \cos\gamma\,\mathbf{n}_1 + \sin\gamma\,\mathbf{n}_2$, we obtain

$$N = n_i\,\sigma_{ij}\,n_j = \sigma_1\cos^2\gamma + \sigma_2\sin^2\gamma = \sigma + \tau\cos 2\gamma \tag{2.32}$$

where

$$\sigma \equiv (\sigma_1 + \sigma_2)/2; \quad \tau \equiv (\sigma_1 - \sigma_2)/2, \quad \sigma_1 \geq \sigma_2 \tag{2.33}$$

The *mean stress* σ is the analog of the pressure in fluid mechanics; it is shown later that the *deviator stress* τ represents the magnitude of the maximum shear stress. The quantities σ and τ are invariants of $\boldsymbol{\sigma}$, as they can be expressed in terms of the principal invariants I_1 and I_2, which were introduced earlier (see (2.29)). The variables σ, τ, and γ are called *Sokolovskii variables*, after Sokolovskii (1965), who used them extensively in his analysis of granular statics.

To determine the shear stress, let **t** be a unit vector chosen so that it is inclined at an angle of 90° measured clockwise from **n** (Fig. 2.9b). The shear stress T exerted across the plane in the direction of **t**, on the material into which **n** points, is given by

$$T \equiv (\mathbf{n} \cdot \boldsymbol{\sigma}) \cdot \mathbf{t} = (\sigma_1 - \sigma_2)\cos\gamma\sin\gamma = \tau\sin 2\gamma \tag{2.34}$$

Equations (2.32) and (2.34) can be elegantly interpreted by constructing a circle with center $(\sigma, 0)$ and radius τ in the N–T plane (Fig. 2.10). The state of stress on a plane whose normal makes an angle γ (measured anticlockwise from the σ_1 axis) can be found by locating a point D on the circle such that $\angle DEF = 2\gamma$. This circle is called the *Mohr's circle for stress*, after Mohr (1882). Figure 2.10 shows that τ is the magnitude of the maximum shear stress, which occurs on two planes, with $\gamma = \pm 45°$.

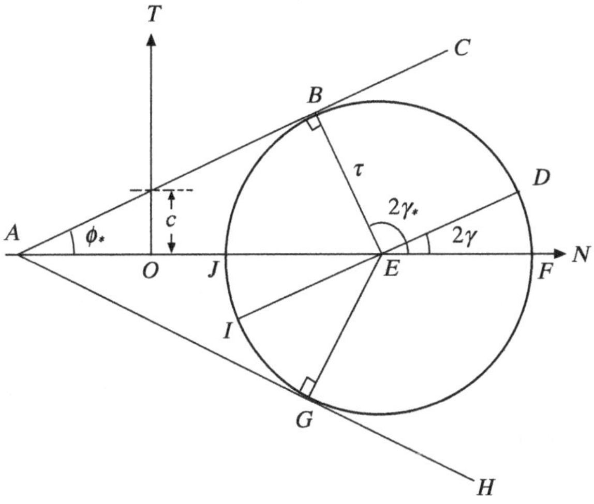

Figure 2.10. The Mohr's circle for the two-dimensional stress tensor. Here T and N are the shear and normal stresses, respectively, acting on a plane whose normal is inclined at an angle γ measured anticlockwise from the σ_1 axis. The origin is at $T = 0$, $N = 0$, and $OF = \sigma_1$, $OJ = \sigma_2$, $OE = \sigma = (\sigma_1 + \sigma_2)/2$, $BE = ED = \tau = (\sigma_1 - \sigma_2)/2$, and σ_1 and σ_2 are the major and minor principal stresses, respectively.

For flow problems, N and T must be related to the components of the stress tensor relative to any chosen coordinate system. Considering a plane whose normal **n** is directed along the x axis (Fig. 2.11), (2.32) and (2.34) imply that

$$N = \sigma_{xx} = \sigma + \tau \cos 2\gamma; \quad T = -\sigma_{xy} = \tau \sin 2\gamma \tag{2.35}$$

With this choice of **n** and **t**, the coordinates of the point D (Fig. 2.10) are $(\sigma_{xx}, -\sigma_{xy})$. Similarly, the coordinates of the point I are $(\sigma_{yy}, \sigma_{yx})$, where

$$\sigma_{yy} = \sigma - \tau \cos 2\gamma; \quad \sigma_{yx} = \sigma_{xy} = -\tau \sin 2\gamma \tag{2.36}$$

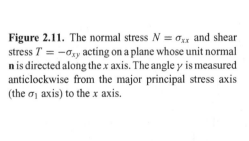

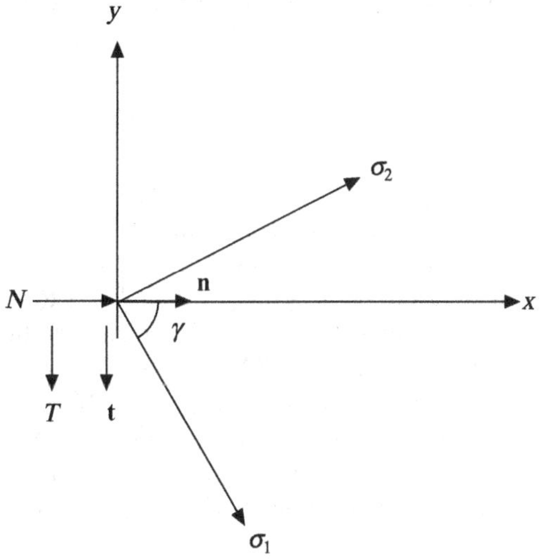

Figure 2.11. The normal stress $N = \sigma_{xx}$ and shear stress $T = -\sigma_{xy}$ acting on a plane whose unit normal **n** is directed along the x axis. The angle γ is measured anticlockwise from the major principal stress axis (the σ_1 axis) to the x axis.

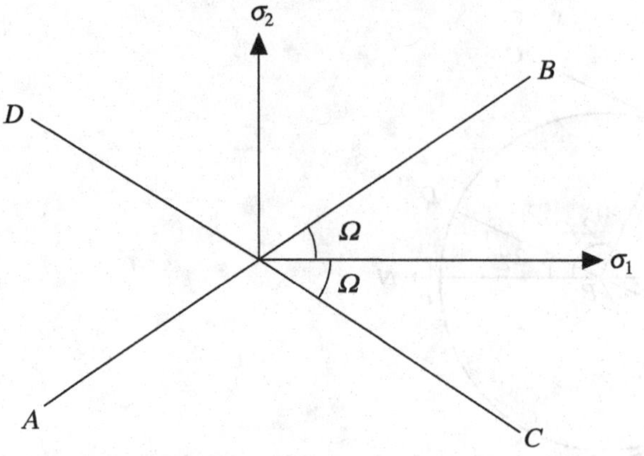

Figure 2.12. The slip planes *AB* and *CD*, which are inclined at angles of $\pm\Omega = \pm[(\pi/4) - (\phi_*/2)]$ relative to the major principal stress axes. Here ϕ_* is the angle of internal friction.

2.8. THE RELATION BETWEEN THE COULOMB AND MOHR–COULOMB YIELD CONDITIONS

In the *N–T* plane, the Coulomb yield condition (2.16) can be represented by the straight lines *ABC* and *AGH* (Fig. 2.10). Consider a Mohr's circle which is tangential to *ABC* and *AGH* at the points *B* and *G*, respectively. It follows from the right-angled triangle *ABE* that the radius τ and the center σ of the circle are related by

$$\tau - (\sigma + c\,\cot\phi_*)\,\sin\phi_* = 0 \tag{2.37}$$

It will now be shown that (2.37) is identical to the Mohr–Coulomb yield condition (2.30). As the coordinates of points *D* and *I* are $(\sigma_{xx}, -\sigma_{xy})$ and $(\sigma_{yy}, \sigma_{xy})$, respectively, the radius *DE* of the circle is given by

$$\tau = \sqrt{((\sigma_{xx} - \sigma_{yy})/2)^2 + {\sigma_{xy}}^2}$$

Similarly, the center of the circle is given by $\sigma = (\sigma_{xx} + \sigma_{yy})/2$. With these expressions for σ and τ, (2.37) reduces to (2.30). Both (2.30) and (2.37) represent the Mohr–Coulomb yield condition. For a cohesionless material, (2.37) reduces to

$$\tau - \sigma\,\sin\phi_* = 0 \tag{2.38}$$

When this yield condition is satisfied, the Coulomb yield condition (2.16) holds only on two planes, called the *slip planes*, which correspond to the points *B* and *G* in Fig. 2.10. As $2\gamma = \pm((\pi/2) + \phi_*)$ for *B* and *G*, respectively, the normals to the slip planes are inclined at angles of $\pm[(\pi/4) + (\phi_*/2)]$ relative to the σ_1 axis. Hence the slip planes are inclined at angles of $\pm[(\pi/4) - (\phi_*/2)]$ relative to the σ_1 axis (Fig. 2.12). The maximum possible value of the ratio $|T|/(N + c\,\cot\phi_*)$ occurs on the slip planes.

The Coulomb yield condition (2.16) permits yielding only on the slip planes, whereas the Mohr–Coulomb yield condition (2.37) permits yielding on *any* plane, provided τ and σ satisfy (2.37). If the Mohr–Coulomb yield condition holds, the Coulomb yield condition is always satisfied on the slip planes. However, the converse is not true, unless additional assumptions are used (see §2.15).

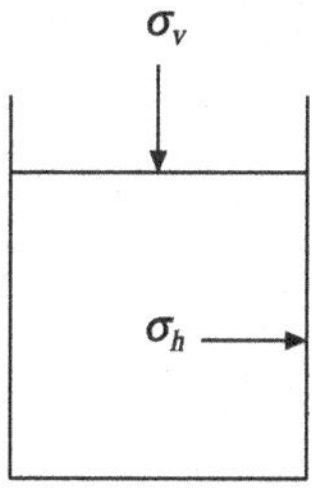

Figure 2.13. Uniaxial compression of a granular material in a cylindrical vessel. Here σ_v is the cross-sectional averaged vertical normal stress and σ_h is the horizontal normal stress.

2.9. ACTIVE AND PASSIVE STATES OF STRESS AND THE VALUE OF THE JANSSEN *K*-FACTOR

As a simple application of the Mohr–Coulomb yield condition (2.37), consider the bin problem discussed in §2.3. If (2.37) holds, (2.35) and (2.36) imply that the ratio $K \equiv \sigma_{xx}/\sigma_{yy}$ is given by

$$K = \frac{1 + \sin\phi_* \cos 2\gamma + (c/\sigma)\cos\phi_* \cos 2\gamma}{1 - \sin\phi_* \cos 2\gamma - (c/\sigma)\cos\phi_* \cos 2\gamma} \tag{2.39}$$

Thus K is not a constant, as assumed by Janssen, but varies depending on the values of c, σ, and γ.

To simplify the analysis, consider the special case of a cohesionless material ($c = 0$). For symmetric solutions, the shear stress σ_{xy} vanishes along the centerline $x = 0$ (Fig. 2.6). Hence (2.35) implies that $\sin 2\gamma = 0$ at $x = 0$. This equation has two roots, (i) $\gamma = \pi/2$ and (ii) $\gamma = 0$, which correspond to the points J and F, respectively, in Fig. 2.10.

The first root $\gamma = \pi/2$ implies that the major principal stress axis is vertical (Fig. 2.11); this is called an *active* state of stress. Conversely, the second root $\gamma = 0$ implies that the major principal stress axis is horizontal; this is called a *passive* state of stress. The notion of active and passive states is due to Rankine (1857). His interpretation differs slightly from the one given earlier; he terms the vertical normal stress generated by gravity as the active stress and the horizontal normal stress which prevents the material from spreading as the passive stress. At the centerline, (2.39) implies that if the yield condition holds, the possible values of K for a cohesionless material are given by (Nedderman, 1992, p. 86)

$$K = K_a = (1 - \sin\phi_*)/(1 + \sin\phi_*); \quad K = K_p = (1 + \sin\phi_*)/(1 - \sin\phi_*) \tag{2.40}$$

where the subscripts a and p denote active and passive states, respectively.

Thus plane problems admit *two* solutions if the yield condition is assumed to be valid. In the case of statics, both are admissible. In flow problems, other considerations may permit one solution to be discarded, as discussed in Chapter 3. With reference to silos, it is often assumed that an active state prevails when they are filled and a passive state prevails when they are allowed to discharge.

If the yield condition does not hold, and the material satisfies a linear elastic constitutive equation (Problem 5.12), it can be shown (Problem 5.13) that

$$K = K_e \equiv \nu_p/(1 - \nu_p) \tag{2.41}$$

where ν_p is a material constant called the Poisson's ratio. For this estimate of K to be valid, it must satisfy

$$K_a < K_e < K_p \tag{2.42}$$

as discussed in §2.11. As noted by Evesque (1997), $K_e < K_a$ for some materials, in which case (2.41) cannot be used.

Let us briefly consider the values of K, calculated using the stresses measured during the uniaxial compression of granular material in a cylindrical vessel (Fig. 2.13). Sawicki

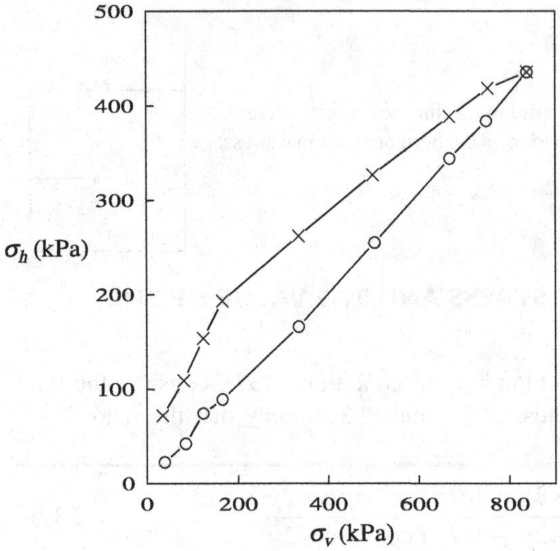

Figure 2.14. Variation of the horizontal normal stress σ_h with the cross-sectional averaged vertical normal stress σ_v at the upper surface of the sample (see Fig. 2.13) for medium dense Sobieszewo sand: $\circ$, loading (increasing values of σ_v); $\times$, unloading (decreasing values of σ_v). The lines are drawn to guide the eye. (Data of Sawicki and Swidzinski, 1998.)

and Swidzinski (1998) have measured the variation of the horizontal normal stress σ_h with the cross-sectional averaged vertical normal stress σ_v during loading (an increase of σ_v) and unloading (a decrease of σ_v). The quantity σ_h is not an average value, but the "local" value measured using a strain gauge fixed to a small region on the circumference of the vessel (Sawicki and Swidzinski, 1995). Further, σ_v is the normal stress at the upper surface of the granular column. Figure 2.14 shows that $\overline{K} \equiv \sigma_h/\sigma_v$ is approximately constant during loading, but not during unloading. For steel balls, lead shot, wheat, and various sands, $\overline{K}$ was in the range 0.3–0.58 during loading (Sawicki and Swidzinski, 1998). For water at rest, the curves for loading and unloading coincide.

2.10. SHEAR TESTS

So far we have discussed the concept of a yield condition, and how it can be used to examine problems of plane statics. For problems of plane flow, additional information relating the deformation of the material to the stress field is needed. Such information can be obtained by conducting tests on the material under controlled conditions. Here we discuss one class of tests that have been widely used in soil mechanics, namely, *shear tests*.

Fig. 2.15 shows a tester called the *Jenike shear cell*, which was developed by Jenike (1961). This is a variant of the *direct shear box*, which is used for testing soil samples. The stress levels employed in the latter are comparable to those exerted on soils by structures such as buildings and retaining walls and are much higher than the levels prevailing in vessels used for the storage and flow of granular materials. For example, in the shear tests reported in Roscoe et al. (1958), the normal stresses were in the range 70–700 kPa (0.7–6.9 atm). Similarly, for the uniaxial compression data shown in Fig. 2.31, the normal stresses are in the range 0–1,500 kPa. On the other hand, the normal stresses measured on the wall of a bunker containing sand at rest are in the range 10–40 kPa (Fig. 2.16). During flow, the normal stress increases, but the maximum value is only around 100 kPa. Most of the vessels used in laboratory-scale experiments are much smaller than the bunker shown in Fig. 2.16, and hence the stresses are also smaller.

With the Jenike shear cell, normal stresses in the range 1–10 kPa can be used. Therefore, the values of some of the material properties can be obtained under conditions that are

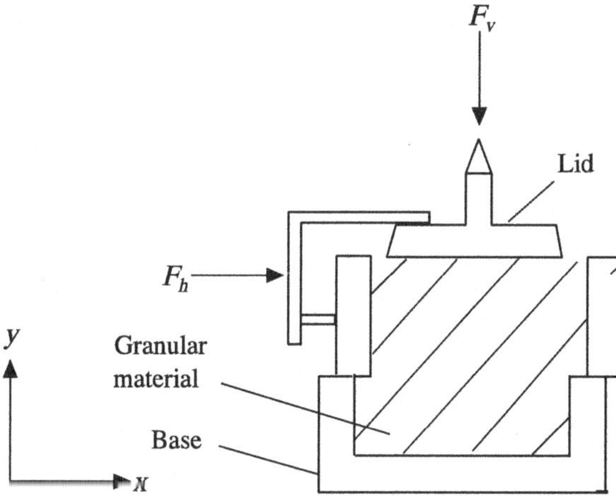

Figure 2.15. The Jenike shear cell. The force F_v is applied to the lid, and the force F_h required to move the upper half of the cell to the right at a constant speed is measured. (Adapted from Fig. 10 of Jenike, 1964b, p. 17.)

more relevant to the design of storage vessels such as bunkers. Subsequently, variants of the Jenike shear cell have been developed (Nedderman, 1992, p. 147; Hsiau and Shieh, 2000; Schwedes, 2000). In particular, the annular shear cell can be used to obtain data at both low and high shear rates.

The Jenike shear cell (Fig. 2.15) consists of a cylindrical cell that is split into two halves. The base is stationary, whereas the shearing ring is free to move in the horizontal direction. The shearing ring is initially mounted eccentrically, as shown in Fig. 2.15. The granular material is poured into the cell and is covered with a lid which is free to move in the vertical direction. A normal load is applied to the lid, and the ring is driven to the right at a constant speed of about 0.04 mm s^{-1} (Jenike, 1964b, p. 35). The shear force exerted by the driving mechanism on the shearing ring is measured. Changes in the height of the lid, and hence in the height of the sample, reflect volume changes during shear.

When sheared as described earlier, granular materials display two distinct types of behavior, depending on the initial density ρ_i of the sample. As discussed earlier, ρ_i is bounded by two values ρ_{min} and $\rho_{max} > \rho_{min}$. A sample is termed *dense* if ρ_i is close to ρ_{max} and *loose* if ρ_i is close to ρ_{min}. A more precise definition will be given shortly. Figure 2.17a shows the results of a strain-controlled shear test, obtained using a shear box.

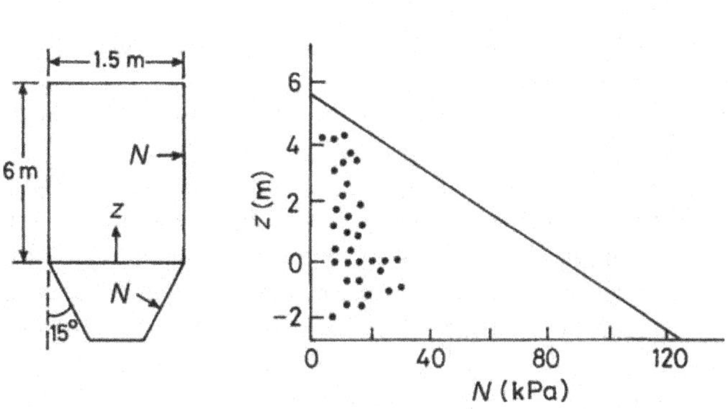

Figure 2.16. The normal stress N on the wall of an axisymmetric bunker filled with sand at rest: •, data of Van Zanten and Mooij (1977); —, hydrostatic profile.

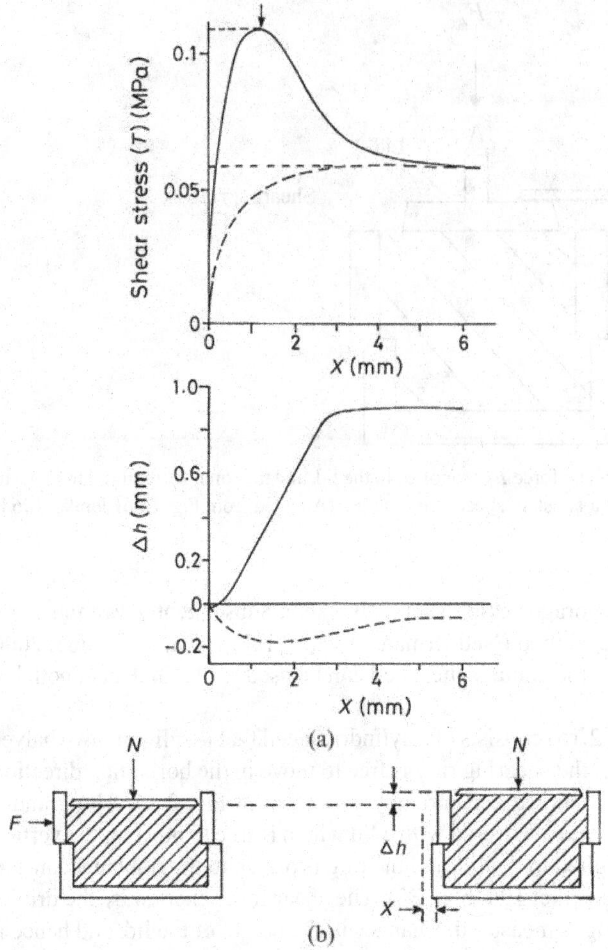

Figure 2.17. (a) Shear stress T versus displacement X and change in sample height Δh versus displacement X curves for sand sheared in a direct shear box under a constant normal stress $N = 100$ kPa: —, dense sample; - - -, loose sample. (b) Schematic diagram of the shear box. Here F is the applied shear force. (Figure 2.17a has been reproduced from Fig. 5.12 of Feda, 1982, with permission from Prof. J. Feda.)

(In such a test, one-half of the box is moved in the horizontal direction at a constant speed relative to the other half.) The Jenike shear cell is expected to give similar results, but these are rarely reported in the form shown in Fig. 2.17a.

The shear stress T versus displacement X curve for the dense sample shows a pronounced peak, whereas the curve for the loose sample is almost monotonic (Fig. 2.17a). However, for large displacements, both curves converge to a common limiting value of the shear stress, denoted by T_c. In the soil mechanics literature, the state corresponding to the peak shear stress is termed a state of *failure* (Atkinson and Bransby, 1982, pp. 179–180).

There are similarities and differences between the shearing of dense granular materials and the sliding of a solid block on a rough surface (Fig. 2.5). In the latter case, the block does not move until the applied force H exceeds a threshold value H_t. Once the block begins to slide, the value of H required for steady sliding is usually less than H_t. If the adhesion term in (2.2) and (2.3) is neglected, this observation implies that the coefficient of kinetic friction is usually less than the coefficient of static friction. In the case of a dense granular material, there is a threshold value for the shear stress, but the material deforms

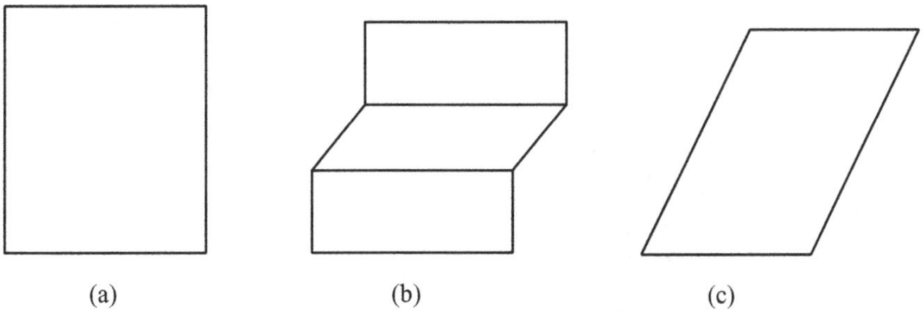

Figure 2.18. Shapes of samples: (a) undeformed sample, (b) dense sample after deformation, and (c) loose sample after deformation. (Adapted from Figs. 7 and 8 of Jackson, 1983.)

even before this is attained. Hence the terms *static friction* and *kinetic friction* cannot be precisely defined for these materials.

The volume change behavior shown in Fig. 2.17a is also interesting. After a small initial compaction or decrease in volume, the dense sample dilates or increases in volume as X increases. In contrast, the loose sample shows a significant compaction, followed by dilation. At the end of the test, the loose sample is denser than it was in the initial state. At large displacements under a constant normal load, both samples deform isochorically, i.e., at constant volume. This state of isochoric deformation was termed a *critical voids ratio state* by Roscoe et al. (1958). Subsequently, this phrase was shortened to *critical state* (Schofield and Wroth, 1968, p. 19). Henceforth, the critical state will be denoted by the subscript "c".

An important aspect of the deformation of the samples must be discussed. Let us first consider dense samples. Figure 2.17a shows that in the post-peak stage of deformation, the shear stress decreases, i.e., the sample becomes progressively weaker, as it is sheared. Therefore, if a thin rupture layer or shear band is initiated at some region, both the shear stress and the density decrease in that region. The adjacent regions, where the density is still close to the original density, can withstand a higher shear stress than the material in the rupture layer. This causes the deformation, and hence the density change, to be confined mainly to the rupture layer, and the material on either side of the layer remains essentially rigid (Fig. 2.18b). For sands, the data of Roscoe (1970), Vardoulakis and Graf (1985), and Vardoulakis et al. (1985) suggest that the thickness of the rupture layer is about 10–20 particle diameters.

The preceding discussion implies that post-peak deformation of dense samples is nonuniform, and boundary displacements do not provide reliable estimates of strains and volume changes. Unfortunately, most of the available deformation data are based on boundary displacements. This points to an important gap in the field, which should receive more attention in the future.

In contrast, the deformation of loose samples is expected to be more homogeneous. Here the shear stress increases with deformation, and hence a rupture layer, if formed, spreads to encompass the whole sample (Fig. 2.18c).

2.10.1. The Critical State

Roscoe et al. (1958) conducted shear tests on sand, glass beads, and steel balls, using a *simple shear apparatus* (Fig. 2.19). This is a refined version of the shear box, with load cells permitting measurement of both shear and normal stresses on the bounding surfaces of the sample. For all the materials tested, the shear stress at a critical state T_c was found to be proportional to the normal stress N (Fig. 2.20a). Similarly, the solids fraction at a critical state ν_c was found to depend on N, as shown in Fig. 2.20b. Thus the locus of critical states

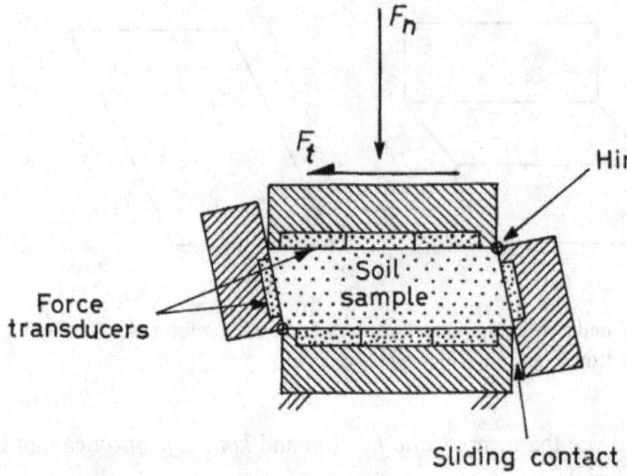

Figure 2.19. The simple shear apparatus. Here F_n and F_t denote the normal and shear forces, respectively, exerted on the lid of the apparatus. Roscoe (1970) notes that shear strains are relatively uniform in the central region. (Reproduced from Fig. 5.12 of Atkinson and Bransby, 1982, p. 84, with permission from Prof. J. H. Atkinson.)

represents a *critical state curve* in a three-dimensional space with Cartesian coordinates (N, ν, T).

For a given value of N, a sample is termed *loose* if $\nu < \nu_c(N)$ and *dense* if $\nu > \nu_c(N)$ (Scott, 1963, p. 327). For materials such as sand and glass beads, it is difficult to obtain samples that are "looser than critical." This is because the weight of even a few layers of particles suffices to compact the samples to a density which is larger than the critical state density. (With the advent of the space shuttle, data from low-gravity shear tests may become

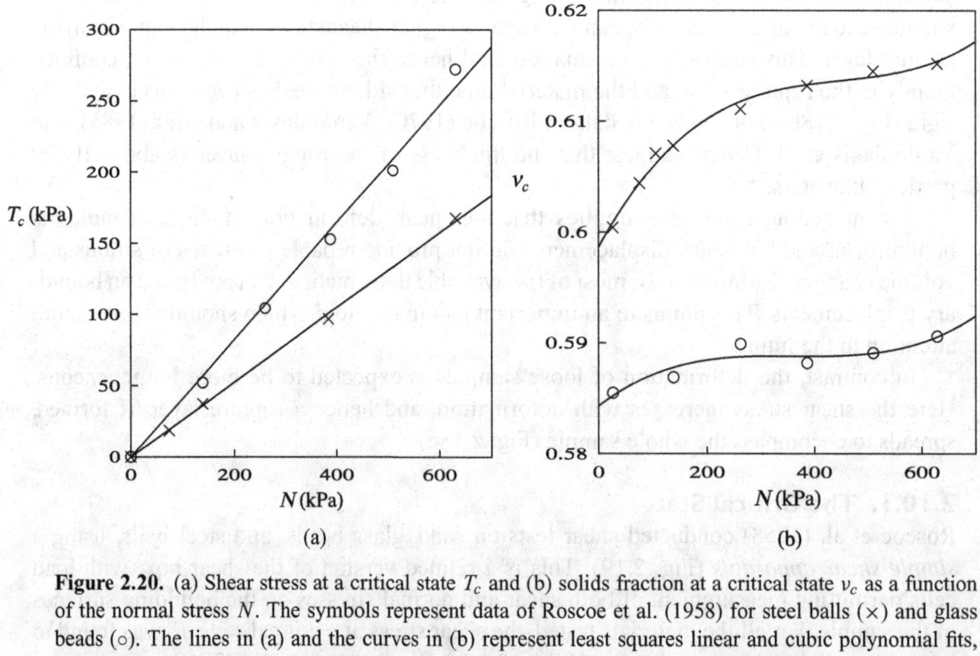

Figure 2.20. (a) Shear stress at a critical state T_c and (b) solids fraction at a critical state ν_c as a function of the normal stress N. The symbols represent data of Roscoe et al. (1958) for steel balls (×) and glass beads (○). The lines in (a) and the curves in (b) represent least squares linear and cubic polynomial fits, respectively, to the data. In (a), if the lines are defined by $T_c = N \tan \phi_c$, we have $\phi_c = 15°$ for steel balls and 22° for glass beads. The data were obtained using the simple shear apparatus.

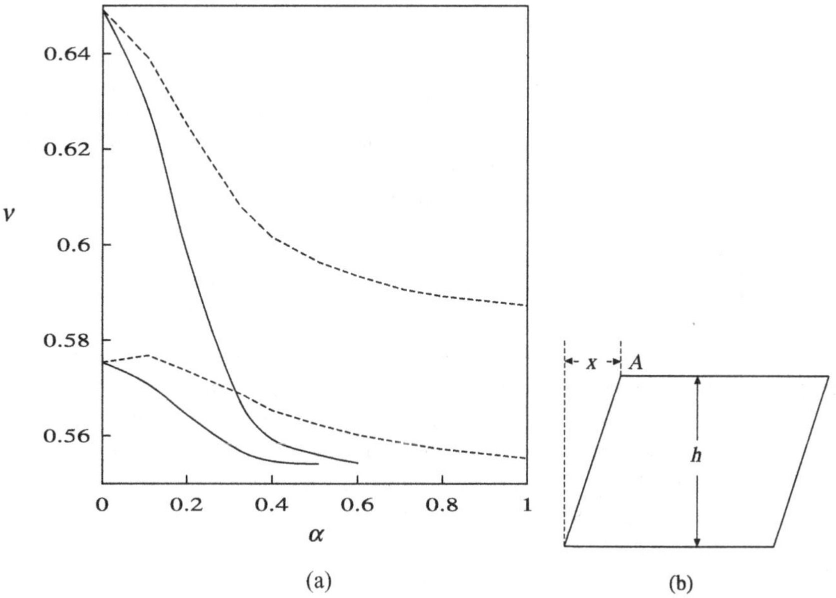

Figure 2.21. (a) Variation of the solids fraction ν with the shear strain α for sand, for two different values of the initial solids fraction: —, local data from γ-ray measurements; - - -, average data from boundary displacements. (Data of Cole, 1967, and Coumoulos, 1968, cited in Roscoe, 1970.) (b) Evaluation of the shear strain $\alpha \equiv \Sigma_i (\Delta x)_i / h_i$, where $(\Delta x)_i$ is the change in the x coordinate of the corner point A during the ith increment of displacement, and h_i is the height of the sample at the beginning of this increment.

available in the future.) For steel balls, it appears that Roscoe et al. (1958) were able to generate loose samples.

Figure 2.21a shows the variation of the solids fraction ν with the shear strain α, for the deformation of sand in the simple shear apparatus. Here the shear strain α is calculated as shown in Fig. 2.21b. (For small displacements, α is approximately equal to the change in the angle between two line elements that are initially orthogonal.) The solids fraction was estimated from boundary displacements (broken curves) and by γ-ray attenuation (solid curves). It is seen that the use of boundary displacements overestimates the solids fraction considerably, particularly for dense samples. The local (γ-ray) measurements show that when samples with different initial densities are sheared under the same normal stress, they attain the same density at the critical state. The recent experiments of Desrues et al. (1996) on cylindrical samples also confirm this result. Therefore it is reasonable to assume that the solids fraction at a critical state is a unique function of the normal stress for each material.

2.10.2. The Hvorslev Surface

Having introduced the concept of a critical state, let us now discuss how dense samples approach it. The post-peak behavior of dense samples is characterized by the *Hvorslev surface*, as explained later.

Following Roscoe et al. (1958), consider a three-dimensional space with Cartesian coordinates representing the normal stress N, the shear stress T, and the solids fraction ν. Thus the "state" of the material is represented by the coordinates $(N,\ \nu,\ T)$. As the shear test progresses, the behavior of the sample can be represented by a curve or *loading path* in this space. By analyzing test data on clays, sand, glass beads, and steel balls, Roscoe et al. (1958) reached the following important conclusion: The post-peak segments of loading paths corresponding to samples with different initial states all lie on a common surface. For example, consider shear tests at a constant normal stress $N = N_1$, on two samples with

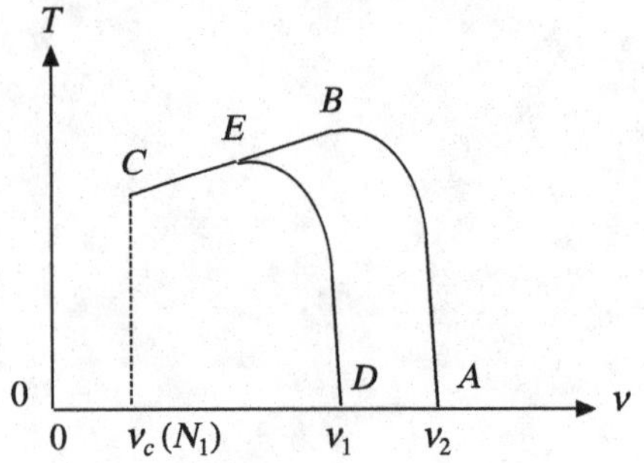

Figure 2.22. Hypothetical loading paths in the shear stress T – solids fraction ν plane for shear tests on dense samples at a constant normal stress $N = N_1$. Here C is the critical state corresponding to N_1.

initial solids fractions $\nu_i = \nu_1$ and $\nu_i = \nu_2$. In the T–ν plane (Fig. 2.22), the preceding result implies that the postpeak segment EC of the loading path for the sample with $\nu_i = \nu_1$ coincides with the segment BEC of the loading path for the sample with $\nu_i = \nu_2$. The curve BEC represents a section of the common surface by the plane $N =$ constant $= N_1$. Following Roscoe et al. (1958), this surface is called the *Hvorslev surface*, after Hvorslev (1937) (cited in Roscoe et al., 1958).

Figure 2.23 shows sections of the Hvorslev surface by planes of constant ν. Figure 2.23a is based on the data of Roscoe et al. (1958) for a cohesive material (clay). They do not report data for tensile normal stresses ($N < 0$), but the curves can extend into this region, as discussed in Molerus (1978). The curves in Fig. 2.23b represent cohesionless materials such as coarse glass beads and steel balls. The segments shown by broken curves denote regions where experimental data are scanty.

It is commonly assumed that states on the Hvorslev surface will cause *plastic* or irrecoverable deformation if the material is allowed to deform. On the other hand, when the state is below the Hvorslev surface, i.e., at a point such as A in Fig. 2.23b, the material is assumed to either remain *rigid* (Jenike, 1961, p. 1; Schofield and Wroth, 1968, p. 106) or deform *elastically* (Schofield and Wroth, 1968, p. 136; Atkinson and Bransby, 1982, pp. 263–265).

The notions of elastic and plastic deformation can be illustrated by considering a body resting on a table. In this state, it has a certain shape and size, which together define its "original" configuration. This changes when forces are applied on its surface, unless the

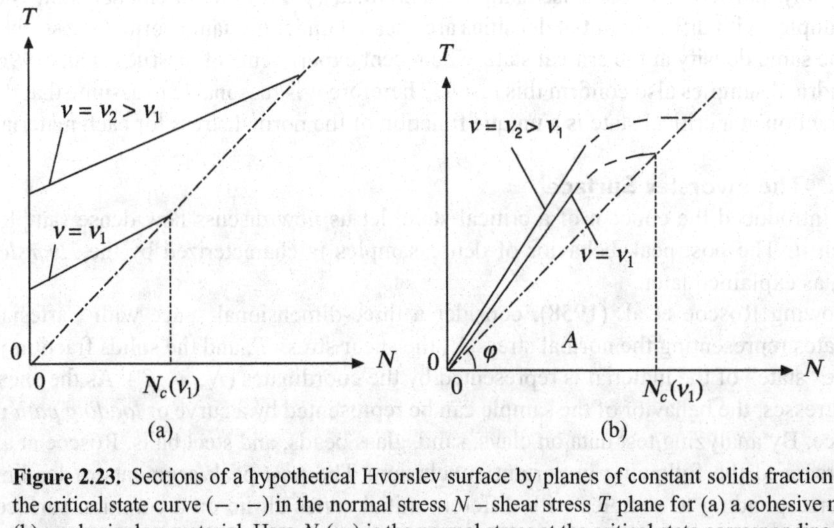

Figure 2.23. Sections of a hypothetical Hvorslev surface by planes of constant solids fraction ν (—), and the critical state curve (– · –) in the normal stress N – shear stress T plane for (a) a cohesive material and (b) a cohesionless material. Here $N_c(\nu_1)$ is the normal stress at the critical state corresponding to ν_1.

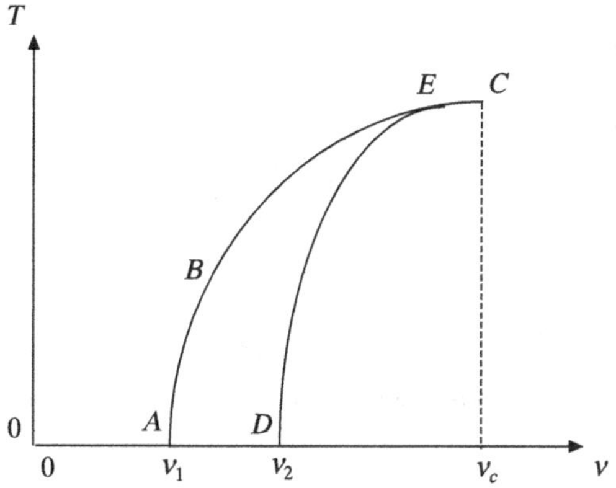

Figure 2.24. Hypothetical loading paths in the shear stress T – solids fraction ν plane for shear tests on loose samples at a constant normal stress $N = N_1$. The point C represents the critical state corresponding to N_1.

body is rigid. If the original configuration is *recovered* when the applied forces are gradually reduced to zero, the deformation is said to be elastic. On the other hand, if the original configuration is not recovered, plastic or irrecoverable deformation is said to have occurred. A simple example is provided by the behavior of a paper clip made of a metal. If it is bent slightly and the applied force is then removed, it recovers its original shape, and hence the deformation is elastic. However, if a large force is applied and then released, the clip does not recover its original shape. In this case, it has deformed plastically.

2.10.3. The Roscoe Surface

Let us now discuss how loose samples approach the critical state. As noted earlier, in terrestrial experiments, it is difficult to obtain samples of cohesionless materials that are "looser than critical." For the broken curves shown in Fig. 2.17a, there does not appear to be any particular point on the shear stress–displacement curve that can be identified with the onset of plastic deformation. The situation is more satisfactorily addressed by triaxial tests, which will be discussed later. Even in this case, most of the available data are confined to clays, which can be readily obtained in states that are looser than critical.

Figure 2.24 shows hypothetical loading paths for loose samples, inferred from triaxial tests on clays. Here ν_1 represents the minimum solids fraction consistent with the specified normal stress, say, $N = N_1$. Such a sample is said to be *normally consolidated*. Following Roscoe et al. (1958), it is postulated that the loading path $ABEC$ represents states of plastic deformation. The loading path terminates at the state C, with coordinates (ν_c, T_c). This represents the critical state for $N = N_1$. Analogous to the Hvorslev surface, which represents the states of plastic deformation of dense samples, it will be assumed that the corresponding states for loose samples all lie on a common surface in N–ν–T space. This is called the *Roscoe surface* (Atkinson and Bransby, 1982, p. 199). The curve $ABEC$ represents a section of this surface by the plane $N =$ constant $= N_1$. Consider a sample whose initial solids fraction ν_2 lies between ν_c and ν_1 (Fig. 2.24). In a test at constant N, the shear stress T rises along the loading path DE. At the point E, the path touches the section $ABEC$ of the Roscoe surface. The preceding assumption implies that the subsequent deformation is represented by movement along the segment EC of the curve $ABEC$. Figure 2.25 shows sections of the Roscoe surface by planes of constant solids fraction. This may be compared with the corresponding figure for the Hvorslev surface (Fig. 2.23).

2.10.4. The Yield Surface

Following the viewpoint adopted by Schofield and Wroth (1968, pp. 109–123) (see also Atkinson and Bransby, 1982, pp. 221–228), we consider both the Roscoe and Hvorslev surfaces as parts of a common surface, which is called the *yield surface*. The yield surface

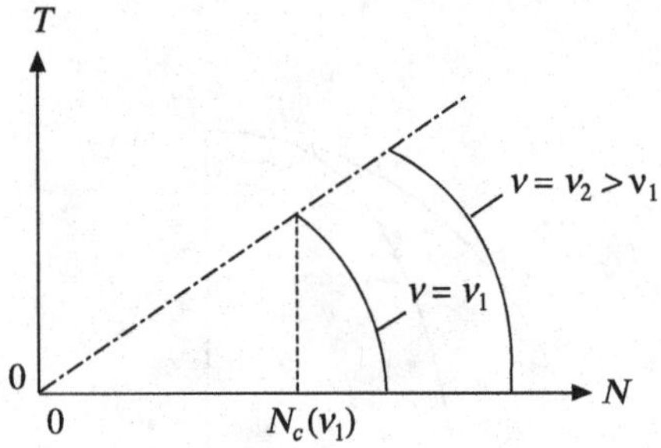

Figure 2.25. Sections of a hypothetical Roscoe surface by planes of constant solids fraction ν (—), and the critical state line (– · –) in the normal stress N – shear stress T plane.

can be represented by an equation of the form

$$T = T(N, \nu) \tag{2.43}$$

For isotropic materials, if a state (N, ν, T) lies on the yield surface, so does the state $(N, \nu, -T)$. Thus the yield surface is *symmetric* about the N–ν plane. However, only the upper half of the yield surface, corresponding to $T \geq 0$, will be shown in most of the figures referred to in this chapter.

Figure 2.26 shows sections of the yield surface by planes of constant solids fraction. These sections are called *yield loci*. Consider a typical yield locus *OCE*. The segment *OC*, which lies on the left of the critical state curve *OCF*, is called the *dilation* branch. Similarly, the segment *CE*, which lies on the right of *OCF*, is called the *compaction* branch.

Data obtained from uniaxial compression tests (§2.13) and isotropic compression tests (§5.1.4) show that plastic deformation occurs even for states lying inside the yield surface. Thus the yield surface is not truly a surface that separates states of elastic and plastic deformation, but rather a bounding or ultimate surface that cannot be crossed. To avoid complicating the analysis considerably, and in keeping with the bulk of the literature on granular flow, the following simplified model is used here: the material is assumed to be rigid for states below the yield surface and is capable of plastic deformation for states on this surface. This defines a *rigid-plastic* material. More refined models that include elastic deformation for states below the yield surface are described in Taylor (1948, pp. 134–151), Atkinson and Bransby (1982, pp. 283–289), Desai and Siriwardane (1984, pp. 282–298, 381–394), Krenk (2000), and in the references listed in Table 5.1. The inclusion of plastic deformation for states below the yield surface is discussed in Lade and Duncan (1975) and Lade (1977).

Consider a shear test at constant $N(= N_A)$ on a rigid-plastic sample with an initial solids fraction ν_3 (Fig. 2.26). The critical state corresponding to $N = N_A$ is given by $T =$

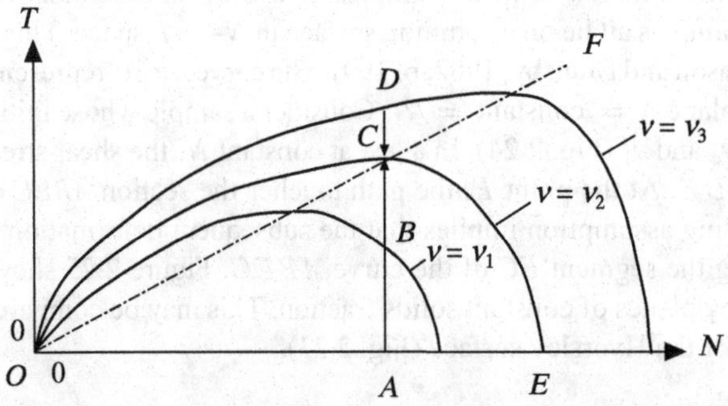

Figure 2.26. Hypothetical yield loci or contours of constant solids fraction ν in the normal stress N – shear stress T plane. The loading paths *ADC* and *ABC* represent the behavior of dense and loose samples, respectively, and *OCF* is the critical state curve. The yield loci are shown for a cohesionless material, with $\nu_1 < \nu_2 < \nu_3$.

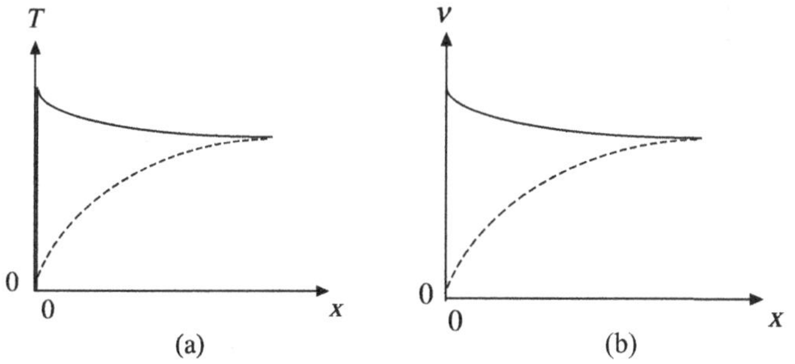

Figure 2.27. (a) Shear stress T versus displacement x and (b) solids fraction ν versus displacement x curves for dense (—) and loose (- - -) rigid-plastic samples sheared under a constant normal stress.

$T_c(N_A)$, $\nu = \nu_2 = \nu_c(N_A)$. As $\nu_3 > \nu_2$, the sample is denser than critical. A rigid-plastic sample remains rigid as the shear stress increases from zero to T_D, where $T_D = T(N_A, \nu_3)$. The sample then yields plastically, with T decreasing from T_D to $T_c(N_A)$ and the solids fraction decreasing from ν_3 to ν_2. Similarly, for a sample with an initial solids fraction $\nu = \nu_1$, the sample remains rigid as T increases from zero to T_B. A further increase in T along the path BC causes the sample to compact. It follows that the shear stress–displacement and solids fraction–displacement curves for rigid-plastic samples will be of the form shown in Fig. 2.27. In the context of triaxial tests, which are discussed in Chapter 5, the Granta-Gravel model of Schofield and Wroth (1968, pp. 109–123) gives rise to curves of this form. The curves shown in Fig. 2.27 may be compared and contrasted with the data shown in Fig. 2.17a.

The key ideas discussed in this section can be summarized as follows:

(i) There is a yield surface in N–ν–T space, which is a combination of the Hvorslev and Roscoe surfaces. Plastic deformation is associated with movement on the yield surface. For states within the yield surface, the material is usually assumed to be either rigid or deforming elastically. Here the former alternative is chosen to simplify the analysis.
(ii) For points lying on the Hvorslev surface, the material dilates as it deforms. For points lying on the Roscoe surface, the material compacts as it deforms. The curve of intersection between the Hvorslev and Roscoe surfaces is called the critical state curve. If the normal stress remains constant and the material is at a critical state, it deforms with constant volume, like an incompressible fluid.
(iii) At present there are very little data about the Roscoe surface for cohesionless materials. Thus it is not possible to construct a complete yield surface for these materials by fitting parameters in model equations to experimental data.

The preceding features (i) and (ii) constitute the *critical state theory* (Schofield and Wroth, 1968, pp. 19–21, 149–151; Atkinson and Bransby, 1982, pp. 184–234). It should be emphasized that this is only one model for the behavior of granular materials. Alternative models, including those that do not involve the concept of a yield surface, are available. Some examples of the latter type of models are discussed in Chapter 5. Most of the results presented in subsequent chapters are based on the critical state model, as it is relatively simple and has been widely used in various contexts.

2.11. YIELD SURFACES IN σ_1–σ_2–ν SPACE

In the previous section, some features of yield surfaces in N–ν–T space have been discussed. For solving flow problems, it is essential to have an expression for the yield surface in a

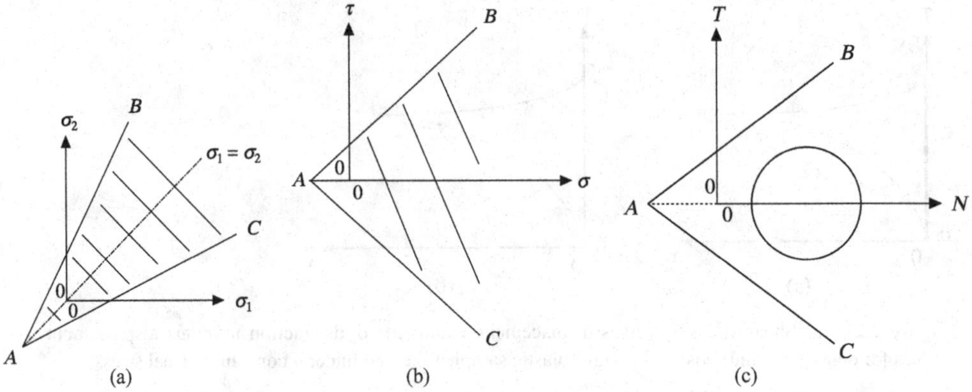

Figure 2.28. The Mohr–Coulomb yield condition is represented by the straight lines AB and AC in the (a) σ_1–σ_2, (b) σ–τ, and (c) N–T planes. Here σ_1 and σ_2 are the principal stresses, $\sigma = (\sigma_1 + \sigma_2)/2$, $\tau = (\sigma_1 - \sigma_2)/2$, and T and N denote the shear and normal stresses acting on any plane. The hatched regions in (a) and (b) show the states of stress which do not satisfy the yield condition. For a point in the hatched region, the Mohr's circle for the stress tensor does not touch the lines AB and AC, as shown in (c).

space with Cartesian coordinates $(\sigma_1, \sigma_2, \nu)$, where σ_1 and σ_2 are the principal stresses of the two-dimensional stress tensor. This space is called the *extended principal stress space*. Test data obtained from the direct shear box and the Jenike tester do not permit the orientation of the principal axes to be evaluated. Hence as discussed in §2.15, additional assumptions must be used to construct yield surfaces in the extended principal stress space. The simple shear apparatus (see Fig. 2.19) is free from this limitation, as it measures the stresses acting on *two* planes. Thus both σ_1 and σ_2 can be determined.

Let us proceed by assuming that it is possible to construct yield surfaces in the extended principal stress space. The yield surfaces can be represented by the equivalent forms

$$F(\sigma_1, \sigma_2, \nu) = 0; \quad f(\sigma, \tau, \nu) = 0 \tag{2.44}$$

The latter form is equivalent to (2.28), as $I_1 = \sigma_1 + \sigma_2 = 2\,\sigma$ and $I_2 = \sigma_1\sigma_2 = \sigma^2 - \tau^2$. As the invariants are symmetric functions of the principal stresses, the functions F and f in (2.44) must share this property. Hence yield surfaces must be symmetric about the plane $\sigma_1 = \sigma_2$ in σ_1–σ_2–ν space and about the plane $\tau = 0$ in the σ–τ–ν space. The latter condition holds only if $\tau \equiv (\sigma_1 - \sigma_2)/2$, without the restriction $\sigma_1 \geq \sigma_2$. On the other hand, if τ is defined by (2.33), we must have $\tau \geq 0$ and the symmetry property of the yield surface is not visually evident in the σ–τ–ν space. Both the definitions of τ will be used here but the choice made will be evident from the context in which τ is used.

In this book, (2.44) will be regarded as the general form of the yield condition for plane flow. The Mohr–Coulomb yield condition (2.37) is an example of (2.44).

For specified values of σ and ν, say ν_* and σ_*, respectively, let τ_* be the value of τ which satisfies the yield condition (2.44). (For isotropic materials, $-\tau_*$ also satisfies the yield condition.) If σ_1 is identified with the major principal stress, we must have $\tau_* = (\sigma_{1*} - \sigma_{2*})/2 \geq 0$. For $\sigma = \sigma_*$ and $\nu = \nu_*$, states of stress with $\tau > \tau_*$ are not permitted within the present framework. Conversely, if $\tau < \tau_*$, the yield condition is not satisfied. Henceforth, the yield function $f(\sigma, \tau, \nu)$ will be defined such that the elastic/rigid region corresponds to $f < 0$. For example, consider the Mohr–Coulomb yield condition (2.38) for a cohesionless material. The elastic/rigid region is shown by the hatched lines in Figs. 2.28a and 2.28b. For states of stress in this region, the Mohr's circle lies within the wedge bounded by the lines AB and AC in Fig. 2.28c. These lines represent the Coulomb yield condition (2.16).

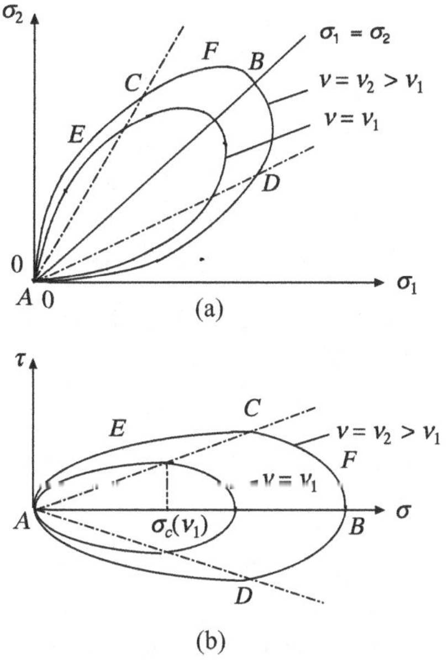

Figure 2.29. Hypothetical yield loci (—) and critical state curves (– · –) in the (a) σ_1–σ_2 and (b) σ–τ planes for a cohesionless material. Here σ_1 and σ_2 are the principal stresses, $\sigma = (\sigma_1 + \sigma_2)/2$, $\tau = (\sigma_1 - \sigma_2)/2$, and AC and AD represent the critical state curves.

In general, there may be spatial regions where the yield condition is satisfied and other regions where it is not satisfied. In the problems discussed here it will be assumed that the yield condition is satisfied at every point in the domain of flow. Even though this approach leads to satisfactory results for some problems, it cannot be justified a priori.

Considering the bin problem discussed in §2.3, suppose that the material is cohesionless and the yield condition is not satisfied along the centerline $x = 0$. As $\sigma_{xy}(x = 0, y) = 0$, (2.35) implies that $\sin 2\gamma = 0$. Hence $\cos 2\gamma = \pm 1$, and (2.35) and (2.36) imply that

$$K_0 \equiv K(x = 0, y) \equiv \frac{\sigma_{xx}(x = 0, y)}{\sigma_{yy}(x = 0, y)} = \frac{\sigma \pm \tau}{\sigma \mp \tau} \tag{2.45}$$

where the upper and lower signs correspond to passive and active states of stress, respectively. For the Mohr–Coulomb yield condition, the elastic/rigid region corresponds to the hatched region in Fig. 2.28b. Hence if $c = 0$, $|\tau/\sigma| < \sin\phi_*$ and

$$K_a < K_0 < K_p \tag{2.46}$$

where K_a and K_p are given by (2.40). Thus the value of K_0 is bounded by the active and passive values. Unfortunately, the range is rather wide. For example, if $\phi_* = 30°$ and $c = 0$, $1/3 < K_0 < 3$. If the yield condition is satisfied, K_0 is equal to either K_a or K_p.

2.12. YIELD LOCI IN THE σ_1–σ_2 AND σ–τ PLANES

For a cohesionless material, Fig. 2.29 shows hypothetical yield loci in the σ_1–σ_2 and σ–τ planes. For the yield locus with $\nu = \nu_2$, AEC and CFB represent the dilation and compaction branches, respectively, and the point C represents the critical state. For a cohesive material, parts of the yield loci are shifted to the region of negative principal stresses (Fig. 2.30). Jenike (1964b, p. 12) states that the critical state curve passes through the origin for many materials. This is supported by the data for Weald clay (Roscoe et al., 1958) and penicillin powder (Williams and Birks, 1965). However, the critical state curves for barytes powder and limestone powder do not pass through the orgin (Molerus, 1978).

Henceforth, it will be assumed that the locus of critical states in the σ–τ plane is a straight line passing through the origin, called the *critical state line*. Linerality of this locus

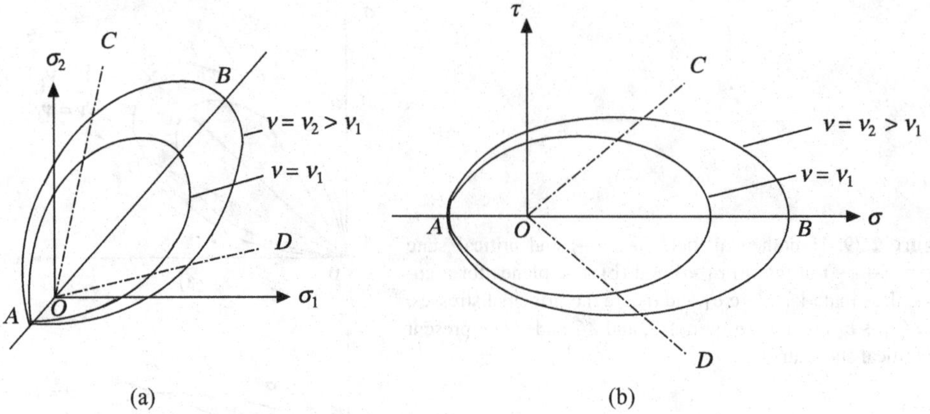

Figure 2.30. Hypothetical yield loci (—) and critical state curves (– · –) in the (a) σ_1–σ_2 and (b) σ–τ planes for a cohesive material. Here σ_1 and σ_2 are the principal stresses, $\sigma = (\sigma_1 + \sigma_2)/2$, $\tau = (\sigma_1 - \sigma_2)/2$, and AC and AD represent the critical state curves.

is not a fundamental requirement but is suggested by the data shown in Fig. 2.20 and also other data reported in Roscoe et al. (1958), Williams and Birks (1965), and Molerus (1978). The critical state line is defined by

$$\tau = \sigma \sin\phi \tag{2.47}$$

where ϕ is a material constant called the *angle of internal friction*. Unfortunately, the same terminology is used in the literature for both ϕ and ϕ_*, where ϕ_* is the angle occurring in the Mohr–Coulomb yield condition (2.37). The two angles differ in general and cannot be related unless additional assumptions are used. For example, if the material satisfies the Mohr–Coulomb yield condition and the deformation is *assumed* to be incompressible, the yield locus coincides with the critical state line. Hence $\phi_* = \phi$. On the other hand, if this yield condition is used in conjunction with the associated flow rule (see §2.14.4), only dilation occurs. Hence there is no critical state and ϕ is not defined, even though ϕ_* is.

To construct the stress–deformation curves sketched in Fig. 2.27, the yield surfaces must be supplemented by constitutive equations relating the stresses to the deformation of the sample. This is discussed below.

2.13. FLOW RULES

At present, there is no universally accepted constitutive equation for slow granular flow. Here we discuss one approach, based on *plasticity theory*, which has been widely used to describe the behavior of soils and metals at large strains.

Consider the uniaxial compression of a sample of sand, which is confined in a rigid cylindrical vessel (Fig. 2.13). (In the civil engineering literature, the tester used for this purpose is called an *oedometer*.) When a compressive normal stress σ_v is applied to the upper surface of the sample, the volume of the sample reduces from its intial value V_0 to a value V.

Fig. 2.31 shows the variation of the cross-sectional averaged normal stress σ_v with the *volumetric strain* $\epsilon_v \equiv (V_0 - V)/V_0$ for sand. At point B, if the sample is *unloaded* by reducing the normal stress, the stress–strain curve follows the path BC. Further, when the sample is *reloaded* from the point C by increasing the stress, the path CD is traced. Thus there is no *unique* relation between the stress and the strain. This rules out the use of elastic

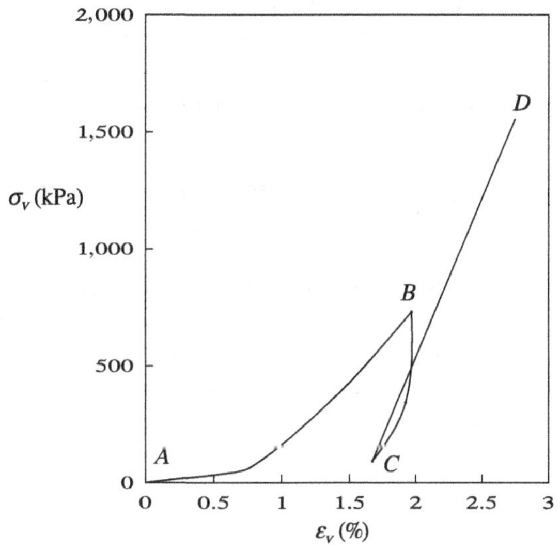

Figure 2.31. Variation of the cross-sectional averaged vertical normal stress σ_v with the volumetric strain $\epsilon_v \equiv (V_0 - V)/V_0$ with during loading (*AB*), unloading (*BC*), and reloading (*CD*) of a sample of sand. Here V_0 and V are the initial and current volume, respectively, of the sample. (Data of Lambe and Whitman, 1969, p. 125.)

constitutive equations, wherein the stress is a function of the strain, to describe granular flow. At the point C, there is a significant strain even though the stress is nearly zero, and the strain at C is less than that at B. Thus the sample has experienced both recoverable or elastic deformation and irrecoverable or plastic deformation. The hysteresis loop shows that energy is dissipated in the unloading–reloading cycle.

As noted by Mendelson (1983, p. 99), the dependence of the strain on the loading history suggests that it may be useful to employ *strain increments*, rather than strains, in constitutive relations describing plastic deformation. The *strain increment* tensor $\mathbf{d\epsilon}$ is defined by (Malvern, 1969, p. 151; Chakrabarthy, 1987, p. 66)

$$\mathrm{d}\boldsymbol{\epsilon} \equiv \mathbf{C}\,\mathrm{d}t \tag{2.48}$$

where

$$\mathbf{C} \equiv -\frac{1}{2}(\nabla\mathbf{v} + (\nabla\mathbf{v})^T) \tag{2.49}$$

is the *rate of deformation* tensor, defined in the compressive sense, and $\mathbf{v}$ is the velocity vector. The components of $\mathbf{C}$ relative to a Cartesian coordinate system are given by

$$C_{ij} = -\frac{1}{2}\left(\frac{\partial v_j}{\partial x_i} + \frac{\partial v_i}{\partial x_j}\right) \tag{2.50}$$

where v_i is the ith component of the velocity vector.

Before proceeding further, a few remarks regarding the physical significance of $\mathbf{C}$ may be helpful. It can be shown (Problem 2.12) that the diagonal components of $\mathbf{C}$ (C_{11}, C_{22}, C_{33}) represent the rate of compression per unit length of line elements that are parallel to the coordinate axes. At any time t, consider a parallelepiped A whose surfaces are parallel to the coordinate planes. Then the diagonal components of $\mathbf{C}$ represent the instantaneous rate of change of *size* of A. The off-diagonal components of $\mathbf{C}$ ($C_{12} = C_{21}$, $C_{13} = C_{31}$, $C_{23} = C_{32}$) are proportional to the rate at which the angle between two initially orthogonal line elements changes due to flow. Hence the off-diagonal components of $\mathbf{C}$ represent the instantaneous rate of change of *shape* of A. If a material element undergoes a local *rigid body* motion, there is no change in shape and size of the element, and hence all components of $\mathbf{C}$ vanish.

In this book, both stresses and strain increments are defined in the compressive sense. For example, the length of a line element parallel to the x_1 axis decreases if $d\epsilon_{11} > 0$. In most of the books on plasticity (see, e.g., Hill, 1950; Mendelson, 1983; Chakrabarthy, 1987), strain increments are defined in the tensile sense.

2.13.1. The Lévy–Mises and the Prandtl–Reuss Equations

Here we discuss some simple flow rules that were originally developed to describe the plastic flow of metals.

The first attempt to formulate plastic constitutive equations was due to de Saint-Venant (1871) (cited in Kachanov, 1974, p. 65), who proposed a relation between the stress tensor and the strain increment tensor for the case of plane deformation. This was extended to three-dimensional deformation fields by Lévy (1871) (cited in Kachanov, 1974, p. 65) and von Mises (1913) (cited in Kachanov, 1974, p. 65). The *Lévy–Mises* equations are given by

$$\mathrm{d}\epsilon_{ij} = s'_{ij}\,\mathrm{d}\lambda, \quad i, j = 1, 3 \tag{2.51}$$

where $\mathrm{d}\lambda$ is a scalar factor of proportionality, s'_{ij} is the component of the *deviatoric* stress tensor, defined by

$$s'_{ij} \equiv \sigma_{ij} - p\,\delta_{ij}, \quad p \equiv \sigma_{ii}/3, \quad i, j = 1, 3 \tag{2.52}$$

p is the pressure and δ_{ij} is the Kronecker delta, defined by (A.6). The factor $\mathrm{d}\lambda$ is not a material parameter but must be determined as a part of the solution. (Even though this chapter is devoted to plane flow, (2.51) is valid for both plane and three-dimensional flows.)

Equation (2.51) is an example of a *flow rule*. In general, a flow rule is a relation between the strain increments (or equivalently, the components of the rate of deformation tensor) and the stresses, which is constructed so that only the *ratios* of the strain increments can be determined explicitly. The Lévy–Mises equations apply to a rigid-plastic material. If the yield condition $\tilde{F}(\boldsymbol{\sigma}, \nu) = 0$ is satisfied at $(\boldsymbol{\sigma}_1, \nu_1)$, *and* $\boldsymbol{\sigma}$ and ν change such that $\mathrm{d}\tilde{F} = 0$, then $\mathrm{d}\lambda \neq 0$ and plastic deformation occurs. In this case, the flow rule (2.51) must be used along with the yield condition to obtain the complete constitutive relation between the stress and the strain increment tensors. Conversely, if either $\tilde{F} < 0$ or $\mathrm{d}\tilde{F} < 0$, we set $\mathrm{d}\lambda = 0$. Equation (2.51) then implies that the material is rigid, and hence the stress field cannot be determined uniquely.

For the case of plane deformation, Prandtl (1924) (cited in Malvern, 1969, p. 339) generalized the Lévy–Mises equations to include elastic effects by assuming that $\mathrm{d}\epsilon_{ij} = \mathrm{d}\epsilon_{ij}{}^{e} + \mathrm{d}\epsilon_{ij}{}^{p}$, where the superscripts "e" and "p" denote elastic and plastic contributions, respectively. The elastic strain increments are evaluated using an incremental form of the linear elastic constitutive equations (5.126), and the plastic strain increments follow from (2.51), with $\mathrm{d}\epsilon_{ij}$ replaced by $\mathrm{d}\epsilon_{ij}{}^{p}$. Subsequently, Reuss (1930) (cited in Malvern, 1969, p. 339) extended these equations to the general case of three-dimensional deformation.

Examples involving the use of the Prandtl–Reuss equations can be found in books on plasticity (see, e.g., Hill, 1950; Chakrabarthy, 1987). The use of strains and strain increments is convenient for problems which do not involve sustained flow. An example of this type is provided by the analysis of spherical shell subjected to an internal pressure p (Chakrabarthy, 1987, pp. 97–102). As p is increased from zero to its steady-state value of p_s, the shell expands elastically. If p_s is large enough, plastic deformation begins at the inner surface of the shell and spreads outward (Problem 5.12). When p is maintained at p_s, there are nonzero elastic and plastic strains, but the *velocity* vanishes at every point in the shell.

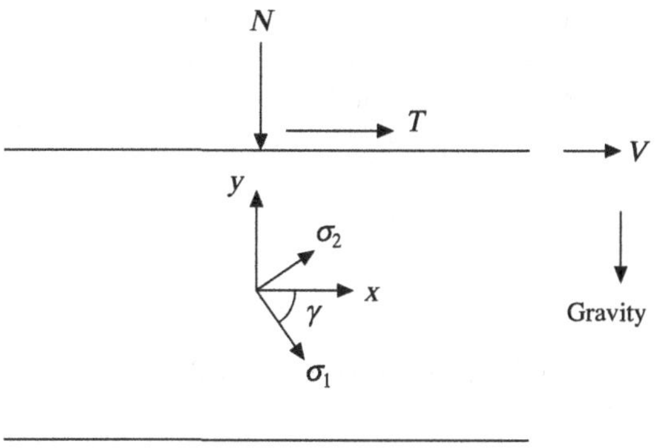

Figure 2.32. Steady, fully developed flow between horizontal plates. The lower plate is stationary relative to the laboratory frame, and the upper plate moves to the right with a constant velocity V. Here σ_1 and σ_2 are the major and minor principal stresses, respectively, and N and T are the normal and shear stresses, respectively, exerted by the upper plate on the granular material.

On the other hand, the velocity is in general nonzero in problems such as extrusion and flow through hoppers. Here the material is subjected to large strains, and elastic strains can be neglected in comparison with plastic strains (Chakrabarthy, 1987, p. 407). Hence a rigid-plastic analysis is permissible, except in regions where the material is either at rest or moving very slowly. Therefore, elastic effects will usually be neglected in this book.

In flowing regions, it is convenient to formulate constitutive equations in terms of velocities and velocity gradients, rather than strains and strain increments. In view of (2.48), the Lévy–Mises equations (2.51) can be rewritten as (Malvern, 1969, p. 337; Shames and Cozzarelli, 1992, p. 291)

$$C_{ij} = \dot{\lambda}\, s'_{ij}, \quad i, j = 1, 3 \tag{2.53}$$

where C_{ij} is a component of the rate of deformation tensor, defined by (2.50), and $\dot{\lambda}$ is a scalar factor of proportionality.

Equation (2.53) is similar in form to the constitutive equation (1.49) for an incompressible Newtonian fluid, which can be rewritten as

$$C_{ij} = \frac{s'_{ij}}{2\,\mu_{\mathrm{f}}}, \quad i, j = 1, 3 \tag{2.54}$$

where μ_{f} is the shear viscosity of the fluid. However, unlike μ_{f}, $\dot{\lambda}$ is not a material parameter. Hence the response of a Lévy–Mises material is quite different from that of an incompressible Newtonian fluid, as illustrated by the following example.

Consider steady, fully developed flow between horizontal plates (Fig. 2.32). The velocity field is of the form $v_x = v_x(y),\ v_y = 0,$ and $v_z = 0$, and the stress components are assumed to depend only on the y coordinate. It is desired to predict the stress and velocity fields for a Newtonian fluid and for a Lévy–Mises material. For both materials, the momentum balances (1.31) reduce to

$$\sigma_{xy} = \text{constant} \tag{2.55}$$

$$\frac{\mathrm{d}\sigma_{yy}}{\mathrm{d}y} = -\rho\, g \tag{2.56}$$

where ρ is the density of the material and g is the acceleration due to gravity.

The Newtonian fluid model (1.49) implies that

$$\sigma_{xx} = \sigma_{yy} = \sigma_{zz} = p_{\mathrm{f}}; \quad \sigma_{xz} = 0 = \sigma_{yz} \tag{2.57}$$

$$\sigma_{xy} = \sigma_{yx} = -\mu_{\mathrm{f}} \frac{\mathrm{d}v_x}{\mathrm{d}y} = \text{constant} \tag{2.58}$$

where p_{f} is the pressure of the fluid. Assuming that the shear viscosity μ_{f} is a constant, the velocity profile is linear in y and the shear stress σ_{xy} is constant across the shear layer.

Now consider the Lévy–Mises material. If the material deforms, the yield condition is satisfied and $\dot{\lambda} \neq 0$. Then the velocity field and the flow rule (2.53) imply that

$$\sigma_{xx} = \sigma_{yy} = \sigma_{zz} \equiv p; \quad \sigma_{xz} = 0 = \sigma_{yz} \tag{2.59}$$

$$\sigma_{xy} = \sigma_{yx} = -\frac{1}{2\,\dot{\lambda}}\frac{\mathrm{d}v_x}{\mathrm{d}y} \tag{2.60}$$

The momentum balances imply that

$$\sigma_{xy} = \sigma_{yx} = -\frac{1}{2\,\dot{\lambda}}\frac{\mathrm{d}v_x}{\mathrm{d}y} = \text{constant} \tag{2.61}$$

$$\frac{\mathrm{d}\sigma_{yy}}{\mathrm{d}y} = \frac{\mathrm{d}p}{\mathrm{d}y} = -\rho_{\mathrm{p}}\nu g \tag{2.62}$$

where ρ_{p} is the particle density and ν is the solids fraction.

Equation (2.61) must be supplemented by a yield condition. Suppose that we use the Mohr–Coulomb yield condition (2.30) for a cohesionless material ($c = 0$). Using (2.59), (2.30) reduces to

$$\sigma_{xy} = \pm p \sin\phi_* \tag{2.63}$$

Equations (2.61) and (2.63) imply that p is a constant, which is inconsistent with (2.62). Thus the combination of the Lévy–Mises equations and the Mohr–Coulomb yield condition does not admit a solution. Conversely, if the yield condition is not satisfied, $\dot{\lambda} = 0$, and the material is rigid. Thus the material cannot be sheared, in contrast to observations. The problem can be resolved by modifying the frictional constitutive equations, as discussed in §5.1.10.

The Lévy–Mises equations are *rate-independent*, i.e., if all the components of the rate of deformation tensor **C** are multiplied by a common factor, the stresses given by the flow rule and the yield condition are unaffected. This is in contrast to a Newtonian fluid, which shows rate-dependent behavior. In order to discriminate between these models, data on the variation of the stresses with **C** are discussed below.

Tardos et al. (1998) measured the torque required to shear granular materials in the annular gap of a cylindrical Couette cell (Fig. 2.1). The outer cylinder is stationary and the inner cylinder rotates about its axis with an angular velocity ω. The torque M required to rotate the inner cylinder is measured, and the nominal shear rate S is calculated as $S = R\,\omega/W$, where R is the radius of the inner cylinder and W is the width of the gap between the cylinders (Fig. 2.1). For glass beads, the dimensionless torque M_* decreases by about 8% as the dimensionless shear rate S_* increases from 0.0026 to 0.24, i.e., as log S_* increases from -2.58 to -0.62 or S increases from 0.5 s^{-1} to 47.5 s^{-1} (Fig. 2.33). The decrease in torque with an increase in the shear rate implies that the glass beads show a *velocity-weakening* behavior, similar to that observed for the sliding of pieces of cardboard relative to each other (Heslot et al., 1994) and for the sliding of blocks of granite (Dieterich, 1981) (cited in Gu et al., 1984). As the uncertainty in the torque measurements is about 1.3%, more detailed measurements are warranted.

The torque is independent of the shear rate for a rate-independent model and increases with the shear rate for a Newtonian fluid. The data suggest that the rate-independent model is preferable to a Newtonian fluid model for the case of slow flow. For $S_* > 0.24$, the torque increases with the shear rate, as shown in Fig. 2.33.

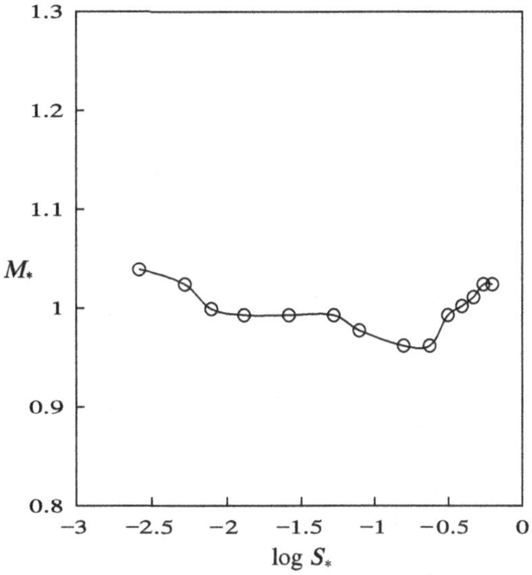

Figure 2.33. Variation of the dimensionless torque M_* with the dimensionless shear rate S_* for shearing of glass beads in a Couette cell (see Fig. 2.1). The torque is scaled by $\pi \rho g R^2 L^2 \sin \phi_*$ and the shear rate by $\sqrt{g/d_p}$. The circles represent the data of Tardos et al. (1998), and the curve is a spline fit drawn to guide the eye. Here ϕ_* is the angle of internal friction, ρ is the density of the granular material, and d_p is the particle diameter. Parameter values: $\phi_* = 28.5°$, $\rho = 1{,}490$ kg m^{-3}, $d_p = 0.25$ mm, $R = 0.125$ m, $L = 0.195$ m, $W = 0.0127$ m. The symbols R, L, and W are defined in Fig. 2.1.

Based on visual observation of the upper free surface of the granular material, Tardos et al. (1998) state that only a part of the material near the inner cylinder sheared at low values of S_*. The thickness of the shearing region increased with S_* and became equal to the gap width at the shear rate corresponding to the minimum in the torque versus shear rate curve. For larger values of S_*, the material sheared vigorously throughout the gap.

At low values of S_*, the actual shear rate may differ considerably from the nominal shear rate. This is because the thickness of the shearing region is less than the gap width, and the velocity of the shearing material may be nonzero at the boundary between shearing and rigid regions.

Albert et al. (1999) have measured the drag force F exerted on a cylindrical rod dipped into a rotating cylinder filled with glass beads (Fig. 2.34a). For velocities in the range 0.1–1.5 mm s^{-1}, the drag force is independent of the velocity (Fig. 2.34b), suggesting rate-independent behavior.

2.13.2. The Coaxiality Condition

The Lévy–Mises equations (2.53) imply that the flow is incompressible, as tr($\mathbf{C}$) $= -\nabla \cdot \mathbf{v} = 0$. They must be modified to allow for density variation during flow, as this is a characteristic feature of granular materials. Let us therefore consider a more general constitutive equation of the form

$$\mathbf{C} = \mathbf{A}(\boldsymbol{\sigma}, \nu) \tag{2.64}$$

where $\mathbf{C}$ is the rate of deformation tensor, $\boldsymbol{\sigma}$ is the stress tensor, and ν is the solids fraction.

For isotropic materials, as shown in Appendix F, (2.64) implies that the principal axes of the stress and rate of deformation tensors are aligned with each other (Serrin, 1959, p. 232; Hunter, 1983, pp. 136–137). This is called the *coaxiality condition*. It does not uniquely determine the relative orientations of the two sets of axes. For example, the major principal stress axis (σ_1 axis) can be aligned with any one of the two axes of the two-dimensional rate of deformation tensor $\mathbf{C}$. We shall assume that the major principal stress axis coincides with

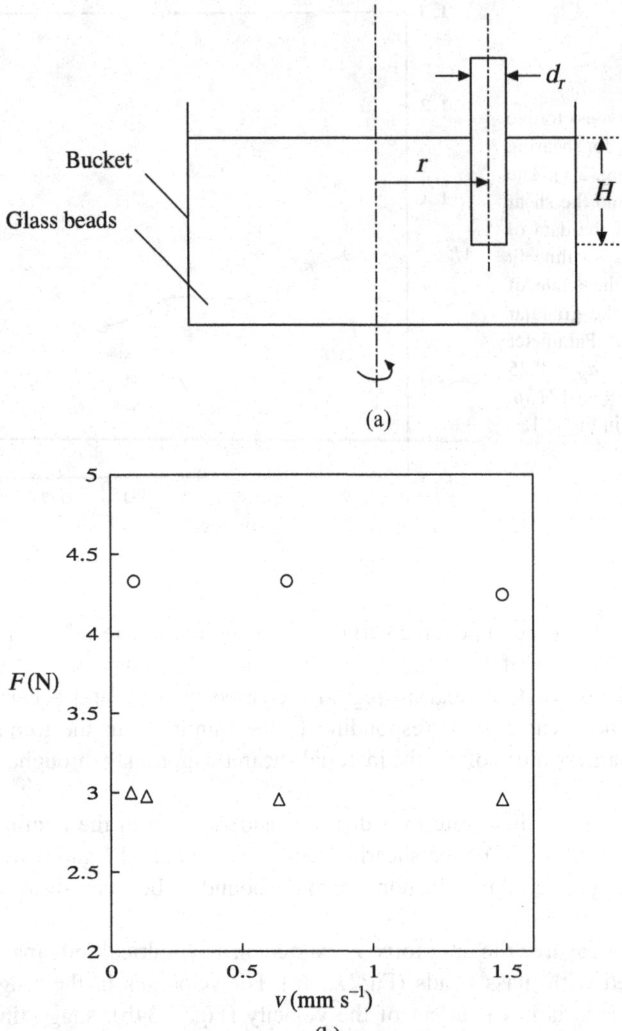

Figure 2.34. (a) A cylindrical rod dipped into a rotating cylindrical bucket filled with glass beads of diameter d_p. (b) Variation of the drag force F with the grain velocity $v = r\,\omega$, where r is distance between the axis of the bucket and the axis of the rod, and ω is the angular velocity of the bucket. Parameter values in mm: diameter and height of the bucket = 255 and 235, respectively, ○, $d_p = 3$, $d_r = 4.7$, $H = 95$; △, $d_p = 3$, $d_r = 19$, $H = 51$. (Data of Albert et al., 1999.)

the major principal axis of **C**. This is intuitively reasonable, as it implies that the maximum rate of compression occurs along the major principal stress axis. Unfortunately, a rigorous justification is not available. Some data related to the coaxiality condition are discussed below.

Figure 2.35 shows data obtained using the simple shear apparatus (Fig. 2.19). The normal stress N exerted on the upper face of the apparatus is kept constant during the test, and the shear stress T is measured as the sample deforms. Figure 2.35a shows the variation of the stress ratio T/N with the shear strain α during loading (curve AB), unloading (curve BC), and reloading (curve CD). In Fig. 2.35b, the angles ψ, ξ, and χ represent the inclinations of the major principal axes of the stress, strain increment, and stress increment tensors, respectively, relative to the vertical.

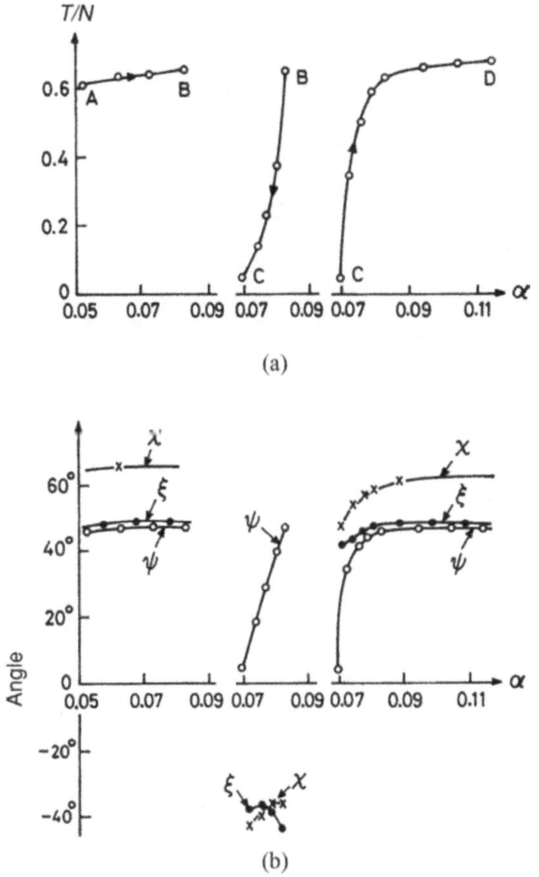

Figure 2.35. (a) Variation of the stress ratio T/N with the shear strain α for the deformation of Leighton Buzzard sand in a simple shear apparatus. Here N (= 131 kPa) and T are the normal and shear stresses, respectively, exerted on the upper surface of the sample, and α is calculated as shown in the caption of Fig. 2.21b. The curves AB, BC, and CD represent loading, unloading, and reloading, respectively. (b) Variation of the inclinations of the major principal axes of the stress tensor (ψ), the strain increment tensor (ξ), and the stress increment tensor (χ) with α. The angles are measured from the vertical. (Reproduced from Roscoe, 1970, with permission from the Institution of Civil Engineers, London.)

During loading, the principal axes of the strain increment and the stress tensors are aligned. In view of (2.48), this implies that the principal axes of the rate of deformation and the stress tensors are aligned, in accord with the plastic constitutive equation (2.64). During unloading, the principal axes of the strain increment and stress increment tensors are aligned. This is consistent with the behavior of an isotropic, linear elastic material, which is described by (5.126). In the advanced stage of reloading, the principal axes of the stress and strain increment tensors are aligned with each other. Thus the material deforms plastically during loading and the advanced stage of reloading, and elastically during unloading.

Even though the models considered here ignore elastic effects, it is likely that such effects may be important in certain parts of the flowing zone. For example, granular materials are often subjected to oscillating stresses when they flow through the transition region of bunkers (§4.1.3), and loading and unloading may occur intermittently.

The experiments of Drescher (1976) also provide some data related to the coaxiality condition. The double-shear apparatus (Fig. 2.36a) was used to study the deformation of crushed glass particles. The particles were filled in the gap between two parallel glass plates and were bounded from above and below by the hinged metal plates ABC and DEF. The lower plate was moved upward at a constant speed, whereas the upper plate was free to move in response to the deformation of the granular material. As the dry material was opaque, it was immersed in castor oil, which has approximately the same refractive index as glass. This rendered the material transparent to light.

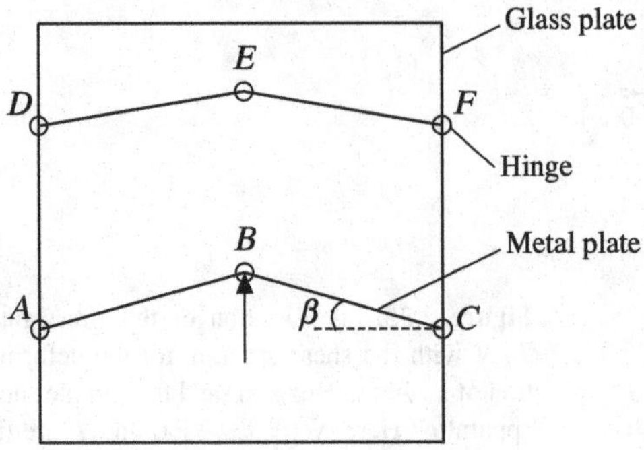

(a)

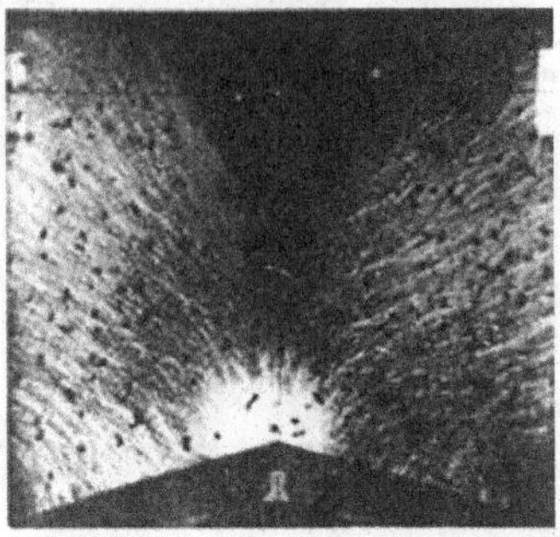

(b)

Figure 2.36. (a) The double-shear apparatus. (b) Stripes produced when the granular material (crushed glass particles) was subjected to stresses and viewed in polarized light. (Figure 2.36a has been adapted from Fig. 2 of Drescher, 1976. Figure 2.36b has been reproduced from Fig. 1b of Drescher, 1976, with permission from the Institution of Civil Engineers, London.)

When this system was subjected to stresses and viewed in circularly polarized light, a pattern of stripes of light was observed (Fig. 2.36b). (The reader is referred to Timoshenko and Goodier (1970, pp. 150–157) for a discussion of the use of polarized light to determine stress fields in photoelastic specimens.) These stripes were assumed to represent the trajectories of the major principal stress axis. It is not clear whether this assumption is applicable when plastic deformation occurs. As the refractive indices of the particles and the oil differed slightly, the particles were visible to some extent. Hence their positions were tracked photographically after small displacements of the lower plate, and the orientations of the principal axes of the strain increment tensor were calculated. With the angle β measured as shown in Fig. 2.36a, the results obtained for a change of 1.5° in β are shown in Fig. 2.37. At several points, the short, solid lines, representing the major principal axis of the strain increment tensor, are not parallel to the major principal stress axis. Thus it appears that coaxiality does not hold in this experiment.

In view of the preceding data, it appears that the extent of deviation from coaxiality may depend on the material used and also on the type of the flow field.

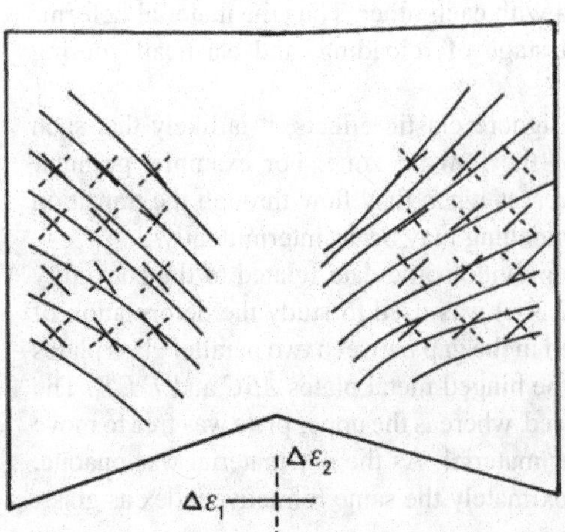

Figure 2.37. Trajectories of the major principal stress axes of the stress tensor (solid curves) and the strain increment tensor (short, solid lines), corresponding to a change of 1.5° in the angle β. Here β is measured as shown in Fig. 2.36a. (Reproduced from Drescher, 1976, with permission from the Institution of Civil Engineers, London.)

2.13.3. The Plastic Potential

Rate independence can be incorporated into (2.64) by rewriting it as (Malvern, 1969, p. 257; Hunter, 1983, pp. 480–481; Nedderman, 1992, p. 251)

$$C_{ij} = \dot{\lambda}\left(\frac{\partial \tilde{G}}{\partial \sigma_{ij}}\right) \tag{2.65}$$

where $\tilde{G}(\boldsymbol{\sigma}, \nu)$ is a scalar function called the *plastic potential*. As before, $\dot{\lambda}$ is a scalar factor of proportionality. The plastic potential was first used by von Mises (1928) (cited in Khan and Huang, 1995, p. 142).

In the three-dimensional case, Hunter (1983, pp. 480–481) has shown that (2.64) can always be rewritten as (2.65), provided that the plastic deformation is (i) incompressible and (ii) does not depend on the mean stress $p = \sigma_{kk}/3$ (see also Problem 5.7). Data suggest that these conditions hold for the plastic deformation of metals, but not for granular materials. It is not known whether a similar proof is available when the conditions (i) and (ii) are relaxed.

For isotropic materials, the plastic potential can be expressed as a function of ν and the invariants of the stress tensor. As the latter are symmetric functions of the principal stresses σ_1 and σ_2, the plastic potential can also be expressed as

$$G = G(\sigma_1, \sigma_2, \nu) \tag{2.66}$$

where G is a *symmetric* function of σ_1 and σ_2. Choosing coordinate axes that are aligned with the principal stress axes, (2.65) can be rewritten as

$$c_i = \dot{\lambda}\left(\frac{\partial G}{\partial \sigma_i}\right), \quad i = 1, 2 \tag{2.67}$$

where c_1 and c_2 are the principal compressive rates of deformation, i.e., the eigenvalues of $\mathbf{C}$. Equation (2.67) can be interpreted geometrically as follows. For a specified value of ν, the equation $G(\sigma_1, \sigma_2, \nu) =$ constant defines a curve called the *plastic potential locus* in the σ_1–σ_2 plane. With a suitable value for the constant, the locus *ABC* can be made to pass through any chosen point B (Fig. 2.38). For movement along *ABC*, we have

$$\mathrm{d}G = 0 = \dot{\lambda}\left(\frac{\partial G}{\partial \sigma_1}\,\mathrm{d}\sigma_1 + \frac{\partial G}{\partial \sigma_2}\,\mathrm{d}\sigma_2\right)$$

or

$$\left(\mathbf{e}_1\frac{\partial G}{\partial \sigma_1} + \mathbf{e}_2\frac{\partial G}{\partial \sigma_2}\right)\cdot(\mathbf{e}_1\,\mathrm{d}\sigma_1 + \mathbf{e}_2\,\mathrm{d}\sigma_2) = 0 \tag{2.68}$$

As $\mathbf{e}_1\,\mathrm{d}\sigma_1 + \mathbf{e}_2\,\mathrm{d}\sigma_2$ is the tangent vector to the curve *ABC*, it follows that the vector $\mathbf{e}_1\,(\partial G/\partial \sigma_1) + \mathbf{e}_2\,(\partial G/\partial \sigma_2)$ is directed along the *normal* to *ABC*.

Hence (2.67) implies that the principal rates of deformation are proportional to the components $\partial G/\partial \sigma_i$ of the normal to the plastic potential locus. Thus the vector $\mathbf{c}$, which has components (c_1, c_2), is directed along the vector *BD* (Fig. 2.38). (In this figure, the principal axes of the stress and rate of deformation tensors have been aligned on account of coaxiality.) The flow rule (2.67) determines only the direction of $\mathbf{c}$. The magnitude of $\mathbf{c}$ must be determined by solving all the governing equations. Once the stress and velocity fields have been obtained, it seems reasonable to ensure that they satisfy the following constraint.

2.13.4. Positive Dissipation

As discussed in §1.5.7, the stress power $\dot{\Phi} = -\boldsymbol{\sigma}^{\mathrm{T}} : (\nabla \mathbf{v})$ represents the rate of interconversion per unit volume of kinetic and internal energies. It is convenient to rewrite the

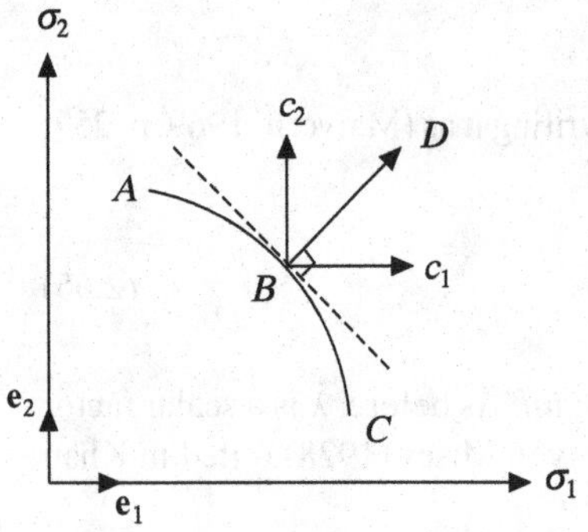

Figure 2.38. The plastic potential locus *ABC* and the normal *BD* to this locus at the point *B*. Here σ_1 and σ_2 are the principal stresses, c_1 and c_2 are the principal compressive rates of deformation, and $\mathbf{e}_1$ and $\mathbf{e}_2$ are unit vectors directed along the principal stress axes. In view of coaxiality, the principal axes of the rate of deformation tensor are aligned with the principal stress axes. At the point *B*, the vector **c**, which has components (c_1, c_2), is directed along the normal *BD* to the plastic potential locus.

expression for $\dot{\Phi}$ as follows. The velocity gradient tensor can be expressed as

$$\nabla \mathbf{v} = -\mathbf{C} + \mathbf{W}$$

where the symmetric rate of deformation tensor **C** and the antisymmetric *vorticity tensor* (or spin tensor) **W** are defined by

$$\mathbf{C} \equiv -\frac{1}{2}(\nabla \mathbf{v} + (\nabla \mathbf{v})^T); \quad \mathbf{W} \equiv \frac{1}{2}(\nabla \mathbf{v} - (\nabla \mathbf{v})^T) \tag{2.69}$$

As $\boldsymbol{\sigma}$ is symmetric and **W** is antisymmetric, (2.69) implies that

$$\dot{\Phi} = -\boldsymbol{\sigma}^{\mathrm{T}} : (\nabla \mathbf{v}) = \boldsymbol{\sigma}^{\mathrm{T}} : \mathbf{C} - \boldsymbol{\sigma}^{\mathrm{T}} : \mathbf{W} = \boldsymbol{\sigma}^{\mathrm{T}} : \mathbf{C} = \boldsymbol{\sigma} : \mathbf{C} \tag{2.70}$$

As elastic effects have been neglected, the deformation is plastic or irrecoverable. Hence it is reasonable to assume that the *dissipation inequality*

$$\dot{\Phi} = \boldsymbol{\sigma} : \mathbf{C} > 0 \tag{2.71}$$

holds whenever the material deforms, i.e., whenever $\mathbf{C} \neq 0$. This is a common assumption in plasticity theory (Hill, 1950, p. 26; Hunter, 1983, p. 482; Chakrabarthy, 1987, p. 73; Shames and Cozzarelli, 1992, p. 296). (For a Newtonian fluid, the Clausius–Duhem inequality (see §1.5) implies that $(\boldsymbol{\sigma} - p_{\mathrm{f}}\mathbf{I}) : \mathbf{C} > 0$, where p_{f} is the fluid pressure. This equation differs from (2.71), as the term involving p_{f} is assumed to represent a reversible contribution to the entropy production rate, and only the irreversible contribution is constrained to be nonnegative.)

In terms of the common principal axes of $\boldsymbol{\sigma}$ and **C**, the inequality (2.71) reduces to

$$\boldsymbol{\sigma}_* \cdot \mathbf{c} = \sigma_i\, c_i > 0, \quad |\mathbf{c}| \neq 0 \tag{2.72}$$

where $\boldsymbol{\sigma}_*$ and **c** are vectors with components (σ_1, σ_2) and (c_1, c_2), respectively, relative to the basis vectors $\mathbf{e}_1$ and $\mathbf{e}_2$ (Fig. 2.38).

2.13.5. Associated and Nonassociated Flow Rules

As both the plastic potential $G(\sigma_1, \sigma_2, \nu)$ and the yield function $F(\sigma_1, \sigma_2, \nu)$ are symmetric functions of the stresses, it is tempting to assume that G and F are identical to within an arbitrary additive or multiplicative constant. With this assumption, the plastic potential flow rule (2.67) reduces to the *associated flow rule*

$$c_i = \dot{\lambda}\frac{\partial F}{\partial \sigma_i}, \quad i = 1, 2 \tag{2.73}$$

Conversely, if F and G are not identical, (2.67) represents a *nonassociated flow rule.*

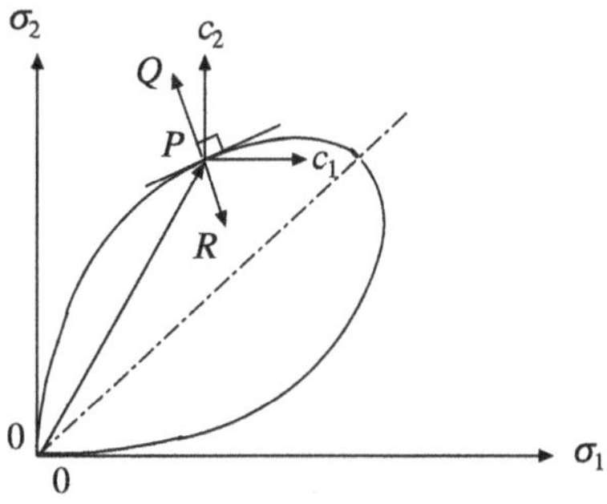

Figure 2.39. An example of a convex yield locus. At a point P on the locus, the vector OP has components (σ_1, σ_2). If the associated flow rule holds, the vector $\mathbf{c} = (c_1, c_2)$ is directed along either the inward normal PR or the outward normal PQ. For consistency with the assumption of positive dissipation, the outward normal should be chosen. Here (σ_1, σ_2) and (c_1, c_2) are the principal stresses and the principal rates of deformation, respectively. The two sets of axes are aligned because of the assumption of coaxiality.

For an associated flow rule, the principal rate of deformation vector $\mathbf{c}$ is directed along the normal to the yield locus (Fig. 2.39). This is called the principle of *normality*. Normality permits either the inward normal PR or the outward normal PQ to be used, but the assumption of positive dissipation (§2.13.4) enforces a choice. For example, consider the special case of a *convex* yield locus, i.e., one for which a straight line joining any two points on the locus is contained within the region enclosed by it (see Fig. 2.39). As $\boldsymbol{\sigma}_* = (\sigma_{1P}, \sigma_{2P}) = OP$ at the point P, and the angle between the vectors OP and PQ is an acute angle, the dissipation inequality (2.72) implies that $\mathbf{c}$ must be directed along the *outward* normal PQ to the yield locus.

The associated flow rule has often been used to describe the plastic deformation of metals (Hunter, 1983, p. 485; Chakrabarthy, 1987, p. 75). The model of Bishop and Hill (1951), which assumes that single crystals deform plastically by sliding along preferred directions, leads to this flow rule when it is applied to the deformation of a polycrystalline material. The data of Lianis and Ford (1957) are also consistent with it, provided the von Mises yield function is used as the plastic potential. (The latter is discussed in Chapter 5.)

For granular materials, deviations from the associated flow rule have been observed in some cases, particularly along the dilation branch of the yield locus (Ko and Scott, 1967; Lade and Duncan, 1973; Lade, 1988). Even though a sound theoretical basis is lacking, the associated flow rule has been used by several investigators to describe the compaction of granular materials (Roscoe and Burland, 1968; Schofield and Wroth, 1968, pp. 106, 140; DiMaggio and Sandler, 1971; Desai and Siriwardane, 1984, pp. 288–295). Examples of nonassociated flow rules are discussed in §2.14.5 and 6.2.3, and some data related to flow rules are presented in §5.1.9.

2.14. EQUATIONS FOR PLANE FLOW

The relevant equations for plane flow will now be derived. In the next two chapters, these equations will be used to examine flow through wedge-shaped hoppers and bunkers.

For isotropic materials, it is convenient to formulate the yield condition and the plastic potential in terms of the principal stresses σ_1 and σ_2 and the principal rates of deformation c_1 and c_2. However, the balance laws involve the stress and velocity components relative to a coordinate system that is chosen to suit the geometry of the problem. In §2.7, the Cartesian components of $\boldsymbol{\sigma}$ have been related to σ_1 and σ_2. A similar approach can be used to relate the Cartesian components of $\mathbf{C}$ to c_1 and c_2, as discussed below.

2.14.1. The Mohr's Circle for the Rate of Deformation Tensor

Referring to Fig. 2.40, we have

$$C \equiv (\mathbf{n} \cdot \mathbf{C}) \cdot \mathbf{n} = c_* + s \cos 2\gamma_*; \quad S \equiv (\mathbf{n} \cdot \mathbf{C}) \cdot \mathbf{t} = s \sin 2\gamma_* \tag{2.74}$$

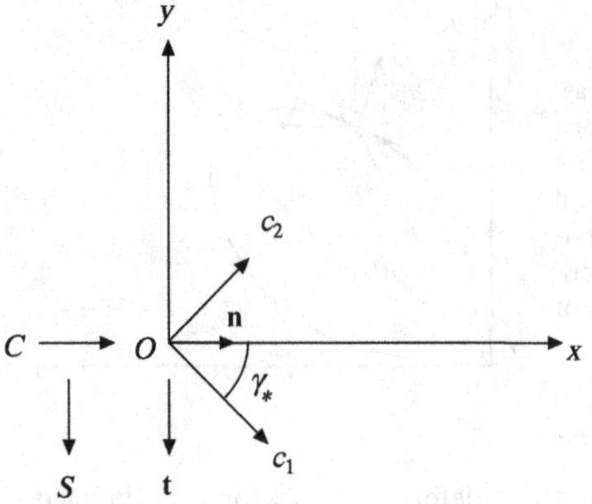

Figure 2.40. Orientation of the principal axes Oc_1 and Oc_2 of the rate of deformation tensor. Here $C = (\mathbf{n} \cdot \mathbf{C}) \cdot \mathbf{n}$ and $S = (\mathbf{n} \cdot \mathbf{C}) \cdot \mathbf{t}$, where $\mathbf{C}$ is the rate of deformation tensor and $\mathbf{n}$ and $\mathbf{t}$ are unit vectors.

where

$$c_* \equiv (c_1 + c_2)/2; \quad s \equiv (c_1 - c_2)/2, \quad c_1 \geq c_2 \tag{2.75}$$

Here c_1 and c_2 are the principal compressive rates of deformation. Equations (2.74) can be represented by a point on a circle in the C–S plane, with center at $(c_*, 0)$ and radius s (Fig. 2.41). This circle is called the *Mohr's circle for the rate of deformation tensor*. The quantity C represents the rate of compression in the direction of $\mathbf{n}$ and S represents the rate of shear in the direction of $\mathbf{t}$ (Fig. 2.40). Figure 2.41 shows that the maximum shear rate occurs on two planes that are inclined at $\pm 45°$ relative to the c_1 axis.

The quantity $c_* = (c_1 + c_2)/2$ can be interpreted as follows. As shown in §2.5.1 and §5.1.1, tr($\mathbf{C}$), the trace of $\mathbf{C}$, is an invariant of $\mathbf{C}$. Hence

$$\nabla \cdot \mathbf{v} = \frac{\partial v_i}{\partial x_i} = -\text{tr}(\mathbf{C}) = -\sum_{i=1}^{3} c_i \tag{2.76}$$

As $c_3 = 0$ for flow parallel to the c_1–c_2 plane, we have

$$\nabla \cdot \mathbf{v} = -(c_1 + c_2) = -2\, c_* \tag{2.77}$$

for plane flow. If $c_* > 0$, (2.77) and the mass balance (1.27) imply that the density of the material increases as it flows.

2.14.2. The Coaxiality Condition

Equations (2.74) will now be used to derive an expression for the coaxiality condition in Cartesian coordinates. In view of coaxiality, Figs. 2.11 and 2.40 imply that $\gamma_* = \gamma$.

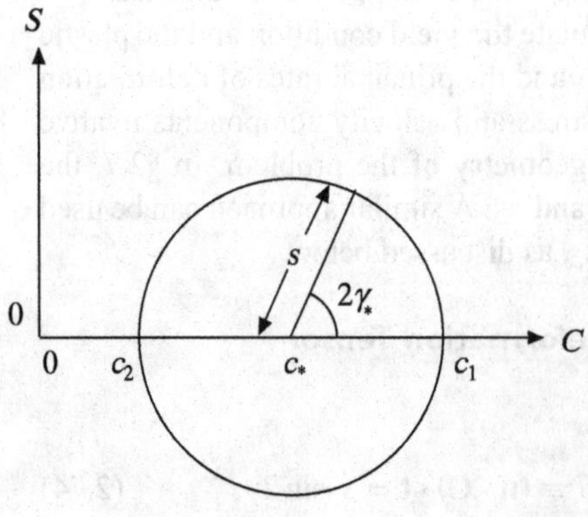

Figure 2.41. The Mohr's circle for the rate of deformation tensor. Here c_1 and c_2 are the principal compressive rates of deformation, $c_* = (c_1 + c_2)/2$, $s = (c_1 - c_2)/2$, and C and S are the rates of compression and shearing, respectively, on a plane whose normal is inclined at an angle γ_* to the c_1 axis (see Fig. 2.40).

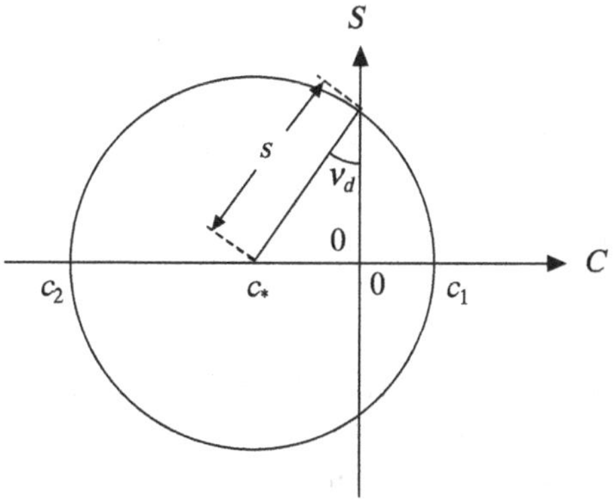

Figure 2.42. Geometric interpretation of the angle of dilation ν_d for the case $|c_*/s| \leq 1$. Here c_1 and c_2 are the principal compressive rates of deformation, $c_* = (c_1 + c_2)/2$, and $s = (c_1 - c_2)/2$.

Identifying **n** with the x direction (Fig. 2.40), it follows from (2.74) that

$$C = C_{xx} = -\frac{\partial v_x}{\partial x} = c_* + s\,\cos 2\gamma$$

$$S = -C_{xy} = \frac{1}{2}\left(\frac{\partial v_y}{\partial x} + \frac{\partial v_x}{\partial y}\right) = s\,\sin 2\gamma \tag{2.78}$$

Similarly,

$$C_{yy} = -\frac{\partial v_y}{\partial y} = c_* - s\,\cos 2\gamma \tag{2.79}$$

Equations (2.78) and (2.79) can be rearranged to obtain

$$\sin 2\gamma\left(\frac{\partial v_y}{\partial y} - \frac{\partial v_x}{\partial x}\right) - \cos 2\gamma\left(\frac{\partial v_y}{\partial x} + \frac{\partial v_x}{\partial y}\right) = 0 \tag{2.80}$$

which is one form of the *coaxiality condition*.

2.14.3. The Flow Rule

The plastic potential flow rule (2.67) can be expressed in terms of the Cartesian components of velocity as follows. Equations (2.78) and (2.79) imply that

$$\left(\frac{c_*}{s}\right)\left(\frac{\partial v_x}{\partial x} - \frac{\partial v_y}{\partial y}\right) - \cos 2\gamma\left(\frac{\partial v_x}{\partial x} + \frac{\partial v_y}{\partial y}\right) = 0 \tag{2.81}$$

which can be rewritten as

$$\sin \nu_d\left(\frac{\partial v_y}{\partial y} - \frac{\partial v_x}{\partial x}\right) - \cos 2\gamma\left(\frac{\partial v_x}{\partial x} + \frac{\partial v_y}{\partial y}\right) = 0 \tag{2.82}$$

where ν_d is the *angle of dilation*, defined by (Hansen, 1958, cited in Roscoe, 1970)

$$\sin \nu_d \equiv -(c_1 + c_2)/(c_1 - c_2) = -c_*/s \tag{2.83}$$

Equation (2.83) is meaningful only when $|c_*/s| \leq 1$. If this inequality holds, Fig. 2.42 shows that the Mohr's circle for the rate of deformation tensor provides a geometric interpretation for ν_d. If $|c_*/s| > 1$, the form (2.81) is preferable to (2.82).

Using (2.67) and (2.75), the ratio c_*/s in (2.81) can be expressed in terms of the stresses to obtain

$$\frac{c_*}{s} = -\sin \nu_d = \frac{\dfrac{\partial G}{\partial \sigma_1} + \dfrac{\partial G}{\partial \sigma_2}}{\dfrac{\partial G}{\partial \sigma_1} - \dfrac{\partial G}{\partial \sigma_2}} \tag{2.84}$$

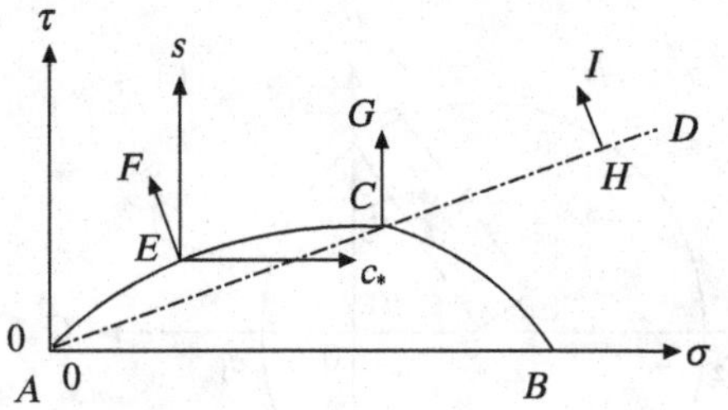

Figure 2.43. A smooth yield locus *ACB* and the critical state line *ACD*. Vectors with components (c_*, s) at the point *E* on the dilation branch and at the critical state *C* are represented by *EF* and *CG*, respectively, and the vector *IH* is normal to *AD*.

where $G(\sigma_1, \sigma_2, \nu)$ is the plastic potential. Equations (2.81) and (2.84) constitute the *flow rule* for plane flow.

For the special case of an associated flow rule, (2.84) reduces to

$$\frac{c_*}{s} = \frac{\dfrac{\partial F}{\partial \sigma_1} + \dfrac{\partial F}{\partial \sigma_2}}{\dfrac{\partial F}{\partial \sigma_1} - \dfrac{\partial F}{\partial \sigma_2}} = \frac{\dfrac{\partial f}{\partial \sigma}}{\dfrac{\partial f}{\partial \tau}} - \left(\frac{\partial \tau}{\partial \sigma}\right)_\nu \tag{2.85}$$

where $F(\sigma_1, \sigma_2, \nu) = 0$ and $f(\sigma, \tau, \nu) = 0$ are equivalent expressions for the yield condition for plane flow. Thus the *associated flow rule* is given by

$$\left(\frac{\partial \tau}{\partial \sigma}\right)\left(\frac{\partial v_y}{\partial y} - \frac{\partial v_x}{\partial x}\right) - \cos 2\gamma \quad \left(\frac{\partial v_x}{\partial x} + \frac{\partial v_y}{\partial y}\right) = 0 \tag{2.86}$$

In this case, the angle of dilation is related to the slope of the yield locus in the τ–σ plane by

$$\sin \nu_d = \left(\frac{\partial \tau}{\partial \sigma}\right)_\nu \tag{2.87}$$

2.14.4. Implications of the Associated Flow Rule

For a cohesionless material, consider a hypothetical convex yield locus *ACB* in the σ–τ plane (Fig. 2.43). Owing to coaxiality, the c_*–s axes are aligned with the σ–τ axes. However, the origin of the c_*–s axes can be located at any point on the yield locus. At a point *E* on the dilation branch of the yield locus, the rate of deformation vector *EF*, which has the components (c_*, s), is directed along the outward normal to the yield locus. As the normal has a negative projection on the c_* axis, the material dilates during flow. Similarly, at the critical state *C*, the normal is perpendicular to the c_* axis, and hence the flow is incompressible.

The critical state line *ACD* (Fig. 2.43) differs from the yield locus *AECB*. This distinction was ignored in an early attempt to apply plasticity theory to soils (Drucker and Prager, 1952). They treated *ACD* as a linear yield locus. Hence (i) the rate of deformation vector *HI* has a negative projection on the c_* axis, implying dilation during flow, (ii) the angle of dilation is a constant, and (iii) the stress power $\sigma_i\, c_i = 2(\sigma\, c_* + \tau\, s)$, which is proportional to the scalar product of the vectors *AH* and *HI*, vanishes. These defects were corrected by Drucker et al. (1957), who suggested the use of a curved yield locus along with a linear critical state curve.

2.14.5. Rowe's Stress–Dilatancy Relation

An alternative to the associated flow rule is provided by the stress–dilatancy relation of Rowe (1962). This relation is discussed below with reference to a sample of a cohesionless material which is of infinite extent in the x_3 direction (Fig. 2.44a) and which does not deform in this direction.

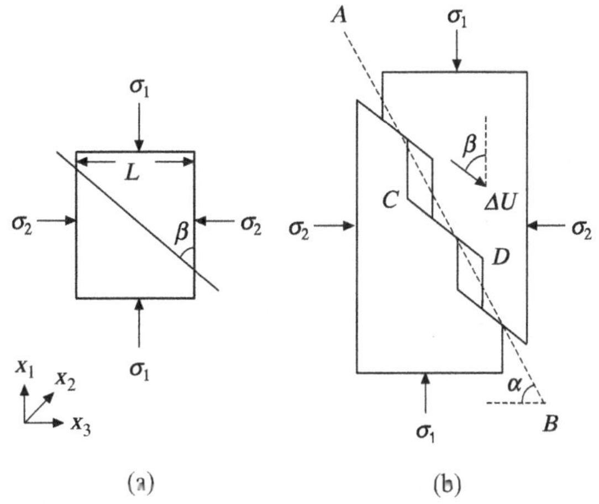

Figure 2.44. (a) Biaxial compression of a sample. Here σ_1 and σ_2 are principal stresses, with $\sigma_1 \geq \sigma_2$. (b) The sawtooth model of Rowe (1962). The line AB represents a hypothetical interlocking plane, and ΔU is the displacement vector corresponding to a small relative displacement of the blocks.

The sample is subjected to the principal stresses σ_1 and $\sigma_2 \leq \sigma_1$ (Fig. 2.44a). If σ_2 is kept constant and σ_1 is increased slowly, the sample will fail by sliding along some direction. Suppose that sliding occurs along a plane inclined at an angle β to the direction of the major principal stress axis (Fig. 2.44a) and that the material on either side of this plane is rigid. There is no change of volume, which is at variance with experimental observations. To overcome this difficulty, Rowe (1962) assumed that sliding occurred along a sawtooth type of plane (Fig. 2.44b). In this case, sliding leads to a volume change. To facilitate the analysis, consider a hypothetical interlocking plane AB (Fig. 2.44b), which is inclined at an angle α to the horizontal and which has the same area as the sawtooth plane.

Let the *energy ratio* E be defined as the ratio of the work done on the sample by the major principal stress σ_1 to the work done by the sample against the minor principal stress σ_2. Rowe related E to the angle of dilation ν_d. As some of his arguments are difficult to follow, the following presentation is partly based on an alternative approach due to de Josselin de Jong (1976).

The upper and lower blocks slide relative to each other along directions that are inclined at an angle β to the vertical (Fig. 2.44b). This angle is not known a priori but will be evaluated later using an assumption. On the plane CD (Fig. 2.44b), the contact forces exerted on the upper block are the normal force F_n and the shear force F_s. Assuming that the material is cohesionless, and the Coulomb yield condition (2.15) holds on this plane, $F_s = F_n \tan\phi_\mu$, where ϕ_μ is a constant. (Rowe identified ϕ_μ with the true angle of friction between the grains and measured it by sliding the grains against a solid block made of the same material.) Resolving F_s and F_n into horizontal and vertical components, the horizontal force H_{CD} and the vertical force V_{CD} exerted across CD on the upper block are found to be related by $V_{CD} = H_{CD} \tan(\phi_\mu + \beta)$. Summing the horizontal and vertical forces acting on all the contact planes, we obtain

$$V = H \tan(\phi_\mu + \beta) \tag{2.88}$$

where V and H are the total vertical and horizontal forces, respectively, exerted on the upper block.

Neglecting inertial effects and body forces, a balance of forces on the upper block (Fig. 2.45) leads to

$$V = \sigma_1 L W; \quad H = \sigma_2 L W \tan\alpha \tag{2.89}$$

where L is the length of the block in the horizontal direction, W is the width of the block in the direction perpendicular to the plane of the figure, and σ_1 and σ_2 are the principal

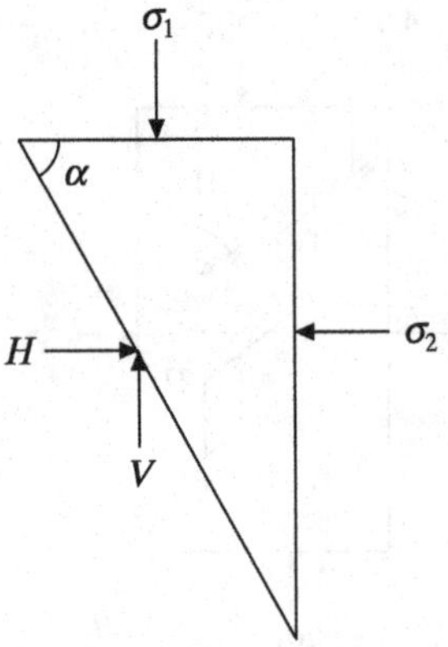

Figure 2.45. The vertical force V and the horizontal force H exerted across the interlocking plane on the upper block.

stresses. Equations (2.88) and (2.89) imply that

$$\sigma_1/\sigma_2 = \tan\alpha\ \tan(\phi_\mu + \beta) \tag{2.90}$$

The energy ratio E can be evaluated as follows. The work done by the major principal stress on the upper block during a small relative displacement of the blocks is given by $\sigma_1\, L W\, \Delta Y$, where ΔY is the vertical displacement. Similarly, the work done by the block against the minor principal stress is given by $\sigma_2\, L\, \tan\alpha\, W\, \Delta X$, where ΔX is the horizontal displacement. As the displacement vector ΔU is inclined at an angle β to the vertical (Fig. 2.44b), we have $\Delta Y = \Delta U\, \cos\beta$ and $\Delta X = \Delta U\, \sin\beta$. Hence

$$E \equiv (\sigma_1\, L W\, \Delta Y)/(\sigma_2\, L \tan\alpha\, W\, \Delta X) = \sigma_1/(\sigma_2\, \tan\alpha\, \tan\beta)$$

or using (2.90)

$$E = \tan(\phi_\mu + \beta)/\tan\beta \tag{2.91}$$

Note that α does not occur in the final expression.

To convert (2.91) to an expression involving the angle of dilation, consider an equivalent expression for E. The mechanical energy balance (1.44) implies that the rate of working of the stresses per unit volume $-\nabla \cdot (\boldsymbol{\sigma} \cdot \mathbf{v})$ is approximately equal to the stress power $\dot{\Phi} = \boldsymbol{\sigma} : \mathbf{C}$, provided the rate of change of kinetic energy and the rate of working of the body forces are negligible. Here $\mathbf{C}$ is the rate of deformation tensor, defined in the compressive sense. In terms of principal axes, $\boldsymbol{\sigma} : \mathbf{C} = \sigma_1 c_1 + \sigma_2 c_2$, where (σ_1, σ_2) and (c_1, c_2) are the principal stresses and the principal rates of deformation, respectively. Hence the rate of working per unit volume by the major principal stress is $\approx \sigma_1\, c_1$, and the rate of working per unit volume by the sample against the minor principal stress is $\approx -\sigma_2\, c_2$. Using these results, (2.91) can be rewritten as

$$E = (\sigma_1 c_1)/(-\sigma_2 c_2) = \tan(\phi_\mu + \beta)/\tan\beta \tag{2.92}$$

or using (2.83)

$$E = (\sigma_1/\sigma_2)\,((1 - \sin\nu_d)/(1 + \sin\nu_d)) = \tan(\phi_\mu + \beta)/\tan\beta \tag{2.93}$$

where ν_d is the angle of dilation.

Equation (2.93) cannot be used directly, as the value of β is unknown a priori. Rowe (1962) chose the value of β so as to minimize E. The basis for this assumption is not clear, but the resulting expression appears to fit data obtained from triaxial tests on dense samples of sand, glass beads, and steel balls. (Triaxial tests are discussed in §5.1.3.) Rowe's

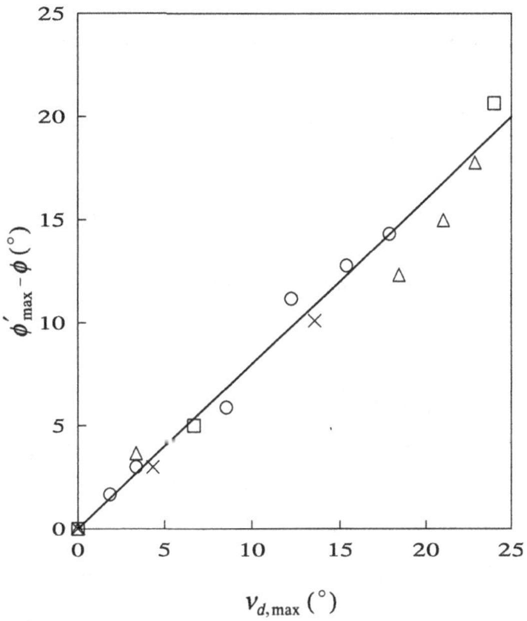

Figure 2.46. Variation of the peak angle of internal friction ϕ'_{max} with the corresponding value of the angle of dilation $\nu_{d,max}$ for various sands: symbols, data cited in Bolton (1986) (○, Brasted river; ×, Welland river; Δ, Leighton Buzzard; □, Monterey no. 0); —, equation (2.98). Here ϕ is the value of ϕ' at a critical state, and ϕ' is defined by (2.95).

assumption leads to $\beta = (\pi/4) - (\phi_\mu/2)$, and hence

$$E = \left(\frac{\sigma_1}{\sigma_2}\right)\left(\frac{1-\sin\nu_d}{1+\sin\nu_d}\right) = \frac{\tan\left(\frac{\pi}{4}+\frac{\phi_\mu}{2}\right)}{\tan\left(\frac{\pi}{4}-\frac{\phi_\mu}{2}\right)}$$

$$= \tan^2\left(\frac{\pi}{4}+\frac{\phi_\mu}{2}\right) = \frac{1+\sin\phi_\mu}{1-\sin\phi_\mu} \tag{2.94}$$

Equation (2.94) represents the *stress–dilatancy relation* between the principal stresses and the angle of dilation. Introducing an angle ϕ' such that

$$\sigma_1/\sigma_2 \equiv (1+\sin\phi')/(1-\sin\phi') \tag{2.95}$$

(2.94) can be solved for $\sin\nu_d$ to obtain

$$\sin\nu_d = (\sin\phi' - \sin\phi_\mu)/(1-\sin\phi'\sin\phi_\mu) \tag{2.96}$$

As mentioned earlier, (2.96) was found to fit data for dense samples. For loose samples, Rowe suggested that (2.96) be modified to

$$\sin\nu_d = (\sin\phi' - \sin\phi_f)/(1-\sin\phi'\sin\phi_f) \tag{2.97}$$

where ϕ_f increases from ϕ_μ at high densities to ϕ at the critical state. Here ϕ is the value of ϕ' at a critical state, where ϕ' is defined by (2.95). This modification robs the stress–dilatancy relation of its predictive power, as the dependence of ϕ_f on ϕ' is not known a priori.

By analogy with (2.97), Bolton (1986) suggested that

$$\nu_d = (\phi' - \phi)/0.8 \tag{2.98}$$

where ϕ' is defined by (2.95). The relation between the two forms is discussed in Problem 2.19. Fig. 2.46 shows that plane strain data for a number of sands can be fitted well by (2.98). Here ϕ'_{max} is the value of ϕ' corresponding to the peak in the σ_1/σ_2 versus compressive strain curves for dense samples (see Fig. 2.17) and $\nu_{d,max}$ is the value of ν_d at this point.

2.14.6. Summary of the Governing Equations for Plane Flow

The forms of the governing equations in Cartesian coordinates are given below for a rigid-plastic material described by a yield condition and the plastic potential flow rule. The particle density ρ_p is assumed to be a constant.

Mass balance

$$\frac{\partial \nu}{\partial t} + \frac{\partial(\nu\, v_x)}{\partial x} + \frac{\partial(\nu\, v_y)}{\partial y} = 0 \tag{2.99}$$

Momentum balance (x component)

$$\rho_p \left(\frac{\partial(\nu\, v_x)}{\partial t} + \frac{\partial(\nu\, v_x v_x)}{\partial x} + \frac{\partial(\nu\, v_y\, v_x)}{\partial y} \right) + \frac{\partial \sigma_{xx}}{\partial x} + \frac{\partial \sigma_{yx}}{\partial y} - \rho_p\, \nu\, g_x = 0 \tag{2.100}$$

Momentum balance (y component)

$$\rho_p \left(\frac{\partial(\nu\, v_y)}{\partial t} + \frac{\partial(\nu\, v_x\, v_y)}{\partial x} + \frac{\partial(\nu\, v_y\, v_y)}{\partial y} \right) + \frac{\partial \sigma_{xy}}{\partial x} + \frac{\partial \sigma_{yy}}{\partial y} - \rho_p\, \nu\, g_y = 0 \tag{2.101}$$

Stress components

$$\sigma_{xx} = \sigma + \tau \cos 2\gamma; \quad \sigma_{xy} = \sigma_{yx} = -\tau \sin 2\gamma; \quad \sigma_{yy} = \sigma - \tau \cos 2\gamma$$
$$\sigma = (\sigma_1 + \sigma_2)/2; \quad \tau = (\sigma_1 - \sigma_2)/2 \tag{2.102}$$

Yield condition

$$F(\sigma_1, \sigma_2, \nu) = 0 \quad \text{or} \quad f(\sigma, \tau, \nu) = 0 \tag{2.103}$$

Coaxiality condition

$$\sin 2\gamma \left(\frac{\partial v_y}{\partial y} - \frac{\partial v_x}{\partial x} \right) - \cos 2\gamma \left(\frac{\partial v_y}{\partial x} + \frac{\partial v_x}{\partial y} \right) = 0 \tag{2.104}$$

Flow rule

$$\sin \nu_d \left(\frac{\partial v_y}{\partial y} - \frac{\partial v_x}{\partial x} \right) - \cos 2\gamma \left(\frac{\partial v_x}{\partial x} + \frac{\partial v_y}{\partial y} \right) = 0 \tag{2.105}$$

Plastic potential flow rule

$$\sin \nu_d \equiv -\left(\frac{c_1 + c_2}{c_1 - c_2} \right) = -\left(\frac{\dfrac{\partial G}{\partial \sigma_1} + \dfrac{\partial G}{\partial \sigma_2}}{\dfrac{\partial G}{\partial \sigma_1} - \dfrac{\partial G}{\partial \sigma_2}} \right) \tag{2.106}$$

Associated flow rule

$$G = F; \quad \sin \nu_d = \left(\frac{\partial \tau}{\partial \sigma} \right)_\nu \tag{2.107}$$

Here g_x and g_y denote the x and y components of the acceleration due to gravity, the angle γ is defined as shown in Fig. 2.11, $G(\sigma_1, \sigma_2, \nu)$ is the plastic potential, σ_1 and σ_2 are the major and minor principal stresses, respectively, and c_1 and c_2 are the major and minor principal rates of deformation, respectively.

2.15. THE RELATION BETWEEN YIELD LOCI IN THE *N–T* AND σ–τ PLANES

Here we discuss the problem of using yield loci in the normal stress – shear stress plane to generate yield loci in the σ–τ plane. An apparatus such as the Jenike shear cell measures the normal stress N and the shear stress T on a single plane. This is not enough to determine the stress tensor for plane flow, as the latter has three components. Through every point in

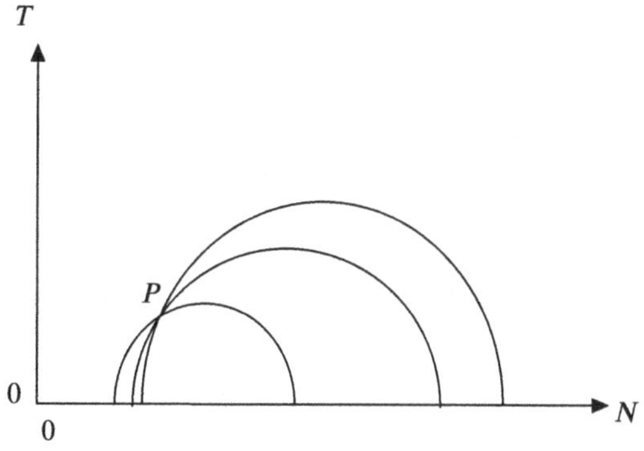

Figure 2.47. Three Mohr's circles passing through a point P in the normal stress N – shear stress T plane.

the N–T plane, an infinite number of Mohr's circles can be drawn (Fig. 2.47). Each Mohr's circle corresponds to a pair of values of σ and τ. Thus the mapping between points in the N–T and σ–τ planes is not unique, unless additional conditions are specified.

It is commonly assumed that the yield locus in the N–T plane is the *envelope* of Mohr's circles (Jenike, 1961, p. 15; Brown and Richards, 1970, p. 51; Desai and Siriwardane, 1984, p. 70). The envelope of a one-parameter family of curves is a curve that is tangential to some member of the family at each point, and different members of the family are tangential to the envelope at different points (Piskunov, 1969, p. 498). This assumption permits the construction of a unique σ–τ yield locus corresponding to the N–T yield locus. For example, consider a cohesionless material that satisfies the Coulomb yield condition $T = N \tan\phi_*$. Referring to Fig. 2.48, the envelope condition implies that the yield condition in the σ–τ plane is given by $\tau = \sigma \sin\phi_*$.

Jenike (1961) has justified the envelope condition on the grounds that states of stress cannot lie above the yield locus in the N–T plane. For an isotropic material, the yield condition must be expressed in terms of the principal stresses σ_1 and σ_2. By construction, states of stress cannot lie above the yield locus in the σ_1–σ_2 plane. However, it is not clear that this statement is equivalent to Jenike's assertion, as the mapping between yield loci in the two planes is not unique. As discussed in Problem 2.15, the following assumptions provide an alternative route to the envelope condition: (i) coaxiality, (ii) the associated flow rule, and (iii) the condition of plane shear, $\partial v_x/\partial x = 0$. (With reference to the Jenike shear cell (Fig. 2.15), the x direction can be taken as the direction of motion of the upper half of the cell.)

2.16. THE DOUBLE-SHEARING MODEL

An alternative model for plane flow, which is not based on the assumption of coaxiality, will now be discussed. This is based on the work on Mandel (1947) (cited in Spencer, 1982),

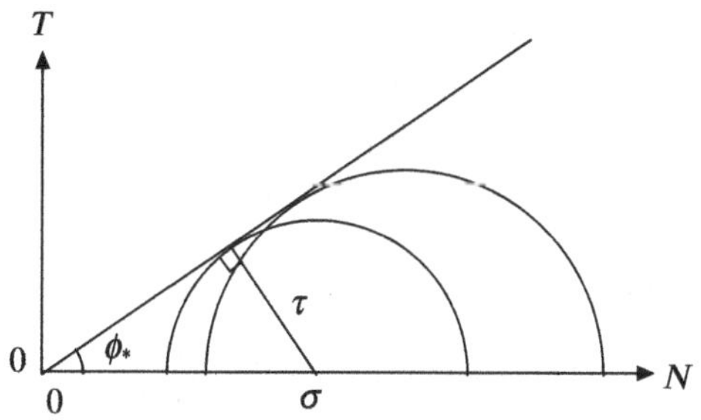

Figure 2.48. The envelope condition for a linear yield locus in the normal stress N – shear stress T plane. Here σ_1 and σ_2 are the principal stresses, $\sigma = (\sigma_1 + \sigma_2)/2$, and $\tau = (\sigma_1 - \sigma_2)/2$.

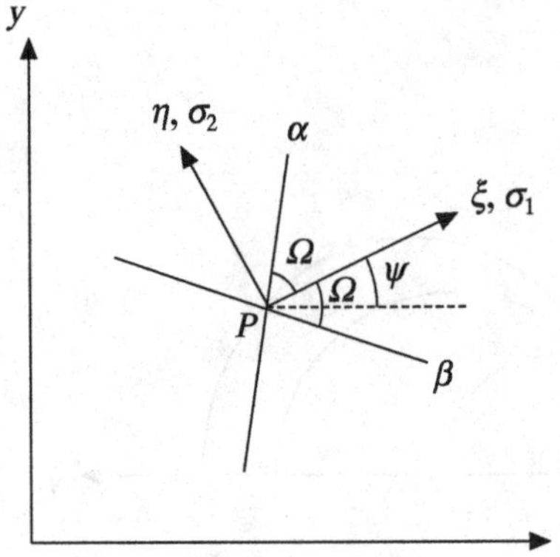

Figure 2.49. A coordinate system with coordinates (x, y), which is fixed relative to the laboratory frame, and a moving coordinate system with coordinates (ξ, η). The origin P of the latter moves with the material, and its axes always coincide with the major and minor principal stress axes, respectively. The slip planes $P\alpha$ and $P\beta$ are inclined at angles of $\pm\Omega = \pm[(\pi/4) - (\phi_*/2)]$ relative to the σ_1 axis.

de Josselin de Jong (1959, 1971, 1977), Spencer (1964, 1982), and Zagainov (1967) (cited in Spencer, 1982). Here we follow the treatment of Spencer (1982).

Consider a material satisfying the Mohr–Coulomb yield condition (2.37). As discussed in §2.8, the Coulomb yield condition (2.16) is satisfied on the slip planes, which are inclined at angles of $\pm\Omega = \pm[(\pi/4) - (\phi_*/2)]$ relative to the major principal stress axis (the σ_1 axis) (see Fig. 2.12). As the orientation of the σ_1 axis can vary with position, the slip planes are not truly planar in general.

The key assumption is that plane deformation of the material can be expressed as a superposition of *shearing deformations* along the slip planes. This can be regarded as an extension of Coulomb's idea that the slip planes separate *rigid* blocks which move relative to each other. Because two families of slip planes are used, the model is called the *double-shearing* model. The notion of a shearing deformation is explained below.

Consider two Cartesian coordinate systems with coordinates (x, y) and (ξ, η) (Fig. 2.49). The former is fixed relative to the laboratory frame. The origin of the latter moves with a velocity that is equal to the velocity of the material particle at that location, and its axes are always aligned with the principal stress axes. Let the slip planes passing through the point P be denoted by $P\alpha$ and $P\beta$. Let $\mathbf{v}'$ denote the velocity measured relative to the ξ–η coordinate system, and s_α and s_β denote distances measured along $P\alpha$ and $P\beta$, respectively. The motion of the material in the neighborhood of P is said to be a shearing deformation along $P\alpha$ if (i) $\partial\mathbf{v}'/\partial s_\alpha = 0$ and (ii) $\partial\mathbf{v}'/\partial s_\beta$ is parallel to $P\alpha$. Similarly, the motion is a shearing deformation along $P\beta$ if (iii) $\partial\mathbf{v}'/\partial s_\beta = 0$ and (iv) $\partial\mathbf{v}'/\partial s_\alpha$ is parallel to $P\beta$. It will be shown shortly that these assumptions imply *incompressible* flow. For a general deformation that involves shearing deformations along both the $\alpha-$ and $\beta-$ lines, the conditions (i) and (iii) are abandoned, and (ii) and (iv) are retained.

The preceding conditions (ii) and (iv) will now be used to derive the governing kinematic equations. Considering (ii) first, and referring to Fig. 2.49, the ξ and η components of $\partial\mathbf{v}'/\partial s_\beta$ are $\partial v'_\xi/\partial s_\beta$ and $\partial v'_\eta/\partial s_\beta$, respectively. Similarly, the ξ and η components of a vector directed along $P\alpha$ are proportional to $\cos\Omega$ and $\sin\Omega$, respectively. Hence (ii) implies that

$$\frac{\partial v'_\xi}{\partial s_\beta} \Big/ \frac{\partial v'_\eta}{\partial s_\beta} = \cos\Omega/\sin\Omega$$

or

$$\sin\Omega\,\frac{\partial v'_\xi}{\partial s_\beta} - \cos\Omega\,\frac{\partial v'_\eta}{\partial s_\beta} = 0 \tag{2.108}$$

As

$$\frac{\partial}{\partial s_\beta} = \cos\Omega\,\frac{\partial}{\partial \xi} - \sin\Omega\,\frac{\partial}{\partial \eta}$$

(2.108) can be rewritten as

$$\sin\Omega\,\cos\Omega\,\frac{\partial v'_\xi}{\partial \xi} - \sin^2\Omega\,\frac{\partial v'_\xi}{\partial \eta} - \cos^2\Omega\,\frac{\partial v'_\eta}{\partial \xi} + \cos\Omega\,\sin\Omega\,\frac{\partial v'_\eta}{\partial \eta} = 0$$

or, using the relation $\Omega = (\pi/4) - (\phi_*/2)$

$$\cos\phi_*\left(\frac{\partial v'_\xi}{\partial \xi} + \frac{\partial v'_\eta}{\partial \eta}\right) + \sin\phi_*\left(\frac{\partial v'_\xi}{\partial \eta} - \frac{\partial v'_\eta}{\partial \xi}\right) - \left(\frac{\partial v'_\xi}{\partial \eta} + \frac{\partial v'_\eta}{\partial \xi}\right) = 0 \tag{2.109}$$

Using a similar procedure, condition (iv) leads to

$$-\cos\phi_*\left(\frac{\partial v'_\xi}{\partial \xi} + \frac{\partial v'_\eta}{\partial \eta}\right) + \sin\phi_*\left(\frac{\partial v'_\xi}{\partial \eta} - \frac{\partial v'_\eta}{\partial \xi}\right) - \left(\frac{\partial v'_\xi}{\partial \eta} + \frac{\partial v'_\eta}{\partial \xi}\right) = 0 \tag{2.110}$$

It is convenient to rewrite (2.109) and (2.110) as follows. Subtracting (2.110) from (2.109), we obtain

$$\frac{\partial v'_\xi}{\partial \xi} + \frac{\partial v'_\eta}{\partial \eta} = 0 \tag{2.111}$$

Hence the present version of the double-shearing model is valid only for incompressible flow. (The extension of this class of models to compressible flow is discussed in de Josselin de Jong (1977), Mehrabadi and Cowin (1978), and Harris (1995).) Adding (2.109) and (2.110), we obtain

$$\sin\phi_*\left(\frac{\partial v'_\xi}{\partial \eta} - \frac{\partial v'_\eta}{\partial \xi}\right) - \left(\frac{\partial v'_\xi}{\partial \eta} + \frac{\partial v'_\eta}{\partial \xi}\right) = 0 \tag{2.112}$$

Equations (2.111) and (2.112) involve velocity components relative to the moving coordinate system. The latter can be expressed in terms of velocity components v_x and v_y relative to the fixed coordinate system as follows. The components of the rate of deformation tensor **C**, defined by (2.49), relative to the ξ–η coordinate system are given by

$$C_{\xi\xi} = -\frac{\partial v'_\xi}{\partial \xi};\quad C_{\xi\eta} = -\frac{1}{2}\left(\frac{\partial v'_\xi}{\partial \eta} + \frac{\partial v'_\eta}{\partial \xi}\right);\quad C_{\eta\eta} = -\frac{\partial v'_\eta}{\partial \eta} \tag{2.113}$$

Similarly, the nonzero component of the vorticity tensor **W**, defined by (2.69), is given by (see (G.34))

$$W_{\xi\eta} = \frac{1}{2}\left(\frac{\partial v'_\eta}{\partial \xi} - \frac{\partial v'_\xi}{\partial \eta}\right) \tag{2.114}$$

Using (2.113) and (2.114), (2.111) and (2.112) can be rewritten as

$$\begin{aligned} C_{\xi\xi} + C_{\eta\eta} &= 0 \\ -\sin\phi_*\quad W_{\xi\eta} + C_{\xi\eta} &= 0 \end{aligned} \tag{2.115}$$

As discussed in Problem G.5, (2.115) can be rewritten in terms of the components of the tensors **C** and **W** relative to the x–y coordinate system to obtain

$$C_{xx} + C_{yy} = 0$$
$$\sin\phi_*\,(-W_{xy} + \dot{\psi}) + \sin\psi\,\cos\psi\,(C_{yy} - C_{xx}) + \cos 2\psi\, C_{xy} = 0 \qquad (2.116)$$

where ψ is the inclination of the σ_1 axis relative to the x axis (Fig. 2.49) and $\dot{\psi} \equiv D\psi/Dt$ is the material derivative of ψ. Using (2.50) and (G.34), (2.116) can be rewritten as

$$\frac{\partial v_x}{\partial x} + \frac{\partial v_y}{\partial y} = 0 \qquad (2.117)$$

$$\sin\phi_* \left(\frac{\partial v_x}{\partial y} - \frac{\partial v_y}{\partial x} + 2\,\dot{\psi}\right) + \sin 2\psi \left(\frac{\partial v_x}{\partial x} - \frac{\partial v_y}{\partial y}\right) - \cos 2\psi \left(\frac{\partial v_y}{\partial x} + \frac{\partial v_x}{\partial y}\right) = 0 \qquad (2.118)$$

The first term in brackets on the left-hand side of (2.118) can be interpreted as follows. As discussed in Problem 5.6

$$\frac{\partial v_x}{\partial y} - \frac{\partial v_y}{\partial x} = -2\,\omega_z$$

where ω_z is the local angular velocity associated with the rigid body rotation of material elements about the z axis. Hence the term in brackets represents twice the difference between the local rate of rotation of the principal stress axes and the local angular velocity.

The governing equations of the double-shearing model consist of the Mohr–Coulomb yield condition (2.37), the linear momentum balances (2.100) and (2.101), the expressions (2.102) for the stress components σ_{xx}, σ_{xy}, and σ_{yy} in terms of σ, τ, and γ, and the kinematic equations (2.117) and (2.118). As seen by comparing Figs. 2.11 and 2.49, $\psi = -\gamma$.

The experiments of Drescher (1976), which were discussed in §2.13.2, provide data regarding the displacement field and the orientation of the principal stress axes. Using these data, Δu_x and Δu_y, the increments of displacement in x and y directions, respectively, can be calculated for small changes in the angle β (Fig. 2.36a). Finally, as $v_x \propto \Delta u_x$ and $v_y \propto \Delta u_y$, the second of (2.118) can be used to calculate $\dot{\psi}$. Results reported in Drescher (1976) show that the calculated values of $\dot{\psi}$ differ in sign from the values of $D\psi/Dt$, whereas the model assumes that $\dot{\psi} = D\psi/Dt$. Similarly, Savage and Lockner (1997) find that the predictions of the double-shearing model do not match their data for the shearing of a layer of sand. Their data were obtained using normal stresses in the range 450–700 MPa; it is not known whether similar trends would be observed at lower stress levels. As noted by Hill and Spencer (1999), more detailed data are required before firm conclusions can be drawn about the validity of the double-shearing model.

A simple application of this model is provided by the following example. Consider steady shear flow between horizontal plates (Fig. 2.32). Assuming that all quantities depend only on the y coordinate and $v_y = 0$, the second of (2.118) reduces to

$$(\sin\phi_* - \cos 2\psi)\,\frac{dv_x}{dy} = 0$$

as $\dot{\psi} = D\psi/Dt = 0$. If the material is shearing, this implies that $\sin\phi_* - \cos 2\psi = 0$ or $\psi = \pm[(\pi/4) - (\phi_*/2)]$. Hence the x direction coincides with a slip plane (see §2.8), and the velocity field cannot be determined uniquely. (As discussed in §2.13.1, the Lévy–Mises model also has the same defect.) Another example is discussed in Problem 2.20.

2.17. SUMMARY

Friction, dilatancy, and the formation of rupture layers are some of the features associated with the statics and flow of granular materials. The effect of friction and cohesion between the material and the wall of a vessel can be modeled by using the wall yield condition. This permits construction of the approximate Janssen solution for the stress field in a bin. The effect of friction and cohesion between adjacent parts of the material can be modeled by using the Coulomb yield condition and its generalization, a yield condition expressed in terms of the invariants of the stress tensor. The invariants are quantities that are unaffected by the rotation and reflection of the coordinate axes and can be obtained from the coefficients of the characteristic equation (2.25).

Shear tests conducted using a constant normal stress show that dense samples dilate and loose samples compact as the material shears. Eventually, all the samples attain a critical state, wherein the solids fraction depends only on the normal stress and is independent of the initial solids fraction. The progress of shear tests can be represented by a curve or loading path in a three-dimensional space with Cartesian coordinates defined by the normal stress, the shear stress, and the solids fraction. The shear stress–displacement curve for a dense sample exhibits a peak. The experiments of Roscoe et al. (1958) suggest that the post-peak segments of loading paths for dense samples with different initial states lie on a common surface. This surface is called the Hvorslev surface. Similarly, the Roscoe surface represents the loading paths of normally consolidated samples, i.e., samples having the lowest solids fraction consistent with the specified normal stress. The yield surface is formed by the union of the Roscoe and the Hvorslev surfaces. It is assumed that plastic or irrecoverable deformation occurs for loading paths traversing the yield surface, and the material is either rigid or deforms elastically for paths lying below this surface.

A flow rule relates the rate of deformation tensor $\mathbf{C}$ to the stress tensor $\boldsymbol{\sigma}$ and the solids fraction ν. Following the approach used in metal plasticity and soil mechanics, it is assumed that the rate of deformation tensor is proportional to the gradient (with respect to the stress components) of a scalar function $G(\boldsymbol{\sigma}, \nu)$. The latter is called the plastic potential. As the proportionality factor is left unspecified, only ratios of the components of $\mathbf{C}$ can be determined using the flow rule. This feature results in rate-independent behavior, i.e., if all components of $\mathbf{C}$ are scaled by a common factor, the stresses are unaffected. Experiments at low shear rates provide some support for rate independence.

In the framework adopted here the constitutive equations for slow granular flow consist of a yield condition and a flow rule.

For isotropic materials, the flow rule implies that the principal axes of the stress and rate of deformation tensors are aligned with each other. This is called the coaxiality condition. Experiments suggest that coaxiality holds in some situations but not in others. The double-shearing model of Spencer (1964) provides an example of a noncoaxial flow rule.

If the plastic potential coincides with the yield function to within arbitrary additive or multiplicative constants, the flow rule is called an associated flow rule; if not, it is called a nonassociated flow rule. A simple alternative to the former is provided by the stress–dilatancy theory of Rowe (1962). However, this theory assumes the validity of the Mohr–Coulomb yield condition.

Overall, it appears that none of the existing flow rules can be used with confidence for all types of slow flows. Further experimental and theoretical work in this direction is essential.

PROBLEMS

2.1. Behavior of wet beach sand

The following phenomenon is often observed while walking on wet sand on the beach. As the foot falls on it, the sand whitens or appears to dry momentarily around the foot. On raising the foot it is seen that the sand which was under the foot becomes wet again. Why does this happen?

This problem has been adapted from Reynolds (1885).

2.2. Forces acting on a block

For the block shown in Fig. 2.5b, explain why the normal force F_n does not pass through the center of mass of the block.

2.3. Forces acting on an hourglass immersed in water

Consider a sealed hourglass containing sand and air, which is in a sealed cylindrical vessel containing water and air (Fig. 2.50a). When the cylinder is inverted, the hourglass remains near the top of the vessel (Fig. 2.50b), until a fair amount of sand has run out of the upper bulb. It then descends to its expected position at the bottom of the cylinder. Explain why this happens.

This problem has been adapted from Feilden (1986).

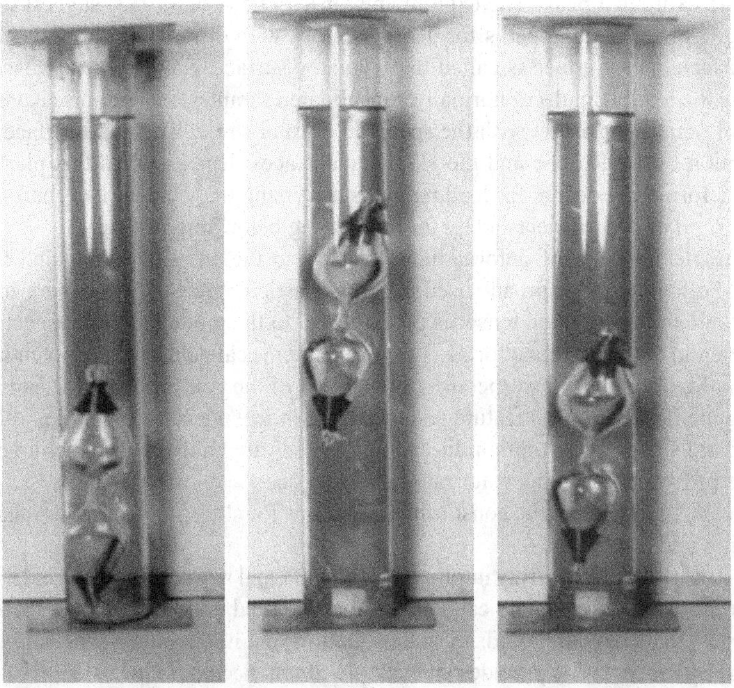

(a) (b) (c)

Figure 2.50. An hourglass containing sand and air, that is immersed in a glass cylinder containing water and air. The photographs show the positions of the hourglass at various times: (a) initial position, (b) 10 s after inverting the cylinder, and (c) 39 s after inverting the cylinder. The inner diameter of the cylinder is 64 mm and its height is 450 mm. The hourglass remains in the position shown in (b) for about 20–30 s, depending on the speed with which the cylinder is inverted. The total time required for the sand to drain from the upper bulb to the lower bulb is about 100–110 s. The setup is a minor modification of the one described by Feilden (1986); it was set up by Ms. R. Jyotsna.

2.4. Stresses exerted on the wall of a hopper

Consider a conical hopper filled with sand at rest (Fig. 2.51). The upper surface of the sand is a spherical cap, which is stress free. The exit orifice is a circle of radius R and is covered by a plate.

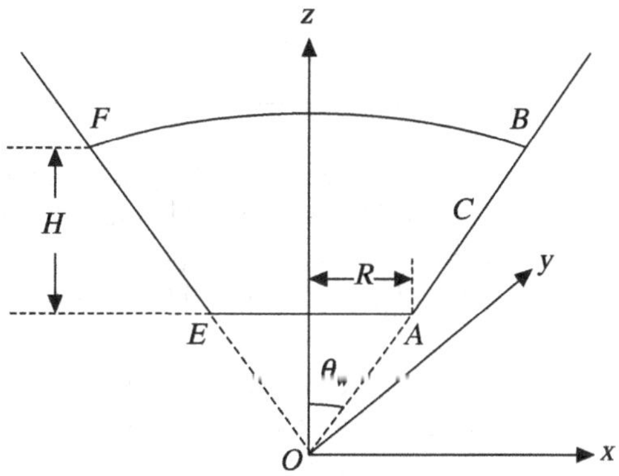

Figure 2.51. A conical hopper filled with sand.

(a) The plane $y = 0$ intersects the hopper wall along the lines AB and EF. At a point C on AB, derive expressions for the normal stress N and the shear stress T exerted on the sand by the hopper wall. Here T is the shear stress exerted in the positive radial direction, where the radial coordinate is measured from the apparent vertex of the hopper. Express the results in terms of stress components referred to (i) the Cartesian coordinate system shown in the figure and (ii) a spherical coordinate system with origin at O (see also Fig. A.4).

(b) Suppose that the stress field is axisymmetric, i.e., the stress components are independent of the ϕ' coordinate (see Fig. A.4). Assuming that $\sigma_{\theta\phi'} = 0$, show that $\sigma_{r\phi'} = 0$. Balance equations in spherical coordinates are given in Appendix A. These results can be used in parts (c)–(f) also.

(c) Derive expressions for the principal stresses at C in terms of stress components relative to the spherical coordinate system.

(d) At C, $\sigma_{rr} = 1$, $\sigma_{\theta\theta} = 2$, and $\sigma_{r\theta} = -2$ kPa. Use this data to examine whether the material is cohesionless.

(e) Assuming that the density of the sand ρ is a constant, derive expressions for the total body F_b and the total contact force F_v exerted in the vertical direction on the whole mass of the sand. The result for the latter should be expressed in terms of the areas of the curved and flat surfaces and the area-averaged values of the stresses. For a stress component N acting in a specified direction on a surface with area A, the area-averaged value of N is defined by

$$< N >= \frac{1}{A} \int N \, \mathrm{d}A$$

where the integral is taken over the area of the surface. Use stress components relative to the spherical coordinate system for the curved surfaces and components relative to the Cartesian coordinate system for the flat bottom surface.

(f) On the hopper wall, the values of the area-averaged stresses are given by $< \sigma >_{\theta\theta} = 1.2$ kPa and $< \sigma_{\theta r} > = -0.5$ kPa. Given that $R = 0.02$ m, $H = 0.3$ m, $\theta_w = 30°$, and density $\rho = 1{,}560$ kg m^{-3}, calculate $< \sigma >_{zz}$, the area-averaged vertical stress exerted by the sand on the bottom plate.

2.5. A block on an inclined plane

Consider a block of mass m, at rest on an inclined plane (Fig. 2.52). Suppose that the relation between the frictional force F and the normal force W is given by a generalization

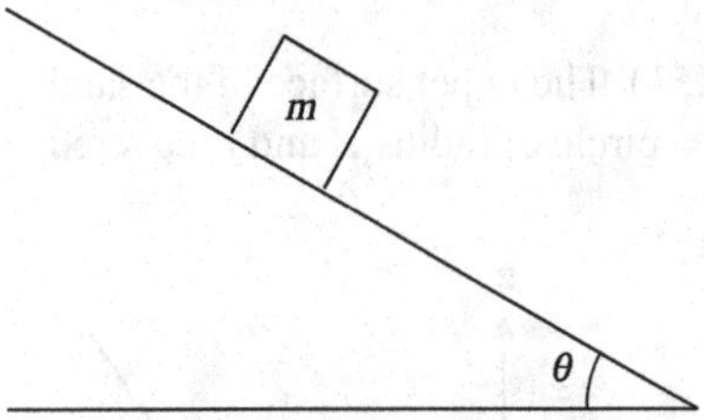

Figure 2.52. A block of mass m on an inclined plane.

of Coulomb's law, i.e., $F = a\,W^n$, where a and n are positive constants. (The data of Tüzün et al. (1988) for single particles sliding on a glass plate suggest that $n \approx 1,\ 0.88$, and 0.78 for mustard seeds, glass beads, and polyethylene pellets, respectively. The corresponding ranges of the normal load are 0.001–0.028 N for glass beads and 0.015–0.15 N for polyethylene pellets.)

(a) Derive an expression for the critical value of θ ($\equiv \theta_c$) required to initiate sliding. For $n < 1$, show that θ_c decreases as the weight of the block increases.

(b) Suppose that the block is replaced by a sphere of mass m. If either Coulomb's law or its generalization holds, and there is a point contact between the sphere and the plane, what is the value of θ_c?

This problem arose out of a discussion with Prof. R. Jackson.

2.6. An application of the Coulomb yield condition to the stability of a slope

(a) Consider a cohesive material that satisfies the Coulomb yield condition (2.15). By examining the equilibrium of the wedge ABC (Fig. 2.53), where AC denotes a possible rupture surface, show that the wedge does not slip downward for any value of α, provided h satisfies a certain inequality involving the density ρ, the acceleration due to gravity g, the cohesion c, the angle of internal friction ϕ_*, and the inclination θ of AB. The slope is of infinite extent in the direction perpendicular to the plane of the figure, and the surfaces AB and BC are traction-free, i.e., the shear and normal stresses vanish on these surfaces. Use this inequality to derive expressions for the critical height h_* of a slope, where h_* is the value of h for which the inequality is satisfied as an equality. If $h_* = 0.2$ m, $\phi_* = 30°$, $\theta = 90°$, and $\rho = 1{,}600$ kg m^{-3}, calculate the value of c.

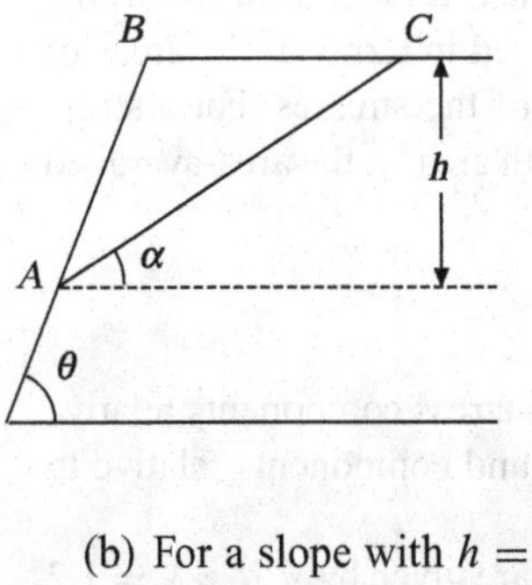

Figure 2.53. A slope AB with a possible rupture surface AC. The surface ABC is stress free.

(b) For a slope with $h = h_* > 0$, consider the limit $c \to 0$. Show that the maximum permissible value of θ is ϕ_*, i.e., the steepest slope for a cohesionless Coulomb material is one which is inclined to the horizontal at the angle of internal friction ϕ_*.

This problem has been adapted from Nedderman (1992, pp. 36–40).

2.7. Eigenvalues and eigenvectors of a symmetric second-order tensor in two dimensions

(a) Considering the two-dimensional stress tensor $\boldsymbol{\sigma}$, and assuming that it is symmetric, show that the eigenvalues σ_1 and σ_2 are real numbers.

(b) Let $\mathbf{n}_i$ denote the eigenvector corresponding to the eigenvalue σ_i, so that

$$\boldsymbol{\sigma} \cdot \mathbf{n}_1 = \sigma_1 \mathbf{n}_1; \quad \boldsymbol{\sigma} \cdot \mathbf{n}_2 = \sigma_2 \mathbf{n}_2 \tag{2.119}$$

Forming the scalar product of the first of (2.119) with $\mathbf{n}_2$, and using index notation, show that $\mathbf{n}_1 \cdot \mathbf{n}_2 = 0$ if $\sigma_1 \neq \sigma_2$. (For a proof of this result in the three-dimensional case, see, e.g., Fung, 1977, pp. 99–101.)

(c) If $\sigma_1 = \sigma_2$, show that $\sigma_{12} = \sigma_{21} = 0$ and $\sigma_1 = \sigma_{11} = \sigma_{22}$, where $\{\sigma_{ij}\}$ are the components of $\boldsymbol{\sigma}$ relative to a Cartesian coordinate system (Fig. 2.8). Hence show that any direction in the x_1–x_2 plane is an eigenvector.

2.8. Invariance of the eigenvalues and the principal invariants under rotation of coordinate axes

Consider two Cartesian coordinate systems, with coordinates (x_1, x_2) and (x_1', x_2'), respectively (Fig. 2.54). Let $\{\sigma_{ij}\}$ and $\{\sigma_{kl}'\}$ denote the components of the two-dimesional stress tensor $\boldsymbol{\sigma}$ relative to these coordinate systems. As shown in Appendix F, the components are related by

$$\sigma_{kl}' = Q_{ki}\, Q_{lj}\, \sigma_{kl} \tag{2.120}$$

where $Q_{ki} \equiv \mathbf{e}_k' \cdot \mathbf{e}_i$, and $\mathbf{e}_i, i = 1, 2$, are unit vectors along the coordinate axes of the unprimed coordinate system. In matrix notation, (2.120) can be written as

$$[\boldsymbol{\sigma}'] = [\mathbf{Q}]\,[\boldsymbol{\sigma}]\,[\mathbf{Q}]^{\mathrm{T}} \tag{2.121}$$

where $[\boldsymbol{\sigma}'] = [\sigma_{kl}']$, $[\mathbf{Q}] = [Q_{ki}]$, and $[\boldsymbol{\sigma}] = [\sigma_{ij}]$.

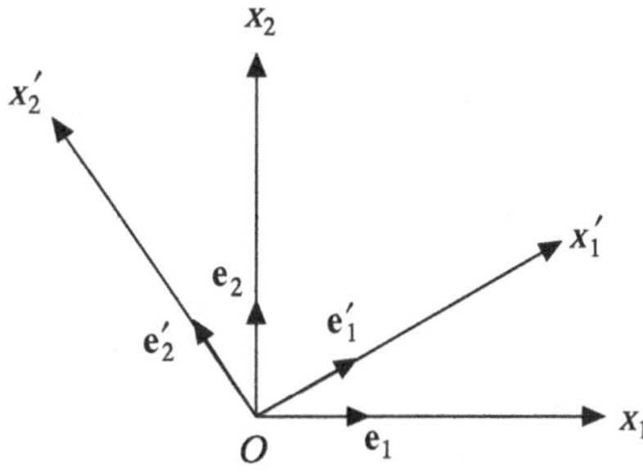

Figure 2.54. Rotation of coordinate axes. Here $\mathbf{e}_i$ and $\mathbf{e}_i', i = 1, 2$, are unit vectors along the axes.

(a) Consider the scalar $A \equiv \det([\boldsymbol{\sigma}'] - \lambda\,[\mathbf{I}])$, where λ is a scalar and $[\mathbf{I}]$ is the unit matrix. Using (2.121) and (A.82), and noting that $[\mathbf{Q}]$ is an orthogonal matrix (see (A.48) and (F.18)), show that $A = \det([\boldsymbol{\sigma}] - \lambda\,[\mathbf{I}])$. Hence $[\boldsymbol{\sigma}]$ and $[\boldsymbol{\sigma}']$ have the same characteristic equation and the same eigenvalues. This result implies that the eigenvalues are unaffected by the rotation of coordinate axes; it also holds for the three-dimensional stress tensor.

(b) Evaluate the eigenvalues (principal stresses) σ_1 and σ_2, and hence show that the principal invariants are given by $I_1 \equiv \mathrm{tr}(\boldsymbol{\sigma}) = \sigma_1 + \sigma_2$ and $I_2 \equiv \det(\boldsymbol{\sigma}) = \sigma_1\sigma_2$. As the invariants depend only on the principal stresses, it follows from the result obtained in part (a) that I_1 and I_2 are also unaffected by the rotation of coordinate axes. This result also holds for the three-dimensional stress tensor, as it can be shown that the principal invariants depend only on the principal stresses $\sigma_i, i = 1, 3$.

Part (a) of this problem has been adapted from Hunter (1983, p. 87).

2.9. Determination of the principal stresses and the principal stress axes

The components of the two-dimensional stress tensor $\boldsymbol{\sigma}$ relative to a Cartesian coordinate system are given by $\sigma_{11} = a$, $\sigma_{12} = \sigma_{21} = 2$, and $\sigma_{22} = 3$ kPa.

(a) If the Mohr–Coulomb yield condition holds, with $\phi_* = 40°$ and $c = 0$, calculate the values of the shear stress T and the normal stress N acting on the slip planes.

(b) Find the normalized eigenvectors $\mathbf{n}_k$, $k = 1, 2$.
(c) Find the orientation of the major principal stress axis relative to the x axis.

2.10. The orientation of the major principal stress axis at the wall of a hopper

(a) Consider a wedge-shaped hopper, with flow parallel to the x–y plane (Fig. 3.13). Assuming that the material is cohesionless, and that it satisfies the Mohr–Coulomb yield condition and the wall yield condition (with $c_w = 0$), derive an expression for γ_w, where $\gamma_w \equiv \gamma'(r, \theta = \theta_w)$. Here (r, θ) denote polar coordinates, with origin at the virtual apex of the hopper, and γ' is the inclination of the σ_1 axis to the circumferential direction. Solve for γ_w (i) algebraically, without using the concept of the Mohr's circle, and (ii) geometrically by using the Mohr's circle. Hint: It may be helpful to express the term $\tan\delta$ in the wall yield condition as $\sin\delta/\cos\delta$.
(b) Calculate the value(s) of γ_w for $\phi_* = 30°$ and $\delta = 20°$.

2.11. The Hvorslev and the Roscoe surfaces

For a clay, the Hvorslev and the Roscoe surfaces in τ–σ–e_* space are given by

$$\tau = 0.595\,\sigma + 561\, e^{-3.59\, e_*}$$
$$2.59 \times 10^{-3}\, \tau^2 = 7.042 \times 10^3\, e^{-7.18\, e_*} + 1.828 \times 10^3\, e^{-3.59\, e_*} - \sigma \tag{2.122}$$

respectively, where $\tau = (\sigma_1 - \sigma_2)/2$ and $\sigma = (\sigma_1 + \sigma_2)/2$, σ_1 and σ_2 are the major and minor principal stresses, respectively, $e_* \equiv (1/\nu) - 1$ is the *voids ratio*, and the stresses have units of kPa.

(a) For $e_* = 0.55$, calculate the values of τ and σ at the critical state and hence the value of the slope $\sin\phi$ of the critical state curve, assuming that the curve is of the form (2.47).
(b) Plot contours of constant e_* in the τ–σ plane, for $e_* = 0.55$ and $e_* = 0.6$.
(c) Plot a contour of constant σ in the τ–e_* plane, for $\sigma = 200$ kPa.

2.12. Interpretation of the components of the rate of deformation tensor

Consider a Cartesian coordinate system with coordinates (x_1, x_2, x_3), and let $\mathbf{C} \equiv -(\nabla\mathbf{v} + (\nabla\mathbf{v})^{\mathrm{T}})/2$ denote the rate of deformation tensor, defined in the compressive sense. Here $\mathbf{v}$ is the velocity vector.

(a) Consider three material points P, P', and P'' (Fig. 2.55). Show that
$$\frac{\mathrm{D}}{\mathrm{D}t}(\Delta x_j'\, \Delta x_j'') = -2C_{lj}\, \Delta x_l'\, \Delta x_j'' \tag{2.123}$$
where $\mathrm{D}/\mathrm{D}t$ denotes the material derivative, repeated indices imply summation, and $\Delta x_j' = x_j' - x_j$ and $\Delta x_j'' = x_j'' - x_j$. Note that $\mathrm{D}x_j/\mathrm{D}t = v_j$, as x_j represents a coordinate of a material point. The points can be assumed to be near each other, so that $\Delta v_k' = v_k' - v_k \approx (\partial v_k/\partial x_j)\,\Delta x_j'$.
(b) Let L' and L'' denote the lengths of the lines PP' and PP'', respectively, and θ the angle between them (Fig. 2.55). Use (2.123) to show that
$$\cos\theta\left(\frac{1}{L'}\frac{\mathrm{D}L'}{\mathrm{D}t} + \frac{1}{L''}\frac{\mathrm{D}L''}{\mathrm{D}t}\right) - \sin\theta\,\frac{\mathrm{D}\theta}{\mathrm{D}t} = -2C_{lj}\,\alpha_l\,\beta_j \tag{2.124}$$
where $\boldsymbol{\alpha}$ and $\boldsymbol{\beta}$ are unit vectors directed along PP' and PP''. Note that the vector PP' is given by $L'\,\boldsymbol{\alpha}$.
(c) Suppose that PP' and PP'' coincide and that PP' is parallel to the x_1 axis. Show that
$$C_{11} = -\frac{1}{L'}\frac{\mathrm{D}L'}{\mathrm{D}t} \tag{2.125}$$
Thus C_{11} is the rate of compression per unit length of a material line element that is parallel to the x_1 axis.

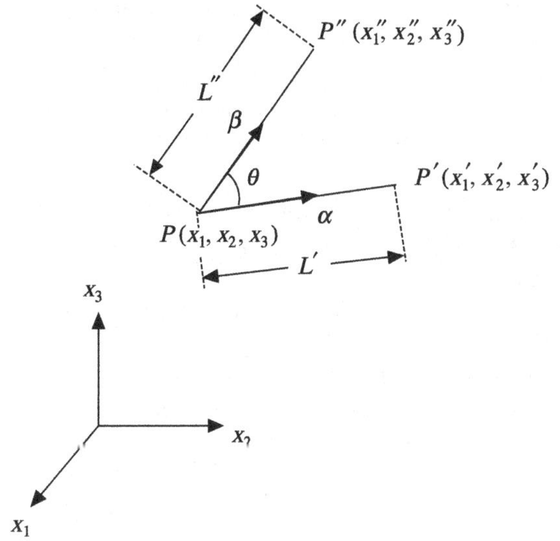

Figure 2.55. Two material line elements PP' and PP''. Here $\boldsymbol{\alpha}$ and $\boldsymbol{\beta}$ are unit vectors.

(d) Suppose that PP' and PP'' are initially orthogonal and PP' is parallel to the x_1 axis. Show that

$$C_{12} = \frac{1}{2}\frac{\mathrm{D}\theta}{\mathrm{D}t} \tag{2.126}$$

Thus the off-diagonal components of **C** are equal to half the rate of increase of the angle between two material line elements that are initially orthogonal.

This problem has been adapted from Prager (1961, p. 63).

2.13. Value of the angle of dilation ν_d for incompressible plane flow

Consider steady, incompressible flow parallel to the x–y plane. If the material is deforming, show that (i) the invariant s, defined by (2.75), is nonzero and (ii) the angle of dilation ν_d, defined by (2.83), vanishes.

2.14. A yield condition and the associated flow rule

The yield condition for a material is given by

$$\tau/\sigma_c = a\,(\sigma/\sigma_c) - b\,(\sigma/\sigma_c)^2, \quad \sigma_1 \geq \sigma_2 \tag{2.127}$$

where $\sigma_c(\nu)$ is the mean stress at a critical state, and a and b are constants.

(a) Does (2.127) represent a cohesionless material or a cohesive material?
(b) The critical state curve is given by $\tau = \sigma_c \sin\phi$, $\sigma = \sigma_c$. If $\phi = 30°$, $\sigma/\sigma_c = 0.8$, and the *associated flow rule* holds, calculate the value of τ/σ_c at this point.

2.15. Relation between yield loci in the N–T and σ–τ planes

Consider plane shear flow between parallel plates, which is characterized by $v_z = 0$, $\partial v_x/\partial x = 0$ (Fig. 2.32). Let N and T denote the normal and shear stresses, respectively, exerted by the upper plate on the material below it. Note that T represents the shear stress exerted in the x direction on a plane whose normal points in the negative y direction by the material above it. Use the associated flow rule to show that the yield locus in the N–T plane is the *envelope* of Mohr's circles whose centers and radii correspond to points on the yield locus in the σ–τ plane. In other words, the Mohr's circle passing through any point on the yield locus is tangential to the locus. Hint: Express T and N in terms of σ, τ, and γ, and use the expression for the Mohr's circle. Mark the point (N, T) and the angle γ carefully on the Mohr's circle.

2.16. Mohr's circle for the rate of deformation tensor for shear between parallel plates Consider the shear flow of a granular material between parallel plates (Fig. 2.32). The lower plate is stationary and the upper plate is driven to the right with a constant speed. The gap between the plates may change due to changes in the solids fraction as the material shears. It can be assumed that the flow is parallel to the x–y plane, and $\partial v_x/\partial x \approx 0$, where v_x is the x component of velocity.

(a) Sketch the Mohr's circle for the rate of deformation tensor for two cases: (i) the material is dilating and (ii) the material is at a critical state. In each case, label the point on the circle that has the coordinates $(C_{xx}, -C_{xy})$.

(b) If coaxiality holds and the material is at a critical state, what is the inclination of the major principal stress axis relative to the x axis?

(c) A "fully rough" wall is one which has a monolayer of the granular material stuck to it. Suppose that the lower plate (Fig. 2.32) is fully rough and the wall yield condition (2.6) holds at this plate. If the assumptions used in part (b) hold, and the critical state curve is given by (2.47), show that the angle of wall friction for a fully rough wall is given by

$$\delta = \tan^{-1}(\sin\phi) \tag{2.128}$$

The measured value of δ for the flow of glass beads past a fully rough wall is within 1° of that predicted by (2.128) (Kaza, 1982, p. 239).

Parts (b) and (c) have been adapted from Kaza (1982, pp. 236–238).

2.17. Arching in a conical hopper

(a) In the normal stress–shear stress plane, the yield locus for $\nu = \nu_1$ is shown by the curve $ABECG$ in Fig. 2.56, where ABE is a straight line, ECG is the arc of a circle, and C represents the critical state for $\nu = \nu_1$. The point C coincides with the top of the circular arc, so that T_C is the maximum value of T along the yield locus. The *unconfined yield strength* f_u is the value of the major principal compressive stress corresponding to the Mohr's circle for stress that passes through the origin of the N–T plane and is tangential to the yield locus at some point P. Let σ_{1C} denote the major principal compressive stress corresponding to the Mohr's circle passing through the critical state C and $\tan\omega$ the slope of a straight line passing through the origin and drawn tangential to this circle. Assuming that the Mohr's circle passing through C is tangential to the yield locus, relate f_u to ω, ϕ_*, and σ_{1C}. It may be helpful to draw separate figures showing the Mohr's circles corresponding to f_u and σ_{1C}.

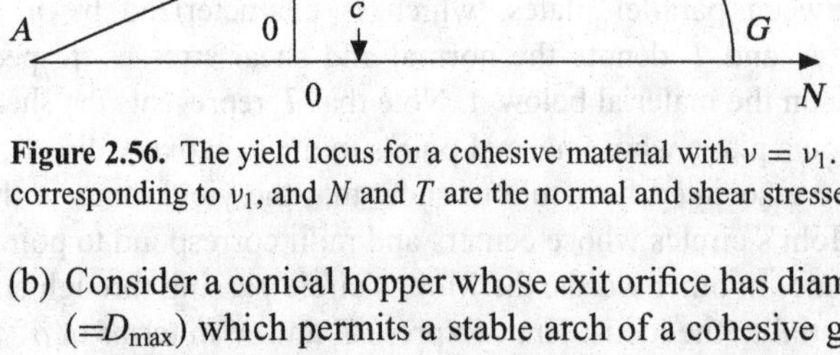

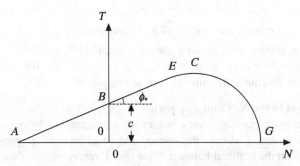

Figure 2.56. The yield locus for a cohesive material with $\nu = \nu_1$. The point C represents the critical state corresponding to ν_1, and N and T are the normal and shear stresses, respectively.

(b) Consider a conical hopper whose exit orifice has diameter D. The maximum value of D ($=D_{max}$) which permits a stable arch of a cohesive granular material can be estimated as follows. Suppose that the material is at rest in the hopper and forms an arch $ABEC$ of constant density $\rho = \rho_p\,\nu_1$ and constant thickness Δz in the z direction (Fig. 2.57).

It can be assumed that $\Delta z \ll D$. Consider the forces acting on the element $FCEH$ (Fig. 2.57), where the lines FC and EH are drawn perpendicular to the traction-free surface AB at the points F and H, respectively. Deduce a relation between D, β, and f_u, where β is the inclination of the tangent to AB at F. Estimate D_{max} by choosing β to maximize the value of D. The length of the line CF can be assumed to be $\approx \Delta z \cos\beta$.

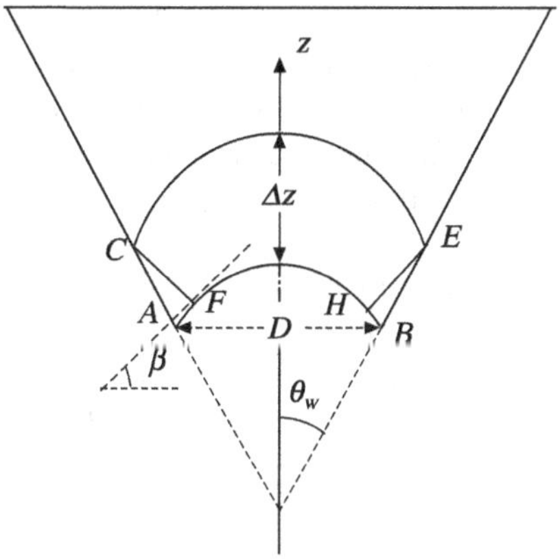

Figure 2.57. An arch of a cohesive material in a hopper.

(c) What happens for (i) $D > D_{max}$ and (ii) $D < D_{max}$?

(d) Suppose that there is material above the upper surface of the arch. Explain qualitatively how this will affect the value of D_{max}.

Parts (a) and (b) of this problem have been adapted from Nedderman (1992, pp. 322–325); the method was suggested by Jenike (1961).

2.18. Rotation of a cylinder of sand

A cylindrical beaker of radius R is filled with sand upto a height h. The beaker is then rotated about its axis with a constant angular velocity ω, causing the free surface to be inclined at an angle β relative to the horizontal (Fig. 2.58a). Note that β may vary with the radial coordinate r. It can be assumed that the sand is at rest relative to the beaker.

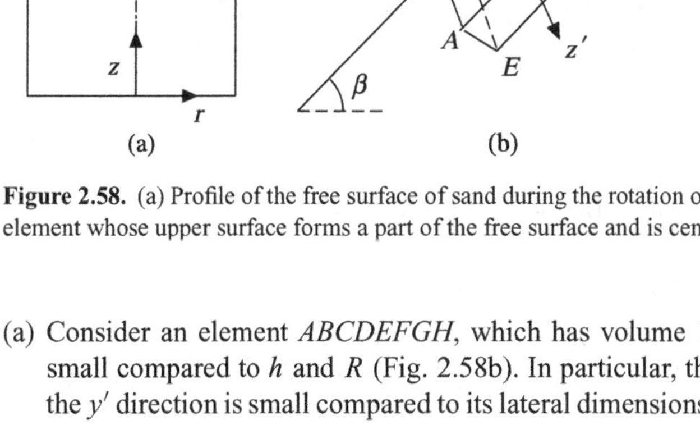

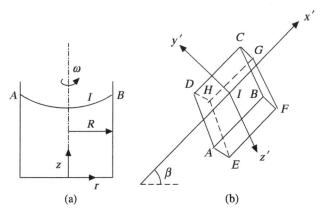

Figure 2.58. (a) Profile of the free surface of sand during the rotation of the cylinder about its axis. (b) An element whose upper surface forms a part of the free surface and is centered at the point I in (a).

(a) Consider an element $ABCDEFGH$, which has volume V and whose dimensions are small compared to h and R (Fig. 2.58b). In particular, the thickness of the element in the y' direction is small compared to its lateral dimensions. The plane $ABCD$ is located on the free surface of the sand, and its center is at the point I, whose coordinates are $(r, z_f(r))$, where $z = z_f(r)$ is the equation for the free surface. The effects of cohesion and fluid drag can be neglected, and the stress field is axisymmetric. Identify the forces acting on the element, in a coordinate frame which is at rest relative to the beaker. Hint: See Problems G.2 and G.4.

(b) Formulate the x' and y' components of the force balance on the element, assuming that the density of the sand is a known constant ρ. Let F_n and F_t represent the normal and shear forces exerted on $EFGH$ by the material below it, with F_t representing the force exerted in the x' direction. The projection of the x' direction at I onto the horizontal plane coincides with the radial direction at I. Derive expressions for F_n and F_t.

(c) Suppose that along $EFGH$, the material is constrained by the Coulomb yield condition $F_t = \alpha\, \mu\, F_n$, where $|\alpha| \le 1$, $\mu \equiv \tan\phi_*$, and ϕ_* is a material constant called the angle of internal friction. Using the results of part (b), formulate a differential equation for the variation of z_f with r.

(d) Express the result obtained in part (c) in dimensionless form, using the dimensionless variables $\eta = z_f/R$, and $\xi = r/R$, and the Froude number $G \equiv (\omega^2 R^2)/(R\, g) = (\omega^2 R)/g$, where g is the acceleration due to gravity.

(e) Starting with a flat, free surface $\eta = \eta_0 = h/R$ at $G = 0$, show that as G is gradually increased, the yield condition is satisfied, i.e., $\alpha = -1$, at some point on the free surface for $G = G_c = \mu$. What is the shape of the free surface for $G < G_c$?

(f) For $G > G_c$, assume that the yield condition is satisfied, for $\xi_c \le \xi \le 1$. Determine the value of ξ_c. Assuming that $\eta(\xi_c) = \eta_0$, integrate the differential equation to obtain $\eta(\xi)$.

(g) Explain whether the solution derived in part (f) is consistent with the assumption of a constant density.

This problem is a slightly modified version of the problem examined by Medina et al. (1995).

2.19. The connection between the stress–dilatancy relation and the relation proposed by Bolton (1986)

Bolton (1986) stated that the stress–dilatancy relation can be written as

$$\frac{\sigma_1}{\sigma_2} = \left(\frac{\sigma_1}{\sigma_2}\right)_c \left(1 - \frac{d\epsilon_v}{d\epsilon_1}\right) \tag{2.129}$$

where the subscript c denotes a critical state, and $d\epsilon_v$ and $d\epsilon_1$ are the volumetric and axial strain increments, respectively, that are related to the components of the rate of deformation tensor by (2.48). Thus $d\epsilon_v = (c_1 + c_2)\, dt$ and $d\epsilon_1 = c_1\, dt$, where c_1 and c_2 are the major and minor principal compressive rates of deformation, respectively.

(a) Show that (2.129) can be rewritten as

$$\sin\nu_d = (\sin\phi' - \sin\phi)/(1 - \sin\phi' \,\sin\phi) \tag{2.130}$$

where ν_d is angle of dilation, defined by (2.83), ϕ' is defined by (2.95), and ϕ is the value of ϕ' at a critical state. Note that (2.130) agrees with Rowe's relation (2.97) only at the critical state.

(b) For small values of ν_d, ϕ', and ϕ, show that (2.130) reduces to

$$\nu_d = \phi' - \phi \tag{2.131}$$

which is qualitatively similar to Bolton's relation (2.98).

2.20. An application of the double-shearing model

(a) Consider a steady, linear velocity field $v_x = k\, y$ and $v_y = 0$, where k is a positive constant. Assuming that $\psi = \psi(t)$, where ψ is the inclination of the major principal stress axis relative to the x axis (Fig. 2.49) and t is the time, show that (2.118) reduces to

$$\frac{\mathrm{d}\psi}{\mathrm{d}t} = \frac{k\,(\cos 2\psi - \sin\phi_*)}{2\,\sin\phi_*} \tag{2.132}$$

(b) Show that (2.132) can be rewritten as

$$\frac{\mathrm{d}\psi}{\mathrm{d}t} = \frac{k\,\sin(\Omega + \psi)\,\sin(\Omega - \psi)}{\sin\phi_*} \tag{2.133}$$

where $\Omega \equiv (\pi/4) - (\phi_*/2)$.

(c) The steady-state solutions of (2.133) are given by $\psi = \psi_s = \pm\Omega$. If the dissipation inequality (2.71) holds, and $0 \le \phi_* \le \pi/2$, show that the permissible steady state is given by $\psi_s = -\Omega$.

(d) Consider the solutions of (2.133) subject to the initial condition $\psi(t = 0) = \psi_0$. By examining the slopes of the trajectories in the ψ–t plane, show that trajectories which satisfy either $|\psi_0| < \Omega$ or $-\pi + \Omega < \psi_0 < -\Omega$ move away from the steady state $\psi_s = -\Omega$. In other words, this steady state is *unstable*. It is not necessary to solve (2.133) to deduce the result.

This problem has been adapted from Spencer (1986).

3

Flow through Hoppers

Hoppers are vessels used for storing granular materials and feeding them into other devices (Fig. 1.5b). The cross sections of wedge-shaped and axisymmetric hoppers are rectangular and circular, respectively (Fig. 3.1).

Though hoppers have been in use for a long time, the mechanics of hopper flow is still poorly understood. There is at present no theory whose predictions are in good agreement with all the experimental observations. Some of the observations are summarized below, and this is followed by a discussion of models for plane flow through wedge-shaped hoppers. Models for axisymmetric hoppers and bunkers are discussed in Chapter 6.

3.1. EXPERIMENTAL OBSERVATIONS

3.1.1. Flow Rate

When coarse materials flow from hoppers, it is generally agreed that the mass flow rate $\dot{M}$ is approximately independent of the head of material H, provided H is greater than a few multiples of the slot width D (Fig. 3.2) (Huber-Burnand, 1829, cited in Herbst and Winterkorn, 1965, p. 3; Hagen, 1852, cited in Wieghardt, 1975; Beverloo et al., 1961; Al-Din and Gunn, 1984; Nedderman, 1992, p. 293). For cylindrical bins whose diameter is $>2D$, Nedderman (1992, p. 293) suggests that the minimum value of H required for head independence is about $2D$.

The above observation is supported by the data shown in Table 3.1. In contrast, the data in Fig. 3.4 show that the flow rate varies with the head of the material for fine materials. Gu et al. (1993) attribute this behavior to the effect of interstitial air pressure gradients.

Some correlations for the mass flow rate $\dot{M}$ will now be discussed. Assuming that $\dot{M}$ is independent of the head of material, it may be expected to depend on the dimensions of the exit slot, the acceleration due to gravity g, the particle diameter d_p, the density ρ of the granular material, and the frictional properties of the granular and wall materials. As noted by Nedderman (1992, p. 293), and verified by experiments (see, e.g., Le Pennec et al., 1998), there is a range of particle diameters d_p for which $\dot{M}$ is independent of d_p. For a circular exit slot or orifice of diameter D, we have $\dot{M} = \dot{M}(D, g, \rho)$.

As the flow is driven by gravity, the velocity scale can be chosen as $\sqrt{gD}$. Hence $\dot{M}$, which is the product of the density, the average vertical velocity, and the cross-sectional area, is given by $\dot{M} \sim \rho \sqrt{g D}\,(\pi D^2/4)$. Thus $\dot{M} \propto D^{5/2}$ for a circular orifice. This result is due to Hagen (1852) (cited in Wieghardt, 1975) and Beverloo et al. (1961).

Based on experiments with sand and various seeds, Beverloo et al. (1961) found that $\dot{M} \propto D^{2.77}$ for sand, and $\dot{M} \propto D^{2.95}$ for spinach seeds. The deviation of the exponent from the expected value of 2.5 was attributed to the existence of an empty zone adjacent to the

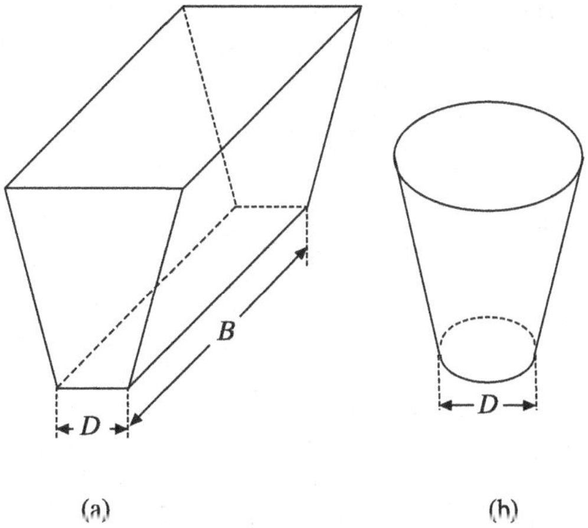

Figure 3.1. Sketches of (a) a wedge-shaped hopper and (b) an axisymmetric or conical hopper.

edge of the orifice (Fig. 3.5). Therefore, they assumed that $\dot{M} \propto (D - k_* \, d_p)^{5/2}$, where d_p is the particle diameter and k_* is a constant. (The use of this scaling had been suggested earlier by Hagen (1852) (cited in Wieghardt, 1975) and Wieghardt (1952).) Thus $(k_* \, d_p)/2$ is the thickness of the empty zone. Even though such a zone has not been observed experimentally, the inclusion of k_* represents an attempt to account for the increased resistance to flow when the dimensions of the particles and the slot become comparable. It was found that $k_* = 2.9$ for sand and $k_* = 1.3$–1.4 for the seeds. Beverloo et al. (1961) proposed the correlation

$$\dot{M} = 0.58 \, \rho \, \sqrt{g} \, (D - 1.4 \, d_p)^{5/2} \tag{3.1}$$

which is similar in form to that proposed by Hagen (1852).

Equation (3.1) predicts the mass flow rates measured by Beverloo et al. (1961) to within 5% for the seeds and to within 12.5% for wheat, sugar, and polystyrene beads. It can also be used to estimate the flow rates of other free-flowing materials. Nedderman (1992, p. 296) recommends that it should be used only for particle diameters in the range 0.4 mm $< d_p < D/6$, where D is the diameter of the exit slot. Above the upper limit, mechanical interlocking of particles is likely to occur at the exit; below the lower limit, drag caused by the interstitial air may affect the flow rate.

The lack of a unique density for a flowing material poses difficulties in using (3.1). Reasonable predictions are obtained by using the value corresponding to a vessel loosely filled with the material (Nedderman, 1992, p. 295).

For the special case $D \gg k_* d_p$, the mean vertical velocity v_e at the hopper exit, as estimated from (3.1), is $0.74 \sqrt{gD}$. This justifies the use of $\sqrt{gD}$ as a velocity scale. This value of v_e is approximately equal to the free-fall velocity of a particle falling from rest through a height $D/4$.

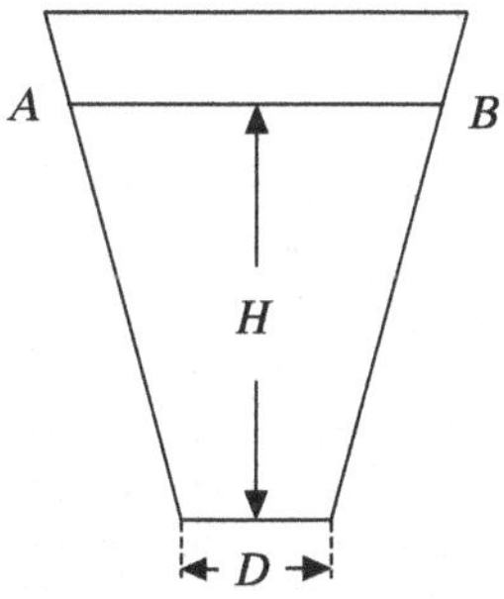

Figure 3.2. Discharge of granular material from a hopper. AB is the free surface of the material, H is the head, and D is either the width of the exit slot for a wedge-shaped hopper, or the diameter of the exit slot for an axisymmetric hopper.

Table 3.1. Variation of the Mass Flow Rate $\dot{M}$ of Glass Beads With the Head H of Material Above the Orifice of a Bunker (Fig. 3.2)*

H (mm)	$\dot{M}$ (g s^{-1})
340	10.5
490	10.1
640	10.7
740	10.5
840	11.0
940	10.2

* The particle diameter is 0.5 mm, and the bulk density is 1,700 kg m^{-3}. Data of Al-Din and Gunn (1984).

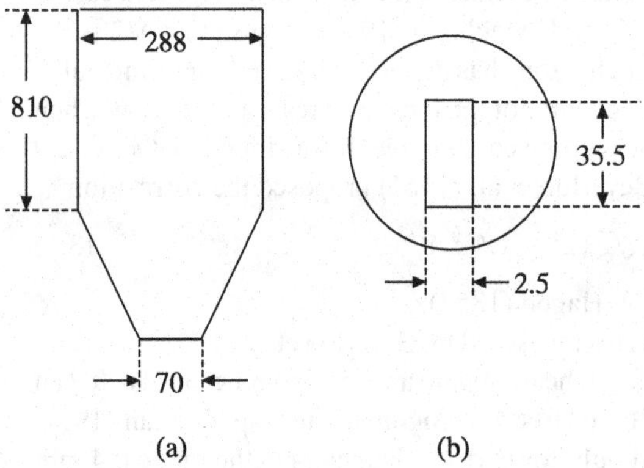

Figure 3.3. (a) Elevation of the axisymmetric bunker used in the experiments of Al-Din and Gunn (1984), (b) plan view of the base. All dimensions are in mm. The data obtained using this bunker are shown in Table 3.1.

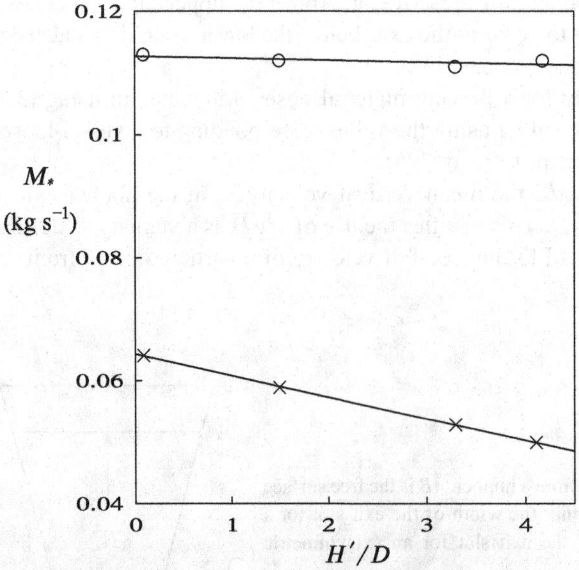

Figure 3.4. Variation of the mass flow rate M_* with the head H' of material above the bin-hopper transition of a bunker (see Fig. 1.5c for a sketch of the bunker): ○, sugar (median diameter = 784 μm); ×, PVC powder (median diameter = 127 μm). The data have been obtained using an axisymmetric bunker with an exit orifice of diameter $D = 20$ mm. (Adapted from Fig. 1a of Gu et al., 1993.)

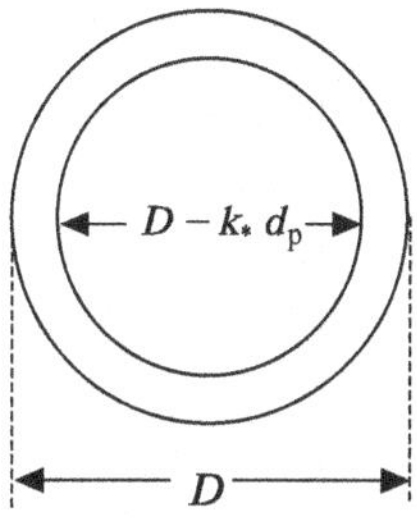

Figure 3.5. Empty zone of thickness $(k_* d_p)/2$ adjacent to the edge of a circular orifice of diameter D. Here d_p is the particle diameter and k_* is a material constant.

Equation (3.1) is based on data obtained using axisymmetric bins. In a bin, there are static "shoulders" of material adjacent to the base, and the material effectively flows in a hopper with curved walls (Fig. 3.6). Therefore, (3.1) can also be used for axisymmetric hoppers and bunkers, provided the walls are not very steep. For example, the measured mass flow rate of kale seeds from a bunker (Fig. 1.5) with $D = 0.034$ m and $\theta_w = 30°$ corresponds to a discharge velocity of $0.78\sqrt{gD}$ (Cleaver and Nedderman, 1993a), which is close to the value estimated from (3.1) ($= 0.74\sqrt{gD}$).

For hoppers with steep walls, (3.1) underestimates the mass flow rate $\dot{M}$ considerably. The effect of the wall angle θ_w on $\dot{M}$ was examined by Rose and Tanaka (1959). Using data obtained from cylindrical bunkers, with θ_w in the range 15°–90°, they found that $\dot{M} \propto F(\theta_w, \beta)$, where

$$F(\theta_w, \beta) \equiv (\tan\theta_w / \tan\beta)^{-0.35}, \quad \theta_w \leq \beta$$
$$\equiv 1, \quad \theta_w > \beta \tag{3.2}$$

where β is the angle between the vertical and the boundary of the stagnant zone in a bin (Fig. 3.6). Nedderman (1992, p. 296) combined (3.2) with the Beverloo correlation (3.1) to obtain

$$\dot{M} = 0.58\,\rho\,\sqrt{g}(D - 1.4\,d_p)^{5/2}\,F(\theta_w, \beta) \tag{3.3}$$

This will be called the *Rose–Tanaka–Beverloo–Nedderman* ($RTBN$) correlation. The measured values of β reported in Brown and Richards (1970, p. 152) for coal, sand, rice, and glass beads are in the range 34–58°. In case the value of β is not known, Nedderman (1992, p. 296) recommends a value of 45°.

The above discussion has been confined to vessels with circular exit slots. For noncircular exit slots, Fowler and Glastonbury (1959) and Brown and Richards (1959) suggested the use of a velocity scale based on the hydraulic diameter D_h. For a slot of area A and perimeter P, $D_h \equiv 4\,A/P$. Using this velocity scale and incorporating the concept of an empty zone, (3.3) can be modified as suggested by Nedderman et al. (1982) to obtain

$$\dot{M} = 0.74\,\rho\,A_*\,\sqrt{g\,D_{h*}}\,F(\theta_w, \beta) \tag{3.4}$$

Figure 3.6. Flow pattern in a bin.

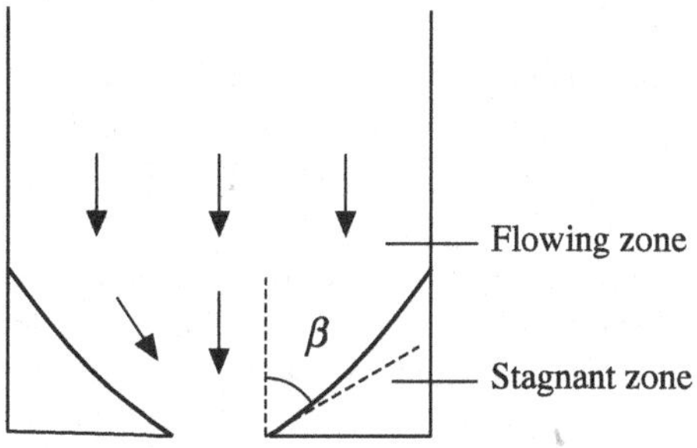

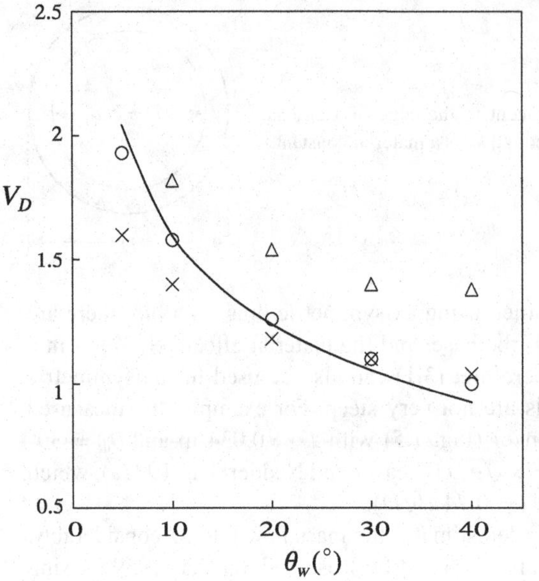

Figure 3.7. Variation of the dimensionless mass flow rate V_D with the wall angle θ_w of a wedge-shaped hopper: –, Rose-Tanaka–Beverloo–Nedderman correlation; Δ, data of Sullivan (1972) (cited in Brennen and Pearce, 1978) for glass beads ($d_p = 0.325$ mm, $B =$ 55.6–68.3 mm, $D = 6.35$ mm, $\rho = 1{,}460$ kg m^{-3}, $\phi = 25°$, $\delta = 17.7°$ for aluminium walls); $\circ$, $\times$, data of Kaza (1982) for glass beads (d_p = 0.85–1.25 mm, $B = 127$ mm, $D = 25.5$–26.5 mm, $\rho = 1{,}630$ kg m^{-3}, $\phi = 32.4°$; $\circ$, $\delta =$ 15.1° for aluminum walls; $\times$, $\delta = 26.9°$ for "fully rough" walls). Here d_p is the particle diameter, B and D are the dimensions of the exit slot, ρ is the density of the beads, ϕ is the angle of internal friction, and δ is the angle of wall friction (see (3.34)). The fully rough walls were formed by sticking a layer of glass beads on Perspex walls.

where $D_{h*} = 4\,A_*/P_*$, and the stars indicate that both the area A and the perimeter P have been corrected for the empty space near the edges of the slot. For a circular exit slot of diameter D, $D_{h*} = D - k_*\,d_p$, and (3.4) reduces to (3.1).

For a rectangular exit slot with $B \gg D \gg d_p$ (Fig. 3.1), (3.4) implies that the mass flow rate scales with $B\,D^{3/2}$, rather than $D^{5/2}$. Further, the mean vertical component of the exit velocity is given by $0.74\sqrt{g\,D_{h*}}F(\theta_w, \beta) = 1.05\sqrt{g\,D}F(\theta_w, \beta)$.

To compare (3.4) with data, it is convenient to introduce a dimensionless *mass flow rate* V_D, defined by

$$V_D \equiv \dot{M}/(\rho\,A\sqrt{g\,D}) \tag{3.5}$$

where A is the area of the exit slot and D is either the diameter of the exit slot for an axisymmetric hopper or the width of the exit slot for a wedge-shaped hopper (Fig. 3.1). (In the latter case, the edge of the exit slot which forms a part of the vertical face is of length D, as shown in Fig. 3.1.) Using (3.5), (3.4) can be rewritten as

$$V_D = 0.74\,F(\theta_w, \beta)\,G(k_* d_p, B, D) \tag{3.6}$$

where F is given by (3.2) and G is a dimensionless function defined by

$$G \equiv \sqrt{2}\,[1 - (k_*\,d_p/D)]\,[1 - (k_*\,d_p/B)]\sqrt{\frac{[1 - (k_*\,d_p/D)]\,[1 - (k_*\,d_p/B)]}{1 + (D/B) - (2\,k_*\,d_p/B)}} \tag{3.7}$$

for a wedge-shaped hopper and by

$$G \equiv [1 - (k_*\,d_p/D)]^{5/2} \tag{3.8}$$

for an axisymmetric hopper.

Figure 3.7 compares the mass flow rate predicted by the $RTBN$ correlation (3.6) with data for the flow of glass beads through wedge-shaped hoppers. The correlation agrees fairly well with the data of Kaza (1982), but it underestimates the data of Sullivan (1972) (cited in Brennen and Pearce, 1978) at large wall angles. Alternative correlations for noncircular orifices, and for nonspherical particles, have been proposed by Al-Din and Gunn (1984).

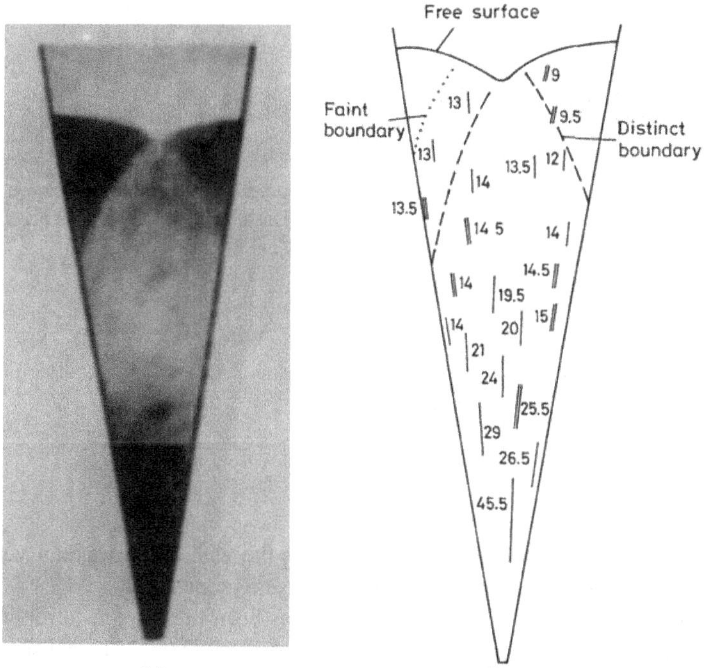

(a) (b)

Figure 3.8. (a) Radiograph of a hopper filled with Leighton Buzzard sand (particle diameter = 0.6–1.2 mm), taken after an increment of flow. The outlet was closed before taking the radiograph. The dark regions near the free surface represent dense material, and the light regions represent dilated material. The dark rectangle which is approximately at a height $H/3$ above the base represents a support used in the experiment. Drescher et al. (1978) note that the arrangement of the X-ray source and films causes the region near the exit to appear denser than it actually is. (b) Displacement of lead markers (particle diameter = 2 mm) after an increment of flow. The numbers show the displacement in mm. Dimensions of the exit slot: 10 mm (in the plane of the figure) $\times$ 60 mm (perpendicular to the plane of the figure), height of the hopper $H = 610$ mm, wall angle $\theta_w = 10^\circ$. (Reproduced from Drescher et al., 1978, with permission from the Institution of Civil Engineers, London.)

The crosses and circles in Fig. 3.7 show that at $\theta_w = 40^\circ$, the mass flow rate increases with the wall roughness. This may be caused by a change in the flow pattern at large wall angles (see Fig. 3.10). It is plausible that the rougher walls cause the material to flow through a channel with an effectively smaller wall angle, leading to a higher mass flow rate than the smoother walls.

3.1.2. Kinematics

The work of Drescher et al. (1978) throws light on flow patterns in a wedge-shaped hopper (Fig. 3.8). The hopper is filled with sand, and the outlet is closed after allowing the sand to discharge for some time. Figure 3.8a shows a radiograph of the hopper, taken after a small increment of discharge. The dark regions adjacent to the free surface of the sand represent dense material, whereas the lighter region below represents material with a lower bulk density. It is implicitly assumed that cessation of flow does not affect the density significantly, but this assumption has not yet been verified.

The displacement of lead spheres, used as tracers, is shown in Fig. 3.8b. Based on the data shown in Fig 3.8, Drescher et al. (1978) proposed the model indicated in Fig. 3.9. They assumed that displacements were parallel to the wall in the feeding zone, and were radially

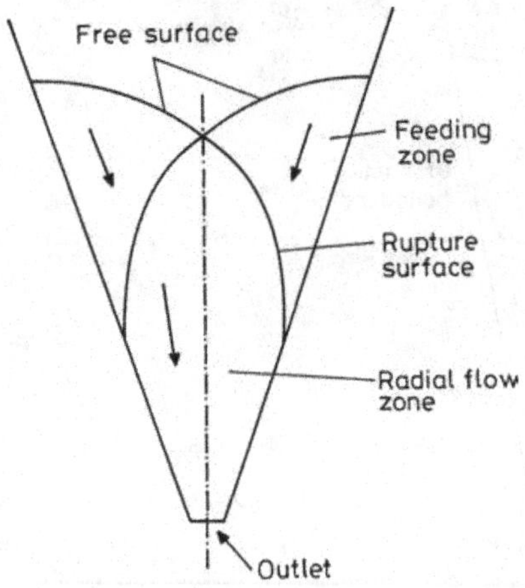

Figure 3.9. Assumed flow pattern based on the data shown in Fig. 3.8. (Reproduced from Drescher et al., 1978, with permission from the Institution of Civil Engineers, London.)

directed in the zone below the rupture surface. However, as the wall angle was not very large, the displacements were approximately radial in the feeding zone also.

The boundary between the dense and dilated material has been idealized as a rupture surface (see also Figs. 2.2 and 2.3), which represents a surface across which the density is discontinuous. If the rupture surface persists during flow, the normal component of velocity must also be discontinuous in order to conserve the mass flux of material. Thus the rupture surface can be regarded as an analog of an oblique shock in gas dynamics. An interesting difference is that the density usually increases as the gas crosses the shock, whereas it usually decreases as the granular material crosses the rupture layer.

In the feeding zone above the rupture surface, the material appears to move like a rigid body (Fig. 3.8b). Hence it is not certain that the yield condition will be satisfied at every point in this zone.

Near the hopper exit, two types of flow patterns have been observed, depending on the inclination of the walls. At a given height above the exit slot, the velocity varies smoothly across the hopper when the walls are steep (Fig. 3.10a). This pattern is called *mass flow*. On the other hand, when the wall angle is large, the flow pattern resembles that in a bin (Fig. 3.6). There is a core of rapidly moving material, bounded by shoulders of material which move very slowly (Fig. 3.10b). This pattern is called *funnel flow* or *core flow*.

As noted by Jenike (1964b, pp. 58–71), the flow rate is relatively steady during mass flow and the material follows a "first in, first out" pattern. On the other hand, if a perishable material flows through a vessel in funnel flow, the material near the walls has a high residence time, and may deteriorate. Funnel flow may be advantageous for abrasive materials, as the stagnant zone prevents contact between the flowing material and the hopper wall, thereby reducing the wear of the wall.

3.1.3. Solids Fraction Profiles

The measurements of Fickie et al. (1989) show that the solids fraction decreases smoothly as the material flows down the hopper (Fig. 3.11a), but the variation is very rapid near the exit slot. Within the hopper, the transverse profiles show little variation (Fig. 3.11b), probably because the walls used were fairly smooth. On the other hand, there is a marked transverse variation below the hopper. The spreading of the jet, as inferred from Fig. 3.11b, is related to its structure. Based on photographs of particle jets issuing from a bin, Darton (1976) suggested that the jets consisted of central regions of dense material, surrounded by expanding clouds of low density.

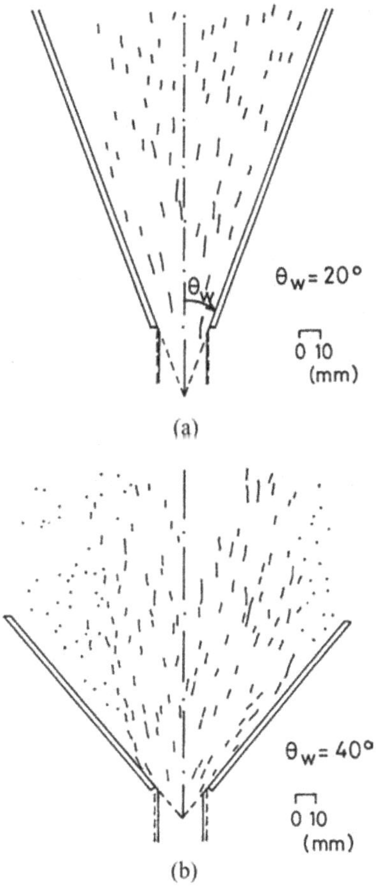

Figure 3.10. Flow patterns in (a) mass flow, and (b) funnel flow. The streaks represent particle paths of coloured glass beads, traced from photographs of glass beads (diameter = 0.85–1.25 mm) flowing through a wedge-shaped hopper with glass faces and aluminum walls. (Source: Kaza, 1982.)

The results shown in Fig. 3.11 are based on a hopper whose exit slot is only about ten particle diameters wide. It would be instructive to examine whether the use of wider slots leads to qualitative changes in the solids fraction profiles and the shape of the particle jet.

3.2. THEORY FOR STEADY, PLANE FLOW

Here we discuss some models for flow through wedge-shaped hoppers, and compare the predictions with data. The models assume that the dependent variables do not vary in the direction normal to the plane of flow, and hence are expected to be valid only for hoppers with $B \gg D$ (Fig. 3.1a).

3.2.1. The Critical State Approximation

The solids fraction data in Fig. 3.11 suggest that ν is approximately constant, except close to the exit. As a first attempt to solve the hopper problem, consider the special case $\nu =$ constant. This assumption will be relaxed in §3.8.

The assumption of constant ν implies that the flow is incompressible, and hence the material is at a critical state. In other words, the state of stress must lie on the critical state curve *ACD* (Fig. 2.43). Assuming that the latter is a straight line, (2.47) implies that

$$\tau = \sigma \sin\phi; \quad \sigma = \sigma_c(\nu) \tag{3.9}$$

where τ and σ are the deviator stress and the mean stress, respectively, defined by (2.33) and ϕ is the angle of internal friction.

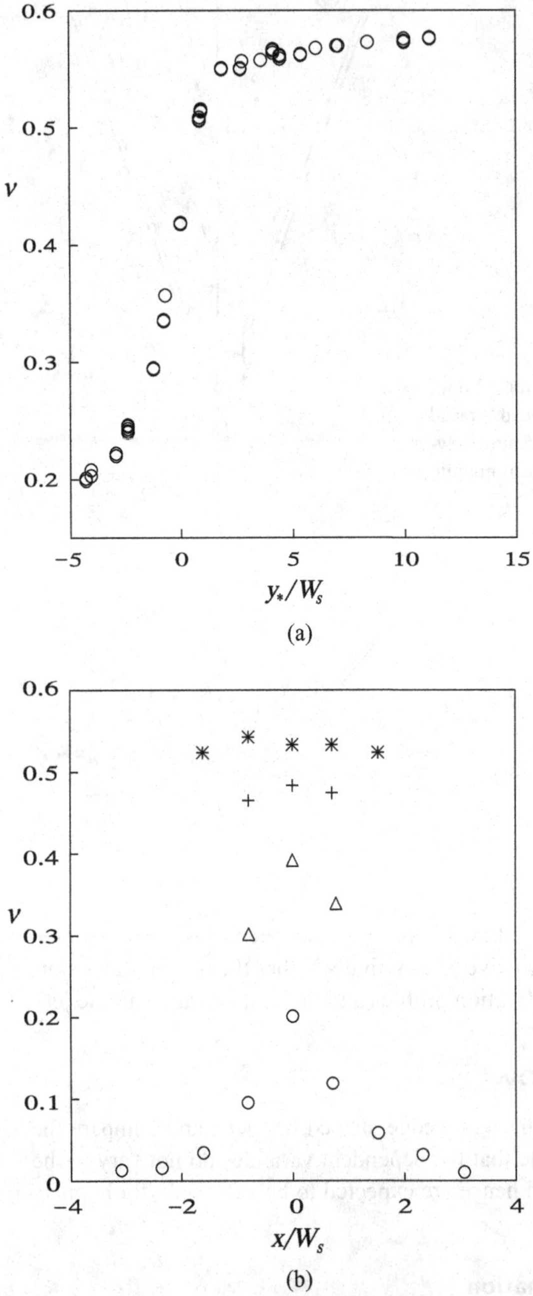

Figure 3.11. Profiles of the solids fraction ν during the discharge of glass beads through a wedge-shaped hopper with glass faces and Plexiglas walls. The data were obtained using a γ-ray densitometer. Here y_* is the vertical coordinate, measured upward from the plane of the exit slot, and W_s is the half-width of this slot. The x coordinate is measured from the centerline, as shown in Fig. 3.13. (a) Profile along the centerline of the hopper, (b) transverse profiles: $y_*/W_s = 2.69$ (*); 0.85 (+); −0.077 (Δ); −4.31 (○). Parameter values: $W_s = 6.5$ mm, wall angle $\theta_w = 23°$, particle density $\rho_p = 2900$ kg m^{-3}, particle diameter $d_p \approx 1$ mm. (Data of Fickie et al., 1989.)

Equation (3.9) and the assumption $\nu =$ constant imply that both τ and σ are constants. This leads to the following difficulty. In view of (2.102) and (3.9), the stress components can be expressed in terms of σ and γ, where γ is the orientation of the major principal stress axis, defined as shown in Fig. 2.11. Thus the governing equations (2.100), (2.101), (2.104), and (2.105) must be solved for σ, γ, and the velocity components v_x and v_y. If σ is a constant, the system of equations becomes overdetermined, and cannot in general be solved.

To resolve this problem, consider the variation of σ_c with ν, which has the qualitative behavior shown in Fig. 3.12. If ν_{drp} is assumed to be very close to ν_{lrp}, small changes in ν

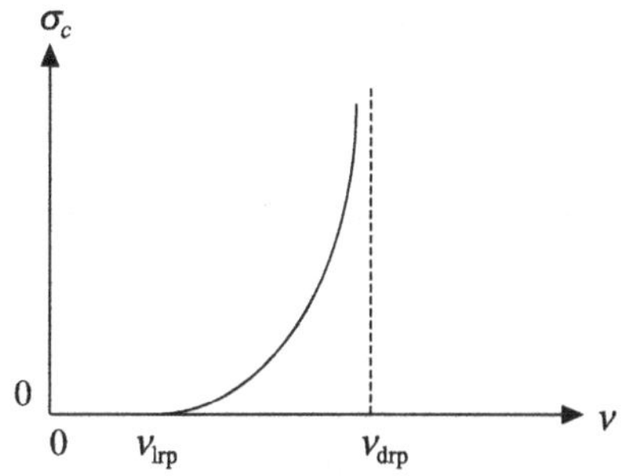

Figure 3.12. Variation of the mean stress at a critical state σ_c with the solids fraction ν. Here ν_{lrp} and ν_{drp} are the solids fractions corresponding to loose random packing and dense random packing, respectively.

will cause large changes in $\sigma = \sigma_c$. Thus σ can vary over a wide range, even though ν is approximately constant. In particular, the second of (3.9) can be discarded. This is called the *critical state approximation* (Jackson, 1983); it is analogous to the case of incompressible fluid flow, where the pressure of the fluid varies but the density remains constant.

From an alternative viewpoint, we can assume that (i) the flow is incompressible, and (ii) the material satisfies the Mohr–Coulomb yield condition for a cohesionless material. Thus

$$\tau = \sigma \sin\phi \tag{3.10}$$

As ν = constant, the angle of dilation ν_d vanishes (see the first of (2.106)). Hence the flow rule (2.105) reduces to the incompressible version of the mass balance (2.99).

3.3. THE SMOOTH WALL, RADIAL GRAVITY (*SWRG*) PROBLEM

An elegant solution due to Savage (1965) is discussed below. It was independently discovered later by Sullivan (1972) and Davidson and Nedderman (1973). Despite the use of numerous simplifying assumptions, it provides valuable insights about the forms of the stress and velocity fields in the hopper.

It is convenient to use cylindrical coordinates, with the origin at the virtual apex of the hopper (Fig. 3.13). For steady, plane flow, the velocity field is given by

$$v_r = v_r(r, \theta); \quad v_\theta = v_\theta(r, \theta); \quad v_z = 0 \tag{3.11}$$

where v_r and v_θ are the radial and circumferential components of velocity, respectively, and the z axis is directed perpendicular to the plane of Fig. 3.13.

Using (3.11), and assuming that (i) the stresses do not vary in the z direction, and (ii) the solids fraction ν is a constant, the balance equations in cylindrical coordinates (see (A.63)–(A.65)) reduce to

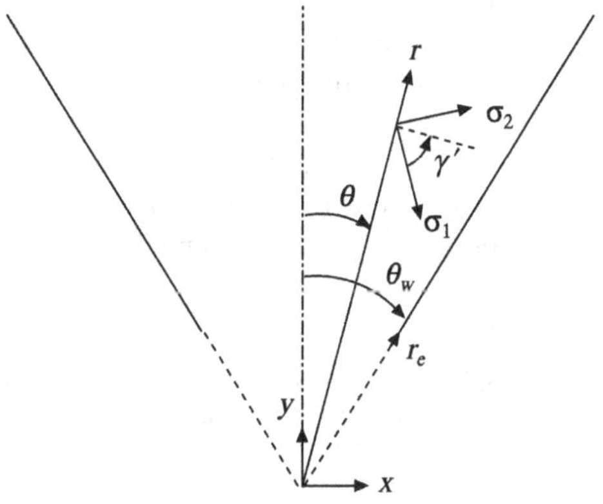

Figure 3.13. Coordinate system for the hopper problem. Here r_e is the radial coordinate corresponding to the hopper exit, and the angle γ' is measured anticlockwise from the major principal stress axis to the circumferential direction.

mass balance

$$\frac{1}{r}\frac{\partial(r\,v_r)}{\partial r} + \frac{1}{r}\frac{\partial v_\theta}{\partial \theta} = 0 \tag{3.12}$$

momentum balance (r component)

$$\rho\left(v_r\frac{\partial v_r}{\partial r} + \frac{v_\theta}{r}\frac{\partial v_r}{\partial \theta} - \frac{v_\theta^2}{r}\right) + \frac{\partial \sigma_{rr}}{\partial r} + \frac{1}{r}\frac{\partial \sigma_{\theta r}}{\partial \theta}$$
$$+\frac{\sigma_{rr}-\sigma_{\theta\theta}}{r} + \rho\, g\, \cos\theta = 0 \tag{3.13}$$

momentum balance (θ component)

$$\rho\left(v_r\frac{\partial v_\theta}{\partial r} + \frac{v_\theta}{r}\frac{\partial v_\theta}{\partial \theta} + \frac{v_r v_\theta}{r}\right) + \frac{\partial \sigma_{r\theta}}{\partial r} + \frac{1}{r}\frac{\partial \sigma_{\theta\theta}}{\partial \theta}$$
$$+\frac{2\sigma_{r\theta}}{r} - \rho\, g\, \sin\theta = 0 \tag{3.14}$$

where $\rho = \rho_p\, \nu$ is the density of the granular material and ρ_p is the particle density, assumed constant. Expressions for the stress components follow from (2.35) and (2.36) when the x and y axes are identified with the circumferential and radial directions, respectively:

$$\sigma_{\theta\theta} = \sigma + \tau\, \cos 2\gamma'; \quad \sigma_{r\theta} = \sigma_{\theta r} = -\tau\, \sin 2\gamma'; \quad \sigma_{rr} = \sigma - \tau\, \cos 2\gamma' \tag{3.15}$$

where

$$\sigma = (\sigma_1 + \sigma_2)/2; \quad \tau = (\sigma_1 - \sigma_2)/2; \quad \sigma_1 \geq \sigma_2 \tag{3.16}$$

σ_1 and σ_2 are the major and minor principal stresses, respectively, and γ' is the orientation of the σ_1 axis relative to the circumferential direction (Fig. 3.13).

As discussed in §3.2.1, the flow rule (2.105) reduces to (3.12). The coaxiality condition can be deduced from (2.78) and (2.79) as follows. Using these equations, identifying the x and y directions with the θ and r directions, respectively, and writing γ as γ', we obtain

$$\sin 2\gamma'\,(C_{\theta\theta} - C_{rr}) + 2\,\cos 2\gamma'\, C_{\theta r} = 0 \tag{3.17}$$

where C_{ij} are the components of the rate of deformation tensor $\mathbf{C}$, defined in the compressive sense (see (2.49)).

Using (A.68), (3.17) reduces to the coaxiality condition

$$\sin 2\gamma'\left(\frac{\partial v_r}{\partial r} - \frac{1}{r}\frac{\partial v_\theta}{\partial \theta} - \frac{v_r}{r}\right) - \cos 2\gamma'\left(\frac{\partial v_\theta}{\partial r} + \frac{1}{r}\frac{\partial v_r}{\partial \theta} - \frac{v_\theta}{r}\right) = 0 \tag{3.18}$$

The above equations can be simplified by assuming that (i) the material is yielding at every point in the hopper, and (ii) the flow is radial, i.e., $v_\theta = 0$. In view of the critical state approximation discussed in §3.2.1, the yield condition can be replaced by (3.10). Further, the mass balance (3.12), together with the assumption of radial flow, implies that $v_r = -f(\theta)/r$. (The minus sign has been introduced for convenience, to ensure that $f(\theta) > 0$ for downward flow.) The velocity field can be further simplified by assuming that the flow is *cylindrically symmetric*, i.e., $f(\theta) = \text{constant} \equiv A$. Thus

$$v_r = -A/r \tag{3.19}$$

Equation (3.19) and the coaxiality condition (3.18) imply that $A \sin 2\gamma' = 0$. As $A \neq 0$ when the material is flowing, we have $\sin 2\gamma' = 0$. There are two possibilities (roots):

$$\text{(i)}\ \gamma' = 0 \quad \text{and} \quad \text{(ii)}\ \gamma' = \pi/2 \tag{3.20}$$

Figure 3.14. Major (—) and minor (- - -) principal stress directions for (a) the passive state ($\gamma' = 0$), and (b) the active state ($\gamma' = \pi/2$) for the smooth wall, radial gravity problem. The major principal axis is inclined at an angle γ' relative to the circumferential direction, as shown in Fig. 3.13.

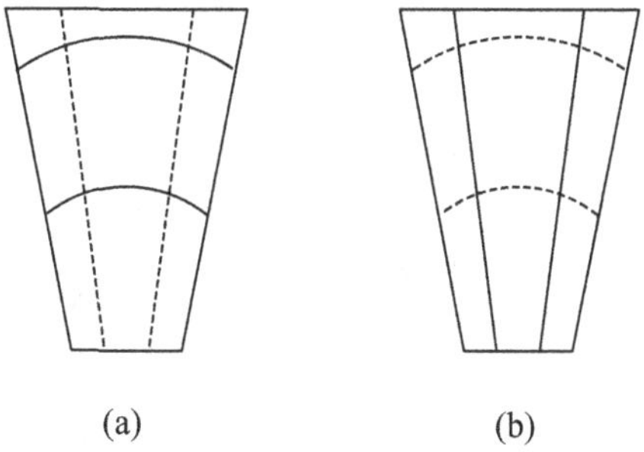

In view of the definition of γ' (Fig. 3.13), the principal stress axes are directed along the radial and circumferential directions (Fig. 3.14). As discussed in §2.9, the roots (i) and (ii) represent the *passive* and *active* states of stress, respectively. At the centerline of the hopper, the major principal stress axis is horizontal in the passive state, and vertical in the active state (Fig. 3.14).

To decide which root should be used, consider the Mohr's circle for the rate of deformation tensor (Fig. 3.15). The passive root corresponds to $C_{rr} = -\partial v_r/\partial r < 0$ (Fig. 3.15a), and the active root to $C_{rr} > 0$ (Fig. 3.15b). Equation (3.19) implies that $C_{rr} < 0$, and hence the passive root $\gamma' = 0$ should be used. From an alternative viewpoint, the condition $C_{rr} < 0$ implies that for steady downward flow,

$$\frac{\mathrm{D}v_r}{\mathrm{D}t} = v_r \frac{\partial v_r}{\partial r} = -v_r\, C_{rr} < 0$$

Hence material elements accelerate on moving down the hopper, in accord with observations on mass flow hoppers. Consider the use of the active root $\gamma' = \pi/2$. This root corresponds to $C_{rr} > 0$ or $\partial v_r/\partial r < 0$ if coaxiality is used. Hence (3.19) implies that $A < 0$, which corresponds to *upward* flow. Conversely, if $A > 0$ and $\gamma' = \pi/2$, coaxiality cannot be satisfied. Further, the mass flow rate does not level off as the head increases.

Returning to the solution for the passive case, there is no shear stress on the hopper wall, as it coincides with a principal stress direction. Thus the assumption of cylindrical symmetry is valid only for a hopper with *smooth* walls.

At this stage, the mass balance and the coaxiality condition have been satisfied, and the momentum balances must be solved. Using (3.10) and (3.20)(i), (3.15) reduces to

$$\sigma_{\theta\theta} = \sigma(1 + \sin\phi); \quad \sigma_{r\theta} = 0; \quad \sigma_{rr} = \sigma(1 - \sin\phi) \tag{3.21}$$

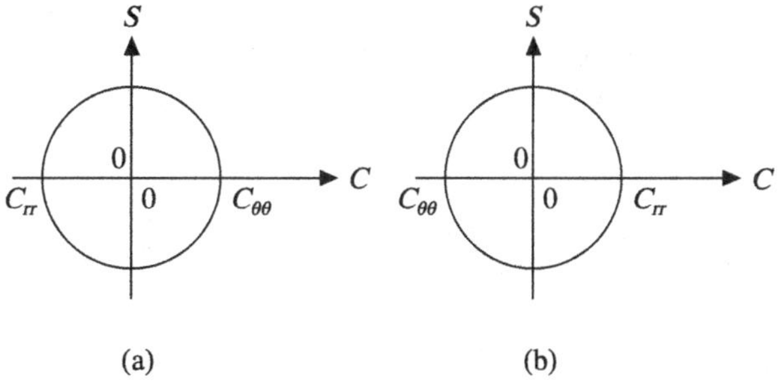

Figure 3.15. Mohr's circle for the rate of deformation tensor, for the case of incompressible flow: (a) $\gamma' = 0$, and (b) $\gamma' = \pi/2$. Here C and S are the rates of compression and shearing, respectively, on a plane whose normal is inclined at an angle γ' measured anticlockwise from the major principal compressive rate of deformation axis. This angle is marked as γ_* in Fig. 2.40.

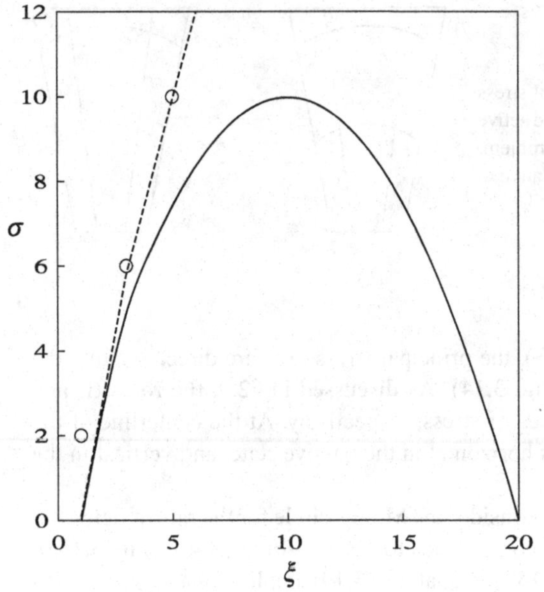

Figure 3.16. Profiles of the mean stress predicted by the smooth wall, radial gravity solution for $\phi = 30°$: —, $\xi_1 = 20$; - - -, $\xi_1 \to \infty$. The circles represent the first term on the right-hand side of (3.25).

Assuming that the stress field is also cylindrically symmetric, i.e., $\sigma = \sigma(r)$, the θ component of the momentum balance (3.14) reduces to $\rho\, g\, \sin\theta = 0$, which can be satisfied only along the centerline $\theta = 0$. To avoid an inconsistency at other locations, it is assumed that gravity is directed radially toward the apparent vertex of the hopper. Then the θ component of gravity vanishes, ensuring that (3.14) is identically satisfied. This assumption is likely to be reasonable when the wall angle θ_w (Fig. 3.13) is small enough to ensure that $\cos\theta_w \approx 1$, $\sin\theta_w \approx 0$.

To determine the mean stress $\sigma(r)$, consider the r component of the momentum balance (3.13), which reduces to

$$\frac{-\rho A^2}{r^3} + (1 - \sin\phi)\frac{\mathrm{d}\sigma}{\mathrm{d}r} - \frac{2\,\sin\phi\,\sigma}{r} + \rho\, g = 0 \qquad (3.22)$$

Equation (3.22) can be rewritten in dimensionless form as

$$\frac{\mathrm{d}\overline{\sigma}}{\mathrm{d}\xi} - \frac{k\,\overline{\sigma}}{\xi} + \frac{1}{1 - \sin\phi} - \frac{\overline{A}^2}{(1 - \sin\phi)\,\xi^3} = 0 \qquad (3.23)$$

where

$$\overline{\sigma} \equiv \sigma/(\rho\, g\, r_e); \quad \xi \equiv r/r_e; \quad k \equiv (2\sin\phi)/(1 - \sin\phi); \quad \overline{A} \equiv A/(\sqrt{g\, r_e{}^3}) \qquad (3.24)$$

Here r_e is the radial coordinate corresponding to the hopper exit (Fig. 3.13). The second, third, and fourth terms in (3.23) represent the effects of friction, gravity, and inertia, respectively.

Equation (3.23) can be integrated to obtain

$$\overline{\sigma} = \frac{1}{1 - \sin\phi}\left(\frac{\xi}{k - 1} - \frac{\overline{A}^2}{(k + 2)\,\xi^2}\right) + c_1\,\xi^k \qquad (3.25)$$

where c_1 is an integration constant.

The constants $\overline{A}$ and c_1 can be determined by specifying two boundary conditions. These conditions follow from the assumption that the flowing material in the hopper is bounded from below and above by *traction-free* surfaces, i.e., surfaces on which the shear and normal stresses vanish. For a cohesionless material, this implies that the mean stress $\overline{\sigma}$

vanishes on such surfaces. As $\overline{\sigma} = \overline{\sigma}(\xi)$, the traction-free surfaces are cylindrical segments, defined by ξ = constant. The lower surface passes through the edges of the exit slot, i.e., $\xi = 1$, and the upper surface is located at $\xi = \xi_1 > 1$. The value of ξ_1 depends on the head of material maintained in the hopper.

The traction-free boundary conditions permit a simple solution to be obtained, but their physical basis is dubious. Suppose it is specified that

$$\overline{\sigma}(\xi = 1) = 0; \quad \overline{\sigma}(\xi = \xi_1) = 0 \tag{3.26}$$

A constant head of material can be maintained in the hopper only by feeding material onto the upper surface $\xi = \xi_1$. Therefore, this surface cannot be truly traction free. However, if the hopper is *deep* i.e., if $\xi_1 \gg 1$, the rate of addition of material will be small. In this case, the assumption that the upper surface $\xi = \xi_1$ is traction-free surface is reasonable. Even though the stresses are expected to decrease toward the hopper outlet, it is difficult to justify the first of (3.26). Indeed, its use leads to certain inconsistencies, as noted later.

Accepting the validity of the boundary conditions (3.26) for the moment, (3.25) and (3.26) imply that

$$\begin{aligned} \overline{A}^2 &= \frac{k+2}{k-1}\left(\frac{1-\xi_1{}^{-(k-1)}}{1-\xi_1{}^{-(k+2)}}\right) \\ c_1 &= -\frac{1}{(1-\sin\phi)(k-1)}\left(\frac{\xi_1{}^{-(k-1)}-\xi_1{}^{-(k+2)}}{1-\xi_1{}^{-(k+2)}}\right) \end{aligned} \tag{3.27}$$

Thus the velocity and stress fields are given by

$$u \equiv v_r/\sqrt{g r_e} = -\overline{A}/\xi$$

and (3.25), respectively. Finally, the mass flow rate $\dot{M}$ is obtained by integrating the mass flux $\rho\, v_r$ across a cylindrical segment to obtain

$$\dot{M} = -2\int_0^{\theta_w} \rho\, v_r\, B\, r\, \mathrm{d}\theta = 2\rho\, B\, r_e\sqrt{g r_e}\,\theta_w\,\overline{A} \tag{3.28}$$

where B is the length of the exit slot in the direction perpendicular to the r–θ plane (Fig. 3.1). Using (3.5), the dimensionless mass flow rate is given by

$$V_D = \frac{2\,\theta_w\,\overline{A}}{(2\,\sin\theta_w)^{3/2}} \tag{3.29}$$

The qualitative behavior of the solution depends on the value of the parameter $k = (2\sin\phi)/(1-\sin\phi)$. For materials with $k > 1 (\phi > 19.5°)$, (3.27) implies that

$$\overline{A} \approx \left(\frac{k+2}{k-1}\right)^{1/2}; \quad \xi_1 \gg 1,\ k > 1 \tag{3.30}$$

Therefore, for a *deep* hopper with $\xi_1 \gg 1$, the mass flow rate V_D attains a value

$$V_D \equiv V_{Da} = \frac{2\,\theta_w}{(2\,\sin\theta_w)^{3/2}}\left(\frac{k+2}{k-1}\right)^{1/2} \tag{3.31}$$

which is independent of the head ξ_1. As discussed in §3.1.1, this behavior is in accord with data for coarse materials.

With $k > 1$, $\xi_1 \gg 1$, and $\xi \ll \xi_1$, (3.25) reduces to

$$\overline{\sigma} \approx \frac{1}{3\,\sin\phi - 1}\left(\xi - \frac{1}{\xi^2}\right) \tag{3.32}$$

This solution is not valid close to the upper free surface $\xi = \xi_1$, where the term $c_1 \xi^k$ in (3.25) becomes important. The first term on the right-hand side of (3.32) represents the combined effect of gravity and internal friction. If $k > 1$ ($\sin\phi > 1/3$), the mean stress increases with ξ, in contrast to the hydrostatic variation of pressure in a body of water at rest. The second term, representing inertial effects, fades as ξ increases. The effect of inertial terms on the stress profile is less than 1% if $1/\xi^3 < 0.01$, i.e., if $\xi > 4.6$.

Consider the effect of the traction-free boundary condition at the upper surface of the fill ($\xi = \xi_1$). For a deep hopper, (3.25) and (3.27) imply that

$$\overline{\sigma} \approx \frac{\xi}{3\,\sin\phi - 1}\left(1 - (\xi/\xi_1)^{k-1}\right), \quad \xi \gg 1, k > 1$$

Thus the effect of the upper boundary condition can be neglected if

$$(\xi/\xi_1)^{k-1} \ll 1 \tag{3.33}$$

For example, if $\phi = 30°$, the profile will be affected by less than 1% if $\xi < 0.01\,\xi_1$. As $\xi \geq 1$, we must have $\xi_1 \gg 100$. This condition is usually violated in laboratory-scale hoppers. However, inclusion of wall roughness in the model causes the "entry" effect to decay much faster than is implied by the inequality (3.33).

The solid curve in Fig. 3.16 shows the stress profile for $\phi = 30°$ and $\xi_1 = 20$. The broken curve shows the profile for a deep hopper, and the circles denote the linear term in (3.25). Comparison of the broken curve and the circles shows that inertial effects decay as ξ increases; in the present case, they are negligible for $\xi > 5$. The boundary condition at the upper surface exerts a significant effect, except close to the hopper exit. In spite of this feature, the asymptotic value of the mass flow rate, given by (3.31), is within 2.5% of the actual value, given by (3.29).

The curve labeled *SWRG* in Fig. 3.17 shows that the predicted mass flow rate increases as the wall angle decreases. To understand this behavior, consider the limit $\theta_w \to 0$, for a fixed value of the width D of the exit slot. As the walls have been assumed to be smooth, they do not exert any shear stress on the granular material. Further, as the effect of air drag has been ignored, there is no force opposing gravity. Hence the mass flow rate becomes unbounded. For $\theta_w > 0$, the normal stress exerted by the wall on the material has a vertical component which opposes gravity. Hence the mass flow rate is expected to be less than that predicted in the limit $\theta_w \to 0$.

For deep hoppers, the mass flow rates predicted by (3.31) are much larger than measured values. For glass beads, the discrepancy is 150–90% of the latter for wall angles in the range 5–40° (Fig. 3.17). It is shown in the next section that the inclusion of wall roughness reduces the discrepancy considerably. The qualitative effect of changes in the wall angle is captured by the model.

The above discussion has been confined to the case $k > 1$ ($\phi > 19.5°$). On the other hand, if $k < 1$, the mass flow rate depends on the head. Indeed, in the limit $k \to 0$ ($\phi \to 0$), the equations for an inviscid fluid are recovered. The discharge rate for a deep hopper is then proportional to the square root of the head, i.e., to $\sqrt{\xi_1 - 1}$. In contrast to the earlier case, the stress distribution is hydrostatic, except close to the exit.

For many materials, $k > 1$ ($\phi > 19.5°$), as shown in Table 3.2. Thus the flow rate of coarse materials will usually be independent of the head, provided the hopper is deep. The limiting value of $\phi = 19.5°$ is based on the assumption of smooth walls; inclusion of wall roughness leads to a smaller limiting value.

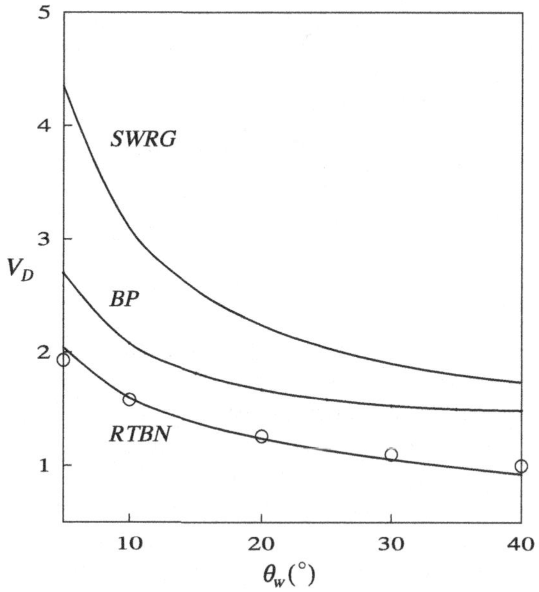

Figure 3.17. Variation of the dimensionless mass flow rate V_D with the wall angle θ_w for a wedge-shaped hopper: $SWRG$, the smooth wall, radial gravity solution (3.31) for a deep hopper; BP, the Brennen–Pearce solution (3.56); $RTBN$, the Rose–Tanaka–Beverloo–Nedderman correlation (3.6); ∘, data of Kaza (1982) for glass beads ($d_p = 0.85$–1.25 mm, $B = 127$ mm, $D = 25.5$–26.5 mm, $\rho = 1630$ kg m^{-3}, $\phi = 32.4°$, $\delta = 15.1°$ (aluminium walls)). Here d_p is the particle diameter, B and D are the dimensions of the exit slot (see Fig. 3.1), ρ is the density of the granular material, ϕ is the angle of internal friction, and δ is the angle of wall friction (see (3.34)).

3.4. THE EFFECT OF WALL ROUGHNESS

When the hopper wall is rough, it exerts a nonzero shear stress on the material flowing past it. The vertical component of this stress opposes gravity, and hence is likely to cause a reduction in the predicted flow rate. Wall roughness is usually incorporated into the model by using the friction boundary condition (2.6). Thus it is assumed that

$$T/N = \text{constant} \equiv \tan\delta \qquad (3.34)$$

where T and N are the shear and normal stresses, respectively, exerted by the wall on the material (Fig. 3.18), and δ is the angle of wall friction.

Equation (3.34) is widely used, but does not have a sound basis. Experiments described in the next chapter show that the wall stresses at some locations tend to oscillate in time, and

Table 3.2. Angles of Internal Friction (ϕ_* or ϕ) for Various Materials*

Material	d_p (mm)	ϕ_* or ϕ (∘)	Reference
Glass beads	0.27	18.2	Brennen and Pearce (1978)
"	0.59	24.3	"
"	1.33	26.8	"
Mustard seeds	2.07	38.2	"
Sand	0.32	30.7	"
"	0.68	30.6	"
"	0.5–2.0	38–40	Benink (1989, p. 19)
Plastic grains	2–3	35–38	"
Soya beans	2–7	34–39	"
Sugar		30–40	Nedderman (1992, p. 25)
Alumina		27–44	"

* The values cited by Brennen and Pearce (1978) and Nedderman (1992) correspond to the angle ϕ_* which occurs in the Mohr–Coulomb yield condition (2.37), and those cited by Benink (1989) correspond to the angle ϕ which occurs in the equation (2.47) for the critical state line.

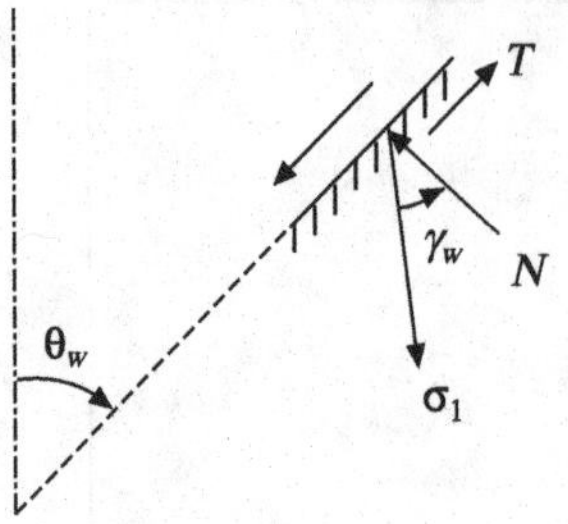

Figure 3.18. Shear stress $T = -\sigma_{\theta r}$ and normal stress $N = \sigma_{\theta\theta}$ exerted by the wall on the material adjacent to it and flowing downward.

hence δ also oscillates. For the analysis of steady state problems, the mean value of δ can be used, provided the amplitude of oscillations is small. More extensive data are needed before a realistic friction boundary condition can be formulated. Another complication should also be noted. Using time-averaged values of T and N, Tüzün et al. (1988) computed δ at various locations along a bin wall. For glass ballotini, δ varied considerably with the time averaged normal stress $\langle N \rangle$ for low values of $\langle N \rangle$ (Fig. 3.19). On the other hand, there was no systematic variation with $\langle N \rangle$ for mustard seeds. Further, for a given value of $\langle N \rangle$, there was considerable scatter in the value of δ. To summarize, δ varies in a complicated manner with time, the normal stress, and other variables. At present, there is no satisfactory model; however, the model proposed by Tüzün et al. (1988) is worth pursuing further.

An alternative friction boundary condition is provided by the rate- and state variable class of models (Dieterich, 1979; Ruina, 1983; Gu et al., 1984; Scholz, 1998), which have been widely used in the geophysics literature. These models permit the frictional shear stress to depend on time and velocity, and hold promise for predicting the observed temporal oscillations of the wall stresses. However, differential equations describing the evolution of one or more "state variables" must be solved along with the mass and momentum balances. This complicates the analysis considerably, and the physical interpretation of the state variables is not clear. These models have not yet been applied to problems of granular flow, except in special cases (Lacombe et al., 2000; Ovarlez and Clément, 2003; Dhoriyani et al., 2006).

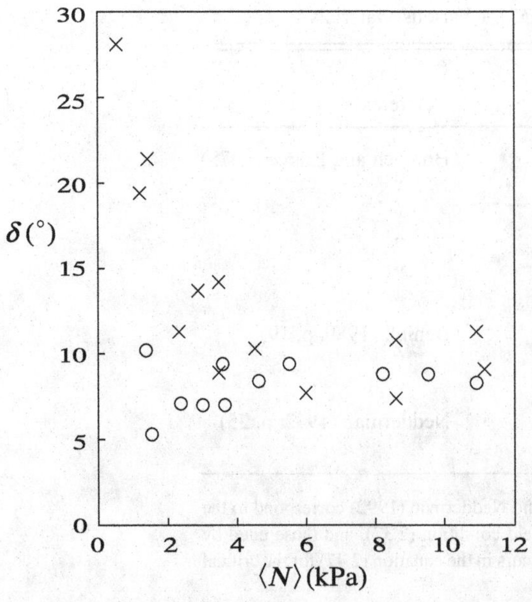

Figure 3.19. Variation of the angle of wall friction δ with the time-averaged normal stress $\langle N \rangle$ exerted on the wall of a bin of rectangular cross section: $\circ$, data for mustard seeds of diameter 2.1–2.4 mm; $\times$, data for glass beads of diameter 1.1–1.3 mm. The bin had glass faces and aluminum walls. (Adapted from Fig. 8 of Tüzün et al., 1988.)

Figure 3.20. The active root γ_{wa} and the passive root γ_{wp} obtained using the wall friction condition (3.35). Here ϕ and δ are the angles of internal friction and wall friction, respectively. The semicircle is drawn tangential to DE.

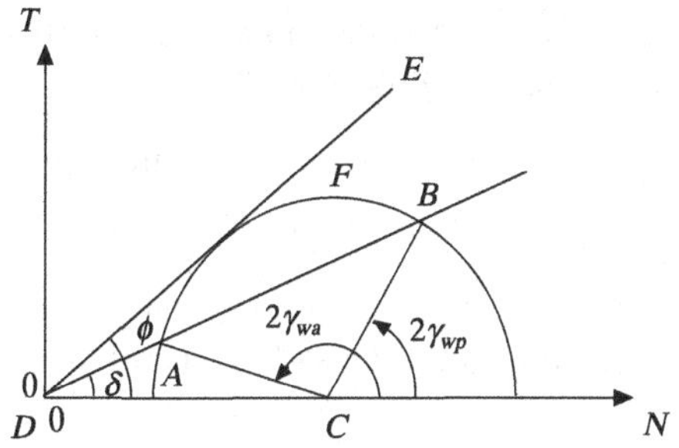

For want of a better alternative, (3.34) will be used here. Equations (3.34) (with $T = -\sigma_{\theta r}$ and $N = \sigma_{\theta\theta}$) and (3.15) imply that

$$\frac{\tau \sin 2\gamma_w}{\sigma + \tau \cos 2\gamma_w} = \tan\delta$$

or, using (3.10)

$$(\sin\phi \sin 2\gamma_w)/(1 + \sin\phi \cos 2\gamma_w) = \tan\delta \tag{3.35}$$

where $\gamma_w = \gamma'(r, \theta_w)$ is the value of γ' at the wall. Equation (3.35) has two roots γ_{wp} and γ_{wa}, as shown in Fig. 3.20. Rewriting $\tan\delta$ as $\sin\delta/\cos\delta$, (3.35) implies that

$$\gamma_{wp} = (\delta + \omega)/2; \quad \gamma_{wa} = (\pi + \delta - \omega)/2 \tag{3.36}$$

where $\sin\omega \equiv \sin\delta/\sin\phi$.

Suppose that the line $T = N \tan\delta$ passes through the point F (Fig. 3.20), which is the highest point on the semicircle. Then (3.10) implies that $T_F = \tau = \sigma \sin\phi = N_F \sin\phi$, or $\tan\delta = \sin\phi$. Hence if $\tan\delta \le \sin\phi$, we have $2\gamma_{wp} \le \pi/2$, $2\gamma_{wa} \ge \pi/2$ (see Fig. 3.20). Using the coaxiality condition, it follows from Fig. 3.21 that $C_{rr} = -\partial v_r/\partial r \le 0$ for $\gamma_w = \gamma_{wp}$ and $C_{rr} \ge 0$ for $\gamma_w = \gamma_{wa}$. As C_{rr} is expected to be ≤ 0 for downward flow, the *passive* root $\gamma_w = \gamma_{wp}$ is appropriate. This is consistent with the choice $\gamma' = 0$ at the centerline (see §3.3).

If $\sin\phi < \tan\delta < \tan\phi$, both γ_{wa} and γ_{wp} imply that C_{rr} is > 0, which is unrealistic for downward flow. However, it appears that $\tan\delta \le \sin\phi$ even for a "fully rough" wall, i.e., a wall covered with a layer of the granular material (Jenike, 1961; Nedderman, 1992, pp. 161–162) (see also Problem 2.16). Hence the case $\tan\delta > \sin\phi$ need not be considered.

Figure 3.21. Mohr's circle for the rate of deformation tensor evaluated at the wall. The circle has been drawn for the case of incompressible flow. Here C and S are the rates of compression and shearing, respectively, on a plane whose normal is inclined at an angle $\gamma' = \gamma_w$ measured anticlockwise from the major principal compressive rate of deformation axis, as shown in Fig. 2.40. In Fig. 2.40, the angle γ' is marked as γ_*.

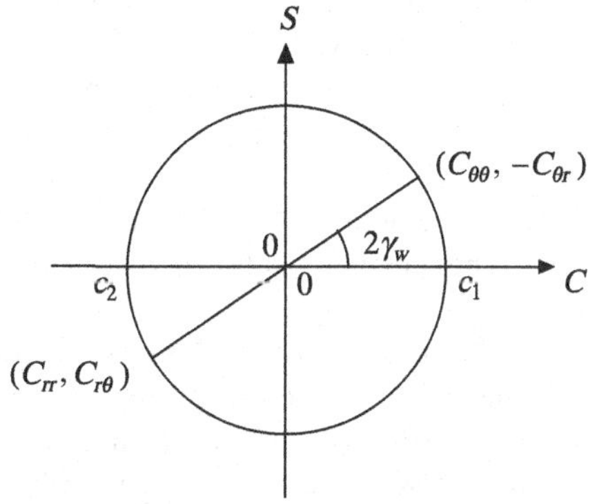

3.5. SOLUTIONS WITH ALLOWANCE FOR ROUGH WALLS AND VERTICAL GRAVITY

Using the critical state approximation, and introducing dimensionless variables

$$\overline{\sigma} = \sigma/(\rho g r_e); \quad \xi = r/r_e; \quad u = v_r/(\sqrt{g r_e}); \quad v = v_\theta/(\sqrt{g r_e}) \tag{3.37}$$

where r_e is the radial coordinate corresponding to the edge of the exit slot (Fig. 3.13), the governing equations (3.12)–(3.14) and (3.18) can be rewritten in dimensionless from as

$$\frac{1}{\xi}\frac{\partial(\xi\, u)}{\partial \xi} + \frac{1}{\xi}\frac{\partial v}{\partial \theta} = 0 \tag{3.38}$$

$$u\frac{\partial u}{\partial \xi} + \frac{v}{\xi}\frac{\partial u}{\partial \theta} - \frac{v^2}{\xi} + \frac{\partial[\overline{\sigma}(1 - \sin\phi\,\cos 2\gamma')]}{\partial \xi} - \frac{1}{\xi}\frac{\partial(\overline{\sigma}\,\sin\phi\,\sin 2\gamma')}{\partial \theta}$$
$$- \frac{2\overline{\sigma}\,\sin\phi\,\cos 2\gamma'}{\xi} + \cos\theta = 0 \tag{3.39}$$

$$u\frac{\partial v}{\partial \xi} + \frac{v}{\xi}\frac{\partial v}{\partial \theta} + \frac{u\,v}{\xi} - \frac{\partial(\overline{\sigma}\,\sin\phi\,\sin 2\gamma')}{\partial \xi} + \frac{1}{\xi}\frac{\partial[\overline{\sigma}\,(1 + \sin\phi\,\cos 2\gamma')]}{\partial \theta}$$
$$- \frac{2\overline{\sigma}\,\sin\phi\,\sin 2\gamma'}{\xi} - \sin\theta = 0 \tag{3.40}$$

$$\sin 2\gamma'\left(\frac{\partial u}{\partial \xi} - \frac{1}{\xi}\frac{\partial v}{\partial \theta} - \frac{u}{\xi}\right) - \cos 2\gamma'\left(\frac{\partial v}{\partial \xi} + \frac{1}{\xi}\frac{\partial u}{\partial \theta} - \frac{v}{\xi}\right) = 0 \tag{3.41}$$

Equations (3.38)–(3.41) constitute a system of four first-order partial differential equations, and hence four boundary conditions must be specified in the ξ and θ directions.

Considering symmetric solutions, the normal component of velocity and the shear stress must vanish along the centerline $\theta = 0$. Hence

$$v(\xi,\,\theta = 0) = 0; \quad \overline{\sigma}_{\theta r}(\xi,\,\theta = 0) = 0 \tag{3.42}$$

The second of (3.42) implies that $\gamma' = 0$ or $\gamma' = \pi/2$. As discussed in §3.4, the passive root $\gamma' = 0$ is appropriate for downward flow. Hence (3.42) can be rewritten as

$$v(\xi,\,\theta = 0) = 0; \quad \gamma'(\xi,\,\theta = 0) = 0 \tag{3.43}$$

Along the wall, the normal component of velocity must vanish. Assuming that the friction boundary condition (3.35) is valid, the wall boundary conditions are given by

$$v(\xi,\,\theta = \theta_w) = 0; \quad \gamma'(\xi,\,\theta = \theta_w) = \text{constant} = \gamma_w \tag{3.44}$$

Additional boundary conditions must be prescribed in the ξ direction. As in the case of the smooth wall, radial gravity problem, it is assumed that the material in the hopper is bounded from above and below by two traction-free surfaces T_1 and T_e (Fig. 3.22). Thus

$$\overline{\sigma} = 0 \text{ along } T_1 \text{ and } T_e \tag{3.45}$$

Two more boundary conditions are needed; these will be specified later.

The remaining subsections describe attempts to solve (3.38)–(3.41).

3.5.1. The Brennen–Pearce Solution

Brennen and Pearce (1978) constructed an approximate solution using expansions in powers of θ. If θ is replaced by $-\theta$ in (3.38)–(3.41), they are unaffected, provided γ' and v are replaced by $-\gamma'$ and $-v$, respectively, and $\overline{\sigma}$ and u are left unchanged. Hence γ' and v are

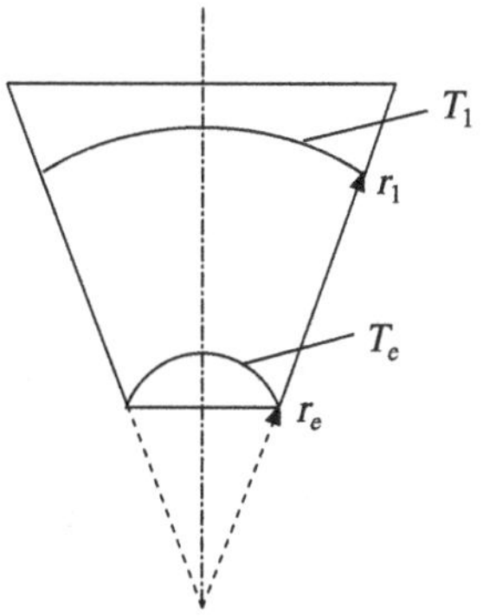

Figure 3.22. The traction-free surfaces T_1 and T_e. Here T_1 represents the upper free surface of the fill, and r_e is the radial coordinate corresponding to the edge of the exit slot.

odd functions of θ, and $\overline{\sigma}$ and u are even functions. This result is also in agreement with our intuitive expectation that the flow must be symmetric about the centerline $\theta = 0$. It is assumed that the dependent variables can be approximated by

$$\overline{\sigma} = \sigma_0(\xi) + \sigma_2(\xi)\theta^2 + \cdots; \quad \gamma' = \gamma_1(\xi)\theta + \gamma_3(\xi)\theta^3 + \cdots;$$
$$u = u_0(\xi) + u_2(\xi)\theta^2 + \cdots; \quad v = v_1(\xi)\theta + v_3(\xi)\theta^3 + \cdots \tag{3.46}$$

where the functions $\sigma_0(\xi)$, $\sigma_2(\xi)$, etc. have to be determined.

Along the centerline $\theta = 0$, the boundary conditions (3.43) are identically satisfied. Along the wall, the boundary conditions (3.44) imply that

$$v_1\,\theta_w + v_3\,\theta_w^3 + \cdots = 0; \quad \gamma_1\,\theta_w + \gamma_3\,\theta_w^3 + \cdots = \gamma_w \tag{3.47}$$

First, consider the special case where terms involving θ^3 and higher powers of θ are omitted are omitted in (3.46). Hence

$$\overline{\sigma} = \sigma_0(\xi) + \sigma_2(\xi)\theta^2; \quad \gamma' = \gamma_1(\xi)\theta$$
$$u = u_0(\xi) + u_2(\xi)\theta^2; \quad v = v_1(\xi)\theta \tag{3.48}$$

Equations (3.47) imply that

$$v_1 = 0; \quad \gamma_1 = \gamma_w/\theta_w \tag{3.49}$$

Hence the flow is radial and γ' varies linearly with θ.

A series solution can be constructed by substituting the expansions (3.46) into (3.38)–(3.41), collecting terms involving like powers of θ, and setting the coefficient of each power of θ to zero. For the finite series (3.48), the coefficients of θ^0 and θ^1 can be solved for u_0, σ_0, u_2, and σ_2, as shown below. The coefficients of θ^2 lead to a different set of equations involving the same variables. As these equations cannot be satisfied, they are discarded.

Using the above procedure, (3.38) and (3.39) imply that

$$u_0 = -A_0/\xi\,; \quad A_0 = \text{constant}$$

$$-\frac{A_0{}^2}{\xi^3} + (1 - \sin\phi)\frac{\mathrm{d}\sigma_0}{\mathrm{d}\xi} - \frac{2\,\sigma_0\,\sin\phi}{\xi}\left(1 + \frac{\gamma_w}{\theta_w}\right) + 1 = 0 \tag{3.50}$$

To derive (3.50), Taylor's series expansions for $\cos 2\gamma'$ and $\sin 2\gamma'$ about $\gamma' = 0$ have been used, and, after substituting for γ' from (3.48), terms involving θ^3 and higher powers of θ have been omitted. Equation (3.50) is identical in form to (3.23), provided the constant k in (3.23) is replaced by

$$k' \equiv \frac{2\,\sin\phi}{1 - \sin\phi}\left(1 + \frac{\gamma_w}{\theta_w}\right) \tag{3.51}$$

and $\overline{A}$ by A_0. Hence the solution is given by

$$\sigma_0 = \frac{1}{1-\sin\phi}\left(\frac{\xi}{k'-1} - \frac{{A_0}^2}{(k'+2)\,\xi^2}\right) + c_1'\,\xi^{k'} \tag{3.52}$$

where c_1' is an integration constant.

Similarly, (3.40) and (3.41) imply that

$$\begin{aligned}\sigma_2 = {} & \frac{\xi}{2\,(1+\sin\phi)}\left(1 + \frac{2\,\sin\phi\,\gamma_w\,(3\,\theta_w + 2\,\gamma_w)}{(1-\sin\phi)\,(k'-1)\,\theta_w^2}\right)\\ & - \frac{2\,\sin\phi\,\gamma_w^2\,{A_0}^2}{(1-\sin^2\phi)(k'+2)\,\theta_w^2}\,\frac{1}{\xi^2}\\ & + \frac{c_1'\,\sin\phi\,\gamma_w((2+k')\,\theta_w + 2\,\gamma_w)}{(1+\sin\phi)\,\theta_w^2}\,\xi^{k'}\\ u_2 = {} & \frac{2\,\gamma_w\,A_0}{\theta_w\,\xi}\end{aligned} \tag{3.53}$$

The integration constants A_0 and c_1' in (3.53) can be determined by using the traction-free boundary conditions (3.45) at the edge of the exit slot ($\xi = 1, \theta = \theta_w$), and at the intersection of the upper traction-free surface with the hopper wall ($\xi = \xi_1, \theta = \theta_w$). Hence we have

$$\sigma_0(\xi = 1) + \sigma_2(\xi = 1)\,\theta_w^2 = 0; \quad \sigma_0(\xi = \xi_1) + \sigma_2(\xi = \xi_1)\,\theta_w^2 = 0 \tag{3.54}$$

The two-term expansion (3.48) is adequate if the second term is small compared to the first term. This requires that $|(\sigma_2(\xi)\,\theta^2)/\sigma_0(\xi)| \ll 1$, and $|(u_2(\xi)\,\theta^2)/u_0(\xi)| \ll 1$. However, (3.54) forces the two terms to be comparable at $\xi = 1$ and at $\xi = \xi_1$. Similar problems arise at the hopper wall when two terms are included in the expansion for the circumferential velocity v. These symptoms suggest that more terms are required near the traction-free surfaces and the hopper wall.

Equations (3.52)–(3.54) imply that

$$\begin{aligned}{A_0}^2 = {} & \frac{k'+2}{k'-1}\left(\frac{1-{\xi_1}^{-(k'-1)}}{1-{\xi_1}^{-(k'+2)}}\right)\\ & \times \frac{2\,(1+\sin\phi) + (3\,\sin\phi - 1)\theta_w^2 + 4\,\sin\phi\,\gamma_w(2\,\theta_w + \gamma_w)}{2\,(1+\sin\phi) + 4\,\sin\phi\,\gamma_w^2}\end{aligned} \tag{3.55}$$

Using (3.5), the dimensionless mass flow rate is given by

$$V_D = \frac{2\,\theta_w\,A_1}{(2\,\sin\theta_w)^{3/2}} \tag{3.56}$$

where

$$A_1 \equiv \frac{1}{\theta_w}\int_0^{\theta_w} u(\xi = 1, \theta)\mathrm{d}\theta = A_0\left(1 - \frac{2\,\gamma_w\,\theta_w}{3}\right) \tag{3.57}$$

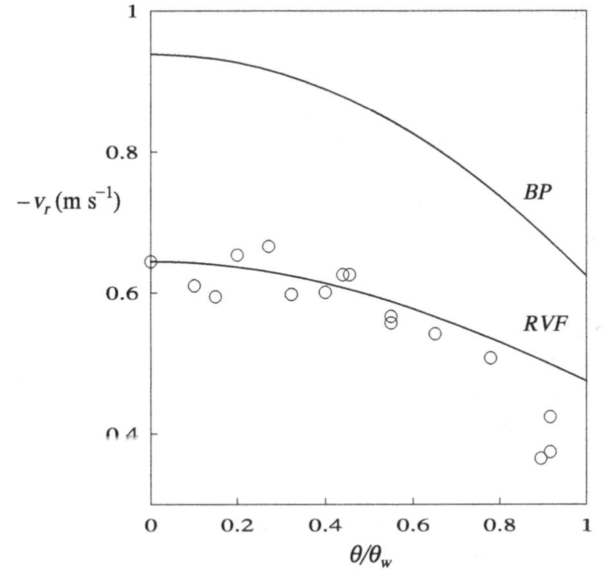

Figure 3.23. Velocity profiles along a circular arc spanning the exit slot of a wedge-shaped hopper: BP, the Brennen–Pearce solution, obtained using (3.37) and (3.48); RVF, the radial velocity field (3.63); ○, data of Bosley et al. (1969) for sand (particle diameter = 0.9 mm). Here θ is the angle measured as shown in Fig. 3.13. Parameter values: width of the exit slot = 32 mm, wall angle of the hopper $\theta_w = 20°$.

Using (3.55) and (3.57), and omitting cubic and higher order terms such as $\theta_w^2\,\gamma_w$ and $\theta_w\,\gamma_w^2$, we obtain

$$A_1{}^2 = \frac{k'+2}{k'-1}\left(\frac{1-\xi_1{}^{-(k'-1)}}{1-\xi_1{}^{-(k'+2)}}\right) \times \left(1+\frac{(9\,\sin\phi-3)\theta_w^2+(16\,\sin\phi-8)\,\theta_w\,\gamma_w}{6\,(1+\sin\phi)}\right) \tag{3.58}$$

It follows from (3.56) and (3.58) that the mass flow rate for a deep hopper with $\xi_1 \gg 1$ is independent of the head of material, provided $k' > 1$.

The above solution can be compared with the smooth wall, radial gravity (*SWRG*) solution and with data. The Brennen–Pearce solution, denoted by the curve labeled BP in Fig. 3.17, predicts a lower mass flow rate than the *SWRG* solution. This is to be expected, as the former includes the effect of wall roughness. The predictions of the Brennen–Pearce solution differ from the measured values by 30–50% of the latter for wall angles in the range 5–40°. The curve labeled $RTBN$ in Fig. 3.17 represents the Rose–Tanaka–Beverloo–Nedderman correlation (3.6), which fits the data better than the Brennen–Pearce solution. The Brennen–Pearce solution overestimates the data of Sullivan (1972) (the triangles in Fig. 3.7) by 10–20% for wall angles in the range 10–40°. In this case, the $RTBN$ correlation underestimates the mass flow rates by 10–30%. At this stage, it can be concluded that the inclusion of wall roughness reduces the mass flow rate significantly, and is a step in the right direction.

Comparison with the velocity data of Bosley et al. (1969) (Fig. 3.23) shows that the solution overestimates radial velocities considerably, but captures qualitatively the variation of the velocity in the circumferential direction. The data shown in this figure were obtained by using a cine camera to track particles moving adjacent to the front face of the hopper. The data given in Bosley et al. (1969) show that the velocity near the face is about 15% lower than that near the mid-plane of the hopper. So a part of the discrepancy between theory and experiment can be attributed to the retarding effect of the front and back faces, which has been ignored in the models for plane flow. More extensive data at various locations within the hopper are needed for a proper assessment of the theory.

The Brennen–Pearce solution represents a significant improvement over the smooth wall, radial gravity solution. Higher order solutions can in principle be constructed by retaining more terms in the expansions (3.46). However, this has not been done so far, probably because a coupled system of nonlinear ordinary differential equations is obtained. An alternative perturbation scheme for the case of steady, incompressible, radial flow is discussed in Meric and Tabarrok (1982).

3.5.2. The Radial Stress and Velocity Fields

Let us consider an alternative approach developed by Jenike (1961), which results in an approximate solution for the stress and velocity fields. This solution is obtained by omitting the inertial terms in the momentum balances, i.e., the terms involving the velocities u and v in (3.39) and (3.40). Hence it is not valid near the hopper exit, where these terms are important. Further, the velocity field is determined only to within a multiplicative constant. Thus the form of the velocity profile can be predicted, but not the mass flow rate.

To examine when the inertial terms can be omitted, consider the mass balance (3.38). This implies that the mean radial velocity is $\propto 1/\xi$. Hence the inertial terms are expected to fade as ξ increases, and can be neglected for $\xi \gg 1$. Jenike (1961) (see also Jenike, 1964a, Johanson, 1964) discovered that the inertia-free momentum balances (i.e., (3.39) and (3.40) with the terms involving velocities omitted) admit a solution of the form

$$\overline{\sigma} = \xi\, b_0(\theta); \quad \gamma' = g_0(\theta) \tag{3.59}$$

where b_0 and g_0 are solutions of the differential equations

$$-\frac{\mathrm{d}}{\mathrm{d}\theta}(b_0 \sin\phi \sin 2g_0) + b_0(1 - 3 \sin\phi \cos 2g_0) + \cos\theta = 0 \tag{3.60}$$

$$\frac{\mathrm{d}}{\mathrm{d}\theta}[b_0\,(1 + \sin\phi \cos 2g_0)] - 3\, b_0 \sin\phi \sin 2g_0 - \sin\theta = 0 \tag{3.61}$$

subject to the boundary conditions

$$g_0(0) = 0\,; \quad g_0(\theta_w) = \gamma_w \tag{3.62}$$

Equations (3.59) represent the *radial stress field*. This terminology has been used by Jenike (1961) because the stress field (3.59) is compatible with a radial velocity field.

The basis for inferring a solution of the form (3.59) has not been indicated in the literature. However, as suggested by E. B. Pitman (1999, private communication), we can attempt to solve the governing equations by separation of variables. Assuming that $\overline{\sigma} = b(\theta)\, h_1(\xi)$, and $\gamma' = g'(\theta)\, h_2(\xi)$, the boundary conditions (3.43) and (3.44) imply that h_2 is a constant. Hence $\gamma' = g_0(\theta)$, and the governing equations can be satisfied if h_1 is proportional to ξ.

Substituting (3.59) into the coaxiality condition (3.41), we find that the mass balance (3.38) and the coaxiality condition admit a solution of the form

$$u = -f_0(\theta)/\xi; \quad v = 0 \tag{3.63}$$

where

$$f_0(\theta) = f_0(0)\, \exp\left(-2 \int_0^{\theta} \tan 2g_0(\theta')\, \mathrm{d}\theta'\right) \tag{3.64}$$

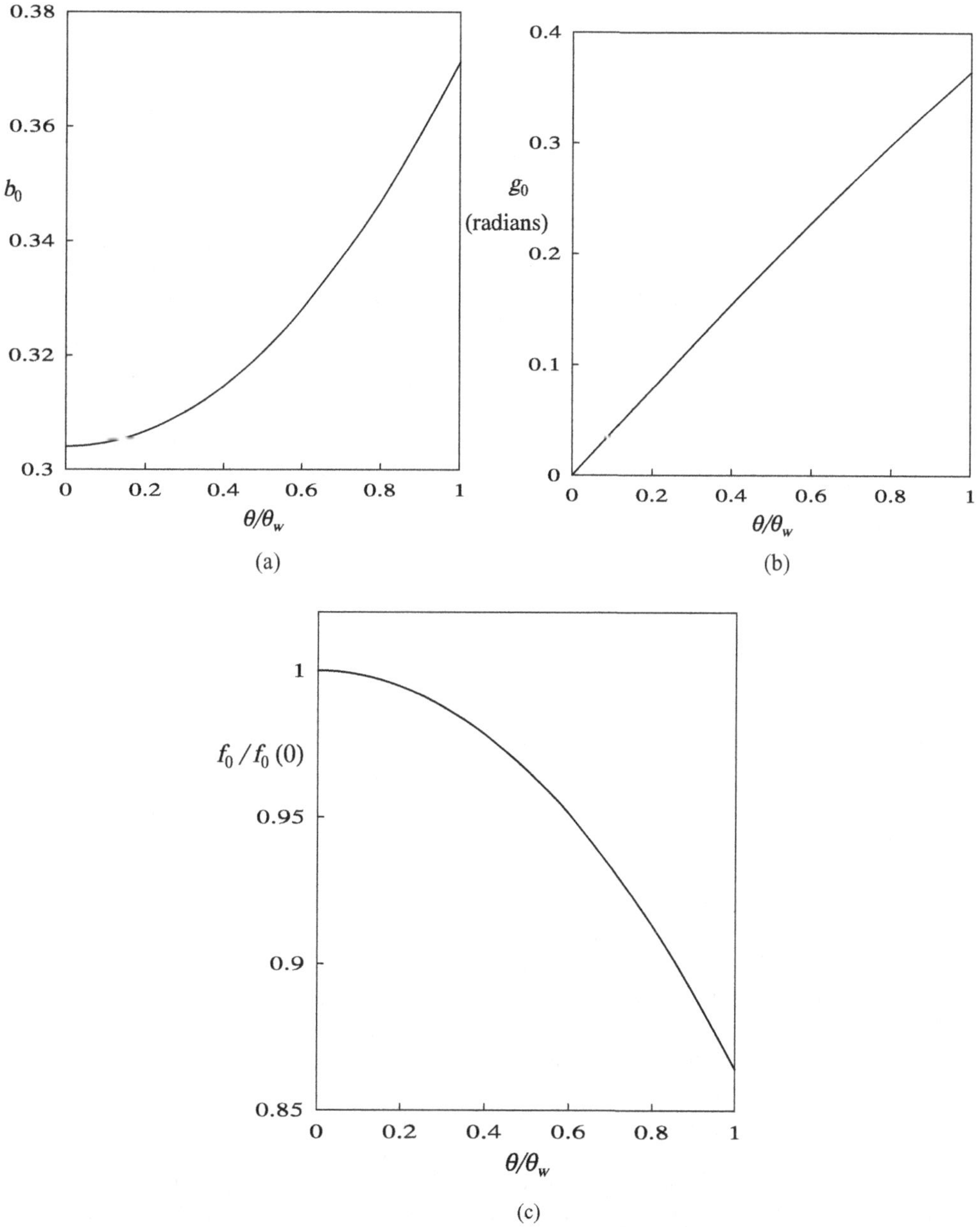

Figure 3.24. Profiles of (a) $b_0(\theta)$, (b) $g_0(\theta)$, and (c) $f_0(\theta)/f_0(0)$. The radial stress and velocity fields are given by $\overline{\sigma} = b_0(\theta)\,\xi$, $\gamma' = g_0(\theta)$, and $u = -f_0(\theta)/\xi$, where $\overline{\sigma} = \sigma/(\rho\, g\, r_e)$, $u = \sqrt{g r_e}$, $\xi = r/r_e$, ρ is the density of the granular material, g is the acceleration due to gravity, and r_e is an arbitrary length scale. Parameter values: $\phi = 35^\circ$, $\delta = 15^\circ$, $\theta_w = 10^\circ$. (Adapted from Kaza and Jackson, 1982b.)

The *radial velocity field* (3.63) is due to Jenike (1961) (see also Jenike, 1964a, Johanson, 1964). As it satisfies the boundary conditions (3.43) and (3.44), $f_0(0)$ is an indeterminate constant. Thus the magnitude of the radial velocity is not determined uniquely, but the shape of the velocity profile is known.

Figures 3.24a–3.24c show the radial stress and velocity fields for a hopper with a wall angle $\theta_w = 10^\circ$, and a material with an angle of internal friction $\phi = 35^\circ$. The value of the angle of wall friction δ is typical of glass beads sliding on an aluminum

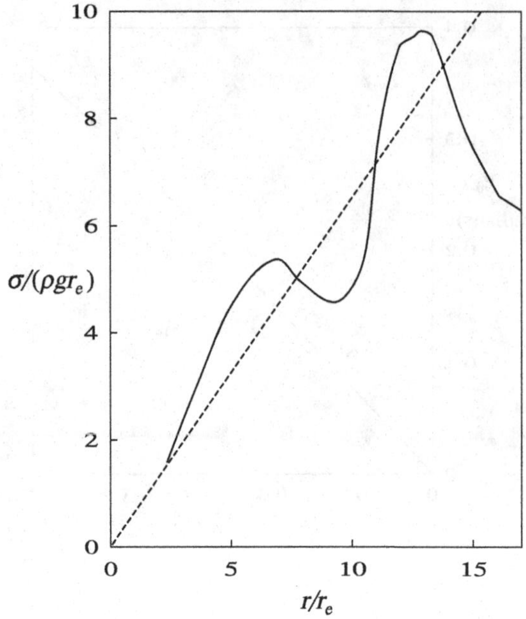

Figure 3.25. Profile of the dimensionless mean stress $\overline{\sigma} \equiv \sigma/(\rho g r_e)$ along the radial line $\theta/\theta_w = 0.88$: - - -, radial stress field; —, data of Handley and Perry (1968) for sand flowing through a wedge-shaped bunker. The normal stresses were measured using a pressure-sensitive radio pill of cylindrical shape. Here ρ is the density of the sand, g is the acceleration due to gravity, and r_e is the radial coordinate corresponding to the hopper exit. Parameter values: $\phi = 35.0°$, $\delta = 20°$, $\theta_w = 25°$, dimensions of the exit slot = 210 mm × 16 mm, $r_e = 18.9$ mm, length of the pill = 20 mm, diameter of the pill = 10 mm, $\rho = 1{,}233$ kg m^{-3}.

wall. Both the functions b_0 and g_0 increase with θ, but this is not always the case. Indeed, for certain ranges of parameter values, the g_0 profile exhibits a maximum. As expected, wall roughness causes the material to move more slowly at the wall than at the centerline.

Even though the radial fields have been widely used (Jenike, 1961; Jenike, 1964a; Johanson, 1964; Kaza and Jackson, 1982b; Michalowski, 1987; Cleaver and Nedderman, 1993b), there have been very few attempts to compare them with data. Consider the experiments of Handley and Perry (1968), who measured the normal stress exerted on the diaphragm of radio pill. The pill was embedded in the granular material and allowed to flow along with it. Experiments were conducted with two orientations of the diaphragm, namely, horizontal and vertical. The average of the normal stresses recorded with these orientations was taken as the mean stress. It was assumed that the pill did not rotate as it moved, but this was not verified. Figure 3.25 shows that the values of $\overline{\sigma}$ given by the radial stress field are comparable to the measured values, but the oscillatory behavior is not captured. The radial velocity field agrees quite well with the data of Bosley et al. (1969) for a wedge-shaped hopper (see the curve labeled RVF in Fig. 3.23), except close to the hopper wall. For a conical hopper, Cleaver and Nedderman (1993b) found that the measured velocity profile was close to that predicted by the radial velocity field. Further, the shape of the velocity profile was independent of the flow rate, in accord with the predictions.

When the inertial terms are omitted in the momentum balances (3.39) and (3.40), it has been observed that steady state solutions of the incompressible frictional equations (3.38)–(3.41) tend to converge to the radial fields as ξ decreases (Jenike, 1964a; Johanson, 1964; Pitman, 1986, 1988). Pitman (1988) used linearized stability analysis to examine the stability of the incompressible radial fields to time-independent perturbations. The governing equations considered were the inertia-free momentum balances, the continuity equation, and the coaxiality condition. The procedure used and the main results obtained are described briefly below.

3.5.3. Linearized Stability Analysis

We look for solutions for the form

$$\overline{\sigma} = \xi\, b_0(\theta) + \overline{\sigma}_1(\xi, \theta); \quad \gamma' = g_0(\theta) + \gamma_1'(\xi, \theta)$$
$$u = -f_0(\theta)/\xi + u_1(\xi, \theta); \quad v = 0 + v_1(\xi, \theta) \tag{3.65}$$

where functions with subscript 1, such as $\overline{\sigma}_1$, represent perturbations to the radial fields. Suppose that nonzero values of the perturbations are prescribed along some curve spanning the hopper, and the governing equations are integrated downward. Intuitively, we can think of θ as a spatial dimension and ξ as the "time" dimension. If the perturbations decay relative to the radial fields as ξ decreases, the latter are said to be downward stable. Conversely, if the perturbations grow, the fields are downward unstable. Many of the available numerical solutions suggest downward stability for certain ranges of parameter values. In particular, downward stability requires that

$$\overline{\sigma}_1/(\xi b_0) \to 0; \quad \gamma_1'/g_0 \to 0; \quad u_1/(f_0/\xi) \to 0; \quad v_1 \to 0, \qquad \text{as } \xi \to 0 \tag{3.66}$$

As the equations for the perturbations are nonlinear, they are linearized about the radial fields, so that a semianalytical solution can be constructed. It is tacitly assumed that the effect of the higher order terms, which are nonlinear in the perturbations, is negligible provided small perturbations are considered. Additional information regarding linearized stability analysis is given in Appendix I.

Introducing a new variable $t' = \ln \xi$, the linearized equations are found to have coefficients which depend only on θ. Hence they admit solutions of the form

$$\overline{\sigma}_1 = \exp(\lambda\, t')\, b_1(\theta) \tag{3.67}$$

where λ is a constant whose value has to be determined. Substituting (3.67) into the linearized equations, and using a suitable difference scheme (Pitman, 1988), a system of linear simultaneous algebraic equations is obtained. These equations admit nontrivial solutions only for certain special values of λ, which are called the *eigenvalues*. For the special case of a smooth-walled hopper, approximate analytical expressions can be constructed for some of the eigenvalues (Problem 3.5).

The inertia-free momentum balances are independent of the velocity field and hence can be solved first to obtain the "stress" eigenvalues λ_s. Let $Re(\lambda_s)$ denote the real part of λ_s. Then, for $Re(\lambda_s) > 1$, the radial field for $\overline{\sigma}$ is downward stable as $\overline{\sigma}_1/(\xi\, b_0) \propto \xi^{\lambda-1} \to 0$ as $\xi \to 0$. On the other hand, for $Re(\lambda_s) > 0$, $\gamma_1'/g_0 \to 0$ as $\xi \to 0$. Thus the radial stress field is downward stable for $Re(\lambda_s) > 1$.

For a specified value of the wall angle θ_w, the eigenvalues depend on the angle of internal friction ϕ and the angle of wall friction δ. In the δ–ϕ plane, the curve corresponding to $Re(\lambda_s) = 1$ is called the *neutral stability* curve. For $\theta_w = 10°$, this is shown by the solid curve in Fig. 3.26. At points such as S, which are on the right of this curve, the analysis predicts that perturbations decay relative to the radial stress field as ξ decreases. Hence the field is downward stable for the parameter values corresponding to such points. Conversely, for parameter values corresponding to points such as U, which are on the left of the neutral stability curve, the field is downward unstable. The significance of the broken curve AB in Fig. 3.26 is explained below.

The t' derivatives of $\overline{\sigma}_1$ and γ_1' in the perturbation equations are multiplied by a factor $(\sin\phi - \cos 2g_0)$. Hence the eigenvalues cannot be determined if

$$\sin\phi = \cos 2g_0 \tag{3.68}$$

For a specified value of ϕ, (3.68) and (3.35) imply a certain value for δ. The curve AB in Fig. 3.26 denotes the locus of these (ϕ, δ) values. Considering the passive case, γ_w

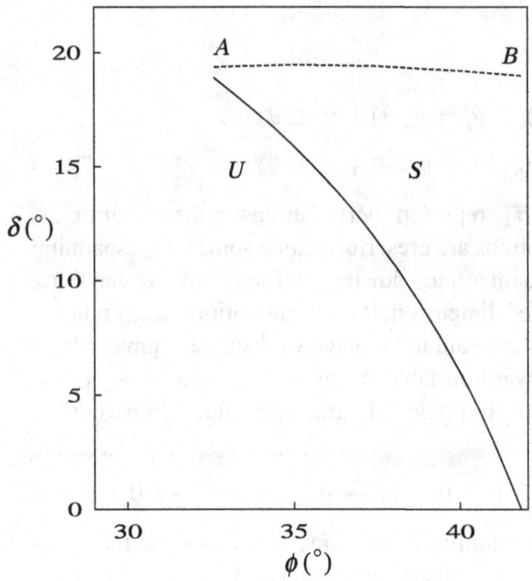

Figure 3.26. The neutral stability curve (—) for the radial stress field: S, downward stable; U, downward unstable. The curve has been generated by Pitman (1988) using a linearized stability analysis. The method used is not applicable in the region above the broken curve AB. Parameter value: $\theta_w = 10°$.

decreases as δ decreases, implying that $\sin\phi - \cos 2g_0 < 0$ in the region below AB. In particular, this inequality holds along the centerline, as $g_0(0) = 0$. Hence solutions can be constructed using the present approach only in the region below AB.

Figure 3.27 shows the results obtained by numerical integration of the governing equations, without linearization. For parameter values corresponding to the stable region of the ϕ–δ plane (for example, the point S in Fig. 3.26), the profile of the mean stress $\overline{\sigma}$ approaches the radial field as ξ decreases (Fig. 3.27a). For parameters values corresponding to the unstable region (for example, the point U in Fig. 3.26), the stress field shows sustained oscillations about the radial field (Fig. 3.27b). It is not clear from the results presented in Fig. 3.27b whether the stress field diverges from the radial field at smaller values of ξ. As ξ decreases, linearized stability analysis predicts that the perturbations become unbounded in the unstable case. However, the analysis is not valid once they become large enough for nonlinear effects to be important.

Consider the velocity field. The linearized analysis predicts that the radial velocity field is always downward stable. Figure 3.28 shows the results obtained by numerical integration of the mass balance and the coaxiality condition. For parameters corresponding to a stable stress field, the velocity field approaches the radial field as ξ decreases (Fig. 3.28a). A similar behavior is observed for parameters corresponding to an unstable stress field (Fig. 3.28b), but it is not certain whether the velocity field eventually converges to the radial field.

Pitman (1988) omitted a term arising from the perturbation to $g_0(\theta)$ (see (3.65)) in the coaxiality condition. This was because he treated γ' as a known function of position as far as the coaxiality condition was concerned, and examined the effect of perturbations to the radial velocity field (E. B. Pitman, private communication 1999).

Inclusion of perturbations to the radial field value of γ' in the coaxiality condition leads to a nonhomogeneous system of equations for the velocity perturbations. If $Re(\lambda_s) > 0$, the nonhomogeneous term decays as ξ decreases, and the stability of the velocity field is effectively governed by a homogeneous system of equations. Hence Pitman's results are applicable. On the other hand, if $Re(\lambda_s) < 0$, the nonhomogeneous term grows as ξ decreases, and may have a destabilizing effect on the velocity field. This issue merits further investigation.

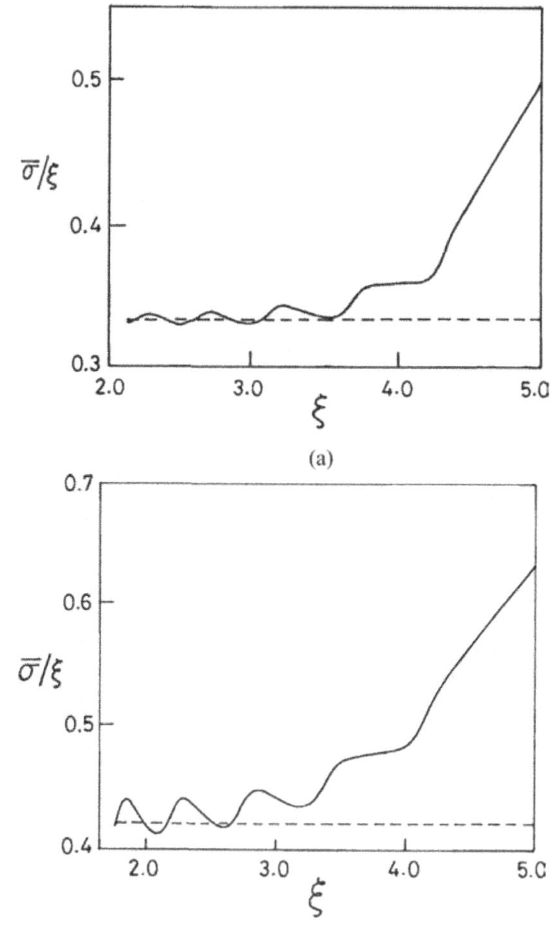

Figure 3.27. Profile of the dimensionless mean stress $\overline{\sigma}/\xi = \sigma/(\rho\, g\, r_e\, \xi)$ along the radial line $\theta = \theta_w = 10°$: —, numerical solution of the inertia-free momentum balances; - - -, radial stress field. Here ρ is the density of the granular material, g is the acceleration due to gravity, $\xi = r/r_e$, and r_e is an arbitrary length scale. Parameter values: (a) $\phi = 40°$, $\delta = 15°$, (b) $\phi = 30°$, $\delta = 15°$. (Reproduced from Pitman, 1988, with permission from the Society of Industrial and Applied Mathematics.)

3.5.4. Downward Integration from the Radial Fields

The governing equations (3.38)–(3.41) are *hyperbolic* (see Appendix C). Hence an initial-boundary value problem can be formulated by specifying "initial conditions" along some noncharacteristic curve Γ which spans the hopper, and boundary conditions (3.43) and (3.44) along the centerline and the wall. (As discussed in Appendix C, a *characteristic* is a curve along which the values of all the dependent variables such as γ' cannot be prescribed arbitrarily. Further, derivatives of these variables in the direction normal to the characteristic cannot be solved for uniquely.) Using the radial fields (3.59) and (3.63) as initial conditions specified along a circular arc, Kaza and Jackson (1982b) integrated the equations downward. The constant $f_0(0)$ in the radial velocity field was chosen as unity. The first of (3.28), and (3.63), (3.64), and the condition $f_0(0) = 1$ imply that the mass flow rate is fixed by the initial conditions. Hence an exit condition must be used to determine the size of the exit slot which corresponds to the specified flow rate. In this context, the constant r_e in (3.37) differs from r_{*e}, the actual value of the radial coordinate corresponding to the hopper exit, and is merely an arbitrary length scale. As the value of r_{*e} is not known a priori in this approach, it cannot be used as the characteristic length scale, but will be determined as a part of the solution.

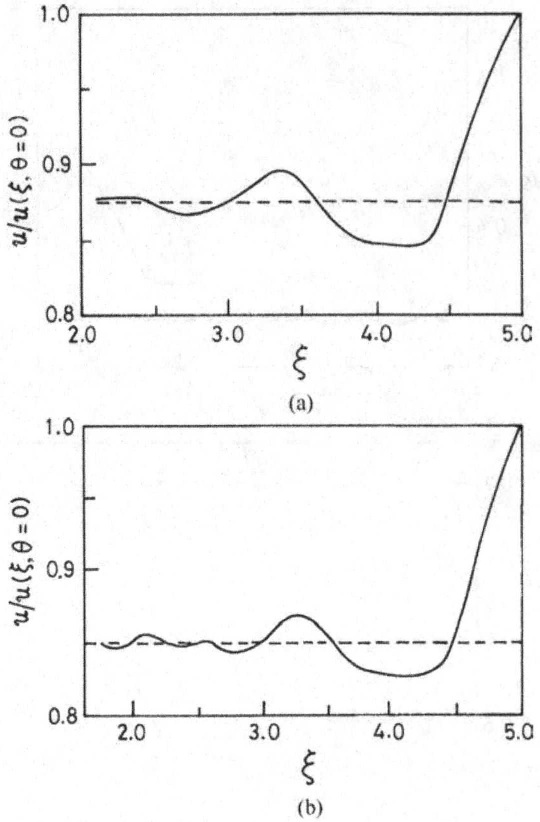

Figure 3.28. Profile of the dimensionless radial velocity $u/u(\xi, \theta = 0)$ along the radial line $\theta = \theta_w = 10°$: —, numerical solution of the mass balance and the coaxiality condition; - - -, radial velocity field. Here $\xi = r/r_e$, $u = v_r/\sqrt{g r_e}$, g is the acceleration due to gravity, and r_e is an arbitrary length scale. Parameter values: (a) $\phi = 40°$, $\delta = 15°$, (b) $\phi = 30°$, $\delta = 15°$. (Reproduced from Pitman, 1988, with permission from the Society of Industrial and Applied Mathematics.)

As discussed in §3.3, it can be assumed that the hopper is bounded from below by a traction-free surface which intersects the walls at $\xi = \xi_{*e} \equiv r_{*e}/r_e$. Hence

$$\overline{\sigma}(\xi = \xi_{*e}, \theta = \theta_w) = 0 \tag{3.69}$$

In principle, the equations can be integrated downward until (3.69) holds, and this condition determines the value of ξ_{*e}. The solid curves in Figs. 3.29 and 3.30 show the results of such an attempt.

Inertial terms cause the mean stress to deviate from the radial field as ξ decreases (Fig. 3.29a). The γ' profile along the radial line $\theta/\theta_w = 0.5$ conforms closely to the radial field value $\gamma' = 0.19$ radians for $\xi > 1.6$ (Fig. 3.29b). For $\xi < 1.6$, oscillations of growing amplitude set in, and it is difficult to construct the solution by numerical integration for $\xi < 1$. Similar oscillations are also observed in the profile of the circumferential velocity (Fig. 3.30b), but not in the profiles of the mean stress and the radial velocity (Figs. 3.29a and 3.30a).

The structure of the momentum balances provides a possible reason for this behavior (Kaza and Jackson, 1982b). Equations (3.39) and (3.40) show that all the derivatives of γ' are multiplied by $\overline{\sigma}$. Hence it becomes difficult to solve for the derivatives of γ' in the limit $\overline{\sigma} \to 0$. From an alternative viewpoint, it may be noted that any curve along which $\overline{\sigma}$ vanishes is a characteristic, as shown in Appendix C. Thus the traction-free curve, if any, is a characteristic, and downward integration from arbitrary initial fields is likely to face problems as this curve is approached. It is not known whether there is any solution which satisfies reasonable initial conditions and remains bounded as $\overline{\sigma} \to 0$.

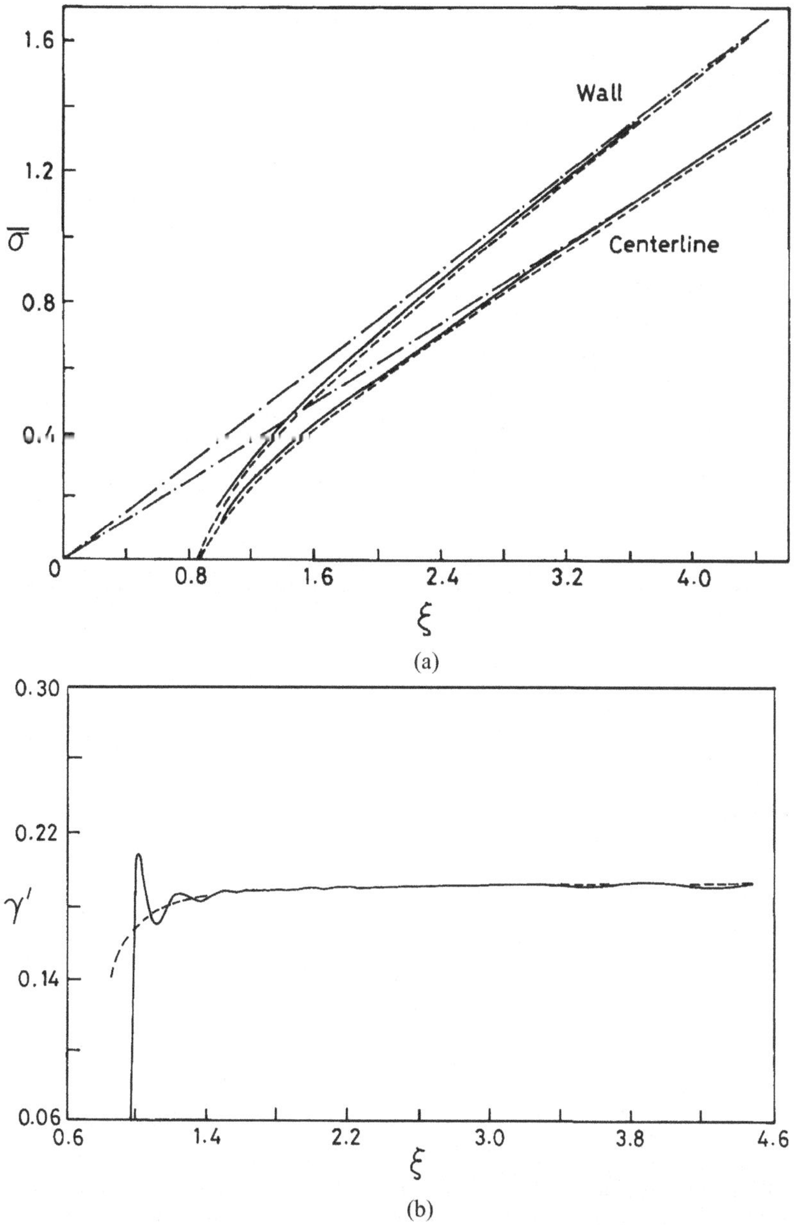

Figure 3.29. Profile of (a) the dimensionless mean stress $\overline{\sigma} \equiv \sigma/(\rho\, g\, r_e)$ along the centerline and the wall, and (b) the orientation γ' (in radians) of the major principal stress axis along a radial line at $\theta/\theta_w = 0.5$, for steady flow through a wedge-shaped hopper: —, numerical solution by downward integration from the circular arc $\xi = r/r_e = 4.5$; - - -, power series solution; — · —, radial stress field. The "initial" conditions are given by (3.59) and (3.63), with $f_0(0) = 1$. Here ρ is the density, g is the acceleration due to gravity, and r_e is an arbitrary length scale. Parameter values: $\phi = 35°$, $\delta = 15°$, $\theta_w = 10°$. (Reprinted from Kaza and Jackson, 1982b, with permission from Elsevier Science.)

3.5.5. The Successive Approximation Procedure

As discussed in the previous section, the numerical integration procedure breaks down when $\overline{\sigma} \to 0$. To overcome this problem, Kaza and Jackson (1982b) constructed an approximate solution to (3.38)–(3.41) as follows. As shown in Table 3.3, the equations are split into two

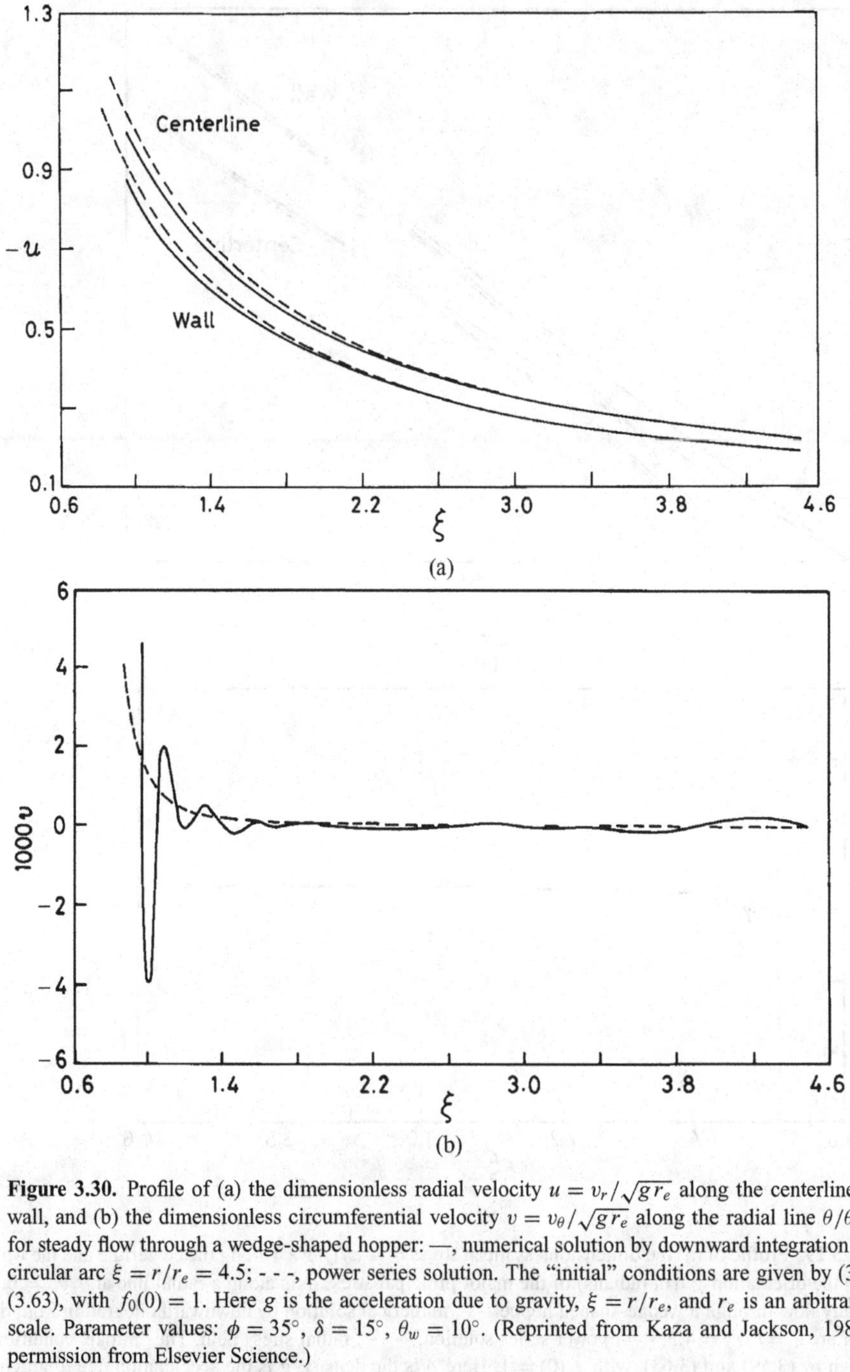

Figure 3.30. Profile of (a) the dimensionless radial velocity $u = v_r/\sqrt{g r_e}$ along the centerline and the wall, and (b) the dimensionless circumferential velocity $v = v_\theta/\sqrt{g r_e}$ along the radial line $\theta/\theta_w = 0.5$, for steady flow through a wedge-shaped hopper: —, numerical solution by downward integration from the circular arc $\xi = r/r_e = 4.5$; - - -, power series solution. The "initial" conditions are given by (3.59) and (3.63), with $f_0(0) = 1$. Here g is the acceleration due to gravity, $\xi = r/r_e$, and r_e is an arbitrary length scale. Parameter values: $\phi = 35°$, $\delta = 15°$, $\theta_w = 10°$. (Reprinted from Kaza and Jackson, 1982b, with permission from Elsevier Science.)

groups, one comprising the momemtum balances, and the other the coaxiality condition and the mass balance. The groups are solved alternately, starting with the momentum balances. When the inertial terms are omitted, the momentum balances admit the radial stress field as a particular solution. The radial field for γ' is then substituted in the coaxiality condition and the second group of equations is solved to obtain the radial velocity field. As discussed in §3.5.4, we set $f_0(0) = 1$ and use (3.69) to determine the size of the exit slot.

Table 3.3. The Successive Approximation Procedure*

Momentum balances		Coaxiality condition and mass balance
$\overline{\sigma}(\xi,\theta) = \xi\, b_0(\theta)$ $\gamma'(\xi,\theta) = g_0(\theta)$		
	↘	
		$u = -f_0(\theta)/\xi$ $v = 0$
	↙	
$\overline{\sigma}(\xi,\theta) = \xi\, b_0(\theta) + b_1(\theta)/\xi^2$ $\gamma'(\xi,\theta) = g_0(\theta) + g_1(\theta)/\xi^3$		
	↘	
		$u = -f_0(\theta)/\xi + f_1(\theta)/\xi^4$ $v = 0 + h_1(\theta)/\xi^4$

* Here $\xi = r/r_e$, $\overline{\sigma} = \sigma/(\rho g r_e)$, $u = v_r/\sqrt{g r_e}$, $v = v_\theta/\sqrt{g r_e}$, ρ is the density, g is the acceleration due to gravity, and r_e is an arbitrary length scale.

The radial fields constitute the zeroth approximation to the solution. To construct the first correction to the stress field, the radial velocity field is substituted in the momentum balances, and it is assumed that

$$\overline{\sigma}(\xi,\theta) = \xi\, b_0(\theta) + \overline{\sigma}_1(\xi,\theta); \quad \gamma'(\xi,\theta) = g_0(\theta) + \gamma_1'(\xi,\theta) \tag{3.70}$$

where $\overline{\sigma}_1$ and γ_1' are small corrections to the radial fields. The momentum balances admit a separable solution of the form

$$\overline{\sigma}_1 = b_1(\theta)/\xi^2; \quad \gamma_1' = g_1(\theta)/\xi^3 \tag{3.71}$$

when they are linearized in $\overline{\sigma}_1$ and γ_1'. The functions b_1 and g_1 are determined by integrating ordinary differential equations in θ.

To construct the first correction to the velocity field, (3.70) is substituted for γ' in (3.41), and it is assumed that

$$u = -(f_0(\theta)/\xi) + u_1(\xi,\theta); \quad v = 0 + v_1(\xi,\theta) \tag{3.72}$$

where u_1 is a small correction to the radial velocity $-f_0/\xi$. Equations (3.38) and (3.41) then admit a solution of the form

$$u_1 = f_1(\theta)/\xi^4; \quad v_1 = h_1(\theta)/\xi^4 \tag{3.73}$$

when they are linearized in u_1 and v_1.

The boundary condition (3.69) and the expansion (3.70) imply that $\xi_{*e}\, b_0(\theta_w) + \overline{\sigma}_1(\xi_{*e},\theta_w) = 0$. Hence the two terms are of the same order of magnitude, suggesting that (3.70) may not be valid near traction-free surfaces. As discussed earlier, the two-term Brennen–Pearce solution (see the first of (3.48)) also faces the same difficulty.

The above procedure generates a solution in inverse cubic powers of ξ, as shown in Table 3.3. This will henceforth be called the power series solution. For $\phi = 35°$, $\theta_w = 10°$, and $\delta = 15°$, four corrections terms appear to adequately represent the $\overline{\sigma}$, γ' and u profiles (Figs. 3.31 and 3.32). For the v profile (Fig. 3.32b), convergence appears to be slow, but the values of v are about a thousand times smaller than the values of the radial velocity u. Thus the streamlines are approximately radial. As gravity acts vertically downward, v is expected to be positive; this conjecture is supported by the results shown in Fig. 3.32b.

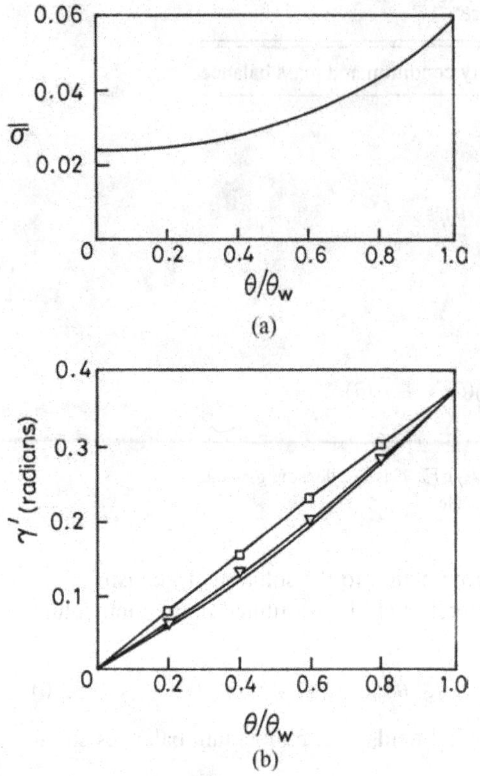

Figure 3.31. Profiles obtained using the power series solution for (a) the dimensionless mean stress $\overline{\sigma} \equiv \sigma/(\rho g r_e)$ $(N = 1)$, and (b) the orientation of the major principal stress axis along the circular arc $\xi = r/r_e = 0.9$, which is near the traction-free curve ($N = 0$ (□), 2 (∇), 4 (—)). Here N denotes the number of corrections to the radial stress and velocity fields, ρ is the density, g is the acceleration due to gravity, and r_e is an arbitrary length scale. Successive approximations to $\overline{\sigma}$ are indistinguishable for $N \geq 1$. Parameter values: $\phi = 35°$, $\delta = 15°$, $\theta_w = 10°$. (Reprinted from Kaza and Jackson, 1982b, with permission from Elsevier Science.)

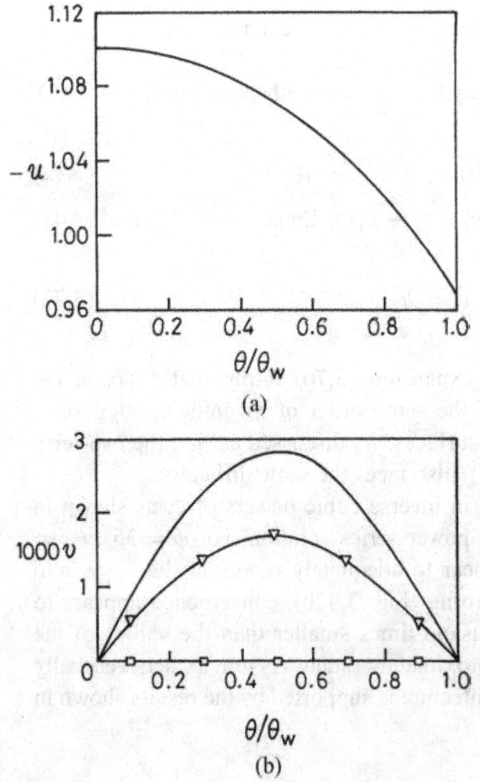

Figure 3.32. Profile obtained using the power series solution for (a) the dimensionless radial velocity $u = v_r/\sqrt{g r_e}$ $(N = 1)$, and (b) the dimensionless circumferential velocity $v = v_\theta/\sqrt{g r_e}$ along the circular arc $\xi = r/r_e = 0.9$, that is near the traction-free curve ($N = 0$ (□), 2 (∇), 4 (—)). Here N denotes the number of corrections to the radial stress and velocity fields, ρ is the density, g is the acceleration due to gravity, and r_e is an arbitrary length scale. Successive approximations to u are indistinguishable for $N \geq 1$. Parameter values: $\phi = 35°$, $\delta = 15°$, $\theta_w = 10°$. (Reprinted from Kaza and Jackson, 1982b, with permission from Elsevier Science.)

The broken curves in Figs. 3.29 and 3.30 show that the power series solution agrees fairly well with the numerical results, except in the region where $\overline{\sigma} \to 0$. It is instructive to compare this solution with the Brennen–Pearce solution. For this purpose, residuals or absolute values of the left-hand sides of (3.38)–(3.41) can be calculated using both the approximate solutions. Consider the results for $\phi = 35°$, $\delta = 15°$, and $\theta_w = 10°$. Along the respective traction-free curves, the mean residuals for the radial component of the momentum balance (3.39) and the coaxiality condition (3.41) are found to be about 3 to 5 times smaller for the power series solution than for the Brennen–Pearce solution (Kaza and Jackson, 1982b). On the other hand, both solutions give comparable residuals for circumferential component of the momentum balance (3.40). The mass balance is identically satisfied by the Brennen-Pearce solution, leading to a vanishing residual; for the power series solution, the residual is ≈ 0.001. Overall, these results suggest that the latter is more accurate than the former.

The dimensionless mass flow rate V_D, defined by (3.5), can be calculated as follows. The dimensionless radial coordinate ξ_{*e} corresponding to the edge of the exit slot can be found using (3.69). For large values of ξ, $u \approx -f_0(\theta)/\xi$ (see (3.72) and (3.73)). Using this result along with the first of (3.28), the mass flow rate is given by

$$\dot{M} = -2\,\rho\,B \int_0^{\theta_w} v_r\,r\,\mathrm{d}\theta = 2\,\rho\,B\,\sqrt{g\,r_e}\,r_e\,\theta_w\,\overline{f}_0$$

where the integration is performed along a circular arc, and

$$\overline{f}_0 \equiv (1/\theta_w) \int_0^{\theta_w} f_0(\theta)\,\mathrm{d}\theta$$

Hence (3.5) implies that

$$V_D = \frac{2\theta_w \overline{f}_0}{(D/r_e)^{3/2}}$$

As

$$\frac{D}{r_e} = \frac{D}{r_{*e}}\frac{r_{*e}}{r_e} = 2\,\sin\theta_w\,\xi_{*e}$$

we obtain

$$V_D = \frac{2\theta_w \overline{f}_0}{(2\xi_{*e}\,\sin\theta_w)^{3/2}} \tag{3.74}$$

The curves labeled PS and BP in Fig. 3.33 represent the dimensionless mass flow rates predicted by the power series solution and the Brennen-Pearce solution, respectively, for deep hoppers. As the power series solution does not converge for large values of θ_w, the upper curve does not extend till $\theta_w = 40°$. On the other hand, the lowest order Brennen–Pearce solution does not face this limitation.

As the power series solution is more accurate than the Brennen–Pearce solution, it is surprising that mass flow rates predicted by the latter are closer to the data. Kaza and Jackson (1982b) suggested that this might be caused by a physical shortcoming of the model, which is discussed below.

3.6. A RE-EXAMINATION OF THE EXIT CONDITION

All the solutions discussed above have been based on the assumptions that (i) the solids fraction is constant, (ii) a traction-free surface Γ spans the exit slot, and (iii) the mean stres

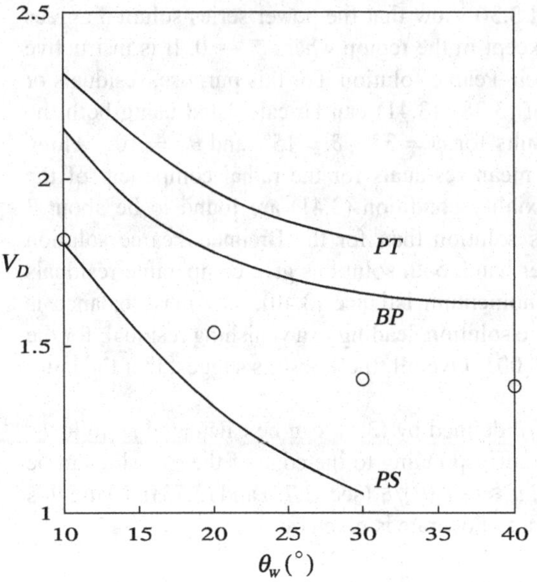

Figure 3.33. Variation of the dimensionless mass flow rate V_D with the wall angle θ_w: PT, power series solution (traction-free exit condition); PS, power series solution (exit shock); BP, Brennen–Pearce solution (traction-free exit condition); o, data of Sullivan (1972) (cited in Brennen and Pearce, 1978). Here V_D is defined by (3.5). Parameter values: $\phi = 25°$, $\delta = 17°$.

stends to zero as Γ is approached from above. As cohesionless materials cannot support tensile stresses, the mean stress must be zero at points below Γ. If the mean stress is positive below Γ, there is no satisfactory reason for terminating the hopper at a traction-free surface. Assuming that the effect of fluid drag is negligible, the material below the traction-free surface must be in free-fall.

Kaza and Jackson (1984) showed that this picture was inconsistent with the mass and momentum balances. Consider a point P on an arc AB of the traction-free curve (Fig. 3.34). Let ρ_1, $\mathbf{u}$, and $\boldsymbol{\sigma}$ represent the density, velocity vector, and stress tensor, respectively, just above AB, and ρ_2, and $\mathbf{v}$ the density and the velocity vector, respectively, just below AB. Choosing a Cartesian coordinate system with the x axis tangential to AB at P, the steady state mass balance and the y component of the momentum balance can be written as

upstream side of AB ($y = 0^-$)

$$(\mathbf{u} \cdot \nabla)\,\rho_1 = -\rho_1 \left(\frac{\partial u_x}{\partial x} + \frac{\partial u_y}{\partial y} \right) \tag{3.75}$$

$$\rho_1 \left(u_x \frac{\partial u_y}{\partial x} + u_y \frac{\partial u_y}{\partial y} \right) = \rho_1\, g\, \cos\alpha - \frac{\partial \sigma_{xy}}{\partial x} - \frac{\partial \sigma_{yy}}{\partial y} \tag{3.76}$$

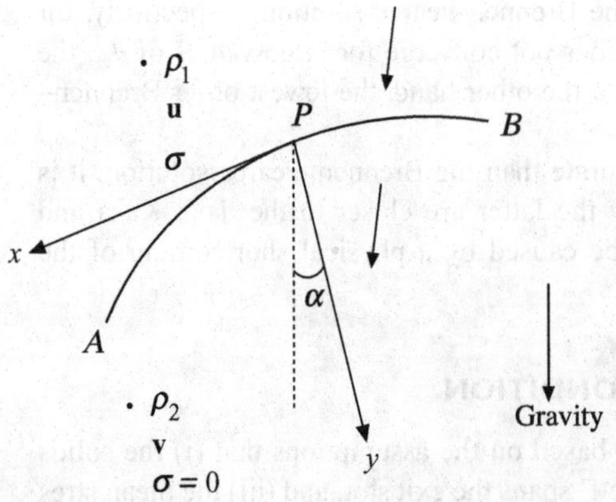

Figure 3.34. An arc AB of a traction-free curve spanning the exit slot of the hopper. The stresses vanish below AB.

where

$$\mathbf{u} \cdot \nabla = u_x \frac{\partial}{\partial x} + u_y \frac{\partial}{\partial y}$$

g is the acceleration due to gravity and α is the inclination of the y axis to the vertical (Fig. 3.34).

downstream side of AB ($y = 0^+$)

$$(\mathbf{v} \cdot \nabla)\, \rho_2 = -\rho_2 \left(\frac{\partial v_x}{\partial x} + \frac{\partial v_y}{\partial y} \right) \tag{3.77}$$

$$\rho_2 \left(v_x \frac{\partial v_y}{\partial x} + v_y \frac{\partial v_y}{\partial y} \right) = \rho_2\, g \cos\alpha \tag{3.78}$$

As there is no stress jump on crossing AB, the density and velocity must be continuous across AB. Hence

$$\rho_2 = \rho_1; \quad \mathbf{u} = \mathbf{v}; \quad \frac{\partial \mathbf{u}}{\partial x} = \frac{\partial \mathbf{v}}{\partial x}; \quad \frac{\partial \sigma}{\partial x} = 0 \tag{3.79}$$

at the point P. Subtracting (3.75) from (3.77), and using (3.79), we obtain

$$(\mathbf{v} \cdot \nabla)\, \rho_2 = (\mathbf{u} \cdot \nabla)\, \rho_1 - \rho_1 \left(\frac{\partial v_y}{\partial y} - \frac{\partial u_y}{\partial y} \right) \tag{3.80}$$

or, using the momentum balances (3.76) and (3.78), and the continuity conditions (3.79)

$$(\mathbf{u} \cdot \nabla)\, \rho_2 = (\mathbf{u} \cdot \nabla)\, \rho_1 - \frac{1}{u_y} \left(\frac{\partial \sigma_{yy}}{\partial y} \right)_{y=0^-} \tag{3.81}$$

As the density is assumed to be constant in the region above AB, (3.81) reduces to

$$[(\mathbf{u} \cdot \nabla)\, \rho_2]_{y=0^+} = -\frac{1}{u_y} \left(\frac{\partial \sigma_{yy}}{\partial y} \right)_{y=0^-} \tag{3.82}$$

The normal stress σ_{yy} tends to zero as y approaches zero from above, and hence

$$\left(\frac{\partial \sigma_{yy}}{\partial y} \right)_{y=0^-} \leq 0$$

The available approximate solutions imply that

$$\left(\frac{\partial \sigma_{yy}}{\partial y} \right)_{y=0^-} < 0 \tag{3.83}$$

and the normal component of velocity u_y must be positive for downward flow (Fig. 3.34). Hence (3.82) and (3.83) imply that $(\mathbf{u} \cdot \nabla)\, \rho_2 > 0$. For steady flow, $(\mathbf{u} \cdot \nabla)\, \rho$ represents the rate of change of density of a material element as it moves along a streamline. As this is positive on the downstream side of the traction-free surface, the material must compact. Physically, compaction occurs because the stress gradient $-\partial \sigma_{yy}/\partial y$ is positive on the upstream side and zero on the downstream side, causing the material to decelerate on crossing the traction-free surface. Compaction implies that frictional stresses must develop, in violation of the assumption of free-fall with zero stresses.

The above difficulty can be avoided if the stress gradient vanishes as the traction-free curve is approached from above. Alternatively, if density variation is permitted and the material dilates sufficiently within the hopper, (3.81) suggests that compaction just below the traction-free curve may be avoided. For the special case of the smooth wall, radial gravity problem, the numerical results of Prakash (1989) show that inclusion of density variation does not prevent compaction.

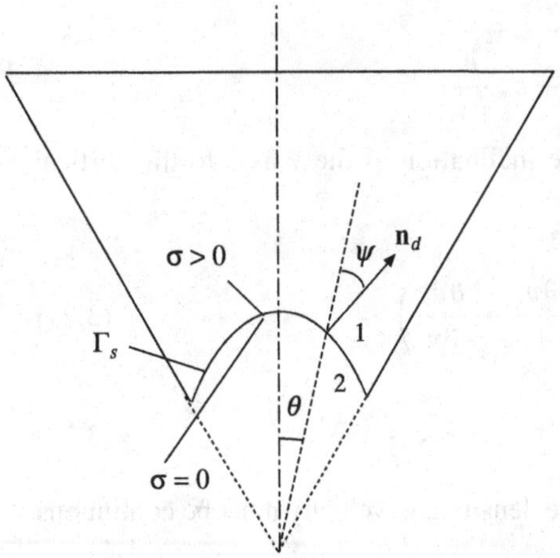

Figure 3.35. An exit shock Γ_s at the bottom of the hopper. Here σ is the mean stress, $\mathbf{n}_d$ is the unit upward normal to Γ_s, and ψ is the angle between the radial direction and $\mathbf{n}_d$.

3.7. AN ALTERNATIVE EXIT CONDITION

In view of the difficulties associated with the traction-free surface, it is natural to seek an alternative exit condition. Kaza and Jackson (1982a) assumed that the hopper terminated at a discontinuity surface or exit shock Γ_s, across which the mean stress σ jumped from a positive value on the upstream side to zero on the downstream side (Fig. 3.35). Thus the problem of integrating the equations until $\sigma = 0$ was avoided, but other problems arose, as explained shortly.

The values of the variables on either side of the shock can be related by using the *jump* mass and momentum balances (Truesdell and Toupin, 1960, pp. 525–527; Courant and Hilbert, 1962, pp. 488–490; Slattery, 1999, pp. 25, 34; see also Appendix D). Let ψ denote the angle measured clockwise from the radial direction to the upward normal to the shock (Fig. 3.35). The dimensionless jump balances are given by (see Problem 3.7)

$$J\{\nu\,(u\,\cos\psi + v\,\sin\psi)\} = 0 \tag{3.84}$$

$$J\{\cos\psi\,(\nu\,u^2 + \sigma'_{rr}) + \sin\psi\,(\nu\,u\,v + \sigma'_{r\theta})\} = 0 \tag{3.85}$$

$$J\{\cos\psi\,(\nu\,u\,v + \sigma'_{r\theta}) + \sin\psi\,(\nu\,v^2 + \sigma'_{\theta\theta})\} = 0 \tag{3.86}$$

where

$$u \equiv v_r/\sqrt{g\,r_e};\quad v \equiv v_\theta/\sqrt{g\,r_e};\quad \sigma'_{rr} \equiv \sigma_{rr}/(\rho_p\,g\,r_e)$$

$$\sigma'_{r\theta} \equiv \sigma_{r\theta}/(\rho_p\,g\,r_e);\quad \sigma'_{\theta\theta} \equiv \sigma_{\theta\theta}/(\rho_p\,g\,r_e) \tag{3.87}$$

ρ_p is the particle density, g is the acceleration due to gravity, r_e is an arbitrary length scale, and $J\{A\} \equiv A_1 - A_2$ is the jump in any variable A on crossing the shock. Here A_1 and A_2 are the values of A on the upstream and downstream sides of the shock, respectively. Equation (3.84) implies that the mass flux is continuous across the shock. Similarly, (3.85) implies the continuity of the sum of the normal stress and the normal component of the momentum flux.

If the upstream values of all the variables are known, and the stresses are assumed to vanish on the downstream side, (3.84)–(3.86) involve four unknowns, namely, the downstream velocities u_2 and v_2, the downstream solids fraction ν_2, and the orientation ψ of the shock. One more assumption is needed for closure, and Kaza and Jackson (1982a) assumed that the downstream velocity vector was vertical, i.e., $u_2\,\cos\theta - v_2\,\sin\theta = 0$ (Fig. 3.35). The assumptions used imply that the material below the shock is in vertical free-fall.

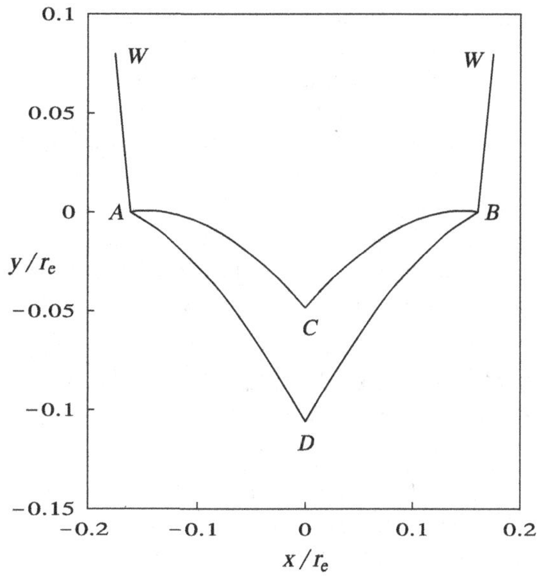

Figure 3.36. The boundary ACB of the domain of determinacy of the exit slot, and the exit shock ADB corresponding to the upper bound on the mass flow rate, for flow through a wedge-shaped hopper. The lines AW and BW represent the hopper walls. The curves were generated using the power series solution. Here r_e is an arbitrary length scale. Parameter values: $\phi = 25°$, $\delta = 15°$, $\theta_w = 10°$. (Adapted from Fig. 4 of Kaza and Jackson, 1982a.)

Using the above assumptions, (3.84)–(3.86) can be solved for v_2, u_2, v_2, and ψ. In cylindrical coordinates, let the shock be defined by $r = r_s(\theta)$. Then $\xi_s(\theta) \equiv r_s/r_e$ can be obtained by integrating

$$\frac{1}{\xi_s}\frac{\mathrm{d}\xi_s}{\mathrm{d}\theta} = -\tan\psi \tag{3.88}$$

subject to the initial condition

$$\xi_s(\theta_w) = \xi_{sw} \tag{3.89}$$

The value of ξ_{sw} can be determined as discussed shortly.

Consider an exit slot of a specified size, or equivalently, a specified value of the constant ξ_{sw} in (3.89). For sufficiently small values of the mass flow rate $\dot{M}$, the mean stress σ is positive at every point on the upstream side of the exit shock. The shock can be constructed for a range of values of $\dot{M}$, and hence the solution is not unique. As $\dot{M}$ increases, the SWRG solution, and also the numerical results for the present problem, suggest that the value of the mean stress σ at any point in the hopper decreases. For a certain value of $\dot{M}$, say, $\dot{M}_u$, σ vanishes at some point Q on the upstream side of the shock. At Q, σ vanishes on both sides of the shock, and hence there is no velocity jump. In view of the assumption that the downstream velocity vector is vertical, the only allowed location for Q, if any, is the intersection of the shock with the centerline of the hopper. As σ must be nonnegative for a cohesionless material, $\dot{M}_u$ represents an upper bound on admissible values of the mass flow rate.

As explained in §3.5.4, direct numerical integration of the governing equations is not feasible in the limit $\sigma \to 0$. However, an approximate solution can be constructed using the power series solution. In this case, $\dot{M}$ is specified (see §3.5.5), and the value of ξ_{sw} in (3.89) is determined by requiring the mean stress σ on the upstream side of the shock to vanish at the centerline $\theta = 0$. The shock constructed using this approach is shown by the curve labeled ADB in Fig. 3.36.

A defect of this solution must be noted. By integrating the equations within the hopper, the values of the variables can be determined in the plane of the exit slot. Then, as the governing equations are hyperbolic (see Appendix C), the solution in the region below this plane can be constructed only within the domain of determinacy of the exit slot, i.e., the region bounded by the exit slot and the characteristics AC and BC which issue from its edges (Fig. 3.36). As the shock ADB lies *outside* the domain of determinacy, it cannot be constructed rigorously, unless additional information is specified along some other boundaries. Probably the inconsistency arises because the approximate solution does not take account of the existence of characteristics.

In spite of this defect, mass flow rates based on the exit shock agree better with the data of Sullivan (1972) (cited in Brennen and Pearce, 1978) than those based on the traction-free surface (see the curves labeled PS and PT in Fig. 3.33). The two predictions bound the data, but it is not known whether this is a general result. Figure 3.33 implies that the mass flow rate is sensitive to the exit condition used.

3.8. THE SMOOTH WALL, RADIAL GRAVITY PROBLEM FOR COMPRESSIBLE FLOW

So far, models for incompressible flow have been discussed. The effect of density variation will now be considered. To simplify the analysis, attention will be confined to the smooth wall, radial gravity problem. The assumptions of smooth walls and radial gravity are relaxed in §4.2.3. This section is based on the work of Prakash and Rao (1988).

Using cylindrical coordinates with origin at the apparent vertex of the hopper, the governing equations for steady, plane compressible flow are given by the mass balance (A.63), which reduces to

$$\frac{1}{r}\frac{\partial(\rho\, r\, v_r)}{\partial r} + \frac{1}{r}\frac{\partial(\rho\, v_\theta)}{\partial \theta} = 0 \tag{3.90}$$

the momentum balances (3.12)–(3.14), the expressions for the stress components (3.15), the coaxiality condition (3.18), and the flow rule.

The flow rule can be deduced from (2.78) and (2.79) as follows. Using these equations, we obtain

$$\cos 2\gamma\,(C_{xx} + C_{yy}) + (c_*/s)(C_{yy} - C_{xx}) = 0 \tag{3.91}$$

where C_{ij} is a component of the rate of deformation tensor $\mathbf{C}$, defined in the compressive sense (see (2.49)). Identifying the x and y directions with the θ and r directions, respectively, and writing γ as γ' (Fig. 3.13), (3.91) can be rewritten as

$$\cos 2\gamma'\,(C_{\theta\theta} + C_{rr}) + (c_*/s)(C_{rr} - C_{\theta\theta}) = 0 \tag{3.92}$$

Using (2.83) and (A.68), (3.92) reduces to

$$\cos 2\gamma'\left(\frac{1}{r}\frac{\partial v_\theta}{\partial \theta} + \frac{v_r}{r} + \frac{\partial v_r}{\partial r}\right) + \sin \nu_d \left(\frac{1}{r}\frac{\partial v_\theta}{\partial \theta} + \frac{v_r}{r} - \frac{\partial v_r}{\partial r}\right) = 0 \tag{3.93}$$

As in the incompressible case (§3.3), it is assumed that (i) the walls are smooth, and (ii) gravity is radially directed toward the apparent vertex of the hopper. Then the equations admit a cylindrically symmetric solution of the form

$$\nu = \nu(r); \quad v_r = v_r(r); \quad v_\theta = 0; \quad \sigma = \sigma(r); \quad \tau = \tau(r) \tag{3.94}$$

The coaxiality condition (3.18) reduces to

$$\sin 2\gamma' \left(\frac{\mathrm{d}v_r}{\mathrm{d}r} - \frac{v_r}{r}\right) = 0$$

Hence either

$$\sin 2\gamma' = 0 \tag{3.95}$$

or

$$\frac{\mathrm{d}v_r}{\mathrm{d}r} - \frac{v_r}{r} = 0 \tag{3.96}$$

Equations (3.96) and (3.90) imply that $v_r \propto r$ and $\nu \propto 1/r^2$. Hence the material compacts as it flows down the hopper, in contrast to the data shown in Fig. 3.11a. We therefore discard the root (3.96) and choose the root (3.95).

The roots of (3.95) are given by

$$\text{(i)}\ \gamma' = 0 \quad \text{or} \quad \text{(ii)}\ \gamma' = \pi/2 \tag{3.97}$$

In the incompressible case, the passive root $\gamma' = 0$ causes material elements to accelerate as they flow down the hopper. This behavior is in accord with the mass balance for cylindrically symmetric flow, and hence this root is chosen. A similar argument applies in the compressible case, provided the material dilates as it flows. However, as the present model does not guarantee dilation, it is difficult to justify *a priori* the choice of the root $\gamma' = 0$. This root will be used here for consistency with the incompressible case.

Using the above assumptions, the θ component of the momentum balance and the coaxiality condition are identically satisfied. Introducing the dimensionless variables

$$\xi = r/r_e; \quad u = v_r/\sqrt{g\,r_e}; \quad \sigma' = \sigma/(\rho_p\, g\, r_e); \quad \tau' = \tau/(\rho_p\, g\, r_e) \tag{3.98}$$

where r_e is the radial coordinate corresponding to the edge of the exit slot (Fig. 3.13), g is the acceleration due to gravity and ρ_p is the particle density, the mass balance, the r component of the momentum balance, and the flow rule reduce to

$$\nu\, \xi\, u = \text{constant} \equiv -V \tag{3.99}$$

$$\nu\, u\, \frac{\mathrm{d}u}{\mathrm{d}\xi} + \frac{\mathrm{d}}{\mathrm{d}\xi}(\sigma' - \tau') - \frac{2\,\tau'}{\xi} + \nu = 0 \tag{3.100}$$

$$(1 - \sin \nu_d)\, \frac{\mathrm{d}u}{\mathrm{d}\xi} + (1 + \sin \nu_d)\, \frac{u}{\xi} = 0 \tag{3.101}$$

Here V is a dimensionless mass flow rate defined by

$$V = \frac{\dot{M}}{2B\,\theta_w\, \rho_p r_e \sqrt{g\, r_e}} \tag{3.102}$$

where $\dot{M}$ is the mass flow rate, and B is the length of the exit slot in the direction perpendicular to the plane of Fig. 3.13.

Expressing the yield condition as $\tau' = \tau'(\sigma', \nu)$, and substituting for u from (3.99) into (3.100) and (3.101), we obtain

$$\left(1 - \frac{\partial \tau'}{\partial \sigma'}\right) \frac{\mathrm{d}\sigma'}{\mathrm{d}\xi} = \left(\frac{1 + \sin \nu_d}{1 - \sin \nu_d}\right) \frac{V^2}{\nu\, \xi^3} + \left(\frac{2 \sin \nu_d}{1 - \sin \nu_d}\right) \frac{\nu}{\xi}\frac{\partial \tau'}{\partial \nu} + \frac{2\,\tau'(\sigma', \nu)}{\xi} - \nu \tag{3.103}$$

$$\frac{\mathrm{d}\nu}{\mathrm{d}\xi} = \left(\frac{2 \sin \nu_d}{1 - \sin \nu_d}\right) \frac{\nu}{\xi} \tag{3.104}$$

Compared to the incompressible case, an additional equation has to be solved to determine ν. Hence three boundary conditions have to be specified for (3.104) and (3.103). Prakash and Rao (1988) examined the solutions of these equations subject to the usual

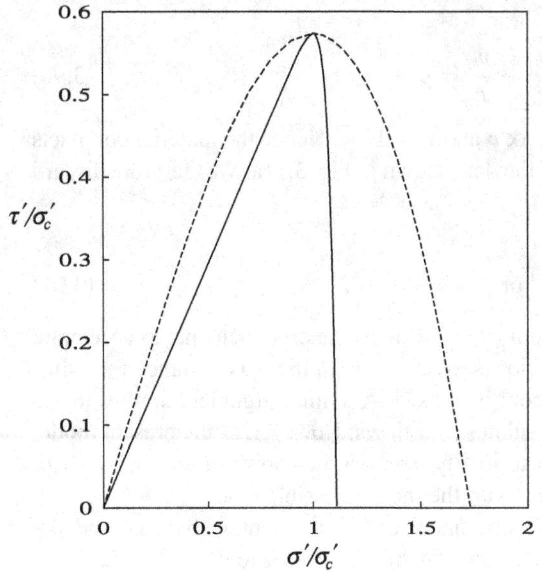

Figure 3.37. Yield loci corresponding to the yield condition (3.107) for $\phi = 35°$: —, $n = 1.03$; - - -, $n = 1.5$.

traction-free boundary conditions

$$\sigma'(\xi_1) = 0; \quad \sigma'(\xi = 1) = 0 \tag{3.105}$$

and

$$\nu(\xi_1) = \nu_1 \tag{3.106}$$

where $\xi_1 \equiv r_1/r_e$ defines the location of the upper traction-free surface. The value of ν_1 is assumed to be within the range of values for poured granular materials. Without recourse to measurements, it is not possible to specify the value of ν_1. Fortunately, except close to the upper traction-free surface, the results are found to be insensitive to the choice of ν_1.

To proceed further, expressions for the yield condition $\tau' = \tau'(\sigma', \nu)$ and the angle of dilation ν_d are required. Here we use the yield condition proposed by Prakash and Rao (1988), along with an associated flow rule. Thus

$$\tau' = \sigma_c'(\nu) \sin\phi \left(n\,\alpha - (n-1)\,(\alpha)^{n/(n-1)}\right); \quad \alpha \equiv \sigma'/\sigma_c'(\nu) \tag{3.107}$$

where $\sigma_c' \equiv \sigma_c/(\rho_p\, g\, r_e)$ is the dimensionless mean stress at a critical state, and ϕ and n are material constants. Jeffries (1993) proposed a similar yield condition in the context of triaxial tests.

Equation (3.107) has two parameters n and ϕ, which can be interpreted as follows. At a critical state, $\alpha = 1$, and hence (3.107) reduces to

$$\tau' = \sigma_c'(\nu) \sin\phi; \quad \sigma' = \sigma_c'(\nu)$$

Thus the slope of the critical state line in the τ'–σ' plane is $\sin\phi$. Similarly, the slope of the yield locus at $\sigma' = 0$ is $n \sin\phi$. Figure 3.37 shows plots of the yield locus for two values of n.

Using (3.107) and the associated flow rule, we have

$$\sin\nu_d = \frac{\partial\tau}{\partial\sigma} = n \sin\phi\,(1 - (\alpha)^{1/(n-1)}) \tag{3.108}$$

For Leighton Buzzard sand, the values of the angle of internal friction ϕ and the parameter n in the yield condition (3.107) can be estimated using the simple shear data of Airey et al. (1985) (Problem 3.6). It is found that n depends on the solids fraction ν.

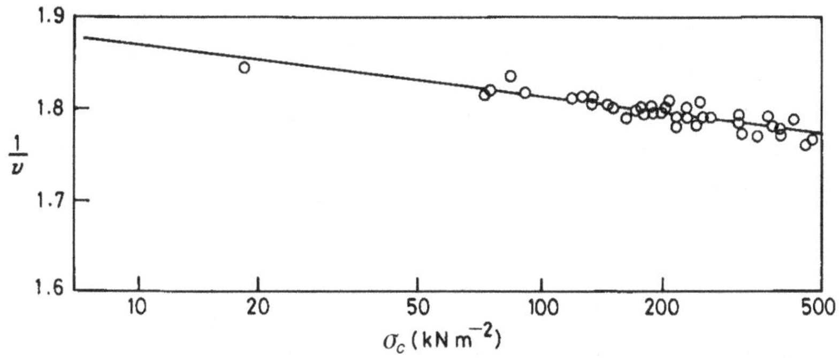

Figure 3.38. Variation of the solids fraction ν with the mean stress at a critical state σ_c for Leighton Buzzard sand: ○, data of Stroud (1971) (cited in Airey et al., 1985), obtained using the simple shear apparatus; —, equation (3.109). (Adapted from Fig. 23 of Airey et al., 1985.)

Prakash and Rao (1991) also estimated the value of n, but their value is lower as they used data which were close to the critical state, and did not allow for the dependence of the data on the normal stress. Hence the values estimated in Problem 3.6 are believed to be more accurate. However, in keeping with the work of Prakash and Rao (1988), results will be presented here only for the case where n is independent of ν. As shown below, the results do not appear to be sensitive to the value n.

An expression for $\sigma_c(\nu)$ must now be specified. Figure 3.38 shows data for Leighton Buzzard sand, obtained using a simple shear apparatus (Fig. 2.19). The data can be fitted by the linear relation

$$1/\nu = \Gamma - \lambda \ln(\sigma_c/p_a) \tag{3.109}$$

where Γ and λ are material constants and $p_a = 101.33$ kPa is the standard value of the atmospheric pressure. Here p_a is used as a convenient nondimensionalizing parameter, as stresses in laboratory-scale hoppers and bunkers rarely exceed 100 kPa. In Fig. 3.38, the smallest value of σ_c is about 20 kPa, which is comparable to the static stress levels in a large bunker (Fig. 2.16). In laboratory-scale hoppers, the stresses are much smaller. For example, the maximum value of the mean stress for the data in Fig. 3.25 is about 2.2 kPa. This illustrates the need for data at low values of the mean stress.

Equations of the form (3.109) are widely used in the soil mechanics literature (see, e.g., Roscoe et al., 1958; Atkinson and Bransby, 1982, p. 190; Airey et al., 1985). Some features of (3.109) will now be discussed. Equation (3.109) implies that σ_c levels off at large values of ν. For typical parameter values, this occurs at solids fractions which exceed the value corresponding to dense random packing, i.e., $\nu \approx 0.64$. Hence the flat portion of the $\sigma_c(\nu)$ curve is rarely sampled. From the viewpoint of flow problems, the low-density limit is also of interest. Equation (3.109) implies that σ_c and all its derivatives vanish as $\nu \to 0$. Owing to a lack of data and theoretical guidelines, the form of $\sigma_c(\nu)$ at low densities cannot be specified with confidence. Alternative expressions for σ_c are discussed below.

Rathbone et al. (1987) use

$$1/\nu = \Gamma - \lambda \ln(\sigma/p_a); \quad \sigma > \sigma_* \tag{3.110}$$

$$= \Gamma_1 - \lambda_1 \sigma; \quad 0 \le \sigma \le \sigma_* \tag{3.111}$$

where Γ, λ, Γ_1, λ_1, and σ_* are material constants. Two of these constants can be determined by requiring that $1/\nu$ and its derivative with respect to σ be continuous at $\sigma = \sigma_*$, and the rest can be determined by fitting (3.111) to data. Equation (3.110) is similar to (3.109), except that σ_c has been replaced by the mean stress σ. With $\sigma = \sigma(\nu)$, (3.103) and (3.104) constitute an overdetermined system. Hence these authors discard the flow rule. Further, if (3.110) is used for all values of σ, the boundary condition (3.105) implies that $\nu(\xi = 1) = 0$. To avoid unrealistic values of ν at the hopper exit, (3.110) is replaced by (3.111) at low values of σ. A few other forms for σ_c are discussed below.

Jenike (1961, p. 12) proposed that $\nu = a\,(b + \sigma)^c$, where σ is the mean stress and a, b, and c are constants. Pitman and Schaeffer (1987) replaced σ by σ_c, and set $b = 0$ in this expression. Johnson et al. (1990) proposed that

$$\sigma_c = a\,(\nu - \nu_{\mathrm{lrp}})^b/(\nu_{\mathrm{drp}} - \nu)^c\,, \quad \nu > \nu_{\mathrm{lrp}}$$
$$= 0, \quad \nu \leq \nu_{\mathrm{lrp}} \tag{3.112}$$

where a, b, and c are material constants. Here $\nu_{\mathrm{lrp}} \approx 0.5$ and $\nu_{\mathrm{drp}} \approx 0.65$ are the solids fractions corresponding to loose random packing and dense random packing, respectively.

At low stress levels, all the expressions listed above must be regarded as tentative. Equation (3.109) will be used here.

The value of the mass flow rate V is determined iteratively by integrating (3.103) and (3.104) from $\xi = \xi_1 > 1$ to $\xi = 1$ and checking whether the second of (3.105) is satisfied to within a desired tolerance. The results of Prakash and Rao (1988) are discussed below, but the parameter values used in Figs. 3.39–3.41 differ slightly from their values.

The stress and solids fraction profiles for two values of n (the parameter in the yield condition (3.107)) differ only near the upper traction-free surface $\xi = \xi_1$ (Fig. 3.39), even though the corresponding yield loci are quite different (Fig. 3.37). This is because the material remains close to the critical state $\sigma'/\sigma_c' = 1$ over most of the hopper (Fig. 3.40). At the critical state, the value of n is immaterial, as $\tau' = \sigma_c'(\nu)\,\sin\phi$ and $\partial\tau'/\partial\sigma' = 0$ for all values of n.

The stress profiles (Fig. 3.39a) are qualitatively similar to the incompressible stress profile (Fig. 3.16), except for a local maximum near the upper traction-free surface for $n = 1.5$.

Near the hopper exit $\xi = 1$, the boundary condition (3.105) forces the mean stress to vanish, whereas σ_c is nonzero in general. Hence $\sigma/\sigma_c < 1$ and the state of stress lies on the dilation branch of the yield locus. This causes the solids fraction to decrease sharply near $\xi = 1$ (Fig. 3.39b). In §3.2.1, the critical state approximation was used to construct a solution to the hopper problem. Figure 3.40 shows that this approximation breaks down near the upper and lower traction-free surfaces.

The density profile is insensitive to the value of $\nu(\xi_1)$, except close to the upper traction-free surface (Fig. 3.41). This suggests the existence of asymptotic density and stress fields, the latter representing a compressible analog of the radial stress field. As discussed in Prakash and Rao (1991) (see also Problem 3.9), approximate expressions for the asymptotic fields can be constructed using perturbation methods. The actual fields approach the asymptotic fields as ξ decreases (Figs. 3.39a and 3.41). As in the incompressible case, the deviation observed near the hopper exit is caused by the increasing influence of the inertial terms.

Figure 3.40 shows that $\sigma'(1) \neq 0$, even though it appears to be equal to zero on the scale of Fig. 3.39a. However, this value of $\sigma'(1)$ can be decreased by using a slightly larger value of the mass flow rate V. For example, $\sigma'(1) = 0.013$ for $V = 0.794$, and is negative for $V = 0.795$. As the material is cohesionless, we set $V = 0.794$.

The effect of dilation on the mass flow rate V will now be discussed. Because of the existence of asymptotic stress and solids fraction fields, the effect of upstream boundary conditions fades on moving down the hopper. Hence V is determined mainly by the solution near the hopper exit. In the exit region, the solids fraction decreases as ξ decreases (Fig. 3.39b), and hence the radial velocity u increases in magnitude to maintain a constant value of V (see (3.99)). In particular, the increase in $|u|$ is more than that required by the

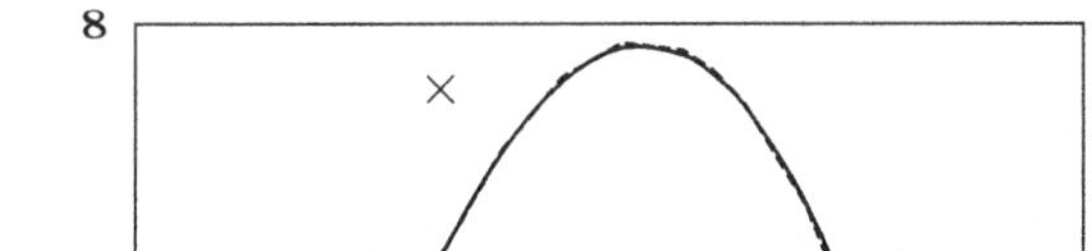

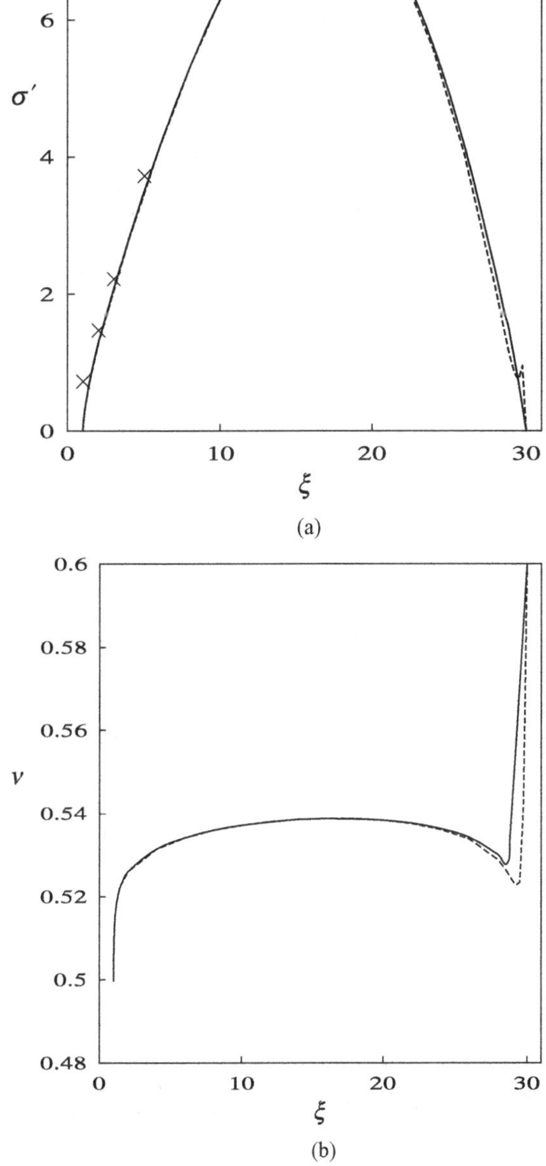

Figure 3.39. Profile of (a) the dimensionless mean stress $\sigma' = \sigma/(\rho_\text{p}\, g\, r_e)$, and (b) the solids fraction ν for the compressible, smooth wall, radial gravity problem: —, $n = 1.03$, $V = 0.794$; - - -, $n = 1.5$, $V = 0.796$; $\times$, asymptotic field for $n = 1.03$. Here ρ_p is the particle density, g is the acceleration due to gravity, $\xi = r/r_e$, and r_e is the radial coordinate corresponding to the edge of the exit slot. The profiles are obtained by downward integration from $\xi_1 = r_1/r_e = 30$, with $\nu(\xi_1) = 0.6$. Parameter values : $\beta = p_a/(\rho_\text{p}\, g\, r_e) = 50$, $\phi = 35°$, $\Gamma = 1.81$, $\lambda = 0.025$.

converging geometry of the hopper. As the inertial terms can be regarded approximately as the product of the density, the square of the velocity, and the reciprocal of a suitable length scale, dilation tends to increase inertial effects. Therefore, compared to the incompressible case, a smaller value of the mass flow rate V suffices to ensure that the exit condition $\sigma'(1) = 0$ is satisfied. Overall, dilation tends to reduce the mass flow rate. For example, let us consider Leighton Buzzard sand, with the parameter values listed in Fig. 3.39a. The compressible mass flow rate V, defined by (3.99), is 0.794, whereas the incompressible mass flow rate $V_i = \nu\,\overline{A}$ (see (3.28) and (3.102)) based on $\nu = \nu(\xi_1) = 0.6$ is 0.998. The

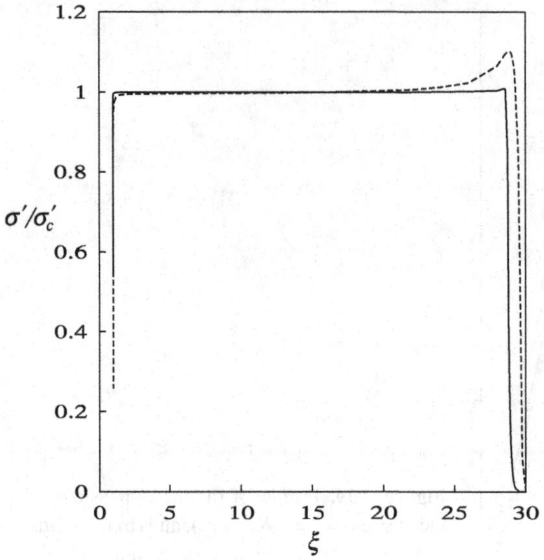

Figure 3.40. Profiles of σ'/σ_c' for the compressible, smooth wall, radial gravity problem: —, $n = 1.03$, $V = 0.794$; - - -, $n = 1.5$, $V = 0.796$. Here $\sigma' = \sigma/(\rho_p g r_e)$ is the dimensionless mean stress, $\sigma_c' = \sigma_c/(\rho_p g r_e)$ is the dimensionless mean stress at a critical state, $\xi = r/r_e$, ρ_p is the particle density, g is the acceleration due to gravity, and r_e is the radial coordinate corresponding to the edge of the exit slot. The profiles are obtained by downward integration from $\xi_1 = r_1/r_e = 30$, with $\nu(\xi_1) = 0.6$. Parameter values are as in Fig. 3.39.

discrepancy can be reduced by choosing a smaller value of ν, but the desired value is not known a priori.

3.9. SUMMARY

Measurements show that the mass flow rate of coarse materials from hoppers is approximately independent of the head of material, provided the head is large compared to the width of the exit slot. For fine materials, interstitial air pressure gradients cause the flow rate to depend on the head. The Rose-Tanaka–Beverloo–Nedderman correlation provides a reasonable estimate of the flow rate of coarse materials.

Two types of flow patterns, called mass flow and funnel flow, are possible. In the former case, material flows throughout the hopper, whereas in the latter case, there are quasi-static

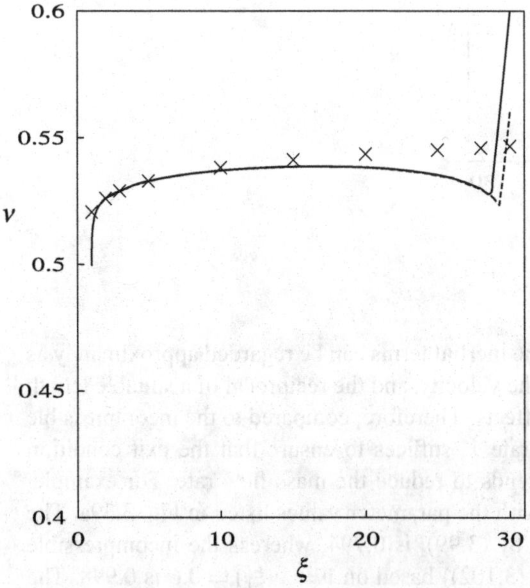

Figure 3.41. Profiles of the solids fraction ν for the compressible smooth wall, radial gravity problem : —, $\nu(\xi_1) = 0.6$; - - -, $\nu(\xi_1) = 0.56$; ; $\times$, asymptotic field. The profiles are obtained by downward integration from $\xi_1 = r_1/r_e = 30$, with $\nu(\xi_1) = 0.6$. Parameter values: $n = 1.03$, $V = 0.794$, the rest as in Fig. 3.39.

regions adjacent to the hopper walls. In general, the preferred flow pattern is mass flow, though funnel flow may be preferred for abrasive materials.

The measurements of Fickie et al. (1989) show that the solids fraction is approximately constant in the upper part of the hopper, but decreases significantly near the exit slot.

Models based on yield conditions and flow rules have been applied to the hopper problem. If the solids fraction ν is assumed to be constant, the flow rule implies that the material is at a critical state. Hence the mean stress σ must also be a constant, whereas the momentum balances require it to vary with position. These conflicting requirements are reconciled by invoking the critical state approximation. This assumes that the mean stress at a critical state σ_c is a strong function of ν, so that very small variations in ν permit large variations in σ_c and hence σ. The assumption of smooth walls and radially directed gravity permit an analytical solution to be constructed. This solution predicts, in accord with observations, that the mass flow rate tends to a constant as the head of material increases. However, the flow rate is overestimated considerably. The discrepancy can be reduced to a large extent by including the effect of wall roughness, as shown by the Brennen–Pearce solution.

The radial stress field is a particular solution of the inertia-free momentum balances. Similarly, the mass balance for incompressible flow and the coaxiality condition admit a particular radial flow solution called the radial velocity field. The linearized stability analysis of Pitman (1988) shows that solutions of the velocity equations tend to converge to the radial velocity field on moving down the hopper. The good agreement between the latter and the measured velocity profile of Bosley et al. (1969) provides some support for the frictional model. For a specified value of the wall angle of the hopper, the results of Pitman (1988) show that solutions of the inertia-free momentum balances tend to the radial stress field for certain ranges of values of the angle of internal friction and the angle of wall friction. Thus the radial stress field is not as robust as the radial velocity field. Comparison of measured and predicted stress profiles shows that the radial stress field provides an order of magnitude estimate of the stresses, but does not capture the spatial oscillations observed by Handley and Perry (1968).

Using the radial fields as upstream conditions, the governing equations for steady flow can be integrated numerically down the hopper. It is often assumed that a traction-free surface, on which the shear and normal stresses vanish, spans the exit slot. However, the latter cannot be approached from above owing to a singularity in the momentum balances. An alternative approach, based on the successive approximation procedure, can be used to construct a series solution. Even though this solution appears to be more accurate than the Brennen–Pearce solution, the discrepancy between predicted and measured mass flow rates is larger in the former case. This anomalous behavior suggests that the traction-free exit condition is perhaps unrealistic. Indeed, it is shown that the existing approximate solutions lead to compaction on the downstream side of the traction-free surface.

Hence an alternative condition is suggested, wherein a discontinuity surface or shock spans the exit slot. Compaction below this surface is avoided by requiring the mean stress to vanish and the velocity vector to be vertically directed on the downstream side of this surface. Based on the limited computational results available for small wall angles, it appears that the mass flow rates obtained using this exit condition agree better with measurements than those obtained using the traction-free surface as an exit condition.

The effect of density variation is examined in the context of the smooth wall, radial gravity problem. The numerical results show that the critical state approximation is valid over most of the hopper, except near the upper and lower traction-free surfaces. The material dilates strongly in the exit region, causing the mass flow rate to be less than the incompressible value.

PROBLEMS

3.1. Stresses exerted on the wall of a hopper

(a) A wedge-shaped hopper is filled with sand. At a point on the hopper wall, the stress tensor has the components $\sigma_{xx} = 8$, $\sigma_{yy} = 4$, and $\sigma_{xy} = -2$ kPa relative to the Cartesian coordinate system shown in Fig. 3.13. Calculate the *normal* stress on the hopper wall, given that the wall angle θ_w of the hopper is 10°.

(b) If the angle of wall friction δ is 20°, is the wall yield condition satisfied?

3.2. Profile of a slip plane in a hopper

A cohesionless granular material satisfying the Mohr–Coulomb yield condition is at rest in a wedge-shaped hopper (Fig. 3.13). The orientation of the major principal stress axis (σ_1 axis) relative to the circumferential direction is given by $\gamma' = (\gamma_w/\theta_w)\,\theta$, where γ_w is a positive constant. It can be assumed that the σ_1 and σ_2 axes lie in the x–y plane.

(a) Does the stress field represent an active state of stress or a passive state?

(b) Derive an expression for the slip plane that passes through a point O on the centerline, and, for $\theta > 0$, lies *below* the horizontal line passing through O. The radial coordinate of O is r_0. If $\phi = 30°$, $\theta_w = 20°$, and $\gamma_w = 10°$, what is the value of r/r_0 at the point where the slip plane intersects the right-hand hopper wall?

3.3. Use of correlations to predict the mass flow rate

Consider the data of Al-Din and Gunn (1984) for the flow of glass beads through a cylindrical bunker with a rectangular exit slot (see Table 3.1 and Fig. 3.3).

(a) Predict the mass flow rate $\dot{M}$ using the correlation (3.4) for two cases: (i) without incorporating an "empty zone" adjacent to the edges of the exit slot, and (ii) incorporating the empty zone, with $k_* = 1.4$. It can be assumed that the factor F in (3.4) is unity. Compare your results with the average value of $\dot{M}$ obtained from the data shown in Table 3.1.

(b) Repeat part (a) using the correlation of Al-Din and Gunn (1984), which is given by

$$\dot{M} = 0.406\, F_o\, F_p\, \rho \sqrt{g}\, A\, ((D - d_p)^{2.5}/D^2)\,(D/d_p)^{0.154}$$

where F_o and F_p are coefficients related to the shapes of the orifice and the particles, respectively, ρ is the density of the granular material, g is the acceleration due to gravity, A is the cross-sectional area of the exit slot, D is the width of the exit slot (Fig. 3.1a) (the smaller dimension), and d_p is the particle diameter. The factors F_o and F_p are given by $F_o = 1.0$ for circular, triangular, and square orifices, $F_o = 1.4$ for rectangular and elliptical orifices with the ratio of the largest to the smallest dimension greater than 1.5, and $F_p = 1.0$ for round particles such as glass beads.

3.4. Discharge of sand from a hopper

Using the result obtained in Problem 1.13b, indicate conditions under which the force exerted by the sand on the hopper wall varies approximately linearly with the time t. The sand begins to flow at $t = 0$.

3.5. Stability of the radial stress field: smooth wall analysis

Consider a wedge-shaped hopper with smooth walls ($\delta = 0$). In the passive case, this condition implies that $\gamma_w = 0$. As $\gamma'(0) = 0$, it can be assumed that γ' is small everywhere, and hence $\sin 2\gamma' \approx 2\gamma'$, $\cos 2\gamma' \approx 1$.

(a) Introducing a new variable $\omega = \overline{\sigma}\,\gamma'$, show that the inertia-free momentum balances reduce to a pair of linear equations in $\overline{\sigma}$ and ω. Here $\overline{\sigma}$ is the dimensionless mean stress, defined by (3.37). Note that the result can be obtained only if the above approximations are used *before* evaluating the derivatives of the stress components in (3.39) and (3.40).

(b) Consider perturbations to the radial stress field, of the form

$$\overline{\sigma} = \xi\, b_0(\theta) + \overline{\sigma}_1(\xi, \theta); \quad \omega = \xi\, \omega_0(\theta) + \omega_1(\xi, \theta)$$

where $\omega_0 = b_0\, g_0$. Substituting these expressions into the governing equations, and considering solutions of the form

$$\overline{\sigma}_1 = \exp(\lambda\, t')\, b_2(\theta); \quad \omega_1 = \exp(\lambda\, t')\, \omega_2(\theta)$$

where $t' \equiv \ln\, \xi$, show that ω_2 satisfies the equation

$$\frac{\mathrm{d}^2\omega_2}{\mathrm{d}\eta^2} = \beta\, \omega_2 \tag{3.113}$$

Here $\eta \equiv \theta/\theta_w$, and β is a constant.

(c) Specify the boundary conditions for ω_2.

(d) Show that (3.113) admits *nontrivial* solutions satisfying the boundary conditions only if

$$\lambda = \lambda_n = \frac{2\, \sin\phi - 1 \pm \sqrt{1 - (n\,\pi/\mu)^2}}{1 - \sin\phi}, \qquad n = 1,\, 2,\, \ldots,\, \infty \tag{3.114}$$

where $\mu \equiv \theta_w/\cos\phi$. (A nontrivial solution is one which does not vanish identically for all values of η.)

(e) Assuming that $\mu < \pi$, show that the radial stress field is downward unstable for $\phi < \sin^{-1}(2/3) = 42^\circ$.

This problem has been adapted from Kaza (1982, pp. 243–246) and Pitman (1988).

3.6. Yield locus for Leighton Buzzard sand

(a) Table 3.4 shows the data of Airey et al. (1985) for Leighton Buzzard sand. Assuming that the critical state line is defined by $\tau = \sigma_c(\nu)\, \sin\,\phi$; $\sigma = \sigma_c(\nu)$, estimate the value of the angle of internal friction ϕ.

(b) Airey et al. (1985) find that $R \equiv |\sigma_{yx}|/\sigma_{yy} = k\, \cot\gamma$, where $k = 0.669$, and γ is the inclination of the major principal stress axis relative to the x axis (Fig. 2.32). Calculate the values of the mean stress σ, the solids fraction ν, and $\alpha \equiv \sigma/\sigma_c$ for the data in the first six rows of Table 3.4. As the values of ν for rows 1–3 do not differ significantly, it can be assumed that the data in these rows correspond to a common value of $\nu(\equiv \nu_1)$. Similarly, the data in rows 4–6 correspond to $\nu = \nu_2$.

Table 3.4. Data of Airey et al. (1985) for Leighton Buzzard Sand, Obtained Using the Simple Shear Apparatus*

No.	e_λ	$\beta \equiv \tau/\sigma$	N (kPa)
1	0.65	0.78	13.79
2	0.689	0.757	48.26
3	0.735	0.739	172.4
4	0.828	0.674	11.72
5	0.9	0.616	82.74
6	0.898	0.639	165.4
7	0.927	0.574	

* Each of the first six rows corresponds to tests conducted under a constant vertical normal stress N, whose value is listed in the fourth column. The last row denotes conditions at a critical state. Here $e_\lambda \equiv e' + \lambda\, \ln\sigma$, where $e' = (1/\nu) - 1$ is the voids ratio, λ is a constant, and σ is the mean stress in kPa. In the e'–σ plane, the critical state line is represented by $e' = H - \lambda\, \ln\sigma_c$, where H is a constant, and σ_c is the mean stress at a critical state. When σ_c has units of kPa, $H = 0.927$, and $\lambda = 0.025$. The values of N are not reported in Airey et al. (1985), but have been provided by Prof. D. M. Wood (private communication, 2000).

(c) Assuming that the yield condition is given by the dimensional equivalent of (3.107), estimate approximately the value of n by using the data for $\nu = \nu_1$. Repeat using the data for $\nu = \nu_2$. Hint: For typical values of n, $n - 1$ is small compared to 1.

3.7. Jump conditions across the exit shock
The mass and linear momentum balances are of the form

$$\frac{\partial F}{\partial t} + \nabla \cdot G + H = 0 \tag{3.115}$$

where F and H are scalars or vectors and G is a vector or a second- order tensor. Consider a discontinuity surface or shock at the hopper exit (Fig. 3.35). For a shock which is stationary relative to the hopper, the jump mass and momentum balances are given by (see Appendix D)

$$J\{\mathbf{n}_d \cdot G\} = 0 \tag{3.116}$$

where $J\{A\} \equiv A_1 - A_2$ is the jump in any variable A on crossing the shock, and $\mathbf{n}_d$ is the unit normal to the shock (Fig. 3.35).

(a) By comparing (3.115) with the differential mass and momentum balances, and using (3.116), show that

$$J\{\mathbf{n}_d \cdot (\rho\, \mathbf{v})\} = 0; \quad J\{\mathbf{n}_d \cdot (\rho\, \mathbf{v}\, \mathbf{v} + \boldsymbol{\sigma})\} = 0 \tag{3.117}$$

(b) Using the components of $\mathbf{v}$ and $\boldsymbol{\sigma}$ relative to the cylindrical coordinate system (Fig. 3.35), show that (3.117) can be rewritten as (3.84)–(3.86), where (3.85) and (3.86) represent the r and θ components of the second of (3.117).

3.8. Solids fraction profile for vertical free-fall below the hopper
Consider a wedge-shaped hopper, through which a granular material flows steadily parallel to the x–y plane (Fig. 3.13).

(a) Derive an expression for the variation of the solids fraction ν with position, in the particle jet *below* the hopper. It can be assumed that (i) the granular material is in *vertical free-fall*, i.e., the contact forces between the particles vanish, (ii) the effect of air drag can be neglected, and (iii) $\nu(x, y = H) = \nu_0(x)$, $v_y(x, y = H) = v_0(x)$, where v_y is the y component of velocity, $y = H$ denotes the plane of the exit slot, and ν_0 and v_0 are known functions of x. Express the result in a dimensionless form, using the half-width of the exit slot ($W_s = D/2$) (Fig. 3.1) as the characteristic length scale.
(b) Using the correlation (3.4) for the mass flow rate $\dot{M}$, estimate $\overline{v}$, the cross-sectional averaged value of $v_0(x)$, assuming that $D \gg k_*\, d_p$, $D = 20$ mm, $B = 60$ mm, and $\theta_w = 20°$. Here B is the length of the exit slot in the z direction (Fig. 3.1).
(c) Estimate approximately the value of $\nu(x = 0, y = H - W_s)$. The parameter values are as in part (b), and $\nu_0(0) = 0.4$.

3.9. Asymptotic fields for compressible smooth wall, radial gravity problem
To derive expressions for the asymptotic fields, consider the momentum balance (3.100) (with the inertial term omitted), the yield condition (3.107), and the flow rule, given by (3.104) and (3.108). Assuming that the parameter λ in (3.109) $\ll 1$, the dependent variables $\nu, \sigma', \sigma_c', \alpha$, and $\sin \nu_d$ can be expanded in powers of λ. For example, $\nu = \nu_0(\xi) + \lambda\, \nu_1(\xi) + \lambda^2\, \nu_2(\xi) + \cdots$. To simplify the representation of the expansions, it is helpful to introduce the order symbol O. A function $F(\xi, \lambda)$ is said to be of *order* λ^n if $\lim_{\lambda \to 0} F/\lambda^n$ is bounded (Bender and Orszag, 1984, p. 318). If this condition holds, we write $F = O(\lambda^n)$. Thus the expansion for ν can be written as $\nu = \nu_0(\xi) + \lambda\, \nu_1(\xi) + O(\lambda^2)$.

(a) Substituting the expansions for ν and σ'_c into (3.109), and setting the coefficients of each power of λ to zero, show that

$$\nu_0 = 1/\Gamma; \quad \nu_1 = (1/\Gamma^2)\ \ln(\sigma'_{c0}/\beta) \tag{3.118}$$

where $\beta = p_a/(\rho_p\, g\, r_e)$, and σ'_{c0} is the first term in the expansion for σ'_c.

(b) Using a similar procedure for the flow rule (3.104), show that $\alpha_0 = 1,\ \sigma'_0 = \sigma'_{c0}$.

(c) Similarly, considering the momentum balance (3.100), show that

$$\sigma'_{c0} = C\,\xi \tag{3.119}$$

where $C \equiv \nu_0/(3\,\sin\phi - 1)$. To derive this result, a first-order ordinary differential equation involving σ'_{c0} has to be solved. It can be assumed that (i) the hopper is deep, and (ii) $3\,\sin\phi > 1$. This permits the term involving the integration constant to be omitted, but the resulting solution is not valid near the upper traction-free surface.

(d) Hence show that

$$\nu = \nu_0\left[1 + \lambda\,\nu_0\,\ln\left(\frac{C\,\xi}{\beta}\right)\right] + O(\lambda^2) \tag{3.120}$$

(e) Setting the coefficient of λ to zero in (3.104), show that

$$\alpha_1 = -n_1/(2\,\Gamma\,n\,\sin\phi)$$

where $n_1 \equiv n - 1$.

(f) Show that

$$\frac{\sigma'}{C\,\xi} = 1 + \lambda\left[\frac{3\alpha_1\,\sin\phi + \nu_0\,(1 - \sin\phi)}{3\,\sin\phi - 1} + \nu_0\,\ln\left(\frac{C\,\xi}{\beta}\right)\right] + O(\lambda^2) \tag{3.121}$$

4

Flow through Wedge-Shaped Bunkers

A bunker is a combination of a bin and a hopper (Fig. 1.5). The abrupt change in geometry at the bin–hopper transition results in a rich variety of flow patterns, and oscillatory wall stresses. These features pose formidable difficulties for modeling bunker flow. At present, there are no satisfactory models which can capture all the observed features.

Some of the experimental observations are summarized below. This is followed by a discussion of models for the bin section, the transition region, and the hopper section.

4.1. EXPERIMENTAL OBSERVATIONS

4.1.1. Flow Regimes

The patterns observed by Nguyen et al. (1980) for the flow of sand through a bunker are shown in Fig. 4.1. For $\theta_w < 40°$, mass flow (type A) occurs regardless of the value of H/W (Fig. 4.2). Here W and H are the half-width of the bunker and the height of the free surface of the material relative to the exit slot, respectively. (For small values of H, the free surface is not flat. In this case, H is the elevation of the point of intersection of the free surface with the bunker wall.) For $\theta_w > 70°$, funnel flow of type B occurs when H/W exceeds a critical value (≈ 3) and funnel flow of type C occurs for a smaller value of H/W. As indicated by the hatched regions in Fig. 4.2, the boundaries between various regimes are not sharply defined. For $\theta_w = 60°$, there is an interesting transition from mass flow to funnel flow, and back again to mass flow as H/W decreases from high values. The experiments of Benink (1989, p. 100) are qualitatively similar, except that the reverse transition from funnel flow to mass flow does not occur as H/W decreases.

When type B flow occurs, the boundary between stagnant and flowing regions is inclined at an angle $(\pi/2) - \beta$ relative to the horizontal (Fig. 4.1). Usually, $(\pi/2) - \beta > \phi$, the angle of internal friction (Table 4.1). Assuming that ϕ approximates the angle of repose, i.e., the inclination of the free surface of a "two-dimensional" heap of cohesionless material relative to the horizontal, it follows that the boundary of the stagnant zone is usually steeper than the free surface of a heap. As the material in the flowing zone exerts forces on this boundary, it is not a traction-free surface.

4.1.2. Kinematics

The initiation of flow is accompanied by the formation of *rupture layers* or *rupture surfaces*. As discussed in Chapters 2 and 3, rupture layers are thin bands of material across which the density and velocity change sharply.

Consider the following experiment. A bunker is filled with a material such as sand. The exit slot is opened, permitting the material to discharge for a short while. The slot is then

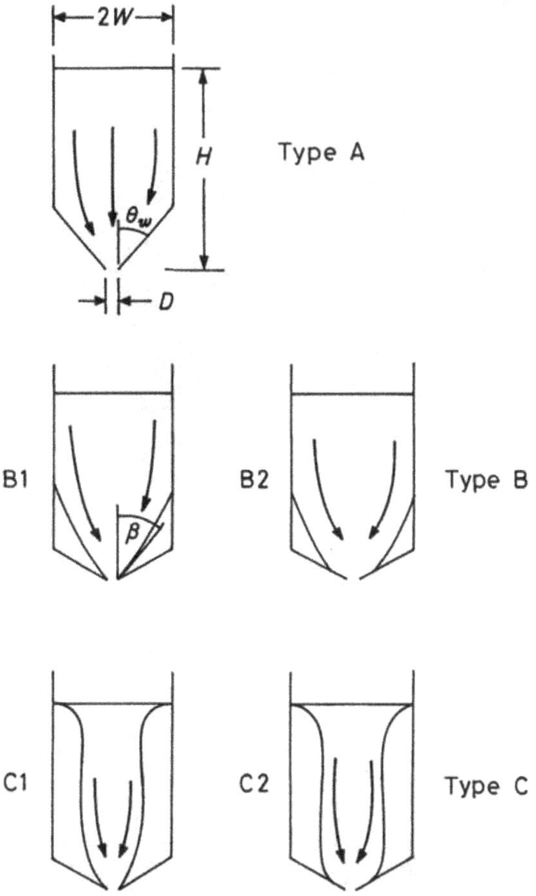

Figure 4.1. Flow regimes observed for the flow of sand through a bunker made of Lucite: type A, mass flow; types B and C, funnel flow. (Reproduced from Nguyen et al., 1980, with permission from the American Society of Mechanical Engineers.)

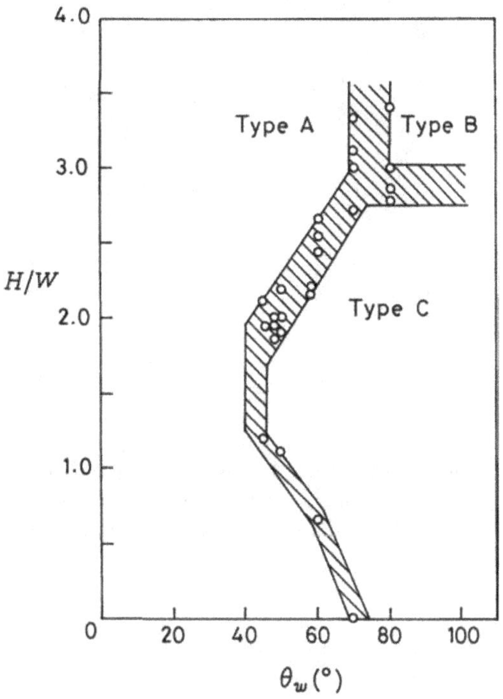

Figure 4.2. Flow regime map for the flow of sand through a bunker made of Lucite: type A, mass flow; types B and C, funnel flow. The flow patterns corresponding to types A, B, and C are shown in Fig. 4.1. The hatched region indicates the boundary between different regimes. Parameter values: particle diameter = 0.5–1.0 mm, distance between the front and back faces = 152 mm. (Reproduced from Nguyen et al., 1980, with permission from the American Society of Mechanical Engineers.)

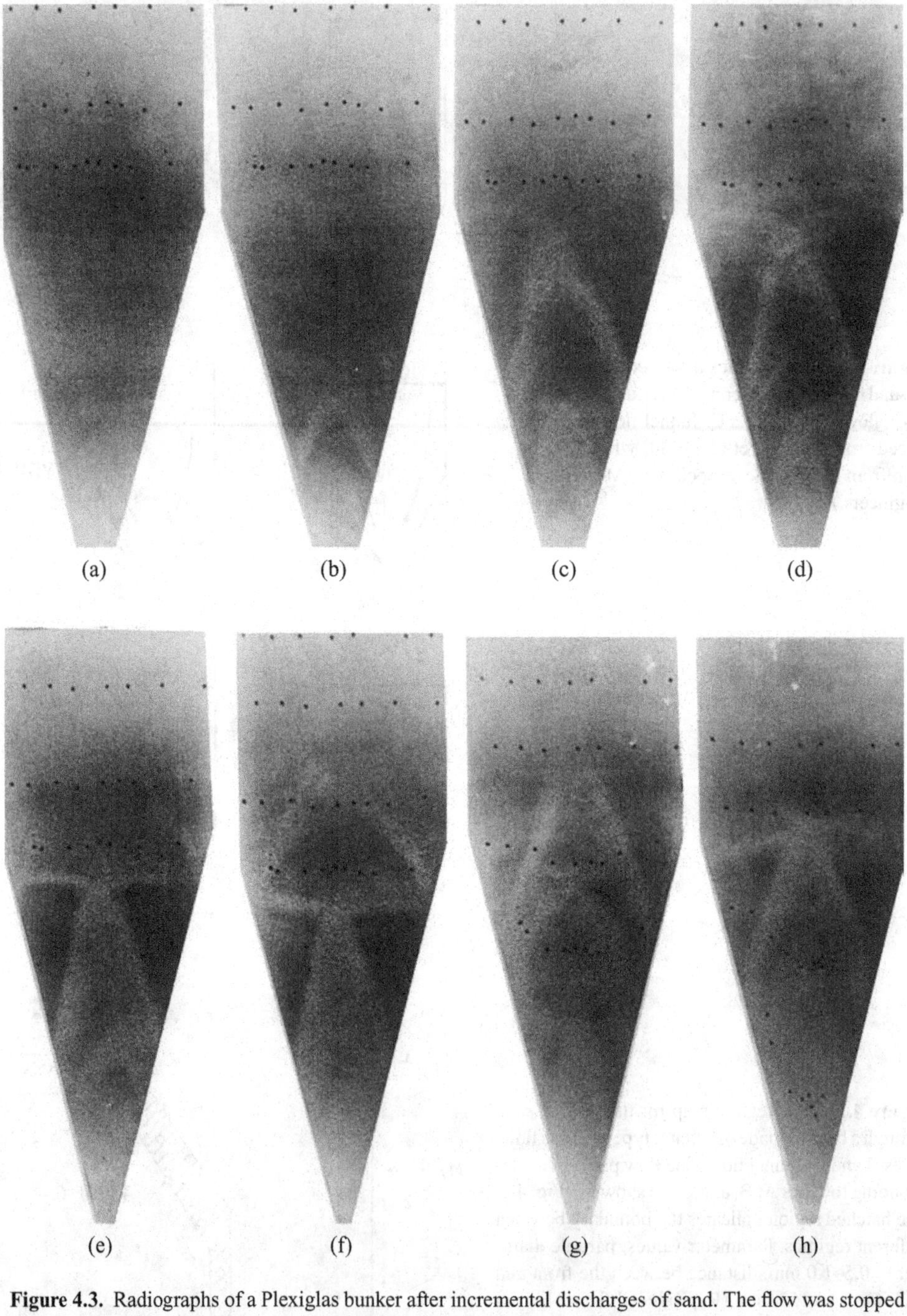

Figure 4.3. Radiographs of a Plexiglas bunker after incremental discharges of sand. The flow was stopped before taking the radiographs. Parameter values: particle diameter = 0.25–1.75 mm, wall angle of the hopper section $\theta_w = 15°$, angle of internal friction $\phi = 32°$, angle of wall friction $\delta = 11°$, width of the bin section ≈300 mm, gap between the faces = 40 mm. The radiographs were taken by A. Drescher. (Reprinted from Michalowski, 1987, with permission from Elsevier Science.)

Table 4.1. Slope of the Boundary of the Stagnant Zone*

Material	ϕ (°)	δ (°)	θ_w (°)	$90 - \beta$ (°)	Reference
Plastic grains	38	14	70	45	Benink (1989, p. 101)
"	"	31	60	45	"
Soya beans	34	29	60	45	"
Sand	37	16	70	40	"
"	"	31	60	40	"
Polystyrene	39	12	80	50	Nguyen et al. (1980)
Glass beads	25	15	"	45	"

* Here ϕ is the angle of internal friction, which is a parameter occurring in the equation for the critical state line (see (4.7)), δ is the angle of wall friction, θ_w is the wall angle of the hopper, and β is the angle between the boundary of the stagnant zone and the vertical, as shown in Fig. 4.1.

closed, and a radiograph is taken using X-rays. By repeating this procedure, the sequence of pictures shown in Fig. 4.3 was generated. Here the dark regions correspond to dense material and the light regions to dilated material. Some rupture layers are visible in the pictures, and they form the basis for the pattern inferred by Michalowski (1987) (Fig. 4.4). The fine structure of the pattern differs from the results of a similar experiment due to Blair-Fish and Bransby (1973) (Fig. 4.5). Perhaps the size of the bunker, the nature of the granular and wall materials, and the method of filling have a role to play.

For some reason, rupture layers do not appear to propagate into the upper part of the bin section. In the advanced stages of flow, the rupture layers have the shapes shown in Fig. 4.4h, but are not stationary. The pattern deduced by Michalowski (1987) for the advanced stage is shown in Fig. 4.6. The solid circle in the figure represents a marker particle. The abrupt change in direction, and hence velocity, of the particle as it crosses the rupture layer is evident. The measured velocity vectors (Fig. 4.7) show that the material moves like a plug in the bin section and approximately radially in the lower part of the hopper. These features

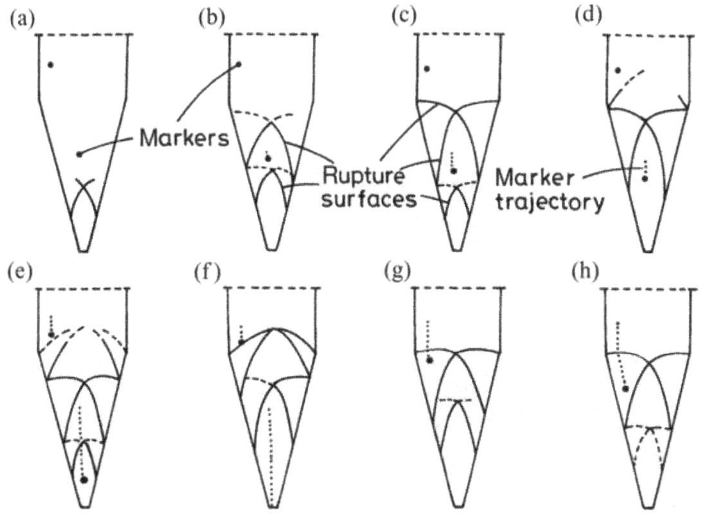

Figure 4.4. Pattern of rupture layers inferred from the radiographs shown in Fig. 4.3. (Reprinted from Michalowski, 1987, with permission from Elsevier Science.)

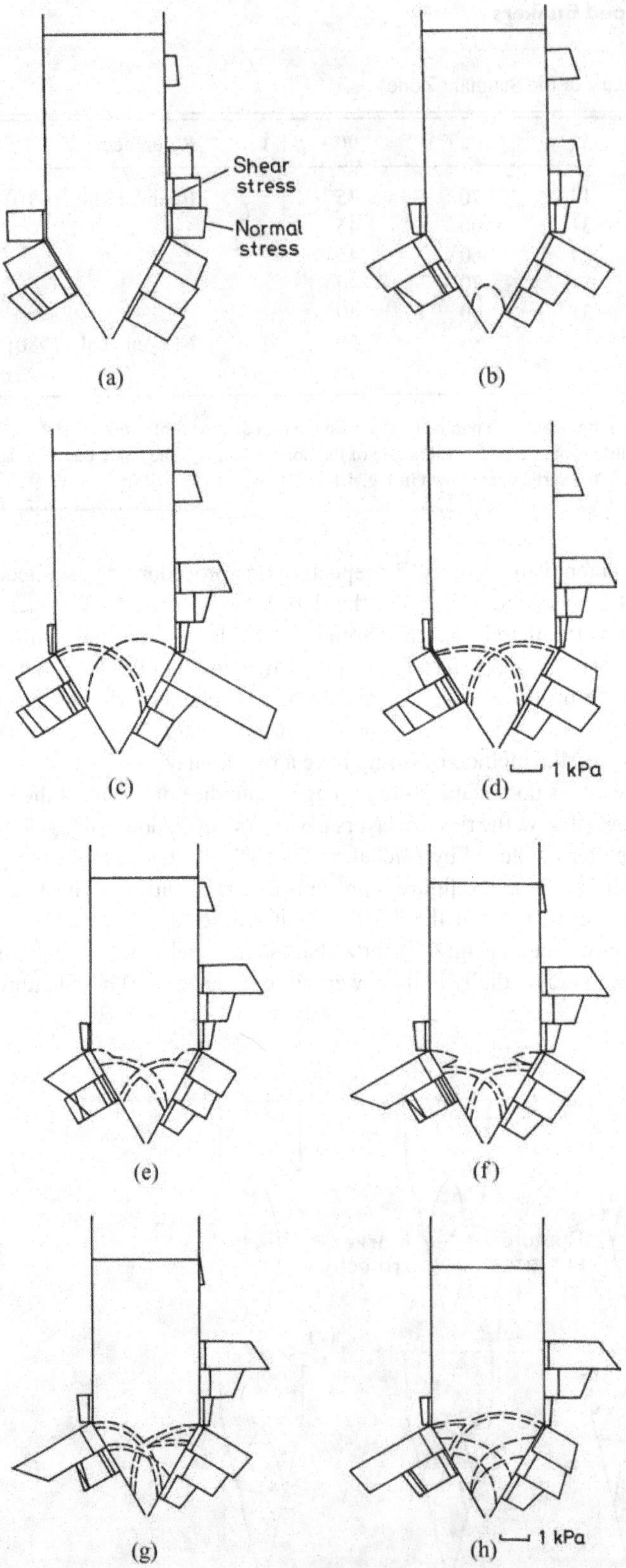

Figure 4.5. Wall stresses and the pattern of rupture layers after intermittent discharge of sand from a Perspex bunker of square cross section. The wall stresses were measured by strain-gauge load cells, and the rupture layers were traced using radiographs. Figure (a) shows the stresses after filling the bunker, and figures b–l show the stresses and positions of the rupture layers after incremental discharges. Parameter values: particle diameter = 0.6–1.2 mm, wall angle of hopper section $\theta_w = 30°$, width of the bin section = 150 mm, slot width = 5 mm. (Reproduced from Blair-Fish and Bransby, 1973, with permission from the American Society of Mechanical Engineers.)

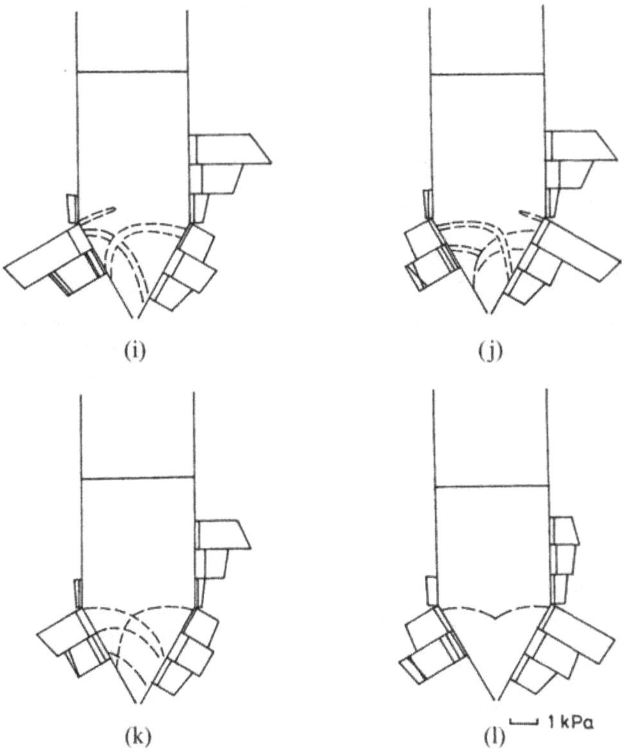

Figure 4.5. *(continued)*

of the velocity field have also been reported by earlier investigators (Bransby et al., 1973; Lee et al., 1974; Bransby and Blair-Fish, 1975).

Thus the transition from vertical flow to quasi-radial flow occurs through a region bounded by the hopper wall and two rupture layers. In this region, velocity vectors are approximately parallel to the hopper wall (Bransby et al., 1973; Lee et al., 1974; Bransby and Blair-Fish, 1975; Drescher et al., 1978). For a bunker with steep hopper walls, as in Fig. 4.7, it is difficult to discriminate between "radial" flow and flow "parallel to the wall." When the walls are shallower, as in Fig. 4.8, the material does appear to move parallel to them. There does not appear to be a simple physical explanation for the observed kinematics. On the other hand, more extensive data are needed before definite statements can be made about the velocity field.

The experiments of Blair-Fish and Bransby (1973), Bransby and Blair-Fish (1974), and Michalowski (1987) suggest that rupture layers are dynamic entities. In the advanced stage of flow, they form at the bin–hopper transition, grow for some time, and then move downward. Meanwhile, new rupture layers form at the transition points, as shown in Fig. 4.9.

Simultaneous monitoring of flow patterns and wall stresses shows that the initiation of rupture layers at the transition is accompanied by a large increase in the normal stress on the hopper wall (Fig. 4.5). The stress then decreases as the rupture layer grows. This process is repeated when the next rupture layer forms. To understand the occurrence of a peak in the normal stress, an analogy with the shear tests discussed in §2.10 may be helpful. When a dense sample is sheared, the shear stress T exhibits a maximum (Fig. 2.17a). In the shear box (Fig. 2.17b), the shear stress is proportional to the force F, which is the normal

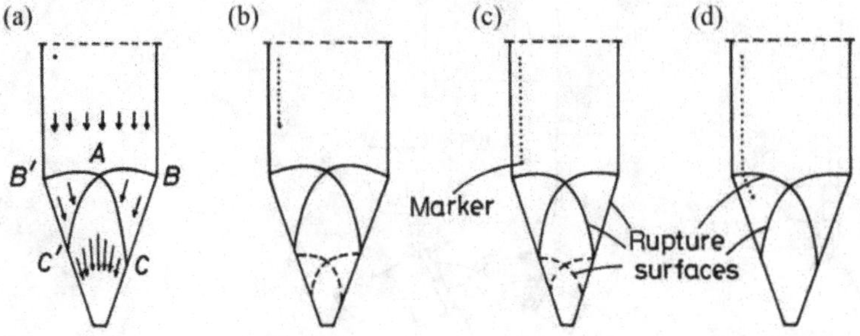

Figure 4.6. Pattern of rupture layers in the advanced stages of the flow of sand through a mass flow bunker having glass faces and wooden walls. Parameter values: wall angle of the hopper section $\theta_w = 20°$, the rest are as in Fig. 4.3. The patterns (a)–(d) were deduced from radiographs taken at different instants of time, but the time interval between them has not been reported. (Reprinted from Michalowski, 1987, with permission from Elsevier Science.)

force acting on the vertical surface of the box. Suppose that the normal stress N exerted by the hopper wall on the material adjacent to it is regarded as an analog of F. Then a dense element of the material which enters the transition region will shear under the action of N and dilate. In this process, N will exhibit a maximum.

If this argument is correct, and the bunker is filled with a loose sample of the material, compaction should occur when the material begins to flow. Such behavior has not been reported in the literature. Perhaps the weight of the material invariably causes it to be in a state which is "denser than critical." Experiments conducted under microgravity conditions may provide valuable information about this point.

The intermittent formation of rupture layers may be responsible for the observed oscillations of the wall stresses in the transition region (Fig. 4.10). The amplitude of the oscillations is much less just above the transition than just below, suggesting that rupture layers do not propagate far into the bin section.

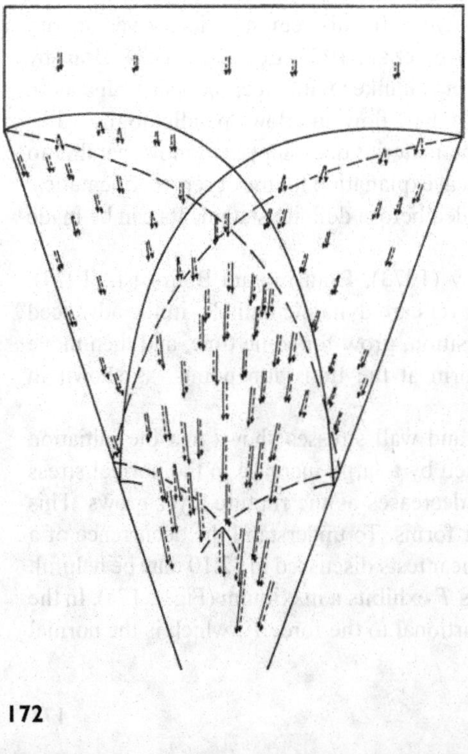

Figure 4.7. Predicted (- - -) and measured (—) velocity vectors and rupture layers for the flow of sand through a bunker. Parameter values: $\theta_w = 20°$, the rest are as in Fig. 4.3. (Reprinted from Michalowski, 1987, with permission from Elsevier Science.)

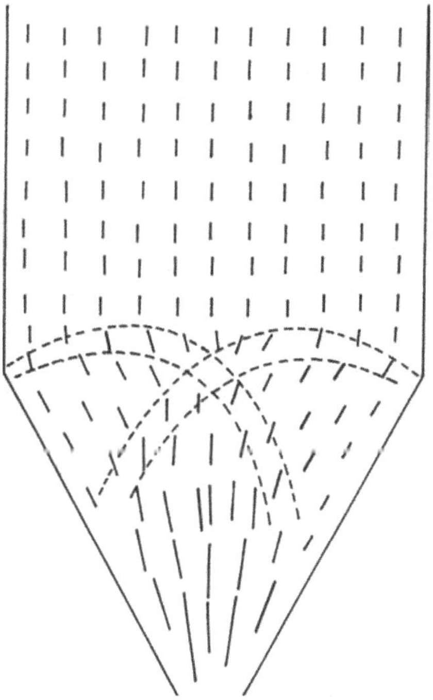

Figure 4.8. Measured displacements (—) of marker particles (lead shot of diameter 2 mm) during the discharge of sand through a bunker having Plexiglas faces. The broken curves indicate the rupture layers. Parameter values: particle diameter = 0.6–8 mm, wall angle of the hopper section $\theta_w = 30°$, dimensions of the exit slot = 76.2 × 5.1 mm. (Reproduced from Lee et al., 1974, with permission from Prof. S. C. Cowin and the Society of Rheology.)

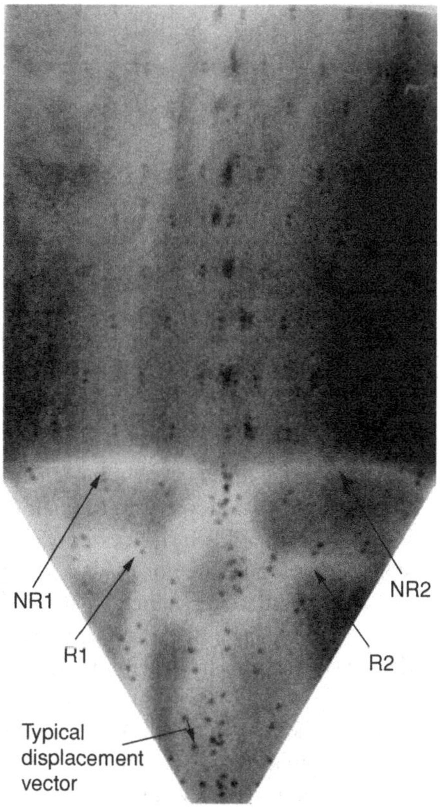

Figure 4.9. Double-exposure radiograph of a bunker containing sand. The old rupture layers R1 and R2, and the new rupture layers NR1 and NR2 are marked in the figure. Parameter values are as in Fig. 4.5. (Reproduced from Blair-Fish and Bransby, 1973, with permission from the American Society of Mechanical Engineers.)

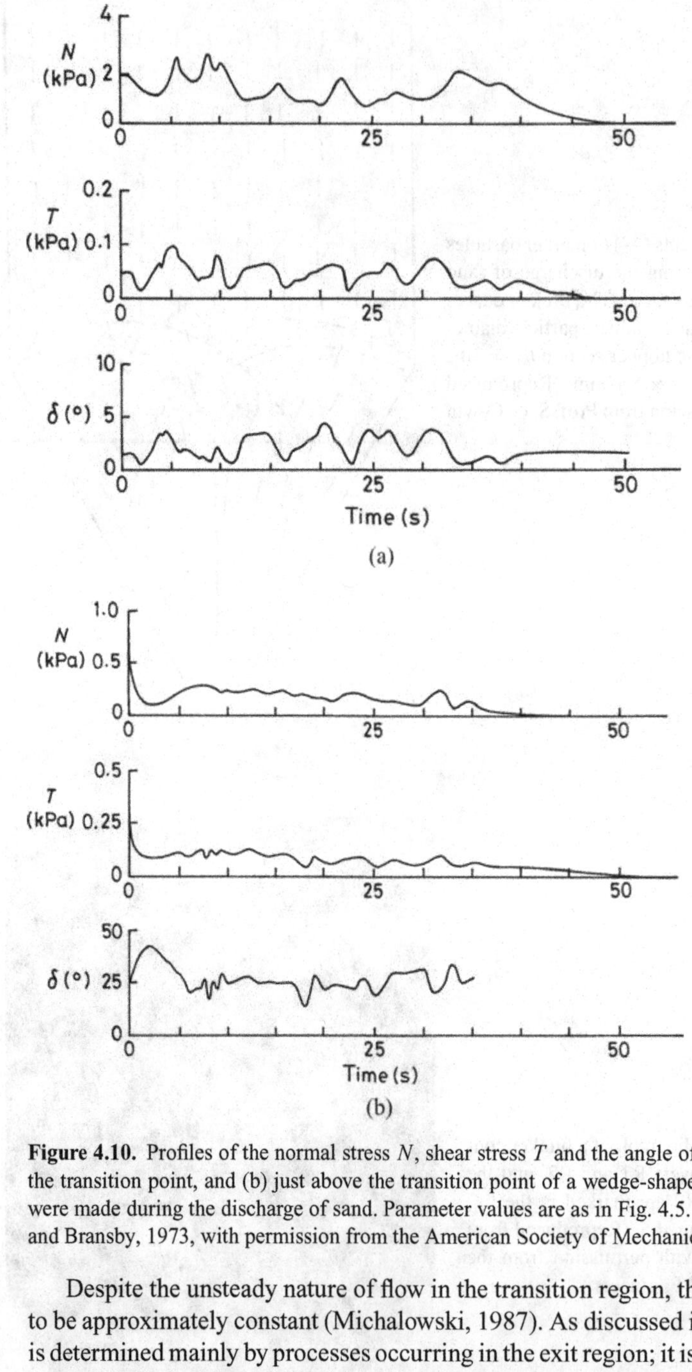

Figure 4.10. Profiles of the normal stress N, shear stress T and the angle of wall friction δ (a) just below the transition point, and (b) just above the transition point of a wedge-shaped bunker. The measurements were made during the discharge of sand. Parameter values are as in Fig. 4.5. (Reproduced from Blair-Fish and Bransby, 1973, with permission from the American Society of Mechanical Engineers.)

Despite the unsteady nature of flow in the transition region, the mass flow rate is found to be approximately constant (Michalowski, 1987). As discussed in Chapter 3, the flow rate is determined mainly by processes occurring in the exit region; it is likely that a quasi-steady state prevails in this region. This conjecture is supported by a decrease in the amplitude of the stress oscillations as the exit is approached (Fig. 4.11).

So far, we have discussed displacement and stress fields. Let us now briefly consider density variation. The density of the material in the hopper section is lower than that in the bin section (Table 4.2). This is caused by the dilation of material on crossing the rupture layers in the transition region. The density in the hopper section is significantly different from, and relatively insensitive to, the initial or poured density. When the initial density is high, dilation is observed in the bin section also.

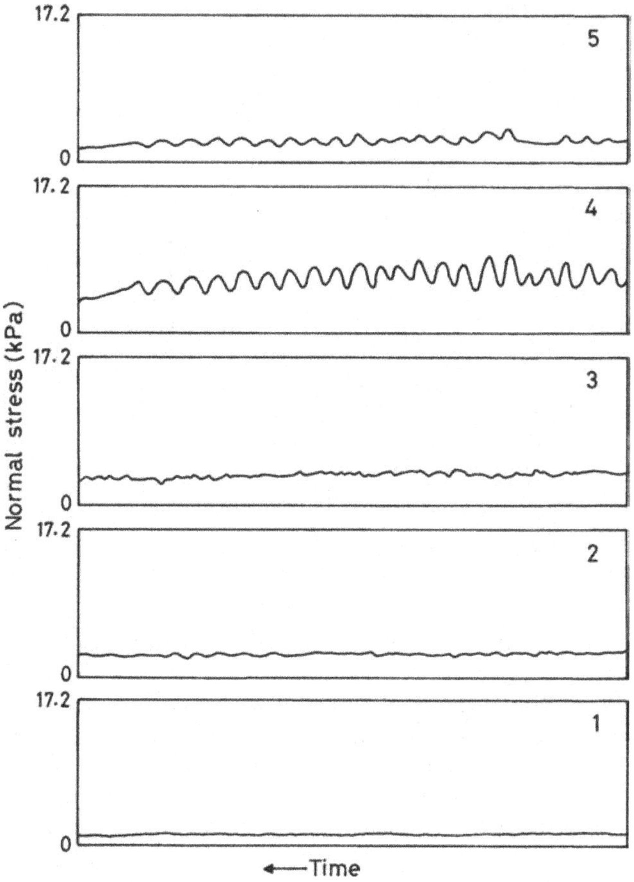

Figure 4.11. Profiles of the normal stress at various locations along the wall for the flow of glass beads through of a Plexiglas bunker. The elevation of the points 1–5 relative to the virtual apex of the hopper section is given in parentheses in mm: 1 (61.3); 2 (125.2); 3 (189.1); 4 (239.2); 5 (253.0). The transition point is at a height of 264 mm above the virtual apex. Parameter values: particle diameter = 0.125–0.175 mm, width of the bin section = 305 mm, width of the exit slot = 50.8 mm, wall angle of the hopper section $\theta_w = 30°$. (Reproduced from Connelly, 1979, with permission from the American Society of Mechanical Engineers.)

The above remarks apply mainly to coarse materials flowing through mass flow bunkers. Most of the data pertain to this case. The reader is referred to Michalowski (1987) for a discussion of rupture layers in funnel flow bunkers.

4.1.3. Wall Stresses

Consider the stress profiles shown in Fig. 4.5. Stresses in the poured material (Fig. 4.5a) are quite different from those prevailing after some amount of material has been discharged (Figs. 4.5b–l). As explained earlier, this is due to the formation of rupture layers and the consequent dilation of material in the hopper section.

During the first three increments of discharge (Fig. 4.5b–d), the stresses in the bin section change, but the underlying mechanism is not understood. As indicated in Table 4.2, the material may have dilated here. In the interval between the fourth and seventh increments of discharge (Figs. 4.5e–h), the stress profile in the bin section does not change much. This suggests that changes caused by the intermittent formation of rupture layers at the bin–hopper transition do not propagate far into the bin section. Thus a "fully developed" stress field prevails after a few increments of discharge. This can also be seen from the dynamic

Table 4.2. Density of Sand in the Bin and the Hopper Sections During the Advanced Stage of Flow through a Bunker with Plexiglas Faces and Wooden Walls*

θ_w (°)	ρ_i	ρ_b	ρ_h
10	1,560	1,560	1,510
"	1,640	1,580	1,520
"	1,730	1,590	1,530
15	1,560	1,560	1,490
25	1,580	1,580	1,480
"	1,740	1,620	1,510

* θ_w = wall angle of the hopper section, ρ_i = initial (filling) density, ρ_b = density in the bin section, measured near the centerline of the bunker and at a height of about 200 mm above the transition point, ρ_h = density in the hopper section, measured near the centerline of the bunker and in the region of approximately radial flow. An ultrasonic technique was used to measure the density, which is listed in units of kg m^{-3}. The standard deviation of the measurements was 1.1–20.7 kg m^{-3}. Parameter values: particle diameter = 0.25–1.75 mm, width of the bin section ≈300 mm, gap between the faces = 40 mm.
Source: Michalowski (1987).

experiments (Fig. 4.10b), where the stresses in this section attain quasi-steady values after about 10 s.

Considering stresses in the transition region, the normal stress exerted on the hopper wall N_h is much larger than that exerted on the bin wall N_b (Figs. 4.5 and 4.10). The data shown in Table 4.3 suggest that N_h/N_b is in the range 2–30 for flow through laboratory scale bunkers. The values of N_h and N_b are rarely measured simultaneously at the transition point. Hence the range given above for N_h/N_b has been obtained by calculating appropriate ratios from the ranges of values reported for N_h and N_b. The data show that the stress field changes considerably in the transition region. Attempts to model this behavior will be discussed later.

Consider the stress oscillations shown in Figs. 4.10 and 4.11. As cohesionless granular materials discharge rapidly, and it is difficult to store enough material in laboratory-scale

Table 4.3. Ratio of the Normal Stress Exerted on the Hopper Wall N_h to the Normal Stress Exerted on the Bin Wall N_b*

Material	δ (°)	ϕ (°)	θ_w (°)	N_h/N_b (Theory)	N_h/N_b (Experiment)	Source of data
Sand	25	30	30	1.9	2.6–29.7	Blair-Fish and Bransby (1973)
"	20	33	14.5	6.1	4.3–5.6	Manjunath (1987, p. 38)
Plastic Grains	11	35	30	9.3	3.8–8.8	Benink (1989, pp. 147–149)
"	11	38	30	12.4	3.8–8.8	"
"	11	35	45	6.4	2.5–8.0	"
"	11	38	45	8.2	2.5–8.0	"
"	11	35	60	4.5	2.1–8.2	"
"	11	38	60	5.5	2.1–8.2	"

* The stresses were measured using load cells mounted near the transition point of the bunker. The numbers in the column labeled "theory" are based on (4.25). Here δ is the angle of wall friction, ϕ is the angle of internal friction (see (4.7)), and θ_w is the wall angle of the hopper section. For plastic grains, the two values of ϕ shown correspond to estimates obtained using the ring shear tester and the Jenike shear cell, respectively.

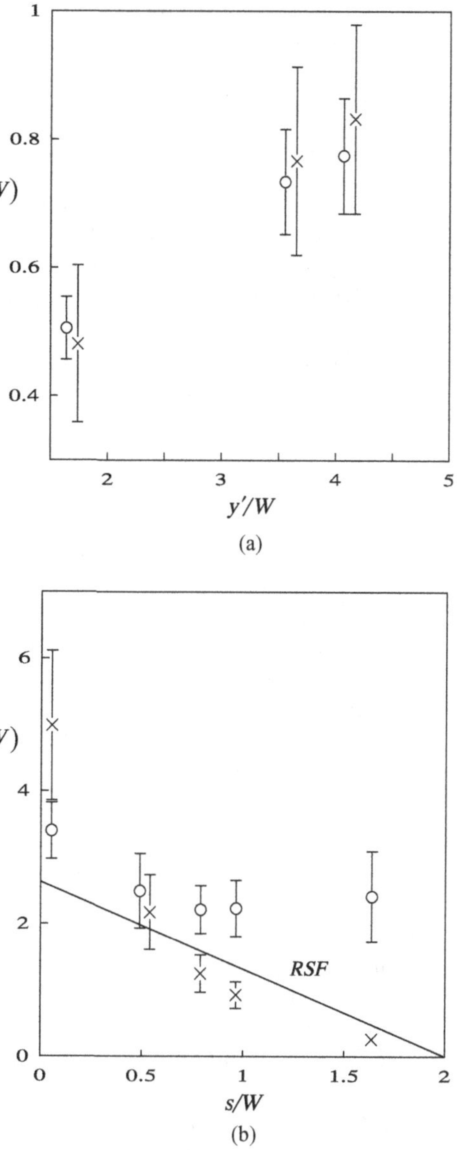

Figure 4.12. Profiles of the dimensionless normal stress $N/(\rho\, g\, W)$ measured along (a) the bin wall, and (b) the hopper wall of a wooden bunker after filling (○), and during flow (×). Here ρ is the density of the granular material, g is the acceleration due to gravity, W is the half-width of the bin section, and s is the distance measured downward along the hopper wall (Fig. 4.26). The transition point corresponds to $s/W = 0$ ($y'/W = 4.35$), and the exit slot to $s/W = 1.94$, where y' is the vertical distance measured downward from the upper surface of the fill (Fig. 4.17). Plastic grains were used as the granular material. The vertical lines show the ranges of the measured stresses. In (a), the abscissas for the circles have been reduced by 0.1 for the sake of clarity. Similarly, in (b), the abscissa for one of the filling stresses has been reduced by 0.05. In (b), the line labeled RSF represents the radial stress field. Parameter values: particle diameter = 2–3 mm, $W = 0.575$ m, height of the bin section = 3 m, initial height of the fill above the transition point = 2.5 m, distance between the front and back faces = 1 m, width of the exit slot = 60 mm, wall angle of the hopper section $\theta_w = 30°$, angle of internal friction $\phi = 36.5°$ (mean of the values obtained using the Jenike tester (38°) and the ring shear tester (35°)), angle of wall friction $\delta = 11°$, diameter of the active face of the load cell = 53 mm. (Adapted from Benink, 1989, p. 147.)

bunkers, the experiments are usually of short duration (Fig. 4.10). If the initial and final transients are omitted from the traces shown in Fig. 4.10, the frequency of oscillations is about 0.2 Hz. Data over a longer duration of flow are required to get a proper estimate of the frequency and amplitude of the oscillations. The data shown in Fig. 4.11 are better suited for this purpose. There is no scale on the time axis, but Connelly (1979) states that the frequency of the stress oscillations at location 5 is 2.5–4 Hz. Figure 4.11 also shows that the amplitude of the oscillations decreases on moving down the hopper.

The angle of wall friction δ (Fig. 4.10) is calculated as $\tan^{-1}(|T|/N)$, where T and N are the measured shear and normal stresses, respectively, exerted by the material on the wall. In the transition region, the value of δ fluctuates. Hence the friction boundary condition (3.34) must be modified in this region, but a satisfactory alternative has yet to be suggested.

Figure 4.12 shows the normal stress measured on the wall of a bunker. The stress increases on moving down the bin section. As indicated earlier, there is a large increase in stress on crossing the transition point, which is located at $y'/W = 4.35$ or $s/W = 0$. Here y' is the vertical coordinate, measured as shown in Fig. 4.17. The mean values of the stresses measured during filling (circles) and flow (crosses), do not differ significantly in the bin section (Fig. 4.12a), in accord with the observations cited earlier. However, the range is larger during flow. In the hopper section, the mean values of the flow stresses are smaller than the mean values of the filling stresses, except just below the transition point (Fig. 4.12b). The results suggest that the stress fields for filling and flow differ significantly in the hopper section, but not in the bin section. Further, the decrease in the flow stresses on moving down the hopper is an indication of the tendency to approach the radial stress field. The latter can be derived as discussed in §3.5.2 and is represented by the line labeled RSF in Fig. 4.12b.

To compare the data of various investigators, obtained using bunkers of rectangular and circular cross sections, it is convenient to introduce a scaled normal stress N', defined by

$$N' = \left(\frac{N_b}{\rho\, g\, W}\right)\left(\frac{P}{A}\right)\tan\delta \tag{4.1}$$

Here N_b is the normal stress exerted on the wall of the bin section, ρ is the density of the granular material, g is the acceleration due to gravity, W is a characteristic length scale and P and A denote the dimensionless perimeter and cross-sectional area, respectively, of the bin section, i.e., the perimeter is scaled by W and the area by W^2. The length W is chosen either as the half-width or the radius of the bin section, depending on whether the cross section is rectangular or circular. The basis for the above scaling is indicated in Problem 4.2.

The symbols in Fig. 4.13 show the measured values of N'. There is considerable scatter in the data, particularly for $\eta' > 3$. In the static case (open symbols), this may be caused by the variation in the solids fraction ν and the structure of the packing from one run to the next. In the dynamic case (filled symbols), temporal oscillations (see Fig. 4.10) also contribute to the scatter. These oscillations are more pronounced near the bin–hopper transition, which corresponds to η' values in the range 2–7 for the data shown in Fig. 4.13. In most cases, the ranges of the static and dynamic stresses at a point tend to overlap, with the maximum dynamic stress often exceeding the maximum static stress.

Benink (1989, p. 66) conducted experiments on the batch discharge of plastic grains through a wedge-shaped bunker. Both normal and shear stresses were measured at one location on the wall of the bin section. The dynamic normal stress had a range of values, with the maximum exceeding the static value, but the dynamic shear stress was slightly lower than the static value. Correspondingly, the dynamic angle of wall friction had a range of values, with the minimum being lower than the static value. More extensive experiments, including simultaneous measurement of stress, density, and velocity fields, are needed before a clear picture emerges.

4.1.4. Bins

A bin is a special case of a bunker, having horizontal hopper walls (Fig. 3.6), in which funnel flow occurs. The flow patterns and wall stresses observed in bins of rectangular cross section are discussed briefly below.

Figure 4.14 shows the pattern of streamlines obtained by photographically tracking marker particles as they flow through the bin. The band labeled "velocity discontinuity" indicates the region where the velocities change sharply. Though radiographs were not taken, the bands may be similar to the rupture layers issuing from the transition point of a bunker (Fig. 4.5). It would be instructive to measure densities in this region to check whether there is significant dilation across the velocity discontinuities.

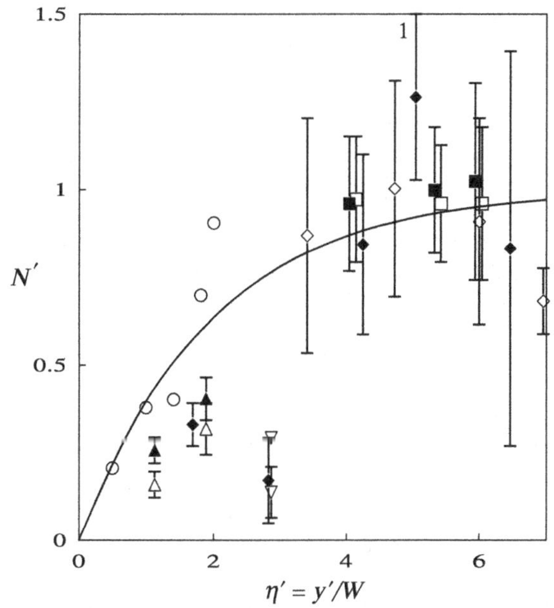

Figure 4.13. Variation of the dimensionless normal stress N' with the dimensionless depth $\eta' = y'/W$ below the upper surface of the fill. Here N' is the normal stress on the wall of the bin section, defined by (4.1), y' is the vertical distance measured downward from the upper surface of the fill (Fig. 4.17), and W is a characteristic length. It is chosen either as the half-width of the bin section for a bunker of rectangular cross section, or as the radius of the bin section for a bunker of circular cross section. The curve shows the Janssen solution (4.2) for $K = 0.36$, $\delta = 35°$. The symbols represent the data of several investigators (see Tables 4.4 and 4.5): open symbols – static stresses measured after filling the bunker, filled symbols – dynamic stresses measured during discharge of the material. In the case of the squares and the triangles, the abscissas for the dynamic stresses have been reduced by 0.1 for the sake of clarity. The vertical lines show the ranges of the measured stresses. For the point marked 1, the actual range of the normal stress is 1.0–3.1.

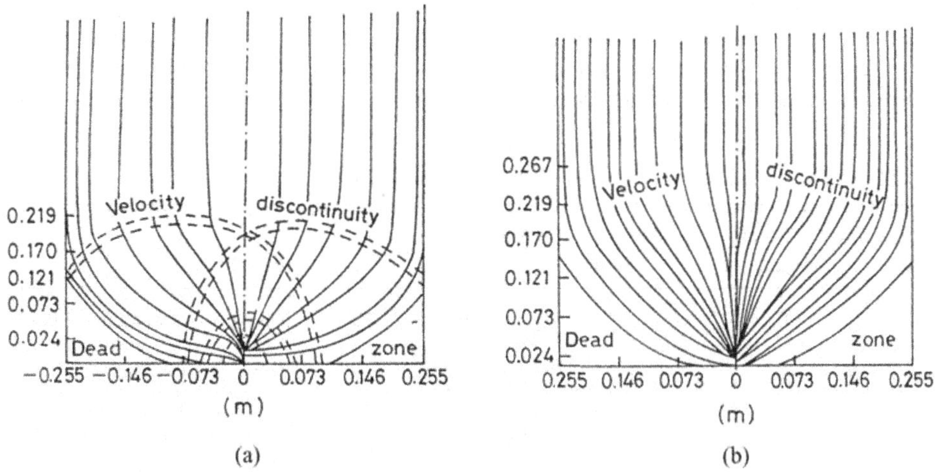

Figure 4.14. Streamlines observed during the flow of mustard seeds through a bin with glass faces and aluminium walls. Parameter values: particle diameter = 2–2.7 mm, width of the bin = 510 mm, distance between the faces = 80 mm, width of the exit slot = 6.4 mm (Fig. 4.14a), 12.7 mm (Fig. 4.14b). (Reprinted from Tüzün and Nedderman, 1985a, with permission from Elsevier Science.)

Table 4.4. Sources of Data on Wall Stresses in Bunkers*

Symbol	Granular material	Wall material	Cross section	W (m)	Reference
▽	Sand	Perspex	Rectangular	0.075	Blair-Fish and Bransby (1973)
□	Iron ore	–	Circular	0.4825	Clague (1973) (cited in Blight, 1986)
◇	Sand	Steel	Circular	0.75	Van Zanten and Mooij (1977)
△	Sand	Perspex	Rectangular	0.265	Manjunath (1987)
○	Sand	Steel	Rectangular	1.0	Jarrett et al. (1995)

* The data is shown in Fig. 4.13, and the material properties are given in Table 4.5.

Consider the stress profiles, which were measured using load cells. Figure 4.15a shows the load cell traces at two locations on the bin wall, which are well above the boundaries of the stagnant zone. The stresses oscillate with a frequency of 0.5–1 Hz, and the amplitude is about 20–25% of the mean value. Tüzün and Nedderman (1985b) attributed these oscillations to the *stick–slip* motion of the particles. A brief description of the phenomenon of stick–slip is given below.

Consider a block resting on a horizontal surface (Fig. 2.5), and suppose that the horizontal force H is applied to the block through a spring. For certain ranges of values of the weight and the applied force, the block may not slide steadily, but in a jerky manner. It alternately "sticks" and "slips" leading to an oscillatory frictional force. Such a stick–slip motion has been observed during the sliding of a layer of granular material relative to a plane surface (Platonov and Poltorak, 1970; Budny, 1979; Nasuno et al., 1997). In the experiments of Tüzün and Nedderman (1985b), it is likely that the observed stress oscillations were due to stick–slip.

For load cells located near the stagnant zone, in addition to oscillations of the stick–slip type, there are slower oscillations with a frequency $\approx$ 0.01–0.02 Hz (Fig. 4.15b). It is tempting to attribute the latter to the intermittent formation of rupture layers. Tüzün and Nedderman (1985b) expressed reservations about this viewpoint, as the velocity field deduced from their photographs appeared to be steady. More experiments are needed to clarify this point.

Figure 4.16 shows profiles of the normal and shear stresses along the bin wall. The circles and triangles represent static stresses measured after filling, and time-averaged flow stresses, respectively. At many locations, data for two runs show a scatter of 10–40% of

Table 4.5. Parameter Values Corresponding to the Data Shown in Fig. 4.13*

Particle density (kg m^{-3})	Solids fraction ν	Density ρ (kg m^{-3})	Angle of wall friction δ (°)	θ_w (°)	Reference
2,650*	0.66	–	9 (static) 25 (dynamic)	30	Blair-Fish and Bransby (1973)
–	–	1,774	28	11	Clague (1973) (cited in Blight, 1986)
–	–	1,525	23.5	15	Van Zanten and Mooij (1977)
–	–	1,520	20	14.5	Manjunath (1987)
2,650*	0.6*	–	35	43	Jarrett et al. (1995)

* Quantities marked with an asterisk are assumed values, as the actual values have not been reported.

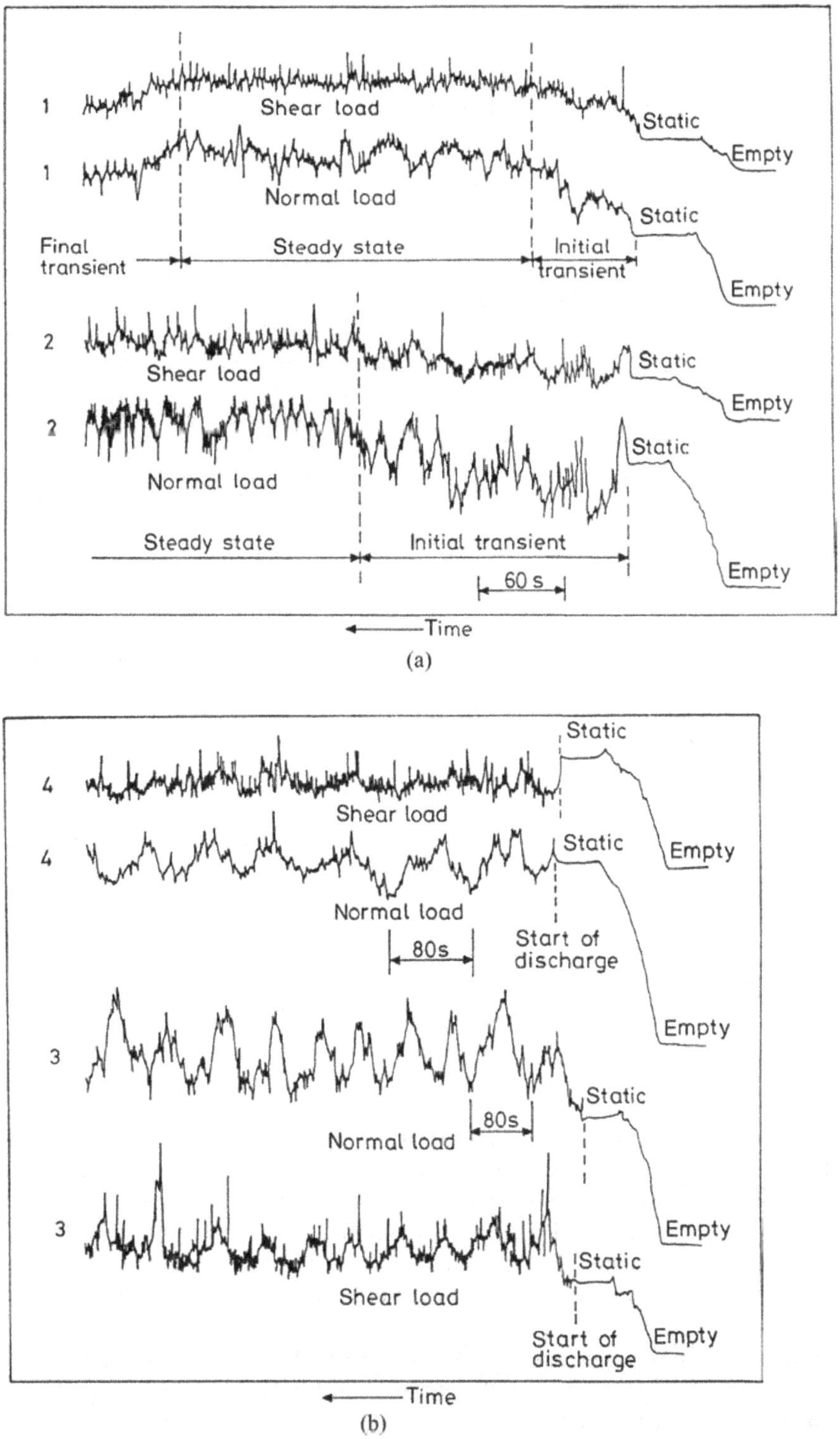

Figure 4.15. Shear and normal stresses exerted on the wall of a bin during the flow of mustard seeds: (a) 1, $H = 1.21$ m; 2, $H = 0.75$m, (b) 3, $H = 0.28$ m; 4, $H = 0.13$ m. The stresses were measured using a load cell with a circular active surface of diameter 28.6 mm. The center of this surface was located at a height H above the base of the bin. The dimensions of the bin are given in Fig. 4.14. (Reprinted from Tüzün and Nedderman, 1985b, with permission from Elsevier Science.)

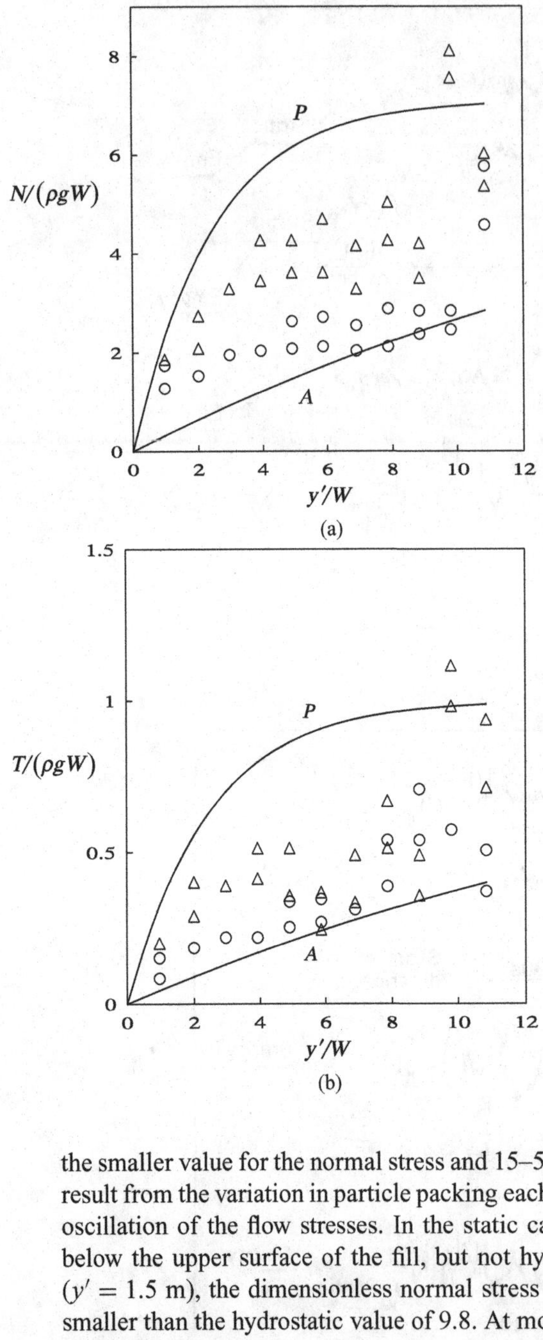

Figure 4.16. Profiles of (a) the dimensionless normal stress $N/(\rho g W)$, and (b) the dimensionless shear stress $T/(\rho g W)$ exerted on the bin wall, measured after filling (○) and during flow (Δ) of mustard seeds through a bin. Here ρ is the density, g is the acceleration due to gravity, W is the half-width of the bin, and y' is the distance measured vertically downward from the top of the bin. Flow stresses are time-averaged values. The curves marked P and A denote the Janssen–Walker solution for the passive and active stress fields, respectively (see Problem 4.1). Parameter values are as given in Fig. 4.14, except that the value of the width of the exit slot has not been reported. (Data of Tüzün and Nedderman, 1985b.)

the smaller value for the normal stress and 15–50% for the shear stress. The difference may result from the variation in particle packing each time the bin is filled, and also the temporal oscillation of the flow stresses. In the static case, the stress profiles increase with depth below the upper surface of the fill, but not hydrostatically. For example, at $y'/W = 9.8$ ($y' = 1.5$ m), the dimensionless normal stress on the bin wall is 2.5–2.8, which is much smaller than the hydrostatic value of 9.8. At most of the locations, flow stresses are higher than static stresses.

To summarize, flow in bins is similar in many respects to flow in mass flow bunkers. It is not yet known whether the stress oscillations near the top of the converging flow region are accompanied by velocity and density fluctuations, as in the case of bunkers.

4.2. MODELS FOR BUNKER FLOW

So far, some of the experimental observations related to stress and velocity fields in bunkers have been presented. Let us now discuss attempts to predict these fields. Two broad classes

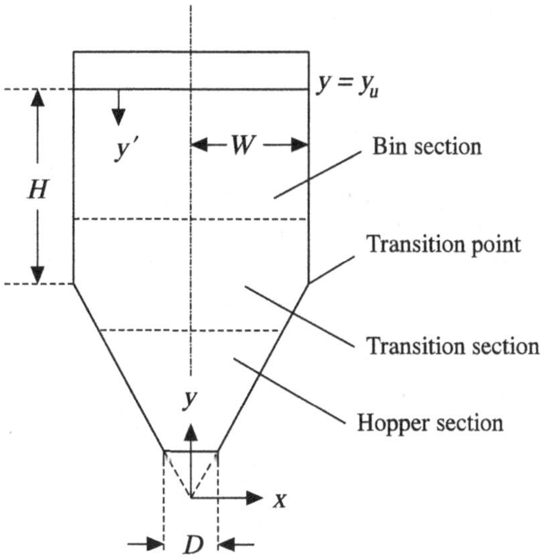

Figure 4.17. The three sections used for the analysis of bunker flow. Here H is the height of the material above the transition point of the bunker.

of models are available: (i) models in which the bunker is divided into the bin, hopper, and transition sections (Fig. 4.17), and each section is analyzed separately, and (ii) models which treat the bunker as a whole. Here we discuss models of type (i), deferring a discussion of models of type (ii) to §6.3.

Though experiments suggest that flow in the transition region is unsteady, only steady state models have been used so far. Further, most of the models have been confined to the determination of either the stresses (Jenike, 1961; Walker, 1966; Jenike and Johanson, 1968; Savage and Yong, 1970; Walters, 1973; Enstad, 1975; Horne and Nedderman, 1978; Wilms and Schwedes, 1985; Drescher, 1991, pp. 86–132) or the kinematics (Jenike, 1961; Drescher et al., 1978; Michalowski, 1987). Consider each section of the bunker in turn.

4.2.1. The Bin Section

Most of the models assume that the material in the bin section moves like a plug. This assumption is supported by the data shown in Figs. 4.7 and 4.8. Careful measurements of the velocity profiles show that there are shear layers adjacent to the bin walls, whose thickness is of the order of 10 particle diameters (Nedderman and Laohakul, 1980; Natarajan et al., 1995). These layers are ignored here.

The assumption of plug flow in the bin section implies that $v_x = 0$, $v_y =$ constant (Fig. 4.17). Hence the coaxiality condition and the flow rule are identically satisfied, and the inertial terms vanish in the momentum balances. Further, the magnitude of the vertical velocity, and the solids fraction in the bin section, are both indeterminate. The stress field can be determined independently of the velocity field, but the solids fraction field must be specified a priori. It is usually assumed that the solids fraction is a known constant. The problem then reduces to the determination of the stress field by integrating the equilibrium equations. Hence the methods described in this section are applicable in the static case also.

Available methods can be grouped into three categories: (a) the method of wedges (Coulomb, 1776; Nedderman, 1992, pp. 47–83), (b) the method of differential slices (Janssen, 1895; Walker, 1966; Jenike and Johanson, 1968; Walters, 1973; Hancock and Nedderman, 1974; Drescher, 1991, pp. 95–132; Nedderman, 1992, pp. 84–126), and (c) the method of characteristics (Johanson, 1964; Savage and Yong, 1970;

Horne and Nedderman, 1976; Drescher, 1991, pp. 86–95; Nedderman, 1992, pp. 190–242). Examples illustrating (b) and (c) are discussed below.

(a) *The Janssen solution*
The Janssen solution has been discussed earlier in §2.3 and Walker's modification is discussed in Problem 4.1. Here we examine certain features of the solution, and compare the predicted normal stresses with measurements. To compare with data for bunkers of rectangular and circular cross sections, the solution must be modified as discussed by Cowin (1977) (see also Problem 4.2). Thus (2.13) must be replaced by

$$V = (1/M') + [V_0 - (1/M')]\, e^{-M'\eta'} \tag{4.2}$$

where $M' \equiv (P/A)\,K\ \tan\delta$, K is the coefficient of earth pressure at rest, $\overline{N} = N_b/(\rho g W)$ is the dimensionless normal stress exerted on the wall of the bin section, δ is the angle of wall friction, A and P are the dimensionless cross-sectional area and perimeter, respectively, $\eta' = y'/W$, y' is the vertical coordinate measured downward from the horizontal upper surface of the material (Fig. 4.17), W is a characteristic length, chosen either as the half-width or the radius of the bin section, depending on whether the cross section is rectangular or circular, and V is the area averaged dimensionless normal stress in the vertical direction, defined by

$$V \equiv (1/A) \int \overline{\sigma}_{y'y'}\, \mathrm{d}A \tag{4.3}$$

In (4.2), $V_0 \equiv V(\eta' = 0)$ is the *surcharge* or average normal stress exerted on the upper surface of the material, and A and P are scaled by W^2 and W, respectively. The stresses are scaled by $\rho g W$, where ρ is the density, assumed constant, and g is the acceleration due to gravity.

As noted earlier, the Janssen solution (4.2) shows that the vertical normal stress V tends to a constant value $V_a = 1/M'$ as the depth η' increases. Hence the normal stress $\overline{N} = K\,V$ exerted on the bin wall also tends to a constant value

$$\overline{N}_a = A/(P\ \tan\delta) \tag{4.4}$$

or using (4.1)

$$N' = 1 \tag{4.5}$$

Similarly, the aymptotic value of the vertical shear stress acting in the downward direction on the bin wall is given by

$$\overline{T}_a = \overline{N}_a\ \tan\delta = A/P \tag{4.6}$$

Further, both $\overline{N}_a$ and $\overline{T}_a$ are independent of the surcharge V_0 acting on the upper surface.

Equation (4.2) cannot be used until a value is specified for the constant M', which in turn depends on the coefficient of earth pressure K and the angle of wall friction δ. It is difficult to estimate K, except in restrictive special cases (see §2.9 and Problem 4.1). Though δ can be estimated from independent experiments, the actual value, as inferred from the measured wall stresses, shows a wide variation. Hence one may wonder about the utility of Janssen's solution. As noted by Nedderman (1992, p. 88), the most reliable elements of the solution are the predicted asymptotic values of the wall stresses, given by (4.4) and (4.6). Any uncertainty in the value of K does not affect the values of $\overline{N}_a$ and $\overline{T}_a$. Further, $\overline{T}_a$ is a purely geometric quantity. In contrast, the vertical normal stress V, and the rate of approach of the stresses to their asymptotic values, depend strongly on K and δ.

By choosing $K = 0.36$, (4.2) can be fitted to a part of the data of Jarrett et al. (1995), as shown by the solid curve and the circles in Fig. 4.13. The data deviate from the curve near the transition region ($\eta' \approx 2$).

The scatter in the data on wall stresses precludes the use of a single value of K for either the static or the dynamic stresses. Even though the predicted asymptotic value of N' is unity (see (4.5)), the data for η' in the range 4–7 show considerable scatter about this limiting value (Fig. 4.13). Therefore, in the absence of reliable estimates of K, ν, and δ, the Janssen solution provides only an order of magnitude estimate of the stresses in the bin section.

Figure 4.16 compares predictions of the Janssen–Walker solution (Walker, 1966; Tüzün and Nedderman, 1985b) (see also Problem 4.1) with the data of Tüzün and Nedderman (1985b) for the flow of mustard seeds through a bin. As mentioned in §3.3, the frictional equations admit two solutions. The major principal stress is nearly vertical in the active case (curve A), and is nearly horizontal in the passive case (curve P). Thus the normal stress on the bin wall is larger in the passive case. Except close to the bottom of the bin, the data lie within the region bounded by the active and passive curves. Hence Tüzün and Nedderman (1985b) suggest that the state of stress in the upper part of the bin is ill defined and the stresses can have any value between the active and passive limits. The indeterminacy of the stresses in this region is consistent with the rigid-body motion in the bin section (Fig. 4.8).

When the bunker is filled, the stresses are reasonably close to the active curve. Near the top of the converging flow zone, the dynamic stresses increase rapidly and attain values which are closer to the passive curve (Fig. 4.16). Thus there may be a switch to the passive state of stress in the converging section. In some cases, the passive values are exceeded, indicating that the passive solution does not represent a true upper bound.

For a bin with a surcharge or applied normal stress V_0 on the upper surface of the granular material, the Janssen solution (4.2) predicts a monotonic variation of the vertical normal stress V with the depth η'. However, recent experiments on quasi-static beds (Vanel et al., 2000; Ovarlez et al., 2003) show that for a range of values of V_0, V exhibits a maximum as η' increases. Models exhibiting such behavior are discussed in Vanel et al. (2000), Ovarlez et al. (2003), and Ovarlez and Clément (2005). However, a simple physical explanation is lacking.

(b) *Solution by the method of characteristics*

The Janssen solution discussed above involves several assumptions. Many of these assumptions can be relaxed, at the expense of greater computational effort, by directly solving the force balances (2.7) and (2.17). As discussed in §3.2.1, we invoke the critical state approximation and assume that the states of stress must lie along the critical state curve. Assuming that the latter is a straight line, the mean stress σ and the deviator stress τ are related by

$$\tau = \sigma \sin\phi \tag{4.7}$$

where ϕ is a material constant called the angle of internal friction. Alternatively, (4.7) can be regarded as the Mohr–Coulomb yield condition for a cohesionless material (see (2.38)). To proceed further, it is assumed that (4.7) holds at every point in the bin section.

Using (2.102) to express the stress components in terms of σ and τ, the balances (2.17) and (2.7) can be written in dimensionless form as

$$\begin{aligned} &\frac{\partial}{\partial \xi}[\overline{\sigma}(1+\sin\phi\,\cos 2\gamma)] + \frac{\partial}{\partial \eta}(-\overline{\sigma}\,\sin\phi\,\sin 2\gamma) = 0 \\ &\frac{\partial}{\partial \xi}(-\overline{\sigma}\,\sin\phi\,\sin 2\gamma) + \frac{\partial}{\partial \eta}[\overline{\sigma}(1-\sin\phi\,\cos 2\gamma)] + 1 = 0 \end{aligned} \tag{4.8}$$

where

$$\xi = x/W; \quad \eta = y/W; \quad \overline{\sigma} = \sigma/(\rho g W); \quad \overline{\tau} = \tau/(\rho g W) \tag{4.9}$$

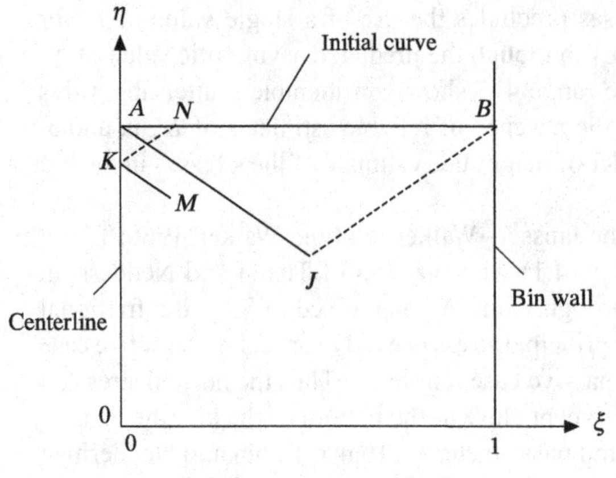

Figure 4.18. Characteristics corresponding to $j = 1$ (the solid lines AJ and KM) and to $j = 2$ (the broken lines BJ and KN). The characteristics are shown as straight lines, but are curved in general. Here $\xi = x/W$, $\eta = y/W$, W is the half-width of the bin section, and the coordinate system is chosen as shown in Fig. 4.17.

W is the half-width of the bin section, γ is angle measured anticlockwise from the major principal stress axis to the x axis (Fig. 2.11), ρ is the density, g is the acceleration due to gravity, and the coordinate system is shown in Fig. 4.17.

Equations (4.8) have to be solved for the mean stress $\overline{\sigma}$ and the angle γ. As shown in Appendix C, these equations are *hyperbolic*, and hence reduce to ordinary differential equations along the characteristic curves defined by

$$\frac{d\xi}{ds_j} = \cos(\gamma \pm \Omega), \quad \frac{d\eta}{ds_j} = -\sin(\gamma \pm \Omega), \quad j = 1, 2 \tag{4.10}$$

where

$$\Omega \equiv (\pi/4) - (\phi/2) \tag{4.11}$$

and s_j is the dimensionless arc length along the jth characteristic. In (4.10), the upper sign corresponds to $j = 1$, and the lower sign to $j = 2$. The *stress characteristics* (4.10) are symmetrically disposed relative to the major principal stress axis, as indicated in Fig. C.2.

In addition to the stress characteristics, any curve along which $\overline{\sigma}$ vanishes is also a characteristic (see Appendix C). This case will be discussed later.

Along each of the stress characteristics, (4.8) reduces to an ordinary differential equation, which is called the *compatibility condition*. The compatibility conditions are given by (see Appendix C)

$$\cos\phi \frac{d\overline{\sigma}}{ds_j} \pm 2\overline{\sigma} \sin\phi \frac{d\gamma}{ds_j} \mp \cos(\gamma \mp \Omega) = 0, \quad j = 1, 2 \tag{4.12}$$

where the upper sign corresponds to $j = 1$, and the lower sign to $j = 2$. Equations (4.12) are nonlinear, and hence have to be integrated numerically along the characteristics. A brief outline of the procedure is given in Appendix C.

The schemes described in §C.1 generate the stress field in the triangular region ABJ (Fig. 4.18). As the solution at J depends only on the values of $\overline{\sigma}$ and γ which are specified along the "initial" curve AB, this curve is said to represent the *domain of dependence* of J.

To continue the solution below AJB, boundary conditions must be specified along the centerline $\xi = 0$ and the bin wall $\xi = 1$. Consider a point K on the centerline. As there are two dependent variables $\overline{\sigma}$ and γ, and only one characteristic passing through K intersects AB, one boundary condition must be specified along $\xi = 0$. Similarly, one boundary condition must be specified along $\xi = 1$.

The boundary conditions are similar to those used for hopper flow (see §3.4 and 3.5). Along $\xi = 0$, the condition of zero shear stress implies that

$$\gamma(\xi = 0, \eta) = 0, \quad \text{passive case;} \quad \gamma(\xi = 0, \eta) = \pi/2, \quad \text{active case} \tag{4.13}$$

Along the wall, the friction boundary condition (3.34) implies that

$$\gamma(\xi = 1, \eta) = \text{constant} \equiv \gamma_w \tag{4.14}$$

where γ_w is obtained by solving (3.35). Equation (3.35) has two roots, the smaller value corresponding to the passive case, and the larger value to the active case.

Using the values of $\overline{\sigma}$ and γ along the initial curve AB, and the boundary conditions (4.13) and (4.14), the solution in the bin section can be constructed by marching in the direction of decreasing η. Two distinct solutions are possible, depending on whether the active or passive roots are used in the boundary conditions. The data shown in Fig. 4.16 suggest that neither of these solutions is truly representative of the material flowing in the bin section.

Consider the solution obtained by Horne and Nedderman (1976) for the active case. They assumed that the upper surface of fill AB was horizontal and

$$\gamma = \pi/2; \quad \sigma = \text{constant} \equiv \sigma_u \quad \text{along} \quad AB \tag{4.15}$$

Equations (4.15) specify the *initial* data, which imply that the shear stress σ_{yx} vanishes on this surface (see (2.102)) and the dimensionless normal stress $\overline{\sigma}_{yy}$ is uniform ($= \overline{\sigma}_u (1 + \sin\phi)$).

A solution satisfying (4.15) is given by (Rankine, 1857)

$$\gamma = \pi/2$$
$$\overline{\sigma} = \overline{\sigma}_u + (\tan\Omega/\cos\phi)(\eta_u - \eta) = \overline{\sigma}_u + (\eta_u - \eta)/(1 + \sin\phi) \tag{4.16}$$

The second of (4.16) follows on substituting $\gamma = \pi/2$ in (4.8). Thus the characteristics are straight lines, and the shear stress $\overline{\sigma}_{xy}$ vanishes. The initial data (4.15) suffice to determine the solution within the triangular *Rankine zone AJB* (Fig. 4.18).

Considering the wall boundary condition (3.35), the active root is given by (see (3.36))

$$\gamma_w \equiv \gamma_{wa} = (\pi + \delta - \omega)/2; \quad \omega \equiv \sin^{-1}(\sin\delta/\sin\phi) \tag{4.17}$$

For the parameter values used by Horne and Nedderman (1976), i.e., $\phi = 30°$ and $\delta = 25°$, we obtain $\gamma_{wa} = 73.6°$. At the corner B (Fig. 4.19a), γ is discontinuous as $\gamma_{wa} \neq \pi/2$, the value of γ in the Rankine zone. To compute the solution in this region, it is helpful to let γ vary smoothly from $\pi/2$ to γ_w along the curve BB' (Fig. 4.19b) (Nedderman, 1992, p. 203). Thus the 1-characteristics form a fan in the angular region $S_1 B S_{1w}$. If BB' is chosen arbitrarily, the value of $\overline{\sigma}$ at any point on BB' cannot be computed. Hence BB' is chosen as the 2-characteristic passing through B. Equation (4.12) implies that the compatibility condition along BB' ($j = 2$) is given by

$$\cos\phi \, \mathrm{d}\overline{\sigma} - 2\overline{\sigma} \sin\phi \, \mathrm{d}\gamma + \cos(\gamma + \Omega)\, \mathrm{d}s_2 = 0 \tag{4.18}$$

In the limit $B' \to B$, $\mathrm{d}s_2 = 0$, and hence (4.18) can be integrated using (4.15) to obtain

$$\overline{\sigma} = \overline{\sigma}_u \exp[2 \tan\phi \, (\gamma - (\pi/2))] \tag{4.19}$$

The Rankine solution (4.16) is valid along BS_1 and (4.19) is valid along BB' (Fig. 4.19b). Hence the solution at a point G in the fan region can be constructed by integrating the compatibility conditions along the 1-characteristic GE and the 2-characteristic GF. Finally, in the angular region $S_{1w}B'D$, the solution can be constructed by using the wall boundary condition $\gamma = \gamma_w$ and the compatibility condition along the 2-characteristics.

Horne and Nedderman (1976) used this procedure to generate the wall shear stress profile shown in Fig. 4.20. The shear stress $\overline{T} = \overline{\sigma}_{xy'}(\xi = 1, \eta')$ exerted in the downward direction by the material on the right-hand wall is nonzero at upper surface of the fill ($\eta' = 0$) as γ changes discontinuously from $\pi/2$ to γ_{wa} at the point O. The pattern of the

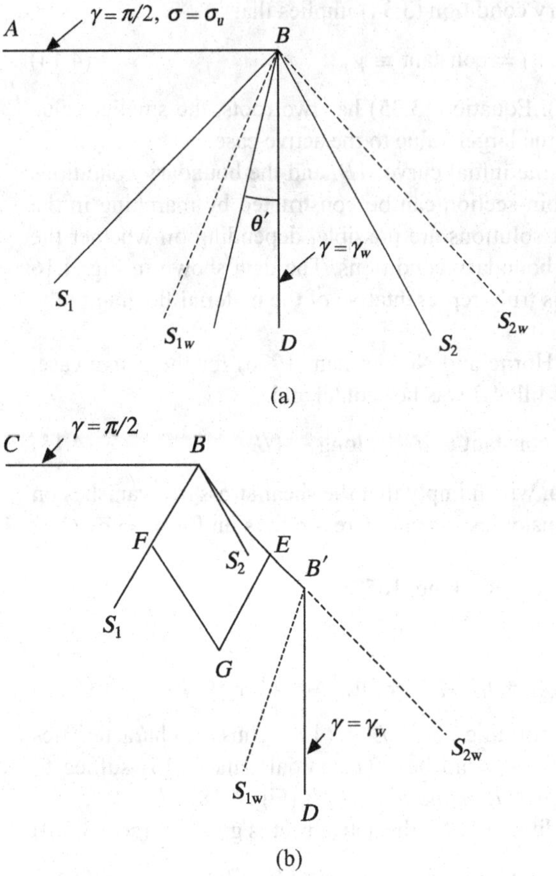

Figure 4.19. (a) Characteristics issuing from the point of intersection of the upper surface of the fill AB and the right-hand bin wall BD. Here BS_1 and BS_2 are the 1- and 2-characteristics, respectively, corresponding to $\gamma = \pi/2$, and BS_{1w} and BS_{2w} are the characteristics corresponding to $\gamma = \gamma_w$. (b) Expanded view of the degenerate $j = 2$ characteristic BB' which connects the upper surface to the wall. The angle γ varies from $\pi/2$ at B to γ_w at B'. Even though the points B and B' are shown as separate points for the sake of clarity, they represent the same physical location.

characteristics shows that the left-hand wall influences the stress profile only below the point C (Fig. 4.20). In the interval CD, $\overline{T}$ decreases as the discontinuous change in γ at the point O' causes $\overline{\sigma}$ to decrease from the specified value $\overline{\sigma}_u$ to a smaller value as the fan-shaped region is traversed (see (4.19)). At large depths, $\overline{T}$ approaches its asymptotic value of unity.

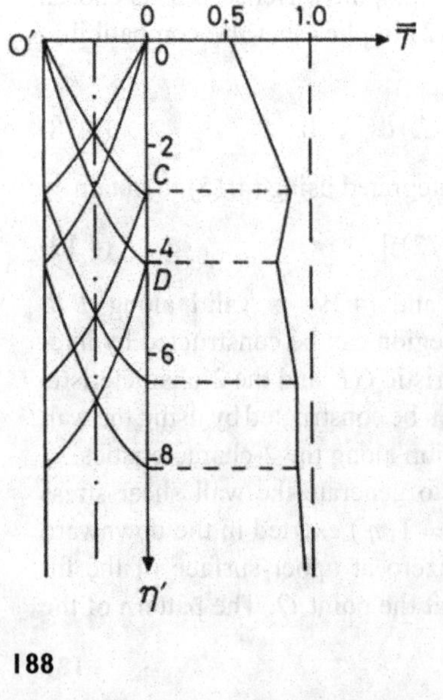

Figure 4.20. Pattern of stress characteristics and the profile of the dimensionless shear stress $\overline{T} = \sigma_{xy'}(\xi = 1, \eta')/(\rho\, g\, W)$ exerted in the downward direction by the material on the right-hand wall of the bin. An active stress field is assumed. Here $\eta' = y'/W$, y' is the vertical distance measured downward from the upper surface of the fill (see Fig. 4.17), W is the half-width of the bin section, ρ is the density of the granular material, and g is the acceleration due to gravity. The broken vertical line $\overline{T} = 1$ represents the asymptotic dimensionless shear stress predicted by the Janssen–Walker solution. Parameter values: $\phi = 30°$, $\delta = 25°$, $\overline{\sigma}_u = 2.67$. The dimensionless normal stress exerted on the upper surface of the fill is $\overline{\sigma}_u(1 + \sin\phi) = 4.0$. (Adapted from Fig. 8 of Horne and Nedderman, 1976.)

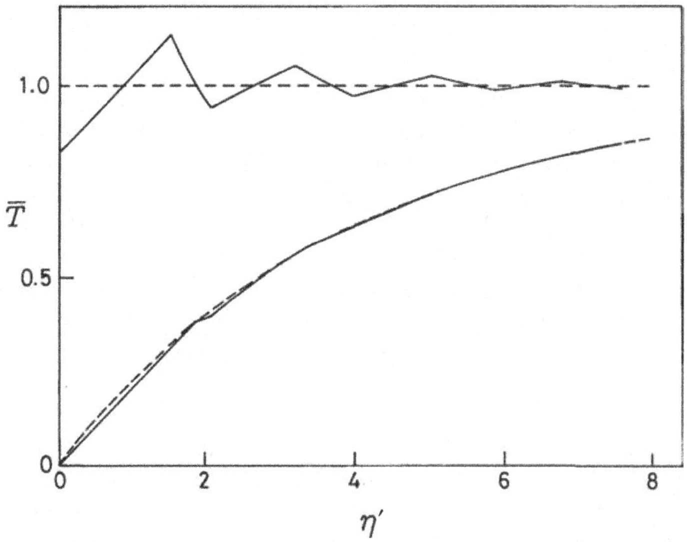

Figure 4.21. Profile of the dimensionless shear stress $\overline{T} = \overline{\sigma}_{xy'}(\xi = 1, \eta')$ exerted in the downward direction by the material on the right-hand wall of the bin: —, method of characteristics; - - -, Janssen–Walker solution. Here $\eta' = y'/W$, y' is the vertical distance measured downward from the upper surface of the fill (Fig. 4.17), W is the half-width of the bin section, ρ is the density of the granular material, and g is the acceleration due to gravity. Parameter values: $\phi = 30°$, $\delta = 20°$. The two upper curves are for $\overline{\sigma}_u = 5.32$. (Adapted from Fig. 9 of Horne and Nedderman, 1976.)

The procedure described above must be modified if the surcharge specified on the upper surface AB (Fig. 4.19a) vanishes, i.e., $\overline{\sigma}_u = 0$. In such a case, (4.19) implies that $\overline{\sigma} = 0$ along the degenerate 2-characteristic BB' (Fig. 4.19b). With $\overline{\sigma} = 0$ along BB', Hancock and Nedderman (1974) found that the wall characteristic $B'S_{1w}$ (Fig. 4.19b) became curved and intersected the characteristic BS_1. Hence the method of characteristics could not be used in this region. A possible reason for this behavior is discussed below.

Any curve along which $\overline{\sigma}$ vanishes is a characteristic (see Appendix C). Hence the value of γ along such a curve cannot be arbitrarily prescribed, but must satisfy a certain condition (Problem C.1). This condition is in general incompatible with the requirement that the curve should be a characteristic of the 2-family. Therefore, it appears that the latter requirement should be relaxed.

Sokolovskii (1965, pp. 237–238) circumvented this difficulty by using the radial stress field in the region S_1BD (Fig. 4.19a). As discussed in §3.5.2, the radial stress field is a particular solution of the force balances. This can be used in the present problem by considering a polar coordinate system with coordinates (r, θ') and origin at B. Then the radial stress field is given by

$$\overline{\sigma} = r_* b_0(\theta'); \quad \gamma = g_0(\theta') + \theta' \tag{4.20}$$

where $r_* = r/W$ is the dimensionless radial coordinate, W is the half-width of the bin section, and the angle θ' is measured clockwise from the bin wall BD to a radial line (Fig. 4.19a). The term θ' occurs in the second of (4.20) as $g_0(\theta')$ is the inclination of the major principal stress axis relative to the circumferential direction. The functions b_0 and g_0 are determined by integrating (3.60) and (3.61) subject to the boundary conditions $g_0(\theta' = 0) = \gamma_w; g_0(\theta' = \Omega) = (\pi/2) - \Omega$.

Using the approaches outlined above, Horne and Nedderman (1976) have constructed stress profiles for the case of zero surcharge ($\overline{\sigma}_u = 0$) and for the case of positive surcharge ($\overline{\sigma}_u = 5.32$). Figure 4.21 shows that the approximate Janssen–Walker solution

(Problem 4.1) agrees well with the rigorous method of characteristics solution for $\overline{\sigma}_u = 0$. On the other hand, for $\overline{\sigma}_u = 5.32$, the solutions differ significantly near the upper surface of the fill.

Surprisingly, there do not appear to have been any attempts to compare stresses predicted by the method of characteristics with measured values. As the experiments are often conducted with zero surcharge, it may be expected that the discrepancy between predicted and measured values is not very different from that obtained with the Janssen solution.

The procedure used for constructing the active stress field in the bin section has been described above. In the passive case, characteristics corresponding to the initial data specified along AB (Fig. 4.19) may overlap with the wall characteristics at B, leading to discontinuous stress profiles (Horne and Nedderman, 1976; Nedderman, 1992, pp. 227–242). The construction of a discontinuous solution is discussed briefly in the next section.

4.2.2. The Transition Region

During the discharge of granular materials from a mass flow bunker, the velocity of the material adjacent to bunker wall is usually nonzero. Therefore, the velocity is discontinuous at the transition point (Fig. 4.17). This may cause a discontinuity in the density, as discussed later. The stress field in the transition region is not well defined, and is unsteady (see Figs. 4.5, 4.10, and 4.11). Steady state analyses have often been based on the assumption of an active state in the bin section and a passive state in the hopper section (Walker, 1966; Bransby and Blair-Fish, 1974; Enstad, 1975; Horne and Nedderman, 1978; Prakash and Rao, 1991; Nedderman, 1992, p. 238); this assumption will be adopted here also. The flow field in the hopper section suggests that a passive state of stress is likely to prevail, provided the coaxiality condition holds (see §3.3). Based on the available data, it is not certain that the *flowing* material is in an active state of stress in the bin section.

The experiments of Perry and Handley (1967) provide some support for the assumption that the state of stress changes as the transition region is traversed. They used a pressure-sensitive radio pill to measure the stresses in sand flowing through a conical bunker. They state that "as the sand approaches the converging section, the axis of the major pressure begins to rotate, the vertical pressure falling as the horizontal radial and circumferential pressures rise."

Some models which have been proposed for the transition region are discussed below.

(a) *Horne and Nedderman (1978)*

Horne and Nedderman (1978) examined the *static* case, with an active state of stress in the bin section and a passive state of stress in the hopper section.

On moving down from the upper surface of the fill, the stress field in the bin section tends to an asymptotic field which is independent of the depth (see §4.1.4). Therefore, for a deep bin with $H/W \gg 1$ (Fig. 4.17), the stress field just above the transition region can be approximated by the asymptotic stress field. The stress field in the transition region can be constructed by using the method of characteristics, as explained below.

The nature of solutions depends on the relative orientations of the bin and hopper characteristics at the transition point. The slopes of the characteristics depend in turn on the values of γ at this point, as shown by (4.10). Owing to the abrupt change of geometry at the transition point, and also the assumed change in the state of stress from active to passive, there is a jump in the value of γ at this point. The magnitude of the jump can be calculated as follows.

Along the bin wall, the wall friction condition (4.14) holds. As an active state of stress is assumed in the bin section, the appropriate value of γ is given by (see (4.17))

$$\gamma_b = (\pi + \delta_b - \omega_b)/2; \quad \omega_b \equiv \sin^{-1}(\sin\delta_b/\sin\phi) \qquad (4.21)$$

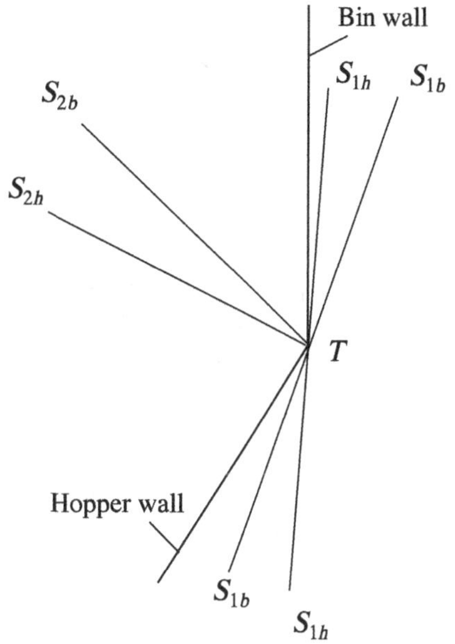

Figure 4.22. Characteristics of the 1-family TS_{1b} and TS_{1h}, and of the 2-family TS_{2b} and TS_{2h}, issuing from the transition point of a bunker. The subscripts b and h denote the bin and hopper sections, respectively. The characteristics are shown as straight lines for ease of exposition. Parameter values: $\theta_w = 30^\circ$, $\phi = 30^\circ$, $\delta_b = \delta_h = 20^\circ$.

where the subscript b denotes the bin section, and δ_b is the angle of wall friction between the material and the bin wall. For a passive state of stress in the hopper (4.14) implies that (see (3.36))

$$\gamma_h' = (\delta_h + \omega_h)/2; \quad \omega_h = \sin^{-1}(\sin\delta_h/\sin\phi) \tag{4.22}$$

Here the subscript h denotes the hopper section, δ_h is the angle of wall friction between the material and the hopper wall, and γ' is the angle measured anticlockwise from the major principal stress axis to the circumferential direction (Fig. 3.12). Hence the angle measured anticlockwise from the major principal stress axis to the horizontal is

$$\gamma_h = \gamma_h' + \theta_w = (\delta_h + \omega_h)/2 + \theta_w \tag{4.23}$$

Bransby and Blair-Fish (1974) noted that there were two cases of interest: (i) $\gamma_h < \gamma_b$ (*steep* hopper wall), and (ii) $\gamma_h > \gamma_b$ (*shallow* hopper wall). These cases are discussed below.

(i) *Steep hopper wall*

Horne and Nedderman (1978) presented results for a material with an angle of internal friction $\phi = 30^\circ$, and a hopper section with a wall angle $\theta_w = 30^\circ$. They also assumed that the angles of wall friction corresponding to the bin and hopper walls were equal, with $\delta_h = \delta_b = 20^\circ$. With these parameter values, we obtain $\omega_h = \omega_b = 43.2^\circ$, $\gamma_b = 78.4^\circ$, $\gamma_h' = 31.6^\circ$, and $\gamma_h = 61.6^\circ$. For these parameter values, the 1-characteristics (i.e., the characteristics corresponding to $j = 1$ in (4.10)) issuing from the transition point T lie outside the bunker, and the 2-characteristics form a fan (Fig. 4.22). Here TS_{1b} and TS_{1h} represent the 1-characteristics corresponding to the bin and hopper sections, respectively.

The solution in the fan region $S_{2b}TS_{2h}$ can be constructed using the procedure described in §4.2.1. Equation (4.19) implies that the variation of the mean stress $\overline{\sigma}$ along the degenerate 1-characteristic through T is given by

$$\overline{\sigma}_h = \overline{\sigma}_b \exp[2\tan\phi\,(\gamma_b - \gamma_h)] \tag{4.24}$$

where all quantities are evaluated at T, and the subscripts b and h refer to the bin and hopper sections, respectively. As $\gamma_h < \gamma_b$ for a steep hopper wall, the mean stress at T shows a step increase from $\overline{\sigma}_b$ to $\overline{\sigma}_h$ on moving downward from the bin section to the hopper section.

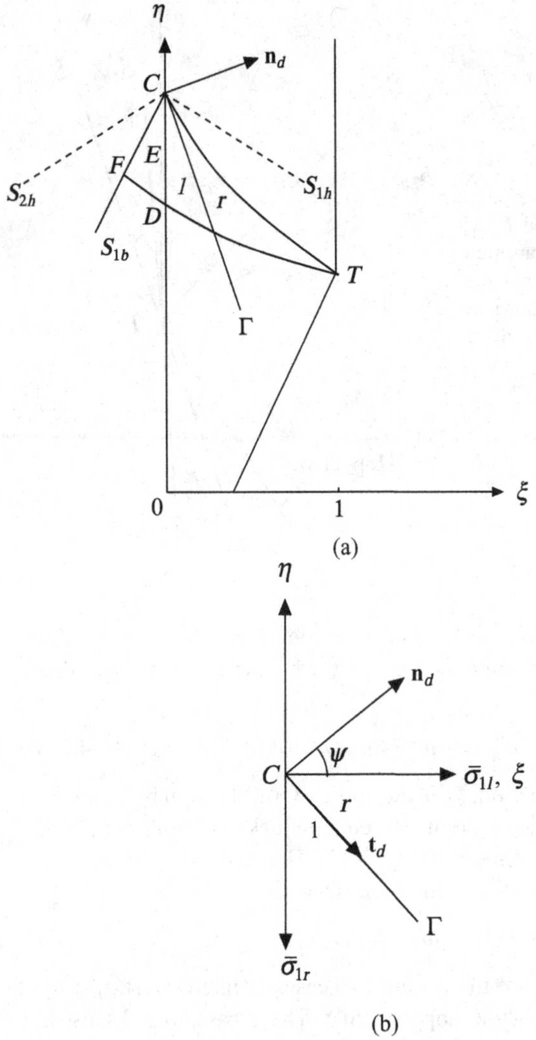

Figure 4.23. (a) Sketch of the discontinuity surface Γ passing through a point C on the centerline of the bunker. The curves CS_{1b} and CT denote the 1- and 2-characteristics corresponding to the bin section, and CS_{1h} and CS_{2h} denote the 1- and 2-characteristics corresponding to the hopper section. Here $\xi = x/W$, $\eta = y/W$, W is the half-width of the bin section, and the x and y coordinates are chosen as shown in Fig. 4.17. (b) Orientations of the discontinuity surface Γ, the major principal stress axis corresponding to the stress state to the left $\overline{\sigma}_{1l}$, and to the right $\overline{\sigma}_{1r}$. Here $\mathbf{n}_d$ and $\mathbf{t}_d$ are the unit normal and unit tangent vector, respectively, to Γ.

Unfortunately, these does not appear to be a simple physical explanation for the increase in $\overline{\sigma}$.

As measurements are confined to wall stresses rather than mean stresses, it is useful to calcuate the ratio of the normal stress on the hopper wall N_h to normal stress on the bin wall N_b. This is given by (see (2.102) and (3.15))

$$N_h/N_b = (\overline{\sigma}_h/\overline{\sigma}_b)(1 + \sin\phi\ \cos 2\gamma_h')/(1 + \sin\phi\ \cos 2\gamma_b) \qquad (4.25)$$

where $\overline{\sigma}_h/\overline{\sigma}_b$ follows from (4.24). For the parameter values used by Horne and Nedderman (1978) ($\phi = 30°$, $\theta_w = 30°$, $\delta_b = \delta_h = 20°$), we obtain $\overline{\sigma}_h/\overline{\sigma}_b = 1.4$ and $N_h/N_b = 3.2$. As discussed in §4.1.3, measurements on laboratory scale bunkers suggest that N_h/N_b is in the range 2–30. Thus the assumption of an active to passive transition produces a jump in the wall stress which is roughly of the same order of magnitude as the measured values.

Let us now discuss some features of the solution in the transition region. The bin characteristic TS_{2b} and the hopper characteristic TS_{2h} issuing from the transition point T (Fig. 4.22) intersect the centerline at the points C and D, respectively, as indicated in Fig. 4.23a. The solution in the region above TC can be constructed using the boundary conditions $\gamma = \pi/2$ along the centerline $\xi = 0$ and $\gamma = \gamma_b$ (see (4.21)) along the bin wall

$\xi = 1$, and values of $\overline{\sigma}$ and γ specified along a suitable curve spanning the bin section. The values of $\overline{\sigma}$ and γ along TC can then be used to construct the solution in the region $TCFDT$ (Fig. 4.23a). Here CF represents the 1-characteristic through C, with $\gamma_C = \pi/2$.

The assumption of an active state of stress in the bin section and a passive state of stress in the hopper section implies that (see (4.13))

$$\gamma(\xi = 0, \eta) = \pi/2, \quad \text{bin section}; \quad \gamma(\xi = 0, \eta) = 0, \quad \text{hopper section} \tag{4.26}$$

At points such as E, which are on the centerline and within the region $TCFDT$ (Fig. 4.23a), γ_E does not satisfy (4.26) in general. To circumvent this difficulty, consider a fan-type solution at C, with a change in the value of γ_C from $\pi/2$ to 0.

Figure 4.23a shows that the 1-characteristic corresponding to the hopper section CS_{1h} projects into the region where the solution has already been determined by the bin characteristics. Hence a fan-type solution cannot be constructed at C, and a discontinuous solution must be considered. Horne and Nedderman (1978) introduce a discontinuity curve Γ passing through C, such that $\gamma(\xi \to 0^+, \eta_C) = \pi/2$ and $\gamma(\xi \to 0^-, \eta_C) = 0$.

The discontinuous solution leads to a curve traversing the bunker, such that $\overline{\sigma}$ and γ are discontinuous across it. The orientation of this curve, and the jumps in the values of $\overline{\sigma}$ and γ are related by the jump momentum balance (D.9). In the static case, this reduces to

$$\mathbf{n}_d \cdot \overline{\sigma}_l - \mathbf{n}_d \cdot \overline{\sigma}_r = 0 \quad \text{on} \quad \Gamma \tag{4.27}$$

where $\mathbf{n}_d$ is the unit normal to Γ, $\overline{\sigma}$ is the dimensionless stress tensor, and the subscripts l and r denote limiting values attained as Γ is approached from the left and the right, respectively (Fig. 4.23a). The construction of a solution for a flowing material is discussed later in this section.

Equation (4.27) implies that the stress vector, and hence the shear and normal stresses, are continuous across Γ. Thus

$$\overline{N}_l = \overline{N}_r; \quad \overline{T}_l = \overline{T}_r \tag{4.28}$$

where the normal stress $\overline{N}$ and the shear stress $\overline{T}$ acting on Γ are given by

$$\overline{N} \equiv (\mathbf{n}_d \cdot \overline{\sigma}) \cdot \mathbf{n}_d; \quad \overline{T} \equiv (\mathbf{n}_d \cdot \overline{\sigma}) \cdot \mathbf{t}_d \tag{4.29}$$

Here $\mathbf{t}_d$ is the unit vector tangential to Γ at C (Fig. 4.23b). Equations (4.28) provide two relations involving $\overline{\sigma}_r$, γ_r, $\overline{\sigma}_l$, γ_l, and ψ, where ψ is the angle measured anticlockwise from the horizontal to the normal $\mathbf{n}_d$ (Fig. 4.23b). As $\gamma_l = 0$, $\gamma_r = \pi/2$, and the value of $\overline{\sigma}_r$ is known from the solution in the bin section, (4.28) can be solved for $\overline{\sigma}_l$ and ψ.

We now describe a graphical solution procedure, based on the Mohr circle for the stress tensor (§2.7). Bransby and Blair-Fish (1974) used this procedure to find the slope of the discontinuity curve at the transition point T (Fig. 4.23a). Horne and Nedderman (1978) note that such a solution cannot be continued into the interior of the bunker, owing to the lack of additional information. Hence the fan-type solution is preferred at T. The construction of a discontinuous solution at the point C on the centerline (Fig. 4.23b) is discussed below.

At C, $\gamma_l = 0$, and $\gamma_r = \pi/2$. Hence the major principal stress axis is horizontal for the state on the left of the discontinuity curve Γ, and vertical for the state on the right of Γ. As the normal to Γ is inclined at an angle ψ to the horizontal (Fig. 4.23b), the angle measured anticlockwise from the σ_1 axis to the normal is ψ for the state on the left, and $(\pi/2) + \psi$ for the state on the right. Hence these states correspond to the points marked l and r, respectively, on the Mohr's circles (Fig. 4.24). The circles are drawn tangential to the line $\overline{T} = \overline{N}\tan\phi$ in order to satisfy the critical state condition (4.7). Equation (4.28) implies that the points l and r must coincide. Hence $\psi = 0$, i.e., the discontinuity is tangential to the centerline, and $\overline{\sigma}_l < \overline{\sigma}_r$.

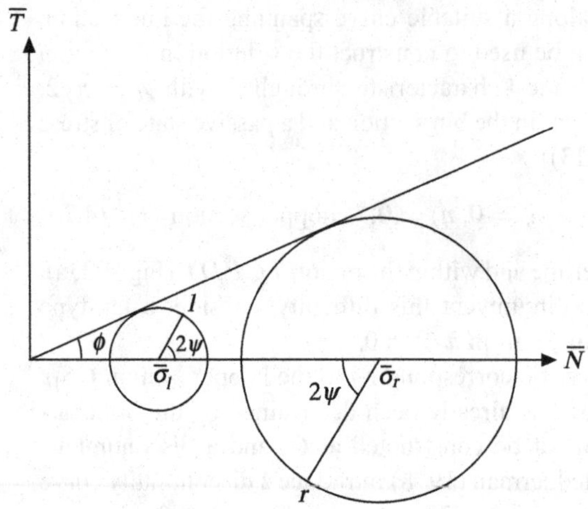

Figure 4.24. Mohr's circles for states of stress to the left l and to the right r of the discontinuity curve Γ at the centerline. Here $\overline{N}$ and $\overline{T}$ are the dimensionless normal and shear stresses, respectively, $\overline{\sigma}$ is the dimensionless mean stress, and ϕ is the angle of internal friction. The stresses are scaled by $\rho\, g\, W$, where ρ is the density of the granular material, g is the acceleration due to gravity, and W is the half-width of the bin section. The normal to Γ is inclined at an angle ψ relative to the horizontal, as shown in Fig. 4.23b.

The discontinuity can be continued downward as follows. Consider a point J, with η_J slightly less than η_C (Fig. 4.25). As the discontinuity is defined by

$$\frac{\mathrm{d}\eta}{\mathrm{d}\xi} = -\cot\psi$$

(see Fig. 4.23b), the modified Euler method (see, e.g., Gerald and Wheatley, 1994, p. 401) implies that

$$\eta_J - \eta_C \approx -\cot\left(\frac{\psi_J + \psi_C}{2}\right)\xi_J$$

Using an assumed value of ψ_J, the fan-type solution for states on the right of the discontinuity CJ, and the jump balances (4.28) at J, the values of ξ_J, $\overline{\sigma}_{Jl}$ and γ_{Jl} can be obtained. Here the subscript l denotes states on the left of CJ. Using the values of $\overline{\sigma}_{Jl}$ and γ_{Jl}, and the 1-characteristic JH (Fig. 4.25), the value of $\overline{\sigma}_H$ can be obtained. Finally, the value of ψ_J can be determined iteratively by requiring the left state at G, obtained by interpolation using the values of C and J, to be consistent with the compatibility condition along the 2-characteristic HG.

The profile of the dimensionless normal stress $\overline{\sigma}_{\theta\theta}$ acting on the hopper wall is shown in Fig. 4.26. At the transition point, the normal stress is $\approx$3 on the bin wall and $\approx$9 on the hopper wall. This discontinuous change is caused by the use of the fan-type solution (4.24) at the transition point. The largest value of $\overline{\sigma}_{\theta\theta}$ does not occur at this point, but at some

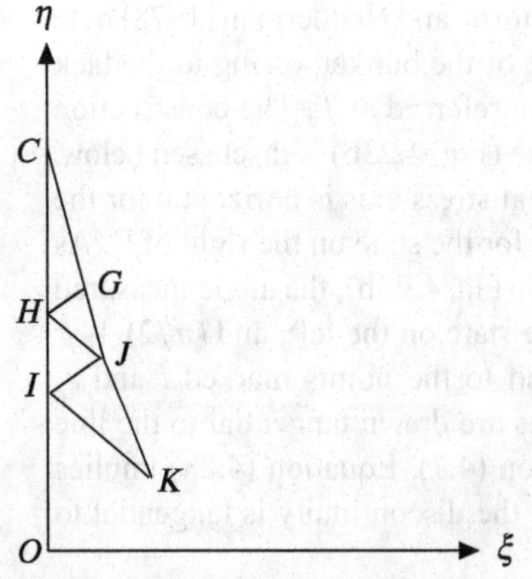

Figure 4.25. Mesh of characteristics used to continue the discontinuity curve $CGJK$ downward from C. Here $\xi = x/W$, $\eta = y/W$, W is the half-width of the bin section, and the x and y coordinates are chosen as shown in Fig. 4.17. The curves JH and KI represent the 1-characteristics, and GH and JI the 2-characteristics.

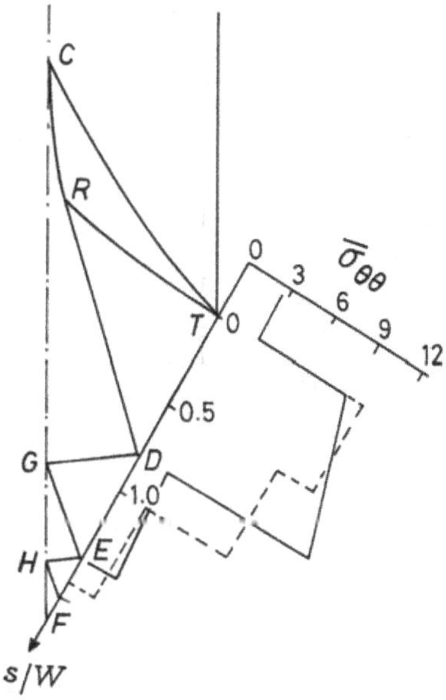

Figure 4.26. Profile of the dimensionless normal stress $\overline{\sigma}_{\theta\theta} = \sigma_{\theta\theta}/(\rho\, g\, W)$ along the wall TF of the hopper section: —, profile predicted by Horne and Nedderman (1978); - - -, profile predicted by averaging the stress over a distance $s/W = 0.5$. Here ρ is the density of the granular material, g is the acceleration due to gravity, W is the half-width of the bin section, and s is the distance measured downward along the hopper wall from the transition point T. The region CTR represents the fan of 2-characteristics issuing from T, and CD, DG, GE, EH, and HF represent discontinuity curves. Parameter values: $\phi = 30°$, $\delta = 20°$, $\theta_w = 30°$. (Adapted from Fig. 7 of Horne and Nedderman, 1978.)

distance below it. It is surprising that there have been no attempts to verify this prediction experimentally.

In this context, consider the stress measurements of Blair-Fish and Bransby (1973). They used a load cell with an active face of size 37.5 mm × 37.5 mm, and a bunker with a bin section of half-width $W = 75$ mm. As the spatial coordinate in Fig. 4.26 is scaled by W, the predicted profile must be averaged over a dimensionless length of 37.5/75 = 0.5 for purposes of comparison with the measurements. The averaged profile is shown by the broken lines in Fig. 4.26. An interesting consequence of the averaging is that the load cell which is closest to the transition registers the maximum normal stress, even though the peak in the actual stress profile occurs at a point which is outside the region occupied by this cell.

(ii) *Shallow hopper wall*

As discussed earlier, this case corresponds to $\gamma_h > \gamma_b$, where the subscripts h and b denote the hopper and the bin sections, respectively. Choosing the same material parameters as in (i) above, i.e., $\phi = 30°$ and $\delta_h = \delta_b = 20°$, we obtain $\gamma_b = 78.4°$ and $\gamma_h = 31.6° + \theta_w$. Thus the limiting value of θ_w is 46.8°.

For $\theta_w > 46.8°$, the 2-characteristics corresponding to the bin and hopper sections overlap, as shown in Fig. 4.27 for $\theta_w = 60°$. Hence the hopper characteristic TS_{2h} projects into a region where the solution has already been determined using the bin characteristics. As a continuous solution cannot be constructed in such cases, a discontinuous solution must be considered, analogous to the situation prevailing at the centerline of a steep walled hopper (Fig. 4.23a).

Horne and Nedderman (1978) note that a discontinuous solution cannot be constructed near the transition point T for the following reason. If the discontinuity curve through T lies in the region BTS_{2b}, it destroys the solution obtained using the bin characteristics. Conversely, if it lies outside this region, the problem caused by the overlapping of 2-characteristics persists. In particular, at the point where the discontinuity intersects the centerline, the value of γ on the downstream side of the discontinuity does not in general satisfy the boundary condition $\gamma(\xi = 0, \eta) = 0$.

The above discussion suggests that a fully plastic solution (i.e., one where the material satisfies the yield condition everywhere) cannot be constructed for a bunker with a shallow

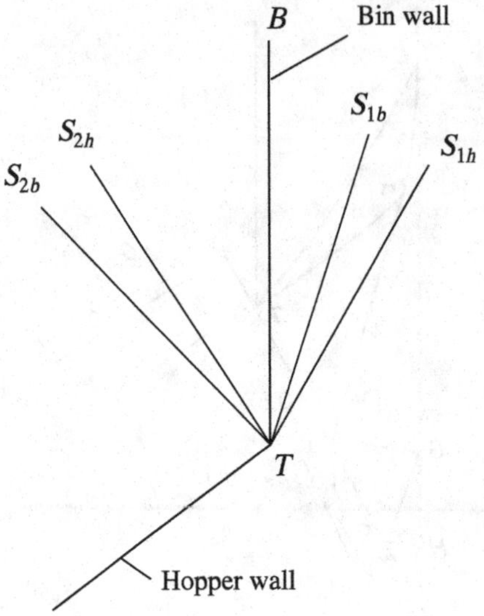

Figure 4.27. Characteristics of the 1-family TS_{1b} and TS_{1h}, and of the 2-family TS_{2b} and TS_{2h}, issuing from the transition point T of a bunker with a shallow hopper section. The subscripts b and h denote the bin and the hopper sections, respectively. Parameter values: $\theta_w = 60°$, $\phi = 30°$, $\delta_b = \delta_h = 20°$.

hopper wall. If the material does not yield, it is unlikely that mass flow can occur near the transition point. As discussed in §4.1.1, shallow hopper walls tend to promote funnel flow rather than mass flow.

Using (4.21) and (4.23), it follows that funnel flow may occur if

$$\theta_w > \frac{1}{2}(\pi + \delta_b - \omega_b - \delta_h - \omega_h) \equiv \theta_c \tag{4.30}$$

where

$$\omega \equiv \sin^{-1}(\sin\delta / \sin\phi) \tag{4.31}$$

and the subscripts b and h denote the bin and hopper sections, respectively. As noted by Nedderman (1992, p. 239), it is likely that mass flow cannot occur if (4.30) is satisfied, but mass flow may not occur even if it is violated. Bransby and Blair-Fish (1974) have also suggested the same criterion, but their arguments are less satisfactory than those of Horne and Nedderman (1978).

The critical wall angle θ_c is compared with the data of Nguyen et al. (1980) and Benink (1989) in Table 4.6. In the first three rows, two values of θ_e, say θ_{e1} and $\theta_{e2} > \theta_{e1}$, are listed for each material. The observed flow patterns are as follows: $\theta_w < \theta_{e1}$ (mass flow), $\theta_{e1} < \theta_w < \theta_{e2}$, (mass flow for small and large values of the head of material H, and funnel flow for intermediate values of H), $\theta_w > \theta_{e2}$ (funnel flow). It is seen that funnel flow occurs for wall angles θ_e which are much less than θ_c, except for the case of rough walls. Hence $\theta_w > \theta_c$ may be a sufficient condition, but not a necessary condition, for the occurrence of funnel flow.

(b) *Michalowski (1987)*

This model is confined to the kinematic aspects. Assuming that the material slips relative to the bunker wall, the velocity is discontinuous at T (Fig. 4.28). The magnitude of the velocity jump can be determined using the coaxiality condition (2.104) and the flow rule (2.105). Michalowski (1987) assumed that (i) the angle of dilation ν_d was a constant, and (ii) the orientation of the major principal stress axis was a *known* continuous function of position. The occurrence of rupture layers suggests that the rate of dilation in the transition region is nonuniform, and hence the angle of dilation may not be constant. Unfortunately, data related to the local values of ν_d at various points in bunkers are not available. As the stresses change rapidly in the transition region, the use of assumption (ii) is also suspect. A

Table 4.6. Predicted Wall Angle of the Hopper Section θ_c for the Transition from Mass Flow to Funnel Flow in a Wedge-Shaped Bunker*

Material	Mean diameter (mm)	ϕ (°)	δ (°)	θ_e (°)	θ_c (°)
Sand	0.5–1.0	31	15	40, 70	60
Polystyrene	0.25–0.39	39	12	35, 60	71
Glass beads	0.325	25	15.3	30, 50	51
River sand	0.5–2.0	38, 40	14	$40^m, 50^i$	67–68
Rape seed	1–2.5	31, 32	17	$30^m, 40^i$	55–57
"	"	"	27	30^i	28–31
Plastic grains	2–3	35, 38	14	$30^m, 40^i$	65–67
Soya beans	2–7	34, 39	12	$40^m, 50^i$	68–71
"	"	"	29	30^i	30–40

* The angle is calculated using (4.30). Here ϕ and δ denote the angles of internal friction and wall friction, respectively, and θ_e is the wall angle of the hopper section used in the experiments. The first three rows represent the data of Nguyen et al. (1980) for a bunker made of Lucite. The last six rows represent the data of Benink (1989, pp. 151–154) for a bunker made of Perspex. Here the superscripts m and i denote mass flow and intermediate flow, respectively. Intermediate flow is a special case of funnel flow, where the flow pattern changes from mass flow to funnel flow when the head of material decreases below a critical value. For the data of Benink (1989), the two values of ϕ shown for each material correspond to estimates obtained using the ring shear tester and the Jenike shear cell, respectively.

redeeming feature of these assumptions is that they permit a relatively simple analysis of the kinematics.

It can be shown (Problem C.3) that the governing equations (2.104) and (2.105) are hyperbolic, and the characteristics of these equations, called the *velocity* characteristics, are inclined at angles of $\pm\,\Omega_v = \pm\,[(\pi/4) - (\nu_d/2)]$ relative to the major principal stress axis. As γ is assumed to be a known function of position, the equations are linear. Hence velocity discontinuities can occur only across velocity characteristics, as shown in Appendix E. This result can also be deduced using the following approach (Shield, 1953).

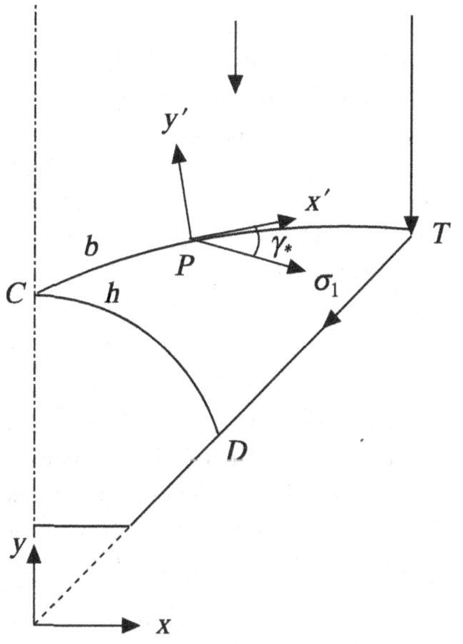

Figure 4.28. Discontinuity curves TC (2-characteristic) and CD (1-characteristic) across which velocity jumps occur. The x' axis is tangential to TC at the point P. The velocity vector is vertical above TC.

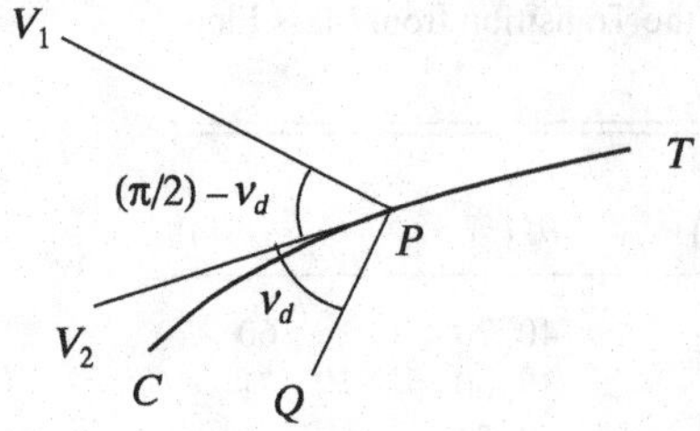

Figure 4.29. The lines PV_1 and PV_2 are tangents to the 1- and 2-characteristics, respectively, at the point P, and hence $\angle V_1 P V_2 = (\pi/2) - \nu_d$. As the discontinuity curve TC coincides with the 2-characteristic, and $\angle V_2 PQ = \nu_d$, the velocity jump vector PQ is normal to PV_1.

Referring to Fig. 4.28, consider a Cartesian coordinate system whose x' axis is tangential to the discontinuity curve TC at the point P. The curve TC can be regarded as an approximation to a thin region across which the velocity field varies sharply. Hence tangential derivatives such as $\partial v_x'/\partial x'$ can be neglected in comparison with normal derivatives such as $\partial v_x'/\partial y'$, where v_x' is the x' component of velocity. Using this approximation, the coaxiality condition (2.104) and the flow rule (2.105) reduce to

$$\sin 2\gamma_* \frac{\partial v_y'}{\partial y'} - \cos 2\gamma_* \frac{\partial v_x'}{\partial y'} = 0 \tag{4.32}$$

$$(\sin \nu_d - \cos 2\gamma_*) \frac{\partial v_y'}{\partial y'} = 0 \tag{4.33}$$

where γ_* is the inclination of the σ_1 axis relative to the x' axis (Fig. 4.28).

Equation (4.33) has two roots: (i) $\cos 2\gamma_* - \sin \nu_d = 0$, or $\gamma_* = \pm [(\pi/4) - (\nu_d/2)]$, and (ii) $\partial v_y'/\partial y' = 0$. The second root is irrelevant, as (4.32) then implies that $\partial v_x'/\partial y' = 0$, i.e., there is no velocity discontinuity. The first root implies that the discontinuity curve coincides with a velocity characteristic (Problem C.3).

Equation (4.32) provides additional information about the velocity jump across TC. As γ_* is assumed to be continuous across TC, (4.32) can be integrated with respect to y' over a thin region spanning TC to obtain

$$v_{xh}' - v_{xb}' = \pm \tan[(\pi/2) - \nu_d](v_{yh}' - v_{yb}') \tag{4.34}$$

where $v_{xh}' = v_x'(x', y' = 0^-)$ and $v_{xb}' = v_x'(x', y' = 0^+)$ (Fig. 4.28). Thus the vector representing the velocity jump, henceforth called the jump vector, is inclined at an angle of ν_d relative to TC. Referring to Fig. 4.29, where the 2-characteristic TC represents the discontinuity curve and PV_1 the 1-characteristic, $\angle V_1 P V_2 = (\pi/2) - \nu_d$, and $\angle V_2 P Q = \nu_d$. Hence the jump vector PQ is normal to the 1-characteristic. An alternative approach is described in Problem E.1.

If the velocity $\mathbf{v}_b$ of the material adjacent to the bin wall is specified at the transition point T, the velocity jump at this point can be determined as follows. Referring to Fig. 4.30, the vertical vector TB represents the velocity $\mathbf{v}_b$. The velocity characteristics are defined by (Problem C.3)

$$\frac{d\eta}{d\xi} = -\tan(\gamma \pm \Omega_v) \tag{4.35}$$

where $\eta = y/W$, $\xi = x/W$, W is the half-width of the bin section, $\Omega_v \equiv (\pi/4) - (\nu_d/2)$, and γ is the inclination of the major principal stress axis relative to the x axis (Fig. E.3). In (4.35), the plus sign refers to the 1-characteristics and the minus sign to the 2-characteristics. As γ is assumed to be a known function of position, the slopes of the characteristics are known a priori. Michalowski (1987) evaluated γ using the radial stress field, i.e.,

$$\gamma = \theta + g_0(\theta) \tag{4.36}$$

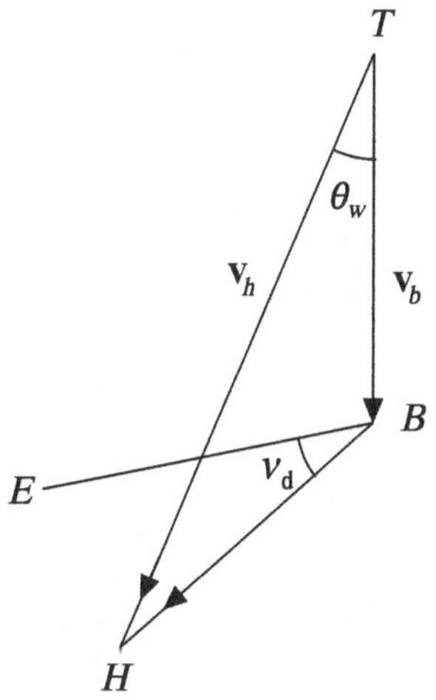

Figure 4.30. Velocity vectors at the transition point T of a bunker. Here $\mathbf{v}_b$ is the velocity of the material adjacent to the bin wall, $\mathbf{v}_h$ is the velocity of the material adjacent to the hopper wall, θ_w is the wall angle of the hopper, ν_d is angle of dilation, assumed constant, and BE is parallel to the tangent to the 2 -velocity characteristic passing through T.

where the angle θ represents the inclination of a radial line to the vertical (Fig. 3.12) , and the function $g_0(\theta)$ is obtained by integrating (3.60) and (3.61). As the stress field in the transition region differs from the radial stress field, it is difficult to justify the use of (4.36).

Assuming that $\nu_d < \pi/2$, the discontinuity curve TC (Fig. 4.28) coincides with the 2-characteristic passing through T. The alternative choice of the 1-characteristic as the discontinuity curve is unsuitable, as discussed later. Let BE denote a line drawn parallel to the tangent to the 2-characteristic at T (Fig. 4.30). The vector $\mathbf{v}_h - \mathbf{v}_b$ representing the velocity jump across the discontinuity is inclined at an angle ν_d relative to BE. Hence the jump vector is directed along BH (Fig. 4.30). The magnitude of the velocity jump can be determined by noting that the vector $\mathbf{v}_h$, which represents the velocity of the material adjacent to the hopper wall, must be parallel to this wall.

To construct the velocity jumps at other points of the discontinuity curve TC (Fig. 4.28), the compatibility conditions must be used. As suggested by Shield (1953), it is convenient to express these in terms of the velocity components (v_1, v_2) which represent the orthogonal projections of the velocity vector onto the characteristics (Fig. C.4). In the other words, v_1 is the velocity component tangential to the 1-characteristic. Let subscripts "b" and "h" denote states on either side of the discontinuity (Fig. 4.28). As shown above (see also Problem E.1), the vector representing the velocity jump across the 2-characteristic is normal to the 1-characteristic (Fig. 4.29). Hence the 1-component of the jump vector vanishes, and $v_{1b} - v_{1h} = 0$. If the discontinuity curve is approached from above, the compatibility condition (C.34) is given by

$$\mathrm{d}v_{2b} + (v_{2b} \tan \nu_d - v_{1b} \sec \nu_d)\, \mathrm{d}\gamma = 0 \tag{4.37}$$

Similarly, if the discontinuity curve is approached from below, the compatibility condition is given by

$$\mathrm{d}v_{2h} + (v_{2h} \tan \nu_d - v_{1h} \sec \nu_d)\, \mathrm{d}\gamma = 0 \tag{4.38}$$

Subtracting (4.38) from (4.37), using the relation $v_{1b} = v_{1h}$, and integrating, we obtain

$$v_{2b} - v_{2h} = (v_{2b} - v_{2h})_T\, e^{(-\tan \nu_d\, (\gamma - \gamma_T))} \tag{4.39}$$

and the subscript T indicates that the variable is evaluated at the transition point. Note that γ has been assumed to be continuous across TC.

The material in the bin section is assumed to be in plug flow, with a downward vertical velocity of magnitude v_b. Hence both v_{1b} and v_{2b} are known, and v_{2h} follows from (4.39). Finally, as $v_{1h} = v_{1b}$, the complete velocity field is known at points on the downstream side

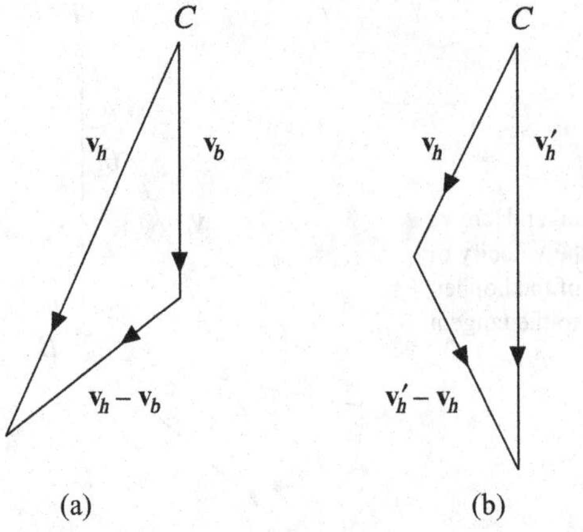

Figure 4.31. Velocity vectors at the point C (see Fig. 4.28) where the discontinuity curves (a) TC and (b) CD intersect the centerline of the bunker. Here $\mathbf{v}_b$ is the velocity vector on the upstream side of TC, i.e., in the bin section, $\mathbf{v}_h$ is the velocity vector on the downstream side of TC, $\mathbf{v}_h - \mathbf{v}_b$ is the velocity jump on crossing TC, $\mathbf{v}'_h$ is the velocity vector on the downstream side of CD, and $\mathbf{v}'_h - \mathbf{v}_h$ is the velocity jump on crossing CD.

of TC (Fig. 4.28). Using this information and the condition of zero normal velocity along the hopper wall, the velocity field can be constructed in the region TCD (Fig. 4.28), where CD is the I-characteristic passing through C.

At C, the velocity vector $\mathbf{v}_h$ on the downstream side of TC is not vertical in general (Fig. 4.31a). Hence another discontinuity curve CD must be introduced (Fig. 4.28), and the magnitude of the velocity jump adjusted so that the downstream velocity vector $\mathbf{v}'_h$ on the downstream side of CD is vertical (Fig. 4.31b).

The above procedure can be used to construct the velocity field in the hopper section. The curves in Fig. 4.32b show the predicted profiles of the radial velocity v_r along the radial line $\theta = -12°$, for two values of the angle of dilation ν_d. Here the angle θ is measured as shown in Fig. 4.32a. The vertical segments of the profiles represent velocity jumps across the discontinuity curves. For $\nu_d > 0$, the material dilates as it flows. Hence at any point in the hopper section, the radial velocity is expected to be larger in magnitude than the radial velocity for incompressible flow ($\nu_d = 0$). This is in accord with the results shown in Fig. 4.32b.

The triangles and filled and open circles in Fig. 4.32b represent data obtained using a stereophotographic technique, at three different times during the advanced stage of flow. The data suggest that the velocity field is fairly steady, even though oscillations in wall stresses have been reported by other investigators (see Fig. 4.10). Therefore, simultaneous measurement of the stress, density, and velocity fields would be valuable. Predicted and measured velocities are of the same order of magnitude (Fig. 4.32), and the overall agreement between theory and experiment is better for the incompressible model ($\nu_d = 0$) than for the compressible model ($\nu_d = 5°$). In contrast to the velocity discontinuities predicted by the model, the triangles suggest that the measured profile is smooth. More data are needed to examine this feature.

Consider the rupture layers predicted by the model. The upper rupture layer FCT (Fig. 4.33) can be regarded as a curve across which the direction of the velocity vector changes from vertical on the upstream side to approximately parallel to the hopper wall on the downstream side. The latter direction persists in the regions FCE and TCD; across the rupture layer ECD, the velocity field changes so that it is radially directed on the downstream side.

The predicted rupture layers FCT and ECD are qualitatively similar to the observed rupture layers FGT and IGH (Fig. 4.33). The discrepancy between theory and experiment is larger in the upper part of the transition region, where the assumption that γ is given by the radial stress field is likely to be in error.

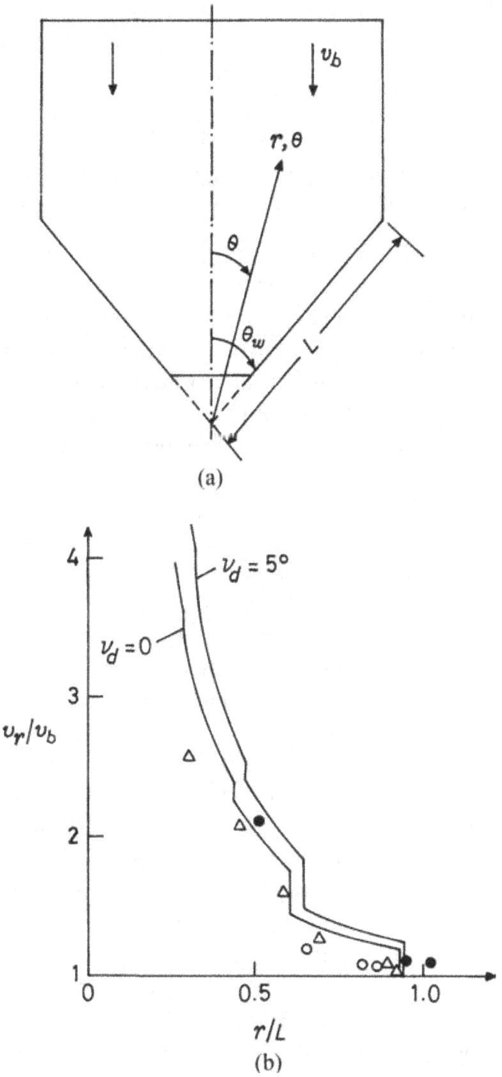

Figure 4.32. (a) Coordinate system for the bunker. Here v_b is the magnitude of the vertical velocity in the bin section, and L is the radial coordinate corresponding to the transition point. (b) Measured (△, ○, •) and predicted (—) profiles of the dimensionless radial velocity v_r/v_b along the radial line $\theta = -12°$ for the flow of sand. Here ν_d is the angle of dilation. The triangles, circles, and dots represent measurements at three instants during the advanced stage of flow. Parameter values: particle diameter = 0.25–1.75 mm, ϕ = angle of internal friction = 32°, δ = angle of wall friction = 11°, $\theta_w = 20°$. (Figure 4.32a has been adapted from Fig. 9 of Michalowski, 1987, and Fig. 4.32b has been reprinted from Michalowski, 1987, with permission from Elsevier Science.)

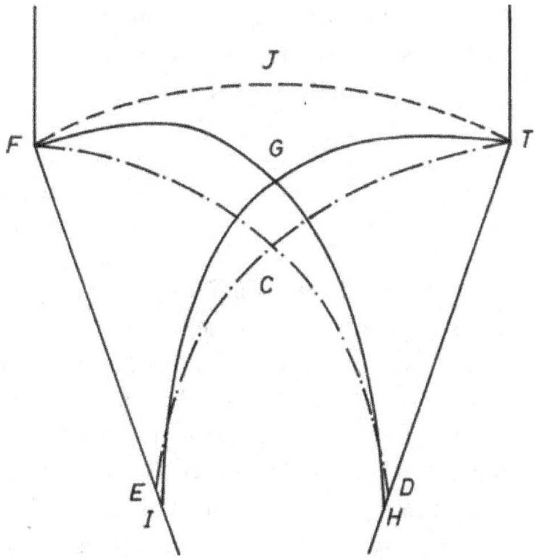

Figure 4.33. Predicted (– · –, - - -) and observed (—) rupture layers in a wedge-shaped bunker: – · –, Michalowski (1987) (parameter values : $\nu_d = 0°$, the rest are as in Fig. 4.32); - - -, Prakash and Rao (1991) (parameter values: $\beta^* = p_a/(\rho_p\, \nu_{max}\, g\, W) = 35$, p_a = 101.33 kPa, $\nu_{max} = 0.74$, $\Gamma = 1.81$, $\lambda = 0.027$, solids fraction in the bin section = 0.577, $v_b/\sqrt{g\,W} = 0.0125$, the rest are as in Fig. 4.32), —, data of Drescher, cited in Michalowski (1987). Here ρ_p is the particle density, g is the acceleration due to gravity, W is the half-width of the bin section, v_b is the magnitude of the vertical velocity in the bin section, and the parameters Γ and λ occur in the expression (3.109). (Reprinted from Prakash and Rao, 1991, with permission of Cambridge University Press.)

In this section, the 2-characteristic passing through the transition point T (Fig. 4.28) has been chosen as the discontinuity curve Γ. On the other hand, if the 1-characteristic through T is chosen as Γ, the solution cannot be continued downward for the following reason. At the point of intersection of Γ with the centerline, another discontinuity curve Γ' must be introduced, as explained above. The curve Γ' must then be chosen as the 2-characteristic, which lies above the 1-characteristic (see Fig. 4.28). As Γ' projects into the region where the solution has already been constructed, this choice is unsuitable.

(c) *Prakash and Rao (1991)*
Here a single discontinuity curve is considered, such that both the velocity and the stress fields are discontinuous across it. This is in contrast to the approaches of Horne and Nedderman (1978) and Michalowski (1987), wherein only one of the two fields is discontinuous.

Using a cylindrical coordinate system with origin at the apparent vertex of the hopper (Fig. 3.12), the governing equations for steady, compressible, plane flow are given by the mass balance (3.12), the momentum balances (3.13) and (3.14), the coaxiality condition (3.18), the associated flow rule ((3.93) and the first of (3.108)), and the yield condition (3.107). As discussed in Appendix E, if the system of equations is hyperbolic and is in the conservation law form (E.1), it admits discontinuous solutions.

Unfortunately, the present system of five equations is not hyperbolic, as the characteristic matrix $\tilde{B}$ (see (C.3)) has only three real roots and three linearly independent eigenvectors (Prakash and Rao, 1991). Further, the coaxiality condition (3.18) and the flow rule (3.93) cannot be expressed in conservation law form. The reader is referred to Schaeffer (1992) for an exploratory attempt to deduce jump conditions for a system which is not in the conservation law form.

In view of the above discussion, it is not clear whether admissible discontinuous solutions can be constructed rigorously. Prakash and Rao (1991) used the jump balances (3.84)–(3.86) (see also Appendix D) associated with the mass and momentum balances, together with three additional assumptions on the downstream side of the discontinuity curve. As suggested by Drescher et al. (1978), it is assumed that (i) the velocity vector is directed radially toward the apparent vertex of the hopper section, and (ii) the material is at a critical state, so that $\sigma = \sigma_c(\nu)$, $\tau = \sigma_c(\nu)\sin\phi$, and the angle of dilation $\nu_d = 0$. Here $\sigma = (\sigma_1 + \sigma_2)/2$ is the mean stress, $\sigma_c(\nu)$ is the mean stress at a critical state, $\tau = (\sigma_1 - \sigma_2)/2$ is the deviator stress and σ_1 and σ_2 are the major and minor principal stresses, respectively. The third assumption will be stated later.

Consider the stress and velocity fields on the upstream side of the discontinuity curve. As discussed in §4.2.1, it is assumed that the material is in plug flow, i.e., $v_x = 0$ and v_y is a known constant, where the coordinate system is chosen as indicated in Fig. 2.6. Assuming that the solids fraction is a known constant, and that the stress field is "fully developed," the momentum balances (2.7) and (2.17) reduce to

$$\frac{d\sigma_{xy}}{dx} = \rho\, g; \quad \frac{d\sigma_{xx}}{dx} = 0 \tag{4.40}$$

Using the boundary conditions (4.13) and (4.14), (4.40) are integrated numerically to obtain the stress field. As in other works cited in §4.2.2, it is assumed that the state of stress is active on the upstream side and passive on the downstream side of the discontinuity curve.

On the downstream side, there are five unknowns, namely, the mean stress σ, the orientation γ' of the major principal stress axis (see Fig. 3.12), the radial velocity v_r, the circumferential velocity v_θ, and the solids fraction ν. In addition, the slope of the discontinuity curve has to be determined. At the centerline ($\theta = 0$) and at the hopper wall ($\theta = \theta_w$), the jump balances, the boundary conditions (3.43) and (3.44), and the assumption that the material is at a critical state, with $\sigma_c(\nu)$ given by (3.109), suffice to determine the downstream values of all the unknowns.

At other points on the discontinuity curve, an additional assumption is needed. It is assumed that $\nu = \nu_0 + (\nu_w - \nu_0)(\theta/\theta_w)^2$, where the subscripts 0 and w denote conditions at the centerline and at the wall, respectively. This form is consistent with the requirement that ν should be an even function of θ for a symmetric solution. As other choices are also possible, this lack of uniqueness is a shortcoming of the approach used.

As shown in Fig. 4.33, the observed rupture layer FGT is bounded by the discontinuity curves predicted by Prakash and Rao (1991) (FJT) and Michalowski (1987) (FCT). Near the centerline, the latter curve agrees better with the observed rupture layer.

This completes our discussion of models for the transition region. Clearly, none of the existing models is very satisfactory, and there is ample scope for refinement. We now consider certain aspects of flow in the hopper section.

4.2.3. The Hopper Section

In Chapter 3, both incompressible solutions for plane flow, and compressible solutions for the smooth wall, radial gravity problem have been discussed. Here the assumptions of smooth walls and radial gravity are relaxed, and the two-dimensional equations for steady compressible flow are considered. The material below has been adapted from Prakash and Rao (1991).

The governing equations (3.107) and (3.12)–(3.18) are integrated downward from the discontinuity curve FJT (Fig. 4.33). As the stresses are sensitive functions of the solids fraction, fine grids have to be used, resulting in long computational times. To overcome this problem, the following approximation is used.

As computations show that the material remains close to a critical state in the upper part of the hopper section, the yield condition (3.107) is approximated by the Mohr–Coulomb yield condition

$$\tau = \sigma \sin\phi \tag{4.41}$$

However, (3.107) is retained as the plastic potential, so that the angle of dilation ν_d is given by (3.108). Hence $\sigma_c(\nu)$ is evaluated using (3.109). The use of (4.41) as the yield condition and (3.107) as the plastic potential is called the *Mohr–Coulomb approximation*. It is interesting to note that the equations based on this approximation are hyperbolic, unlike the complete equations. With this approximation, τ does not depend explicitly on ν, and the derivative $\partial\tau/\partial\nu$, whose magnitude necessitates a fine grid when the complete equations are used, is eliminated. Hence the computational time is reduced significantly.

Figure 4.34 shows that the profiles of the mean stress σ^* obtained by integrating the complete equations, and the equations based on the Mohr–Coulomb approximation, agree well for $\xi^* \leq 0.3$. Here ξ^* is defined by

$$\xi^* \equiv 1 - \frac{(r/W)}{a\,(1 + m\,(\theta/\theta_w)^2)} \tag{4.42}$$

where r is the radial coordinate (Fig. 4.32a), and W is the half-width of the bin section. It is found that the discontinuity curve FJT (Fig. 4.33) can be fitted by an expression of the form $r/W = a\,(1 + m\,(\theta/\theta_w)^2)$, where a is a constant, and $m = 1/(a\,\sin\theta_w) - 1$. Thus $\xi^* = 0$ along FJT.

Using the Mohr–Coulomb approximation, the governing equations can be integrated downward until a specified exit condition is satisfied, thus determining the location of the hopper exit. Two possible exit conditions, one based on the traction-free curve (§3.5.4) and another based on the free-fall curve (§3.7) were discussed earlier. A modified version of the free-fall curve is used here as explained below.

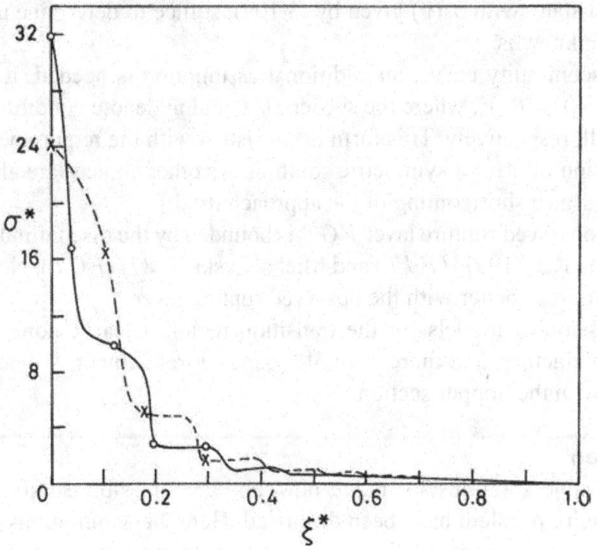

Figure 4.34. Profiles of the dimensionless mean stress $\sigma^* = \sigma/(\rho_p\, \nu_{\max}\, g\, W)$ along the centerline (—,o) and the wall (- - -, ×) of the hopper section : o, × complete equations; —, - - -, Mohr–Coulomb approximation. The variable ξ^* is a dimensionless coordinate, defined by (4.42). Parameter values: $\phi = 37°$, $\delta_b = \delta_h = 16°$, $\theta_w = 10°$, $\nu_{\max} = 0.74$, $\beta^* = p_a/(\rho_p\, \nu_{\max}\, g\, W) = 5.25$, $p_a = 101.33$ kPa, $\nu_{\max} = 0.74$, $\Gamma = 1.81$, $\lambda = 0.027$, $n = 1.05$, solids fraction in the bin section = 0.61, $v_b/\sqrt{g\,W} = 0.0125$, $a = 5.881$. Here the subscripts b and h denote the bin and hopper sections, respectively, ρ_p is the particle density, g is the acceleration due to gravity, W is the half-width of the bin section, Γ and λ are parameters occurring in (3.109), and n is a parameter occurring in (3.107). (Reprinted from Prakash and Rao, 1991, with permission of Cambridge University Press.)

Let ADB denote a curve passing through the edges of the exit slot of the hopper (Fig. 4.35). Assuming that the equations can be integrated from the discontinuity curve FJT till ADB, the values of the dependent variables are known along ADB. The solution can be continued below ADB using the method of characteristics (§4.2.1), but only within the region ACB (Fig. 4.35). Here AC and BC represent the "innermost" characteristics

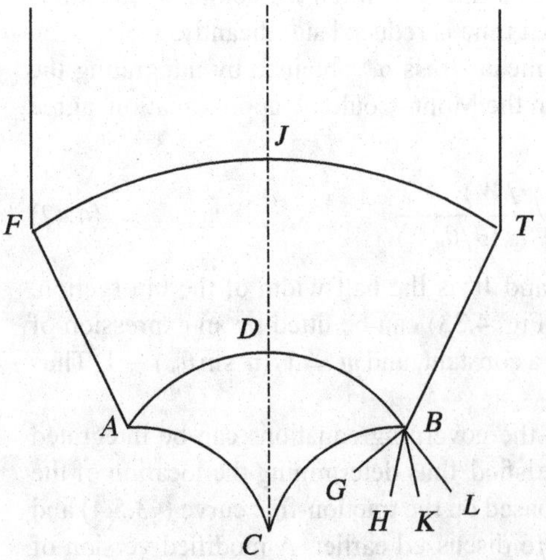

Figure 4.35. Domain of determinacy $ACBDA$ for downward integration from the curve ADB, using the Mohr–Coulomb approximation. Here AC and BC represent the innermost characteristics passing through A and B, respectively. The lines BG, BH, BK, and BL represent tangents to the other characteristics passing through B. The discontinuity curve FJT passes through the transition points F and T. In the work of Prakash and Rao (1991), the curve ADB is chosen as a contour of constant ξ^*, where ξ^* is defined by (4.42).

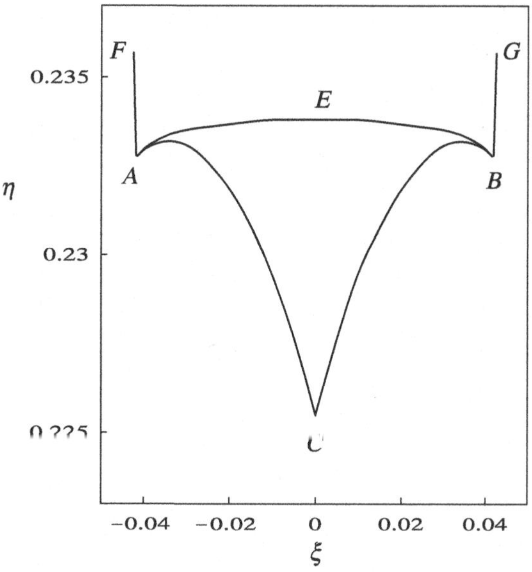

Figure 4.36. Free-fall curve AEB passing through the edges A and B of the exit slot. The lines AF and BG represent the walls of the hopper section. The characteristics AC and BC define the domain of determinacy of the data specified along the curve ADB (Fig. 4.35). Here $\xi = x/W$, $\eta = y/W$, W is the half-width of the bin section, and the origin of the coordinate system is at the apparent vertex of the hopper, as shown in Fig. 4.17. Parameter values are as in Fig. 4.34. (Adapted from Fig. 20 of Prakash and Rao, 1991.)

passing through the edges A and B, respectively. The region bounded by the curves ACB and ADB represents the *domain of determinacy* of ADB.

The free-fall curve must lie above the curve ACB, so that the values of the variables are known on the upstream side of the former. Computations show that this condition is satisfied when the magnitude v_b of the velocity in the bin section (Fig. 4.32) does not exceed a critical value v_{bc}. For $v_b = v_{bc}$, the free-fall curve at B (Fig. 4.35) is tangential to the characteristic BC. This tangency condition serves to fix the value of v_b. It is found that the free-fall curve for $v_b = v_{bc}$ is above the curve ACB formed by the innermost characteristics (Fig. 4.36), except at the end points A and B, where it is tangential.

This behavior is an improvement over the results shown in Fig. 3.36, where the free-fall curve lies below the domain of determinacy of the exit slot. The present results should be viewed with caution, as they are based on limited range of parameter values. Further, as solutions can also be constructed for any value of $v_b < v_{bc}$, the mass flow rate is not uniquely determined by the model. However, $v_b = v_{bc}$ represents an upper bound on the mass flow rate.

The results shown in Figs. 4.34 and 4.36 are for a hopper section with a wall angle $\theta_w = 10°$. For a larger wall angle of $15°$, the equations could not be integrated downward till the exit slot as steep gradients developed and the characteristics tended to converge. To circumvent the problem, the values of the dependent variables on the downstream side of the discontinuity curve FJT (Fig. 4.35) are specified using certain asymptotic fields instead of the jump balances (Prakash and Rao, 1991). These fields are constructed using a modified version of the approach discussed in Problem 3.9, with allowance for wall roughness. Using the asymptotic fields as initial conditions, the governing equations can be integrated downward for a wider range of wall angles.

Using the above procedure, mass flow rate $\dot{M}$ can be evaluated. It is convenient to use a dimensionless mass flow rate V_D', defined by a slight modification of (3.5):

$$V_D' \equiv \frac{\dot{M}}{\rho_p \, v_{max} \, B \, D \sqrt{gD}} \tag{4.43}$$

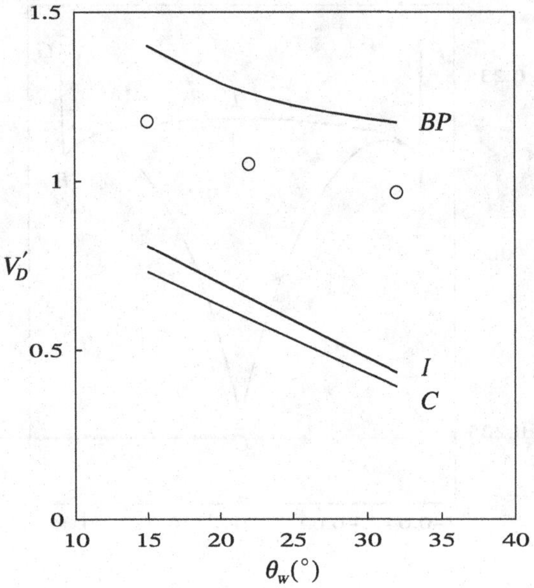

Figure 4.37. Variation of the dimensionless mass flow rate V_D', defined by (4.43), with the wall angle θ_w of the hopper section: curves labeled C and I, compressible and incompressible solutions, respectively, for the free-fall exit condition; curve labeled BP, incompressible Brennen–Pearce solution based on the traction-free exit condition; o, data of Nguyen et al. (1980) for the flow of sand through a wedge-shaped bunker. Parameter values: $\phi = 31°$, $\delta_b = \delta_h = 15°$, $\beta^* \equiv p_a/(\rho_p \, \nu_{max} \, g \, W) = 35$, particle density $\rho_p = 2670$ kg m^{-3}, $\nu_{max} = 0.74$, $a = 3.96$ ($\theta_w = 15°$), 2.747 ($\theta_w = 22°$), 1.934 ($\theta_w = 32°$), the rest as in Fig. 4.34.

Here ρ_p is the particle density, $\nu_{max} = 0.74$ is the solids fraction corresponding to the regular close packing of uniform spheres, B and D are the dimensions of the exit slot (Fig. 3.1), and g is the acceleration due to gravity.

The curve labeled C in Fig. 4.37 shows that the predicted mass flow rates underestimate the measured values for sand (the circles) by about 40–60%. It is surprising that the discrepancy between theory and experiment is slightly less for the incompressible model (the curve labeled I), even though it is expected to be less realistic than the compressible model. Further, the density measurements of Bosley et al. (1969) and Fickie et al. (1989) (Fig. 3.11a) suggest that the material dilates as it approaches the exit slot. Hence the poorer performance of the compressible model may be due to defects in (i) the proposed exit condition, or (ii) the constitutive equations for compressible flow.

The curve labeled BP in Fig. 4.37 represents the predictions of the incompressible Brennen–Pearce solution (§3.5.1), based on the traction-free exit condition (3.45). In contrast to the results based on the free-fall exit condition, this solution overestimates the mass flow rate. Of the three solutions shown in Fig. 4.37, the discrepancy between predicted and measured values is the least for the Brennen–Pearce solution.

Based on the results shown in Figs. 3.33 and 4.37, it appears that the actual flow rate is bounded by predictions based on the traction-free and the free-fall exit conditions. However, it should be noted that the traction-free curve is realized only when approximate solutions such as the Brennen–Pearce or power series solutions (§3.5.1 and 3.5.5) are used. As discussed in §3.5.4, it has not been possible to approach the traction-free curve from above by numerical integration of the governing equation.

Consider the profiles of the solids fraction ν. The solid curve in Fig. 4.38 shows the predicted profile in the hopper, and the vertical line represents the jump in solids fraction on crossing the free-fall curve. The broken curve represents the prediction based on the assumption of vertical free fall. Within the hopper, predicted and measured values agree well for $y_*/W_s > 2$. Near the exit slot, the variation of ν is stronger than that predicted by the model. Surprisingly, in the particle jet below the hopper ($y_* < 0$), there is good agreement between predicted and measured values. More extensive comparisons between

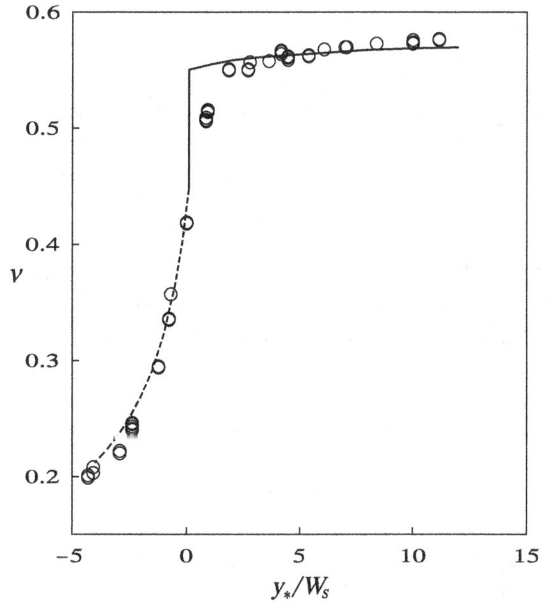

Figure 4.38. Profile of the solids fraction ν along the centerline of a hopper: —, Mohr–Coulomb approximation; - - -, solution for vertical free-fall; ○, data of Fickie et al. (1989) for the flow of glass beads through a wedge-shaped hopper. Here y_* is the vertical coordinate, measured upward from the plane of the exit slot, and W_s is the half-width of this slot. Parameter values: $\phi = 32.4°$, $\delta_h = 15.1°$, $\theta_w = 23°$, $W_s = 6.5$ mm, $\beta^* = p_a/(\rho_p\, \nu_{max}\, g\, W) = 118.6$, $\nu_{max} = 0.74$, $p_a = 101.33$ kPa, particle density $\rho_p = 2900$ kg m^{-3}, particle diameter $d_p \approx 1$ mm, $\Gamma = 1.62$, $\lambda = 0.027$, $a = 2.559$, the rest as in Fig. 4.34. Here Γ and λ are parameters occurring in (3.109). As there was no bin section in the experiments of Fickie et al. (1989), W was chosen as the half-width of the hopper at the highest location for which the density data were reported. (Adapted from Fig. 24 of Prakash and Rao, 1991.)

theory and experiment are required to ascertain whether this is a fortuitous occurrence, or an indication that free-fall conditions do prevail.

The measured transverse profiles (the symbols in Fig. 4.39) show that the particle jet spreads as y_* decreases. On the other hand, the assumption of vertical free-fall constrains the predicted width of the jet to be constant. Apart from this feature, the predicted and measured profiles are qualitatively similar, with a mild variation in the hopper, and a stronger variation in the jet.

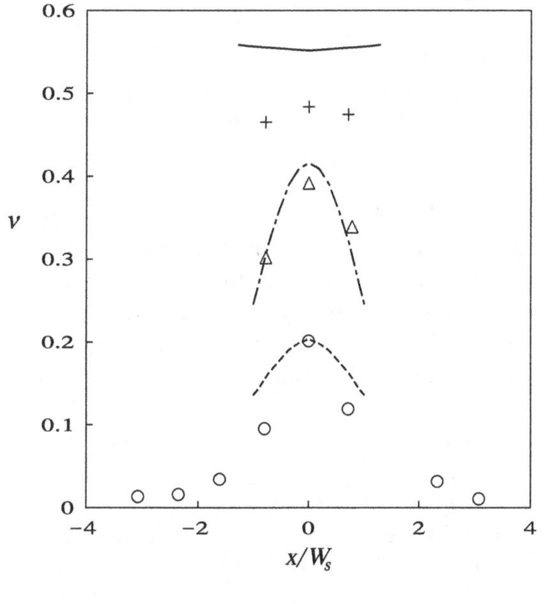

Figure 4.39. Transverse profiles of the solids fraction ν: —, - · -, - - -, Mohr–Coulomb approximation; +, Δ, ○, data of Fickie et al. (1989) for the flow of glass beads through a wedge-shaped hopper. Here y_* is the vertical coordinate, measured upward from the plane of the exit slot, and W_s is the half-width of this slot. The x and y coordinates are defined as shown in Fig. 4.17. Parameter values: $y_*/W_s = 0.85$ (—, +), -0.077 (- · -, Δ), -4.31 (- - -, ○), the rest are as in Fig. 4.38. (Adapted from Fig. 25 of Prakash and Rao, 1991.)

4.3. SUMMARY

Both experimental observations and models for flow through bunkers have been discussed in this chapter. There are many interesting aspects which are not well understood, and none of the available models can predict all the observed features. Hence there is ample scope for further experimental and theoretical work. The material presented in this chapter is summarized below.

(i) *Experimental observations*

As shown in Fig. 4.1, there are several types of flow patterns, depending on the dimensions of the bunker, the head of material, and the properties of the granular and wall materials. At present there is no model which can accurately predict the pattern associated with a given set of parameter values.

When the material begins to discharge from a bunker, rupture layers, or thin regions across which the density and velocity change sharply, form at the exit slot and propagate upward. The data of Michalowski (1987) suggest that the rupture layers do not propagate into the bin section. Experiments involving the simultaneous measurement of density and velocity fields are required to verify this conjecture.

During continuous flow, the pattern of rupture layers appears to be as shown in Fig. 4.6. The rupture layers form at the transition point, grow for some time, and then move downward. This is believed to be a dynamic process, so that the formation and shedding of rupture layers occurs intermittently. This intermittency may be responsible for the temporal stress oscillations observed in the transition region. The amplitude of the oscillations decreases on moving away from this region, leading to quasi-steady stress and velocity fields in the bin section and in the exit region of the hopper section.

In mass flow bunkers, the velocity field appears to consist of three zones: (i) plug flow in the bin section, (ii) approximately radial flow in the hopper section, and (iii) flow parallel to the hopper wall in the transition region.

The density in the hopper section is less than that in the bin section, and is independent of the initial or poured density. Thus the material dilates across a rupture layer or layers as it moves downward from the bin section.

There is a lot of scatter in the data on wall stresses, and the ranges of the static and flowing (dynamic) normal stresses overlap. In the bin section, the maximum values of the dynamic normal stresses exceed static values. The measurements of Benink (1989) suggest that the dynamic shear stresses do not exceed static values. Hence in the bin section, the increase in the dynamic normal stress relative to the static value could possibly be caused by a reduction in the angle of wall friction during flow. There is a need for careful measurements of the shear and normal stresses to check this conjecture.

At the transition point (Fig. 4.17), some of the data suggest that the ratio N_h/N_b of the normal stress on the hopper wall to the normal stress on the bin wall is in the range 2–30. It has often been assumed that this increase in normal stress arises from a switch in the state of stress from an active state in the bin section to a passive state in the hopper section. Using this assumption, the value predicted by Horne and Nedderman (1978) for N_h/N_b is found to be roughly of the same order of magnitude as the observed values. Unfortunately, detailed comparisons of predicted and measured stress profiles are lacking.

In bins, which are special cases of funnel flow bunkers, the measurements of Tüzün and Nedderman (1985a) and Tüzün and Nedderman (1985b) suggest that the velocity field in the transition region is steady, even though the wall stresses oscillate in time. This intriguing feature, and also the similarities and differences between mass-flow bunkers and bins, deserve further experimental investigation.

(ii) *Models*
The main features of the steady-state models are as follows. In the bin section, the Janssen solution provides simple estimates of the stresses. It requires the values of the angle of wall friction δ and the coefficient of earth pressure at rest K. At present, there are no estimates for K which can be used with confidence. The variation of the observed wall stresses with depth below the upper surface of the fill is roughly captured by the Janssen solution (Fig. 4.13). The latter provides an order of magnitude estimate of the stresses, but it cannot be used to establish either lower or upper bounds.

A rigorous alternative to the Janssen solution is provided by the analysis of Horne and Nedderman (1976) for the static case. It is assumed that the density is constant and the Mohr–Coulomb yield condition holds. The resulting force balances are found to constitute a hyperbolic system, which can be integrated numerically using the method of characteristics. In the active case, a saw-tooth type of profile is obtained for the wall stresses (Fig. 4.20). The solution tends to approach the Janssen solution as the depth below the upper surface of the fill increases. These solutions are also valid for the case of steady, fully developed flow, as the momentum balances then reduce to force balances. It is difficult to assess the utility of the method of characteristics solution, as detailed comparisons between predicted and measured stresses are not available.

For the transition region, Horne and Nedderman (1978) have discussed the static case, based on the assumption of (i) an active state in the bin section, and (ii) a passive state in the hopper section. Even though assumption (i) is widely used in the literature, it lacks direct experimental support. Assumption (ii) seems reasonable from the kinematic viewpoint (see §3.3).

For small wall angles (steep hopper walls), the 2-characteristics form a fan at the transition point (Fig. 4.22). At the point where the upper boundary of the fan intersects the centerline, a discontinuity must be introduced in order to satisfy the condition of zero shear stress. Hence the stress field is discontinuous. The profile of the normal stress on the hopper wall (Fig. 4.26) is qualitatively similar to the profile observed by Blair-Fish and Bransby (1973) (Fig. 4.5). It should be noted that the observed profile is time-dependent, whereas the model assumes steady state behavior.

For large wall angles (shallow hopper walls), a solution which satisfies the yield condition everywhere cannot be constructed. Bransby and Blair-Fish (1974) and Horne and Nedderman (1978) have suggested that funnel flow is likely to occur in such cases. The theory provides an estimate of the critical wall angle θ_c, such that if $\theta_w > \theta_c$, funnel flow is likely to occur. Comparison with data (Table 4.6) shows that the theory overestimates the wall angle at which the transition from mass flow to funnel flow occurs.

Michalowski (1987) examined the kinematic aspects of the transition region, assuming that the stress field was a known continuous function of position, and that the angle of dilation was a constant. With these assumptions, the coaxiality condition and the flow rule constitute a linear hyperbolic system. Hence velocity discontinuities can occur only along velocity characteristics. The discontinuities arise because of the change in geometry at the bin–hopper transition. Predicted and measured velocity profiles along a radial line are comparable (Fig. 4.32), but the latter is smoother.

Prakash and Rao (1991) use a single discontinuity curve across which the state of stress changes from active to passive, and the direction of the velocity field from vertical to radial. As the coaxiality condition and the flow rule are not in conservation law form, ad hoc assumptions about the downstream values of the variables are needed in order to construct the discontinuity curve. The observed rupture layer (for the transition from vertical flow to flow parallel to the hopper wall) lies between the discontinuity curves predicted by

Michalowski (1987) and Prakash and Rao (1991) (Fig. 4.33). Near the centerline, the former solution provides a better representation.

Some aspects of compressible flow in the hopper section have been examined. When the associated flow rule is used, the governing equations are not hyperbolic. Hyperbolicity can be restored by using the Mohr–Coulomb approximation, which is equivalent to a nonassociated flow rule. This is because the yield condition is given by (4.41), whereas the plastic potential is given by (3.107). For small wall angles, the equations can be integrated downward from the upper discontinuity curve till the exit slot. For wall angles $\geq 15°$, Prakash and Rao (1991) found that steep fronts formed, making it difficult to integrate the equations. This is an unresolved problem which merits attention. To circumvent the above problem, asymptotic fields have been used as upstream conditions. This permits the equations to be integrated for a wider range of wall angles than in the earlier case.

Using the free-fall exit condition, it is found that the mass flow rates are underestimated by 40–60% when predictions are compared with the data of Nguyen et al. (1980). It is surprising that the incompressible model performs slightly better, in spite of experimental evidence for strong density variation near the exit slot.

Along the centerline of the hopper section, the predicted variation of the solids fraction is weaker than that observed by Fickie et al. (1989) (Fig. 4.38). In the particle jet below the exit slot, the agreement between predicted and measured profiles is good. Transverse profiles of the solids fraction are qualitatively similar to the observed profiles, except for the spreading of the jet in the latter case (Fig. 4.39).

Overall, it appears that flow in the exit region is still poorly understood.

PROBLEMS

4.1. The Janssen–Walker solution for the stress field in a bin

Consider a bin of rectangular cross section (Fig. 2.6), and assume that it is of infinite extent in the z direction. Walker (1966) modified the Janssen solution discussed in §2.3 as follows. The starting point is (2.11):

$$\frac{\mathrm{d}V}{\mathrm{d}\eta} = 1 - \overline{N}\tan\delta \tag{4.44}$$

Here $\eta = y/W$, $V = \int_0^1 \overline{\sigma}_{yy}\,\mathrm{d}\xi$ is the cross-sectional averaged normal stress in the vertical direction, $\xi = x/W$, and $\overline{N} = \overline{\sigma}_{xx}(\xi = 1, \eta)$ is the dimensionless normal stress acting on the wall. Walker (1966) assumed that

$$\overline{N} = K\,\overline{\sigma}_{yy}(\xi = 1, \eta) = K\,F\,V \tag{4.45}$$

where K is the coefficient of earth pressure at rest, and F is a distribution factor. (The Janssen solution assumes that $F = 1$.)

(a) Consider a cohesionless material satisfying the Mohr–Coulomb yield condition. If the friction boundary condition (2.10) holds, show that (i) K is a constant, and (ii) the active and passive values of K are given by

$$K_a = \frac{1 - \sin\phi_* \cos(\delta - \psi)}{1 + \sin\phi_* \cos(\delta - \psi)}; \quad K_p = \frac{1 + \sin\phi_* \cos(\delta + \psi)}{1 - \sin\phi_* \cos(\delta + \psi)} \tag{4.46}$$

where $\sin\psi \equiv \sin\delta/\sin\phi_*$. As noted in §3.4, $\tan\delta \leq \sin\phi_*$, and hence $\sin\psi \leq 1$.

(b) Assuming that F is a constant, derive an expression for V.

(c) At large values of the depth η, assume that the steady state stress and velocity fields are "fully developed," i.e., independent of η. For the special cases where the material is either at rest, or the velocity field is vertically directed ($v_x = 0$) and fully developed, the stress field can be determined by solving the force balances. Using (2.10) and considering solutions which are symmetric about $\xi = 0$, show that

$$\overline{\sigma}_{xx} = 1/\tan\delta\,; \quad \overline{\sigma}_{xy} = \xi \tag{4.47}$$

(d) Using (4.47) and the yield condition (2.30), show that

$$\overline{\sigma}_{yy} = \overline{\sigma}_{xx}\,\frac{1 + \sin^2\phi_* \underline{+} 2\,\sin\phi_*\,\sqrt{1-(\beta\,\xi)^2}}{\cos^2\phi_*} \tag{4.48}$$

where $\beta \equiv \tan\delta/\tan\phi_*$. In (4.48), identify the expressions which correspond to active and passive stress fields, respectively.

(e) Using the results of parts (c) and (d), show that

$$F = \frac{1 + \sin^2\phi_* \underline{+} 2\,\sin\phi_*\,\cos\omega}{1 + \sin^2\phi_* \underline{+}\,\sin\phi_*\,[(\omega/\sin\omega) + \cos\omega]} \tag{4.49}$$

where $\sin\omega \equiv \beta$.

(f) The solution obtained in part (b) shows that the normal stress exerted on the wall attains a limiting value N_a as $\eta \rightarrow \infty$. If the upper surface of the fill is traction free, estimate the "penetration depth" η_*, i.e., the value of η at which the normal stress attains 95% of its asymptotic value. Using the value of K at the wall, calculate the active and passive values of η_* and the asymptotic values of the normal and shear stresses exerted on the wall. The parameter values correspond to Fig. 4.16, i.e., $\phi_* = 30°$, $\delta = 8°$. Generate the active and passive curves shown in Fig. 4.16.

(g) Indicate the shortcomings of the Janssen–Walker analysis.

This problem has been adapted from Walker (1966), Tüzün and Nedderman (1985b), and Nedderman (1992, pp. 93–98).

4.2. The Janssen solution for the stress field in a bin of arbitrary cross section
Consider a granular material at rest in a bin.

(a) Show that the integral momentum balance (1.29) reduces to

$$\int_S \mathbf{n}_* \cdot \boldsymbol{\sigma}\,\mathrm{d}S = \int_V \rho\,\mathbf{b}\,\mathrm{d}V \tag{4.50}$$

where V is a stationary volume with a bounding surface S, $\mathbf{n}_*$ is the unit outward normal to S, ρ is the density, $\mathbf{b}$ is the body force per unit mass, and the stress tensor $\boldsymbol{\sigma}$ is defined in the compressive sense.

(b) Identifying V with the slice bounded by the bin walls and two horizontal planes $y = y_1$ and $y = y_2 = y_1 + \Delta y$ (Fig. 4.40), and considering the limit $\Delta y \rightarrow 0$, show that (4.50) can be rewritten as

$$\frac{\mathrm{d}}{\mathrm{d}y} < \mathbf{e}_y \cdot \boldsymbol{\sigma} > + \frac{P'}{A'}\,(\mathbf{n}_* \cdot \boldsymbol{\sigma})_w = \rho\,\mathbf{b} \tag{4.51}$$

where $\mathbf{e}_y$ is the unit vector in the y direction (Fig. 4.40), A' is the cross-sectional area of the slice, and P' is the perimeter corresponding to the cross section. The angular

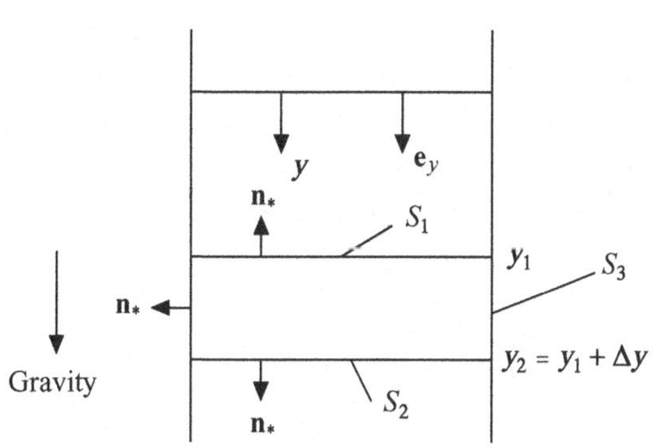

Figure 4.40. A slice of thickness Δy and cross-sectional area A'. In the horizontal plane, the perimeter of the slice is P'. Here $\mathbf{n}_*$ is the unit outward normal to the slice.

brackets denote the cross-sectional averaged value, defined by

$$< \mathbf{e}_y \cdot \boldsymbol{\sigma} > \equiv \frac{1}{A'} \int_{A'} \mathbf{e}_y \cdot \boldsymbol{\sigma} \, \mathrm{d}A' \tag{4.52}$$

and $(\mathbf{n}_* \cdot \boldsymbol{\sigma})_w$ is the perimeter averaged stress vector, evaluated at the wall. It is defined by

$$(\mathbf{n}_* \cdot \boldsymbol{\sigma})_w = \frac{1}{P'} \oint_{P'} \mathbf{n} \cdot \boldsymbol{\sigma} \, \mathrm{d}s \tag{4.53}$$

where s is the arc length measured along the perimeter.

(c) Introducing the assumptions used in Janssen's analysis (see §2.3), and using the dimensionless variables

$$\eta = y/W; \quad \overline{\sigma} = \sigma/(\rho \, g \, W); \quad P = P'/W; \quad A = A'/W^2 \tag{4.54}$$

where W is a characteristic length, show that (4.51) can be integrated to obtain

$$V = (1/M') + [V_0 - (1/M')] \, e^{-M' \eta} \tag{4.55}$$

Here $M' \equiv (P/A) \, K \tan \delta$, $K = \overline{N}/V$, $\overline{N}$ is the dimensionless normal stress exerted on the wall of the bin section, δ is the angle of wall friction, V is the area averaged dimensionless normal stress in the vertical direction, defined by (4.3) and $V_0 \equiv V(\eta = 0)$ is the average normal stress exerted on the upper surface of the material.

(d) The cross section of a bin is a square of side $2\,W$. The bin is filled with a granular material whose parameters are $\phi_* = 35°$, $\delta = 25°$, and particle density $\rho_\mathrm{p} = 2600$ kg m^{-3}. Using the Janssen solution, estimate the value of η at which the normal stress on the bin wall attains 95% of its asymptotic value for (i) a passive stress field, and (ii) an active stress field. The values of K can be approximated by (4.46). Assume that the upper surface is stress free.

(e) The resolution of a strain gauge based load cell used to measure normal stresses is 0.3 kPa, and its range is 0–6 kPa. Can it be used to measure the wall stresses for the bin described in part (d), if (i) $W = 15$ mm, and (ii) $W = 150$ mm? The solids fraction is 0.61.

4.3. Alternative derivation of the Rankine solution

Assuming that $\gamma = \pi/2$ in a region, derive the second of (4.16) by integrating the characteristic equations (4.10) and the compatibility condition (4.12).

5

Theory for Slow Three-Dimensional Flow

Most of the available continuum equations can be broadly classified into two groups: (i) equations involving a yield condition, and (ii) equations which do not involve a yield condition. Group (i) includes the *rigid-plastic*, *elastoplastic*, *viscoplastic*, *elastic-viscoplastic*, *polar elastoplastic*, *gradient plastic* equations and equations based on fluctuations in the rate of deformation tensor **D**, and group (ii) includes the *hypoelastic*, *hypoplastic*, *viscoelastic*, *polar hypoplastic*, and the *Goodman–Cowin* class of equations. Examples of rigid-plastic, hypoelastic, and hypoplastic equations are discussed below, and the model based on fluctuations in the rate of deformation tensor is discussed in §10.3. References to the other classes of equations are listed in Table 5.1.

5.1. CONSTITUTIVE EQUATIONS INVOLVING A YIELD CONDITION

5.1.1. The Yield Condition

The yield condition and the flow rule can be formulated by generalizing the corresponding equations for plane flow. As in Chapter 2, attention will be confined to isotropic materials. First consider the yield condition (2.18), which now takes the form

$$\tilde{F}(\boldsymbol{\sigma}, \nu) = 0 \tag{5.1}$$

where $\boldsymbol{\sigma}$ is the three-dimensional stress tensor and ν is the solids fraction. For isotropic materials, (5.1) can be expressed in terms of ν and the principal invariants of $\boldsymbol{\sigma}$. The latter are defined by (Truesdell and Noll, 1965, p. 23; Slattery, 1999, p. 41)

$$\begin{aligned}
I_1 &= \mathrm{tr}(\boldsymbol{\sigma}) = \sigma_{ii} = \sigma_1 + \sigma_2 + \sigma_3 = \sigma_{xx} + \sigma_{yy} + \sigma_{zz} \\
I_2 &= \left[(\mathrm{tr}(\boldsymbol{\sigma}))^2 - \mathrm{tr}(\boldsymbol{\sigma}^2)\right]/2 = \left[I_1^2 - \mathrm{tr}(\boldsymbol{\sigma}^2)\right]/2 \\
&= \left(\sigma_{ii}\,\sigma_{jj} - \sigma_{ij}\,\sigma_{ji}\right)/2 = \sigma_1\,\sigma_2 + \sigma_2\,\sigma_3 + \sigma_3\,\sigma_1 \\
&= \sigma_{xx}\,\sigma_{yy} + \sigma_{yy}\,\sigma_{zz} + \sigma_{zz}\,\sigma_{xx} - (\sigma_{xy}\,\sigma_{yx} + \sigma_{yz}\,\sigma_{zy} + \sigma_{zx}\,\sigma_{xz}) \\
I_3 &= \det(\boldsymbol{\sigma}) = \left[(\mathrm{tr}(\boldsymbol{\sigma}))^3 - 3\,\mathrm{tr}(\boldsymbol{\sigma})\,\mathrm{tr}(\boldsymbol{\sigma}^2) + 2\,\mathrm{tr}(\boldsymbol{\sigma}^3)\right]/6 \\
&= \left[-2\,I_1^3 + 6\,I_1\,I_2 + 2\,\mathrm{tr}(\boldsymbol{\sigma}^3)\right]/6 \\
&= \left(\sigma_{ii}\,\sigma_{jj}\,\sigma_{kk} - 3\,\sigma_{ii}\,\sigma_{jk}\,\sigma_{kj} + 2\,\sigma_{ij}\,\sigma_{jk}\,\sigma_{ki}\right)/6 \\
&= \sigma_1\,\sigma_2\,\sigma_3 \\
&= \sigma_{xx}\,\sigma_{yy}\,\sigma_{zz} + \sigma_{xy}\,\sigma_{yz}\,\sigma_{zx} + \sigma_{xz}\,\sigma_{yx}\,\sigma_{zy} \\
&\quad - (\sigma_{xx}\,\sigma_{yz}\,\sigma_{zy} + \sigma_{yy}\,\sigma_{zx}\,\sigma_{xz} + \sigma_{zz}\,\sigma_{xy}\,\sigma_{yx})
\end{aligned} \tag{5.2}$$

Table 5.1. References Using Various Types of Constitutive Equations for Flowing Granular Materials

Type of Equation	References
Elastoplastic	Wieckowski (1994), von Wolffersdorf (1996), Anand and Gu (2000)
Gradient plastic	Vardoulakis and Aifantis (1991)
Polar elastoplastic	Mühlhaus (1986), Tejchman and Wu (1993)
Fluctuations in the rate of deformation tensor	Savage (1998), Tardos et al. (2003)
Viscoplastic	Karlsson et al. (1998), Elaskar et al. (2000)
Elastic-viscoplastic	Desai and Zhang (1987)
Viscoelastic	Zhang and Rauenzahn (1997)
Polar hypoplastic	Tejchman and Bauer (1996), Tejchman and Gudehus (2001)
Goodman–Cowin	Goodman and Cowin (1971), Rajagopal and Massoudi (1990), Kumar et al. (2003)

where the repeated indices imply summation, σ_i, $i = 1, 3$ are the principal stresses, σ_{xx} is the x component of the normal stress relative to a Cartesian coordinate system, and $\mathrm{tr}(\boldsymbol{\sigma})$ denotes the sum of the diagonal components of $\boldsymbol{\sigma}$. In some books, the principal invariants are called the invariants (Fung, 1977, p. 99; Hunter, 1983, pp. 87–88). As discussed in §2.5.1, the principal invariants are coefficients of the characteristic equation (2.24), which now takes the form

$$\det([\boldsymbol{\sigma} - a\,\mathbf{I}]) = -a^3 + I_1\,a^2 - I_2\,a + I_3 = 0 \tag{5.3}$$

Thus (5.1) can be replaced by

$$F_*(I_1, I_2, I_3, \nu) = 0 \tag{5.4}$$

or equivalently,

$$F(\sigma_1, \sigma_2, \sigma_3, \nu) = 0 \tag{5.5}$$

where F is a symmetric function of the principal stresses.

The geometric representation of yield conditions is facilitated by considering a three-dimensional Euclidean space with Cartesian coordinates σ_1, σ_2, and σ_3 (Fig. 5.1). This is called the *Haigh–Westergaard* or *principal stress* space (Haigh, 1920; Westergaard, 1920; cited in Yu, 2002). States of stress are represented by points in this space. In particular, cohesionless materials, which cannot support tensile normal stresses, are represented by points

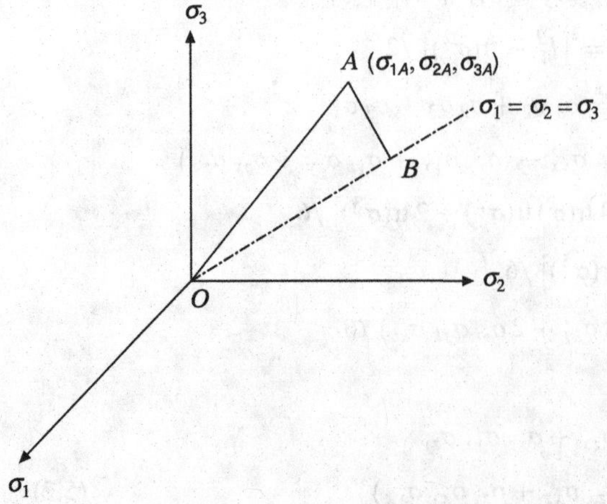

Figure 5.1. The principal stress space. Here OB is the space diagonal and AB is perpendicular to OB. The length of OB is $\sqrt{3}\,p_A$ and the length of AB is $\sqrt{2/3}\,q_A$, where p and q are defined by (5.6).

in the nonnegative octant $\sigma_i \geq 0$, $i = 1, 3$. The yield condition (5.5) can be represented in this space by surfaces of constant solids fraction ν, which are called *yield surfaces*.

Before discussing the forms of specific yield surfaces, it is convenient to introduce two invariants p and q of the stress tensor related to the principal invariants by

$$\begin{aligned} p &\equiv I_1/3 = (\mathrm{tr}\,\boldsymbol{\sigma})/3 = (\sigma_{ii})/3 = (\sigma_1 + \sigma_2 + \sigma_3)/3 \\ &= (\sigma_{xx} + \sigma_{yy} + \sigma_{zz})/3 \\ q &\equiv (I_1{}^2 - 3\,I_2)^{1/2} = \frac{1}{\sqrt{2}}\left[3\,\sigma_{ij}\,\sigma_{ji} - \sigma_{ii}\,\sigma_{jj}\right]^{1/2} \\ &= \frac{1}{\sqrt{2}}[(\sigma_1 - \sigma_2)^2 + (\sigma_2 - \sigma_3)^2 + (\sigma_3 - \sigma_1)^2]^{1/2} \\ &= \frac{1}{\sqrt{2}}[(\sigma_{xx} - \sigma_{yy})^2 + (\sigma_{yy} - \sigma_{zz})^2 + (\sigma_{zz} - \sigma_{xx})^2 \\ &\quad + 6(\sigma_{yx}\,\sigma_{xy} + \sigma_{zx}\,\sigma_{xz} + \sigma_{yz}\,\sigma_{zy})]^{1/2} \end{aligned} \tag{5.6}$$

where the repeated indices imply summation. The variable p is called the *mean stress*; it is the three-dimensional analog of the mean stress σ for plane flow (see (2.33)). The variable q can be interpreted as follows. The magnitude of the maximum shear stress acting on a plane whose normal is parallel to the σ_1–σ_2 plane is $(\sigma_1$–$\sigma_2)/2$ (see Fig. 2.10). Thus q is proportional the root mean square value of the maximum shear stresses associated with the σ_1–σ_2, σ_2–σ_3, and σ_3–σ_1 planes (Nedderman, 1992, p. 133).

Some of the yield conditions in the literature are formulated in terms of an invariant J_2, which is related to q as follows. Using (5.2) and (5.6),

$$\begin{aligned} q^2/3 &= (I_1^2 - 3\,I_2)/3 = (9p^2 - (27/2)p^2 + (3/2)\mathrm{tr}(\boldsymbol{\sigma}^2))/3 \\ &= (\mathrm{tr}(\boldsymbol{\sigma}^2) - 3\,p^2)/2 = (\boldsymbol{\sigma} : \boldsymbol{\sigma} - 3\,p^2)/2 \end{aligned} \tag{5.7}$$

It is conveninent to introduce the deviatoric stress tensor $\mathbf{s}'$ defined by

$$\mathbf{s}' \equiv \boldsymbol{\sigma} - p\,\mathbf{I} \tag{5.8}$$

where $\mathbf{I}$ is the unit tensor. As $\mathbf{s}' : \mathbf{I} = \mathbf{I} : \mathbf{s}' = \mathrm{tr}(\mathbf{s}') = 0$, we have $\boldsymbol{\sigma} : \boldsymbol{\sigma} = (\mathbf{s}' + p\,\mathbf{I}) : (\mathbf{s}' + p\,\mathbf{I}) = \mathbf{s}' : \mathbf{s}' + 3\,p^2$. (Note that $\mathbf{I} : \mathbf{I} = 3$ in a three-dimensional space.) Hence (5.7) reduces to

$$q^2/3 = (\mathbf{s}' : \mathbf{s}')/2 = (s'_{ij}\,s'_{ji})/2 = -I'_2 \equiv J_2 \tag{5.9}$$

where I'_2 is the second invariant of $\mathbf{s}'$.

The variables p and q can be interpreted geometrically as follows. Consider a point A in principal stress space, with coordinates $(\sigma_{1A}, \sigma_{2A}, \sigma_{3A})$, and let AB represent a line which passes through A and is perpendicular to the space diagonal (Fig. 5.1). It can be shown (Problem 5.1) that the length of the line OB is $\sqrt{3}\,p_A$, and the length of AB is $\sqrt{2/3}\,q_A$.

Consider some examples of three-dimensional yield conditions. The classical *von Mises* yield condition (von Mises, 1913, cited in Khan and Huang, 1995, p. 94) is given by

$$q - \sigma_0 = 0; \quad \sigma_0 = \text{constant} > 0 \tag{5.10}$$

Even though (5.10) is usually attributed to von Mises, Khan and Huang (1995, p. 95) note that it was first proposed by Huber (1904). For a state of uniaxial compression or tension, given by $\sigma_1 \neq 0$, $\sigma_2 = 0$, and $\sigma_3 = 0$, (5.6) and (5.10) imply that $q = |\sigma_1| = \sigma_0$. Hence σ_0 is the magnitude of the yield stress in uniaxial compression or tension. As q is proportional to the distance measured perpendicular to the space diagonal (Fig. 5.1), (5.10) represents a circular cylinder in principal stress space (Fig. 5.2a). Thus yielding is independent of the

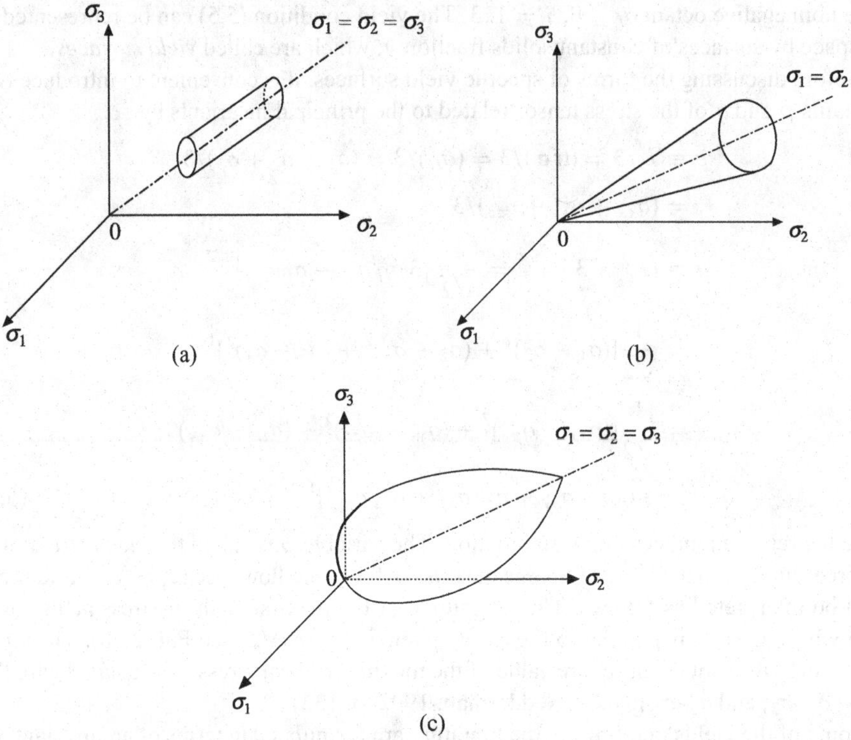

Figure 5.2. Yield surfaces corresponding to the (a) von Mises (see (5.10)), (b) Drucker–Prager (see (5.13)) and (c) Granta gravel (see (5.14)) yield conditions.

value of the mean stress p, which is proportional to the distance measured along the space diagonal. This behavior is appropriate for metals, but not for granular materials. The defect can be remedied by replacing (5.10) with (Schleicher, 1925, 1926; von Mises, 1926)

$$q - E(p) = 0 \tag{5.11}$$

where $E(p)$ is a specified function of p. This can be further generalized to

$$q - H(p, \nu) = 0 \tag{5.12}$$

where ν is the solids fraction. Equation (5.12) will be called the *extended von Mises* yield condition.

Depending on the form chosen for $H(p, \nu)$, yield surfaces can be either open or closed. For example, Drucker and Prager (1952) proposed the yield condition

$$q - k' p = 0, k' = \text{constant} > 0 \tag{5.13}$$

which corresponds to an open cone in principal stress space (Fig. 5.2b). On the other hand, the yield condition for the the Granta-gravel model (Schofield and Wroth, 1968, p. 111) can be represented by a one-parameter family of closed yield surfaces in principal stress space (Fig. 5.2c). This yield condition is defined by

$$(q/p_c) - M(p/p_c)\,[1 - \ln(p/p_c)] = 0 \tag{5.14}$$

where $p_c = p_c(\nu)$ is the mean stress at a critical state, and M is a constant defining the slope of the critical state line $q = Mp = Mp_c$. A part of this yield surface is inaccessible for cohesionless materials, as it lies in the region where some of the principal stresses are negative Schofield and Wroth (1968, p. 132) (see also Problem 5.2).

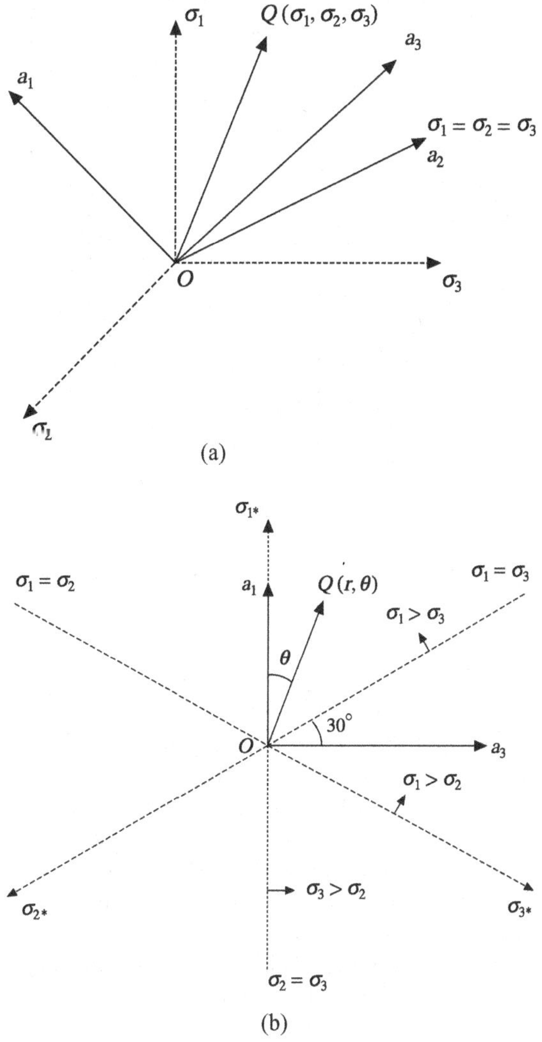

Figure 5.3. (a) The Cartesian coordinate system (a_1, a_2, a_3). The a_2 axis is aligned with the space diagonal, and the a_1 and a_3 axes are in the Π-plane $\sigma_1 + \sigma_2 + \sigma_3 = 0$. The a_1 axis coincides with the projection of the σ_1 axis onto this plane, and the a_1, a_2, and a_3 axes form a right-handed coordinate system. (b) The a_3–a_1 plane. Here $O\sigma_{1*}$, $O\sigma_{2*}$, and $O\sigma_{3*}$ represent projections of the σ_1, σ_2, and σ_3 axes onto this plane. The length of the vector OQ is r.

Before considering other forms for yield surfaces, it is helpful to introduce some terminology. The *octahedral* plane $\sigma_1 + \sigma_2 + \sigma_3 =$ constant is a plane perpendicular to the space diagonal. In particular, the octahedral plane passing through the origin is called the Π-plane (Mendelson, 1983, p. 81), and is defined by $\sigma_1 + \sigma_2 + \sigma_3 = 0$. The intersection of the yield surface with an octahedral plane generates a curve, called a *cross section*. To examine the shapes of cross sections, it is helpful to use new coordinates (a_1, a_2, a_3) instead of the principal stresses, as explained below.

Consider a point Q in principal stress space, with coordinates $(\sigma_1, \sigma_2, \sigma_3)$ (Fig. 5.3a). Let us introduce new coordinates (a_1, a_2, a_3), such that the a_2 axis coincides with the space diagonal, and the a_1 and a_3 axes lie in the Π-plane (Fig. 5.3a). The a_1 axis coincides with the projection of the σ_1 axis onto the Π-plane, and the a_3 axis is chosen so that the a_1, a_2, and a_3 axes form a right-handed coordinate system. The two sets of coordinates are related by (Schofield and Wroth, 1968, p. 41) (see also Problem 5.3)

$$
\begin{aligned}
& a_1 = (2\,\sigma_1 - \sigma_2 - \sigma_3)/\sqrt{6}; \; a_2 = (\sigma_1 + \sigma_2 + \sigma_3)/\sqrt{3} = p\,\sqrt{3} \\
& a_3 = (\sigma_3 - \sigma_2)/\sqrt{2} \\
& \sigma_1 = (\sqrt{6}\,a_1 + \sqrt{3}\,a_2)/3; \; \sigma_2 = (2\sqrt{3}\,a_2 - \sqrt{6}\,a_1 - 3\sqrt{2}\,a_3)/6 \\
& \sigma_3 = (2\sqrt{3}\,a_2 + 3\,\sqrt{2}\,a_3 - \sqrt{6}\,a_1)/6
\end{aligned}
\tag{5.15}
$$

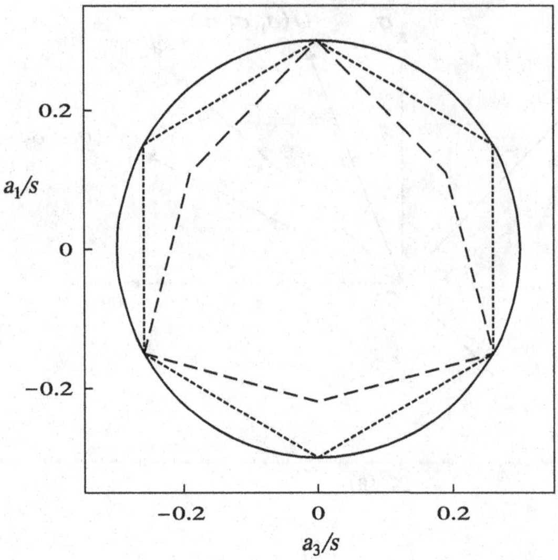

Figure 5.4. Cross sections of some yield surfaces, i.e., intersections of the yield surfaces with the octahedral plane $\sigma_1 + \sigma_2 + \sigma_3 =$ constant $= s > 0$: —, von Mises (see (5.10)) or extended von Mises (see (5.11)); - - -, Tresca (see (5.17)) or extended Tresca (see (5.18)); – – –, Mohr–Coulomb (see (5.19)).

Using (5.15), it follows that the invariant q, defined by (5.6), is related to a_1 and a_3 by

$$q = \sqrt{3/2}\,\sqrt{a_1^2 + a_3^2} \tag{5.16}$$

Thus each cross section is defined by an equation of the form $F(a_1, a_2, a_3) = 0$, where a_2 is a specified constant. Hence it can be represented by a curve in the a_3–a_1 plane (the Π-plane) (Fig. 5.3b). The broken lines $O\sigma_{1*}$, $O\sigma_{2*}$, and $O\sigma_{3*}$ in Fig. 5.3b represent the projections of the σ_1, σ_2, and σ_3 axes onto the a_3–a_1 plane. Along each of these lines, two of the principal stresses are equal, as indicated in the figure.

The yield conditions introduced above have circular cross sections. For example, the von Mises yield condition (5.10) corresponds to a circle of radius $\sqrt{2/3}\sigma_0$ in the a_3–a_1 plane. Similarly, the intersection of the Drucker–Prager yield condition (5.13) with the plane $\sigma_1 + \sigma_2 + \sigma_3 =$ constant $\equiv s$ is a circle of radius $r = \sqrt{2}\,k'\,s/(3\sqrt{3})$ (Fig. 5.4). Some examples of yield conditions with noncircular cross sections will now be discussed.

The *Tresca* yield condition (Tresca, 1864, cited in Khan and Huang, 1995, p. 96) is given by

$$\sigma_i - \sigma_l = \text{constant} \equiv \sigma_0, \quad i, j, l = 1, 3, \quad \sigma_i \geq \sigma_j \geq \sigma_l \tag{5.17}$$

where the indices i, j, l have distinct values. If $\sigma_2 = 0 = \sigma_3$, (5.17) implies that $\sigma_1 = \sigma_0$. Hence σ_0 represents the magnitude of the yield stress in uniaxial tension or compression. If $\sigma_1 \geq \sigma_2 \geq \sigma_3$, (5.17) implies that $\sigma_1 - \sigma_3 = \sigma_0$. Hence the maximum shear stress in the σ_1–σ_3 plane is equal to $\sigma_0/2$.

The cross section corresponding to the Tresca yield condition is a regular hexagon (Problem 5.3). As in the case of the von Mises yield condition, yielding is unaffected by the value of the mean stress, i.e., the equation for the cross section is independent of the coordinate a_2, defined by (5.15). Hence the yield surface is cylinder of hexagonal cross section (Fig. 5.5a). Further, yielding is unaffected by the value of the intermediate principal stress σ_j.

The *extended Tresca* yield condition is defined by

$$\sigma_i - \sigma_l - k'\,p = 0, \quad i, j, l = 1, 3, \quad \sigma_i \geq \sigma_j \geq \sigma_l \tag{5.18}$$

where k' is a positive constant and the indices i, j, l have distinct values. The intersection of (5.18) with the plane $\sigma_1 + \sigma_2 + \sigma_3 =$ constant $\equiv s$ is a regular hexagon of side $\sqrt{2}\,k'\,s/(3\sqrt{3})$ (Fig. 5.4).

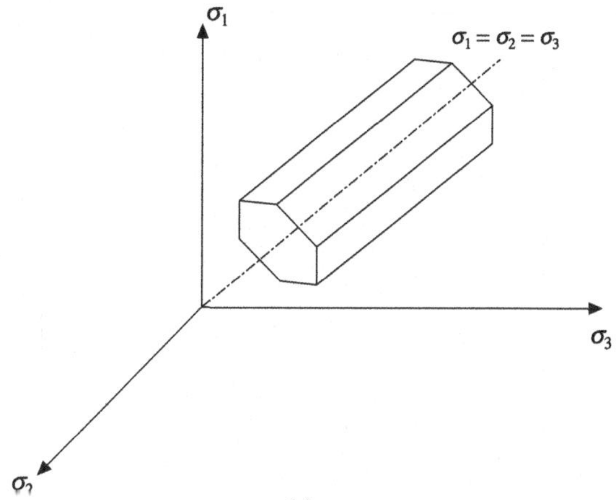

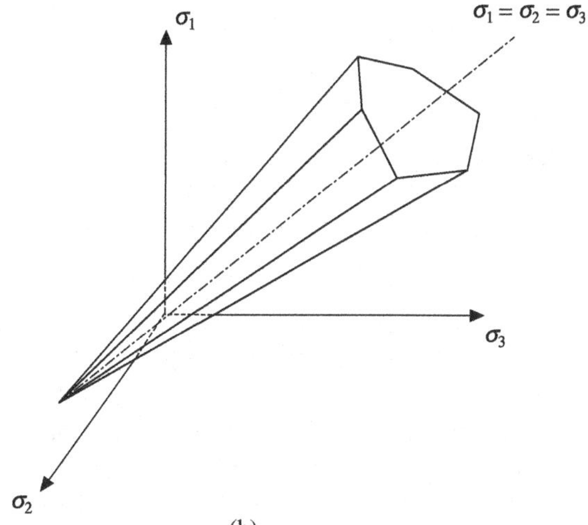

Figure 5.5. Yield surfaces corresponding to the (a) Tresca (see (5.17)) and (b) Mohr–Coulomb (see (5.19)) yield conditions.

A more realistic yield condition for granular materials is given by the *Mohr–Coulomb* yield condition (2.37) (Bishop, 1966; Feda, 1982, p. 264)

$$(\sigma_i - \sigma_l)/2 - ((\sigma_i + \sigma_l)/2)\sin\phi_* - c\cos\phi_* = 0, \quad i, j, l = 1, 3 \tag{5.19}$$

where c and ϕ_* are material constants called the *cohesion* and the *angle of internal friction*, respectively, $\sigma_i \geq \sigma_j \geq \sigma_l$, and the indices i, j, l have distinct values. Thus yielding is unaffected by the value of the intermediate principal stress. It can be shown (Problem 5.3) that the cross section is an irregular hexagon (Fig. 5.4) whose size increases with the mean stress p. The yield surface is a cone with vertex at a point on the space diagonal with coordinates $\sigma_1 = \sigma_2 = \sigma_3 = -c\cot\phi_*$ (Fig. 5.5b).

Jenike and Shield (1959) have constructed a closed yield surface, which for $\nu = \nu_1$ is the union of the Mohr–Coulomb cone and the octahedral plane defined by $p = p_c(\nu_1)$. Here $p_c(\nu)$ is the mean stress at a critical state. Thus the yield surface is a cone with a flat "base." The size of the cone varies with the density in a manner prescribed by the function $p_c(\nu)$.

Lade and Duncan (1975) proposed a yield condition of the form

$$I_1^3 - k_1(\nu)\, I_3 = 0 \tag{5.20}$$

where I_1 and I_3 are invariants of the stress tensor (see (5.2)). As shown in Fig. 5.6, the cross section resembles a smoother version of the Mohr–Coulomb cross section. Equation (5.20)

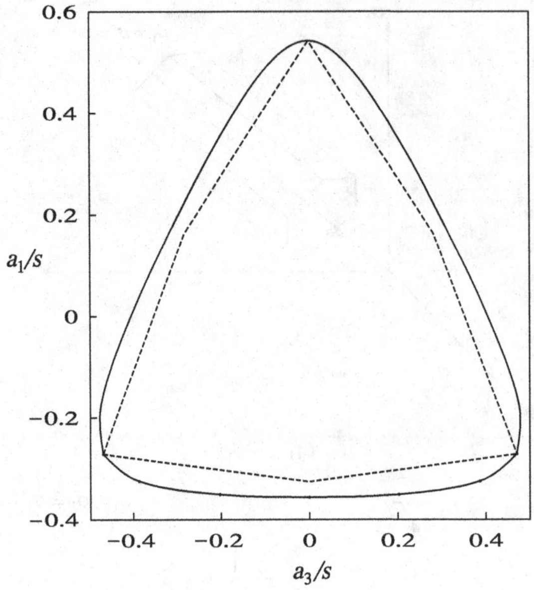

Figure 5.6. Cross sections corresponding to the Lade–Duncan yield condition (5.20) for $k_1 = 103$ (—), and the Mohr–Coulomb yield condition (5.19) (- - -). The cross sections represent intersections of the yield conditions with the octahedral plane $\sigma_1 + \sigma_2 + \sigma_3 = \text{constant} = s > 0$. The parameter values have been chosen so that three of the vertices of the cross section coincide with the corresponding points on the cross section of the Lade–Duncan yield condition.

leads to a family of open yield surfaces in the sense that they intersect the space diagonal in principal stress space only at the origin.

Krenk (2000) proposed a yield condition of the form

$$I_3^3 - p^3\,(p/p_*)^n = 0 \tag{5.21}$$

where I_3 is an invariant of the stress tensor, defined by (5.2), $p = \text{tr}(\boldsymbol{\sigma})$, and n is a positive constant. Equation (5.21) represents a family of closed yield surfaces which intersect the space diagonal at $p = 0$ and at $p = p_*$. In the framework of critical state models, p_* can be regarded as a function of the solids fraction ν. Krenk generalized this by allowing p_* to depend on the work associated with the plastic deformation of the material.

Other expressions for smooth yield surfaces are discussed in Lade (1977), van Eekelen (1980), Desai and Siriwardane (1984, pp. 373–380), and Matsuoka et al. (1989). The article by Yu (2002) contains a large number of references on yield conditions for granular and other materials.

5.1.2. Symmetry Considerations

The yielding of isotropic materials must be unaffected by the manner in which the principal stress axes are labeled. Therefore, if $(\sigma_1, \sigma_2, \sigma_3)$ is a point on the cross section, so must $(\sigma_1, \sigma_3, \sigma_2)$. With reference to the Π-plane (Fig. 5.3b), interchanging the values of σ_2 and σ_3 is equivalent to changing the sign of a_3 while keeping the value of a_1 unchanged (see (5.15)). Hence the cross section must be symmetric about the a_1 axis, which coincides with the projection $O\sigma_{1*}$ of the σ_1 axis onto the Π-plane. Similarly, it must be symmetric about the projections $O\sigma_{2*}$ and $O\sigma_{3*}$ of the σ_2 and σ_3 axes onto this plane. Hence the cross section must be composed of six identical segments. An example is provided by the cross section for the Mohr–Coulomb yield condition (Fig. 5.4).

5.1.3. Conventional Triaxial Tests

In Chapter 2, some features of data obtained using the direct shear box and the Jenike cell have been discussed. For plane flow, the data lead to a yield condition of the form (2.44), or equivalently to

$$H(\sigma_1, \sigma_2, \nu) = 0 \tag{5.22}$$

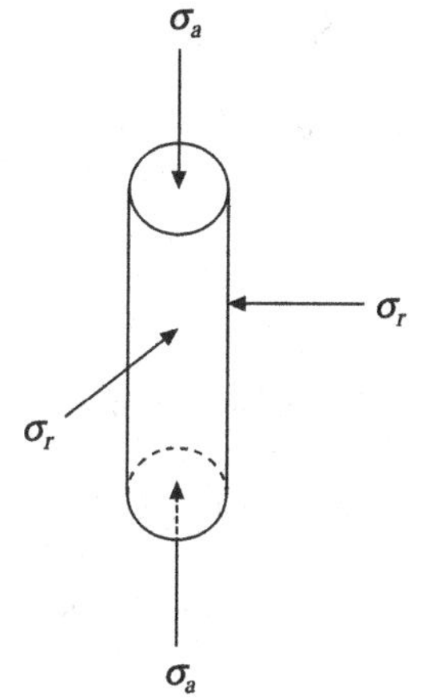

Figure 5.7. Stresses exerted on a cylindrical sample in the triaxial test. Here σ_a is the axial normal stress, and σ_r is the radial normal stress.

where σ_1 and σ_2 are the principal stresses in the plane of flow. Here we consider *triaxial* tests, which provide information regarding the form of the yield condition (5.5) for three-dimensional problems.

In a triaxial test, cylindrical specimens of the material are subjected to an axial normal stress σ_a and a radial normal stress σ_r (Fig. 5.7). The sample is confined by a rubber membrane or sleeve, and subjected to a radial stress by the water in the cell (Fig. 5.8). (The membrane permits samples to be deformed at constant volume, if desired, and also prevents cylindrical samples formed from cohesionless materials such as sand and glass beads from falling apart.) As the tests are performed at slow speeds of axial displacement of the loading ram, the flow of cell water generates negligible shear stresses on the membrane and on the loading cap on top of the sample. Therefore, it can be assumed that the stresses σ_a and σ_r (Fig. 5.7) are principal stresses. Tests are usually conducted with saturated samples, where the gaps between particles are filled with water. The porous filter which supports the base of the sample (Fig. 5.8) permits water to flow into or out of the sample as the test progresses. This provides a convenient method for measuring volume changes of the sample, and also

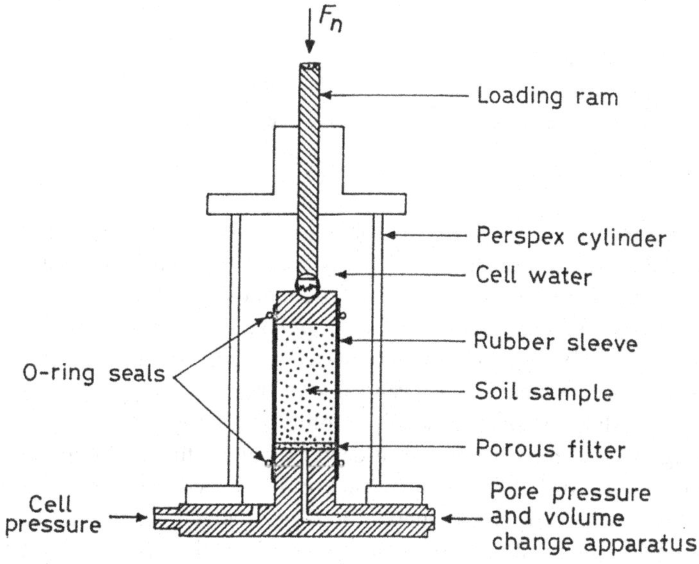

Figure 5.8. The triaxial tester. (Reproduced from Fig. 5.6 of Atkinson and Bransby, 1982, p. 74, with permission from Prof. J. H. Atkinson.)

for monitoring the *pore pressure*, or the pressure in excess of the atmospheric pressure which is exerted by the water on the particles.

Central to the analysis of tests on saturated samples is the concept of an *effective stress*, which was introduced by Terzaghi (1923, cited in Lade and De Boer, 1997). Defining the effective principal stresses σ_i' by

$$\sigma_i' = \sigma_i - u_w, \quad i = 1, 3 \tag{5.23}$$

where u_w denotes the pore pressure, Terzaghi (1936) postulated that the deformation of soils is governed by the effective stresses σ_i', rather than by the total stresses σ_i. (Lade and De Boer (1997) note that the concept of an effective stress was independently postulated by Fillunger (1936).) The following statement taken from Atkinson and Bransby (1982, p. 25) is a fair assessment of this postulate: "Detailed theoretical and experimental examinations of the principle of effective stress have been published by Skempton (1960) and Bishop et al. (1965), and no conclusive evidence has yet been found which invalidates Terzaghi's hypothesis, at least for saturated soils at normal levels of engineering stresses." Lade and De Boer (1997) examined data for sands and found that the principle was satisfactory except at very high stresses.

Accepting the validity of the principle of effective stress, let us now consider triaxial tests. Two types of tests are commonly conducted (Atkinson and Bransby, 1982, pp. 70–78; Bardet, 1997, pp. 443–460). These are *compression* tests, in which $\sigma_a = \sigma_1 > \sigma_2 = \sigma_3 = \sigma_r$, and *extension* tests, in which $\sigma_a = \sigma_3 < \sigma_1 = \sigma_2 = \sigma_r$. Here the principal stresses are numbered so that $\sigma_1 \geq \sigma_2 \geq \sigma_3$, and σ_1, σ_2, and σ_3 are called the major, intermediate, and minor principal stresses, respectively. A limiting case of these two types of tests is given by the *isotropic compression* test, where $\sigma_a = \sigma_r$. The tests can be further classified into *drained* and *undrained* tests. In drained tests, the pore water in the sample is allowed to drain freely as the test proceeds. Thus the volume of the sample changes as dictated by the deformation, and the pore pressure of water u_w is zero. In contrast, drainage of pore water is prevented in undrained tests. Here $u_w \neq 0$ in general, but the volume of the sample is forced to remain approximately constant. This is due to the incompressibility of water at the stress levels usually employed. The reader is referred to Bardet (1997, pp. 443–460) for a detailed discussion of test procedures.

5.1.4. Isotropic Compression Tests

Here all the principal stresses are equal, and hence (5.6) implies that $p = \sigma_1$ and $q = 0$. As discussed earlier, only effective stresses are relevant. The effective mean stress is given by

$$p' = p - u_w = \sigma_1 - u_w \tag{5.24}$$

where u_w is the pore pressure. Figure 5.9 shows data obtained using a sample of kaolin clay. It is common practice (Atkinson and Bransby, 1982, p. 126) to plot test data in the $1/\nu$–ln p' plane. Starting at point A, if p' is increased monotonically, the solids fraction increases along the curve AB, which is called the *loading, normal consolidation*, or *isotropic compression* curve. Atkinson and Bransby (1982, p. 128) note that states on the right of the normal consolidation curve cannot be attained. For a given value of the effective mean stress p', this curve represents the smallest possible value of the solids fraction ν.

At the point B (Fig. 5.9), if p' is decreased, ν decreases along the *unloading* curve BCD, rather than along BA. At D, if p' is increased, ν increases along the *reloading* curve DEF. The profile of ν for isotropic compression is qualitatively similar to that for one-dimensional compression (Fig. 2.31). As discussed in §2.13, the occurrence of the hysteresis loop during unloading and reloading shows that a part of the deformation is plastic or irrecoverable.

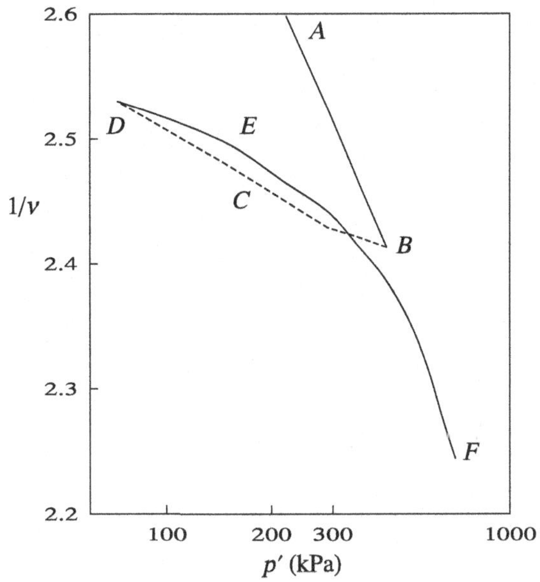

Figure 5.9. Isotropic compression test on kaolin clay. Here ν is the solids fraction, and p' is the effective mean stress, defined by (5.24) and (5.6). (Data of Roscoe and Burland, 1968.)

As shown by the segment AB, the normal consolidation curve can often be approximated by a straight line in the $1/\nu$–ln p' plane. At low values of p', even though $1/\nu$ cannot vary linearly with ln p', the appropriate form is not known owing to a lack of data and theoretical guidelines.

For sand, the data reported in Atkinson and Bransby (1982, p. 132) suggest that the unloading–reloading lines are almost horizontal. Similar results were reported for glass beads by Heertjes et al. (1978). However, in the latter case, the smallest value of p' was 7 kPa. Data at smaller values of p' are needed to understand the variation of ν with p' at low stress levels.

So far, we have discussed the behavior of samples in isotropic compression, where there are no shear stresses. Consider compression tests with $\sigma_a > \sigma_r$.

5.1.5. Compression Tests

Figure 5.10 shows the results of a compression test on sand. Here the radial principal stress $\sigma_r = \sigma_2 = \sigma_3$ is kept constant and the axial principal stress $\sigma_a = \sigma_1$ is measured while the sample is compressed axially at a constant speed. As $\sigma_2 = \sigma_3$, (5.6) implies that $p = (\sigma_1 + 2\,\sigma_3)/3$ and $q = |\sigma_1 - \sigma_3|$. As the test is of the drained type, the pore pressure u_w vanishes and hence the effective and actual stresses are equal. Thus $p' = p$ and $q' = q$. The results are qualitatively similar to those obtained using the shear box (Fig. 2.17). Quantitative comparison of data from these two types of tests is difficult, as the vertical normal stress is a constant in the test based on the shear box, but not in the triaxial test.

Consider the behavior of dense samples. Let q'_* and p'_* denote the values of q' and p', respectively, which correspond to the peak in the q' versus axial strain curve (Fig. 5.10a). Figure 5.11 shows data for Weald clay, obtained using drained and undrained tests on samples with different initial densities. The data collapse onto a single line in the q'_*/p'_e–p'_*/p'_e plane. Here $p'_e(\nu)$ is the equivalent effective mean stress, computed from the normal consolidation curve (see the curve AB in Fig. 5.9) for any specified value of ν. As discussed in Chapter 2, the solid line can be regarded as a section of the normalized Hvorslev surface $q'_*/p'_e = f(p'_*/p'_e)$.

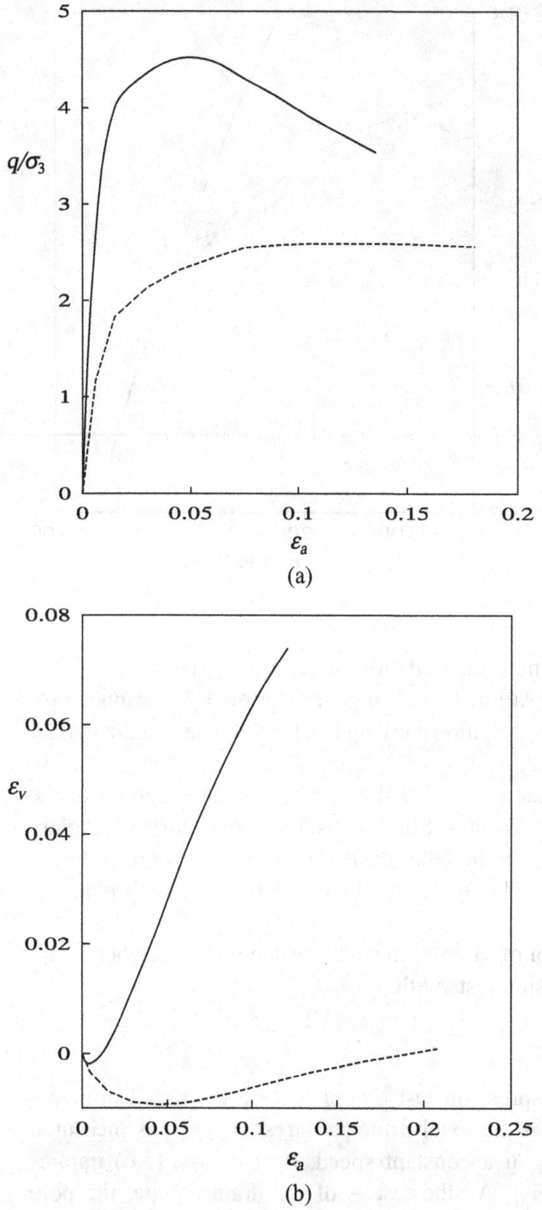

Figure 5.10. Drained triaxial compression test: —, dense sample (initial value of the solids fraction $\nu_i = 0.623$); - - -, loose sample ($\nu_i = 0.545$). (a) Variation of $q/\sigma_3 = (\sigma_1 - \sigma_3)/\sigma_3$ with the axial strain $\epsilon_a = \Delta L/L_i$, (b) variation of the volumetric strain $\epsilon_v = \Delta V/V_i$ with the axial strain. Here L_i and V_i are the initial length and volume, respectively of the cylindrical sample. The radial normal stress is $\sigma_r = \sigma_3 = 207$ kPa and the rate of change of ϵ_a is 0.03% per second. (Adapted from Taylor, 1948, p. 335.)

For loose samples, the solid curves in Fig. 5.12 represent loading paths for undrained tests. Hence the solids fraction ν is a constant along each curve. The broken curves in Fig. 5.12 represent contours of constant water content, or equivalently, contours of constant solids fraction, obtained from drained tests ($u_w = 0$, ν variable). This method of presenting data is due to Rendulic (1936, cited in Atkinson and Bransby, 1982, p. 199). The similarity between the shapes of the full and broken curves is striking. If it is assumed that the loading paths of normally consolidated samples (samples with initial states on the line AB in Fig. 5.9) represent states of plastic deformation, then the curves in Fig. 5.12 represent yield loci corresponding to the Roscoe surface. As the yield loci are similar, it should be possible to collapse them onto a single curve, as shown in Fig. 5.13 for kaolin clay. The data can be regarded as a section of the normalized Roscoe surface.

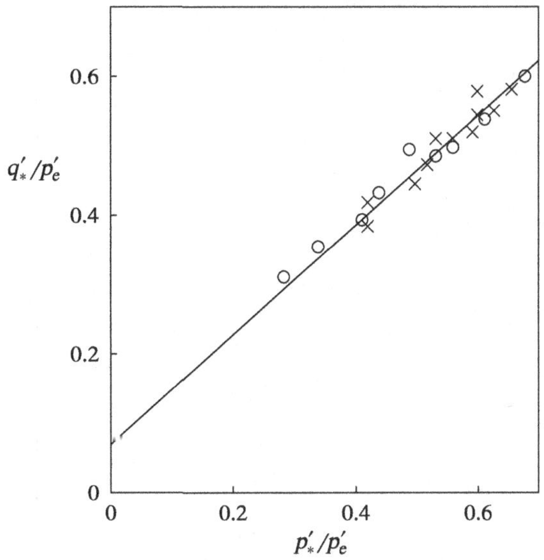

Figure 5.11. States of peak shear stress for drained (o) and undrained (x) tests on dense samples of Weald clay. The straight line fitted to the data can be regarded as a section of the Hvorslev surface by planes of constant solids fraction. Data of Parry (1960). (Reproduced from Fig. 11.4 of Atkinson and Bransby, 1982, p. 213, with permission from Prof. J. H. Atkinson.)

5.1.6. Cubical Triaxial Tests

The triaxial tests described above generate data only in the plane $\sigma_2 = \sigma_3$. Data in other regions of principal stress space can be generated using *cubical* or *true* triaxial tests. In these tests, a cubical sample of the material is subjected to independently variable normal stresses along three orthogonal directions (see, e.g., Ko and Scott, 1967; Lade and Duncan, 1973; Desai and Siriwardane, 1984, pp. 61–62; Kamath and Puri, 1997). Thus stress states which do not lie on the $\sigma_2 = \sigma_3$ plane can also be sampled. Data obtained using such tests are presented below.

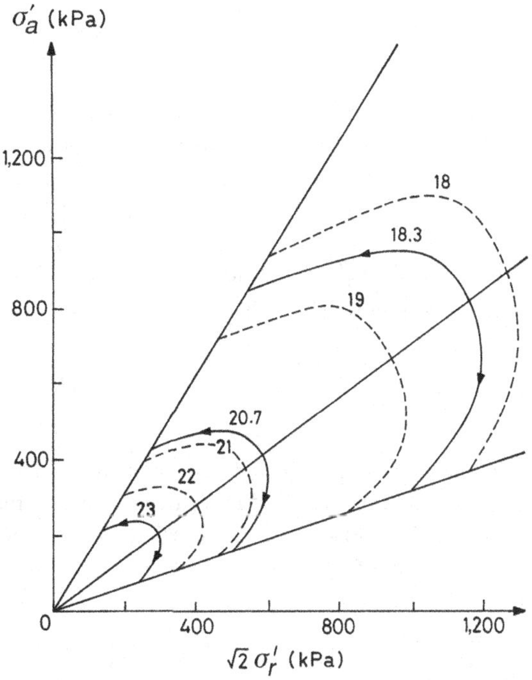

Figure 5.12. Compression and extension tests on normally consolidated samples of Weald clay: —, loading paths in undrained tests, - - -, contours of constant water content w (expressed as a percentage), from drained tests. The initial states of such samples lie on the normal consolidation or isotropic compression line AB in Fig. 5.9. Here σ'_a and σ'_r are the axial and radial effective principal stresses, respectively. The water content is defined by $w \equiv [\rho_w(1-\nu)]/(\rho_p\,\nu)$, where ρ_w and ρ_p are the densities of water and the grains, respectively, and ν is the solids fraction. (Reproduced from Fig. 9 of Henkel, 1960, with permission from the American Society of Civil Engineers.)

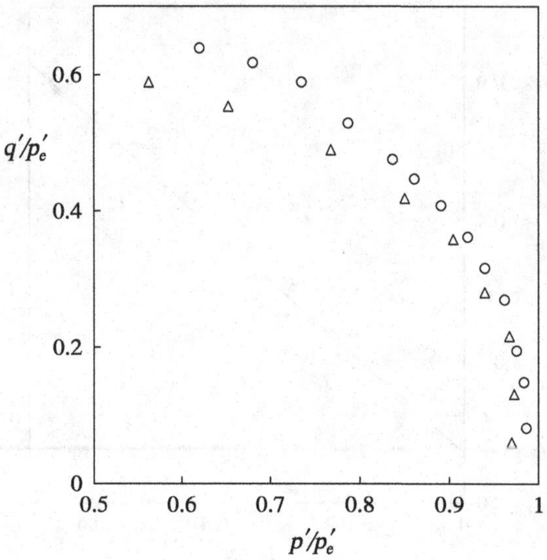

Figure 5.13. Drained (○) and undrained (Δ) compression tests on normally consolidated samples of kaolin clay. The quantity $p_e'(\nu)$ is the effective mean stress, obtained from the normal consolidation curve (a curve such as AB in Fig. 5.9) for any value of ν. Data of Balasubramaniam (1969). (Reproduced from Fig. 10.24 of Atkinson and Bransby, 1982, p. 204, with permission from Prof. J. H. Atkinson.)

5.1.7. Comparison of Yield Conditions with Data

Figure 5.14 shows that data for two sands can be fitted fairly well by the Lade–Duncan yield condition (5.20). The Mohr–Coulomb yield condition also fits the data, except near the corners. Clearly, the extended von Mises and the extended Tresca yield conditions are not satisfactory for these data.

Thornton (2000) simulated the deformation of a collection of spheres using the discrete element method. For radial loading paths in the Π-plane, he found that the locus of peak values of the deviator stress σ_1–σ_3 could be fitted well by the Lade–Duncan yield condition.

5.1.8. Flow Rules

For three-dimensional flow, the flow rule (2.67) for plane flow is replaced by

$$c_i = \dot{\lambda}\left(\frac{\partial G}{\partial \sigma_i}\right), \quad i = 1, 3 \tag{5.25}$$

where c_i, $i = 1, 3$ are the principal compressive rates of deformation, i.e., the eigenvalues of the rate of deformation tensor **C**. The latter is defined in the compressive sense by

$$\mathbf{C} \equiv -\frac{1}{2}(\nabla\mathbf{v} + (\nabla\mathbf{v})^T) \tag{5.26}$$

Here **v** is the velocity vector, and $G(\sigma_1, \sigma_2, \sigma_3, \nu)$ is the plastic potential. As before, G is a symmetric function of the principal stresses. In particular, the *associated* flow rule is given by

$$c_i = \dot{\lambda}\,\frac{\partial F}{\partial \sigma_i}, \quad i = 1, 3 \tag{5.27}$$

where $F(\sigma_1, \sigma_2, \sigma_3, \nu) = 0$ represents the yield condition. In some applications, it is convenient to express the flow rules in terms of the components C_{ij} and σ_{ij} of the rate of deformation tensor and the stress tensor, respectively, relative to any chosen coordinate system. Thus (5.25) is written as

$$C_{ij} = \dot{\lambda}\left(\frac{\partial \tilde{G}}{\partial \sigma_{ij}}\right), \quad i, j = 1, 3 \tag{5.28}$$

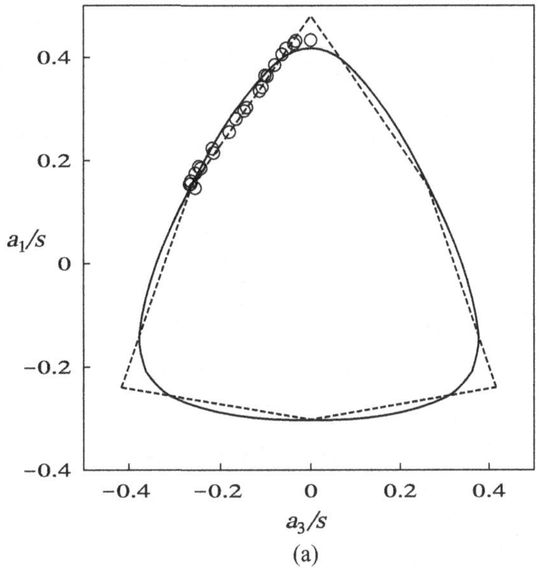

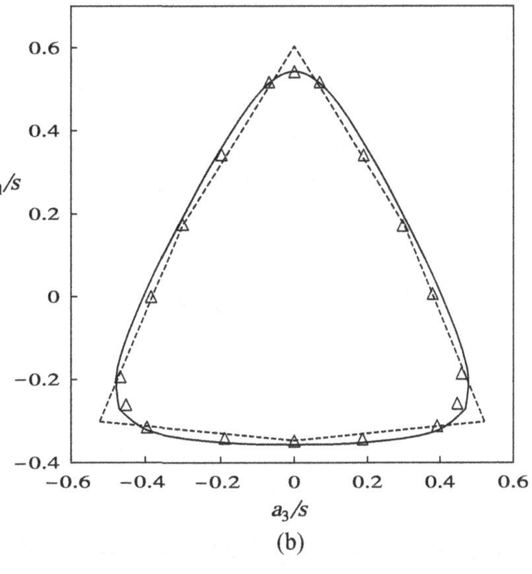

Figure 5.14. Cross sections of yield surfaces, that is intersections of the yield surfaces with the octahedral plane $\sigma_1 + \sigma_2 + \sigma_3 =$ constant $= s > 0$: (a) ∘, data of Green and Bishop (1969) for Ham river sand; —, the Lade–Duncan yield condition (5.20) with $k_1 = 56$; - - -, the Mohr–Coulomb yield condition (5.19) with $\phi_* = 43°$, $c = 0$, (b) △, data of Lade and Duncan (1973) for Monterey sand; —, the Lade–Duncan yield condition with $k_1 = 103$; - - -, the Mohr–Coulomb yield condition with $\phi_* = 54°$, $c = 0$. The data were obtained using cubical triaxial tests.

where $\tilde{G}(\boldsymbol{\sigma}, \nu)$ is the plastic potential. Similarly, the associated flow rule (5.27) is written as

$$C_{ij} = \dot{\lambda} \left(\frac{\partial \tilde{F}}{\partial \sigma_{ij}} \right), \quad i, j = 1, 3 \tag{5.29}$$

where $\tilde{F}(\boldsymbol{\sigma}, \nu) = 0$ represents the yield condition. As discussed in §2.13.2, (5.29) implies that the principal axes of $\mathbf{C}$ and $\boldsymbol{\sigma}$ are aligned with each other. For the special case of the associated flow rule based on the extended von Mises yield condition (5.12), the flow rule (5.29) can be rewritten in a form which resembles the constitutive relation for a Newtonian fluid (Problem 5.4). However, the "shear viscosity" is not a constant, but depends on an invariant of the rate of deformation tensor.

As discussed in §2.13.1, if the yield condition (5.1) is satisfied for $\boldsymbol{\sigma} = \boldsymbol{\sigma}_*$ and $\nu = \nu_*$, i.e., $\tilde{F}(\boldsymbol{\sigma}_*, \nu_*) = 0$, *and* $\boldsymbol{\sigma}$ and ν change such that $\mathrm{d}\tilde{F} = 0$, we assume that $\dot{\lambda} \neq 0$ and plastic deformation occurs. In this case the flow rule (5.28) must be supplemented by the yield

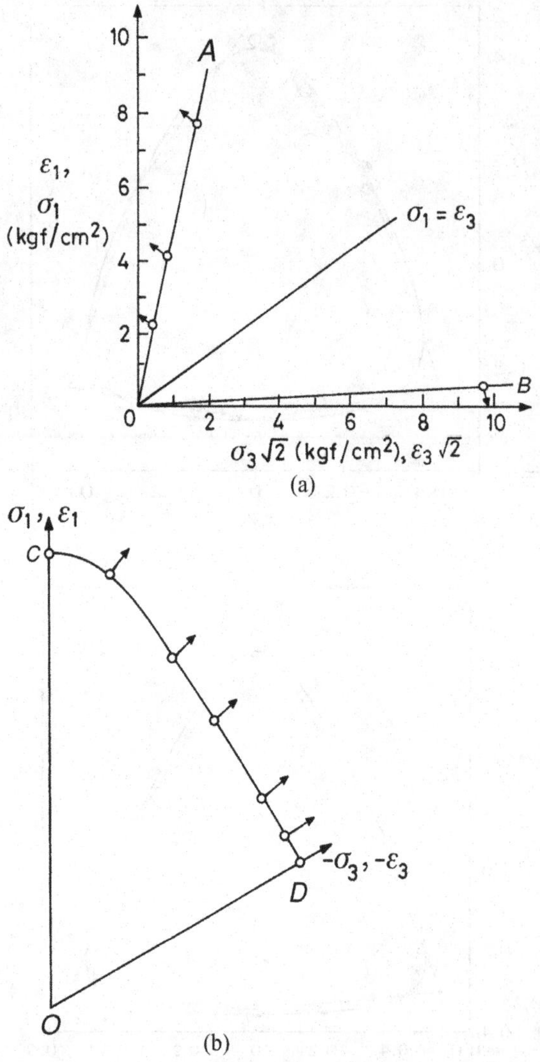

Figure 5.15. (a) Intersection of the yield surface for Monterey sand (solids fraction $= 0.64$) with (a) the triaxial plane $\sigma_2 = \sigma_3$, and (b) an octahedral plane $\sigma_1 + \sigma_2 + \sigma_3 =$ constant. The short arrows represent the directions of the plastic strain increment vectors. The quantities ϵ_1 and ϵ_3 represent the major and minor principal strain increments, respectively. In (a), *OA* represents the yield locus for triaxial compression, and *OB* the yield locus for triaxial extension. (Reproduced from Fig. 8 and 9 of Lade and Duncan, 1973, with permission from the American Society of Civil Engineers.)

condition. Conversely, if either $\tilde{F} < 0$ or $\mathrm{d}\tilde{F} < 0$, we set $\dot{\lambda} = 0$. The flow rule (5.28) then implies that the material remains rigid, and hence the stress field is indeterminate.

5.1.9. Data Related to Flow Rules

Figure 5.15a shows the intersection of the yield surfaces with the triaxial plane $\sigma_2 = \sigma_3$ for Monterey sand. Assuming that the principal axes of the stress and strain increment tensors are aligned with each other, the horizontal axis in Fig. 5.15 represents $\epsilon_3 \sqrt{3}$, where ϵ_3 is the minor principal strain increment, and the vertical axis represents the major principal strain increment ϵ_1. The directions of the observed plastic strain increment vectors are not normal to the yield surfaces, and hence the associated flow rule does not hold. Instead, Lade and Duncan (1975) have used a plastic potential of the form

$$I_1^3 - k_2 I_3 = 0 \tag{5.30}$$

where I_1 and I_3 are invariants of the stress tensor (see (5.2)), and k_2 is a function of the parameter k_1 occuring in the yield condition (5.20). Thus the plastic potential and the yield condition are similar in form, but are not identical. When the plastic strain increment vectors are projected onto an octahedral plane, they are seen to be approximately normal to the cross section of the yield surface, except at a few points (Fig. 5.15b). This is an example of *deviatoric normality*, i.e., normality holds approximately in the planes which are perpendicular to the space diagonal, and nonassociative behavior is caused only by changes in the mean stress. Desai and Siriwardane (1984, p. 381) state that many geological materials exhibit deviatoric normality.

Compaction has often been modeled using the associated flow rule. For an "artificial soil," which consisted of a mixture of clay, sand, and a small amount of mineral oil, Desai and Siriwardane (1984, p. 319) found that the plastic strain increment vectors were approximately normal to the compaction branches of the yield loci. Hence associative behavior can be used to describe the compaction of this material. Similarly, Gerogiannopoulos and Brown (1978) found that the dilatancy of certain rocks could be described by the associated flow rule.

The associated flow rule, and also some of the nonassociated flow rules, implicitly assume that coaxiality holds. Data obtained from triaxial and other tests on sands and clays can in some cases be fitted well by such models (DiMaggio and Sandler, 1971; Lade and Duncan, 1975; Desai and Salami, 1987; Anand and Gu, 2000). However, in many cases, the design of the testing equipment is such that coaxiality is automatically enforced. To check the assumption of coaxiality, data obtained using devices which are free from this constraint are required.

In the following two sections, some simple applications of the rigid-plastic equations are discussed.

5.1.10. Steady, Fully Developed Flow of a Rigid-Plastic Material

The shearing of a material satisfying the Mohr–Coulomb yield condition and the Lévy-Mises flow rule was discussed in §2.13.1. Here we consider the more general case of a material satisfying the yield condition (5.1) and the flow rule (5.28). For steady, fully developed flow between horizontal plates (Fig. 2.32), the velocity field is given by

$$v_x = v_x(y); \quad v_y = 0; \quad v_z = 0 \tag{5.31}$$

All components of the rate of deformation tensor $\mathbf{C}$ vanish, except

$$C_{xy} = C_{yx} = -\frac{1}{2}\frac{\partial v_x}{\partial y}$$

Hence (5.28) implies that

$$\frac{\partial \tilde{G}}{\partial \sigma_{xx}} = 0 = \frac{\partial \tilde{G}}{\partial \sigma_{yy}} = \frac{\partial \tilde{G}}{\partial \sigma_{zz}} = \frac{\partial \tilde{G}}{\partial \sigma_{xz}} = \frac{\partial \tilde{G}}{\partial \sigma_{yz}} \tag{5.32}$$

where $\tilde{G}(\boldsymbol{\sigma}, \nu)$ is the plastic potential, and it has been assumed that $\dot{\lambda} \neq 0$. Equations (5.32) and the yield condition (5.1) can in principle be solved for all the stresses in terms of the solids fraction ν to obtain

$$\sigma_{ij} = \sigma_{ij}(\nu) \tag{5.33}$$

The momentum balances for steady, fully developed flow are given by

$$\sigma_{yx} = \text{constant}; \quad \frac{d\sigma_{yy}}{dy} = -\rho_p\, \nu\, g; \quad \sigma_{yz} = \text{constant} \tag{5.34}$$

where ρ_p is the particle density and g is the acceleration due to gravity. Equations (5.33) and the first of (5.34) imply that ν is a constant, but this is inconsistent with the second of (5.34). Thus as in the case of the Lévy-Mises material, we must abandon the assumption that $\dot{\lambda} \neq 0$. Hence $\dot{\lambda} = 0$, which implies that the material is rigid and cannot be sheared.

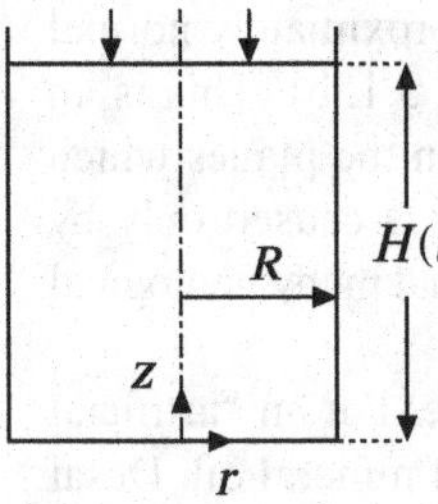

Figure 5.16. One-dimensional deformation of a material in a rigid cylindrical vessel of radius R. At time t, the upper surface of the material is at a height $H(t)$ above the base of the vessel.

The inability of rigid-plastic, hypoelastic, and hypoplastic models (see §5.2.3) to predict the velocity profile in steady, fully developed flow has been attributed (Mühlhaus, 1986, cited in Mühlhaus and Vardoulakis, 1987; Tejchman and Gudehus, 2001) to the absence of a material length scale in the constitutive equations. A length scale such as the particle diameter can be incorporated into the rigid-plastic and hypoplastic models by introducing additional field variables, as discussed in Mühlhaus and Vardoulakis (1987), Tejchman and Wu (1993), Tejchman and Bauer (1996), and Tejchman and Gudehus (2001). Applications of the modified equations to flow between parallel plates, filling and discharge from silos, and fully developed flow through a vertical channel are discussed in Tejchman and Wu (1993), (Tejchman 1998a, 1998b), and Mohan et al. (1999, 2002).

The Goodman–Cowin class of equations (Goodman and Cowin, 1971; Rajagopal and Massoudi, 1990; Wang and Hutter, 1999; Kumar et al., 2003) is linear in the rate of deformation tensor and involves the spatial gradient of the solids fraction. It also leads to nontrivial velocity profiles for steady, simple shear flow. Unlike the equations mentioned in the previous paragraph, these equations are rate dependent.

5.1.11. One-Dimensional Deformation of a Rigid-Plastic Material

Even though the rigid-plastic model cannot be used for steady, fully developed flow, there are other problems which can be examined using this model. Some examples are provided by flow through a hopper (discussed in Chapters 3 and 6), and one-dimensional deformation. The latter will be discussed here. Consider the one-dimensional motion of a rigid-plastic material in the z direction (Fig. 5.16). The material is confined in a rigid cylindrical vessel, and friction between the material and the wall is neglected.

Using a cylindrical coordinate system with the origin at the base of the vessel, consider a velocity field of the form

$$v_r = 0 = v_\theta; \quad v_z = v_z(z, t) \tag{5.35}$$

where t is the time. Assuming that a uniform normal stress is applied at the horizontal upper surface $z = H(t)$ (Fig. 5.16), the stress field is assumed to be of the form

$$\sigma_{rr} = \sigma_{\theta\theta} = \sigma_{\theta\theta}(z, t); \quad \sigma_{zz} = \sigma_{zz}(z, t) \tag{5.36}$$

All components of the rate of deformation tensor **C** vanish, except

$$C_{zz} = -\frac{\partial v_z}{\partial z}$$

Hence (5.28) implies that

$$\frac{\partial \tilde{G}}{\partial \sigma_{rr}} = \frac{\partial \tilde{G}}{\partial \sigma_{\theta\theta}} = \frac{\partial \tilde{G}}{\partial \sigma_{r\theta}} = \frac{\partial \tilde{G}}{\partial \sigma_{rz}} = \frac{\partial \tilde{G}}{\partial \sigma_{\theta z}} = 0 \tag{5.37}$$

where $\tilde{G}(\boldsymbol{\sigma}, \nu)$ is the plastic potential, and it has been assumed that $\dot{\lambda} \neq 0$. Equations (5.37) and the yield condition (5.1) can in principle be solved for all the stresses in terms of the

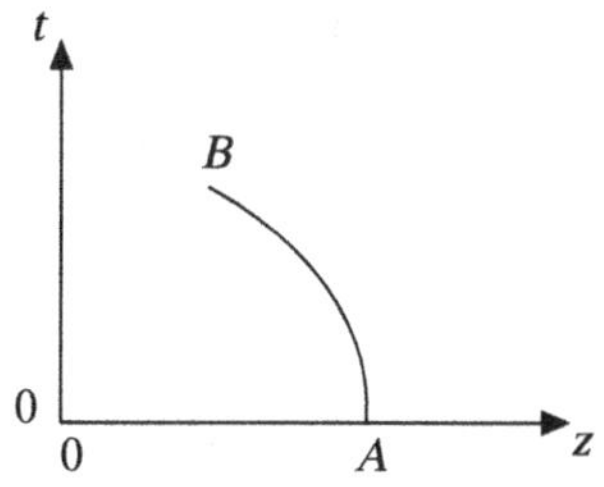

Figure 5.17. A particle path AB for a material undergoing one-dimensional compression.

solids fraction ν to obtain (Problem 5.10)

$$\sigma_{ij} = \sigma_{ij}(\nu) \tag{5.38}$$

The velocity field (5.35) implies that the principal axes of **C** are aligned with the coordinate directions. As the principal axes of **C** and $\boldsymbol{\sigma}$ are aligned with each other, we have

$$\sigma_{rz} = 0; \quad \sigma_{\theta z} = 0; \quad \sigma_{r\theta} = 0 \tag{5.39}$$

At this stage, all the stresses are known as functions of ν, whereas the results of uniaxial tests on granular materials are often presented in terms of the axial *engineering* strain ξ_e, or equivalently, the volumetric engineering strain, as shown in Fig. 2.31. The axial engineering strain is defined as the change in length relative to the original length of the sample, or, in a differential form, as

$$\mathrm{d}\xi_e \equiv -\mathrm{d}L/L_0 \tag{5.40}$$

where L is the current length of the sample, and L_0 is its original length.

To relate ξ_e to ν, it is convenient to introduce a new variable ξ defined by (Davis and Mullenger, 1978)

$$\frac{\mathrm{D}\xi}{\mathrm{D}t} = \frac{\partial \xi}{\partial t} + v_z \frac{\partial \xi}{\partial z} \equiv C_{zz} = -\frac{\partial v_z}{\partial z} \tag{5.41}$$

where $\mathrm{D}/\mathrm{D}t$ is the material derivative. This variable can be interpreted as follows. As discussed in Problem 2.12, $D_{zz} = -C_{zz}$ represents the instantaneous rate of stretching per unit length of a line element which is parallel to the z axis. If the length of the element at time t is L, (2.125) and (5.41) imply that

$$\frac{\mathrm{D}\xi}{\mathrm{D}t} = C_{zz} = -\frac{1}{L}\frac{\mathrm{D}L}{\mathrm{D}t} \tag{5.42}$$

The material derivative $\mathrm{D}\xi/\mathrm{D}t$ represents the rate of change of ξ along a *particle path*, which is defined by

$$\frac{\mathrm{d}z}{\mathrm{d}t} = v_z \tag{5.43}$$

Along any particle path such as the curve AB in Fig. 5.17, (5.42) can be rewritten as

$$\mathrm{d}\xi = -\mathrm{d}L/L \tag{5.44}$$

Thus ξ represents the *natural* strain in the axial direction (Chakrabarthy, 1987, p. 3). Note that ξ increases when the element is compressed.

Using the initial conditions $\xi = 0$ and $\xi_e = 0$ at $L = L_0$, (5.40) and (5.44) imply that $\xi_e = 1 - \exp(-\xi)$. Thus $\xi_e \approx \xi$ at small strains. For example, $\xi_e = 0.18$ when $\xi = 0.2$.

The mass balance can be written as

$$\frac{\mathrm{D}\nu}{\mathrm{D}t} = \nu\,\mathrm{tr}(\mathbf{C}) = \nu\, C_{zz} = -\nu \frac{\partial v_z}{\partial z} \tag{5.45}$$

Using (5.41), (5.45) reduces to

$$\frac{\mathrm{D}\nu}{\mathrm{D}t} = \nu\frac{\mathrm{D}\xi}{\mathrm{D}t} \tag{5.46}$$

or

$$\frac{\mathrm{d}\nu}{\mathrm{d}\xi} = \nu \tag{5.47}$$

along a particle path. Hence (5.47) can be integrated to obtain $\nu = \nu_A\,\mathrm{e}^{\xi}$, where ν_A is the value of ν at the point A (Fig. 5.17), and ξ_A has been chosen as zero for convenience. Hence (5.38) can be rewritten as

$$\sigma_{ij} = \sigma_{ij}(\xi) \approx \sigma_{ij}(\xi_e) \tag{5.48}$$

The above treatment has glossed over a subtle point. The particle paths cannot be determined unless the velocity field $v_z(z, t)$ is known, and the latter must be obtained by solving the z component of the momentum balance

$$\rho_\mathrm{p}\,\nu\left(\frac{\partial v_z}{\partial t} + v_z\frac{\partial v_z}{\partial z}\right) + \frac{\partial \sigma_{zz}}{\partial z} + \rho_\mathrm{p}\,\nu\,g = 0 \tag{5.49}$$

along with (5.41). Let us consider the case where the material is initially at rest, i.e.,

$$v_z(z, t = 0) = 0, \quad 0 \le z \le H_0 \tag{5.50}$$

where H_0 is the intial height of the sample. For $t > 0$, the upper surface $z = H(t)$ is compressed with a known speed $V(t)$, so that

$$v_z(z = H(t), t) = \frac{\mathrm{d}H}{\mathrm{d}t} = -V(t); \quad H(0) = H_0 \tag{5.51}$$

In view of the intial condition (5.50), it is assumed that $V(0) = 0$.

Equations (5.41), (5.46), (5.49), and (5.38) represent a system of coupled partial differential equations which must be solved numerically for ξ, ν, and v_z. An approximate solution can be found by assuming that the speed of compression is slow, so that the inertial terms can be neglected in the momentum balance. If the body force is also neglected, (5.49) implies that the axial normal stress σ_{zz} is independent of position. Equation (5.48) then implies that $\xi = \xi(t)$, as assumed by Davis and Mullenger (1978). It follows from (5.41) that

$$v_z = -\frac{\mathrm{d}\xi}{\mathrm{d}t}z + f(t) \tag{5.52}$$

However, as the axial velocity vanishes at the bottom of the vessel ($z = 0$), we have $f(t) = 0$. Thus

$$v_z = -\frac{\mathrm{d}\xi}{\mathrm{d}t}z \tag{5.53}$$

Equations (5.53), (5.43), and (5.51) imply that

$$\frac{d\xi}{dt} = -\frac{v_z}{z} = -\frac{1}{z}\frac{\mathrm{d}z}{\mathrm{d}t} = -\frac{v_z(z = H, t)}{H(t)} = \frac{V(t)}{H_0 - \int_0^t V(t')\,\mathrm{d}t'} \tag{5.54}$$

Equation (5.54) can be integrated using the initial condition $\xi(0) = 0$, $z(0) = z_0$ to obtain

$$\xi = \ln\left(\frac{H_0}{H_0 - \int_0^t V(t')\,\mathrm{d}t'}\right); \quad \frac{z}{z_0} = 1 - \frac{1}{H_0}\int_0^t V(t')\mathrm{d}t' \tag{5.55}$$

This provides an expression for the particle path $z = z(t, z_0)$.

5.2. CONSTITUTIVE EQUATIONS THAT DO NOT INVOLVE A YIELD CONDITION

5.2.1. Hypoelastic and Hypoplastic Models

Consider a constitutive equation of the form

$$\frac{\mathrm{D}\boldsymbol{\sigma}}{\mathrm{D}t} = \mathbf{F}_1(\boldsymbol{\sigma}, \nabla\mathbf{v}, \nu) \tag{5.56}$$

where $\mathrm{D}/\mathrm{D}t$ is the material derivative defined by

$$\frac{\mathrm{D}}{\mathrm{D}t} = \frac{\partial}{\partial t} + \mathbf{v}\cdot\nabla \tag{5.57}$$

$\mathbf{v}$ is the velocity vector, and ν is the solids fraction. By requiring (5.56) to satisfy the principle of material frame indifference (see Appendix G), Noll (1955) has deduced an admissible constitutive equation as follows.

As discussed in Appendix G, this principle requires constitutive equations to be unaffected by a change of frame. It is convenient to view a change of frame as a time-dependent transformation of position vectors, defined by (G.25) and a translation of the time axis, defined by (G.27). Using the superscript * to denote a change of frame, (5.56) must take the form

$$\frac{\mathrm{D}\boldsymbol{\sigma}^*}{\mathrm{D}t^*} = \mathbf{F}_1(\boldsymbol{\sigma}^*, \nabla^*\mathbf{v}^*, \nu^*) \tag{5.58}$$

at every point in the medium if it is to be frame indifferent. Using (G.29), (5.58) can be rewritten as

$$\mathbf{Q}\frac{\mathrm{D}\boldsymbol{\sigma}}{\mathrm{D}t}\mathbf{Q}^{\mathrm{T}} + \dot{\mathbf{Q}}\,\boldsymbol{\sigma}\,\mathbf{Q}^{\mathrm{T}} + \mathbf{Q}\,\boldsymbol{\sigma}\,\dot{\mathbf{Q}}^{\mathrm{T}}$$
$$= F_1(\mathbf{Q}\,\boldsymbol{\sigma}\,\mathbf{Q}^{\mathrm{T}}, \mathbf{Q}\,(\nabla\mathbf{v})\,\mathbf{Q}^{\mathrm{T}} + \mathbf{Q}\,\dot{\mathbf{Q}}^{\mathrm{T}}, \nu) \tag{5.59}$$

Here $\mathbf{Q}(t)$ is an arbitrary time-dependent orthogonal tensor (see §A.7.7), which satisfies

$$\mathbf{Q}\,\mathbf{Q}^{\mathrm{T}} = \mathbf{Q}^{\mathrm{T}}\,\mathbf{Q} = \mathbf{I} \tag{5.60}$$

where $\mathbf{I}$ is the unit tensor, and

$$\dot{\mathbf{Q}} \equiv \frac{\mathrm{D}\mathbf{Q}}{\mathrm{D}t} = \frac{\mathrm{d}\mathbf{Q}}{\mathrm{d}t} \tag{5.61}$$

Evaluating (5.59) at some time t and at some point B, let us choose $\mathbf{Q}(t) = \mathbf{I}$. Then (5.60) implies that $\dot{\mathbf{Q}}(t) = -\dot{\mathbf{Q}}^{\mathrm{T}}(t)$, and hence $\dot{\mathbf{Q}}(t)$ is an antisymmetric tensor. Let $\dot{\mathbf{Q}}(t)$ be chosen as $\mathbf{W}_B$, where

$$\mathbf{W}_B \equiv \frac{1}{2}((\nabla\mathbf{v})_B - (\nabla\mathbf{v})_B{}^{\mathrm{T}}) \tag{5.62}$$

is the vorticity tensor evaluated at the point B and at time t.

Substituting the above choices for $\mathbf{Q}$ and $\dot{\mathbf{Q}}$ into (5.59), and noting that $\mathbf{W}_B^{\mathrm{T}} = -\mathbf{W}_B$, we obtain

$$\left(\frac{\mathrm{D}\boldsymbol{\sigma}}{\mathrm{D}t}\right)_B + \mathbf{W}_B\,\boldsymbol{\sigma}_B - \boldsymbol{\sigma}_B\,\mathbf{W}_B = \mathbf{F}_1(\boldsymbol{\sigma}_B, (\nabla\mathbf{v})_B - \mathbf{W}_B, \nu_B)$$
$$= \mathbf{F}(\boldsymbol{\sigma}_B, \mathbf{C}_B, \nu_B) \tag{5.63}$$

where $\mathbf{C} \equiv -(\nabla\mathbf{v} - \mathbf{W})$ is the rate of deformation tensor, defined in the compressive sense.

Introducing the *Jaumann* or *corotational* derivative of $\boldsymbol{\sigma}$, defined by Jaumann (1911, cited in Prager, 1961, p. 155)

$$\overset{\circ}{\boldsymbol{\sigma}} \equiv \frac{\mathrm{D}\boldsymbol{\sigma}}{\mathrm{D}t} + \mathbf{W}\boldsymbol{\sigma} - \boldsymbol{\sigma}\mathbf{W} \tag{5.64}$$

(5.63) takes the form

$$(\overset{\circ}{\boldsymbol{\sigma}})_B = \mathbf{F}(\boldsymbol{\sigma}_B, \mathbf{C}_B, \nu_B) \tag{5.65}$$

As the point B has been chosen arbitrarily, (5.65) must hold at every point. Hence the subscript B can be omitted, leading to the form

$$\overset{\circ}{\boldsymbol{\sigma}} = \mathbf{F}(\boldsymbol{\sigma}, \mathbf{C}, \nu) \tag{5.66}$$

A few remarks about (5.66) are in order. The choice

$$\mathbf{Q}(t) = \mathbf{I}; \quad \dot{\mathbf{Q}}(t) = \mathbf{W}_B \tag{5.67}$$

can be interpreted as follows. Equations (5.67) and (G.29) imply that $\mathbf{W}_B^* = 0$. As the nonzero elements of $\mathbf{W}$ represent the angular velocity $\boldsymbol{\omega}$ of the material (Problem 5.6), the change of frame (5.67) nullifies the rotation of the material by rotating position vectors in the opposite direction with an angular velocity $-\boldsymbol{\omega}$. This is perhaps easier to understand if we view a change of frame as a translation and rotation of a reference frame, say the starred frame, relative to another reference frame, say the unstarred frame (see §G.1). In this case, the condition $\mathbf{W}_B^* = 0$ implies that the local angular velocity of the material relative to the starred frame vanishes. In other words, if $\boldsymbol{\omega}_B$ is the local angular velocity relative to the unstarred frame of an infinitesimal element centered at B, the starred frame rotates relative to the unstarred frame with an angular velocity $\boldsymbol{\omega}_B$. The material derivative of $\boldsymbol{\sigma}$ relative to the starred frame takes the form of the Jaumann derivative of $\boldsymbol{\sigma}$ when it is expressed in terms of the unstarred frame. Equivalently, the time derivative relative to a frame which translates with the velocity of the material, and rotates with the local angular velocity of the material takes the form of the Jaumann derivative when it is expressed in terms of the unstarred frame (Bird et al., 1977, p. 321). Such a frame is said to be a *corotational* frame.

Instead of (5.67), we could have chosen $\mathbf{Q}(t) = \mathbf{I}$; $\dot{\mathbf{Q}}(t) = -\mathbf{W}_B$, but this does not lead to any simplification of (5.56).

It can be shown that the Jaumann derivative of $\boldsymbol{\sigma}$ is a frame-indifferent tensor, or equivalently, a frame-indifferent *stress rate* (Problem G.1). Many other frame-indifferent stress rates have been proposed in the literature (see, e.g., Truesdell, 1955; Prager, 1961, p. 156; Reinhardt and Dubey, 1996; Xiao et al., 1997), and can be used in place of the Jaumann derivative in (5.66). At present, it is not clear which of these rates is most suitable for granular materials.

Equation (5.66) has been used by several workers to describe the deformation of granular materials (Desai, 1972, cited in Davis and Mullenger, 1978; Stutz, 1973, cited in Davis and Mullenger (1978); Romano, 1974; Davis and Mullenger, 1978; Kolymbas, 1987, cited in Kolymbas, 1991; Kolymbas, 1991; Chambon et al., 1994; Gudehus, 1996; Bauer, 1996; Herle and Gudehus, 1999). A drawback of these models is the lack of an appealing physical basis.

Variants of (5.66), obtained by adding a term proportional to the Jaumann derivative of the rate of deformation tensor $\mathbf{C}$ on the right-hand side, have also been used to examine problems of sustained flow, such as flow through bins (Eibl and Rombach, 1988; Eibl, 1997) (see §6.3).

5.2.2. Some Features of (5.66)

The principle of material frame indifference will now be used to develop an explicit expression for (5.66). Under a change of frame, (5.66) takes the form

$$\overset{\circ}{\boldsymbol{\sigma}}^{*} = \mathbf{F}(\boldsymbol{\sigma}^{*}, \mathbf{C}^{*}, \nu^{*}) \tag{5.68}$$

Using (G.39), (G.33), and (5.66), (5.68) can be rewritten as

$$\mathbf{Q}\,\overset{\circ}{\boldsymbol{\sigma}}\,\mathbf{Q}^{\mathrm{T}} = \mathbf{F}(\mathbf{Q}\boldsymbol{\sigma}\mathbf{Q}^{\mathrm{T}}, \mathbf{Q}\,\mathbf{C}\,\mathbf{Q}^{\mathrm{T}}, \nu)$$

or

$$\mathbf{F}(\boldsymbol{\sigma}, \mathbf{C}, \nu) = \mathbf{Q}^{\mathrm{T}}\,\mathbf{F}(\mathbf{Q}\boldsymbol{\sigma}\mathbf{Q}^{\mathrm{T}}, \mathbf{Q}\,\mathbf{C}\,\mathbf{Q}^{\mathrm{T}}, \nu)\,\mathbf{Q} \tag{5.69}$$

Thus **F** is an *isotropic* tensor valued function of its arguments.

Hence it can be shown (Rivlin and Ericksen, 1955; Pennisi and Trovato, 1987) that (5.66) can be expressed as

$$\begin{aligned}\overset{\circ}{\boldsymbol{\sigma}} &= \mathbf{F}(\boldsymbol{\sigma}, \mathbf{C}, \nu)\\ &= a_0\,\mathbf{I} + a_1\,\mathbf{C} + a_2\,\boldsymbol{\sigma} + a_3\,\mathbf{C}^2\\ &\quad + a_4\,\boldsymbol{\sigma}^2 + a_5\,(\mathbf{C}\,\boldsymbol{\sigma} + \boldsymbol{\sigma}\,\mathbf{C}) + a_6\,(\mathbf{C}^2\,\boldsymbol{\sigma} + \boldsymbol{\sigma}\,\mathbf{C}^2)\\ &\quad + a_7\,(\mathbf{C}\,\boldsymbol{\sigma}^2 + \boldsymbol{\sigma}^2\,\mathbf{C})\end{aligned} \tag{5.70}$$

where

$$\begin{aligned}a_i &= a_i(\mathrm{tr}(\mathbf{C}), \mathrm{tr}(\boldsymbol{\sigma}), \mathrm{tr}(\mathbf{C}^2), \mathrm{tr}(\boldsymbol{\sigma}^2), \mathrm{tr}(\mathbf{C}^3), \mathrm{tr}(\boldsymbol{\sigma}^3),\\ &\quad \mathrm{tr}(\boldsymbol{\sigma}\,\mathbf{C}), \mathrm{tr}(\boldsymbol{\sigma}^2\,\mathbf{C}), \mathrm{tr}(\boldsymbol{\sigma}\,\mathbf{C}^2), \mathrm{tr}(\boldsymbol{\sigma}^2\,\mathbf{C}^2)), \qquad i = 0, 7\end{aligned} \tag{5.71}$$

and the arguments of the $\{a_i\}$ are the *joint invariants* of **C** and $\boldsymbol{\sigma}$. Equation (5.70) represents a *hypoelastic* material if it is a *linear* function of **C**, and a *hypoplastic* material if it is a *nonlinear* function of **C** (Kolymbas, 1991). Hypoelastic materials are also called *rate-type* materials (Truesdell and Noll, 1965, p. 96).

An example of a hypoplastic material is given by (Romano, 1974; Davis and Mullenger, 1978)

$$\begin{aligned}\overset{\circ}{\boldsymbol{\sigma}} &= a_0\,\mathbf{I} + a_1\,\mathbf{C} + a_2\,\boldsymbol{\sigma}\\ &= (h_1\,\mathrm{tr}(\mathbf{C}) - h_2\,\mathrm{tr}(\boldsymbol{\sigma}\,\mathbf{C}))\,\mathbf{I} + h_3\,\mathbf{C}\\ &\quad -(h_4\,\mathrm{tr}(\mathbf{C}) - h_5\,\mathrm{tr}(\boldsymbol{\sigma}\,\mathbf{C}))\,\boldsymbol{\sigma}\end{aligned} \tag{5.72}$$

where $h_i, i = 1, 5$ are functions of the density and the invariants of $\boldsymbol{\sigma}$. Equation (5.72) can be obtained from (5.70) by setting $a_i = 0,\ i = 3\text{–}7$.

An example of a hypoplastic material is given by (Kolymbas, 1991)

$$\begin{aligned}\overset{\circ}{\boldsymbol{\sigma}} &= a_0\,\mathbf{I} + a_2\,\boldsymbol{\sigma} + a_4\,\boldsymbol{\sigma}^2 + a_5\,(\mathbf{C}\,\boldsymbol{\sigma} + \boldsymbol{\sigma}\,\mathbf{C})\\ &= b_1\,\mathrm{tr}(\boldsymbol{\sigma}\,\mathbf{C})\,\mathbf{I} + b_2\,||\mathbf{C}||\,\boldsymbol{\sigma}\\ &\quad + b_3\,(||\mathbf{C}||/\mathrm{tr}(\boldsymbol{\sigma}))\,\boldsymbol{\sigma}^2 + b_4\,(\mathbf{C}\,\boldsymbol{\sigma} + \boldsymbol{\sigma}\,\mathbf{C})\end{aligned} \tag{5.73}$$

where $b_i, i = 1, 4$ are dimensionless constants, and

$$||\mathbf{C}|| \equiv \sqrt{\mathrm{tr}(\mathbf{C}^2)} \tag{5.74}$$

is the *norm* of **C**. Equation (5.73) can be obtained from (5.70) by setting a_1, a_3, a_6, and $a_7 = 0$.

As shown in §5.2.4, hypoplastic models reqire different sets of parameters to describe loading and unloading. In contrast, as discussed in Bauer (1996), hypoplastic models can describe both these processes using a single set of parameters.

For steady flow, (5.72) and (5.73) are homogeneous functions of degree 1 in **C** and **W**. Hence if all the velocities are multiplied by a common factor, these equations are unaffected. In other words, they are *rate-independent* equations.

Two examples involving these constitutive equations are discussed below.

5.2.3. Steady, Fully Developed Flow of a Hypoelastic Material

Let us apply (5.72) to the steady, fully developed flow of a granular material between horizontal plates (Fig. 2.32). Assuming that

$$v_x = v_x(y); \quad v_y = 0; \quad v_z = 0 \tag{5.75}$$

(5.64) and (5.72) imply that

$$\overset{\circ}{\sigma}_{xx} = 2\, W_{xy}\, \sigma_{yx} = -2\, \sigma_{xy}\, C_{yx}\, (h_2 - h_5\, \sigma_{xx}) \tag{5.76}$$

As σ and **C** are symmetric, and (5.75) implies that $W_{xy} = C_{xy}$, (5.76) reduces to

$$2\, \sigma_{xy}\, C_{xy}\, (h_2 - h_5\, \sigma_{xx} + 1) = 0$$

or, assuming that $\sigma_{xy}\, C_{xy} \neq 0$

$$\sigma_{xx} = (h_2 + 1)/h_5 \tag{5.77}$$

Similarly,

$$\sigma_{yy} = (h_2 - 1)/h_5; \quad \sigma_{zz} = h_2/h_5 \tag{5.78}$$

$$\sigma_{xy}{}^2 = -(h_3\, h_5 + 2)/(2 h_5{}^2); \quad \sigma_{xz} = 0; \quad \sigma_{yz} = 0 \tag{5.79}$$

Equation (5.79) holds provided $C_{xy} = C_{yx} \neq 0$, and $h_3\, h_5 + 2 < 0$.

For the velocity field given by (5.75), the forms assumed by Davis and Mullenger (1978) for h_2, h_3, and h_5 imply that they are all functions of the mean stress $p \equiv (\sigma_{xx} + \sigma_{yy} + \sigma_{zz})/3$. Hence (5.77)–(5.79) permit all the stress components to be expressed as functions of p.

The first of the momentum balances (5.34) implies that p is a constant, and hence the second of (5.34) cannot be satisfied. Further, the velocity field is indeterminate. The inconsistency arises because of the assumption that $C_{xy} \neq 0$. On the other hand, if it is assumed that $C_{xy} = 0$, the material is *rigid* and hence cannot be sheared. Equation (5.72) is identically satisfied, but the stress field is indeterminate. A similar problem is encountered with the Lévy-Mises equations (see §2.13.1), and also with the more general plastic potential flow rule (5.28).

5.2.4. One-Dimensional Deformation of a Hypoelastic Material

Even though the hypoelastic and hypoplastic models cannot be used for steady, fully developed flow, there are other problems which can be examined using these models. For example, consider the one-dimensional compression of a hypoelastic material in the $-z$ direction (Fig. 5.16). The treatment below is adapted from Davis and Mullenger (1978).

The material is confined in a rigid cylindrical vessel, and friction between the material and the walls is neglected. (In §5.1.11, this problem has been discussed for the case of a rigid-plastic material.) The constitutive equation (5.72) will be used here with the velocity and stress fields assumed to be given by (5.35) and (5.36).

The velocity field (5.35) implies that all components of the vorticity tensor **W** vanish (see (5.62)), and hence the Jaumann derivative reduces to the material derivative. As the only nonzero component of $\mathbf{C}$ is C_{zz}, the quantities tr($\mathbf{C}$) and tr($\boldsymbol{\sigma}\mathbf{C}$), which occur in (5.72), are given by tr($\mathbf{C}$) $= C_{zz}$, and tr($\boldsymbol{\sigma}\mathbf{C}$) $= \sigma_{zz}\, C_{zz}$.

Equation (5.72) reduces to

$$\frac{\mathrm{D}\sigma_{rr}}{\mathrm{D}t} = (h_1 - h_2\,\sigma_{zz} - (h_4 - h_5\,\sigma_{zz})\,\sigma_{rr})\,C_{zz}$$

$$\frac{\mathrm{D}\sigma_{zz}}{\mathrm{D}t} = (h_1 - h_2\,\sigma_{zz} + h_3 - (h_4 - h_5\,\sigma_{zz})\,\sigma_{zz})\,C_{zz} \tag{5.80}$$

where it is assumed that (Davis and Mullenger, 1978)

$$h_1 = h_1(p, \nu); \quad h_2 = h_2(p); \quad h_3 = h_3(p); \quad h_4 = h_4(p, \nu) = h_5\, p \tag{5.81}$$

Here ν is the solids fraction, and the mean stress p is defined by $p \equiv$ tr($\boldsymbol{\sigma}$) $= (2\,\sigma_{rr} + \sigma_{zz})/3$.

Introducing the deviator stress q defined by

$$q \equiv \sigma_{zz} - \sigma_{rr} \tag{5.82}$$

(5.80) can be rewritten as

$$\frac{\mathrm{D}p}{\mathrm{D}t} = (h_1 - h_2\,\sigma_{zz} + (h_3/3) - (h_4 - h_5\,\sigma_{zz})\,p)\,C_{zz}$$

$$\frac{\mathrm{D}q}{\mathrm{D}t} = (h_3 - (h_4 - h_5\,\sigma_{zz})\,q)\,C_{zz} \tag{5.83}$$

As in §5.1.11, it is convenient to replace C_{zz} by the natural strain ξ, defined by (5.41). The first of equations (5.83) can then be rewritten as

$$\frac{\mathrm{D}p}{\mathrm{D}t} = (h_1 - h_2\,\sigma_{zz} + (h_3/3) - (h_4 - h_5\,\sigma_{zz})\,p)\,\frac{\mathrm{D}\xi}{\mathrm{D}t} \tag{5.84}$$

Along the particle path (5.43), (5.84) reduces to

$$\frac{\mathrm{d}p}{\mathrm{d}\xi} = h_1 - h_2\,\sigma_{zz} + (h_3/3) - (h_4 - h_5\,\sigma_{zz})\,p \tag{5.85}$$

where $\sigma_{zz} = (3\,p + 2\,q)/3$. Similarly, the second of (5.83) takes the form

$$\frac{\mathrm{d}q}{\mathrm{d}\xi} = h_3 - (h_4 - h_5\,\sigma_{zz})\,q \tag{5.86}$$

As some of the h_i depend on ν, (5.85) and (5.86) must be supplemented by the mass balance, which is given by (5.47).

Equations (5.85), (5.86), and (5.47) can be integrated to obtain p, q, and ν as functions of the axial natural strain ξ. To illustrate the results obtained, let us consider specific forms for the functions h_i. Davis and Mullenger (1978) assume that

$$h_1 + (h_3/3) = h_2\, p + (\kappa(p, \nu)\, p)/(\rho_\mathrm{p}\, \nu); \; h_4 = h_5\, p$$

$$h_2 = 2\, b\, \mu_*\, p/M^2; \; h_3 = 2\,\mu_* = 2\,(\mu_0 + \mu_1\, p)$$

$$h_4 = h_2\,(\nu_c/\nu)^{\gamma_*}; \quad M = M_0\,(1 - \exp(-\beta\, p))$$

$$\gamma_* = \gamma_0\, p \tag{5.87}$$

where ρ_p is the particle density, and b, μ_0, μ_1, M_0, γ_0, and β are constants. The function ν_c in (5.87) represents the solids fraction at a critical state. In the ln p–$1/\nu$ plane, the critical state line is assumed to be parallel to the normal consolidation line (5.90). Some data in support of this assumption are given in Roscoe et al. (1958). Hence

$$1/\nu_c = (1/\nu_{c0}) - (\rho_\mathrm{p}/\kappa_l)\, \ln(p/p_0) \tag{5.88}$$

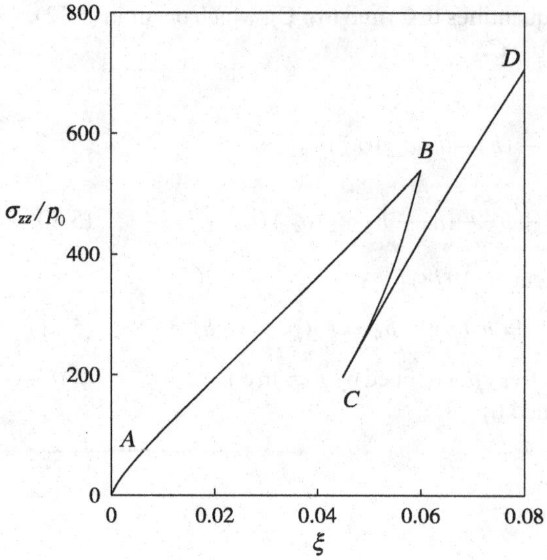

Figure 5.18. Variation of the dimensionless axial normal stress σ_{zz}/p_0 with the axial natural strain ξ, defined by (5.41), for the one-dimensional deformation of a hypoelastic material. Parameter values: $p_0 = 6895$ Pa, $\rho_p\, \nu_{c0} = 1448\,\mathrm{kg\,m^{-3}}$, $\tilde{\kappa}_l = 22.9$, $\tilde{\kappa}_u = 40.0$, $b_u = 2$, $\tilde{\mu}_0 = 4900$, $\mu_1 = 12.5$, $\tilde{M}_0 = 3400$, $\nu_{c0}/\nu_{nc0} = 1.02$, $\gamma_0\, p_0 = 0.05$, $\lambda = 1.5$, $\beta\, p_0 = 4.5 \times 10^{-4}$. The initial conditions used are $\nu_{c0}/\nu = 0.81$, $\tilde{p} = 1.0$, and $\tilde{q} = 0$ at $\xi = 0$.

where ν_{c0} is the value of ν corresponding to a reference mean stress $p = p_0$. The rationale for choosing some of the above forms for h_i is indicated in Problem 5.15.

The function $\kappa(p, \nu)$ is given by

$$\begin{aligned} \kappa &= \kappa_l\,(p_{nc}/p)^{\lambda}\,, \quad \frac{\mathrm{D}\xi}{\mathrm{D}t} > 0 \\ &= \kappa_u, \quad \frac{\mathrm{D}\xi}{\mathrm{D}t} \leq 0 \end{aligned} \tag{5.89}$$

where κ_l, κ_u, and λ are constants. The reason for using two different functions for κ is explained below. The variable p_{nc} in (5.89) represents the mean stress at a state of normal consolidation, i.e., a state where the stress tensor is given by $\boldsymbol{\sigma} = p\,\mathbf{I}$ and the material has the lowest solids fraction consistent with the specified value of p. As discussed in §5.1.4, a linear relation can be assumed between $\ln p_{nc}$ and $1/\nu$. Thus

$$1/\nu = (1/\nu_{nc0}) - (\rho_p/\kappa_l)\,\ln(p_{nc}/p_0) \tag{5.90}$$

where the subscript nc denotes a state of normal consolidation, and ν_{nc0} is the value of ν corresponding to a reference mean stress $p_{nc} = p_0$.

Suppose that (5.85), (5.86), and (5.47) are integrated from $\xi = 0$ to some value of ξ, say $\xi_B > 0$ (Fig. 5.18). The material has been compressed to a natural strain of ξ_B, and the process is termed *loading*. If the material is now *unloaded* by reducing the axial stress, or equivalently, by reducing p, the material expands along the curve BA. The loading and unloading curves coincide as the derivatives $\frac{\mathrm{d}p}{\mathrm{d}\xi}$ and $\frac{\mathrm{d}q}{\mathrm{d}\xi}$ are unique functions of p and ν. However, experiments suggest that materials often unload along a different curve such as BC in Fig. 5.18. Davis and Mullenger (1978) incorporate this behavior in the model by assuming that the functions h_i differ for loading and unloading. In particular, the constants κ_l and κ_u in (5.89) correspond to *loading* ($\mathrm{D}\xi/\mathrm{D}t > 0$) and *unloading* ($\mathrm{D}\xi/\mathrm{D}t \leq 0$), respectively. For more general deformations, it is assumed that $\kappa = \kappa_l\,(p_{nc}/p)^{\lambda}$ for $\mathrm{tr}(\boldsymbol{\sigma}\,\mathbf{C}) > 0$ (loading) and $\kappa = \kappa_u$ for $\mathrm{tr}(\boldsymbol{\sigma}\,\mathbf{C}) \leq 0$ (unloading). Similarly, the constant b in (5.87) has two values, given by

$$\begin{aligned} b &= -1, \quad \frac{\mathrm{D}\xi}{\mathrm{D}t} > 0 \\ &= b_u\,, \quad \frac{\mathrm{D}\xi}{\mathrm{D}t} \leq 0 \end{aligned} \tag{5.91}$$

where b_u is a constant in the range 0–4. For more general deformations, it is assumed that $b = -1$ for $\mathrm{tr}(\mathbf{s}'\mathbf{C}') > 0$ and $b = b_u$ for $\mathrm{tr}(\mathbf{s}'\mathbf{C}') \leq 0$, where $\mathbf{s}'$ and $\mathbf{C}'$ are the deviatoric stress and rate of deformation tensors, defined by

$$\mathbf{s}' \equiv \boldsymbol{\sigma} - (\mathrm{tr}(\boldsymbol{\sigma})/3)\,\mathbf{I} = \boldsymbol{\sigma} - p\,\mathbf{I}; \quad \mathbf{C}' \equiv \mathbf{C} - (\mathrm{tr}(\mathbf{C})/3)\,\mathbf{I} \tag{5.92}$$

and $\mathbf{I}$ is the unit tensor.

Using (5.87)–(5.88), and introducing dimensionless variables

$$\tilde{p} = p/p_0; \tilde{q} = q/p_0; \quad \tilde{M} = M/p_0; \quad \tilde{\mu}_* = \mu_*/p_0$$
$$\tilde{\nu} = \nu/\nu_{c0}; \quad \tilde{\kappa} = \kappa/(\rho_p\,\nu_{c0}) \tag{5.93}$$

(5.85) and (5.86) can be rewritten as

$$\frac{d\tilde{p}}{d\xi} = \frac{\tilde{\kappa}\,\tilde{p}}{\tilde{\nu}} - \frac{4\,b\,\tilde{\mu}_*\,\tilde{p}\,\tilde{q}}{3\,\tilde{M}^2}\left[1 - \left(\frac{\nu_c}{\nu}\right)^{\gamma_*}\right]$$
$$\frac{d\tilde{q}}{d\xi} = 2\,\tilde{\mu}_*\left[1 + \frac{2\,b\,\tilde{q}^2}{3\,\tilde{M}^2}\left(\frac{\nu_c}{\nu}\right)^{\gamma_*}\right] \tag{5.94}$$

Davis and Mullenger (1978) estimated all the parameters occurring in (5.94), except b_u and κ_u (see (5.89) and (5.91)), using triaxial data for Pyramid dam rockfill. Using these parameter values, and assumed values for b_u and κ_u, (5.94) and (5.47) were integrated numerically to obtain the variation of σ_{zz} with ξ (Fig. 5.18). The profile is qualitatively similar to the data shown in Fig. 2.31, except for the absence of a hysteresis loop. (As the vessel is rigid, the volumetric strain is proportional to the axial engineering strain, which is approximately equal to the axial natural strain, except at large strains.) Davis and Mullenger (1978) presented a profile with a hysteresis loop, but the parameter values were reported.

It remains to determine the velocity field, and to relate ξ to z and t. This requires the solution of the coupled equations (5.41), (5.46), and (5.49). An approximate solution can be constructed for the special case of slow deformation and negligible body forces, as discussed in §5.1.11.

5.3. SUMMARY

Rigid-plastic equations involve a yield condition and a flow rule. The yield condition is a scalar relation between the components of the stress tensor $\boldsymbol{\sigma}$ and the solids fraction ν, and the flow rule relates the stresses to the components of the rate of deformation tensor. For isotropic materials, the yield condition depends on ν and the invariants of $\boldsymbol{\sigma}$.

Data related to three-dimensional yield conditions can be obtained using triaxial tests. In a conventional triaxial test, a cylindrical sample saturated with water is immersed in a cell filled with water, and is subjected to an axial load. Thus two of the principal stresses are equal to the normal stress in the radial direction, and the third principal stress is equal to the normal stress in the axial direction.

Isotropic compression tests form a special case of triaxial tests, wherein all the principal stresses are equal. Data from such tests can be used to construct the normal consolidation curve, which relates the solids fraction of the material in its loosest state to the mean stress for a monotonic increase in the latter. The loading and unloading curves do not usually coincide, showing that a part of the deformation is plastic or irrecoverable.

In triaxial compression tests, the axial normal stress exceeds the radial normal stress. The behavior of samples in these tests is qualitatively similar to the behavior in shear tests, which were discussed in Chapter 2. Data for Weald clay reveal the shapes of the Hvorslev and Roscoe surfaces for this material. In this case, the yield loci (contours of constant solids fraction) are geometrically similar.

The flow rule requires the components of the rate of deformation tensor to be proportional to the gradient (with respect to the stress components) of a scalar function called the plastic potential. Further, it implies that the principal axes of the stress tensor $\boldsymbol{\sigma}$ and the rate of deformation tensor $\mathbf{C}$ are aligned with each other. In a space where the principal axes of $\mathbf{C}$ are aligned with the principal axes of $\boldsymbol{\sigma}$, the vector $\mathbf{c}$ representing the principal rates of deformation is directed along the normal to the plastic potential surface. If the plastic potential coincides with the yield function, an associated flow rule results, and $\mathbf{c}$ is directed along the normal to the yield surface. Data for Monterey sand show that the associated flow rule does not hold for this material. However, Desai and Siriwardane (1984, p. 381) note that for many materials, normality holds in planes perpendicular to the space diagonal. In such cases, nonassociative behavior is caused only by changes in the mean stress.

The hypoelastic and hypoplastic equations do not involve the concept of a yield condition, but assume that a suitable stress rate depends on the stress tensor $\boldsymbol{\sigma}$, the rate of deformation tensor $\mathbf{C}$, and the solids fraction. Hypoelastic equations are linear functions of $\mathbf{C}$, whereas hypoplastic equations are nonlinear functions of $\mathbf{C}$.

It is shown that one-dimensional deformation can be modeled using both rigid-plastic and hypoelastic equations. However, for the case of steady shear flow between horizontal plates, both these types of equations fail to predict the velocity profile. This defect is believed to be caused by the absence of a material length scale in the constitutive equations, and can be remedied by using alternative equations suggested by Mühlhaus and Vardoulakis (1987) and others.

PROBLEMS

5.1. Expressions for the distances measured along and perpendicular to the space diagonal

Consider a point A with coordinates $(\sigma_{1A}, \sigma_{2A}, \sigma_{3A})$ in the principal stress space (Fig. 5.1), and let the line AB be drawn perpendicular to the space diagonal OB, which is defined by $\sigma_1 = \sigma_2 = \sigma_3$.

(a) Show that the coordinates $(\sigma_1, \sigma_2, \sigma_3)$ of any point on AB satisfy the equation

$$\sigma_1 + \sigma_2 + \sigma_3 = \text{constant} = \sigma_{1A} + \sigma_{2A} + \sigma_{3A} \equiv 3\, p_A \tag{5.95}$$

(b) Hence show that l_{BO}, the length of BO, and l_{AB}, the length of AB, are given by

$$\begin{aligned} l_{BO} &= \sqrt{3}\, p_A = (\sigma_{1A} + \sigma_{2A} + \sigma_{3A})/\sqrt{3} \\ l_{AB} &= ((\sigma_{1A} - \sigma_{2A})^2 + (\sigma_{2A} - \sigma_{3A})^2 + (\sigma_{3A}\sigma_{1A})^2)^{1/2}/\sqrt{3} \\ &= \sqrt{2/3}\, q_A \end{aligned} \tag{5.96}$$

where q is defined by (5.6).

5.2. A feature of the yield locus for Granta-gravel

Consider the p–q plane, where p and q are defined by (5.6).

(a) Show that if two of the principal stresses vanish, $q = 3\,|p|$.
(b) For $p \geq 0$, show that one or two of the principal stresses must be negative if $q > 3\,p$.
(c) In the (q/p_c)–(p/p_c) plane, sketch the yield locus corresponding to the Granta-gravel yield condition (5.14). Hence show that a segment near the origin is not accessible for cohesionless materials.

5.3. Cross sections corresponding to the Tresca and Mohr–Coulomb yield surfaces

(a) Formulate expressions for the position vector OQ and for unit vectors along the a_1, a_2, and a_3 axes (Fig. 5.3a). Using these expressions, derive (5.15).

(b) Show that the cross section corresponding to the Tresca yield condition is a regular hexagon of side $\sqrt{2/3}\,\sigma_0$, where σ_0 is the magnitude of the yield stress in uniaxial compression or tension.

(c) Show that the intersection of the Mohr–Coulomb yield surface with the octahedral plane $\sigma_1 + \sigma_2 + \sigma_3 = \text{constant} \equiv k$ is an irregular hexagon. The hexagon is irregular if all of its corners are not equidistant from the origin of the a_1–a_3 plane. Show that the yield surface intersects the space diagonal at $p = -c\,\cot\phi_*$.

5.4. An alternative formulation of the associated flow rule for the extended von Mises yield condition

For the extended von Mises yield condition (5.12), the associated flow rule is given by

$$C_{ij} = \dot{\lambda}\left(\frac{\partial \tilde{G}}{\partial \sigma_{ij}}\right), \quad i, j = 1, 3 \tag{5.97}$$

where $\tilde{G} = q - H(p, \nu)$ is the plastic potential, and $\{C_{ij}\}$ are the components of the rate of deformation tensor, defined in the compressive sense.

(a) Using (5.9), (5.12), and (5.97), show that

$$C_{ij} = \dot{\lambda}\left(\frac{3\,s'_{ij}}{2\,H} - \frac{1}{3}\frac{\partial H}{\partial p}\,\delta_{ij}\right) \tag{5.98}$$

where s'_{ij} is a component of the deviatoric stress tensor, defined by (5.92). Caution: Do not set $s'_{ij} = s'_{ji}$ in (5.9) before differentiating q with respect to σ_{ij}.

(b) Using (5.98), show that

$$\mathbf{C}' = \frac{3\,\dot{\lambda}\,\mathbf{s}'}{2\,H} = \sqrt{\frac{3\,\mathbf{C}' : \mathbf{C}'}{2}}\,\frac{\mathbf{s}'}{H}$$

$$\mathrm{tr}(\mathbf{C}) = -\sqrt{\frac{2\,\mathbf{C}' : \mathbf{C}'}{3}}\,\frac{\partial H}{\partial p} \tag{5.99}$$

where $\mathbf{C}'$ and $\mathbf{s}'$ are the deviatoric stress and rate of deformation tensors, defined by (5.92). Comparison of the first of (5.99) with the constitutive relation for an incompressible Newtonian fluid (2.54) shows that the "shear viscosity" μ for the present model of the granular material is $\mu = H/\sqrt{6\,\mathbf{C}' : \mathbf{C}'}$.

This problem has been adapted from Schaeffer (1987) and S. Sundaresan (private communication, 1998).

5.5. Form of the Drucker–Prager yield condition for plane flow

Consider flow parallel to the x–y plane. If all quantities are independent of the z coordinate, and the Lévy-Mises flow rule (2.53) is applicable, show that the Drucker–Prager yield condition (5.13) reduces to the Mohr–Coulomb yield condition $\tau = \sigma\,\sin\phi_*$, provied $k' \equiv \sqrt{3}\,\sin\phi_*$. Here the deviator stress τ and the mean stress σ are defined by (2.33).

5.6. Relation between the local angular velocity of the material and the vorticity tensor

(a) Consider two points A and B with position vectors $r_A = x_i\,\mathbf{e}_i$ and $r_B = (x_i + \Delta x_i)\,\mathbf{e}_i$, respectively, where the $\mathbf{e}_i$ are the basis vectors for a Cartesian coordinate system. Let $\mathbf{v}_A = v_{i\,A}\,\mathbf{e}_i$ denote the linear velocity at point A. Expanding $\mathbf{v}_B$ in a Taylor's series about A and assuming that B is near A, show that

$$\Delta v_i \equiv v_{i\,B} - v_{i\,A} \approx -(W_{ij} + C_{ij})\,\Delta x_j$$

where C_{ij} and W_{ij} denote the components of the rate of deformation tensor and the vorticity tensor, respectively. These are defined by

$$C_{ij} = -\frac{1}{2}\left(\frac{\partial v_j}{\partial x_i} + \frac{\partial v_i}{\partial x_j}\right); \quad W_{ij} = \frac{1}{2}\left(\frac{\partial v_j}{\partial x_i} - \frac{\partial v_i}{\partial x_j}\right) \tag{5.100}$$

Introducing a vector $\boldsymbol{\omega} = \omega_k \mathbf{e}_k$, defined by

$$W_{ij} \equiv \epsilon_{ijk}\,\omega_k \tag{5.101}$$

where ϵ_{ijk} is the permutation symbol defined by (A.23), show that the contribution to Δv_i from W_{ij} is given by

$$\Delta v_{i*} \equiv -W_{ij}\,\Delta x_j = (\,\boldsymbol{\omega} \times \Delta\mathbf{r})_i \tag{5.102}$$

where $\Delta\mathbf{r} \equiv \mathbf{r}_B - \mathbf{r}_A$ and $(\,\boldsymbol{\omega} \times \Delta\mathbf{r})_i$ is the ith component of $\boldsymbol{\omega} \times \Delta\mathbf{r}$. In view of (5.102), Δv_{i*} can be regarded as the linear velocity of B relative to A, caused by rigid body rotation about A with an angular velocity $\boldsymbol{\omega}$. This problem has been adapted from Batchelor (1967, p. 81).

(b) Using (5.101), (A.62), and (A.87), show that

$$-\,\boldsymbol{\epsilon} : \mathbf{W} = \nabla \times \mathbf{v} = 2\,\boldsymbol{\omega} \tag{5.103}$$

where $\boldsymbol{\epsilon}$ is alternating tensor defined by (A.22) and $\mathbf{W}$ is the vorticity tensor whose components are defined by (5.100). Hence the vorticity $\nabla \times \mathbf{v}$ represents twice the local angular velocity of the material.

(c) Using (5.103), show that

$$\omega_3 = W_{12} = \frac{1}{2}\left(\frac{\partial v_2}{\partial x_1} - \frac{\partial v_1}{\partial x_2}\right)$$

5.7. Derivation of the plastic potential flow rule for the deformation of metals
Consider a flow rule of the form

$$\mathbf{C} = \dot{\lambda}\,\mathbf{H}(\boldsymbol{\sigma}, \nu) \tag{5.104}$$

for an isotropic material, where $\mathbf{C}$ is the rate of deformation tensor, defined by (5.26), $\boldsymbol{\sigma}$ is the symmetric stress tensor, ν is the solids fraction, and $\dot{\lambda}$ is a scalar factor of proportionality. For metals, data suggest that plastic deformation is (i) not affected by the value of the mean stress, and (ii) is incompressible. In view of the condition (i), (5.104) can be rewritten as

$$\mathbf{C} = \dot{\lambda}\,\mathbf{H}(\mathbf{s}', \nu) \tag{5.105}$$

where $\mathbf{s}' \equiv \boldsymbol{\sigma} - p\,\mathbf{I}$ is the deviatoric stress tensor, and $p \equiv \mathrm{tr}(\boldsymbol{\sigma})/3$ is the mean stress. For an isotropic material, it can be shown that (5.105) can be rewritten in component form as (see, e.g., Serrin, 1959; Hunter, 1983, pp. 137–139, and also Problem 5.8)

$$C_{mn} = \dot{\lambda}(\alpha\,\delta_{mn} + \beta\,s'_{mn} + \gamma\,s'_{mk}\,s'_{kn}) \tag{5.106}$$

where δ_{mn} is the Kronecker delta, α, β, and γ are functions of ν and the principal invariants I'_2 and I'_3 of $\mathbf{s}'$, and $s'_{mn} = \sigma_{mn} - (1/3)\,\sigma_{kk}\,\delta_{mn}$. (Note that $I'_1 \equiv \mathrm{tr}(\mathbf{s}') = 0$ by construction.)

(a) For incompressible flow, show that (5.106) can be rewritten as

$$C_{mn} = \dot{\lambda}[\beta\,s'_{mn} + \gamma\,(s'_{mk}\,s'_{kn} - (1/3)\,s'_{rq}\,s'_{qr}\,\delta_{mn})] \tag{5.107}$$

where β and γ are functions of ν, I'_2 and I'_3.

(b) Using (5.2), derive expressions for I'_2 and I'_3. Hence show that

$$\frac{\partial I'_2}{\partial s'_{ij}} = -s'_{ij};\quad \frac{\partial I'_3}{\partial s'_{ij}} = I'_2\,\delta_{ij} + s'_{jr}\,s'_{ri} \tag{5.108}$$

Caution: Do not set $s'_{ij} = s'_{ji}$ before differentiating I'_2 and I'_3 with respect to s'_{ij}.

Using (5.108), show that

$$\frac{\partial I'_2}{\partial \sigma_{mn}} = -s'_{mn};\quad \frac{\partial I'_3}{\partial \sigma_{mn}} = s'_{mr}\,s'_{rn} - (1/3)\,s'_{ir}\,s'_{ri}\,\delta_{mn} \tag{5.109}$$

Using (5.109), show that (5.107) reduces to

$$C_{mn} = \dot{\lambda}\left(-\beta\,\frac{\partial I'_2}{\partial \sigma_{mn}} + \gamma\,\frac{\partial I'_3}{\partial \sigma_{mn}}\right) \tag{5.110}$$

(c) Hunter (1983, p. 481) notes that if β and γ are differentiable functions of I_2' and I_3', an integrating factor $\psi(I_2', I_3')$ can be found such that $-\psi\,\beta$ and $\psi\,\gamma$ can be expressed as the partial derivatives of another function $G_*(I_2', I_3')$. Thus

$$-\psi\,\beta = \frac{\partial G_*}{\partial I_2'}; \quad \psi\,\gamma = \frac{\partial G_*}{\partial I_3'} \tag{5.111}$$

Using (5.111), show that (5.110) can be rewritten as

$$C_{mn} = \dot{\Lambda}\,\frac{\partial G_*}{\partial \sigma_{mn}}$$

where $\dot{\Lambda}$ is a scalar factor of proportionality.

This problem has been adapted from Hunter (1983, p. 480–481).

5.8. Representation of the constitutive equation for an isotropic tensor valued function of a second order tensor and a scalar

For an isotropic material, consider a constitutive equation of the form

$$\mathbf{C} = \mathbf{H}(\boldsymbol{\sigma}, \nu) \tag{5.112}$$

where $\mathbf{C}$ is the rate of deformation tensor, $\boldsymbol{\sigma}$ is the stress tensor, and ν is the solids fraction. As discussed in Appendix F, the principal axes of $\mathbf{C}$ and $\boldsymbol{\sigma}$ are aligned with each other. Relative to the common principal axes of $\mathbf{C}$ and $\boldsymbol{\sigma}$, (5.112) can be rewritten as

$$c_1 = h(\sigma_1, \sigma_2, \sigma_3, \nu) \equiv h_{123} \tag{5.113}$$

where h is a scalar function of its arguments. However, as the material is isotropic, (5.113) also holds for permutations of the indices 1, 2, and 3, i.e.,

$$\begin{aligned} c_2 &= h(\sigma_2, \sigma_3, \sigma_1, \nu) = h_{231} \\ c_3 &= h(\sigma_3, \sigma_1, \sigma_2, \nu) = h_{312} \end{aligned} \tag{5.114}$$

On account of isotropy, the value of h is unchanged when the second and third subscripts are interchanged. Thus

$$h_{123} = h_{132} \tag{5.115}$$

Suppose that

$$c_1 = h_{123} = \alpha + \beta\,\sigma_1 + \gamma\,\sigma_1^2 \tag{5.116}$$

where α, β, and γ are functions of ν and the three principal stresses, σ_i, $i = 1, 3$. Note that (5.116) is in general *not* a polynomial in σ_1.

Using permutations of indices 1, 2, and 3, (5.116) and (5.113) imply that

$$\begin{aligned} c_2 &= h_{231} = \alpha + \beta\,\sigma_2 + \gamma\,\sigma_2^2 \\ c_3 &= h_{312} = \alpha + \beta\,\sigma_3 + \gamma\,\sigma_3^2 \end{aligned} \tag{5.117}$$

Relative to the common principal axes of $\mathbf{C}$ and $\boldsymbol{\sigma}$, the diagonal elements of $\mathbf{C}$ are given by c_1, c_2, and c_3, and the off-diagonal elements vanish. Hence (5.116) and (5.117) are equivalent to

$$\mathbf{C} = \alpha\,\mathbf{I} + \beta\,\boldsymbol{\sigma} + \gamma\,\boldsymbol{\sigma}^2 \tag{5.118}$$

As this is a tensor equation, it is valid in all coordinate systems.

(a) Show that (5.116) and (5.117) can always be solved for α, β, and γ provided the principal stresses are distinct.

(b) If $\sigma_1 \neq \sigma_2 = \sigma_3$, show that (5.116) and (5.117) can be solved for α, β, and γ, provided h is differentiable. (If h is not required to be differentiable, $\mathbf{C}$ can be expressed as $\alpha\,\mathbf{I} + \beta\,\boldsymbol{\sigma}$, which is a special case of (5.118).)

(c) Show that α, β, and γ are unaffected by interchanging the subscripts 1 and 2 in (5.113) and (5.114). Similar results hold for interchange of any two subscripts. Hence α, β, and

γ are symmetric functions of the principal stresses and can be expressed in terms of the principal invariants of $\boldsymbol{\sigma}$.

This problem has been adapted from Hunter (1983, p. 136–139).

5.9. The critical state line for modified Cam Clay

The modified Cam clay yield condition is given by (Desai and Siriwardane, 1984, p. 292) $q^2 = M^2\, p\,(p_0(\nu) - p)$, which is an ellipse in the p–q plane. Here M is a constant, and $p = p_0(\nu)$ is the normal consolidation or isotropic compression curve in the p–ν plane (see §5.1.4). Using the associated flow rule, show that the critical state line is given by $q = Mp$. Relate the mean stress at a critical state p_c to p_0.

5.10. Expressions for the components of the stress tensor in terms of ν for one-dimensional deformation

Consider the one-dimensional compression of a granular material in a cylinder with frictionless walls (Fig. 5.16). The material satisfies a yield condition of the form $q - H(p, \nu) = 0$, where p and q are defined by (5.6), and the associated flow rule holds.

(a) Using the velocity field (5.35), show that

$$\sigma_{rr} = \sigma_{\theta\theta} = p - (H/3); \quad \sigma_{zz} = p + (2\,H/3); \quad \partial H/\partial p = -3/2$$

Hence all the stress components can in priciple be expressed in terms of ν.

(b) For Weald clay, the mean stress p_c at a critical state is given by

$$p_c(\nu) = \exp\left(\frac{\Gamma\,\nu - 1}{\lambda\,\nu}\right)$$

with $\Gamma = 2.06$ and $\lambda = 0.093$, when p_c is expressed in kPa (Atkinson and Bransby, 1982, pp. 189–190). The parameter M in the yield condition for modified Cam clay has the value $M \approx 0.9$ (see Problem 5.9). Let ξ denote the natural strain defined by (5.41). If $\xi = 0$ at $\nu = 0.6$, determine the values of the normal stress σ_{zz} at $\xi = 0$ and $\xi = 0.1$.

5.11. Determination of the yield locus for plane flow from the three-dimensional yield surface

The Drucker–Prager yield condition for a cohesive material can be written as

$$\sqrt{J_2} = a\,p + b \tag{5.119}$$

where J_2 is the second invariant of the deviatoric stress tensor (see (5.9) and (5.6)), p is the mean stress, defined by (5.6), and a and b are constants.

(a) As suggested by Bauer and Wu (1995), it is convenient to introduce new variables

$$\sigma_{i*} = \sigma_i + (b/a); \quad p_* = p + (b/a), \quad i = 1, 3 \tag{5.120}$$

where σ_i, $i = 1, 3$ are the principal stresses. Show that (5.119) can be rewritten as

$$F \equiv \sqrt{J_2} - a\,p_* = 0 \tag{5.121}$$

where J_2 is now a function of σ_{1*}, σ_{2*}, and σ_{3*}.

(b) Interpret the change of variables geometrically.

(c) Consider flow parallel to the x–y plane, and assume that all quantities are independent of the z coordinate. Assuming that coaxiality holds, show that one of the principal stress axes, say the σ_3 axis, is parallel to the z axis. Hence the σ_1 and σ_2 axes lie in the x–y plane.

(d) Using the associated flow rule (5.27), show that

$$\sigma_{3*} = h\,\sigma_* \tag{5.122}$$

where

$$h \equiv \frac{1 + (2\,a^2)/3}{1 - (a^2/3)}; \quad \sigma_* \equiv \frac{\sigma_{1*} + \sigma_{2*}}{2} \tag{5.123}$$

(e) Using (5.122), show that the yield condition (5.119) reduces to

$$\tau_* \equiv (\sigma_{1*} - \sigma_{2*})/2 = \sigma_* \sin\phi_* \tag{5.124}$$

where

$$\sin\phi_* \equiv \frac{a}{\sqrt{1-(a^2/3)}}$$

(f) Hence show that the yield condition for plane flow is given by the Mohr–Coulomb yield condition (2.37), i.e.,

$$\tau \equiv (\sigma_1 - \sigma_2)/2 = \sigma \sin\phi_* + c \cos\phi_* \tag{5.125}$$

where

$$\sigma \equiv \frac{\sigma_1+\sigma_2}{2}; \quad c = \frac{b}{\sqrt{1-(4\,a^2/3)}}$$

(g) Using the data reported for Weald clay in Fig. 15 of Roscoe et al. (1958), we have $a =$ 0.55, and $b =$81 kPa. Calculate the values of the angle of internal friction ϕ_* and the cohesion c.

5.12. Elastic solution for a spherical shell

Consider a spherical shell of inner radius r_0 and outer radius r_1. The material of the shell satisfies the constitutive equation for an *isotropic, linear elastic material*, which is given by (Timoshenko and Goodier, 1970, pp. 7–10; Hunter, 1983, p. 369)

$$E\,\epsilon_{*ij} = (1+\nu_p)\,\sigma_{ij} - \nu_p\,\sigma_{kk}\,\delta_{ij}; \quad i,j = 1,3 \tag{5.126}$$

where $\{\epsilon_{*ij}\}$ are the components of the Eulerian infinitesimal strain tensor $\boldsymbol{\epsilon}_*$, defined by in the compressive sense by

$$\boldsymbol{\epsilon}_* \equiv -\frac{1}{2}\left(\nabla\mathbf{u} + (\nabla\mathbf{u})^{\mathrm{T}}\right) \tag{5.127}$$

$\mathbf{u}$ is the displacement vector (Fig. 1.27), ν_p and E are material constants called the *Poisson's ratio* and *Young's modulus*, respectively, and δ_{ij} is the Kronecker delta.

(a) Using a spherical coordinate system with the origin at the center of mass of the shell, it can be assumed that the principal stress directions coincide with the coordinate directions. The outer surface of the shell is *traction-free*, i.e., the shear and normal stresses vanish on this surface, and the inner surface is subjected to a uniform pressure p_0. Assuming that the shell is at rest, body forces can be ignored, and that all quantities depend only on the radial coordinate, formulate the steady state momentum balances. Balances in spherical coordinates are given in Appendix A.

(b) Using the expressions for $\epsilon_{*\theta\theta}$ and ϵ_{*rr} (see (A.78)), show that $A \equiv \sigma_{rr} - \sigma_{\theta\theta}$ can be expressed in terms of the derivatives of the stresses.

(c) Using the result obtained in part (b), and the radial component of the momentum balance, derive expressions for the dimensionless stresses $\tilde{\sigma}_{rr}(\xi) \equiv \sigma_{rr}/p_0$ and $\tilde{\sigma}_{\theta\theta}(\xi) \equiv \sigma_{\theta\theta}/p_0$. Here $\xi \equiv r/r_1$.

(d) Examine whether the normal stresses are tensile or compressive.

(e) If the Tresca yield condition (5.17) holds when the material yields, show that yielding occurs first at the inner surface $\xi = \xi_0 \equiv r_0/r_1$. Hence determine the value of p_0 required to cause yielding at the inner surface.

This problem has been adapted from Chakrabarthy (1987, pp. 306–308).

5.13. Estimate of the coefficient of earth pressure using elasticity theory

This problem deals with the use of a linear elastic model to estimate the coefficient of earth pressure K in a cylindrical bin. A granular material is at rest in a cylindrical bin of radius R (Fig. 5.16). For an isotropic, linear elastic material, the constitutive equation is given by (5.126).

(a) Consider cylindrical coordinate system with origin at the base of the bin (Fig. 5.16). It can be assumed that (i) all displacements and stresses are independent of the circumferential coordinate θ, and (ii) the wall $r = R$ is rigid, so that $u_r(r = R, z) = 0$. Here u_r is the radial component of the displacement vector (Fig. 1.27). In view of assumption (ii), we can look for a solution with $u_r(r, z) = 0$. Using this information along with (5.126) and (A.70), show that $\sigma_{\theta\theta} = \sigma_{rr} = K_e\, \sigma_{zz}$, where $K_e \equiv \nu_p/(1 - \nu_p)$ and ν_p is the Poisson's ratio of the material. Hence the coefficient of earth pressure is given by

$$K \equiv \sigma_{rr}/\sigma_{zz} = K_e = \nu_p/(1 - \nu_p)$$

(b) In view of assumption (i), it can be assumed that $u_\theta = 0$. Hence show that $\sigma_{r\theta} = 0 = \sigma_{\theta z}$.

(c) Using the results of parts (a) and (b), and assuming that the density $\rho \approx$ constant, show that the momentum balances can be manipulated to obtain a single second order partial differential equation for σ_{zz}.

This problem has been adapted from Ovarlez et al. (2003).

5.14. Shearing of a granular material in a cylindrical Couette cell

Schaeffer (1987) has proposed the following constitutive equation for the slow three-dimensional flow of cohesionless granular materials:

$$\boldsymbol{\sigma} = p \left(\mathbf{I} + \sqrt{2}\, \sin\phi_* \frac{\mathbf{C}}{|\mathbf{C}|} \right) \tag{5.128}$$

Here $\mathbf{I}$ is the unit tensor, $3\,p$ is the trace of the stress tensor $\boldsymbol{\sigma}$ (see (A.18)), $\mathbf{C}$ is the rate of deformation tensor, defined in the compressive sense, $|\mathbf{C}| \equiv (\mathbf{C} : \mathbf{C})^{1/2} = (C_{ij}\, C_{ji})^{1/2}$, and ϕ_* is a material constant.

(a) At locations where $|\mathbf{C}| \neq 0$, what is the constraint imposed by (5.128) on the velocity field?

(b) Suppose that a granular material is sheared slowly in a cylindrical Couette cell whose axis is vertical (Fig. 2.1). The inner cylinder rotates about its axis in an anticlockwise direction and the outer cylinder is stationary. If the Couette gap W is much smaller than the radius R of the inner cylinder, the actual flow field can be approximated by flow between vertical flat plates. Assuming that the velocity field is steady, unidirectional, and does not vary in the vertical direction, use (5.128) to express the stress components in terms of C_{ij}. Hence show that the velocity field is *indeterminate* for the case of plane flow.

(c) Assuming that the density ρ is a constant, the free surface F of the material is horizontal, and the stresses do not vary in the direction of flow, show that p varies linearly with the depth below F. The cell is filled to a height L.

(d) Estimate the torque T required to rotate the inner cylinder at an angular velocity ω.

(e) Instead of considering flow between flat plates, examine the flow field in a cylindrical Couette cell (Fig. 2.1). Using cylindrical coordinates with the origin fixed at some point on the axis of the cell, assume that the circumferential velocity v_θ is the only nonzero component of velocity, the velocity field is fully developed, i.e., v_θ is independent of the z coordinate, and the stress fields are axisymmetric. Show that the use of (5.128) along with the r and θ components of the momentum balance leads to an inconsistency. Balance equations in cylindrical coordinates, and also expressions for the C_{ij}, are given in Appendix A.

(f) Suppose that (5.128) is replaced by

$$\boldsymbol{\sigma} = p \left(\mathbf{I} - \sqrt{2}\, \sin\phi_* \frac{\mathbf{C}}{|\mathbf{C}|} \right) \tag{5.129}$$

Examine whether (5.129) is an acceptable constitutive equation.

Parts (a)–(d) have been adapted from Tardos et al. (1998).

5.15. Forms for the functions in the hypoelastic model of Davis and Mullenger (1978)

(a) Using the trace of (5.72), show that

$$\frac{\mathrm{D}p}{\mathrm{D}t} = (h_1 + (h_3/3) - h_2\, p - h_4\, p + h_5\, p^2)\,\mathrm{tr}(\mathbf{C}) + (h_5\, p - h_2)\,\mathrm{tr}(\mathbf{s}'\,\mathbf{C}')$$

$$\overset{\circ}{\mathbf{s}}{}' = h_3\,\mathbf{C}' + [(h_5\, p - h_4)\,\mathrm{tr}(\mathbf{C}) + h_5\,\mathrm{tr}(\mathbf{s}'\,\mathbf{C}')]\,\mathbf{s}' \qquad (5.130)$$

where $p = \mathrm{tr}(\boldsymbol{\sigma})/3$, $\mathbf{s}'$ and $\mathbf{C}'$ are the deviatoric stress and the rate of deformation tensors defined by (5.92), and $\overset{\circ}{\mathbf{s}}{}'$ is the Jaumann derivative of $\mathbf{s}'$. If the response of $\mathbf{s}'$, i.e., $\overset{\circ}{\mathbf{s}}{}'$, is to be independent of tr(**C**), as suggested by Romano (1974), the second of (5.130) implies that

$$h_5\, p = h_4 \qquad (5.131)$$

(b) For the special case of normal consolidation or *isotropic loading* ($\mathbf{s}' = 0,\ \mathbf{C}' = 0$), it is found that for some soils, $\ln p$ can be expressed as a linear function of the specific volume $w \equiv 1/\rho = 1/(\rho_p\, \nu)$, where ρ is the density. Using the first of (5.130), (5.131), and (5.45), show that this observation implies that

$$h_1 + (h_3/3) - h_2\, p = \kappa\, p/(\rho_p \nu) \qquad (5.132)$$

where κ is a constant. Using (5.131) and (5.132), show that the first of (5.130) can be integrated to obtain (5.90). To allow for states which do not lie on the normal consolidation curve (5.90), κ is assumed to be given by (5.89).

(c) Suppose the mean stress p is kept constant during deformation. The material will eventually reach a critical state or state of isochoric deformation, which implies that $\mathrm{D}\nu/\mathrm{D}t = 0$. However, as deviatoric deformation can occur at a critical state, $\mathbf{C}' \neq 0$ in general. Show that at a critical state

$$h_4 = h_5\, p = h_2 \qquad (5.133)$$

(d) Using index notation, and assuming that $\boldsymbol{\sigma}$ is a symmetric tensor, show that

$$\mathrm{tr}(\mathbf{s}'\,\overset{\circ}{\mathbf{s}}{}') = \frac{1}{2}\frac{\mathrm{D}}{\mathrm{D}t}[\mathrm{tr}(\mathbf{s}'^2)] \qquad (5.134)$$

where $\mathrm{tr}(\mathbf{s}'\,\overset{\circ}{\mathbf{s}}{}')$ denotes the trace of the product of the two tensors $\mathbf{s}'$ and $\overset{\circ}{\mathbf{s}}{}'$ (see §A.7.2).

(e) A critical state is defined by

$$\nu = \nu_c(p); \quad \mathrm{tr}(\mathbf{s}'^2) = [M(p)]^2 \qquad (5.135)$$

where M is a function of p. (Even though there are no yield surfaces in hypoelastic models, the second of (5.135) is equivalent to requring that the stresses at critical states satisfy an extended von Mises yield condition.) Using (5.135), the second of (5.130), and (5.134), show that

$$\frac{\mathrm{D}}{\mathrm{D}t}(M^2) = 2\,(h_3 + h_5 M^2)\,\mathrm{tr}(\mathbf{s}'\,\mathbf{C}')$$

As $M = M(p)$, and p does not change at a critical state, we have

$$h_5 = -h_3/M^2 \qquad (5.136)$$

(f) For deviatoric loading defined by $\mathrm{tr}(\mathbf{s}'\mathbf{C}') > 0$, and $\mathrm{D}p/\mathrm{D}t = 0$, dense samples with $\nu > \nu_c$ must dilate, and loose samples with $\nu < \nu_c$ must compact. Here ν_c is the solids fraction at a critical state. Use the first of (5.130) to show that $h_4 - h_2$ must have the same sign as $\nu - \nu_c$. Using this result along with (5.133), Davis and Mullenger (1978) have assumed the forms given in (5.87) for h_4 and h_2.

(g) For an isotropic, linear elastic material, the constitutive relation for small displacement gradients is given by (5.126). Show that the deviatoric stress tensor $\mathbf{s}' \equiv \boldsymbol{\sigma} - (\mathrm{tr}(\boldsymbol{\sigma})/3)\,\mathbf{I}$ is given by

$$\mathbf{s}' = 2\,G\,\boldsymbol{\epsilon}'_* \qquad (5.137)$$

where $G \equiv E/(2(1+\nu_p))$ is a constant called the *shear modulus*, $\boldsymbol{\epsilon}'_*$ is a deviatoric strain tensor, defined by $\boldsymbol{\epsilon}'_* = \boldsymbol{\epsilon}_* - (\mathrm{tr}(\boldsymbol{\epsilon}_*)/3)\,\mathbf{I}$, and $\boldsymbol{\epsilon}_*$ is the Eulerian infinitesimal strain tensor, defined by (5.127). Here tr denotes the trace of a tensor.

(h) Let the *displacement* vector be defined by $\mathbf{u} \equiv \mathbf{x} - \mathbf{X}$, where $\mathbf{x}$ and $\mathbf{X}$ are position vectors corresponding to a material point in the reference and current configurations, respectively (Fig. 1.27). Using the small displacement gradient approximation, i.e., assuming

$$\left|\frac{\partial u_i}{\partial X_j}\right| \ll 1, \quad \text{for all} \quad i, j \tag{5.138}$$

use Cartesian tensor notation to show that

$$\frac{\partial u_i}{\partial X_j} \approx \frac{\partial u_i}{\partial x_j}$$
$$\frac{\mathrm{D}}{\mathrm{D}t}\left(\frac{\partial u_i}{\partial x_j}\right) \equiv \frac{\partial}{\partial t}\left(\frac{\partial u_i}{\partial x_j}\right)\bigg|_{\mathbf{X}} \approx \frac{\partial v_i}{\partial x_j} \tag{5.139}$$

Using (5.138) and (5.139), show that

$$\overset{\circ}{s}{}' \approx 2\,G\,\mathbf{C}' \tag{5.140}$$

where $\mathbf{C}' \equiv \mathbf{C} - (\mathrm{tr}(\mathbf{C})/3)\,\mathbf{I}$. If the second of (5.130) is approximated by

$$\overset{\circ}{\mathbf{s}}{}' \equiv h_3\,\mathbf{C}' \tag{5.141}$$

comparison of (5.140) and (5.141) shows that $h_3 \approx 2\,G$. Hence (5.87) implies that $\mu_* \approx G$. This problem has been adapted from Davis and Mullenger (1978) and Hunter (1983, p. 366).

6

Flow through Axisymmetric Hoppers and Bunkers

In this chapter, experimental observations and models relevant to axisymmetric flow will be discussed. Plane and axisymmetric flows have both similarities and differences, which will be indicated.

6.1. EXPERIMENTAL OBSERVATIONS

6.1.1. Flow Rate

As discussed in §3.1.1, the mass flow rate $\dot{M}$ for a hopper with a circular exit orifice is proportional to $\rho_{\rm p}\, \nu\, \sqrt{g}\, D^{5/2}$, where $\rho_{\rm p}$ is the particle density, ν is the solids fraction, g is the acceleration due to gravity, and D is the diameter of the exit orifice. To compare the flow rates obtained using wedge-shaped and conical hoppers, let

$$V_D \equiv \frac{\dot{M}}{\rho_{\rm p}\, \nu\, A\, \sqrt{g\, D}} \tag{6.1}$$

denote a dimensionless mass flow rate. Here A is the cross-sectional area of the orifice, and, for a wedge-shaped hopper, D is the width of the exit slot (Fig. 3.1). The solids fraction ν used in (6.1) should be the cross-sectional averaged value at the hopper exit. As this value is rarely measured, investigators have often used other estimates, such as the mean solids fraction of a static column of material. Figure 6.1 shows that for a given wall angle, the mass flow rates for conical hoppers (the open symbols) are significantly lower than those for wedge-shaped hoppers (the filled symbols). For the data cited in the latter case, the length B of the exit slot is much larger than its width D (Fig. 3.1). Thus two of the walls are vertical and are far apart. Because of these features, the wedge-shaped hopper may offer less resistance to flow than the conical hopper, leading to a higher flow rate in the former case. The data of Zeininger and Brennen (1985) for glass beads flowing through conical hoppers (Fig. 6.1) show that the V_D values for fine beads (the open inverted triangles) are higher than those for coarse beads (the open triangles). The mass flow rates for fine beads are usually lower than those for coarse beads, as air drag impedes the flow of the former (Crewdson et al., 1977). Zeininger and Brennen (1985) attribute the anomalous behavior to the higher value of the ratio $D/d_{\rm p}$ in the former case (see Table 6.2), as the frictional properties of both types of beads are similar. This is in accord with the correlation of Beverloo et al. (1961) (see (3.1)), which suggests that V_D is proportional to $(D - k_*\, d_{\rm p})^{5/2}$, where k_* is a positive constant.

6.1.2. Velocity Profiles

Here we discuss the measurements of Cleaver and Nedderman (1993a) who used a tracer technique. The hopper was fitted with a guide tube that permitted the introduction of a

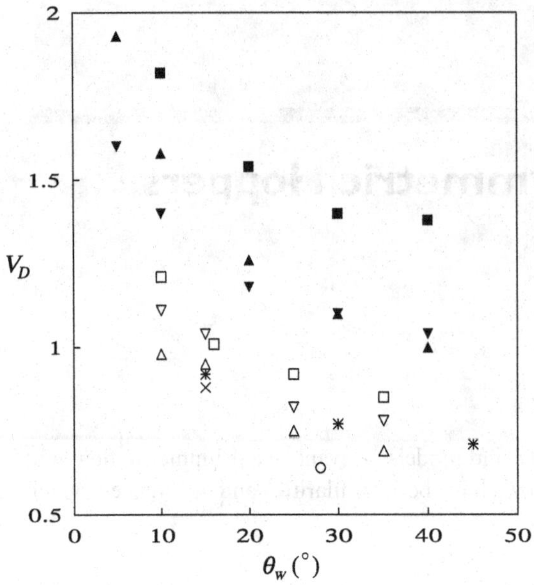

Figure 6.1. Variation of the dimensionless mass flow rate V_D (defined by (6.1)) with the wall angle θ_w: open symbols – conical hopers, filled symbols – wedge-shaped hoppers. The materials used and other parameter values are listed in Tables 6.1 and 6.2.

tracer particle at any chosen position in the mid plane of the hopper. A stop–start procedure was used, wherein the hopper was filled with material and allowed to discharge for about 30 s. This permitted the density field to change from its static state to a flowing state. The flow was then stopped momentarily, while the tracer was introduced through the guide tube. The tube was then removed and the material was allowed to discharge.

A ceramic tracer particle of diameter 4 mm was used. It was chosen to be of a larger size than the granular material, so that it could be captured using a sieve when the material discharged from the hopper. The residence time T_R, i.e., the time taken by the tracer to

Table 6.1. Sources of Data on Mass Flow Rates of Granular Materials Flowing Through Hoppers*

Reference	Granular material	Wall material	Orifice diameter/ slot width D (mm)	Symbol
(a) *Conical hoppers*				
Nguyen et al. (1979)	Sand	Aluminum	20.3–33	□
Zeininger and Brennen (1985)	Glass beads (coarse)	Sheet metal	28.3–32.7	△
Zeininger and Brennen (1985)	Glass beads (fine)	Sheet metal	28.3–32.7	▽
Nedderman (1988)	Kale seed	Aluminum	35	○
Gu et al. (1993)	Sugar	Galvanized steel	44.5	×
Deming and Mehring (1929)	Potassium nitrate pellets		10.1–15.1	*
(b) *Wedge-shaped hoppers*				
Brennen and Pearce (1978)	Glass beads	Aluminum	6.35	■
Kaza (1982)	Glass beads	Aluminum	25.5–26.5	▲
Kaza (1982)	Glass beads	Wall covered with a layer of glass beads	"	▼

* The data are plotted in Fig. 6.1, and the material properties are given in Table 6.2.

Table 6.2. Properties of the Granular Materials Used to Generate the Data Shown in Fig. 6.1*

Reference	Particle diameter d_p (mm)	Angle of internal friction $\phi(\circ)$	Angle of wall friction $\delta(\circ)$	D/d_p	Density $\rho_p\, \nu$ (kg m^{-3})
(a) *Conical hoppers*					
Nguyen et al. (1979)	0.5–1.0	31	24.5	25–63	1500
Zeininger and Brennen (1985)	1.39			20.4–23.5	
Zeininger and Brennen (1985)	0.51			55.5–64.1	
Nedderman (1988)	1.7	27–29		20.6	720
Gu et al. (1993)	0.78	43		57	850
Deming and Mehring (1929)	0.37–0.485	24		20.7–40.8	1244
(b) *Wedge-shaped hoppers*					
Brennen and Pearce (1978)	0.325	25	17	19.5	
Kaza (1982)	0.85–1.25	32.4	15.1	20.4–31.2	1630
Kaza (1982)	"	"	26.5	"	"

* Here D is either the orifice diameter or the width of the exit slot.

reach the hopper exit, was measured. This can be used to construct the velocity field as follows.

Using a spherical coordinate system with origin at the apparent vertex of the hopper (Fig. 6.2), and assuming that the flow is radial and incompressible, the steady state mass balance (A.71) reduces to

$$v_r = -f(\theta)/r^2 \tag{6.2}$$

where v_r is the radial component of velocity, $f(\theta)$ is a function whose form is determined as described later, r is the radial coordinate, θ is the polar angle (Fig. 6.2), and the minus sign has been introduced so that $f > 0$ for downward flow. The data are found to be consistent with the assumptions of incompressible, radial flow, except close to the exit slot and the upper surface of the fill. Near the exit slot, density variation is important; near the upper surface, the flow is not radial. Assuming that the tracer moves with the surrounding material, its radial coordinate R varies with the time t as

$$\frac{\mathrm{d}R}{\mathrm{d}t} = v_r = -\frac{f(\theta)}{R^2} \tag{6.3}$$

or

$$T_R = (R_0{}^3 - R_e{}^3)/(3\, f(\theta)) \tag{6.4}$$

where (R_0, θ) are the initial coordinates of the tracer, and (R_e, θ) are the coordinates corresponding to the point at which the tracer leaves the hopper. By measuring T_R and plotting it versus $R_0{}^3$, the function $f(\theta)$ can be determined. Except close to the centerline, the magnitude of the radial velocity decreases as θ increases (Fig. 6.3). This reflects the retarding effect of wall friction. If the flow is radial and incompressible, the function $f(\theta)$ in (6.2) should be proportional to the mass flow rate. This feature is consistent with the data for two orifice diameters (Fig. 6.3), where the data for the smaller orifice have been scaled by the ratio of mass flow rates of the larger to the smaller orifice.

6.1.3. Density Profiles

Here we discuss the measurements of Hosseini-Ashrafi and Tüzün (1993), who used a γ-ray attenuation technique coupled with a computer-aided tomographic scanner to measure the cross-sectional averaged solids fraction ν_a in horizontal "slices" of a bunker. A binary

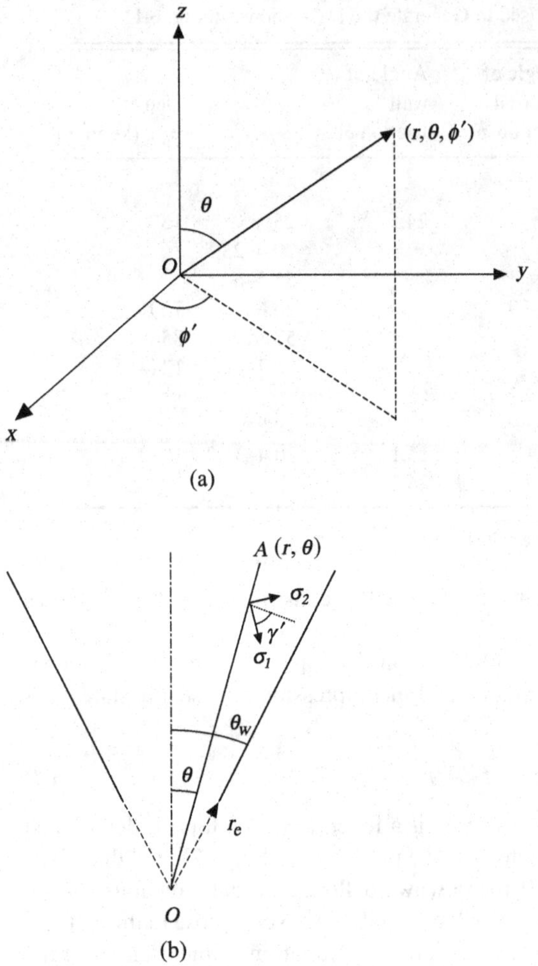

Figure 6.2. (a) Spherical coordinate system, (b) conical hopper with wall angle θ_w. In (b), σ_1 and σ_2 are the major and minor principal stresses, respectively in the r–θ plane, and γ' is the angle measured anticlockwise from the σ_1 axis to the circumferential direction.

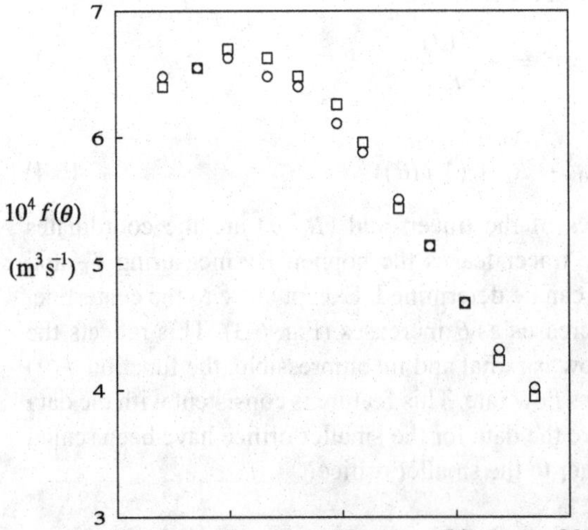

Figure 6.3. Variation of the velocity function $f(\theta) \equiv -r^2\, v_r$ with the polar angle θ for kale seeds flowing through a hopper with aluminium walls. Here v_r is the radial component of velocity, r is the radial coordinate measured from the virtual apex of the hopper, and θ is the angle measured from the centerline (see Fig. 6.2b). Parameter values: wall angle of the hopper $\theta_w = 30°$, diameter of the exit orifice $D = 34$ mm (□), 17 mm (○). The velocity function for $D = 17$ mm has been scaled by the ratio of mass flow rates of the larger to the smaller orifice. (Adapted from Fig. 12 of Cleaver and Nedderman, 1993a.)

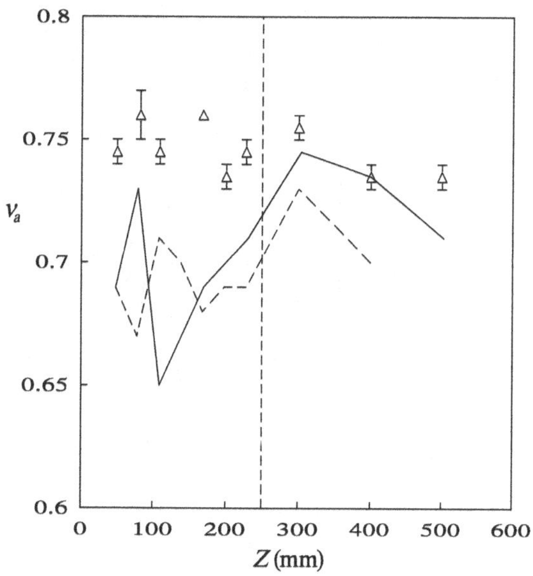

Figure 6.4. Profiles of the cross-sectional averaged solids fraction ν_a for the flow of a binary mixture of poly(methyl methacrylate) spheres of two size ranges (0.85–1.0 mm (mass fraction = 0.75), and 0.125–0.212 mm) through a Perspex bunker: Δ, static bed; —, after 5 s of discharge; - - -, after 60 s of discharge. Here z is the vertical coordinate measured upward from the plane of the exit slot, and the broken vertical line represents the bin–hopper transition. Parameter values: wall angle of the hopper section $\theta_w = 10°$, inner diameter of the bin section = 96 mm, diameter of the exit orifice = 10 mm. (Adapted from Fig. 12 of Hosseini-Ashrafi and Tüzün, 1993.)

mixture of poly(methyl methacrylate) spheres of two size ranges was used to generate the data shown in Fig. 6.4.

In the case of a static bed (the triangles), ν_a is slightly higher in the hopper section than in the bin section. The vertical bars represent the uncertainty in ν_a values, typically about ± 0.005. The solid curve (Fig. 6.4) shows the solids fraction profile measured after allowing the material to discharge for 5.5 s, and then stopping the flow. For the sake of clarity, the error bars have been suppressed here but they are typically about ± 0.01. Comparison of the triangles with the solid curve shows that in the hopper section, the "dynamic" values of ν_a are lower than the static values by about 3–15% of the latter. This may be caused by the dilation of the material during flow. It has been implicitly assumed that stopping the flow does not significantly change the ν_a profile. This can be checked only if the ν_a profile is measured during flow.

The broken curve in Fig. 6.4 shows the profile obtained after the material has been allowed to discharge for 60 s. Comparison of the broken and solid curves suggests that density waves propagate through the bunker as the material flows. Further experiments are needed to ascertain whether this occurs only during the initial stages of flow, or also during the advanced "fully developed" stage.

6.1.4. Flow Patterns

As there is not much information about flow patterns in axisymmetric hoppers, we discuss some observations related to bins. McCabe (1974) studied the flow of sand through a transparent cylindrical bin. Using colored sand (as marker particles) and radioactive tracers, he inferred the flow patterns shown in Fig. 6.5. The initial state corresponds to the silo filled with sand, and the exit slot closed (Fig. 6.5a). When the slot is opened (Fig. 6.5b), a front AB is formed which initially separates static material above it from the flowing material below. McCabe (1974) refers to the region below AB as the free fall zone, but it is not certain that the material is truly in free fall. The zones CDE and GFH represent stagnant material. As the boundaries of these zones were not determined by direct visual observation, it is not certain that CD and GF are straight lines.

As flow progresses, the front AB moves upward and then remains stationary at the position shown in Fig. 6.5c. Simultaneously, another front KL forms and propagates upward. The material is in plug flow above KL, and in converging flow below it. Hence

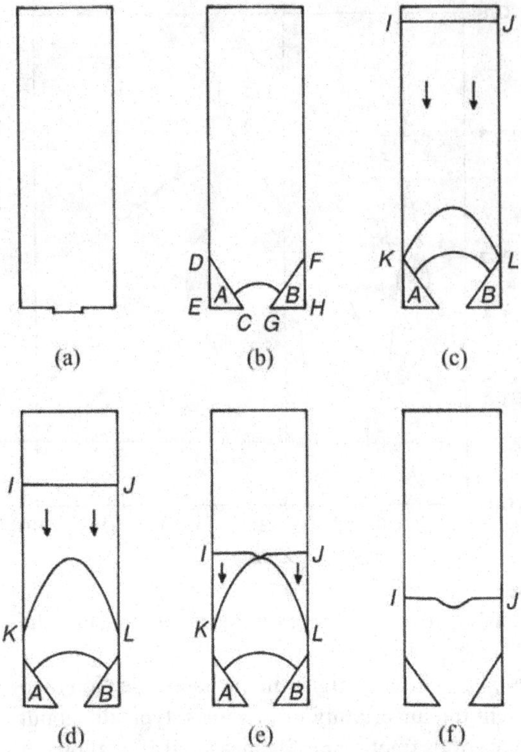

Figure 6.5. Flow patterns observed during the batch discharge of sand from a cylindrical bin. Here IJ represents the free surface of the sand, and the curves KL and AB represent the upper and lower boundaries of the zone of dynamic equilibrium. Above KL, the material is in plug flow; below AB, the vertical velocity increases rapidly. The zones CDE and GFH (Fig. 6.5b) represent stagnant material. (Adapted from Fig. 19 of McCabe, 1974.)

the free surface IJ begins to descend. Further discharge causes KL to move upward and stabilize at the position shown in Fig. 6.5d. McCabe (1974) refers to the region bounded by the curves KL and AB as the zone of dynamic equilibrium. Eventually, a crater or depression is observed when the free surface IJ just touches the apex of KJ (Fig. 6.5e). The crater becomes deeper as IJ descends below this level (Fig. 6.5f).

Consider a special case where the head of material in the bin is maintained approximately constant by feeding material from above. If the upper surface IJ is maintained above the apex of the curve KL, the flow pattern is as shown in Fig. 6.5d. Figure 6.6 shows the streamlines and vertical velocities corresponding to tracers released at different radial positions. (The tangent drawn to a streamline at any point is parallel to the instantaneous velocity vector of the material (Batchelor, 1967, p. 72).) The streamlines remain vertical over a substantial part of the zone of dynamic equilibrium, but the magnitude of the vertical velocity changes after the tracers cross the upper boundary KL. Thus the density of the material in the upper part of this zone differs from the density in the plug flow zone. Indeed, the velocity profiles shown in Fig. 6.6b suggest that the material near the axis of the bin dilates after entering this zone. The stagnant zones at the bottom of the bin cause the streamlines to bend as the tracers approach the lower boundary AB of the zone of dynamic equilibrium. Below AB, the velocity increases rapidly; however, the acceleration (as inferred from Fig. 6.6) is well below that corresponding to free fall.

It is instructive to compare the patterns shown in Fig. 6.5 with those observed by Watson (1993) (cited in Nedderman, 1995). Watson used a hemi-cylindrical bunker whose flat face was made of glass. The flow patterns observed during the batch discharge of polypropylene pellets are sketched in Fig. 6.7. When the exit slot is opened, an interface separating stagnant material above from flowing material below is formed (Fig. 6.7a). The interface propagates upward until it touches the free surface (Figs. 6.7b,c). The free surface then begins to descend, and the shape of the interface changes as shown in Figs. 6.7d–h.

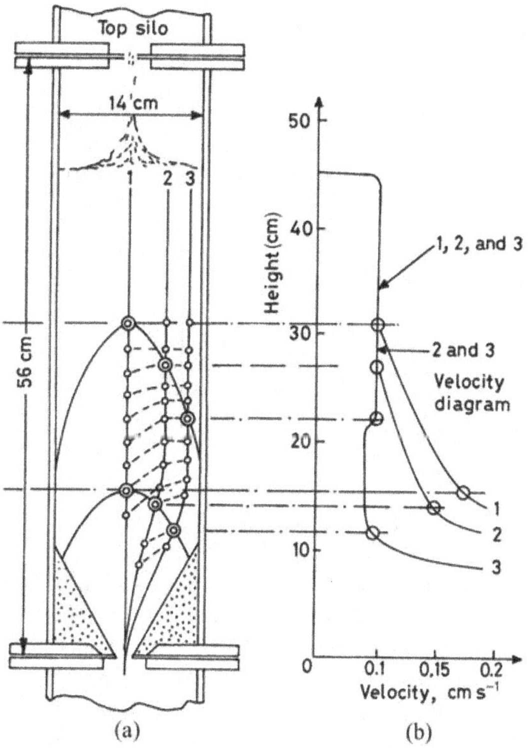

Figure 6.6. (a) Streamlines, and (b) profiles of the vertical velocity corresponding to tracers released at the positions 1, 2, and 3 in figure (a). (Reproduced from Fig. 17 of McCabe, 1974, with permission from the Institution of Civil Engineers, London.)

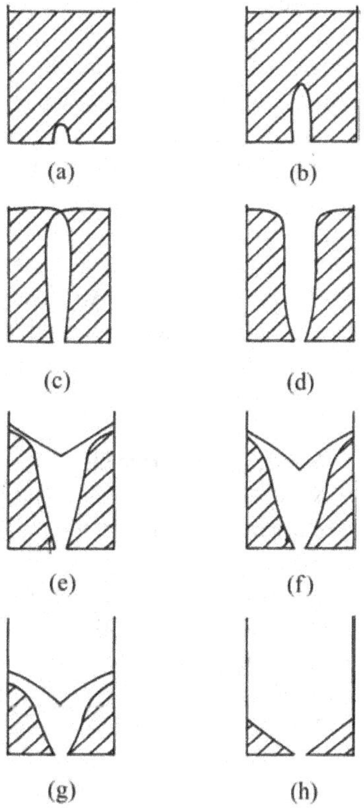

Figure 6.7. Flow patterns observed during the batch discharge of polypropylene pellets from a hemicylindrical bin with a semicircular orifice. The curved surface was made of steel, and the flat surface was made of glass. The pellets were cylinders of diameter 5 mm and length 0.5–4 mm. Parameter values: bin diameter = 650 mm, orifice diameter = 65 mm, initial height of the free surface relative to the base of the bin = 940 mm. Data of Watson (1993). (Reprinted from Fig. 1 of Nedderman, 1995, with permission from Elsevier Science.)

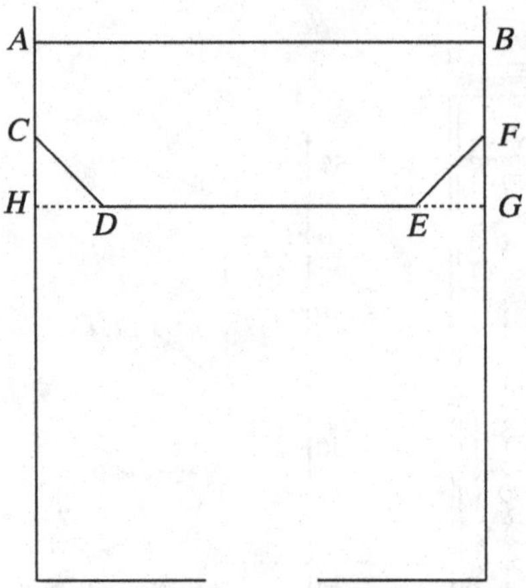

Figure 6.8. Formation of "banks" FE and CD due to the velocity gradient near the wall of the bin. Here AB is the initial location of the upper surface of the fill, and $CDEF$ represents the shape of the upper surface during flow.

The pattern shown in Fig. 6.5d, with a region of plug flow above a converging flow zone, is missing in the experiments of Watson (1993)(Fig. 6.7). This difference can probably be reconciled by McCabe's observation that the former pattern occurs only when the free surface is above the apex of the curve KL (Fig. 6.5d). For the flow of sand with a size range 2–5 mm through a bin of diameter $D_b = 460$ mm, the height of the apex of KL relative to the base of the bin was about 2.2 D_b (McCabe, 1974). In the experiments of Watson (1993), the maximum height of the free surface relative to the base of the bin was 1.4 D_b. Even though the granular materials and the diameters of the bins differed in the two sets of experiments, it is possible that the value of H in Watson's experiments was not large enough for the pattern shown in Fig. 6.5d to be observed.

Returning to the pattern shown in Fig. 6.5d, let us consider the velocity profile in the "plug flow" region above KL. Though the experiments of McCabe (1974) suggest that the material is in plug flow, measurements in channels of rectangular cross section show that there are thin shear layers adjacent to the walls (Nedderman and Laohakul, 1980; Natarajan et al., 1995). Within the shear layer, the velocity increases in magnitude from its value at the wall to the plug velocity at the edge of the layer. The velocity gradient causes an initially flat free surface AB (Fig. 6.8) to assume the form $CDEF$ during flow. The region FEG is called a "bank" (Munch-Andersen, 1987), and its presence can be readily inferred by visual observation. Munch-Andersen (1987) found that the thickness of the bank or shear layer was approximately independent of the bin diameter. In his experiments, it was about 20 mm for barely and 5 mm for wheat. As the dimensions of the two types of grains are comparable, the difference in the thickness may be caused by differences in material properties.

6.1.5. Stress Profiles

Most of the recent measurements have been confined to bins. Let us first discuss the measurements of Munch-Andersen and Askegaard (1993), who studied the flow of Leighton Buzzard sand through a cylindrical bin. Both shear and normal stresses acting on the bin wall were measured using strain gauge based load cells. The dimensions of the active face of the load cell were 100 mm × 100 mm, which were large compared to the particle diameter (0.5–1 mm) (J. Munch-Andersen, private communication 1997), but small compared to the height of the silo (4.9 m).

Each of the panels in Fig. 6.9 shows the variation of the normal stress with time, at a particular location on the wall. The height of the load cell above the base of the bin H is given on the right-hand side of the figure. During flow, the stress fluctuates at all the locations, but the amplitude of fluctuations is more in the upper part of the silo. As in the case of bins of rectangular cross section (Fig. 4.15), there are rapid oscillations

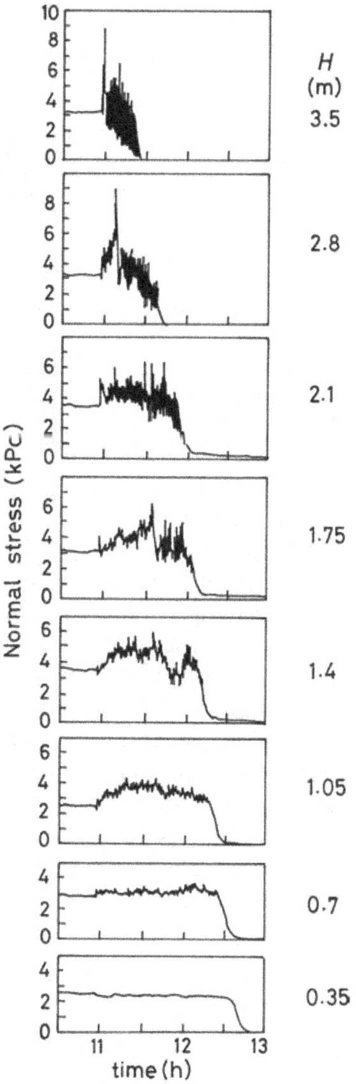

Figure 6.9. Profiles of the normal stress at various locations on the wall of a cylindrical bin. Discharge of sand began at 11.00 hours. The centres of the load cells are aligned vertically, and the height H of each cell above the base of the silo is given on the right of the panel. Silo dimensions: inner diameter = 700 mm, height = 4.9 m. The initial height of fill was about 4.3 m. (Reprinted from Fig. 6 of Munch-Andersen and Askegaard, 1993, with permission from Dr. J. Munch-Andersen.)

superimposed on slower ones. The rapid oscillations are usually attributed to stick-slip flow (Richards, 1977; Tüzün and Nedderman, 1985a); the reason for the slower oscillations is not known. It should be noted that the mode of filling influences the normal stress profiles. If the sand is introduced into the bin through a device which distributes it across the cross section of the bin, the stress profile is as shown in Fig. 6.10. Alternatively, the sand can be introduced through a tube at the top of the bin, leading to the profile shown in Fig. 6.9 (see the panel corresponding to a height of 2.1 m). The two profiles are quite different, with the distributed inlet (Fig. 6.10) leading to lower peak and average stresses. Perhaps the distributed inlet causes the sand to shear more than in the case of the "point" (tube) inlet. Thus the material may be closer to a critical state in the former case, and hence the changes in density and stress may be less pronounced.

Though Munch-Andersen and Askegaard (1993) do not present profiles of the shear stresses exerted on the wall during flow, they state that the mean shear stresses remain close to the values obtained after filling. This behavior can be understood by noting that in the region of fully developed flow, the shear stress supports the weight of the material. Hence

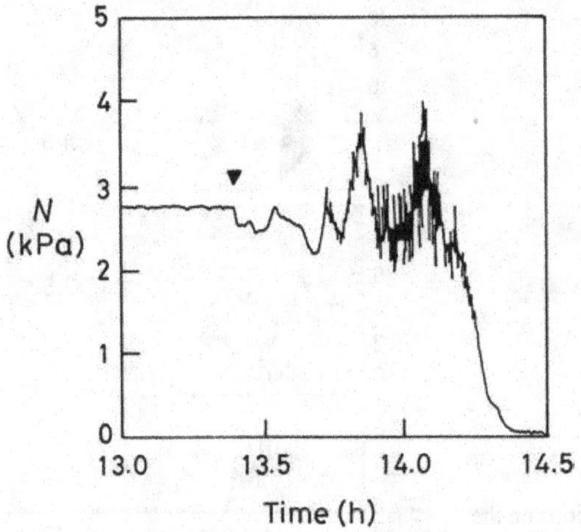

Figure 6.10. Profile of the normal stress N for a load cell located at a height of 2.1 m above the base of the bin. The triangle indicates the start of discharge. A distributed inlet was used to fill the silo. Parameter values are as in Fig. 6.9. (Reproduced from Fig. 9 of Munch-Andersen and Askegaard, 1993, with permission from Dr. J. Munch-Andersen.)

if static and flowing densities do not differ significantly, the mean shear stress during flow will also remain close to its static value. On the other hand, the mean normal stress during flow tends to be higher than the static value at several locations (Fig. 6.9). This implies that the mean angle of wall friction decreases during flow. More data on bins of different aspect ratios are needed to examine whether this is a general result.

Consider the effect of the bin diameter on the normal stress profiles. Figure 6.11a shows the profile at a particular location on the wall of a concrete bin. The corresponding profile at a geometrically similar position in a 1/10th scale model is shown in Fig. 6.11b. The stresses measured at the end of the filling period, and also the peak stresses measured during discharge, scale approximately with the bin diameter. This is in accord with the frictional theory discussed in Chapter 4.

As noted by Munch-Andersen (1987), the normal stress measured during flow is higher than the static value for the model (Fig. 6.11b), whereas it fluctuates about the static value

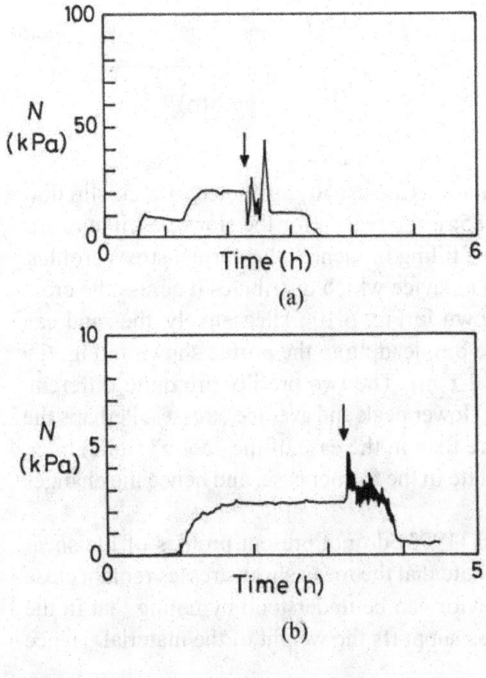

Figure 6.11. Profiles of the normal stress N measured on the wall of (a) a full-scale bin made of concrete, and (b) a 1/10th scale model made of Araldite, during the flow of barley. The arrows indicate the start of discharge. (Reproduced from Fig. 1 of Munch-Andersen, 1987, with permission from The Institution of Engineers, Australia.)

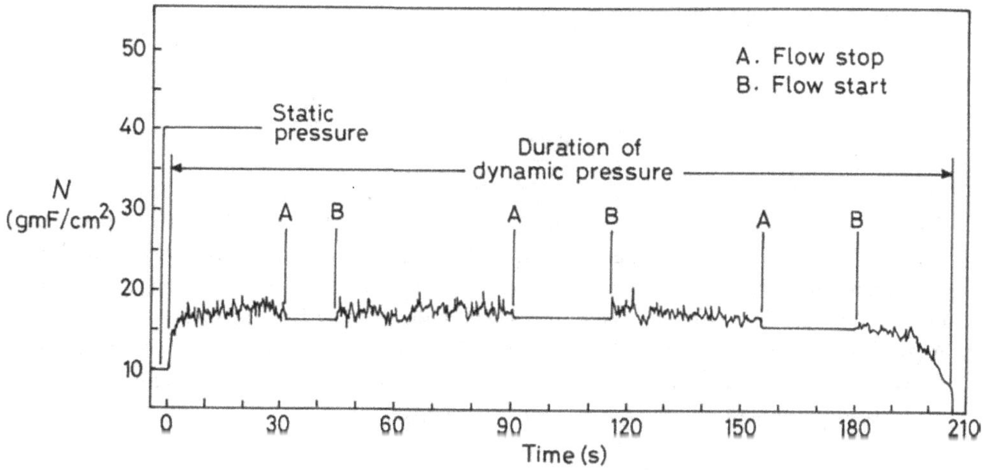

Figure 6.12. Profile of the normal stress N measured on the wall of a cylindrical bunker during the discharge of sand. The material is at rest between points A and B. The normal stress was measured using a pressure-sensitive radio pill, located 164 mm above the plane of the exit orifice. Parameter values: particle diameter = 0.3–0.6 mm, diameter of the exit orifice = 26 mm, wall angle of the hopper section = 20°. (Reprinted from Fig. 14 of Rao and Venkateswarlu, 1974, with permission from Elsevier Science.)

for the full-scale bin (Fig. 6.11a). Although the silos are geometrically similar, the materials of construction differ in the two cases. Further, the maximum deviation of the cross section from a circle was about 1% of the radius (Munch-Andersen, 1987). Hence the absolute size of the wall imperfections is much larger in the full-scale bin. These factors may all contribute to the observed differences in the profiles.

A subtle point regarding static and dynamic stresses needs clarification. The static stresses shown in Figs. 6.9 and 6.10 are the values measured after filling the bin. Although the dynamic stresses measured during flow often exceed the static stresses, this is not the case for "stop–start" operation (Fig. 6.12). When the flow is stopped (the points marked A in Fig. 6.12), the normal stress does not revert to the static value, but remains close to the dynamic value. When the flow is resumed (the points marked B in Fig. 6.12), the normal stress oscillates about the value corresponding to point A. Thus the dynamic stress does not increase significantly, unlike the first time the flow is started after filling the bunker. This behavior suggests that (i) the state of the material immediately after filling differs from the state attained during flow, and (ii) the latter state is preserved to some extent even when the flow is stopped. Further experiments are required to verify this conjecture.

6.2. THEORY FOR STEADY, AXISYMMETRIC FLOW THROUGH A HOPPER

It is convenient to use a spherical coordinate system with origin at the virtual apex of the hopper (Fig. 6.2). For axisymmetric flow, all the variables depend only on r and θ, and the velocity field is given by

$$v_r = v_r(r, \theta); \quad v_\theta = v_\theta(r, \theta); \quad v_{\phi'} = 0 \tag{6.5}$$

Introducing the rate of deformation tensor, defined in the compressive sense by

$$\mathbf{C} \equiv -\frac{1}{2}\,(\nabla \mathbf{v} + (\nabla \mathbf{v})^{\mathrm{T}}) \tag{6.6}$$

(6.5) and (A.76) imply that

$$C_{r\phi'} = -\frac{1}{2r\,\sin\theta}\frac{\partial v_r}{\partial \phi'} = 0; \quad C_{\theta\phi'} = -\frac{1}{2r\,\sin\theta}\frac{\partial v_\theta}{\partial \phi'} = 0$$

$$C_{\phi'\phi'} = -\frac{(v_r + v_\theta\,\cot\theta)}{r} \tag{6.7}$$

As $C_{r\phi'} = 0$, $C_{\theta\phi'} = 0$, and $C_{\phi'\phi'} \neq 0$ in general, the ϕ' direction coincides with a principal axis of **C**. As discussed in Appendix F, if **C** depends only on the stress tensor $\boldsymbol{\sigma}$ and the solids fraction ν, the principal axes of σ and **C** are aligned with each other. Hence one of the principal axes of $\boldsymbol{\sigma}$, say the σ_3 axis coincides with the ϕ' direction, and $\sigma_{\theta\phi'} = 0,\ \sigma_{r\phi'} = 0$. As the principal axes are mutually orthogonal, the σ_1 and σ_2 axes must lie in the r–θ plane. Consider the mass and momentum balances. The ϕ' component of the momentum balance is identically satisfied, and the other balances are given by

mass balance

$$\frac{\partial\rho}{\partial t} + \frac{1}{r^2}\frac{\partial(\rho\, r^2\, v_r)}{\partial r} + \frac{1}{r\,\sin\theta}\frac{\partial(\rho\, v_\theta\,\sin\theta)}{\partial\theta} = 0 \tag{6.8}$$

momentum balance (r component)

$$\rho\left(\frac{\partial v_r}{\partial t} + v_r\frac{\partial v_r}{\partial r} + \frac{v_\theta}{r}\frac{\partial v_r}{\partial\theta} - \frac{v_\theta^2}{r}\right) + \frac{\partial\sigma_{rr}}{\partial r} + \frac{1}{r}\frac{\partial\sigma_{\theta r}}{\partial\theta}$$
$$+\frac{2\,\sigma_{rr} - \sigma_{\theta\theta} - \sigma_{\phi'\phi'} + \sigma_{r\theta}\,\cot\theta}{r} + \rho\, g\,\cos\theta = 0 \tag{6.9}$$

momentum balance (θ component)

$$\rho\left(\frac{\partial v_\theta}{\partial t} + v_r\frac{\partial v_\theta}{\partial r} + \frac{v_\theta}{r}\frac{\partial v_\theta}{\partial\theta} + \frac{v_r\, v_\theta}{r}\right) + \frac{\partial\sigma_{r\theta}}{\partial r} + \frac{1}{r}\frac{\partial\sigma_{\theta\theta}}{\partial\theta}$$
$$+\frac{3\,\sigma_{r\theta} + (\sigma_{\theta\theta} - \sigma_{\phi'\phi'})\,\cot\theta}{r} - \rho\, g\,\sin\theta = 0 \tag{6.10}$$

where $\rho = \rho_p\,\nu$ is the density of the granular material, ρ_p is the particle density, ν is the volume fraction of solids, v_r and v_θ are the velocity components in the r and θ directions, respectively, and g is the acceleration due to gravity. In (6.9) and (6.10), the stress components are taken to be defined in the compressive sense. As in the case of flow through wedge-shaped hoppers and bunkers (Chapters 4 and 5), (6.8)–(6.10) must be supplemented by constitutive equations. Using the plasticity model developed in Chapter 2, the velocity equations for axisymmetric flow are given by the coaxiality condition and the flow rule. Following the procedure used in §3.3 and §3.8 to deduce these equations in cylindrical coordinates from the corresponding equations in Cartesian coordinates, we obtain, in spherical coordinates

coaxiality condition

$$\sin 2\gamma'\left(\frac{1}{r}\frac{\partial v_\theta}{\partial\theta} + \frac{v_r}{r} - \frac{\partial v_r}{\partial r}\right) + \cos 2\gamma'\left(\frac{1}{r}\frac{\partial v_r}{\partial\theta} + \frac{\partial v_\theta}{\partial r} - \frac{v_\theta}{r}\right) = 0 \tag{6.11}$$

flow rule

$$\cos 2\gamma'\left(\frac{1}{r}\frac{\partial v_\theta}{\partial\theta} + \frac{v_r}{r} + \frac{\partial v_r}{\partial r}\right) + \sin\nu_d\left(\frac{1}{r}\frac{\partial v_\theta}{\partial\theta} + \frac{v_r}{r} - \frac{\partial v_r}{\partial r}\right) = 0 \tag{6.12}$$

where ν_d is the angle of dilation, defined by

$$\sin \nu_d = -(c_1 + c_2)/(c_1 - c_2) \tag{6.13}$$

and c_1 and c_2 are the major and minor principal compressive rates of deformation, respectively, in the r–θ plane.

Finally, the stress components $\sigma_{rr}, \sigma_{r\theta}$, and $\sigma_{\theta\theta}$ can be expressed in terms of the Sokolovskii variables $\sigma \equiv (\sigma_1 + \sigma_2)/2$, $\tau \equiv (\sigma_1 - \sigma_2)/2$, and the angle γ' (Fig. 6.2b), to obtain

$$\sigma_{\theta\theta} = \sigma + \tau \cos 2\gamma'; \quad \sigma_{r\theta} = \sigma_{\theta r} = -\tau \sin 2\gamma'; \quad \sigma_{rr} = \sigma - \tau \cos 2\gamma' \tag{6.14}$$

An interesting difference between the balances for plane and axisymmetric flows is that an additional stress component $\sigma_{\phi'\phi'}$ occurs in the latter case. Hence the constitutive equations for plane flow ((2.103)–(2.105)) cannot be used without additional assumptions. Some assumptions, and also an alternative approach based on a three-dimensional flow rule, are described below.

6.2.1. The Haar–von Karman Hypothesis

In their paper on three-dimensional stress and deformation fields in static media, Haar and von Karman (1909) considered a special case where two of the principal stresses were equal. This assumption is called the *Haar–von Karman hypothesis*, and can be used in the present problem as discussed below.

The principal stresses are given by σ_1, σ_2, and $\sigma_3 = \sigma_{\phi'\phi'}$, with σ_1 chosen to be $\geq \sigma_2$. Thus there are three possible ways of implementing the hypothesis: (i) $\sigma_1 = \sigma_2$, (ii) $\sigma_1 = \sigma_3 = \sigma_{\phi'\phi'}$, and (iii) $\sigma_2 = \sigma_3 = \sigma_{\phi'\phi'}$. Case (i) can be ruled out as follows. For a hopper with rough walls, the shear stress $\sigma_{r\theta}(r, \theta = \theta_w)$ at the wall is nonzero in general. As the σ_1 and σ_2 axes lie in the r–θ plane, a nonzero value of $\sigma_{r\theta}$ implies that $\sigma_1 \neq \sigma_2$. Consider Cases (ii) and (iii). At the centerline, the condition of zero shear stress implies that $\sigma_{r\theta} = 0$. Assuming that $\tau \neq 0$, it follows from (6.14) that $\gamma' = 0$ or $\pi/2$. Thus one of the principal stress axes must be vertical. Further, the other two axes must lie in the horizontal plane, and must be equal on account of symmetry. Hence at the centerline, Case (ii) corresponds to a passive state of stress, with the σ_1 axis horizontal, and Case (iii) to an active state of stress, with the σ_1 axis vertical. As two of the principal stresses are equal at the centerline, this result provides a justification for the Haar–von Karman hypothesis. At other locations with $\theta \neq 0$, the basis for the hypothesis is not clear. Nevertheless, it has been widely used (Savage, 1967; Williams, 1977; Nguyen et al., 1979; Cleaver and Nedderman, 1993b).

As discussed in §3.3, the passive state of stress is usually assumed to be appropriate for downward flow through the hopper. However, it appears difficult to rigorously establish this result for the case of three-dimensional hopper flow. Here we shall assume that

$$\sigma_1 = \sigma_3 = \sigma_{\phi'\phi'} \tag{6.15}$$

Assuming that $\sigma_{\phi'\phi'} = \sigma_1$, the number of dependent variables is equal to that for plane flow. With $\sigma_1 = \sigma_3$, the yield condition $F(\sigma_1, \sigma_2, \sigma_3, \nu) = 0$ can be rewritten as $f(\sigma, \tau, \nu) = 0$. Assuming that the yield condition can be solved for τ as a function of σ and ν, there are five dependent variables σ, γ', v_r, v_θ, and ν. Most of the analyses have been confined to incompressible flow, in which case the mass and momentum balances (6.8)–(6.10) must be supplemented by one more equation. Jenike (1961, 1964a), Nguyen et al. (1979), and Cleaver and Nedderman (1993b) used the coaxiality condition (6.11) as the additional equation.

Some solutions obtained using (6.8)–(6.11), (6.14), the assumption $\sigma_3 = \sigma_{\phi'\phi'} = \sigma_1$, and the critical state approximation $\tau = \sigma \sin\phi$ (see §3.2.1) are discussed below. As noted in Chapter 3, the latter approximation is equivalent to assuming that the material satisfies the Mohr–Coulomb yield condition for a cohesionless material.

(a) *The smooth wall, radial gravity (SWRG) problem for incompressible flow*
Proceeding as in §3.3, we obtain (Savage, 1965; Davidson and Nedderman, 1973)

$$\overline{\sigma} = \frac{1}{1-\sin\phi}\left(\frac{\xi}{2k-1} - \frac{\overline{A}^2}{(k+2)\xi^4}\right) + c_1\,\xi^{2k} \tag{6.16}$$

where

$$\overline{\sigma} \equiv \sigma/(\rho\, g\, r_e); \quad \xi \equiv r/r_e; \quad k \equiv (2\sin\phi)/(1-\sin\phi) \tag{6.17}$$

c_1 is an integration constant, and r_e is the radial coordinate corresponding to the hopper exit (Fig. 6.2b). The dimensionless constant $\overline{A}$ in (6.16) is related to the radial component of velocity v_r by

$$u \equiv v_r/\sqrt{g\, r_e} = -\overline{A}/\xi^2 \tag{6.18}$$

The constants $\overline{A}$ and c_1 can be determined using the traction-free boundary conditions (3.26) to obtain

$$\overline{A}^2 = \left(\frac{k+2}{2k-1}\right)\left(\frac{1-\xi_1{}^{-(2k-1)}}{1-\xi_1{}^{-(2k+4)}}\right)$$

$$c_1 = -\frac{1}{(1-\sin\phi)(2k-1)}\left(\frac{\xi_1{}^{-(2k-1)} - \xi_1{}^{-(2k+4)}}{1-\xi_1{}^{-(2k+4)}}\right) \tag{6.19}$$

where $\xi_1 \equiv r_1/r_e$ denotes the radial coordinate corresponding to the upper surface of the fill.

It is instructive to compare the SWRG solutions for conical and wedge-shaped hoppers. For typical values of k, comparison of the stress profiles (6.16) and (3.25) shows that both the inertial term and the term representing the effect of the upper boundary condition decay more rapidly in the axisymmetric case than in the case of plane flow. Using (6.18) and (6.19), we obtain

$$V_D = \frac{8(1-\cos\theta_w)}{(2\sin\theta_w)^{5/2}}\,\overline{A} \tag{6.20}$$

For the special case of a deep hopper with $\xi_1 \gg 1$, (6.19) and (6.20) imply that

$$V_D \approx \frac{8(1-\cos\theta_w)}{(2\sin\theta_w)^{5/2}}\left(\frac{k+2}{2k-1}\right)^{1/2}, \quad \xi_1 \gg 1 \tag{6.21}$$

Using (6.21) and (3.31), the ratio of the mass flow rates for conical and wedge-shaped hoppers is given by

$$\frac{V_{D,\text{conical}}}{V_{D,\text{wedge}}} = \frac{2(1-\cos\theta_w)}{\theta_w\,\sin\theta_w}\left(\frac{k-1}{2k-1}\right)^{1/2}, \quad \xi_1 \gg 1 \tag{6.22}$$

provided the same value of ϕ is used in (6.17) and (3.24). For a typical value of $\phi = 30°$, $V_{D,\text{conical}}/V_{D,\text{wedge}}$ increases slightly from 0.58 to 0.60 as θ_w increases from 10° to 40°. Thus conical hoppers have lower mass flow rates than wedge-shaped hoppers, in accord with the data shown in Fig. 6.1. For $\theta_w = 10°$, the data for coarse glass beads (see the open and filled triangles in Fig. 6.1) imply that $V_{D,\text{conical}}/V_{D,\text{wedge}} = 0.62$, which is close to the SWRG estimate of 0.58.

(b) *The solution of Nguyen et al. (1979)*
This is an approximate solution which accounts for wall roughness and vertical gravity, using the approach proposed by Brennen and Pearce (1978) for a wedge-shaped hopper (see §3.5.1).

Omitting terms involving cubes and higher powers of θ, the solution is given by

$$u = -\frac{\overline{A}}{\xi^2}\left(1 - \frac{3\gamma_w}{\theta_w}\theta^2\right); \quad v \equiv \frac{v_\theta}{\sqrt{g\,r_e}} = 0$$

$$\gamma' = \gamma_w \left(\frac{\theta}{\theta_w}\right); \quad \overline{\sigma} = \overline{\sigma}_o(\xi) + \overline{\sigma}_2(\xi)\left(\frac{\theta}{\theta_w}\right)^2 \tag{6.23}$$

where

$$\overline{A}^2 = \left(\frac{k'+2}{2\,k'-1}\right)\left(1 + \frac{\theta_w\,(\sin\phi\,(14\,\gamma_w + 5\,\theta_w) - \theta_w)}{2\,(1+\sin\phi)}\right) \times \left(\frac{1-\xi_1{}^{-(2\,k'-1)}}{1-\xi_1{}^{-(2\,k'+4)}}\right)$$

$$k' = \frac{2\,\sin\phi}{1-\sin\phi}\left(1 + \frac{\gamma_w}{\theta_w}\right)$$

$$\overline{\sigma}_0(\xi) = \frac{1}{1-\sin\phi}\left(\frac{\xi}{2\,k'-1} - \frac{\overline{A}^2}{(k'+2)\,\xi^4}\right) + c_1\,\xi^{2\,k'}$$

$$\overline{\sigma}_2(\xi) = \left(\frac{\sin\phi\,\gamma_w\,(4\,\theta_w + 3\gamma_w)}{\cos^2\phi\,(2\,k'-1)} + \frac{\theta_w{}^2}{2\,(1+\sin\phi)}\right)\xi$$
$$+\frac{\sin\phi\,\gamma_w\,(\theta_w - 3\,\gamma_w)\,\overline{A}^2}{\cos^2\phi\,(k'+2)\,\xi^4}$$
$$+\frac{\sin\phi\,\gamma_w\,((2k'+3)\,\theta_w + 3\gamma_w)}{1+\sin\phi}\,c_1\,\xi^{2\,k'}$$

$$c_1 = \frac{-1}{(1-\sin\phi)\,(2\,k'-1)}$$
$$\times\left[1 + \frac{\theta_w}{2\,(1+\sin\phi)}\,(\sin\phi\,(\gamma_w\,(6-4\,k') + 5\,\theta_w) - \theta_w)\right] \tag{6.24}$$

Using the above expressions and (6.1), the dimensionless mass flow rate is given by

$$V_D = \frac{8\,E(\gamma_w, \theta_w)\,\overline{A}}{(2\,\sin\theta_w)^{5/2}} \tag{6.25}$$

where

$$E(\gamma_w,\,\theta_w) \equiv (1-\cos\theta_w) \times \left[1 - \frac{3\,\gamma_w}{\theta_w\,(1-\cos\theta_w)}\,(2\,\theta_w\,\sin\theta_w + (2-\theta_w{}^2)\,\cos\theta_w - 2)\right]$$

and $\overline{A}$ follows from the first of (6.24).

(c) *Comparison of measured and predicted mass flow rates*
The curves labeled *SWRG*, *RTBN*, and *N* in Fig. 6.13 show the V_D values obtained by using the *SWRG* analysis (see (6.20)), the Rose–Tanaka–Beverloo–Nedderman correlation (*RTBN*) (3.3), and the analysis of Nguyen et al. (1979) (see (6.25)), respectively. The

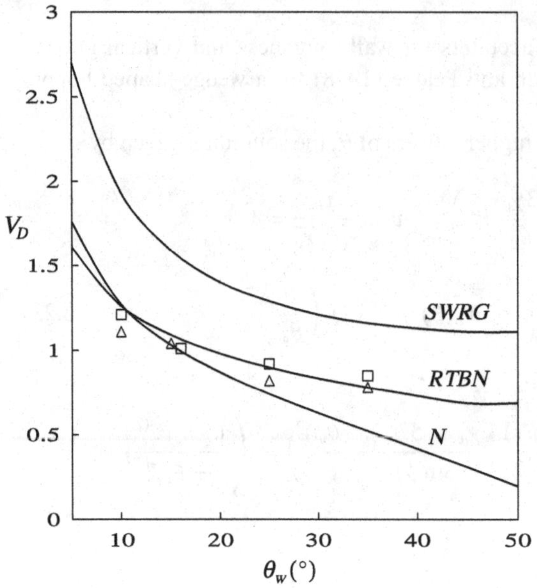

Figure 6.13. Variation of the dimensionless mass flow rate V_D (defined by (6.1)) with the wall angle θ_w for flow through a conical hopper: *SWRG*, the smooth wall, radial gravity estimate (6.21) for a deep hopper; *RTBN*, the Rose–Tanaka–Beverloo–Nedderman correlation (3.3) with $\beta = 45°$; *N*, the estimate of Nguyen et al. (1979) (6.25); □, data of Nguyen et al. (1979) for sand ($d_p = 0.69$ mm, $\phi = 31°$, $\delta = 24.5°$, $D/d_p = 29$–48); △, data of Zeininger and Brennen (1985) for glass beads ($d_p = 0.51$ mm, $D/d_p = 56$–64). The value of d_p is required only for the Rose–Tananka–Beverloo–Nedderman correlation. The curves are shown for $\phi = 31°$, $\delta = 24.5°$, $D/d_p = 48$.

calculations are done for parameters appropriate to the data of Nguyen et al. (1979) (the squares in Fig. 6.13). The *SWRG* analysis overestimates the mass flow rates considerably; the error is 60–36% for wall angles in the range 10–35°. At $\theta_w = 10°$, the error obtained with the analysis of Nguyen et al. (1979) is only 5%. However, at $\theta_w = 35°$, the error is 38%, which is comparable to error obtained with the *SWRG* analysis. As the former solution omits cubes and higher powers of θ, it is likely to give inaccurate results for large values of θ_w. The *RTBN* correlation (3.3) performs surprisingly well; it gives estimates which are within 2–8% of the measured values. The Beverloo correlation (3.1) gives $V_D = 0.69$, which grossly underestimates the mass flow rate at small wall angles. The triangles in Fig. 6.13 represent the data of Zeininger and Brennen (1985) for glass beads. In this case also, the *RTBN* correlation performs well. The data cannot be compared with the predictions of the *SWRG* model and the model of Nguyen et al. (1979)) as the values of ϕ and δ have not been reported by Zeininger and Brennen (1985).

6.2.2. The Radial Stress and Velocity Fields for the Mohr–Coulomb Yield Condition and the Haar–von Karman Hypothesis

If coaxiality holds, the axisymmetric velocity field (6.5) implies that two of the principal stresses, say, σ_1 and σ_2, are in the r–θ plane. As discussed in §6.2.1, the Haar–von Karman hypothesis can be applied to the hopper problem by requiring that $\sigma_3 = \sigma_{\phi'\phi'} = \sigma_1 \geq \sigma_2$. Using the Mohr–Coulomb yield condition (3.10), the stress components are given by (see (6.14) and (6.15))

$$\begin{aligned} \sigma_{rr} &= \sigma\,(1 - \sin\phi\ \cos 2\gamma'); \quad & \sigma_{\theta\theta} &= \sigma\,(1 + \sin\phi\ \cos 2\gamma') \\ \sigma_{r\theta} &= -\sigma\ \sin\phi\ \sin 2\gamma'; \quad & \sigma_{\phi'\phi'} &= \sigma_1 = \sigma\,(1 + \sin\phi) \end{aligned} \tag{6.26}$$

where $\sigma \equiv (\sigma_1 + \sigma_2)/2$ is the mean stress in the r–θ plane, and the angle γ' is defined in Fig. 6.2b.

Consider the radial stress and velocity fields for the special case of incompressible flow. These fields are analogs of the radial fields discussed in §3.5.2 for flow through wedge-shaped hoppers. Referring to Fig. 6.2b and assuming that the flow is radial and incompressible, the mass balance (6.8) implies that

$$v_r = -f(\theta)/r^2; \quad v_\theta = 0; \quad v_{\phi'} = 0 \tag{6.27}$$

The radial velocity field (6.27) implies that the inertial terms in the θ component of the momentum balance vanish identically, and that the inertial terms in the r component of the momentum balance are proportional to $f(\theta)^2/r^5$. Hence for large values of r, the inertial terms can be omitted in comparison with the body force terms.

As in the case of plane flow (see §3.5.2), the inertia-free momentum balances admit a particular solution of the form (Jenike, 1961, 1964a; Cleaver and Nedderman, 1993b)

$$\sigma_* \equiv \sigma/(\rho\, g\, r_e) = b_*(\theta)\, \xi\,; \quad \gamma' = g_*(\theta) \tag{6.28}$$

where $\rho = \rho_p\, \nu$ is the particle density, g is the acceleration due to gravity, r_e is the radial coordinate corresponding to the edge of the exit slot (Fig. 6.2b), and $\xi \equiv r/r_e$. The functions $b_*(\theta)$ and $g_*(\theta)$ are determined by integrating the following differential equations, which are obtained by substituting (6.28) into the inertia-free momentum balances

$$\frac{\mathrm{d}}{\mathrm{d}\theta}(b_*\, \sin\phi\, \sin 2g_*) = b_*\,[1 - \sin\phi\,(4\, \cos 2g_* + \sin 2g_*\, \cot\theta + 1)] + \cos\theta$$

$$\frac{\mathrm{d}}{\mathrm{d}\theta}[b_*\,(1 + \sin\phi_*\, \cos 2g_*)] = b_*\, \sin\phi\,[4\, \sin 2g_* + \cot\theta\,(1 - \cos 2g_*)] \tag{6.29}$$

For a passive solution, the boundary conditions are given by

$$g_*(0) = 0; \quad g_*(\theta_w) = \gamma_w \tag{6.30}$$

where γ_w is given by the passive root γ_{wp} (see (3.36)).

Using the g_* profile and the coaxiality condition (6.11), the incompressible radial velocity field is given by (6.27), with

$$f(\theta) = f(0)\exp\left(-3\int_0^\theta \tan(2\, g_*(\omega))\, \mathrm{d}\omega\right) \tag{6.31}$$

An interesting implication of (6.31) is discussed below.

It is instructive to consider the initial value problem defined by (6.29) and the initial conditions

$$g_*(0) = 0; \; b_*(0) = b_{*0} \tag{6.32}$$

where b_{*0} is a positive number. As shown in Fig. 6.14, $g_*(\theta) = \pi/4$ at some value of $\theta = \theta_c$ for small values of $b_{*0} \equiv b_*(\theta = 0)$. The integrand in (6.31) becomes unbounded at $\omega = \theta_c$, and it can be shown that the integral is also unbounded (Problem 6.2). Hence $v_r(r, \theta_c) = f(\theta_c)/r^2 = 0$, as noted by Jenike (1964a). It is tempting to assume that the material is at rest for $\theta > \theta_c$, even though this is not the only possibility. If this assumption is accepted, the radial velocity field predicts the occurrence of funnel flow or core flow for $\theta_w > \theta_c$. In such a case, the boundary between stagnant and flowing regions is given by the radial line $\theta = \theta_c$. The dependence of θ_c on the material parameters is discussed below.

The value of the constant b_{*0} can be determined iteratively to ensure that the second of (6.30) is satisfied at $\theta = \theta_w$. Alternatively, for each value of b_{*0} and δ, the equations can be integrated from $\theta = 0$ until at some value of θ, say θ_*, $g_*(\theta_*) = \gamma_w$. The stress field so determined corresponds to a hopper with $\theta_w = \theta_*$. For a fixed value of the angle of internal friction ϕ, the maximum value of g_*, i.e., $g_{\max}$, depends only on b_{*0}. Hence the condition $g_{\max} = \pi/4$ at $\theta = \theta_c$ fixes the value of b_{*0}, implying that θ_c depends only on ϕ. In particular, it does not depend on the angle of wall friction δ. For $\phi = 30°$, the curve labeled 2 in Fig. 6.14 shows that $\theta_c \approx 12°$.

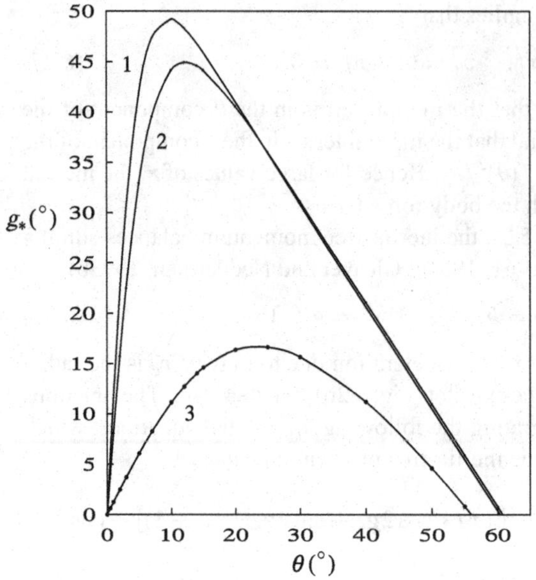

Figure 6.14. Profile of $g_*(\theta)$ for the radial stress field based on the Haar–von Karman hypothesis and the Mohr–Coulomb yield condition and (see (6.29)). Parameter values: $\phi = 30°$, $b_{*0} \equiv b_*(\theta = 0) = 0.04$ (curve 1), 0.056 (curve 2), and 0.25 (curve 3).

6.2.3. The Drucker–Prager Yield Condition and Levy's Flow Rule

The analyses described in the previous two subsections were based on the Haar–von Karman hypothesis. Here we describe an alternative approach (Jenike, 1987; Cleaver and Nedderman, 1993b) which does not use this hypothesis.

Cleaver and Nedderman (1993b) constructed radial stress and velocity fields for the special case of incompressible radial flow. They used the Drucker–Prager yield condition (5.13) and the Levy–Mises flow rule (2.53). For this yield condition, the latter is an example of a nonassociated flow rule which implies incompressible flow. On the other hand, an associated flow rule implies dilation.

The Levy–Mises relations are given by

$$C_{ij} = \dot{\lambda}\, s'_{ij}, \quad i, j = 1, 3 \tag{6.33}$$

where C_{ij} is a component of the rate of deformation tensor, defined in the compressive sense, $\dot{\lambda}$ is a proportionality factor, and s'_{ij} is a component of the deviatoric stress tensor, defined by

$$s'_{ij} = \sigma_{ij} - p\,\delta_{ij} \tag{6.34}$$

Here p is the mean stress defined by

$$p \equiv \text{tr}(\sigma)/3 = \sigma_{kk}/3 \tag{6.35}$$

and the repeated indices imply summation. Using (A.76) and the velocity field (6.27), we obtain

$$C_{\theta\theta} = C_{\phi'\phi'} = -C_{rr}/2 = f/r^3$$
$$C_{r\theta} = f'/(2r^3); \quad C_{r\phi'} = 0 = C_{\theta\phi'} \tag{6.36}$$

where

$$f' \equiv \frac{\mathrm{d}f}{\mathrm{d}\theta} \tag{6.37}$$

Equations (6.36) and (6.33) imply that

$$s'_{\theta\theta} = s'_{\phi'\phi'} = -s'_{rr}/2$$
$$s'_{r\theta}/s'_{\theta\theta} = f'/(2f); \quad s'_{r\phi'} = 0; \; s'_{\theta\phi'} = 0 \tag{6.38}$$

To proceed further, consider the Drucker–Prager yield condition (5.13), which can be rewritten using (5.9) as

$$\frac{1}{2} s'_{ij} s'_{ji} = p^2 \sin^2 \phi \tag{6.39}$$

where ϕ is a material constant called the angle of internal friction. Jenike (1987) refers to (6.39) as the *conical yield function*, as it represents a cone in principal stress space. Henceforth, (6.39) will be called the conical yield condition. For the special case of plane flow, if the Levy–Mises flow rule (6.33) is used, (6.39) reduces to the critical state approximation (2.47) (Problem 5.5). The latter is also equivalent to the Mohr–Coulomb yield condition (2.38) for a cohesionless material.

Using (6.38), (6.39) reduces to

$$s'_{\theta\theta} = \pm (p \sin\phi)/\sqrt{3 + (f'/2f)^2} \tag{6.40}$$

Following Cleaver and Nedderman (1993b), it is convenient to introduce an angle γ_*, defined by

$$\sqrt{3}\tan 2\gamma_* \equiv -f'/(2f) \tag{6.41}$$

so that (6.40) can be rewritten as

$$s'_{\theta\theta} = \pm (p \sin\phi \; \cos 2\gamma_*)/\sqrt{3}$$

Thus the stress components are given by

$$\sigma_{\theta\theta} = p \left[1 \pm \frac{\sin\phi \; \cos 2\gamma_*}{\sqrt{3}}\right] = \sigma_{\phi'\phi'}$$
$$\sigma_{rr} = p \left[1 \mp \frac{2 \sin\phi \; \cos 2\gamma_*}{\sqrt{3}}\right]$$
$$\sigma_{r\theta} = \mp p \sin\phi \; \sin 2\gamma_* = \sigma_{\theta r} \tag{6.42}$$

As in the case of wedge-shaped hoppers (see §3.3), the passive state of stress is appropriate for downward flow. This state corresponds to the upper signs in (6.42).

Unlike the Haar–von Karman hypothesis (§6.2.1), which leads to $\sigma_{\phi'\phi'} = \sigma_1$, the present approach yields $\sigma_{\phi'\phi'} = \sigma_{\theta\theta}$. Thus $\sigma_{\phi'\phi'} \neq \sigma_1$ in general, even though the equality holds at the centerline.

At this stage, all the stress components have been expressed in terms of the mean stress p and the angle γ_*. Both p and γ_* can be determined by substituting (6.27) and (6.42) into the momentum balances (6.9) and (6.10), and integrating them. Cleaver and Nedderman (1993b) constructed an approximate solution as follows.

As discussed in §6.2.2 , the inertia-free momentum balances admit a particular solution of the form

$$\overline{p} \equiv p/(\rho g r_e) = b_*(\theta)\xi\,; \quad \gamma_* = g_*(\theta) \tag{6.43}$$

where $\xi \equiv r/r_e$ and r_e is the radial coordinate corresponding to the edge of the exit orifice (Fig. 6.2b). Equations (6.43) represent a *radial stress field.*

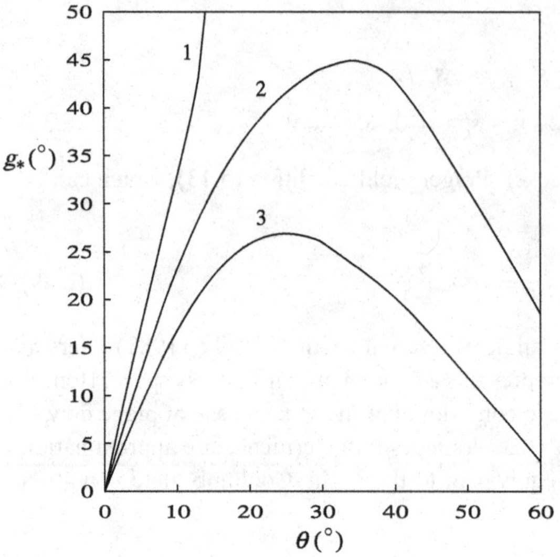

Figure 6.15. Profile of $g_*(\theta)$ for the radial stress field based on the conical yield condition and Lévy's flow rule (see (6.43)). Parameter values: $\phi = 30°$, $b_{*0} = 0.12$ (curve 1), 0.1559 (curve 2), 0.2 (curve 3).

Substituting (6.42) and (6.43) into (6.9) and (6.10), and omitting the inertial terms, we obtain

$$\frac{\mathrm{d}}{\mathrm{d}\theta}(b_* \sin\phi \ \sin 2g_*) = b_* (1 - 8 \sin\hat{\phi} \cos 2g_*) - b_* \sin\phi \ \sin 2g_* \cot\theta + \cos\theta$$

$$\frac{\mathrm{d}}{\mathrm{d}\theta}(b_* (1 + \sin\hat{\phi} \cos 2g_*)) = 4 b_* \sin\phi \ \sin 2g_* + \sin\theta \tag{6.44}$$

where $\sin\hat{\phi} \equiv (\sin\phi)/\sqrt{3}$. Equations (6.44) can be integrated subject to the boundary conditions (3.42) and (3.34)

$$\sigma_{\theta r}(r, \theta = 0) = 0$$

$$-\sigma_{\theta r}(r, \theta_w)/\sigma_{\theta\theta}(r, \theta_w) = \text{constant} \equiv \tan\delta \tag{6.45}$$

For the passive state of stress given by (6.42), (6.45), and (3.36) imply that

$$g_*(0) = 0\,;\ g_*(\theta_w) = (\hat{\delta} + \sin^{-1}(\sin\hat{\delta}/\sin\hat{\phi}))/2 \tag{6.46}$$

where $\tan\hat{\delta} \equiv (\tan\delta)/\sqrt{3}$.

Once $g_*(\theta)$ is known, the velocity field can be obtained by integrating (6.41) to obtain

$$f(\theta) = f(0) \exp\left(-\sqrt{12} \int_0^\theta \tan(2\, g_*(\omega))\, \mathrm{d}\omega\right) \tag{6.47}$$

The radial stress field is determined by solving (6.44) subject to the boundary conditions (6.46). Consider an equivalent initial value problem obtained by replacing (6.46) with

$$g_*(0) = 0; \quad b_*(0) \equiv b_{*0} \tag{6.48}$$

where b_{*0} is a positive number. As shown in Fig. 6.15, $g_*(\theta) = \pi/4$ at some value of $\theta = \theta_c$ for small values of $b_{*0} \equiv b_*(\theta = 0)$. Then, as discussed in §6.2.2, the radial velocity field (6.47) predicts the occurrence of funnel flow or core flow for $\theta = \theta_c < \theta_w$, whenever $g_*(\theta) = \pi/4$. The dependence of θ_c on the material parameters is discussed below.

For $\phi = 30°$, the g_* profile which has $g_{\max} = \pi/4$ is shown by the curve labeled 2 in Fig. 6.15. Thus $\theta_c \approx 34°$. This is much larger than the corresponding value for the analysis based on the Mohr–Coulomb yield conditon and the Haar–von Karman hypothesis

($\theta_c = 12°$; see the curve labeled 2 in Fig. 6.14). Thus the latter analysis predicts a narrower flowing zone in funnel flow, and, according to Jenike (1987), is less realistic. Results presented by Cleaver and Nedderman (1993b) show that the difference between the two analyses arises mainly due to the choice of expressions for $\sigma_{\phi'\phi'}$, i.e., $\sigma_{\phi'\phi'} = \sigma_1$ when the Haar–von Karman hypothesis is used, and $\sigma_{\phi'\phi'} = \sigma_{\theta\theta}$ when Lévy's flow rule is used. Indeed, the Mohr–Coulomb yield condition together with the assumption $\sigma_{\phi'\phi'} = \sigma_{\theta\theta}$ leads to a $g_*(\theta)$ profile which is close to the curve labeled 2 in Fig. 6.15.

Suppose that $\delta = 15°$. Then the second of (6.46) implies that $g_*(\theta_w) = 20.4°$. Hence it follows from the curve labeled 2 in Fig. 6.15 that there are two roots for this equation, namely, $\theta_w \approx 8°$, and $\theta_w \approx 59°$. The first root corresponds to mass flow through a hopper with steep walls ($\theta_w < \theta_c$), and the second root corresponds to funnel flow ($\theta_w > \theta_c$).

For $\phi = 30°$ and $\delta = 15°$, if $\theta_w < 59°$, i.e., $b_{*0} > 0.1559$, the g_* profile lies below the curve labeled 2 in Fig. 6.15, as shown, for example, by the curve labeled 3. In such a case, $g_{max} < \pi/4$, and hence mass flow occurs. For $\theta_w = 59°$, there is an abrupt transition to funnel flow, and the present analysis implies that the material is stagnant in the region $\theta_c = 34° \leq \theta \leq \theta_w = 59°$. For $b_{*0} < 0.1559$, the value corresponding to the curve labeled 2, $g_* = \pi/4$ at some value of $\theta < \theta_c = 34°$, as shown by the curve labeled 1. Thus $\theta = \theta_c$ represents the widest core region in funnel flow. This line of reasoning is due to Cleaver and Nedderman (1993b) and Verghese and Nedderman (1993).

6.2.4. Comparison of Predicted and Measured Velocity Profiles

Here we compare predicted velocity profiles with the measurements of Cleaver and Nedderman (1993a) for the flow of polypropylene granules through a hopper lined with sandpaper. The angle of internal friction ϕ and the angle of wall friction δ were determined using an annular shear cell. Three tests were conducted to determine ϕ, and the minimum value ϕ_{min} and the maximum value ϕ_{max} were reported for each test. The mean values of ϕ_{min} and ϕ_{max} were 30.4° and 37.1°, respectively. The mean value of δ was 28.5°.

Figure 6.16 shows the profiles predicted by (i) the conical yield condition and Levy's flow rule, and (ii) the Mohr–Columb yield condition and the Haar–von Karman hypothesis. For each model, there are two curves corresponding to the two values of ϕ indicated above. Let us first consider model (i). For the curve labeled 1, which corresponds to $\phi = 30.4°$, and $\delta = 28.5°$, g_* exceeds $\pi/4$ for $\theta > 26°$. In this region, the radial velocity field predicts an unrealistic velocity profile, with $|f(\theta)|$ increasing as θ increases. Hence as discussed in §6.2.3, it is assumed that $f(\theta) = 0$ for $\theta > 26°$. (Cleaver and Nedderman (1993b) use a smaller value of δ, so that the maximum value of g_* is $\pi/4$. The resulting curve does not differ significantly from the curve labeled 1.) Similarly, for the curve labeled 2, g_* is slightly greater than $\pi/4$ for $21° < \theta < 25°$. Hence it is assumed that $f(\theta) = 0$ for $\theta > 21°$.

Consider model (ii), i.e., the Mohr–Coulomb yield condition and the Haar–von Karman hypothesis. As discussed in greater detail in the next section, the radial stress field does not exist for $\phi = 30.4°$, $\delta = 28.5°$ and for $\phi = 37.1°$, $\delta = 28.5°$. This is because $g_*(\theta_w)$ does not match the value required by the boundary condition (6.45) for any value of b_{*0} (see Fig. 6.14). In these cases, the velocity profile is obtained by using the g_* profile which leads to $g_{max} = \pi/4$, as discussed in §6.2.3. Figure 6.16 shows that the conical yield condition and Levy's flow rule give better results than the Mohr–Coulomb yield condition and the Haar–von Karman hypothesis. Cleaver and Nedderman (1993b) note that this conclusion is also applicable to their experiments with spherical kale seeds.

6.2.5. Criteria for Mass Flow

As discussed in §3.1.2, two types of flow patterns occur in hoppers, namely mass flow and funnel flow. In many applications, the former is preferred. Hence it is of interest to determine the maximum wall angle of the hopper which permits mass flow to occur for

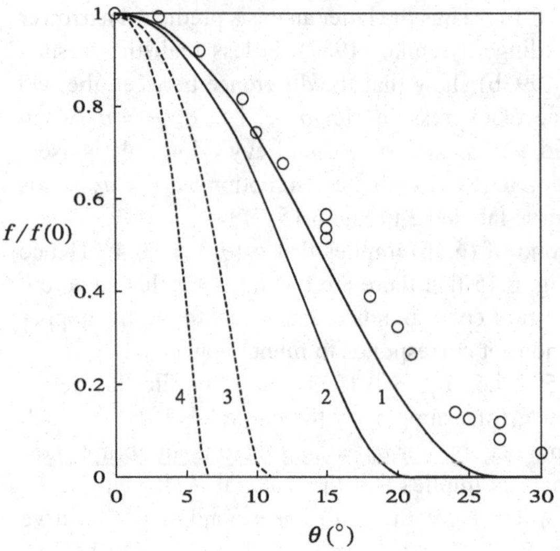

Figure 6.16. Profiles of the dimensionless radial velocity for the flow of polypropylene granules through a hopper lined with sandpaper: o, experiment; — (1, $\phi = 30.4°$, 2, $\phi = 37.1°$), radial velocity field based on the conical yield condition and Lévy's flow rule (see (6.47)); --- (3, $\phi = 30.4°$, 4, $\phi = 37.1°$), radial velocity field based on the Mohr–Coulomb yield condition and the Haar–von Karman hypothesis (see (6.31)). The granules are cylinders of diameter 3 mm and length 3 mm. Parameter values: angle of wall friction $\delta = 28.5°$, wall angle of the hopper $\theta_w = 30°$, $b_{*0} = 0.14975$ (curve 1), 0.0963 (curve 2), 0.053 (curve 3), and 0.024 (curve 4).

a given material. A few a priori criteria which can be used without solving the complete governing equations are discussed below.

(a) *Criterion based on the radial stress field*
Consider the governing equations (6.29), which determine the radial stress field for the special case of the Mohr–Coulomb yield condition and the Haar–von Karman hypothesis. Suppose that the equations are integrated from $\theta = 0$ using the initial conditions (6.32). Choosing any value of θ as θ_w, the corresponding value of δ can be obtained from the value of $g_*(\theta)$ by using (3.35). Thus

$$\tan\delta = (\sin\phi \;\sin 2g_*(\theta))/(1 + \sin\phi \;\cos 2g_*(\theta)) \qquad (6.49)$$

For a fixed value of ϕ, the above procedure generates a one-parameter family of curves $\delta = \delta(\theta_w; b_{*0})$ in the θ_w–δ plane (Fig. 6.17). For any chosen value of δ, computations show that radial field solutions satisfying the initial condition $g_{*0} = 0$ exist only for $0 \le \theta_w \le \theta_{wc}$ (Jenike, 1961, 1964a). Thus there is a limiting curve $\theta_w = \theta_{wc}(\delta)$ in the θ_w–δ plane, such that solutions do not exist for $\theta_w > \theta_{wc}(\delta)$.

For $\phi = 30°$, Fig. 6.17 shows that the limiting curve can be approximated fairly well by the curve labeled 1, which corresponds to $b_{*0} = 5 \times 10^{-4}$. The latter can be regarded as the curve obtained in the limit $b_{*0} \to 0$, and agrees well with the following expression, which is attributed to Jenike (1961) by Benink (1989, p. 87)

$$2\,\theta_w = \pi - \cos^{-1}[(1 - \sin\phi)/(2\,\sin\phi)] - 2\,\gamma_{wp} \qquad (6.50)$$

Here γ_{wp}, the passive value of g_* which satisfies (6.49), is given by the first of (3.36). The basis for (6.50) is not clear. Further, the right-hand side of (6.50) is not defined for $\phi < \sin^{-1}(1/3) \approx 19.5°$. Substituting for γ_{wp} from (3.36), (6.50) can be rewritten in the more useful form

$$2\,\theta_w = \pi - \cos^{-1}[(1 - \sin\phi)/(2\,\sin\phi)] - \delta - \sin^{-1}(\sin\delta/\sin\phi) \qquad (6.51)$$

A minor point should be noted regarding Fig. 6.17. The curves for $b_{*0} = 5 \times 10^{-4}$ and $b_{*0} = 0.056$ cross at $\theta \approx 18°$. Even though the values of δ, and hence of g_{*0}, are the same for both the curves at the intersection point, the slopes of the curves differ as the values of b_{*0} differ at this point.

As the radial stress field does not exist for $\theta_w > \theta_{wc}$, Jenike (1961) suggested that mass flow cannot occur in such cases. As more general solutions can in principle be constructed

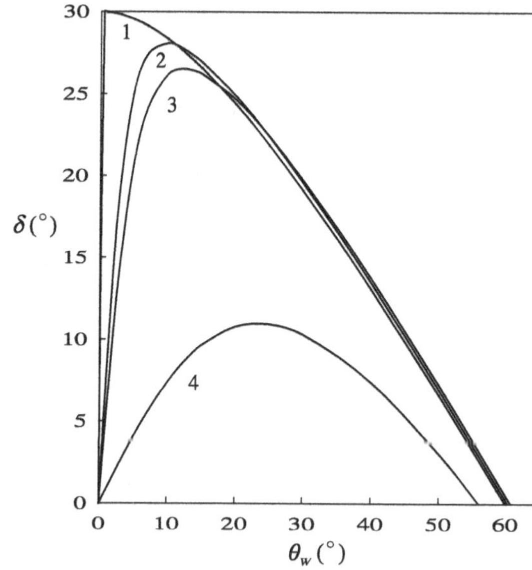

Figure 6.17. Variation of the angle of wall friction δ (calculated using (6.49)) with the wall angle θ_w, based on the radial stress field for the Mohr–Coulomb yield condition and the Haar–von Karman hypothesis. Parameter values: $\phi = 30°$, $b_{*0} = 5 \times 10^{-4}$ (curve 1), 0.04 (curve 2), 0.056 (curve 3), 0.25 (curve 4).

for $\theta_w > \theta_{wc}$, it is not clear that the existence of the radial stress field is a *necessary* condition for mass flow.

If equations corresponding to the conical yield condition and Lévy's flow rule (§6.2.3) are used, computations suggest that the radial stress field exists for all values of θ_w (see Fig. 6.15). Hence there is no analog of (6.50) in this case. Similarly, in the case of wedge-shaped hoppers, there is no limiting curve in the θ_w–δ plane.

(b) *Criterion based on the radial velocity field*

Let us first consider the equations corresponding to the Mohr–Coulomb yield condition and the Haar–von Karman hypothesis. As discussed in §6.2.2, the radial velocity vanishes at any value of θ, say, θ_c, for which $g_*(\theta) = \pi/4$. Hence Jenike (1987) suggested that mass flow cannot occur if this condition is satisfied. The condition can be translated into an equivalent one involving the material and geometric parameters as follows (Nedderman, 1992, p. 264).

As noted earlier, at any value of θ, if we set $\theta_w = \theta$, the value of δ corresponding to $g_*(\theta)$ is given by (6.49). Now $\delta = \delta(\theta_w; b_{*0})$, where b_{*0} is a parameter occurring in the initial conditions (6.32). For typical values of b_{*0}, δ exhibits a maximum $\delta_{\max}$ as θ_w increases from 0 (Fig. 6.17). A suitable choice of b_{*0} ensures that $\tan\delta_{\max} = \sin\phi$, or equivalently, $g_* = \pi/4$ (see (6.49)). For $\phi = 30°$, the appropriate value of b_{*0} is 0.056, which leads to the curve labeled 2 in Fig. 6.14 and the curve labeled 3 in Fig. 6.17. Considering the latter curve, the segment on the *right* of the maximum represents the boundary for mass flow, as every pair of δ and θ_w values on this segment implies that $g_* = \pi/4$ at some value of $\theta \leq \theta_w$. The segment of the curve on the left of the maximum does not satisfy this condition on g_* and hence is irrelevant. It is replaced by $\delta = \tan^{-1}\sin\phi$, as states above this line imply that $g_* = \pi/4$ at some value of θ.

Using a similar approach, the $\delta(\theta_w)$ curve based on the radial velocity field for the conical yield condition and Lévy's flow rule can be constructed.

(c) *Comparison of the criteria*

Figure 6.18 shows the $\delta(\theta_w)$ curves obtained by using the criteria discussed in (a) and (b) above. Each curve represents the boundary between mass flow and funnel flow, with mass flow occurring in the region *below* it. When the Mohr–Coulomb yield condition and the Haar–von Karman hypothesis is used, the curves based on the radial stress field and the

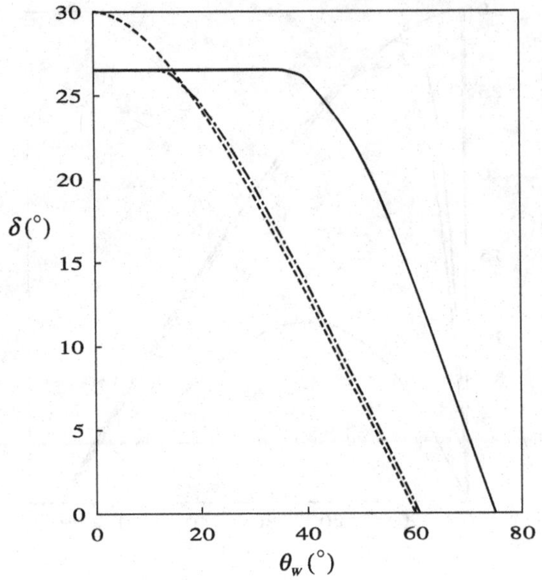

Figure 6.18. Variation of the angle of wall friction δ with the wall angle of the hopper θ_w at the boundary between mass flow and funnel flow: - - -, based on the radial stress field for the Mohr–Coulomb yield condition and Haar–von Karman hypothesis; - · -, based on the radial velocity field for the Mohr–Coulomb yield condition and Haar–von Karman hypothesis; —, based on the radial velocity field for the conical yield condition and Lévy's flow rule. For each curve, mass flow is assumed to occur in the region below it. Parameter values: $\phi = 30°$, $b_{*0} = 10^{-4}$ (- - -), 0.056 (- · -), 0.1559 (—).

radial velocity field do not differ significantly except for small values of θ_w. For a given value of $\delta < \tan^{-1}(\sin\phi)$, mass flow is permitted for a larger range of wall angles when the conical yield condition and Lévy's flow rule is used, than in the other case. These criteria are useful for preliminary design, but do not have a sound basis.

Figure 6.19 compares these criteria with the data of Johanson (1964) for the flow of iron concentrate through conical hoppers. Each hopper was assembled from two metal sheets of semicircular cross section. Horizontal layers of chalk powder were used as marker particles. After allowing the solid to flow for some time, the flow was stopped and one half of the hopper was removed. Based on the observed deformation of the marker layers, Johanson (1964) noted that material flowed along the wall in some cases (the circles in Fig. 6.19), but not in others (the crosses in Fig. 6.19). Thus the circles and crosses can be regarded as representing mass flow and funnel flow, respectively. The data are consistent

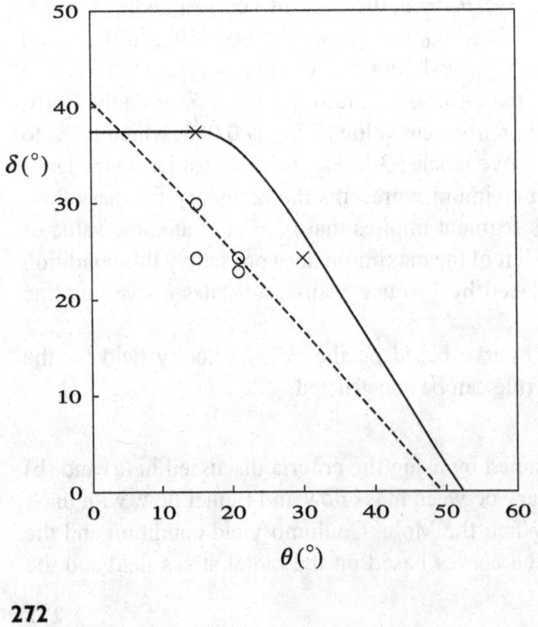

Figure 6.19. Comparison of the criteria for mass flow (curves) with data of Johanson (1964) (symbols) for the flow of iron concentrate: - - -, criterion based on the radial stress field corresponding to the Mohr–Coulomb yield condition and the Haar–von Karman hypothesis; —, criterion based on the radial velocity field corresponding to the conical yield condition and Lévy's flow rule; ∘, ×, data from experiments where the material adjacent to the wall was either stationary (×), or flowing (∘). Parameter values: $\phi = 50°$, $b_{*0} = 5 \times 10^{-4}$ (- - -), 0.0532 (—).

with the broken curve in Fig. 6.19, which represents the criterion based on the radial stress field for the Mohr–Coulomb yield condition and the Haar–von Karman hypothesis. On the other hand, the criterion based on the conical yield condition and Levy's flow rule predicts mass flow even for the data point corresponding to $\theta_w = 30°$ (Fig. 6.19), whereas funnel flow occurs in this case. However, the point lies very close to the boundary for mass flow. More extensive data are needed to assess the utility of these criteria. With the advent of noninvasive techniques such as positron emission particle tracking (see, e.g., Stewart et al., 2001), it is hoped that accurate data will become available shortly.

6.3. A HYBRID HYPOPLASTIC-VISCOUS MODEL

In §6.2, models based on yield conditions and flow rules have been used to examine flow through hoppers and bunkers. Here an alternative approach based on a hypoplastic model (see §5.2) will be discussed. Eibl and coworkers (see, e.g., Eibl and Rombach, 1988; Rombach and Eibl, 1988; Eibl, 1997) have used a modified version of a hypoplastic model. They assume that the Jaumann derivative of $\boldsymbol{\sigma}$ (see §5.2.1) is given by the sum of a rate-independent "frictional" contribution which is appropriate for slow flow, and a rate-dependent "viscous" contribution which is appropriate for rapid flow.

The rationale for this approach is as follows. At points which are far above the exit slot of the bunker, the solids fraction is high and the shear rates are low. Here momentum transfer is likely to occur mainly by sliding and rolling of the grains relative to each other. Hence as discussed in §2.13.1, it is preferable to use a rate-independent model in this region. Near the exit slot, the solids fraction is low (Fig. 3.11a), and the shear rates are likely to be much higher than those prevailing far above. Here collisional transfer of momentum is likely to be important, and hence it may be desirable to use a rate-dependent constitutive equation in this region. As discussed in §3.5–3.7, and §4.2, attempts to use rate-independent plasticity models for the entire bunker lead to severe difficulties in constructing a solution. Similarly, it appears that if the viscous terms are omitted in the present model and a purely hypoplastic model is used, it is difficult to solve the governing equations numerically (G. Rombach, private communication, 2002).

Eibl and Rombach (1988) assume that

$$\overset{\circ}{\boldsymbol{\sigma}} = \overset{\circ}{\boldsymbol{\sigma}}{}^{\mathrm{f}} + \overset{\circ}{\boldsymbol{\sigma}}{}^{\mathrm{v}} \tag{6.52}$$

where $\overset{\circ}{\boldsymbol{\sigma}}$ is the Jaumann derivative of $\boldsymbol{\sigma}$, defined by (5.64), and the superscripts f and v denote the frictional and viscous contributions, respectively. The frictional contribution to stress is given by (Kolymbas, 1991) (see (5.73))

$$\overset{\circ}{\boldsymbol{\sigma}}{}^{\mathrm{f}} = b_1\,\mathrm{tr}(\boldsymbol{\sigma}^{\mathrm{f}}\,\mathbf{C})\,\mathbf{I} + b_2\,||\mathbf{C}||\,\boldsymbol{\sigma}^{\mathrm{f}} + b_3\,(||\mathbf{C}||/\mathrm{tr}(\boldsymbol{\sigma}^{\mathrm{f}}))\,\boldsymbol{\sigma}^{\mathrm{f}^2} + b_4\,(\mathbf{C}\,\boldsymbol{\sigma}^{\mathrm{f}} + \boldsymbol{\sigma}^{\mathrm{f}}\,\mathbf{C}) \tag{6.53}$$

where $b_i, i = 1, 4$ are dimensionless constants, $\mathbf{C}$ is the rate of deformation tensor, defined in the compressive sense, and

$$||\mathbf{C}|| \equiv \sqrt{\mathrm{tr}(\mathbf{C}^2)} = \sqrt{C_{ij}\,C_{ji}} \tag{6.54}$$

is the *norm* of $\mathbf{C}$. The viscous part is assumed to be given by

$$\overset{\circ}{\boldsymbol{\sigma}}{}^{\mathrm{v}} = 2\mu\,\overset{\circ}{\mathbf{C}}{}' \tag{6.55}$$

where μ is the shear viscosity, and $\mathbf{C}' = \mathbf{C} - (\mathrm{tr}(\mathbf{C})\,\mathbf{I})/3$ is the deviatoric rate of deformation tensor. If μ is a constant, (6.55) can be obtained by taking the Jaumann derivative of the constitutive equation (2.54) for an incompressible Newtonian fluid.

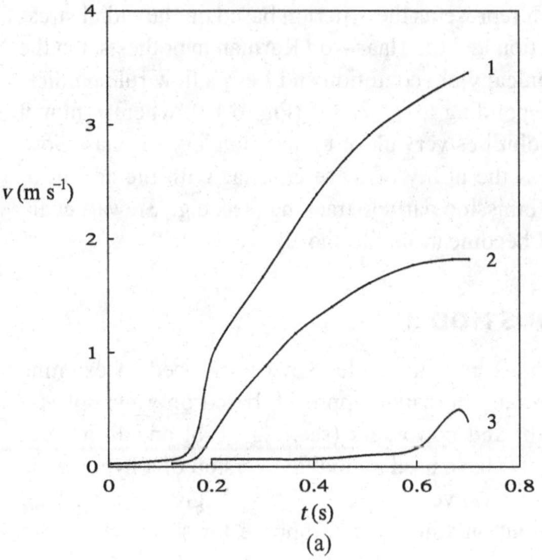

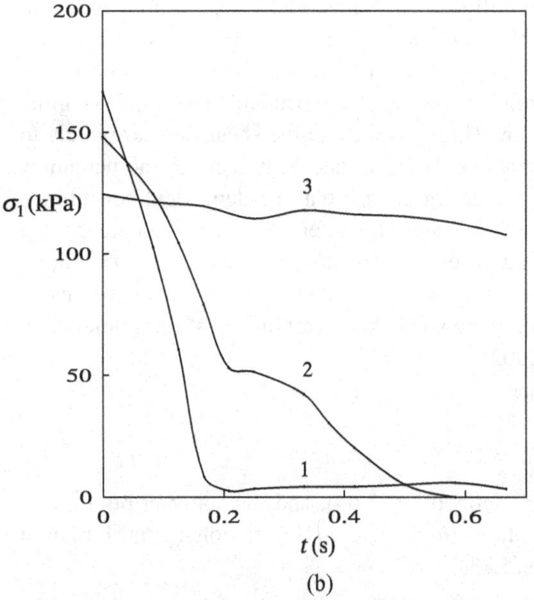

Figure 6.20. Profiles of (a) the downward vertical velocity v and (b) the major principal stress σ_1 predicted by a hybrid hypoplastic-viscous model for the flow of sand through a cylindrical bin. The profiles represent quantities at three points which are on the centerline of the bin, and at dimensionless heights $z/H = 0$ (curve 1), 0.047 (curve 2), 0.13 (curve 3). Here z denotes the height above the base, and H is the height of the bin. Parameter values: diameter of the bin = 5.6 m, diameter of the exit slot = 2.4 m, H = 28 m, constants in the constitutive equation (6.53): $b_1 = -200.0$, $b_2 = -47.9$, $b_3 = 36.6$, $b_4 = -252.7$ (G. Rombach, private communication 2002), angle of wall friction = 26.5°, shear viscosity μ = 1.0 kPa s. (Adapted from Fig. 5 of Eibl and Rombach, 1988.)

In their paper, Eibl and Rombach (1988) give an expression for μ which depends on the trace of $\mathbf{C}^2$. However, it appears that μ was treated as a constant in the computations (G. Rombach, private communication, 2007).

Eibl and Rombach (1988) have examined the batch discharge of sand from a cylindrical bin using (6.52) along with the mass and momentum balances. A finite element method was used to solve the governing equations. The authors model the deformation of the wall and the base of the bin also, treating it as a linear elastic material.

The granular material accelerates as it flows through the converging zone near the exit slot (Fig. 6.20a). The profiles of the major principal stress σ_1 (Fig. 6.20b) show that the stresses at level 3 are not significantly affected by the discharge of the material. On the other hand, the stress falls to very low values near the exit slot. Results reported in Eibl and Rombach (1988) show that when the bin is filled, the major principal stress is nearly vertical except near the walls. Thus an active state of stress prevails over most of the bin.

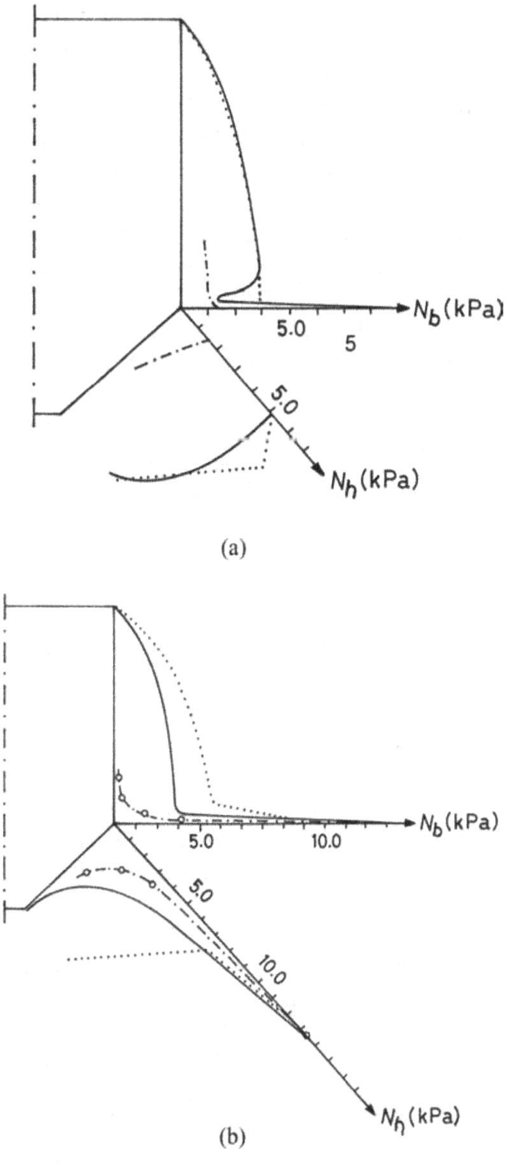

Figure 6.21. Profile of a cylindrical bunker (only the right-half of the bunker is shown), and the profiles of the normal stresses N_b and N_h on the walls of the bin and hopper sections, respectively (a) after filling, and (b) after the material has been allowed to discharge for 0.8 s: —, hypoplastic-viscous model (see (6.52), (6.53), and (6.55)); ..., DIN standard, – · –, – ○ –, data of Perry and Jangda (1970). Parameter values: diameter of the bunker = 0.292 m, height of the cylindrical section = 0.914 m, diameter of the exit slot = 35 mm, wall angle of the hopper section = 20°. The parameters b_1–b_4, have the same values as shown in the caption to Fig. 6.20. The value of the shear viscosity μ is not reported in Rombach and Eibl (1988), but it is likely that $\mu \approx 1.0$ kPa s (G. Rombach, private communication, 2007). (Reproduced from Figs. 11 and 14 of Rombach and Eibl, 1988, with permission from Prof. G. Rombach.)

When the material begins to flow, this orientation is retained in the upper part of the bin. However, it changes in the converging zone near the exit slot, and is horizontal near the centerline. Thus a passive state of stress prevails near the centerline of the converging zone.

The above model has also been used to examine flow through cylindrical bunkers, and bins of rectangular cross section with eccentric outlets. Consider the results of Rombach and Eibl (1988) for a cylindrical bunker. The process of filling the bunker was simulated by increasing the body force from zero in small increments, and solving the unsteady equations after every increment (Eibl and Rombach, 1988). However, the results of Keiter and Rombach (2001) suggest that predicted stress profiles are more realistic if the material is filled layer by layer.

At the end of the filling process, the profile of the normal stress on the bunker wall is as shown in Fig. 6.21a. On moving down from the free surface, the normal stress increases as expected from the Janssen solution (see Fig. 4.13). However, just above the bin–hopper transition, there is a sharp minimum in the normal stress. This is believed to be an artifact of the discretization elements used in the finite element procedure (G. Rombach, private communication, 2002). In the hopper section, the normal stress decreases on moving downward. If a plasticity model is used, the decrease can be anticipated by the existence of a radial stress field (see §6.2.2). The dotted curve in Fig. 6.21a shows the profile given by the German Standard DIN 1055/6. In the bin section, it agrees fairly well with the predictions of the model, except near the transition region.

The broken curve in Fig. 6.21a shows the normal stress measured using a pressure-sensitive radio pill fixed at various locations on the wall (Perry and Jangda, 1970). The predicted stresses are of the same order of magnitude as the measured values, but differ significantly from the latter.

The solid curves in Fig. 6.21b show the predicted stress profiles after the material has been allowed to discharge for 0.8 s. Comparison of the profiles in Figs. 6.21a and 6.21b shows that the dynamic normal stress in the upper part of the bin section is approximately equal to the static value at the same location. However, near the bin–hopper transition, dynamic stresses are significantly larger than static values. The reason for this behavior is not known. In the hopper section, the results of Rombach and Eibl (1988) show that the major principal stress axis is nearly vertical when the material is filled, and it becomes nearly horizontal during flow. Thus the state of stress in the hopper changes from active to passive when the material begins to flow. This is in accord with the experimental observations of Perry and Handley (1967), which were discussed in §4.2.2. The change in the state of stress may be responsible for the occurrence of the large dynamic stresses in the upper part of the hopper section. The dynamic stresses predicted by the model differ significantly from both the German standard and the data of Perry and Jangda (1970).

If a plasticity model is used for the bunker problem, the active and passive regions must be identified a priori (see §4.2.2). The present model is more attractive in this regard, as such an identification is not required. It would be interesting to compare the predictions of the plastic and hypoplastic models, and also to examine the importance of including the rate-dependent contribution to the stress tensor.

Though this model has been used extensively by Eibl and coworkers, there are very few comparisons between predictions and experimental data, particularly for flow problems. Attempts to fill this gap would be worthwhile, as this is one of the few models for which solutions have been reported in complex geometries. An alternative class of hybrid models is discussed in Chapter 10.

6.4. THE KINEMATIC MODEL FOR BATCH DISCHARGE FROM A BIN

Here we describe a simple model for the incompressible velocity field. As suggested by the name, the model is confined to the kinematic aspects only, and it does not predict the stress field. It was proposed by Litwiniszyn (1956) (cited in Tüzün, 1979). Subsequently, Mullins (1972) and Nedderman and Tüzün (1979) deduced the same equations using slightly different ideas.

The model of Nedderman and Tüzün (1979) is constructed as follows. Consider a cylindrical coordinate system, with origin at the base of the bin, and two particles F and G whose centers of mass are in a horizontal plane at time t (Fig. 6.22). For downward flow, suppose that particle F has a larger vertical velocity $v' \equiv -v_z$ than particle G. Consider a particle E, which is above F and G. It is assumed that the radial velocity of E is directed such that it moves toward F. Using a continuum model, this assumption can be expressed as

$$v_r = B\,\frac{\partial v'}{\partial r} = -B\,\frac{\partial v_z}{\partial r} \tag{6.56}$$

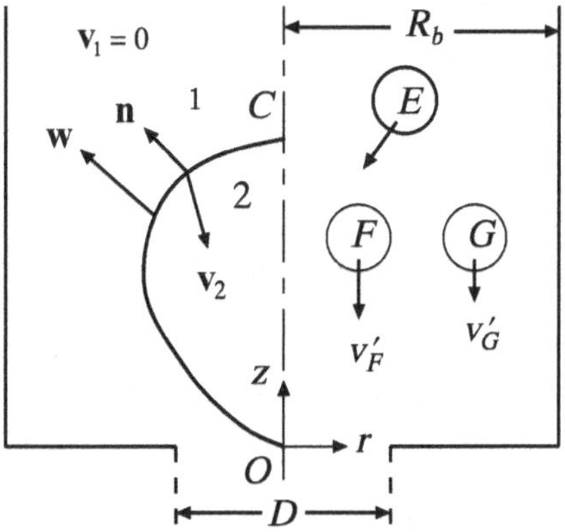

Figure 6.22. The kinematic model of Nedderman and Tüzün (1979). Here OC represents the boundary of the stagnant zone at time t_b. The material is stationary above OC and has velocity $\mathbf{v}_2$ on the downstream side of OC. The boundary moves with speed w in the direction of the unit normal $\mathbf{n}$. The symbol v'_F denotes the downward vertical velocity of the particle F. If $v'_F > v'_G$, the particle E moves toward F.

where v_r is the radial component of velocity, B is a proportionality factor having the dimensions of length. At present there is no predictive model for B, and hence its value is determined by fitting the predicted velocity profiles to data. Assuming that the flow is steady, radial, and axisymmetric, the mass balance (A.63) reduces to

$$\frac{1}{r}\frac{\partial(r\,v_r)}{\partial r} + \frac{\partial v_z}{\partial z} = 0 \tag{6.57}$$

Treating B as a constant and substituting (6.56) in (6.57), we obtain

$$\frac{\partial v_z}{\partial z} = \frac{B}{r}\frac{\partial}{\partial r}\left(r\,\frac{\partial v_z}{\partial r}\right) \tag{6.58}$$

Nedderman (1995) examined the batch discharge of material from a bin using (6.58). His analysis is discussed below.

Consider a boundary OC which separates the stationary material above from the moving material below (Fig. 6.22). The boundary moves with speed w in the direction of the unit normal $\mathbf{n}$. The jump mass balance (Appendix D) implies that the mass flux is continuous across the boundary, i.e., $\rho_1\, w = \rho_2\,(w - \mathbf{n}\cdot\mathbf{v}_2)$. Here the subscripts 1 and 2 denote points on the upstream and downstream sides of the boundary, respectively, and ρ is the density. Solving for w, we obtain

$$w = \frac{\rho_2\,(-\mathbf{n}\cdot\mathbf{v}_2)}{\rho_1 - \rho_2} \tag{6.59}$$

For downward flow, $-\mathbf{n}\cdot\mathbf{v}_2 > 0$ (see Fig. 6.22). Further, if the boundary is to move upward into the bin, as in Fig. 6.22, (6.59) implies that $\rho_1 > \rho_2$. Hence the material must dilate on crossing the boundary.

To proceed further, it is assumed that (i) $\rho_1 - \rho_2 = \text{constant} \equiv \Delta\,\rho$, and (ii) the normal $\mathbf{n}$ to the boundary is antiparallel to the downstream velocity vector $\mathbf{v}_2$ (Fig.6.22), i.e., $\mathbf{n} = -\mathbf{v}_2/|\mathbf{v}_2|$, where $|\mathbf{v}_2|$ denotes the magnitude of $\mathbf{v}_2$. In other words, the boundary is locally perpendicular to the downstream streamline. These assumptions cannot be justified a priori, but their use permits a simple solution to be obtained. In view of the assumption (ii), (6.59) implies that the velocity of the boundary is given by

$$\mathbf{w} = w\,\mathbf{n} = -\rho_2\,\mathbf{v}_2/(\Delta\,\rho) \tag{6.60}$$

To proceed further, it is assumed that the downstream velocity field $\mathbf{v}_2$ is given by the solution of the kinematic model (6.58). If z is regarded as a variable resembling time, (6.58) represents an "unsteady" one-dimensional *diffusion equation* for the vertical velocity v_z. For the special case of a point sink at the origin ($r = 0$, $z = 0$) of a medium which is unbounded in the x and y directions, the solution of (6.58) is (Carslaw and Jaeger, 1959, p. 259) (see also Problem 6.3)

$$v \equiv -v_z = \frac{\dot{Q}}{4\pi B z} \exp\left(\frac{-r^2}{4 B z}\right) \tag{6.61}$$

where $\dot{Q}$ is the volumetric flow rate. There are two sources of error in applying (6.61) to the bin: (i) the exit orifice is not a point sink but a circle of diameter D (Fig. 6.22), and (ii) the bin wall imposes the boundary condition of zero normal velocity ($v_r(r = R_b, z) = 0$), which is violated by (6.61). Hence the velocity field (6.61) is not applicable at points which are near either the exit orifice or the bin wall. Despite these limitations, Nedderman's approach leads to an elegant solution which appears to have the correct qualitative behavior.

Let (r_b, z_b) denote the coordinates of a point on the boundary OC (Fig. 6.22). Using (6.60), (6.61), and (6.56), the components of $\mathbf{w}_2$ are given by

$$\begin{aligned} w_z &= \frac{\mathrm{d}z_b}{\mathrm{d}t} = \frac{\dot{Q}}{4\pi B z_b} \exp\left(-\frac{r_b^2}{4 B z_b}\right) \frac{\rho_2}{\Delta\rho} \\ w_r &= \frac{\mathrm{d}r_b}{\mathrm{d}t} = -B\,\frac{\rho_2}{\Delta\rho}\frac{\partial v}{\partial r} = \frac{\dot{Q}\, r_b}{8\pi B z_b^2} \exp\left(\frac{-r_b^2}{4 B z_b}\right) \frac{\rho_2}{\Delta\rho} \end{aligned} \tag{6.62}$$

Hence the location of the boundary is given by

$$\frac{\mathrm{d}r_b}{\mathrm{d}z_b} = \frac{r_b}{2 z_b}$$

or

$$r_b^2 / z_b = \text{constant} \tag{6.63}$$

Using (6.63) and the initial condition $z_b = 0$ at $t = 0$, the first of (6.62) can be integrated to from $t = 0$ to any time $t = t_b$ to obtain

$$t_b = \left(\frac{2\pi B \Delta\rho}{\dot{Q} \rho_2}\right) z_b^2 \exp\left(\frac{r_b^2}{4 B z_b}\right) \tag{6.64}$$

Equation (6.64) represents the profile of the moving boundary at the time t_b. The profiles for different values of t_b pass through the origin, as they are based on the solution (6.61) for a point sink.

There is one inconsistency in the analysis desribed above. The diffusion equation (6.58) does not admit discontinuous solutions, whereas the velocity field has been assumed to be discontinuous across the moving boundary. The construction of a model which avoids this inconsistency merits attention.

The curves in Fig. 6.23 show the predicted profiles of the boundary at various times t_b. The size of the flowing zone increases with t_b, in accord with the flow patterns shown in Fig. 6.7. The predicted profiles are in reasonable agreement with the data of Watson (1993) (the symbols in Fig. 6.23). Though the experimental bin had a semicircular cross section, the nature of the velocity field may be such that the assumption of axial symmetry does not cause significant errors.

For large values of t_b, the boundary of the stagnant zone, as predicted by (6.64), intersects the free surface of the material at some point F (Fig. 6.24). For $r > r_F$, it is assumed that the free surface is conical, and inclined to the horizontal at the angle of repose β_r. This is

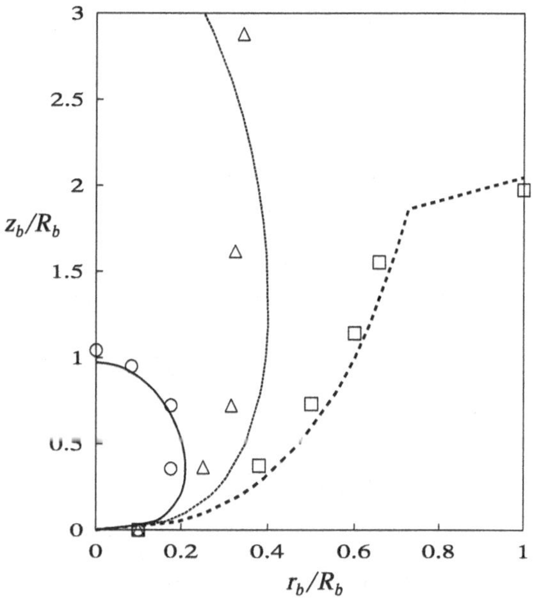

Figure 6.23. Profiles of the boundary of the stagnant zone at various times: t_b (in s) = 0.3 (—, ○); 4 (- - -, △); 109.8 (– – –,□). The curves are plots of (6.64) and the symbols represent the data of Watson (1993) for the flow of polypropylene pellets through a hemi-cylindrical bin. Parameter values: $B = 5$ mm, $\Delta\rho/\rho_2 = 0.1$, $\dot{Q} = 1.05 \times 10^{-3}$ m^3 s^{-1}, angle of repose $\beta_r = 34°$, radius of the bin $R_b = 0.325$ m, the rest are as in Fig. 6.7.

roughly in accord with the data of Watson (1993), which suggest that $\beta_r \approx 34°$. Using the procedure described in Problem 6.1, it is found that the boundary predicted by (6.64) for $t_b = 109.8$ s extends till $\xi = r/R_b = 0.72$ (Fig. 6.23).

Overall, the kinematic model provides a good description of the initiation of flow in the experiments of Watson (1993). It would be desirable to compare model predictions with data for bins of other aspect ratios and other ratios of the orifice diameter to the bin diameter.

In slender bins, when the head of material is large compared to the bin diameter, the stagnant boundary eventually intersects the bin wall. A quasisteady velocity field is then attained, and (6.61) is not applicable. The case of steady flow through a cylindrical bin has been discussed by Tüzün (1979) and Graham et al. (1987). They solved (6.58) subject to the condition of zero normal velocity at the bin wall. For example, Graham et al. (1987) calculated the residence time or time taken for a marker particle released at a specified position to reach the exit. It was found that predicted and measured residence times agreed well, provided a suitable value was chosen for the kinematic constant B.

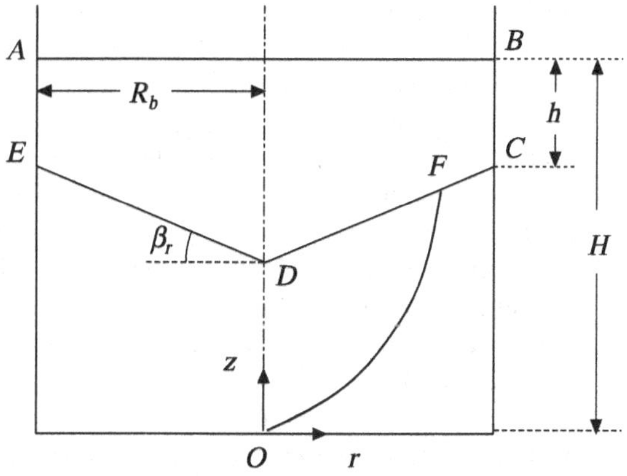

Figure 6.24. The free surface FC and the stagnant boundary OF at some time t_b during the batch discharge of material from a cylindrical bin of radius R_b. AB represents the free surface at the intial time $t_b = 0$.

6.5. SUMMARY

As in the case of wedge-shaped bunkers, there are many interesting aspects of flow through axisymmetric hoppers and bunkers.

For hoppers, both theory and experiment suggest that the dimensionless mass flow rate is larger for a wedge-shaped hopper than for a conical hopper of the same wall angle. This may be due to the larger resistance to flow offered by the walls in the latter case.

Density profiles measured in a bunker suggest that the initiation of flow is accompanied by dilation, particularly in the hopper section. There is also some evidence for the propagation of density waves through the bunker, but more detailed experiments are needed to verify this feature.

The experiments of McCabe (1974) and Watson (1993) show that flow is initiated by the propagation a front separating stationary and moving material. Based on the limited data available, it appears that the flow patterns in tall and short bins differ significantly. In the former case, a quasi-steady flow pattern is eventually established, with the material in the upper part of the bin in plug flow. This pattern changes to converging flow across a curved surface (Fig. 6.5d). In short bins, the boundary of the stagnant zone propagates upward until it touches the free surface. The latter then descends, but there is no region where plug flow occurs (Fig. 6.7). At present, there is no satisfactory criterion for predicting the type of flow pattern which can occur in a bin of a specified aspect ratio.

Considering stress measurements, the wall stresses show both slow and rapid temporal oscillations when the material flows. The rapid oscillations are usually attributed to stick-slip flow, but a convincing explanation for the slow oscillations is not yet available. The dynamic stress profile is affected by the mode of filling the bin. For example, the data of Munch-Andersen and Askegaard (1993) suggest that distributing the feed across the cross section of the bin leads to lower normal stresses than the introduction of the feed in a small region near the axis of the bin.

The shear stresses measured during flow are close to the static values measured after filling the bin, but the normal stresses tend to exceed static values. One possibility is that the angle of wall friction has a lower value during flow than in the static case. On the other hand, the experiments of Rao and Venkateswarlu (1974) suggest that static and dynamic normal stresses do not differ significantly if the static values are measured after allowing the material to flow for some time. When the flow is stopped, the density is perhaps closer to the density of the flowing material than to that attained after filling the bin. Simultaneous measurement of the stress and density fields would be valuable.

The flow through axisymmetric bunkers has been examined using three classes of models, namely, models involving a yield condition and a flow rule, a hypoplastic model modified to include rate-dependent behavior, and a kinematic model. Consider the models involving a yield condition and a flow rule. For steady axisymmetric flow, two of the principal stresses axes, say, the σ_1 and σ_2 axes, lie in the r–θ plane. If it is assumed that σ_1 and σ_2 satisfy the Mohr–Coulomb yield condition, additional information must be specified about $\sigma_3 = \sigma_{\phi'\phi'}$, as the latter occurs in the radial and circumferential components of the momentum balance. An assumption which is often used is the Haar–von Karman hypothesis, which assumes that $\sigma_{\phi'\phi'} = \sigma_1$. This hypothesis is difficult to justify, but its use permits the methods developed for plane flow to be readily extended to axisymmetric flow.

Using the Mohr–Coulomb yield condition and the Haar–von Karman hypothesis, the smooth wall, radial gravity problem has been examined. The predicted variation of the mass flow rate with the wall angle θ_w is qualitatively in accord with data. However, quantitative agreement is poor. The solution of Nguyen et al. (1979), which includes the effect of wall roughness, gives good estimates of the flow rate at small wall angles. The discrepancy between predicted and measured flow rates increases at large wall angles, suggesting that more terms may have to be included in the series solution. For the data shown in Fig. 6.13, the modified Beverloo correlation performs better than either of the two models. Data for a wider range of materials and wall angles are needed before definite conclusions can be drawn.

An alternative approach is based on the conical or Drucker–Prager yield condition and Lévy's flow rule. This approach is more general than that based on the Mohr–Coulomb yield condition and the Haar–von Karman hypothesis, and, unlike the latter, can in principle be used for non-axisymmetric flows also.

Cleaver and Nedderman (1993b) used both these approaches to predict the velocity field for the special case of steady, incompressible radial flow. Comparison with data (Fig. 6.16) shows that the conical yield condition and Lévy's flow rule lead to a more realistic velocity profile than the Mohr–Coulomb yield condition and the Haar–von Karman hypothesis.

When the Mohr–Coulomb yield condition and the Haar–von Karman hypothesis are used, the radial stress field exists only for certain ranges of parameter values. For a given value of ϕ, this corresponds to the region below a curve in the δ–θ_w plane. Jenike (1961) postulated that mass flow cannot occur in the region above this curve. An alternative criterion for mass flow is based on the radial velocity field. The radial velocity vanishes at any value of θ, say, θ_c, for which $g_*(\theta_c) = \pi/4$. It is assumed that the material is at rest for $\theta \geq \theta_c$, and hence funnel flow occurs. For a given value of ϕ, this condition is shown to be equivalent to a curve in the δ–θ_w plane, and mass flow cannot occur in the region above this curve. For $\phi = 30°$, the curves based on the radial stress and velocity fields do not differ significantly, except for small values of θ_w.

A criterion based on the radial velocity field can also be developed for the conical yield condition and Lévy's flow rule. For a given value of $\delta < \tan^{-1}(\sin\phi)$, this criterion predicts mass flow for a larger range of wall angles than the criterion based on the Mohr–Coulomb yield condition and the Haar–von Karman hypothesis. Comparison with the data of Johanson (1964) shows that the latter criterion predicts the flow patterns correctly. On the other hand, the former criterion predicts mass flow for a point which corresponds to funnel flow in the experiments. However, the point is very close to the limiting curve in the δ–θ_w plane. More data are needed for a proper assessment of these criteria.

Eibl and coworkers modified the hypoplastic model of Kolymbas (1991) (see §5.2.1) to include a rate-dependent contribution to the stress tensor. This model has been used to examine unsteady flow through bins and bunkers. For flow through a bin, the results of Eibl and Rombach (1988) show that the major principal stress does not vary significantly with time in the upper region, but it decreases to low values near the exit slot. Along the centerline, the stress field changes from active to passive as the exit slot is approached. For a bunker, the results of Rombach and Eibl (1988) show that the predicted wall stresses differ significantly from the measured values. However, both are of the same order of magnitude. At present, it is not clear how the value of the shear viscosity μ in (6.55) should be chosen. Overall, this class of models is promising, but needs refinement.

Finally, the batch discharge of material from a bin has been examined using a kinematic model. The position of the boundary between stagnant and moving material is predicted by solving a diffusion equation for the vertical velocity in conjunction with the jump mass balance. In spite of many simplifying assumptions, the model predicts the position of the boundary fairly well (Fig. 6.23). Further work is needed to develop a theory which can provide (i) an a priori estimate of the kinematic constant, and (ii) a constitutive relation between stress and deformation. The latter relation should be such that it reduces to either (6.56), or some approximation to it, under certain conditions.

PROBLEMS

6.1. Profile of the free surface for the batch discharge of material from a bin
Consider the batch discharge of material from a bin of radius R_b (Fig. 6.24). Assume that (i) the material is initially filled to a height H above the base, (ii) the volumetric flow rate $\dot{Q}$ is a constant, (iii) shortly after discharge begins, the free surface attains a conical shape, and is inclined at an angle β_r to the horizontal.

(a) At time t_b, determine the depth h at which the edge of the free surface intersects the wall. As the free surface is flat at $t = 0$, it can be assumed that t_b is large compared to the time required for the free surface to aquire a conical shape.
(b) For the parameters listed in the caption of Fig. 6.23, determine the radial and axial coordinates of the point F (Fig. 6.24) for t_b =109.8 s and β_r =34°. The initial height of the free surface is 0.94 m. The value of t_b corresponds to one of the curves in Fig. 6.23.

This problem has been adapted from Nedderman (1995).

6.2. A feature of the incompressible radial velocity field
As discussed in §6.2.2 and 6.2.3, the radial velocity field is given by $v_r = f(\theta)/r^2$, with f given by an expression of the form

$$f(\theta) = f(0)\,\exp\left(-b\int_0^{\theta} \tan(2\,g_*(\omega))\,\mathrm{d}\omega\right) \tag{6.65}$$

where b is a positive constant. Assuming that g_* increases monotonically from 0 to $\pi/4$ as ω increases from 0 to θ_c, show that $f(\theta_c) = 0$. Hint: approximate the integrand in (6.65) by a function which is a lower bound and which can be integrated in a closed form.

6.3. Derivation of the solution to the diffusion equation (6.58)
Using Cartesian coordinates, the analog of (6.56) is

$$v_x = B\,\frac{\partial v}{\partial x}; \quad v_y = B\,\frac{\partial v}{\partial y} \tag{6.66}$$

where v_x is the x component of velocity, $v \equiv -v_z$, and gravity acts in the $-z$ direction, and B is a constant.

(a) Using the mass balance for steady incompressible flow, show that the kinematic model (6.66) leads to

$$\frac{\partial v}{\partial z} = B\left(\frac{\partial^2 v}{\partial x^2} + \frac{\partial^2 v}{\partial y^2}\right) \tag{6.67}$$

(b) Expressing the coordinates (x, y, z) in terms of cylindrical coordinates (r, θ, z) (Fig. A.3), and assuming that the velocity field is axisymmetric, i.e., $v = v(r, z)$, show that (6.67) reduces to (6.58). Hence any axisymmetric solution of (6.67) satisfies (6.58).
(c) For a medium which is unbounded in the x and y directions, (6.67) can be solved using the two-dimensional Fourier transform, as follows. Let $\mathbf{r}$ denote a position vector in the x–y plane. The two-dimensional Fourier transform of the variable $v = v(\mathbf{r}, z)$ is defined by (McQuarrie, 2003, p. 397)

$$\hat{v}(\mathbf{k}, z) \equiv \frac{1}{2\pi}\int_{-\infty}^{\infty} e^{i\mathbf{k}\cdot\mathbf{r}}\, v(\mathbf{r}, z)\,\mathrm{d}\mathbf{r} \tag{6.68}$$

where $\mathbf{k}$ is a wavenumber vector with components (k_x, k_y), $i \equiv \sqrt{-1}$, $\mathbf{k}\cdot\mathbf{r} = k_x\,x + k_y\,y$, and

$$\int_{-\infty}^{\infty} F(\mathbf{r}, z)\,\mathrm{d}\mathbf{r} \equiv \int_{-\infty}^{\infty}\int_{-\infty}^{\infty} F(x, y, z)\,\mathrm{d}x\,\mathrm{d}y$$

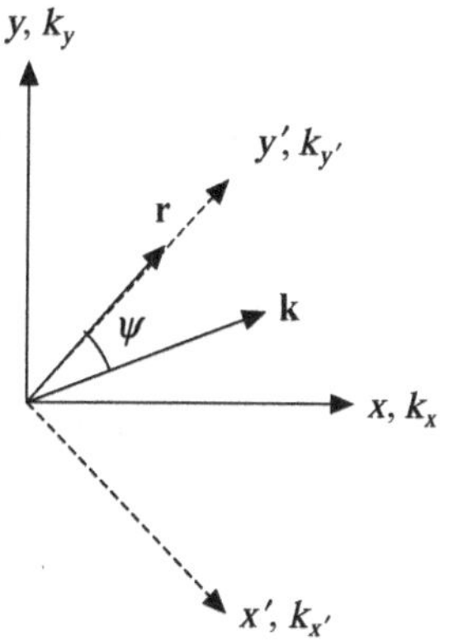

Figure 6.25. The k'_y axis of the k'_x–k'_y coordinate system is aligned with the position vector **r**. Here **k** represents the wavenumber vector.

for any function $F(\mathbf{r}, z)$. The two-dimensional inverse Fourier transform of the variable $\hat{v}(\mathbf{k}, z)$ is defined by (McQuarrie, 2003, p. 397)

$$v(\mathbf{r}, z) \equiv \frac{1}{2\pi} \int_{-\infty}^{\infty} e^{-i\mathbf{k}\cdot\mathbf{r}}\, \hat{v}(\mathbf{k}, z)\, d\mathbf{k} \tag{6.69}$$

where

$$\int_{-\infty}^{\infty} \hat{F}(\mathbf{k}, z)\, d\mathbf{k} \equiv \int_{-\infty}^{\infty} \int_{-\infty}^{\infty} F(k_x, k_y, z)\, dk_x\, dk_y$$

Taking the Fourier transform of (6.67) and using integration by parts, show that

$$\frac{\partial \hat{v}}{\partial z} = -B\, k^2\, \hat{v} \tag{6.70}$$

where $k \equiv |\mathbf{k}| = \sqrt{k_x^2 + k_y^2}$. It can be assumed that v and its first derivatives with respect to x and y vanish in the limit $|x| \to \infty$ and in the limit $|y| \to \infty$. Hence

$$\hat{v}(\mathbf{k}, z) = \hat{v}_0\, e^{-B\, k^2\, z} \tag{6.71}$$

where $\hat{v}_0 \equiv \hat{v}(\mathbf{k}, 0)$.

(d) Choosing new Cartesian coordinates (x', y'), such that the y' axis coincides with the direction of **r** (Fig. 6.25), and taking the inverse Fourier transform of (6.71) show that

$$v(r, z) = \frac{\hat{v}_0}{2\pi} \int_0^{\infty} \int_0^{2\pi} e^{-(B\, k^2\, z + i\, k\, r\, \cos\psi)}\, k\, d\psi\, dk \tag{6.72}$$

where $r = |\mathbf{r}| = \sqrt{x^2 + y^2} = \sqrt{x'^2 + y'^2}$. The integral representation of the Bessel function of the first kind of order zero is given by (McQuarrie, 2003, p. 397, Arfken and Weber, 2001, pp. 674–675)

$$J_0(w) = \frac{1}{2\pi} \int_0^{2\pi} e^{i\, w\, \cos\psi}\, d\psi$$

As $J_0(w)$ is an even function of w, (6.72) can be rewritten as

$$v(r, z) = \hat{v}_0 \int_0^{\infty} e^{-B\, k^2\, z}\, J_0(k\, r) k\, dk \tag{6.73}$$

(e) Using the series expansion for $J_0(w)$, it can be shown that (Lebedev, 1972, p. 132)

$$\int_0^{\infty} e^{-a^2\, w^2}\, J_0(b\, w)\, w\, dw = \frac{e^{-b^2/(4\, a^2)}}{2\, a^2}, \quad a > 0,\ b > 0 \tag{6.74}$$

where a and b are constants. Using (6.74), show that

$$v(r, z) = \frac{\hat{v}_0}{2\, B\, z}\, e^{-r^2/(4\, B\, z)} \tag{6.75}$$

(f) For a point sink at the origin of the coordinate system, the "initial" condition is

$$v(\mathbf{r}, 0) = \dot{Q}\,\delta(\mathbf{r}) \tag{6.76}$$

where $\dot{Q}$ is a constant and $\delta(\mathbf{r})$ is the two-dimensional Dirac delta function, defined by

$$\int_{-\infty}^{\infty} f(\mathbf{r})\,\delta(\mathbf{r})\,\mathrm{d}\mathbf{r} = \int_{-\infty}^{\infty}\int_{-\infty}^{\infty} f(x, y)\,\delta(x, y)\mathrm{d}x\;\mathrm{d}y = f(0, 0)$$

for any function $f(\mathbf{r}) \equiv f(x, y)$. Hence show that $\dot{Q}$ represents the volumteric flow rate.

(g) Show that (6.75) reduces to (6.61) when $\hat{v}_0$ is expressed in terms of $\dot{Q}$.

Parts (c) and (d) of this problem have been adapted from McQuarrie (2003, pp. 388, 397).

7

Theory for Rapid Flow of Smooth, Inelastic Particles

7.1. PRELIMINARIES AND SCALING

A characteristic feature of slow, quasi-static flow of granular materials, considered in the previous chapters, is the rate independence of the stress. As discussed earlier, in the slow flow regime, grains are in abiding contact and friction is the dominant mechanism for generating shear forces. In this chapter we consider the contrasting regime of rapid flow, in which grains are in continuous fluctuational motion, and come into contact only during very brief collisions. We shall see that the stress in this regime of flow is rate dependent; indeed, the stress varies as the square of the shear rate for shear flow with a spatially uniform shear rate.

Our physical picture of rapid flow is that of grains in a state of constant agitation, with interactions between them occurring only through *instantaneous* collisions, as shown in Fig. 7.1. To simplify our analysis, we shall assume uniformity in size and shape, and consider the granular material to be composed of smooth spheres. By this, we mean that there is no tangential force exerted by one sphere on the other at the point of contact. This picture is identical to that of molecules in a gas, which is why granular materials in this state are often referred to as "granular gases" in the literature. However, there is a fundamental and crucial difference between a granular material and a gas: collisions or interactions between molecules are *elastic*, i.e., the net energy of a colliding pair is conserved, but collisions between grains are inelastic. Although a pair of fluid molecules may exchange energy between translational, rotational, and vibrational modes during a collision, in a state of equilibrium the principle of equipartition of energy holds that energy is divided equally among all the active modes of energy storage; on average, molecules possess $\frac{1}{2}k_B T$ of thermal energy for each degree of freedom that is excited at the given temperature, where k_B is the Boltzmann constant and T is the temperature of the gas. In grain collisions, a part of the translational kinetic energy is irrecoverably lost during a collision; it is transformed to thermal energy of the molecules that make up the grains. Because the motion associated with thermal energy is incoherent, it cannot be transformed back into coherent motion of the entire grain. Thus molecules in a gas exhibit fluctuational motion even at static equilibrium, while the fluctuational motion of grains in a granular material decays with time, unless energy is continually fed by external agitation or flow. As a result, the energy in a granular material cascades from the mean motion to the fluctuational motion of the grains, and finally, by the inelasticity of collisions, into the thermal energy of the molecules comprising the grains.

The picture we have invoked above, of smooth spherical particles in a state of continuous agitation, interacting with each other through instantaneous collisions, is certainly a gross simplification of real situations of granular flow. In practice, collisions are not instantaneous,

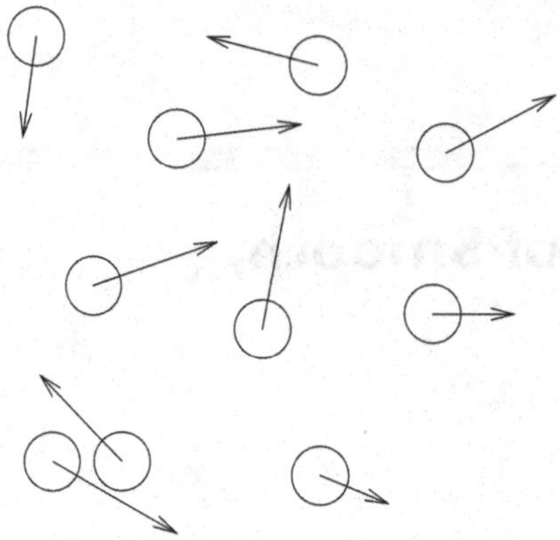

Figure 7.1. Physical picture of rapid granular flow: the grains are always in fluctuational motion. Momentum is transmitted by streaming of grains and by instantaneous collisions between grains.

particles may be rough and angular, and they may form clusters which break up and form intermittently. However, this is clearly too complicated a picture, and we make the above simplifications to allow analytical progress. Nevertheless, we shall see that useful information can be gained from this approach. We shall indicate later the improvements that can be made to account for the effects that have been ignored here: a brief discussion on mixtures comprising particles of two sizes is given in §7.5, and theories for rough particles are described in some detail in Chapter 9.

We shall assume that air drag is small compared to other forces (see §1.7) and can therefore be neglected. Theories that account for interactions with the interstitial fluid are beyond the scope of this book, but can be found in the books of Fan and Zhu (1998) and Jackson (2000).

The dependence of the stress on the shear rate is easily determined by dimensional analysis. Consider a granular material comprising rigid inelastic spheres subjected to a steady, uniform shear rate $\dot{\gamma}$. (The terms "inelastic spheres" and "inelastic particles" are used in this book to refer to particles whose collisions are inelastic.) Momentum is transported by two mechanisms: particle streaming, which refers to the transport of momentum when particles move from one location to another, and collisions. In both mechanisms, it is the inertia of the particles and the difference between their velocities that determine the momentum flux: the former depends on the intrinsic density ρ_p and diameter d_p of the particles, and the latter is $\sim \dot{\gamma}\, d_p$, the difference in the flow velocity over a length of d_p. There is no other velocity scale in the problem. Consequently, we may express the normal stress N on a given plane as

$$N = \alpha \rho_p^a \, d_p^b \, \dot{\gamma}^c \tag{7.1}$$

where α is a factor that depends on all the dimensionless variables and parameters in the problem. The exponents a, b, and c can be determined by matching the dimensions of the left and right hand sides, and we get

$$N \sim \alpha \rho_p \, d_p^2 \, \dot{\gamma}^2 \tag{7.2}$$

The above argument is also valid for every other component of the stress tensor. Thus the stress varies as the square of the shear rate, as mentioned earlier. Experimental measurements of the stresses in rapid shear have been made in many studies, starting from Bagnold (1954); a sample of the data is shown in Fig. 7.2. The data of some studies show a roughly quadratic increase of the shear and normal stresses with the shear rate, but a slower increase is seen in some studies, particularly at low shear rates.

Given our earlier statement that particle interactions in a granular and molecular gases are quite similar, one might wonder why the above scaling for the stress does not hold for the

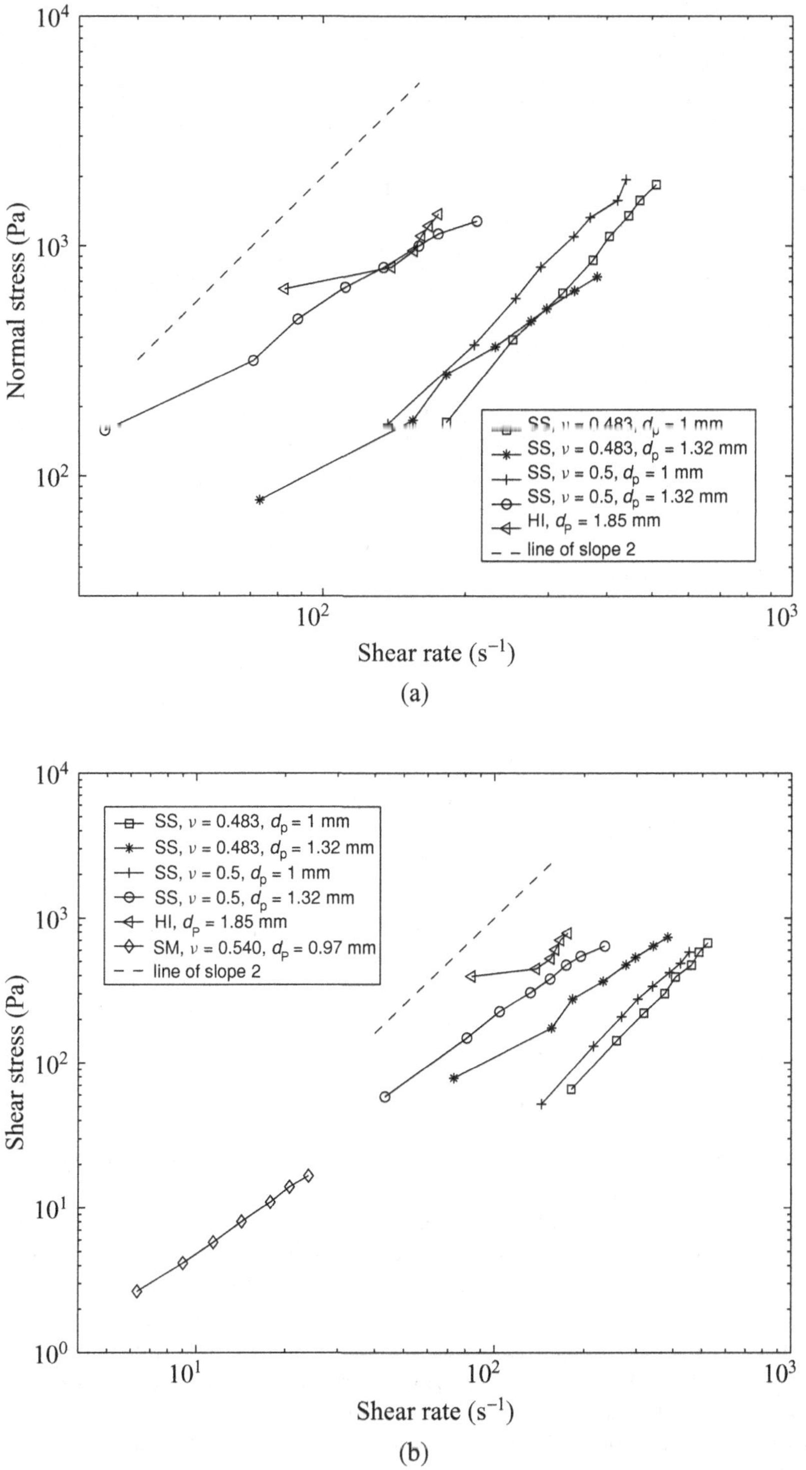

Figure 7.2. Experimental data of the (a) normal and (b) shear stresses as a function of the shear rate for model granular materials in the rapid flow regime. In both the plots, dashed lines of slope 2, representing quadratic variation, are given for reference. Note the roughly quadratic dependence on the shear rate. Data are from Savage and Sayed (1984), Hanes and Inman (1985), and Savage and McKeown (1983) (labeled SS, HI, and SM, respectively, in the legend).

latter. It does not for a very simple reason: for a molecular gas, there is an additional velocity scale, namely, the mean fluctuation velocity of the molecules $w \equiv (k_B T/m_p)^{1/2}$, where m_p is the mass of a molecule. This leads to an additional dimensionless factor $(\dot{\gamma} d_p/w)^{\ell}$ in the expression for N in (7.2), where ℓ is an arbitrary dimensionless constant. This fact alone explains almost all the differences between molecular and granular gases. In a granular gas, w is not independent of $\dot{\gamma}\, d_p$, as fluctuations do not exist in the absence of external forcing. These issues are discussed in more detail in §7.3.4.

In the above analysis we have implicitly assumed that the flow is unidirectional and that variations in the velocity are also in a single direction, so the scalar shear rate $\dot{\gamma}$ fully characterizes the flow field. In a general flow field in three spatial dimensions, there are nine independent components of the velocity gradient tensor; clearly, dimensional analysis cannot throw light on the form of the tensor relation between the stress and the velocity gradient. It also does not yield information on the functional dependence of the factor α on dimensionless variables such as the solids fraction, and dimensionless parameters such as the inelasticity of particle collisions (defined in the following section). To determine the stress tensor or any other bulk property, we must allow the most general form of variation in the state of the granular material, and average the relevant particle properties over a large collection of particles. For this, we first require a description of how particle properties change during a collision, or a collision model.

7.1.1. Model for Inelastic Collisions

The inelasticity of collisions is a result of plastic deformation occurring within the particles during collision, as a result of which the kinetic energy of the particles is dissipated as heat, i.e., vibrational modes of the molecules in the particles. The energy dissipated during plastic deformation depends on the elastic and plastic properties of the material forming the particles, their size, density, and the relative impact velocity before collision. However, most studies have used a simplified model for the inelasticity of collisions. To formulate this model, we consider particles 1 and 2 in Fig. 7.3, whose translation velocities before collision are $\mathbf{c}_1$ and $\mathbf{c}$, respectively. After collision, their velocities become $\mathbf{c}_1'$ and $\mathbf{c}'$, respectively. There is plastic deformation during each collision, but the total plastic strain is so small that the shape of the particles is assumed to remain unaltered; this is a reasonable assumption, as anyone who has played with marbles or billiard balls would attest to. If $\mathbf{g} \equiv \mathbf{c}_1 - \mathbf{c}$ is the translation velocity of particle 1 relative to particle 2, the collision model relates the component of $\mathbf{g}$ normal to the surface at the point of contact before and after collision as

$$(\mathbf{k}\cdot\mathbf{g}') = -e_p\,(\mathbf{k}\cdot\mathbf{g}) \tag{7.3}$$

where the prime denotes a postcollision property, and $\mathbf{k}$ is the unit vector along the line joining the center of particle 1 to the center of particle 2. The parameter e_p is the *coefficient of restitution*, and can take any value in the range $0 \le e_p \le 1$. When $e_p = 1$, the collision is elastic, and the kinetic energy of the particle pair is conserved during collision (see (7.7) below). Although it is evident from experiments (Goldsmith, 1960; Mangwandi et al., 2007) and theoretical analyses (Johnson, 1987a, pp. 361–369; Stronge, 2000, Chapter 6) that e_p depends on the magnitude of the relative velocity $|\mathbf{g}|$, we shall use the model with a constant e_p, because it is simple and adequate for the purpose of illustrating the main consequences of inelastic particle collisions. Additionally, the particles are assumed to be smooth, so that their translation velocities in any direction orthogonal to $\mathbf{k}$, and their angular velocities about their centers of mass, remain unchanged after collision. The consequences of particle roughness are addressed in Chapter 9. The balance of linear momentum for the two particles is

$$\mathbf{c}_1' = \mathbf{c}_1 - \mathbf{J}, \quad \mathbf{c}' = \mathbf{c} + \mathbf{J} \tag{7.4}$$

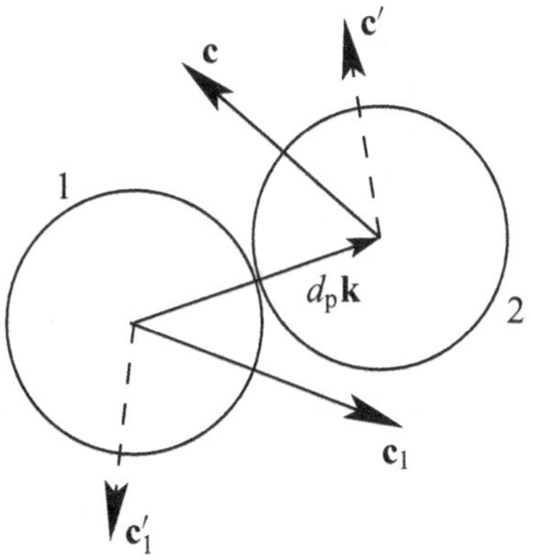

Figure 7.3. A collision between two smooth, inelastic spheres 1 and 2 of diameter d_p. Their translational velocities are $\mathbf{c}_1$ and $\mathbf{c}$, respectively, before collision, and $\mathbf{c}'_1$ and $\mathbf{c}'$, respectively, after collision. The unit vector along the line joining the center of particle 1 to the center of particle 2 is $\mathbf{k}$.

where $\mathbf{J}$ is the impulse per unit mass on particle 2 during collision. From (7.3) and (7.4) we find that $\mathbf{k}\cdot\mathbf{J} = (1+e_p)\mathbf{k}\cdot\mathbf{g}/2$; the tangential impulse $(\mathbf{k}\times\mathbf{J})\times\mathbf{k}$ is zero for smooth particles, and therefore

$$\mathbf{J} = \eta_1(\mathbf{k}\cdot\mathbf{g})\mathbf{k} \tag{7.5}$$

where $\eta_1 \equiv (1+e_p)/2$. Substitution of the above in (7.4) gives the postcollision velocities in terms of the precollision velocities,

$$\mathbf{c}'_1 = \mathbf{c}_1 - \eta_1(\mathbf{k}\cdot\mathbf{g})\mathbf{k}, \quad \mathbf{c}' = \mathbf{c} + \eta_1(\mathbf{k}\cdot\mathbf{g})\mathbf{k} \tag{7.6}$$

The change in kinetic energy E of the pair during the collision is

$$\Delta E = \frac{1}{2}m_p(c'^2 + c_1'^2 - c^2 - c_1^2) = -\frac{m_p}{4}(1-e_p^2)(\mathbf{k}\cdot\mathbf{g})^2 \tag{7.7}$$

where $c^2 = \mathbf{c}\cdot\mathbf{c}$, and similarly for the other velocities. Thus the kinetic energy lost during the collision is proportional to $(1-e_p^2)$.

7.1.2. Hydrodynamic Description of Rapid Granular Flows

Following the pioneering work of Bagnold (1954), several studies attempted to formulate hydrodynamic equations for rapid granular flow. McTigue (1978) generalized the arguments of Bagnold and proposed a constitutive relation for the stress for general deformations, that reduced to the square-rate dependence for homogeneous simple shear. Ogawa (1978) made the important advance of recognizing that the kinetic energy of the random motion of particles, which he called the "quasithermal" energy, determines the transport properties of the material, and realized that a balance equation for this energy is required. This idea was formalized in many subsequent studies (Haff, 1983; Jenkins and Savage, 1983; Lun et al., 1984; Jenkins and Richman, 1985), and the fluctuational kinetic energy is sometimes referred to as the *pseudothermal energy*, by analogy with the thermal energy of molecules in fluids. As an extension of the same analogy, one-third of the pseudothermal energy per unit mass is referred to as the *grain temperature*.

Our physical picture of grain interactions in rapid granular flow, i.e., instantaneous binary collisions,† coincides with that of a dense gas. The latter has been studied using the mathematical machinery of the kinetic theory of gases, and hydrodynamic equations and constitutive relations have been derived for the transport of mass, momentum and energy. It is therefore appropriate to revisit the kinetic theory of dense gases and modify it to account

† What we mean by "instantaneous" collisions is that time of contact between particles during collisions is much smaller than the time of flight between collisions. In this situation, the probability of more than two particles participating in a collision is negligibly small, and we may take all collisions to be binary.

for the inelasticity of particle collisions. This task has been accomplished to varying levels of approximation in many studies over the past two decades, and we shall describe the details shortly.

Before getting into the details of kinetic theory, it is useful to first use our knowledge of molecular fluids to derive a heuristic description of rapid granular flow. Such an analysis was first put forward by Haff (1983), which we discuss in §7.2. Focusing on the physics of the problem and keeping the mathematical analysis simple, we propose the hydrodynamic equations and derive the necessary closures to within undetermined multiplicative constants.

We shall follow this by the more rigorous kinetic theory-based approach to derive the equations of motion and the constitutive relations in §7.3.

7.2. HEURISTIC HYDRODYNAMIC THEORY FOR HIGH-DENSITY FLOWS

Consider a collection of particles of uniform diamater d_p and mass m_p, which are in continuous fluctuational motion. We restrict attention to high enough solids fraction, so that the mean free path of particles (the average distance traversed by particles between successive collisions), s, is small compared to the particle diameter, i.e., $s \ll d_p$. This is not a necessary assumption of the theory, but it is useful because it permits the simplification of constant density (equal to that at dense random packing), thereby making the analysis simpler. It is important to note that though s is small, its variations in space and time are important; indeed, we shall see that s determines the transport properties of the material.

Denoting the particle velocity by $\mathbf{c}$, we define the following field variables: the mean velocity $\mathbf{v} \equiv \langle \mathbf{c} \rangle$, and the root-mean-square velocity fluctuation $w \equiv \langle |\mathbf{c} - \mathbf{v}|^2 \rangle^{1/2}$. The angle brackets represent an average over a large enough number of particles in a region and during a time interval, so that the averages are smooth functions of space and time. However, it is assumed that the size of the region and the time interval are much smaller than the macroscopic length and time scale scales, respectively, that are of interest. This separation of length and time scales is essential for a continuum mechanical treatment; a discussion of its applicability for granular flows is given in §7.3.1. As in fluids, we expect w to determine the transport properties of the granular material: for instance, a simple calculation shows the viscosity of an ideal gas to be $\sim \rho\, w\, s$ (Reif, 1985, pp. 473–475), where ρ is the density. However, in contrast to fluids, w is not a property of the state of thermodynamic equilibrium of the granular material, but is a result of the applied forcing, such as shear or vibration. As a result, the equation for the balance of fluctuational kinetic energy must necessarily be coupled to the balances of mass and momentum.

The hydrodynamic equations that govern the flow are the balances of mass, momentum, and fluctuational kinetic energy. The mass and momentum balances were introduced in §1.5.3 and §1.5.5. For constant density, they assume the forms

$$\nabla \cdot \mathbf{v} = 0 \tag{7.8}$$

$$\rho \frac{\mathrm{D}\mathbf{v}}{\mathrm{D}t} = -\nabla \cdot \boldsymbol{\sigma} + \rho\, \mathbf{b} \tag{7.9}$$

The stress tensor $\boldsymbol{\sigma}$ is defined in the compressive sense (see §1.5.4), as in the previous chapters.

The balance of fluctuational kinetic energy is formulated by separating the internal energy $\hat{U}$ (see §1.5.7) into the translational kinetic energy of the entire grain (i.e., energy due to the coherent motion of the molecules in a grain) and the internal energy due to incoherent motion of the molecules, or the true thermal energy. In other words

$$\hat{U} = \frac{1}{2} w^2 + \hat{U}' \tag{7.10}$$

where $\hat{U}'$ is the thermal energy per unit mass. Similarly, the energy flux $\mathbf{q}$ is the sum of the flux of pseudothermal energy $\mathbf{q}_{pt}$ and the flux of true thermal energy $\mathbf{q}'$. It was shown in §1.5.7 that mechanical energy is converted to internal energy at the rate equal to the stress power $\dot{\Phi} \equiv -\boldsymbol{\sigma}^{T}:\nabla\mathbf{v}$. In our model for rapid granular flows, we shall assume that mechanical energy is first converted to fluctuational kinetic energy, which then is converted to thermal energy by inelastic collisions at the rate of Γ. This is the energy cascade referred to in §7.1. As a result, the balance of internal energy (1.45) may be separated into the two equations

$$\rho\frac{D}{Dt}(\frac{1}{2}w^2) = -\nabla\cdot\mathbf{q}_{pt} + \dot{\Phi} - \Gamma \tag{7.11}$$

$$\rho\frac{D\hat{U}'}{Dt} = -\nabla\cdot\mathbf{q}' + \Gamma \tag{7.12}$$

which represent the balance of pseudothermal energy and true thermal energy, respectively. If changes in the true temperature do not affect the properties of the particles (such as their size, shape, and coefficient of restitution), the balance of thermal energy is not of consequence. We shall assume this to be the case, and henceforth not concern ourselves with the balance of thermal energy (7.12). For the sake of convenience, we shall henceforth drop the subscript "pt" from $\mathbf{q}_{pt}$.

To close the above set of equations, constitutive relations are required for $\boldsymbol{\sigma}$, $\mathbf{q}$, and Γ, and in determining these we follow an intuitive, rather than rigorous approach. The more rigorous analysis in §7.3, based on the kinetic theory of gases, yields constitutive relations of precisely the form given here. We assume the form

$$\boldsymbol{\sigma} = p\,\mathbf{I} - 2\mu\mathbf{D} \tag{7.13}$$

for the stress, where $\mathbf{D} \equiv -\mathbf{C}$ (see (2.49)) is the rate of deformation tensor defined in the tensile sense. Further, we assume a Fourier relation for the pseudothermal energy flux

$$\mathbf{q} = -\kappa\nabla(\frac{1}{2}\rho\,w^2) \tag{7.14}$$

Here p is the pressure, $\mathbf{I}$ is the identity tensor, μ is the viscosity, and κ is the conductivity of pseudothermal energy. (Note that the symbol μ is used for the viscosity in this and the following chapters, whereas in earlier chapters it was used for the coefficient of friction.) Though the stress appears to have the form corresponding to a Newtonian fluid, it will soon be clear that this is not the case: we shall see that μ is not a constant, but depends on the nature of the flow through s and w. Similarly, the pseudothermal conductivity κ is not a constant, as it too depends on s and w.

The properties p, μ, κ, and Γ must be determined from a consideration of the details of particle interactions. To determine μ, we consider a dense granular material subjected to a constant shear rate (Fig. 7.4). On average, the momentum transferred by a particle in one layer to another particle in a layer below it during collision is $\sim m_p\Delta v_x$. Hence the momentum transferred per collision by the lower layer to the upper is $\sim -m_p\Delta v_x$. The frequency of collisions between particles in the two layers scales as w/s, and there are $n(d_p + s) \approx n\,d_p$ particles per unit area in each layer, where n is the number density of particles. The shear stress σ_{yx} on a surface of constant y is the flux of x momentum in the y direction across the surface. Considering such a surface between the upper and lower layers, and recognizing that the shear stress is the product of the momentum impulse per collision, the frequency of collisions, and the number of particles per unit area in each layer, we obtain

$$\sigma_{yx} = -a_2 m_p \Delta v_x \frac{w}{s} n d_p \tag{7.15}$$

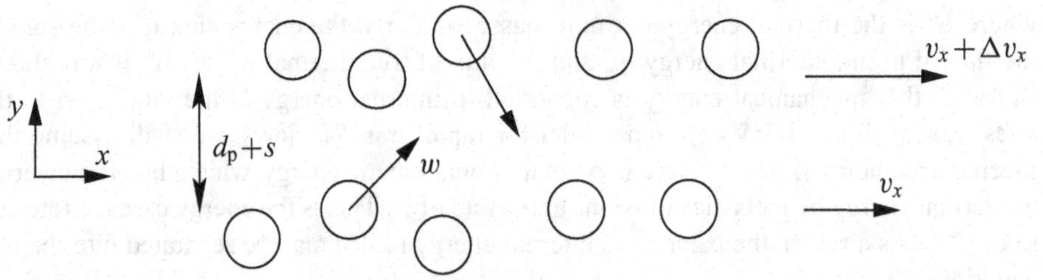

Figure 7.4. Rapid shear of a dense granular material. The gap between the particles is exaggerated for clarity.

The constant a_2 is introduced because the above is not a proper statistical average: the momentum impulse varies according to the velocities of the particles before collision, and so does the frequency of collision. To compute the shear stress due to collisions correctly, one must weight the momentum flux arising from a collision between particles with given precollision velocities with the probability of finding a pair with those velocities, and average over all possible precollision velocities. At this point, we sacrifice rigour for an analysis that is simple but provides physical insight, and write the flux as a product of the *mean* momentum impulse, the *mean* collision frequency, and the number density. The constant a_2 is the scale factor that relates this to the correct statistical average.

Identifying $\Delta v_x/(d_p+s) \approx \Delta v_x/d_p$ as the shear rate dv_x/dy, and the product nm_p as the density ρ, the shear stress becomes $\sigma_{yx} = -a_2\rho\, d_p^2\,(w/s)\,dv_x/dy$, whence the viscosity is

$$\mu = a_2\rho\, d_p^2\,\frac{w}{s} \tag{7.16}$$

The pressure p is isotropic, i.e., it is the same in all directions. It does not arise from the transfer of momentum due to coherent motion of the particles, but from their fluctuational velocities, which is $\sim m_p w$ per collision. Using the same arguments as above, the pressure is a product of $m_p w$, w/s and nd_p, hence

$$p = a_1\rho\, w^2\,\frac{d_p}{s} \tag{7.17}$$

where a_1 is a dimensionless constant. By similar reasoning, the pseudothermal conductivity and the rate of energy dissipation per unit volume are found to be (Haff, 1983)

$$\kappa = a_3\rho\, d_p^2\,\frac{w}{s} \tag{7.18}$$

$$\Gamma = a_4(1-e_p^2)\rho\,\frac{w^3}{s} \tag{7.19}$$

The O(1) dimensionless constants a_1–a_4 cannot be determined from this heuristic analysis. However, they may be determined by comparing (7.16)–(7.19) with the high density limit of the results obtained from kinetic theory in §7.3.13 (see Problem 7.1).

Equations (7.8), (7.9), and (7.11) are now closed, and we are in a position to consider their solution for a few simple flow problems.

7.2.1. Application to Uniform Plane Shear

Consider the steady shearing of a granular material confined between plane parallel walls (Fig. 7.5) in the absence of gravity and any other body force; shear is achieved by moving one wall relative to the other at constant speed. In general, conditions at the wall differ from those in the bulk, and consequently the shear rate and other variables vary with distance from the walls. However, we assume here that the shear rate $dv_x/dy \equiv \dot{\gamma}$ is spatially uniform.

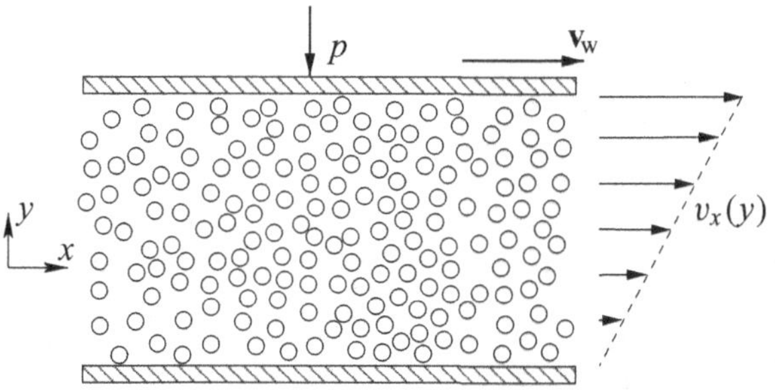

Figure 7.5. Schematic diagram of uniform plane shear: shear is generated by moving the upper wall at a constant velocity v_W while keeping the lower wall stationary. It is assumed that the material shears uniformly, i.e., dv_x/dy is constant.

It is shown in §8.2 that the uniformly shearing state arises for a particular set of boundary conditions, though not physically realistic. Nevertheless, it is useful to consider this case here in order to elucidate some features of the heuristic theory, namely the dependence of the stress and the fluctuation velocity on the shear rate and the inelasticity.

One may wonder how shear stress is transmitted by the walls to the granular material if the particles are smooth. It is possible if the walls are not perfectly flat, but have undulations (see §8.1.2); momentum in the x direction is transmitted to particles striking the fore of the undulations. At this stage, we shall not concern ourselves with the mechanics of stress transmission at the walls – this is done in §8.1, where boundary conditions at solid walls are derived – rather, we take it that stress is indeed transmitted, and study its effects on the granular material far from the boundaries.

When the shear rate is everywhere equal, it is possible to have a situation where there is no spatial gradient in w. This is because the pseudothermal energy generated by the stress work can be *locally* dissipated by inelastic collisions between particles. This is unlike a fluid composed of elastic particles, where the thermal energy generated by the stress work has to be *conducted* away to the boundaries, which requires a gradient in w. In a granular gas, the state of uniform w can be thought of as the simplest solution: other solutions, with spatial variation of w and s, arise as instabilities of this solution, which eventually attain steady nonuniform states, as discussed in §8.4.2.

When w is spatially uniform and the flow is fully developed, (7.17) and the y component of the momentum balance (7.9),

$$\frac{\mathrm{d}p}{\mathrm{d}y} = 0 \tag{7.20}$$

imply that s too is spatially uniform. Thus $\dot{\gamma}$, w, and s are all constant in steady uniform plane shear flow. The value of s is set by the gap maintained between the two walls. If the gap before shearing commences is H, the dilation of the material that accompanies shearing causes the gap to increase to $H + \Delta H$ when steady shearing is attained. The number of particles spanning the distance between the walls before shearing is $\approx H/d_\mathrm{p}$, and the number when steady shearing is attained is $\approx (H + \Delta H)/(d_\mathrm{p} + s)$. Since these must be equal, we obtain $s \approx d_\mathrm{p}(\Delta H/H)$.

The value of w is obtained from the balance of fluctuational energy (7.11),

$$-\sigma_{yx}\dot{\gamma} - \Gamma = 0 \tag{7.21}$$

which after substituting from (7.13), (7.16), and (7.19), gives

$$w^2 = \frac{a_2\, d_p^2\, \dot{\gamma}^2}{a_4(1-e_p^2)} \tag{7.22}$$

Another solution of (7.21) is $w = 0$, which is unacceptable because the normal and shear stresses at the walls vanish. Thus for a granular material of given properties, the mean fluctuation speed is determined by the shear rate. Using the expression for w from above, the shear stress becomes

$$\sigma_{yx} = -\left(\frac{a_2^3}{a_4(1-e_p^2)}\right)^{1/2} \rho\, \dot{\gamma}^2\, d_p^3/s \tag{7.23}$$

which agrees with our expectation that the stress must vary as the square of the shear rate.

Next, to understand the time scale for the relaxation of hydrodynamic quantities, let us consider a sudden change of the shear rate, though still spatially uniform, from $\dot{\gamma}$ to $\dot{\gamma}_1$. The interparticle distance s remains unchanged, because the distance between the walls is kept constant. The fluctuational energy balance now assumes the form

$$\rho\, \frac{\mathrm{d}}{\mathrm{d}t}(\frac{1}{2}w^2) = a_2\, \rho\, d_p^2\, \frac{w}{s}\dot{\gamma}_1^2 - a_4(1-e_p^2)\rho\, \frac{w^3}{s} \tag{7.24}$$

with the initial condition

$$w(0) = \dot{\gamma}\left(\frac{a_2 d_p^2}{a_4(1-e_p^2)}\right)^{1/2}$$

The solution is

$$\frac{w(t)}{w_1} = \frac{(\dot{\gamma}_1+\dot{\gamma})/(\dot{\gamma}_1-\dot{\gamma}) - e^{-t/\tau}}{(\dot{\gamma}_1+\dot{\gamma})/(\dot{\gamma}_1-\dot{\gamma}) + e^{-t/\tau}} \tag{7.25}$$

where

$$\tau \equiv \frac{1}{2a_4(1-e_p^2)}\frac{s}{w_1} \tag{7.26}$$

is the time scale for w to relax to its new steady state value

$$w_1 \equiv d_p\, \dot{\gamma}_1 (\frac{a_2}{a_4(1-e_p^2)})^{1/2}$$

Thus the time scale to attain a steady state is $\sim s/w_1$, i.e., roughly the average time between two successive collisions of a particle, unless e_p is close to unity. When $e_p \to 1$, we see from (7.22) and (7.26) that $\tau \sim (1-e_p^2)^{-1/2}$: it becomes very large.

The expression for the relaxation time in (7.26) is not applicable in the precise limit $e_p = 1$, because in this limit w at a steady state is determined by a balance between the stress work and the gradient of the flux $\mathrm{d}q_y/\mathrm{d}y$ (rather than between the stress work and the dissipation rate Γ); here it is easily seen that $\tau \sim s/w\,(H/d_p)^2$ (see Problem 7.2). Indeed, this is the scaling for the relaxation time of all variables that are *conserved* in particle interactions, namely, the mass, momentum, and energy, and hydrodynamic equations are normally written only for such variables. In granular gases, energy is not conserved, and therefore w relaxes very quickly, unless e_p is close to unity. When e_p is not close to unity, w is not a separate hydrodynamic variable, rather it is "enslaved" to the $\mathbf{v}$ and s fields; one may determine w from (7.22) using the instantaneous, local value of $\dot{\gamma}$, even if the flow is spatially nonuniform and time dependent. These issues are discussed further in §7.3.6.

7.3. KINETIC THEORY FOR A GRANULAR GAS OF SMOOTH INELASTIC PARTICLES

As discussed earlier, the mechanism of interaction between particles in the regime of rapid flow, through collisions of short duration, is much the same as in a gas. The kinetic theory for dense gases has been well developed, and the concepts and methods used there can be put to use for analyzing the behavior of granular gases. Our discussion of kinetic theory is rather brief, as it is a classical subject and detailed descriptions can be found in several books (Chapman and Cowling, 1964; Huang, 1985; Reif, 1985). A good historical summary of kinetic theory can be found in Chapman and Cowling (1964, p. 380). Here we focus on the extension of kinetic theory to granular gases consisting of inelastic particles. We derive the hydrodynamic equations and the necessary closure (or constitutive) relations for rapid granular flows.

A large part of this section is devoted to the derivation of the hydrodynamic equations, or the equations of motion, that govern rapid granular flow, and of the constitutive relations for the fluxes and sinks of certain quantities that occur in the hydrodynamic equations. The reader who is not interested in the derivation, but only in the final result may proceed directly to the relevant section. The equations of motion are given in §7.3.6, and the constitutive relations in §7.3.13. It is however important that the reader recognizes the assumptions made in the derivation, and the ensuing limitations of the theory; a discussion on these may be found in §7.3.4, §7.3.6, and §7.3.7.

The theory is applied to simple flow problems in Chapter 8.

7.3.1. Statistical Preliminaries

Consider a collection of identical spherical particles of mass m_p which are in a state of continuous motion. Since we consider the particles to be smooth and frictionless, the angular velocity of each particle (about its own axes) remains unchanged, no matter what other changes the system undergoes. Hence the angular velocities are irrelevant in our analysis, and the six-dimensional space of position and velocity, or phase space, completely determines the state of all the particles. Collisions between particles, or between particles and the walls of the container, cause their velocities $\mathbf{c}$ to change and fluctuate about the local mean. If they are inelastic, their motion will decay unless maintained by external forcing such as shear or vibration of the container; if the external forcing is discontinued, the particles will lose energy during collisions and ultimately all fluctuational motion will cease.

If there are a large number of particles in the collection, their spatial and velocity distribution can be characterized by a smoothly varying distribution function $f^{(1)}(t, \mathbf{r}, \mathbf{c})$, such that the number of particles at time t in a differential volume $d\mathbf{r}$ and a differential velocity range $d\mathbf{c}$ is

$$f^{(1)}(t, \mathbf{r}, \mathbf{c})\, d\mathbf{r}\, d\mathbf{c}$$

In a Cartesian coordinate system (x,y,z), the volume element $d\mathbf{r}$ is $dx\, dy\, dz$ and the volume in velocity space $d\mathbf{c}$ is $dc_x dc_y dc_z$. The assumption of a large number of particles is satisfied in molecular systems; in a gas under normal conditions of temperature and pressure, there are $\sim 10^7$ molecules in a volume of one cubic micrometer, which is sufficient spatial resolution for most practical purposes. For a granular material comprising macroscopic particles, it would seem that we can only achieve a spatial resolution that is much larger than the particle size. However, we could wait for a sufficiently long period of time until a large enough number of particles have passed through the phase volume $d\mathbf{r}\, d\mathbf{c}$; this would mean that $f^{(1)}$ is defined not instantaneously but only in a time-averaged sense. This is not a problem in molecular gases, where the mean free time $\tau_c \sim s/w$ (i.e., the average time between successive collisions of a particle) is $\sim 10^{-12}$–10^{-14} s. For a sheared granular gas, assuming a fluctuation velocity of 1 cm/s and a mean interparticle spacing of 0.1 cm, $\tau_c \sim 0.1$ s; we can only expect to resolve temporal changes which have a much slower variation, and in some cases a meaningful description of transient phenomena may be in doubt. Thus the separation between microscopic and macroscopic length or time scales, which is so clearly present in molecular systems, is often difficult to achieve in granular materials. Nevertheless, we assume that such a separation is present and proceed with the analysis, reserving further comment on this issue till the end of this chapter.

From the definition of $f^{(1)}$, it follows that the number density of particles is

$$n(t, \mathbf{r}) = \int f^{(1)}(t, \mathbf{r}, \mathbf{c})\, \mathrm{d}\mathbf{c} \tag{7.27}$$

Implicit in all integrals over velocity space, such as the one above, is that the limits of integration are $-\infty$ and ∞ for c_x, c_y, and c_z. The bulk density is $\rho \equiv n\, m_\mathrm{p}$. We often find it convenient to write the density in terms of the solids fraction ν as $\rho \equiv \rho_\mathrm{p}\nu$.

The average $\langle\psi\rangle$ of any particle property ψ is defined as

$$n(t, \mathbf{r})\, \langle\psi\rangle(t, \mathbf{r}) = \int \psi f^{(1)}(t, \mathbf{r}, \mathbf{c})\, \mathrm{d}\mathbf{c} \tag{7.28}$$

Thus the mean velocity $\mathbf{v}$ is given by

$$n(t, \mathbf{r})\, \mathbf{v}(t, \mathbf{r}) = \int \mathbf{c} f^{(1)}(t, \mathbf{r}, \mathbf{c})\, \mathrm{d}\mathbf{c} \tag{7.29}$$

and the grain temperature, which is two-thirds of the mean fluctuational kinetic energy per unit mass, by

$$\frac{3}{2} n(t, \mathbf{r})\, T(t, \mathbf{r}) = \frac{1}{2} \int C^2 f^{(1)}(t, \mathbf{r}, \mathbf{c})\, \mathrm{d}\mathbf{c} \tag{7.30}$$

Here

$$\mathbf{C} \equiv \mathbf{c} - \mathbf{v} \tag{7.31}$$

is the fluctuation, or peculiar, velocity of the particles, and $C \equiv |\mathbf{C}|$ its magnitude. (We have adopted the notation in this and the succeeding chapters that if $\mathbf{b}$ is a vector, then b is its magnitude.) Note that this definition of T differs from the classical definition (which gives the temperature in Kelvin) by a factor $k_\mathrm{B}/m_\mathrm{p}$, where k_B is the Boltzmann constant. The temperature is related to the root-mean-square speed fluctuation w defined in the high density theory as $3T = w^2$.

We shall henceforth refer to $f^{(1)}$ as the *singlet* distribution function, as it characterizes the spatial and velocity distribution of individual particles. It is also useful to define the distribution function for the positions and velocities of a pair of particles, $f^{(2)}(t, \mathbf{r}_1, \mathbf{c}_1, \mathbf{r}_2, \mathbf{c}_2)$, where the indices 1 and 2 refer to the two particles. At time t, the number of particle pairs with one located within the volume $\mathrm{d}\mathbf{r}_1$ and the velocity interval $\mathrm{d}\mathbf{c}_1$, and the other within $\mathrm{d}\mathbf{r}_2$ and $\mathrm{d}\mathbf{c}_2$ is

$$f^{(2)}(t, \mathbf{r}_1, \mathbf{c}_1, \mathbf{r}_2, \mathbf{c}_2)\, \mathrm{d}\mathbf{r}_1\, \mathrm{d}\mathbf{c}_1\, \mathrm{d}\mathbf{r}_2\, \mathrm{d}\mathbf{c}_2$$

It will become clear in the following section that this function is necessary to determine changes caused by particle collisions. By extending (7.27), we define the two particle position distribution function

$$n^{(2)}(t, \mathbf{r}_1, \mathbf{r}_2) = \iint f^{(2)}(t, \mathbf{r}_1, \mathbf{c}_1, \mathbf{r}_2, \mathbf{c}_2)\, \mathrm{d}\mathbf{c}_1\, \mathrm{d}\mathbf{c}_2 \tag{7.32}$$

so that the number of particle pairs with one particle in the volume $\mathbf{dr}_1$, centered at $\mathbf{r}_1$, and the other in $\mathbf{dr}_2$, centered at $\mathbf{r}_2$, is $n^{(2)}(t, \mathbf{r}_1, \mathbf{r}_2)\mathbf{dr}_1\, \mathbf{dr}_2$.

In the following sections, we will derive the Boltzmann equation, which governs the evolution of $f^{(1)}$, and determine from it the Maxwell–Boltzmann velocity distribution for the equilibrium state of a gas of elastic particles. We will see that knowledge of $f^{(1)}$ gives us all the equilibrium properties of the gas. For nonequilibrium states, we will see in §7.3.6 that $f^{(1)}$ is required to determine the constitutive relations for the fluxes and sources that appear in the equations of motion.

7.3.2. The Evolution of $f^{(1)}$

In the absence of collisions, a particle at position $\mathbf{r}$ having velocity $\mathbf{c}$ will be found after an infinitesimal time interval $\mathrm{d}t$ at $\tilde{\mathbf{r}} = \mathbf{r} + \mathbf{c}\,\mathrm{d}t$ with a velocity $\tilde{\mathbf{c}} = \mathbf{c} + \mathbf{b}\,\mathrm{d}t$. Here $\mathbf{b}$ is the external body force per unit mass; we assume that it may be a function of t and $\mathbf{r}$, but not of $\mathbf{c}$. Note that this assumption precludes the presence of a drag force exerted on the particles by an interstitial fluid. For a collection of particles initially in the phase-space volume $\mathbf{dr}\,\mathbf{dc}$ centered at $(\mathbf{r}, \mathbf{c})$, this statement may be written as

$$f^{(1)}(t + \mathrm{d}t, \tilde{\mathbf{r}}, \tilde{\mathbf{c}})\, d\tilde{\mathbf{r}}\, d\tilde{\mathbf{c}} = f^{(1)}(t, \mathbf{r}, \mathbf{c})\, \mathbf{dr}\, \mathbf{dc}$$

The volume in phase space at time $t + \mathrm{d}t$ is related to that at time t by $\mathbf{d}\tilde{\mathbf{r}}\, \mathbf{d}\tilde{\mathbf{c}} = |\mathcal{J}|\, \mathbf{dr}\, \mathbf{dc}$, where $\mathcal{J}$ is the determinant of the Jacobian (see §H.3) for the transformation $(\tilde{\mathbf{r}}, \tilde{\mathbf{c}}) \longrightarrow (\mathbf{r}, \mathbf{c})$. It is straightforward to show that $|\mathcal{J}| = 1$, hence the volume is conserved. To account for collisions, we write the net increase in the number of particles in the volume $\mathbf{d}\tilde{\mathbf{r}}\, \mathbf{d}\tilde{\mathbf{c}}$ due to collisions during the time interval $\mathrm{d}t$ as $\dot{f}_{\text{coll}}\, \mathbf{dr}\, \mathbf{dc}\, \mathrm{d}t$. Expanding $f^{(1)}(t + \mathrm{d}t, \tilde{\mathbf{r}}, \tilde{\mathbf{c}})$ in a Taylor series about $(t, \mathbf{r}, \mathbf{c})$ and dividing the equation by $\mathrm{d}t$, we get the *Boltzmann equation* governing the evolution of $f^{(1)}$ (Chapman and Cowling, 1964, pp. 46–47; Huang, 1985, pp. 57–58; Reif, 1985, p. 509),

$$\frac{\partial f^{(1)}}{\partial t} + \mathbf{c} \cdot \boldsymbol{\nabla} f^{(1)} + \mathbf{b} \cdot \boldsymbol{\nabla}_c f^{(1)} = \dot{f}_{\text{coll}} \tag{7.33}$$

Here $\boldsymbol{\nabla}$ and $\boldsymbol{\nabla}_c$ represent gradients with respect to the position and velocity coordinates, respectively; thus the second and third terms on the left-hand side of (7.33) represent advection of $f^{(1)}$ in position and velocity space, respectively. The term on the right-hand side represents the rate at which the distribution function changes due to collisions.

To determine $\dot{f}_{\text{coll}}$, it is usually assumed that only binary collisions between particles need to be considered. This is correct when the contact time during collisions is much smaller than the mean free time between collisions, so that collisions between three or more particles are improbable. Referring to Fig. 7.6, our task is to determine the rate at which collisions will change the number of particles in the differential phase-space volume $\mathbf{dr}\,\mathbf{dc}$ around $(\mathbf{r}, \mathbf{c})$ during the time interval $\mathrm{d}t$. If the precollision velocity of particle 2 is within $\mathbf{dc}$, a collision with particle 1 will knock it out of the volume. If the precollision velocity of particle 2 is outside $\mathbf{dc}$, it can come into the velocity range after collision with another particle. Following Chapman and Cowling (1964), we refer to the former type of collisions as "direct collisions," and the latter type as "inverse collisions."

We first determine the rate at which particles *leave* $\mathbf{dr}\,\mathbf{dc}$ by direct collisions. Considering particle 1 in Fig. 7.6, if collisional contact with particle 2 is to occur on an infinitesimal element of its surface defined by the solid angle $\mathbf{dk}^{\dagger}$ around $\mathbf{k}$, then its center at the instant of contact must be in a small region around $\mathbf{r}_1 = \mathbf{r} - \mathbf{k}\, d_{\mathrm{p}}$. During a time interval $\mathrm{d}t$ before collision, the center of particle 1 must reside in a cylinder of length $g\,\mathrm{d}t$ and cross-section

† The solid angle subtended by a surface S at a point is the projected area of the surface on a sphere of unit radius centered at that point.

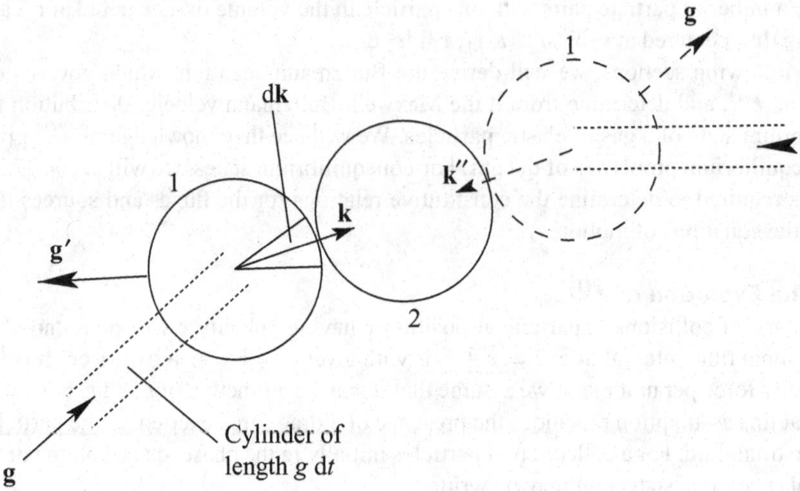

Figure 7.6. The velocity and trajectory of particle 1 before and after it collides with particle 2, in a reference frame that moves with particle 2. The position vector of particle 2 in the laboratory reference frame is **r**, and **k** is the unit vector pointing toward the center of particle 2 from the center of particle 1. The dashed circle represents particle 1 in the "inverse" collision.

area $d_p^2\,\mathrm{d}\mathbf{k}\,\mathbf{k}\cdot\mathbf{g}/g$. Here $\mathbf{g}$ is the relative velocity of particle 1 with respect to particle 2 and $g \equiv |\mathbf{g}|$ is its magnitude. Thus the volume in which particle 1 resides in the time interval $\mathrm{d}t$ before collision is $\mathrm{d}\mathbf{r}_1 = d_p^2\,\mathrm{d}\mathbf{k}\,\mathbf{k}\cdot\mathbf{g}\,\mathrm{d}t$, and it has a velocity in the range $\mathrm{d}\mathbf{c}_1$. Particle 2 resides in the volume $\mathrm{d}\mathbf{r}$ and has a velocity in the range $\mathrm{d}\mathbf{c}$. Hence the number of collisions that deplete particles from $\mathrm{d}\mathbf{r}\,\mathrm{d}\mathbf{c}$ in the time interval $\mathrm{d}t$ is

$$f^{(2)}(t, \mathbf{r} - d_p\mathbf{k}, \mathbf{c}_1, \mathbf{r}, \mathbf{c})\, d_p^2\, \mathbf{k}\cdot\mathbf{g}\, \mathrm{d}\mathbf{k}\, \mathrm{d}\mathbf{c}_1\, \mathrm{d}\mathbf{r}\, \mathrm{d}\mathbf{c}\, \mathrm{d}t \tag{7.34}$$

with the constraint $\mathbf{k}\cdot\mathbf{g} > 0$ so that only impending collisions are counted.

We next determine the rate at which particles *enter* the range $\mathrm{d}\mathbf{r}\,\mathrm{d}\mathbf{c}$ by inverse collisions. We consider collisions in which particles 1 and 2 have precollision velocities $\mathbf{c}_1''$ and $\mathbf{c}''$, respectively (see Fig. 7.6), and leave with velocities $\mathbf{c}_1$ and $\mathbf{c}$, respectively, after collision. The accumulation of particles in $\mathrm{d}\mathbf{r}\,\mathrm{d}\mathbf{c}$ around $(\mathbf{r}, \mathbf{c})$ in the time interval $\mathrm{d}t$ is

$$f^{(2)}(t, \mathbf{r} - d_p\mathbf{k}'', \mathbf{c}_1'', \mathbf{r}, \mathbf{c}'')\, \mathrm{d}\mathbf{r}\, \mathrm{d}\mathbf{c}''\, d_p^2\, \mathbf{k}''\cdot\mathbf{g}''\, \mathrm{d}\mathbf{k}''\, \mathrm{d}\mathbf{c}_1''\, \mathrm{d}t \tag{7.35}$$

From the collision rule (7.3), we see that

$$\mathbf{k}''\cdot\mathbf{g}'' = -\frac{1}{e_p}\,\mathbf{k}''\cdot\mathbf{g} \tag{7.36}$$

The constraint to be enforced for counting only impending collisions is therefore $\mathbf{k}''\cdot\mathbf{g} < 0$. It is convenient to choose

$$\mathbf{k}'' = -\mathbf{k} \tag{7.37}$$

This is in no way restrictive, as we will integrate over all $\mathbf{k}$ (see below). Consequently, (7.4) and (7.5) give $\mathbf{c}_1''$ and $\mathbf{c}''$ in terms of $\mathbf{c}_1$ and $\mathbf{c}$ as

$$\mathbf{c}_1'' = \mathbf{c}_1 - \frac{\eta_1}{e_p}(\mathbf{k}\cdot\mathbf{g})\mathbf{k}, \quad \mathbf{c}'' = \mathbf{c} + \frac{\eta_1}{e_p}(\mathbf{k}\cdot\mathbf{g})\mathbf{k}. \tag{7.38}$$

In other words, for every direct collision with precollision velocities $\mathbf{c}_1$ and $\mathbf{c}$ and center-to-center unit vector $\mathbf{k}$, there is an inverse collision with precollision velocities given by (7.38) and center-to-center unit vector $-\mathbf{k}$. The volumes in velocity space before and after collision are related by $\mathbf{dc''}\,\mathbf{dc}_1'' = |\mathcal{J}|\mathbf{dc}\,\mathbf{dc}_1$ where $\mathcal{J}$ is the determinant of the Jacobian (see §H.3) for the transformation $(\mathbf{c},''\,\mathbf{c}_1'') \longrightarrow (\mathbf{c}, \mathbf{c}_1)$. This transformation is defined by (7.38), and it is easily verified that $|\mathcal{J}| = 1/e_\text{p}$. Substituting this and (7.36) in (7.35), and subtracting (7.34) from the result, we get the net gain of particles within $\mathbf{dr}\,\mathbf{dc}$ due to collisions with particles whose velocity before or after collision is near $\mathbf{c}_1$,

$$\mathbf{dr}\,\mathbf{dc}\,\mathrm{d}t\, d_\text{p}^2 \left[\frac{1}{e_\text{p}^2} f^{(2)}(t, \mathbf{r} + d_\text{p}\mathbf{k}, \mathbf{c}_1'', \mathbf{r}, \mathbf{c}'') - f^{(2)}(t, \mathbf{r} - d_\text{p}\mathbf{k}, \mathbf{c}_1, \mathbf{r}, \mathbf{c})\right] \mathbf{k}\cdot\mathbf{g}\,\mathbf{dk}\,\mathbf{dc}_1$$

Integrating the above over the entire surface of particle 1, i.e., over all $\mathbf{k}$, and over all velocities $\mathbf{c}_1$ yields the net gain of particles $\dot{f}_\text{coll}\,\mathbf{dr}\,\mathbf{dc}\,\mathrm{d}t$, whence the Boltzmann equation assumes the form

$$\frac{\partial f^{(1)}}{\partial t} + \mathbf{c}\cdot\nabla f^{(1)} + \mathbf{b}\cdot\nabla_c f^{(1)} = d_\text{p}^2 \int_{\mathbf{k}\cdot\mathbf{g}>0} \left[\frac{1}{e_\text{p}^2} f^{(2)}(t, \mathbf{r} + d_\text{p}\mathbf{k}, \mathbf{c}_1'', \mathbf{r}, \mathbf{c}'') - f^{(2)}(t, \mathbf{r} - d_\text{p}\mathbf{k}, \mathbf{c}_1, \mathbf{r}, \mathbf{c})\right] \mathbf{k}\cdot\mathbf{g}\,\mathbf{dk}\,\mathbf{dc}_1 \tag{7.39}$$

It now remains to determine the two-particle distribution function $f^{(2)}$: if it can be written in terms of the singlet distribution function $f^{(1)}$, (7.39) is closed and its solution can, in principle, be obtained. The closure that is usually used is the one proposed by Boltzmann (1872) (see also Boltzmann, 1995),

$$f^{(2)}(t, \mathbf{r}_1, \mathbf{c}_1, \mathbf{r}, \mathbf{c}) = f^{(1)}(t, \mathbf{r}_1, \mathbf{c}_1)\, f^{(1)}(t, \mathbf{r}, \mathbf{c}) \tag{7.40}$$

This relation follows from the assumption of *molecular chaos*, which means that the velocities of two particles just prior to collision are uncorrelated. It is expected to be quite accurate for dilute gases. For dense gases, a correction was suggested by Enskog (1922) (see Chapman and Cowling (1964, pp. 273–274) for an elaboration), of the form

$$f^{(2)}(t, \mathbf{r}_1, \mathbf{c}_1, \mathbf{r}, \mathbf{c}) = g(t, \mathbf{r}_1, \mathbf{r}) f^{(1)}(t, \mathbf{r}_1, \mathbf{c}_1)\, f^{(1)}(t, \mathbf{r}, \mathbf{c}) \tag{7.41}$$

where $g(t, \mathbf{r}_1, \mathbf{r}) \equiv n^{(2)}(t, \mathbf{r}_1 \mid \mathbf{r})/n$ is the pair distribution function. In other words, while the velocities of the particles coming into contact are uncorrelated, there is some spatial correlation which is determined by $g(t, \mathbf{r}_1, \mathbf{r})$. Here $n^{(2)}(t,\ \mathbf{r}_1 \mid \mathbf{r})$ is the conditional pair distribution, defined such that $n^{(2)}(t,\ \mathbf{r}_1 \mid \mathbf{r})\ \mathbf{d}r_1$ is the number of particles in the volume $\mathbf{d}r_1$ given that a particle is present at $\mathbf{r}$. With this closure, the final form of the Boltzmann equation for smooth, inelastic spheres is†

$$\frac{\partial f}{\partial t} + \mathbf{c}\cdot\nabla f + \mathbf{b}\cdot\nabla_c f = d_\text{p}^2 \int_{\mathbf{k}\cdot\mathbf{g}>0} \left[\frac{1}{e_\text{p}^2} f_1''(t, \mathbf{r} + d_\text{p}\mathbf{k}) f''(t, \mathbf{r})\, g(t, \mathbf{r} + d_\text{p}\mathbf{k}, \mathbf{r}) - f_1(t, \mathbf{r} - d_\text{p}\mathbf{k}) f(t, \mathbf{r})\, g(t, \mathbf{r} - d_\text{p}\mathbf{k}, \mathbf{r})\right] \mathbf{k}\cdot\mathbf{g}\,\mathbf{dk}\,\mathbf{dc}_1 \tag{7.42}$$

where we have used the notation $f_i(t, \mathbf{r}) \equiv f^{(1)}(t, \mathbf{r}, \mathbf{c}_i)$ and $f_i''(t, \mathbf{r}) \equiv f^{(1)}(t, \mathbf{r}, \mathbf{c}_i'')$. Henceforth, we shall drop the superscript $^{(1)}$ on the singlet-distribution function for the sake of convenience.

† We call this the Boltzmann equation, though classically the phrase is used for a dilute gas of elastic particles, i.e., (7.42) with $g(t, \mathbf{r}_1, \mathbf{r}) = 1$ and $e_\text{p} = 1$.

7.3.3. The Equilibrium Distribution Function

When particle collisions are elastic ($e_p = 1$), as in a molecular gas, and if there are no externally imposed gradients and forces, the total kinetic energy of the system does not change with time. In this situation of thermodynamic equilibrium, we expect no temporal or spatial variations in the state of the material; f will therefore be independent of time and position. As a result, all terms on the left-hand side of (7.42) vanish. Moreover, there is no preferred direction and hence the pair distribution function can depend only on the scalar distance between the particles, i.e., $g(\mathbf{r} + d_p\mathbf{k}, \mathbf{r}) = g(d_p)$ for any $\mathbf{k}$. The Boltzmann equation (7.42) then reduces to

$$0 = g(d_p)\, d_p^2 \int_{\mathbf{k}\cdot\mathbf{g} > 0} (f_1'' f'' - f_1 f)\, \mathbf{k}\cdot\mathbf{g}\, \mathrm{d}\mathbf{k}\, \mathrm{d}\mathbf{c}_1 \tag{7.43}$$

It is clear that

$$f_1'' f'' - f_1 f = 0 \tag{7.44}$$

is a sufficient condition for the solution of (7.43); Boltzmann's H theorem shows that it is also a necessary condition. The proof of this theorem can be found in Chapman and Cowling (1964, pp. 69–71), Huang (1985, pp. 68–70), and most other treatises on kinetic theory. This implies that

$$\ln f_1'' + \ln f'' = \ln f_1 + \ln f$$

i.e., the logarithm of the singlet distribution of a pair of particles is conserved during their collision. It follows that $\ln f$ can be expressed as a linear combination of all the functions of $\mathbf{c}$ that are conserved during a collision. For smooth spheres, these are the momentum, energy, and a constant; hence

$$\ln f = \alpha_1 + \boldsymbol{\alpha}_2\cdot\mathbf{c} + \alpha_3\, c^2 \tag{7.45}$$

The constants α_1, $\boldsymbol{\alpha}_2$, and α_3 may be related to the mean properties of the fluid, defined by (7.27), (7.29), and (7.30). The algebra is eased considerably if we rewrite (7.45) as

$$\ln f = \alpha_1' + \boldsymbol{\alpha}_2'\cdot(\mathbf{c} - \mathbf{v}) + \alpha_3' |\mathbf{c} - \mathbf{v}|^2 \tag{7.46}$$

Equation (7.46) is simply a result of transforming to a reference frame that is moving with the mean velocity $\mathbf{v}$ of the particles. In this reference frame, the value of f for velocity $(\mathbf{c} - \mathbf{v})$ must be the same as that for $-(\mathbf{c} - \mathbf{v})$, as this is simply equivalent to reversing the directions of the coordinate axes. It then follows that $\boldsymbol{\alpha}_2' = 0$. Also $\alpha_3' \le 0$ as f must decay as the magnitude of the velocity becomes large. Application of (7.27) and (7.30) yields α_1' and α_3', and the resulting expression for the velocity distribution function is

$$f(\mathbf{c}) = \frac{n}{(2\pi T)^{3/2}} \exp\left[-\frac{|\mathbf{c} - \mathbf{v}|^2}{2T}\right] \tag{7.47}$$

This is the Maxwell–Boltzmann distribution for the equilibrium state of a gas, from which one can derive all thermodynamic properties of interest.

In the presence of a conservative external force, $\mathbf{b}$ may be expressed as a gradient of a scalar potential field, i.e., $\mathbf{b} = -\boldsymbol{\nabla}\phi(\mathbf{r})$. It is a straightforward exercise (Chapman and Cowling, 1964, pp. 77–80) to show that the distribution function remains as in (7.47), but with the number density varying with position as

$$n(\mathbf{r}) = n_0 e^{-\phi(\mathbf{r})/T} \tag{7.48}$$

Although we have derived (7.47) for a gas whose molecules undergo binary interactions and satisfy molecular chaos, the distribution in fact has greater validity. It also holds for liquids and solids, where molecular interactions are far stronger and their motion strongly correlated. This can be shown through the more general principles of statistical mechanics see, e.g., Huang (1985, p. 139)).

7.3.4. The Departure from Equilibrium

When the gas is not at equilibrium, the left-hand side of (7.42) does not vanish and the distribution is therefore no longer given by (7.47). Departure from equilibrium arises most commonly from spatial gradients of the hydrodynamic variables $\mathbf{v}$, T, and ν. For a small departure from equilibrium, the distribution function can be expressed as a series expansion in the gradients $\nabla\mathbf{v}$, ∇T, and $\nabla\nu$. The departure from equilibrium is small if the gradients are small (as explained more precisely below), which is the case for molecular fluids. Exact solutions of (7.42) in nonequilibrium conditions are not known, but methods for approximate solution have been developed, one of which is discussed in §7.3.7.

If the particles are inelastic, collisions dissipate kinetic energy until all fluctuational motion ceases. The equilibrium state for the distribution function then, no matter what the initial state and how small the inelasticity, is the trivial one of a Dirac delta function centered at the mean velocity $\mathbf{v}$. External forcing is therefore necessary to sustain fluctuational motion, and hence the system is inherently far from equibrium. The forcing may be in the form of a vibrating boundary, which directly imparts fluctuational kinetic energy to the particles adjacent to it, a body force, or relative motion of the boundaries, which result in the conversion of kinetic energy from the mean flow into fluctuational modes. Except in some contrived circumstances, spatial gradients of the hydrodynamic variables are usually present. Therefore the transport properties of a granular material, when it is forced in some manner, are very different from its properties at equilibrium; we do not normally speak of the viscosity of a grain heap, but there is a well-defined viscosity when it is subjected to steady shear.

The state of equilibrium in a gas is characterized by its mean velocity, temperature, and number density (or, equivalently, the pressure), and the molecular velocities are distributed according to the Maxwell–Boltzmann distribution. Even when the gas is not in equilibrium, the velocity distribution usually remains close to that at equilibrium. For instance, in shear flow the departure from equilibrium can be estimated by the ratio of the velocity difference between nearby particles due to the imposed shear to the mean fluctuational speed at equilibrium

$$\delta \equiv \frac{\dot{\gamma} s}{\sqrt{T}} \tag{7.49}$$

where $\dot{\gamma}$ is the imposed shear rate and s is the mean free path. Using the equilibrium estimate for the mean free path $s = 1/(\sqrt{2}\pi n d_p^2)$ (Chapman and Cowling, 1964, p. 91), we find δ to be very small for values of $\dot{\gamma}$ encountered normally. For example, for H_2 gas at NTP ("normal temperature and pressure," referring to $\frac{m}{k_B}T = 300\,\mathrm{K}$, $p = 10^5\,\mathrm{Nm}^{-2}$)

$$T = 300\,\frac{k_B}{m} = 300\,\frac{1.38\times10^{-23}}{3.347\times10^{-27}} = 1.234\times10^{6}\,\mathrm{m}^2\ \mathrm{s}^{-2}$$

and the number density may calculated from the ideal gas equation of state

$$n = \frac{p}{mT} = \frac{1\times10^{5}}{3.347\times10^{-27}\times1.234\times10^{6}} = 2.42\times10^{25}\,\mathrm{m}^{-3}$$

Using a molecular diameter of 1.365×10^{-10} m, we get a mean free path of 1.247×10^{-7} m. Substituting the above in (7.49), we get $\delta\sim10^{-10}$ for $\dot{\gamma} = 1\,\mathrm{s}^{-1}$ and $\delta\sim10^{-5}$ for $\dot{\gamma} = 10^5\,\mathrm{s}^{-1}$.

Following our discussion above, it is not useful to assess the departure from equilibrium for a granular gas comprising inelastic particles, because its forced state bears no resemblance to its equilibrium state. We can however ask what its departure is from the equilibrium state of a gas of elastic particles *at the same temperature*, i.e., what is the shear rate required to maintain a granular gas at the same temperature as a gas of elastic particles? The parameter δ in (7.49) is then a measure of the deviation of a granular gas from a gas of elastic particles. Considering steady uniform shear of a granular material, we recall from §7.2.1 that the temperature varies as

$$T \sim s^2 \dot{\gamma}^2/(1 - e_{\text{p}}^2)$$

and hence $\delta \sim \sqrt{1 - e_{\text{p}}^2}$. This is small when $e_{\text{p}} \to 1$, i.e., when the particles are nearly elastic. However, the temperature diverges as $e_{\text{p}} \to 1$, unless we simultaneously decrease $\dot{\gamma}$ such that the ratio $\dot{\gamma}/\sqrt{1 - e_{\text{p}}^2}$ is kept constant. This idea is implicit in all the kinetic theories of granular flows, but was first enunciated by Sela et al. (1996). For a general flow, with spatial and temporal variation of the hydrodynamic variables, this ratio need not remain constant, but both quantities must be small (Sela and Goldhirsch, 1998) for the state to be near equilibrium. Thus the classical procedure of solving for the nonequilibrium distribution function as a small perturbation of the Maxwell–Boltzmann distribution (see §7.3.7) is correct only in the limit $(1 - e_{\text{p}}^2) \ll 1$ *and* $\dot{\gamma} \ll \sqrt{T}/s$, but keeping the ratio $\dot{\gamma}/\sqrt{1 - e_{\text{p}}^2}$ finite.[†] We shall return to this idea while determining the velocity distribution function in §7.3.7.

In the following section we derive the Maxwell transport equations, from which the necessity for determining the nonequilibrium distribution function will become clear.

7.3.5. Maxwell Transport Equation

The transport equation for any property of a gas may be derived from the Boltzmann equation. Let ψ be any particle property that is not explicitly a function of $\mathbf{r}$ and t. For smooth spheres, a particle's velocity alone fully specifies its state, hence $\psi \equiv \psi(\mathbf{c})$. We obtain the balance equation for $\langle \psi \rangle$ by multiplying (7.42) with ψ and integrating the result over velocity space

$$\int \psi \left[\frac{\partial f}{\partial t} + \mathbf{c} \cdot \nabla f + \mathbf{b} \cdot \nabla_c f \right] \mathrm{d}\mathbf{c} = \int \psi \dot{f}_{\text{coll}} \, \mathrm{d}\mathbf{c} \tag{7.50}$$

Since t, $\mathbf{r}$, and $\mathbf{c}$ are independent variables, we may write the first term within square brackets as $\partial/\partial t(\psi f)$, and the second term as $\nabla \cdot (\psi \mathbf{c} f)$. Interchanging the order of the derivatives and integrals for these two terms, and using the divergence theorem (1.81) for the third term, we get

$$\frac{\partial}{\partial t} \int \psi f \, \mathrm{d}\mathbf{c} + \nabla \cdot \int \psi \, \mathbf{c} f \, \mathrm{d}\mathbf{c} + \int_{c \to \infty} \psi f \mathbf{b} \cdot \mathrm{d}\mathbf{S} - \int f \mathbf{b} \cdot \nabla_c \psi \, \mathrm{d}\mathbf{c} = \int \psi \dot{f}_{\text{coll}} \, \mathrm{d}\mathbf{c} \tag{7.51}$$

The integral in the third term above is over the surface in velocity space on which the magnitude of $\mathbf{c}$ is infinitely large; this term vanishes because f must decay rapidly as $c \to \infty$ for the average of ψ to be meaningful. Using the definition of $\langle \psi \rangle$ (see (7.28)) in (7.51), we obtain

$$\frac{\partial}{\partial t}(n \langle \psi \rangle) + \nabla \cdot (n \langle \mathbf{c} \psi \rangle) = n \langle \mathbf{b} \cdot \nabla_c \psi \rangle + \dot{\psi}_{\text{coll}} \tag{7.52}$$

[†] Nevertheless, we assume $\dot{\gamma}$ to be large enough that the parameter R_*, defined in (2.1), is ~ 1 so that the granular material is in the rapid flow regime.

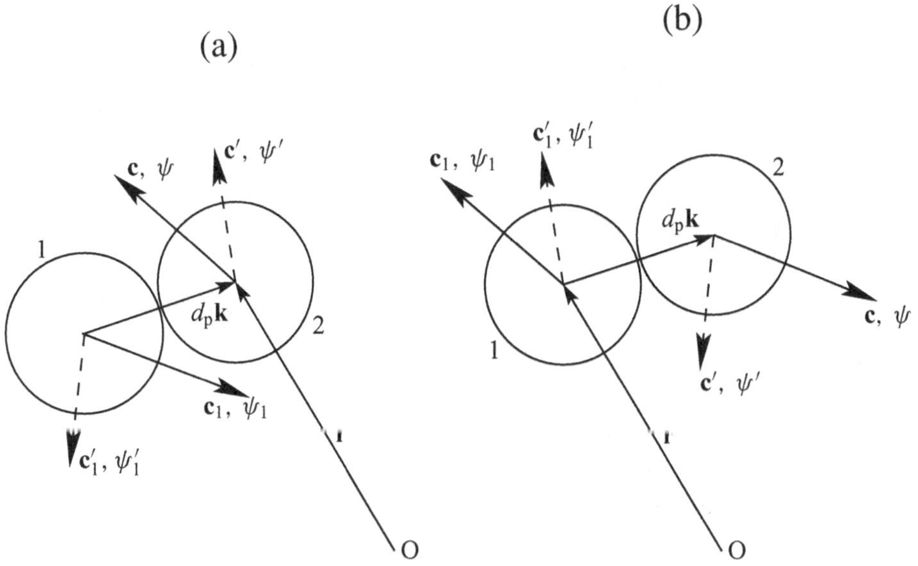

Figure 7.7. The collisional change of the property ψ on a sample particle located in a volume $\mathbf{dr}$ centered at $\mathbf{r}$. In panel (a), the sample particle is the one labeled 2, and in panel (b) it is the one labeled 1.

where $\dot{\psi}_{\text{coll}}$ is the rate of change of $\langle\psi\rangle$ per unit volume by collisions, given by the right-hand side of (7.50). Equation (7.52) is the transport equation for the field $\langle\psi\rangle$, first derived by Maxwell (1867).

Expression for $\dot{\psi}_{\text{coll}}$

We construct an expression for $\dot{\psi}_{\text{coll}}$ by following a procedure similar to that adopted in §7.3.2 for determining $\dot{f}_{\text{coll}}$. We start by determining the change in ψ produced on a sample particle due to collisions during an infinitesimal time interval dt. Consider the sample particle 2 in Fig. 7.7a, located within a volume element $\mathbf{dr}$ centered at $\mathbf{r}$ and having a velocity in the range $\mathbf{dc}$ centered at $\mathbf{c}$. Particle 1, having a velocity in the range $\mathbf{dc}_1$ centered at $\mathbf{c}_1$, collides with the sample particle, with the point of contact lying within a differential surface of the former defined by the solid angle $\mathbf{dk}$. The position of particle 1 at the instant of collision is $\mathbf{r}_1 = \mathbf{r} - \mathbf{k}d_p$. Following the reasoning in §7.3.2, the change in ψ within the volume $\mathbf{dr}$ in the time interval dt is

$$(\psi' - \psi) f^{(2)}(t, \mathbf{r} - d_p\mathbf{k}, \mathbf{c}_1, \mathbf{r}, \mathbf{c}) d_p^2\, \mathbf{k}\cdot\mathbf{g}\, \mathbf{dk}\, \mathbf{dc}_1\, \mathbf{dr}\, \mathbf{dc}\, dt \tag{7.53}$$

the prime indicating postcollision properties. As in (7.34), we impose the constraint $\mathbf{k}\cdot\mathbf{g} > 0$, so as to count only impending collisions. The collisional rate of change of ψ per unit volume is the above expression divided by $\mathbf{dr}\, dt$, and integrated over all possible velocities of the two particles

$$\dot{\psi}_{\text{coll}} = d_p^2 \int_{\mathbf{k}\cdot\mathbf{g} > 0} (\psi' - \psi) f^{(2)}(t, \mathbf{r} - d_p\mathbf{k}, \mathbf{c}_1, \mathbf{r}, \mathbf{c})\, \mathbf{k}\cdot\mathbf{g}\, \mathbf{dk}\, \mathbf{dc}_1\, \mathbf{dc} \tag{7.54}$$

A more symmetric form may be obtained if we consider the collision between a particle at $\mathbf{r}$ with velocity $\mathbf{c}_1$ and another at $\mathbf{r}_1 = \mathbf{r} + \mathbf{k}d_p$ with velocity $\mathbf{c}$ (see Fig. 7.7b), i.e., if we swap the roles of particles 1 and 2 in Fig. 7.7a. The rate of change of ψ per unit volume is now

$$\dot{\psi}_{\text{coll}} = d_p^2 \int\limits_{\mathbf{k}\cdot\mathbf{g}>0} (\psi_1' - \psi_1) f^{(2)}(t, \mathbf{r}, \mathbf{c}_1, \mathbf{r} + d_p\mathbf{k}, \mathbf{c})\, \mathbf{k}\cdot\mathbf{g}\, d\mathbf{k}\, d\mathbf{c}_1\, d\mathbf{c} \tag{7.55}$$

where we have used ψ_i to denote $\psi(\mathbf{c}_i)$. We substitute the Taylor expansion

$$f^{(2)}(t, \mathbf{r} - \mathbf{k}d_p, \mathbf{c}_1, \mathbf{r}, \mathbf{c}) = \left(1 - d_p\mathbf{k}\cdot\nabla + \frac{1}{2!}(d_p\mathbf{k}\cdot\nabla)^2 - \frac{1}{3!}(d_p\mathbf{k}\cdot\nabla)^3 + \cdots\right) f^{(2)}(t, \mathbf{r}, \mathbf{c}_1, \mathbf{r} + \mathbf{k}d_p, \mathbf{c}) \tag{7.56}$$

in (7.55). Upon taking the average of (7.54) and (7.55) and simplifying, the expression for $\dot{\psi}_{\text{coll}}$ takes the form (Jenkins and Savage, 1983)

$$\dot{\psi}_{\text{coll}} = -\nabla\cdot\boldsymbol{\theta} + \chi \tag{7.57}$$

where $\boldsymbol{\theta}$ is the flux of $\langle\psi\rangle$ due to particle collisions,

$$\boldsymbol{\theta} = \frac{d_p^3}{2} \int\limits_{\mathbf{k}\cdot\mathbf{g}>0} \mathbf{k}\,(\psi' - \psi)\,\mathbf{k}\cdot\mathbf{g} \left(1 - \frac{1}{2!}(d_p\mathbf{k}\cdot\nabla) + \cdots\right) \times f^{(2)}(t, \mathbf{r}, \mathbf{c}_1, \mathbf{r} + d_p\mathbf{k}, \mathbf{c})\, d\mathbf{k}\, d\mathbf{c}_1\, d\mathbf{c} \tag{7.58}$$

and χ is the collisional source of $\langle\psi\rangle$,

$$\chi = \frac{d_p^2}{2} \int\limits_{\mathbf{k}\cdot\mathbf{g}>0} (\psi' + \psi_1' - \psi - \psi_1)\, \mathbf{k}\cdot\mathbf{g}\, f^{(2)}(t, \mathbf{r}, \mathbf{c}_1, \mathbf{r} + d_p\mathbf{k}, \mathbf{c})\, d\mathbf{k}\, d\mathbf{c}_1\, d\mathbf{c} \tag{7.59}$$

We now invoke the assumption of molecular chaos with Enskog's correction (7.41) to write $f^{(2)}$ in terms of the singlet distribution $f^{(1)} \equiv f$. For the pair distribution function for two particles at contact $g(\mathbf{r} + d_p\mathbf{k}, \mathbf{r})$, we use the value *at equilibrium* for a gas of elastic particles, which is isotropic. This is an approximation, for one expects the inelasticity of the particles and the anisotropy of the forcing, such as in shear flow, to affect the pair distribution. A discussion on the incorporation of anisotropy of the pair distribution function in the theory is given in §7.4. The equilibrium pair distribution at contact depends only on the local density; we evaluate it for the density (or, equivalently, the solids fraction ν) at the point of contact

$$g(\mathbf{r}+\mathbf{k}\,d_p, \mathbf{r}) = g_0[\nu(\mathbf{r}+\mathbf{k}\,d_p/2)] \tag{7.60}$$

As ν approaches ν_{drp}, the solids fraction at dense random packing, the probability of a particle being in contact with a neighbor approaches unity. Since g_0/V is a probability density, where V is the total volume of the gas, $g_0(\nu)$ must diverge as $\nu \to \nu_{\text{drp}}$. Carnahan and Starling (1969) proposed the functional form

$$g_0(\nu) = \frac{2 - \nu}{2\,(1 - \nu)^3} \tag{7.61}$$

which is commonly used for hard sphere fluids. However, it is not applicable at high densities, as it diverges only as $\nu \to 1$. Ogawa et al. (1980) proposed a relation that diverges as $\nu \to \nu_{\text{drp}}$,

$$g_0(\nu) = \frac{1}{\left[1 - (\nu/\nu_{\text{drp}})^{1/3}\right]} \tag{7.62}$$

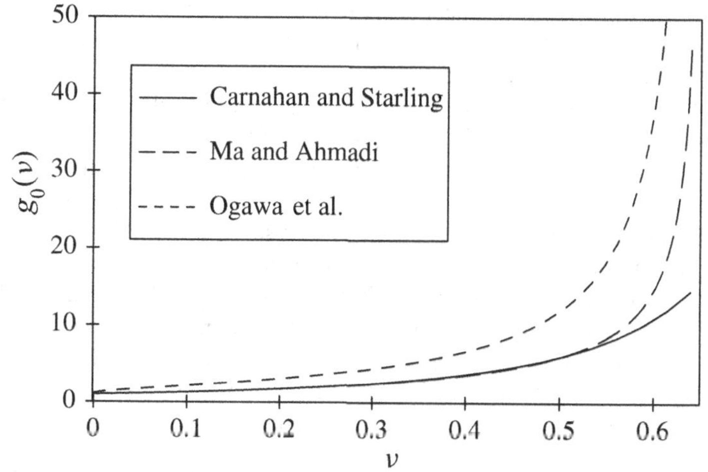

Figure 7.8. Some proposed forms for the radial distribution at contact for hard-sphere fluids. The solids fraction at maximum packing, ν_{drp}, is taken as 0.65.

Ma and Ahmadi (1986) have proposed, on the basis of data obtained from computer simulations, another form that diverges as $\nu \to \nu_{\text{drp}}$,

$$g_0(\nu) = \frac{(1 + 2.5\,\nu + 4.59\,\nu^2 + 4.51\,\nu^3)}{[1 - (\nu/\nu_{\text{drp}})^3]^{0.68}} \tag{7.63}$$

Plots of the above expressions for $g_0(\nu)$ are shown in Fig. 7.8.

With the assumption of molecular chaos, (7.58) and (7.59) become

$$\boldsymbol{\theta} = \frac{d_p^3}{2} \int_{\mathbf{k}\cdot\mathbf{g} > 0} \mathbf{k}\,(\psi' - \psi)\,\mathbf{k}\cdot\mathbf{g} \left[1 - \frac{1}{2!}(d_p\mathbf{k}\cdot\nabla) + \cdots\right]$$
$$\times \left[g_0[\nu(\mathbf{r}{+}d_p\mathbf{k}/2)] f(t, \mathbf{r}, \mathbf{c}_1) f(t, \mathbf{r} + d_p\mathbf{k}, \mathbf{c})\right] \,\mathrm{d}\mathbf{k}\,\mathrm{d}\mathbf{c}\,\mathrm{d}\mathbf{c}_1 \tag{7.64}$$

and

$$\chi = \frac{d_p^2}{2} \int_{\mathbf{k}\cdot\mathbf{g} > 0} (\psi' + \psi_1' - \psi - \psi_1)\,\mathbf{k}\cdot\mathbf{g}\, g_0[\nu(\mathbf{r}{+}d_p\mathbf{k}/2)]$$
$$\times f(t, \mathbf{r}, \mathbf{c}_1) f(t, \mathbf{r} + d_p\mathbf{k}, \mathbf{c})\,\mathrm{d}\mathbf{k}\,\mathrm{d}\mathbf{c}\,\mathrm{d}\mathbf{c}_1 \tag{7.65}$$

The hydrodynamic equations of motion for the continuum are then obtained by substituting the appropriate quantities for ψ, as elaborated below.

7.3.6. The Equations of Motion

If ψ is a property that is conserved during collisions, the source of ψ from collisions, given by (7.65), is zero. Ignoring, for the moment, external forces acting on the particles, the transport equation for ψ (7.52) assumes the form

$$\frac{\partial}{\partial t}(n\langle\psi\rangle) = -\nabla\cdot(n\langle\mathbf{c}\,\psi\rangle) - \nabla\cdot\boldsymbol{\theta} \tag{7.66}$$

All the terms on the right-hand side of the above equation involve spatial derivatives; this implies that the time scale τ for the change of $\langle\psi\rangle$ increases with the length scale H of its spatial variation. This property is indeed required of hydrodynamic variables, for hydrodynamics is the description of phenomena whose length scale is much greater than the microscopic length scale (which here is the mean free path s), and times scale is much greater than the microscopic time scale (here the mean free time τ_c) (Chaikin and Lubensky, 1997, p. 418). The presence of a conservative external force does not alter this result, because it simply balances the hydrostatic pressure; if the momentum balance is written in terms of the pressure deviation from the hydrostatic, it reverts to the form of (7.66).

If ψ is not a conserved property, the source term $\chi(\psi)$ also appears on the right-hand side of (7.66). Since it is proportional to the rate of particle collisions, it scales as $\langle\psi\rangle/\tau_c$. The other terms, being gradients of fluxes, are smaller by a factor of s/H, and hence the balance in (7.66) is between the left-hand side and $\chi(\psi)$. This sets τ_c as the time scale for variations of $\langle\psi\rangle$ to relax, i.e., it relaxes rapidly to a local equilibrium, and is therefore not a hydrodynamic variable. For this reason, hydrodynamic equations are normally written only for conserved variables such as the mass, momentum, and energy densities (for elastic particles). The angular momentum is another conserved variable, but for smooth particles no exchange of angular momentum takes place during a collision, and hence its conservation is implicitly satisfied. Consequently, it is necessary to write the hydrodynamic equation for the angular momentum only when the particles are rough (see §9.2).

There are some situations where the relaxation time for a nonconserved variable increases with some parameter, and become large enough to be comparable to macroscopic time scales; in such instances, the temporal behavior of $\langle\psi\rangle$ is like that of a hydrodynamic variable. Consequently, a hydrodynamic description is sometimes extended to nonconserved quantities, such as an order parameter field near a second-order phase transition (Chaikin and Lubensky, 1997, pp. 419, 464–465), because the relaxation time τ gets very large as the phase transition is approached. For a granular gas, inelastic collisions render the particle energy a nonconserved variable, and the time scale for relaxation of the average particle energy scales as τ_c. However, in the limit $e_p \to 1$, the time scale grows as $(1 - e_p^2)^{-1/2}$ (see §7.2.1) and one can treat energy as a hydrodynamic variable. Having already made the assumption in §7.3.4 that e_p is close to unity, we proceed to determine the normal hydrodynamic equations, viz. the balances of mass, momentum, and energy, for a granular gas.

Substituting $\psi = m_p$ in (7.52), we see that $\boldsymbol{\theta}$ and χ vanish, and we get the mass balance, or the equation of continuity

$$\frac{\partial \rho}{\partial t} + \nabla\cdot(\rho \mathbf{v}) = 0 \tag{7.67}$$

which is often written in the alternative form

$$\frac{\mathrm{D}\rho}{\mathrm{D}t} = -\rho \nabla\cdot\mathbf{v} \tag{7.68}$$

Next, setting $\psi = m_p\mathbf{c}$ gives the linear momentum balance

$$\frac{\partial(\rho \mathbf{v})}{\partial t} + \nabla\cdot(\rho \mathbf{v}\mathbf{v}) = -\nabla\cdot\boldsymbol{\sigma} + \rho\,\mathbf{b} \tag{7.69}$$

where

$$\boldsymbol{\sigma} \equiv \rho\langle\mathbf{C}\mathbf{C}\rangle + \boldsymbol{\theta}(m_p\mathbf{c}) \tag{7.70}$$

is the stress tensor of the granular material, and $\mathbf{C}$ is the peculiar velocity defined by (7.31). The first term on the right-hand side of (7.70) is the streaming contribution, i.e., the momentum flux due to the movement of particles from one place to another, and the second is the collisional contribution. The continuity equation may be used to reduce (7.69) to the more familiar form

$$\rho \frac{\mathrm{D}\mathbf{v}}{\mathrm{D}t} = -\nabla \boldsymbol{\cdot} \sigma + \rho\, \mathbf{b} \tag{7.71}$$

Finally, we obtain the energy balance by substituting $\psi = \frac{1}{2} m_{\mathrm{p}} c^2$ in (7.52). After some manipulation, we get[†]

$$\frac{3}{2}\rho \frac{\mathrm{D}T}{\mathrm{D}t} + \rho \frac{\mathrm{D}}{\mathrm{D}t}\left(\frac{1}{2}v^2\right) = -\nabla \boldsymbol{\cdot} \mathbf{q} - \Gamma + \rho\, \mathbf{b} \boldsymbol{\cdot} \mathbf{v} - \nabla \boldsymbol{\cdot} (\sigma \boldsymbol{\cdot} \mathbf{v}) \tag{7.72}$$

where

$$\mathbf{q} = \frac{1}{2}\rho \langle \mathbf{C}C^2 \rangle + \boldsymbol{\theta}\left(\frac{1}{2} m_{\mathrm{p}} C^2\right) \tag{7.73}$$

$$\Gamma = -\chi\left(\frac{1}{2} m_{\mathrm{p}} C^2\right) \tag{7.74}$$

are the flux and the volumetric dissipation rate, respectively, of the fluctuational kinetic energy. The above is the balance of total kinetic energy, which includes the energy due to mean motion of the particles. A more useful form is obtained by subtracting from it the balance of *mechanical energy* (i.e., energy due to mean flow), obtained by forming the inner (dot) product of $\mathbf{v}$ with (7.71). This gives the fluctuational energy balance

$$\frac{3}{2}\rho \frac{\mathrm{D}T}{\mathrm{D}t} = -\nabla \boldsymbol{\cdot} \mathbf{q} - \boldsymbol{\sigma}^{\mathrm{T}} : \nabla \mathbf{v} - \Gamma \tag{7.75}$$

Equations (7.68), (7.71), and (7.75) are called the "equations of motion." They were derived in §1.5.3–§1.5.7 for a continuum, without any reference to the nature of the particles it is made of. The Maxwell transport equation yields the same equations, and also gives expressions for $\boldsymbol{\sigma}$, $\mathbf{q}$, and Γ in terms of the microscopic properties of the particles and their interactions.

To close the equations, relations for the fluxes $\boldsymbol{\sigma}$ and $\mathbf{q}$, and the energy sink Γ must be determined in terms of the hydrodynamic variables ρ, $\mathbf{v}$, and T. These are called the constitutive relations. To compute the integrals in (7.58) and (7.59), knowledge of the distribution function $f(t, \mathbf{r}, \mathbf{c})$ away from equilibrium is required, and determining this is the subject of the following section.

7.3.7. The Chapman–Enskog Expansion

The nonlinearity of the collision integral makes an exact analytical solution of (7.42) difficult. However, several approximate methods of solution exist, of which the most systematic and mathematically rigorous is the method of Chapman (1916) and Enskog (1917). They independently arrived at a perturbative solution for nonuniform gases (i.e., gases away of equilibrium). Though their approaches were quite different, their results were identical. Extension of the kinetic theory of nonuniform gases to systems of inelastic particles was attempted in several studies (Jenkins and Savage, 1983; Lun et al., 1984; Jenkins and Richman, 1985).

In all approximate methods, advantage is taken of the presence of small parameters in the problem; these are the inelasticity of collisions

$$\epsilon \equiv (1 - e_{\mathrm{p}}^2) \tag{7.76}$$

† To get (7.72), we have used the identities $\boldsymbol{\theta}(\frac{1}{2}m_{\mathrm{p}}c^2) = \boldsymbol{\theta}(\frac{1}{2}m_{\mathrm{p}}C^2) + \mathbf{v}\boldsymbol{\cdot}\boldsymbol{\theta}(m_{\mathrm{p}}C)$, and $\chi(\frac{1}{2}m_{\mathrm{p}}c^2) = \chi(\frac{1}{2}m_{\mathrm{p}}C^2)$, which follow from the linearity of $\boldsymbol{\theta}(\psi)$ in its argument and the conservation of linear momentum of the colliding particles.

and the ratio of microscopic to macroscopic length scales. The latter is the inverse of the *Knudsen number Kn*, defined as

$$Kn \equiv \frac{H}{s} = \sqrt{2}\pi n d_p^2 H = 6\sqrt{2}\,\nu \frac{H}{d_p} \tag{7.77}$$

where H is a characteristic macroscopic length scale, such as the length over which there is variation of ν, $\mathbf{v}$, or T. For instance, for hydrogen gas at NTP (see §7.3.4) flowing through a capillary of radius 1 mm, $Kn \approx 10^4$. At equilibrium there is no spatial variation in $\mathbf{v}$ or T, or alternatively, the length scale over which there is appreciable spatial variation of these quantities diverges, and hence $Kn = \infty$. Therefore $Kn \gg 1$ implies a small departure from equilibrium. In problems where spatial uniformity is assumed, such as in homogeneous cooling, Kn is not of relevance and ϵ is the only small parameter. However these situations are difficult, if not impossible, to achieve physically, for energy is usually imparted through boundaries, thereby causing spatial gradients in the hydrodynamic variables. As a result, Kn and ϵ play a role in most problems.

For elastic particles, the classical method for determining f is the Chapman–Enskog expansion (Chapman and Cowling, 1964; Huang, 1985; Reif, 1985), in which f is determined as a perturbation expansion about the *local* Maxwell–Boltzmann distribution, i.e.,

$$f^0 = \frac{n(\mathbf{r},t)}{[2\pi T(\mathbf{r},t)]^{3/2}} \exp\left[-\frac{|\mathbf{c}-\mathbf{v}(\mathbf{r},t)|^2}{2T(\mathbf{r},t)}\right] \tag{7.78}$$

where $n(\mathbf{r},t)$, $\mathbf{v}(\mathbf{r},t)$, and $T(\mathbf{r},t)$ are the *local* number density, mean velocity, and temperature, respectively. For the local Maxwellian, it is easily verified that the collision integral in (7.42) vanishes, but the left-hand side does not. The Chapman–Enskog method corrects this in a systematic manner.

The Chapman–Enskog procedure becomes clear if we first scale the variables in the following manner:

$$\hat{\mathbf{r}} = \mathbf{r}/H, \quad \hat{\mathbf{c}} = \mathbf{c}/v_s, \quad \hat{t} = t v_s/H, \quad \hat{\mathbf{b}} = \mathbf{b}H/v_s^2, \quad \hat{f} = f v_s^3/n_s \tag{7.79}$$

where v_s and n_s set the scale for the particle velocity and the number density, and H is the macroscopic length scale over which there is appreciable variation in the hydrodynamic variables. The scale for the mean free path is $s_s \equiv 1/(\pi\sqrt{2}\,n_s d_p^2)$, and for convenience we define

$$K \equiv \pi\sqrt{2}\,Kn^{-1} = \frac{\pi d_p}{6\nu_s H} \tag{7.80}$$

as the small parameter. Here $\nu_s \equiv n_s \pi d_p^3/6$ is the volume fraction corresponding to n_s. As stated in §7.3.4, we are interested in the slow and gradual variation of f, caused by the gradients of the hydrodynamic variables. Indeed, the Chapman–Enskog method assumes that spatial and temporal variation of f is *only* through that of the hydrodynamic variables n, $\mathbf{v}$ and T, which vary over macroscopic length and time scales. Any explicit variation of f with time or space will decay rapidly, in a time scale of the mean free time (τ_c) (Huang, 1985, p. 124). Therefore, we use the macroscopic length scale H to scale t and $\mathbf{r}$ (rather than the mean free path s). Using the above, (7.42) can be reduced to the dimensionless form

$$\begin{aligned} \hat{f}K\left(\frac{\hat{\mathrm{D}}}{\mathrm{D}\hat{t}} + \hat{\mathbf{C}}\cdot\hat{\nabla} + \hat{\mathbf{b}}\cdot\hat{\nabla}_c\right)\ln\hat{f} = \int_{\mathbf{k}\cdot\hat{\mathbf{g}}>0} \Bigg[&\frac{1}{1-\epsilon}\hat{f}''(\hat{\mathbf{r}})\hat{f}_1''\left(\hat{\mathbf{r}} + \frac{6\nu_s}{\pi}K\mathbf{k}\right) \\ &\times g_0\left[\nu\left(\hat{\mathbf{r}} + \frac{3\nu_s}{\pi}K\mathbf{k}\right)\right] - \hat{f}(\hat{\mathbf{r}})\hat{f}_1\left(\hat{\mathbf{r}} - \frac{6\nu_s}{\pi}K\mathbf{k}\right) \\ &\times g_0\left[\nu\left(\hat{\mathbf{r}} - \frac{3\nu_s}{\pi}K\mathbf{k}\right)\right]\Bigg]\,\mathbf{k}\cdot\hat{\mathbf{g}}\,\mathrm{d}\mathbf{k}\,\mathrm{d}\hat{\mathbf{c}}_1 \end{aligned} \tag{7.81}$$

where $\hat{D}/D\hat{t} \equiv \partial/\partial\hat{t} + \hat{\mathbf{v}}\cdot\hat{\nabla}$ is the dimensionless material derivative, and $\hat{\mathbf{C}} \equiv \mathbf{C}/v_s$ and $\hat{\mathbf{g}} \equiv \mathbf{g}/v_s$ are the dimensionless peculiar and relative velocities, respectively.

In the limit $K \to 0$ and $\epsilon \to 0$ we recover the Maxwell–Boltzmann distribution

$$\hat{f}^0 = \frac{\nu/\nu_s}{(2\pi\hat{T})^{3/2}}\, e^{-\hat{C}^2/(2\hat{T})} \tag{7.82}$$

To determine the deviation of $\hat{f}$ from the Maxwellian, we set

$$\hat{f} = \hat{f}^0(1+\Phi), \;\; \Phi \ll 1 \tag{7.83}$$

and assume a perturbation expansion for Φ of the form

$$\Phi = K\,\Phi_K + \epsilon\,\Phi_\epsilon + K^2\,\Phi_{KK} + K\epsilon\,\Phi_{K\epsilon} + \epsilon^2\,\Phi_{\epsilon\epsilon} + \cdots \tag{7.84}$$

The classical method of Chapman and Enskog is for elastic particles, and hence Φ is expanded as a perturbation series in K alone. For inelastic particles, ϵ is another parameter that measures the departure from equilibrium. Recall from §7.3.4 that the departure for inelastic particles is measured from the equilibrium state of a system of *elastic* particles having the same temperature; henceforth, we shall adopt this definition of the term "equilibrium." We saw in §7.3.4 that for steady uniform shear, $\sqrt{\epsilon} \sim K$ if the system is to remain close to equilibrium. For a general flow field, the shear rate, density, and temperature may vary in space and time, and ϵ and K must be considered as independent small parameters. Hence the perturbation expansion is in powers of both K and ϵ (Sela and Goldhirsch, 1998).

From (7.78), it is clear that the local density, mean velocity, and temperature are velocity moments of f^0, i.e.,

$$n = \int f^0\, d\mathbf{c}\,, \quad n\mathbf{v} = \int \mathbf{c} f^0\, d\mathbf{c}\,, \quad 3nT = \int C^2 f^0\, d\mathbf{c} \tag{7.85}$$

Therefore, it follows from (7.27)–(7.30) and (7.83) that the corresponding moments of Φ must vanish, i.e.,

$$\int \hat{f}^0 \Phi\, \psi\, d\mathbf{c} = 0 \tag{7.86}$$

for $\psi = 1$, $\hat{\mathbf{c}}$, and $\hat{C}^2$. It is convenient to satisfy (7.86) at every order in the perturbation expansion.

Expressing (7.70) in terms of the dimensionless variables defined in (7.79), we obtain the following expression for the dimensionless stress $\hat{\boldsymbol{\sigma}} \equiv \boldsymbol{\sigma}/(\rho_p v_s^2)$,

$$\begin{aligned}\hat{\boldsymbol{\sigma}} = \nu_s \int \hat{\mathbf{C}}\hat{\mathbf{C}} \hat{f}(\hat{t}, \hat{\mathbf{r}}, \hat{\mathbf{c}})\, d\hat{\mathbf{c}} + \frac{3\nu_s^2}{\pi} \int_{\mathbf{k}\cdot\hat{\mathbf{g}} > 0} \mathbf{k}\,(\hat{\mathbf{c}}' - \hat{\mathbf{c}})\,\mathbf{k}\cdot\hat{\mathbf{g}} \\ \times \left[1 - \frac{6\nu_s K}{\pi}\mathbf{k}\cdot\hat{\nabla} + \cdots\right] g_0[\nu(\hat{\mathbf{r}} + \frac{3\nu_s K}{\pi}\mathbf{k})]\, \hat{f}(\hat{t}, \hat{\mathbf{r}}, \hat{\mathbf{c}}_1) \\ \times \hat{f}(\hat{t}, \hat{\mathbf{r}} + \frac{6\nu_s K}{\pi}\mathbf{k}, \hat{\mathbf{c}})\, d\mathbf{k}\, d\hat{\mathbf{c}}\, d\hat{\mathbf{c}}_1\end{aligned} \tag{7.87}$$

Similarly, (7.73) may be written in terms of the dimensionless variables to yield the following expression for the dimensionless fluctuational energy flux $\hat{\mathbf{q}} \equiv \mathbf{q}/(\rho_p v_s^3)$,

$$\hat{\mathbf{q}} = \frac{\nu_s}{2}\int \hat{\mathbf{C}}\hat{C}^2 \hat{f}(\hat{t},\hat{\mathbf{r}},\hat{\mathbf{c}})\,\mathrm{d}\hat{\mathbf{c}} + \frac{3\nu_s^2}{2\pi}\int_{\mathbf{k}\cdot\hat{\mathbf{g}}>0} \mathbf{k}\,(\hat{C}'^2 - \hat{C}^2)\,\mathbf{k}\cdot\hat{\mathbf{g}}$$

$$\left[1 - \frac{6\nu_s K}{\pi}\mathbf{k}\cdot\hat{\nabla} + \cdots\right] g_0[\nu(\hat{\mathbf{r}} + \frac{3\nu_s K}{\pi}\mathbf{k})]\,\hat{f}(\hat{t},\hat{\mathbf{r}},\hat{\mathbf{c}}_1)$$

$$\times\, \hat{f}(\hat{t},\hat{\mathbf{r}} + \frac{6\nu_s K}{\pi}\mathbf{k},\hat{\mathbf{c}})\,\mathrm{d}\mathbf{k}\,\mathrm{d}\hat{\mathbf{c}}\,\mathrm{d}\hat{\mathbf{c}}_1 \qquad (7.88)$$

In (7.87) and (7.88), the first terms represent the streaming contributions, and the second terms the collisional contributions. When the expansion for Φ in powers of K and ϵ is substituted in the above expressions, we obtain similar expansions for $\hat{\boldsymbol{\sigma}}$ and $\hat{\mathbf{q}}$,

$$\hat{\boldsymbol{\sigma}} = \hat{\boldsymbol{\sigma}}_0 + K\hat{\boldsymbol{\sigma}}_K + \epsilon\,\hat{\boldsymbol{\sigma}}_\epsilon + \cdots \qquad (7.89)$$

$$\hat{\mathbf{q}} = \hat{\mathbf{q}}_0 + K\hat{\mathbf{q}}_K + \epsilon\,\hat{\mathbf{q}}_\epsilon + \cdots \qquad (7.90)$$

An expression for the dimensionless dissipation rate $\hat{\Gamma} \equiv \Gamma d_p/(\rho_p v_s^3)$ can be derived from (7.74) in a similar manner,

$$\hat{\Gamma} = \frac{3\nu_s^2}{2\pi}\int_{\mathbf{k}\cdot\hat{\mathbf{g}}>0} (\hat{C}'^2 + \hat{C}_1'^2 - \hat{C}^2 - \hat{C}_1^2)\,\mathbf{k}\cdot\hat{\mathbf{g}}\,g_0[\nu(\hat{\mathbf{r}} + \frac{3\nu_s K}{\pi}\mathbf{k})]$$

$$\times\,\hat{f}(t,\hat{\mathbf{r}},\hat{\mathbf{c}}_1)\,\hat{f}(t,\hat{\mathbf{r}} + \frac{6\nu_s K}{\pi}\mathbf{k},\hat{\mathbf{c}})\,\mathrm{d}\mathbf{k}\,\mathrm{d}\hat{\mathbf{c}}\,\mathrm{d}\hat{\mathbf{c}}_1 \qquad (7.91)$$

We see from (7.7) that $\hat{C}'^2 + \hat{C}_1'^2 - \hat{C}^2 - \hat{C}_1^2$ is proportional to $(1 - e_p^2)$, hence there are no O(1) or O(K) terms in the expansion for $\hat{\Gamma}$. (Here and henceforth, an O(1) term refers to a term of O($\epsilon^0 K^0$).) Therefore, it has the expansion

$$\hat{\Gamma} = \epsilon\,\hat{\Gamma}_\epsilon + \epsilon K\,\hat{\Gamma}_{\epsilon K} + \epsilon^2\,\hat{\Gamma}_{\epsilon\epsilon} + \cdots \qquad (7.92)$$

We can, in principle, determine $\hat{\boldsymbol{\sigma}}$, $\hat{\mathbf{q}}$, and $\hat{\Gamma}$ to all orders in K and ϵ by determining the complete expansion of f, but in what follows we shall stop at the first correction to their equilibrium values. This is a good approximation if K and ϵ are small. For instance, stopping at the O(K) contribution to the stress for a gas, for which K is indeed very small under normal conditions, yields the Navier–Stokes equations. However, certain aspects of the behavior do not manifest at the Navier–Stokes order, and one has to go to higher order. For example, normal stress differences† appear first at the Burnett order, i.e., at O(K^2) (Goldhirsch and Sela, 1996). The procedure for evaluating the higher order contributions is straightforward, though algebraically more tedious; Sela and Goldhirsch (1998) determined the constitutive relations for a *dilute* granular gas by considering only the streaming contributions to $\boldsymbol{\sigma}$ and $\mathbf{q}$ up to O($K^2\epsilon^2$), and Kumaran (2004) has determined the stress to O(K^2).

7.3.8. Constitutive Relations at Leading Order

The leading terms in (7.89)–(7.92) are determined by simply using the unperturbed distribution function f^0 in place of f.

† Normal stress differences are defined for viscometric flows, i.e., flows for which the directions of velocity (x), velocity gradient (y) and vorticity (z) are everywhere mutually orthogonal. Plane simple shear is an example of viscometric flow. The first normal stress difference is defined as $N_1 \equiv \sigma_{xx} - \sigma_{yy}$ and the second normal stress difference as $N_2 \equiv \sigma_{yy} - \sigma_{zz}$.

Using (7.4) to substitute $\hat{\mathbf{C}}' - \hat{\mathbf{C}} = (\mathbf{k}\cdot\hat{\mathbf{g}})\,\mathbf{k}$ in (7.87), we get

$$\hat{\boldsymbol{\sigma}}_0 = \nu_s \int \hat{f}^0 \hat{\mathbf{C}}\hat{\mathbf{C}}\, d\hat{\mathbf{C}} + \frac{3}{\pi}\nu_s^2 g_0 \int_{\mathbf{k}\cdot\hat{\mathbf{g}}>0} \mathbf{k}\mathbf{k}\,(\mathbf{k}\cdot\hat{\mathbf{g}})^2\, \hat{f}_1^0 \hat{f}^0\, d\mathbf{k}\, d\hat{\mathbf{C}}\, d\hat{\mathbf{C}}_1 \tag{7.93}$$

The evaluation of these and other integrals in this chapter is accomplished using the results in Appendix H. Here we use (H.4) to effect the integration over $\mathbf{k}$ in the second integral, giving

$$\hat{\boldsymbol{\sigma}}_0 = \nu_s \int \hat{f}^0 \hat{\mathbf{C}}\hat{\mathbf{C}}\, d\hat{\mathbf{C}} + \frac{2}{5}\nu_s^2 g_0 \int (2\hat{\mathbf{g}}\hat{\mathbf{g}} + \hat{g}^2\mathbf{I}) \hat{f}_1^0 \hat{f}^0\, d\hat{\mathbf{C}}\, d\hat{\mathbf{C}}_1 \tag{7.94}$$

where $\mathbf{I}$ is the identity tensor. Substituting $\hat{\mathbf{g}} = \hat{\mathbf{C}}_1 - \hat{\mathbf{C}}$, and evaluating the two integrals using (H.18) and (H.19), we get

$$\hat{\boldsymbol{\sigma}}_0 = \left(1 + 4\nu g_0\right) \nu \hat{T}\,\mathbf{I} \tag{7.95}$$

Thus $\hat{\boldsymbol{\sigma}}_0$ is the isotropic pressure determined by the local solids fraction and temperature. The first term within the parenthesis is the streaming contribution, and the second term is the contribution from particle collisions.

The integrands for the streaming and collisional contributions to $\hat{\mathbf{q}}$ at O(1) order are odd functions of $\hat{\mathbf{C}}$, and hence (H.18) implies that $\hat{\mathbf{q}}_0 = 0$.

There is no O(1) term in Γ, as indicated in (7.92). The O(ϵ) contribution to Γ is determined from (7.91) by using $\hat{f} = \hat{f}^0$,

$$\hat{\Gamma}_\epsilon = -\frac{3\nu_s^2}{4\pi} g_0 \int_{\mathbf{k}\cdot\hat{\mathbf{g}}>0} (\mathbf{k}\cdot\hat{\mathbf{g}})^3\, \hat{f}_1^0 \hat{f}^0\, d\mathbf{k}\, d\hat{\mathbf{C}}\, d\hat{\mathbf{C}}_1 \tag{7.96}$$

Integrating over $\mathbf{k}$ using (H.3), we get

$$\hat{\Gamma}_\epsilon = -\frac{3\nu_s^2}{8} g_0 \int \hat{g}^3 \hat{f}_1^0 \hat{f}^0\, d\hat{\mathbf{C}}\, d\hat{\mathbf{C}}_1 \tag{7.97}$$

To evaluate this integral, we transform the variables of integration to $\hat{\mathbf{G}} \equiv (\hat{\mathbf{C}}_1 + \hat{\mathbf{C}})/2$ and $\hat{\mathbf{g}} = \hat{\mathbf{C}}_1 - \hat{\mathbf{C}}$. It is easily verified that the determinant of the Jacobian for this transformation is unity (see §H.3), and hence $d\hat{\mathbf{C}}\, d\hat{\mathbf{C}}_1 = d\hat{\mathbf{G}}\, d\hat{\mathbf{g}}$. Using (H.23), we get

$$\hat{\Gamma}_\epsilon = \frac{12}{\pi^{1/2}} \nu^2 g_0\, \hat{T}^{3/2} \tag{7.98}$$

7.3.9. Distribution Function at O(*K*)

We now turn to determining Φ_K. This correction to f is solely due to the gradients in the hydrodynamic variables, and not due to particle inelasticity. In other words, it is the same whether we consider granular or molecular gases. The equation governing Φ_K is obtained by retaining only terms of O(K) in (7.81).

Since the left-hand side (LHS) of (7.81) is already multiplied by K, its O(K) part LHS_K is obtained by setting $\hat{f} = \hat{f}^0$ and $\epsilon = 0$. This yields

$$\text{LHS}_K = \hat{f}^0 \left[\frac{1}{\nu}\frac{\hat{D}\nu}{D\hat{t}} - \left(\frac{3}{2\hat{T}} - \frac{\hat{C}^2}{2\hat{T}^2}\right)\frac{\hat{D}\hat{T}}{D\hat{t}} + \frac{1}{\hat{T}}\hat{\mathbf{C}}\hat{\mathbf{C}}{:}\hat{\nabla}\hat{\mathbf{v}}\right]$$
$$+ \hat{f}^0 \hat{\mathbf{C}}\cdot \left[\frac{1}{\nu}\hat{\nabla}\nu - \left(\frac{3}{2\hat{T}} - \frac{\hat{C}^2}{2\hat{T}^2}\right)\hat{\nabla}\hat{T} + \frac{1}{\hat{T}}\left(\frac{\hat{D}\hat{\mathbf{v}}}{D\hat{t}} - \hat{\mathbf{b}}\right)\right] \tag{7.99}$$

Here only the O(1) parts of the material derivatives must be retained, which are determined in (7.110).

To determine the $O(K)$ terms on the right-hand side (RHS) of (7.81), we set $\epsilon = 0$ and expand f and g_0 in a Taylor series about $\hat{\mathbf{r}}$. Discarding terms of $O(K^2)$ and higher, we get

$$\begin{aligned}\mathrm{RHS} = g_0 &\int_{\mathbf{k}\cdot\hat{\mathbf{g}}>0} \left[\hat{f}''\hat{f}_1'' - \hat{f}\hat{f}_1\right] \mathbf{k}\cdot\hat{\mathbf{g}}\, \mathrm{d}\mathbf{k}\, \mathrm{d}\hat{\mathbf{c}}_1 \\ &+ \frac{6\nu_s}{\pi} K g_0 \int_{\mathbf{k}\cdot\hat{\mathbf{g}}>0} \mathbf{k}\cdot\left[\hat{f}''\hat{\nabla}\hat{f}_1'' + \hat{f}\hat{\nabla}\hat{f}_1\right] \mathbf{k}\cdot\hat{\mathbf{g}}\, \mathrm{d}\mathbf{k}\, \mathrm{d}\hat{\mathbf{c}}_1 \\ &+ \frac{3\nu_s}{\pi} K \int_{\mathbf{k}\cdot\hat{\mathbf{g}}>0} \left[\hat{f}''\hat{f}_1'' + \hat{f}\hat{f}_1\right] \mathbf{k}\cdot\hat{\nabla} g_0\, \mathbf{k}\cdot\hat{\mathbf{g}}\, \mathrm{d}\mathbf{k}\, \mathrm{d}\hat{\mathbf{c}}_1 \end{aligned} \tag{7.100}$$

All the variables are now evaluated at $\hat{\mathbf{r}}$. The second and third terms above are already multiplied by K, hence their $O(K)$ parts are obtained by substituting $\hat{f}^0$ for $\hat{f}$. In the first term, we use the expansion for $\hat{f}$ from (7.83) and (7.84) and retain only the $O(K)$ terms. Using the identity $\hat{f}^{0''}\hat{f}_1^{0''} = \hat{f}^0\hat{f}_1^0$ (which follows from (7.44)), the $O(K)$ part of (7.100) reduces to

$$\begin{aligned}\mathrm{RHS}_K = g_0 &\int_{\mathbf{k}\cdot\hat{\mathbf{g}}>0} \hat{f}^0\hat{f}_1^0 \left(\Phi_K'' + \Phi_{K1}'' - \Phi_K - \Phi_{K1}\right) \mathbf{k}\cdot\hat{\mathbf{g}}\, \mathrm{d}\mathbf{k}\, \mathrm{d}\hat{\mathbf{c}}_1 \\ &+ \frac{6\nu_s}{\pi} g_0 \int_{\mathbf{k}\cdot\hat{\mathbf{g}}>0} \hat{f}^0\hat{f}_1^0\, \mathbf{k}\cdot\hat{\nabla}\ln(\hat{f}_1^0\hat{f}_1^{0''})\, \mathbf{k}\cdot\hat{\mathbf{g}}\, \mathrm{d}\mathbf{k}\, \mathrm{d}\hat{\mathbf{c}}_1 \\ &+ \frac{6\nu_s}{\pi} \int_{\mathbf{k}\cdot\hat{\mathbf{g}}>0} \hat{f}^0\hat{f}_1^0\, \mathbf{k}\cdot\hat{\nabla} g_0\, \mathbf{k}\cdot\hat{\mathbf{g}}\, \mathrm{d}\mathbf{k}\, \mathrm{d}\hat{\mathbf{c}}_1 \end{aligned} \tag{7.101}$$

where we have used the notation $\Phi_{Ki} \equiv \Phi_K(\mathbf{c}_i)$ and $\Phi_{Ki}'' \equiv \Phi_K(\mathbf{c}_i'')$. The integration over $\mathbf{k}$ in the third term above can be carried out using (H.2) to yield

$$4\nu_s \int \hat{\mathbf{g}}\cdot\hat{\nabla} g_0\, \hat{f}^0\hat{f}_1^0\, \mathrm{d}\hat{\mathbf{c}}_1 \tag{7.102}$$

which upon integration over $\hat{\mathbf{c}}_1$ using (H.23) gives

$$-4\nu\hat{f}^0\hat{\mathbf{C}}\cdot\hat{\nabla} g_0 \tag{7.103}$$

The second term in (7.101), on substituting

$$\ln\hat{f}^0 = -\ln[\nu_s(2\pi)^{3/2}] + \ln\nu - \frac{3}{2}\ln\hat{T} - \frac{\hat{C}^2}{2\hat{T}} \tag{7.104}$$

evaluates to

$$\begin{aligned}\frac{6\nu_s}{\pi} g_0 &\int_{\mathbf{k}\cdot\hat{\mathbf{g}}>0} \mathbf{k}\cdot\left[\frac{2}{\nu}\hat{\nabla}\nu - \frac{3}{\hat{T}}\hat{\nabla}\hat{T} + \frac{1}{2\hat{T}^2}\hat{\nabla}\hat{T}(\hat{C}_1^2 + \hat{C}_1'^2)\right. \\ &\left. + \frac{1}{\hat{T}}\hat{\nabla}\hat{\mathbf{v}}\cdot(\hat{\mathbf{C}}_1 + \hat{\mathbf{C}}_1')\right] \hat{f}^0\hat{f}_1^0\, \mathbf{k}\cdot\hat{\mathbf{g}}\, \mathrm{d}\mathbf{k}\, \mathrm{d}\hat{\mathbf{C}}_1 \end{aligned} \tag{7.105}$$

Note that $\hat{\mathbf{C}}_1'$ depends on $\mathbf{k}$ through (7.6), with e_p set to unity at this order of the perturbation expansion. We then carry out the integration over $\mathbf{k}$ using (H.2) to get

$$4\nu_s g_0 \int \Big\{ \left[\frac{2}{\nu}\hat{\nabla}\nu - \frac{3}{\hat{T}}\hat{\nabla}\hat{T}\right]\cdot\hat{\mathbf{g}} + \frac{1}{2\hat{T}^2}\hat{\nabla}\hat{T}\cdot\Big[2\hat{C}_1^2\hat{\mathbf{g}} - \frac{2}{5}\left(2\hat{\mathbf{g}}\,\hat{\mathbf{g}}\cdot\hat{\mathbf{C}}_1 + \hat{g}^2\hat{\mathbf{C}}_1\right) + \frac{3}{5}\hat{g}^2\hat{\mathbf{g}}\Big] + \frac{1}{\hat{T}}\hat{\nabla}\hat{\mathbf{v}}:\left[2\hat{\mathbf{C}}_1\hat{\mathbf{g}} - \frac{2}{5}\hat{\mathbf{g}}\hat{\mathbf{g}}\right] - \frac{1}{5\hat{T}}\hat{g}^2\hat{\nabla}\cdot\hat{\mathbf{v}}\Big\} \hat{f}^0\hat{f}_1^0 \mathrm{d}\hat{\mathbf{C}}_1 \qquad (7.106)$$

Finally, substituting $\hat{\mathbf{g}} = \hat{\mathbf{C}}_1 - \hat{\mathbf{C}}$ and integrating over $\hat{\mathbf{C}}_1$ gives

$$-4\nu g_0 \hat{f}^0 \Big[\frac{2}{\nu}\hat{\nabla}\nu\cdot\hat{\mathbf{C}} + \frac{1}{\hat{T}}\hat{\nabla}\hat{T}\cdot\hat{\mathbf{C}}\left(\frac{3\hat{C}^2}{10\hat{T}} - \frac{1}{2}\right) + \frac{2}{5\hat{T}}\hat{\mathbf{C}}\hat{\mathbf{C}}:\hat{\nabla}\hat{\mathbf{v}} - (1 - \frac{\hat{C}^2}{5\hat{T}})\hat{\nabla}\cdot\hat{\mathbf{v}}\Big] \qquad (7.107)$$

Substituting the expressions for the second and third terms of (7.101) from (7.107) and (7.103), respectively, into (7.101), we get

$$\mathrm{RHS}_K = g_0 \int_{\mathbf{k}\cdot\hat{\mathbf{g}}>0} \hat{f}^0\hat{f}_1^0\left(\Phi_K' + \Phi_{K1}' - \Phi_K - \Phi_{K1}\right)\mathbf{k}\cdot\hat{\mathbf{g}}\,\mathrm{d}\mathbf{k}\,\mathrm{d}\hat{\mathbf{c}}_1 - 4\nu g_0 \hat{f}^0\Big[\frac{2}{\nu}\hat{\nabla}\nu\cdot\hat{\mathbf{C}} + \frac{1}{\hat{T}}\hat{\nabla}\hat{T}\cdot\hat{\mathbf{C}}\left(\frac{3\hat{C}^2}{10\hat{T}} - \frac{1}{2}\right) + \frac{2}{5\hat{T}}\hat{\mathbf{C}}\hat{\mathbf{C}}:\hat{\nabla}\hat{\mathbf{v}} - (1 - \frac{\hat{C}^2}{5\hat{T}})\hat{\nabla}\cdot\hat{\mathbf{v}}\Big] - 4\nu\hat{f}^0\hat{\mathbf{C}}\cdot\hat{\nabla}g_0 \qquad (7.108)$$

Equating (7.99) and (7.108) and rearranging, we obtain

$$g_0 \int_{\mathbf{k}\cdot\hat{\mathbf{g}}>0} \hat{f}^0\hat{f}_1^0\left(\Phi_K' + \Phi_{K1}' - \Phi_K - \Phi_{K1}\right)\mathbf{k}\cdot\hat{\mathbf{g}}\,\mathrm{d}\mathbf{k}\,\mathrm{d}\hat{\mathbf{c}}_1 = \hat{f}^0\Big\{\frac{1}{\nu}\frac{\hat{\mathrm{D}}\nu}{\mathrm{D}\hat{t}} + \left(\frac{\hat{C}^2}{2\hat{T}} - \frac{3}{2}\right)\frac{1}{\hat{T}}\frac{\hat{\mathrm{D}}\hat{T}}{\mathrm{D}\hat{t}}\Big\} + \hat{f}^0\hat{\mathbf{C}}\cdot\Big\{\frac{1}{\hat{T}}\left(\frac{\hat{\mathrm{D}}\hat{\mathbf{v}}}{\mathrm{D}\hat{t}} - \hat{\mathbf{b}}\right) + (1 + 8\nu g_0)\frac{1}{\nu}\hat{\nabla}\nu + \Big[\left(\frac{\hat{C}^2}{2\hat{T}} - \frac{3}{2}\right) + 4\nu g_0\left(\frac{3\hat{C}^2}{10\hat{T}} - \frac{1}{2}\right)\Big]\frac{1}{\hat{T}}\hat{\nabla}\hat{T} + 4\nu\hat{\nabla}g_0\Big\} + \hat{f}^0\left[1 + \frac{8}{5}\nu g_0\right]\frac{1}{\hat{T}}\hat{\mathbf{C}}\hat{\mathbf{C}}:\hat{\nabla}\hat{\mathbf{v}} - 4\nu g_0\hat{f}^0\left(1 - \frac{\hat{C}^2}{5\hat{T}}\right)\hat{\nabla}\cdot\hat{\mathbf{v}} \qquad (7.109)$$

As noted earlier, we must determine the material derivatives of ν, $\hat{\mathbf{v}}$ and $\hat{T}$ at O(1). This is accomplished by using $\hat{\mathbf{q}}_0 = 0$, $\hat{\Gamma}_0 = 0$, and (7.95) for $\hat{\boldsymbol{\sigma}}_0$ in the balance equations (7.68),

(7.71) and (7.75). The result is

$$\frac{1}{\nu}\left(\frac{\hat{\mathrm{D}}\nu}{\mathrm{D}\hat{t}}\right)_0 = -\hat{\nabla}\cdot\hat{\mathbf{v}}$$

$$\frac{1}{\hat{T}}\left(\frac{\hat{\mathrm{D}}\hat{T}}{\mathrm{D}\hat{t}}\right)_0 = -\frac{2}{3\nu\hat{T}}\hat{\sigma}_0^{\mathrm{T}}:\hat{\nabla}\hat{\mathbf{v}} = -\frac{2}{3}(1+4\nu g_0)\hat{\nabla}\cdot\hat{\mathbf{v}}$$

$$\frac{1}{\hat{T}}\left(\frac{\hat{\mathrm{D}}\hat{\mathbf{v}}}{\mathrm{D}\hat{t}} - \hat{\mathbf{b}}\right)_0 = -\frac{1}{\nu\hat{T}}\hat{\nabla}\cdot\hat{\sigma}_0 = -(1+4\nu g_0)\frac{1}{\hat{T}}\hat{\nabla}\hat{T}$$
$$-\frac{1}{\nu}(1+8\nu g_0)\hat{\nabla}\nu - 4\nu\hat{\nabla}g_0 \qquad (7.110)$$

Substituting the above in (7.109) and introducing, for notational convenience, the rescaled velocities

$$\boldsymbol{\xi} \equiv \hat{\mathbf{C}}/(2\hat{T})^{1/2}, \quad \boldsymbol{\zeta} \equiv \hat{\mathbf{g}}/(2\hat{T})^{1/2} \qquad (7.111)$$

we obtain the Boltzmann equation at O(K)

$$\mathcal{L}(\Phi_K) = -\frac{\nu_s}{\nu g_0}\left[\left(1+\frac{12}{5}\nu g_0\right)\left(\xi^2 - \frac{5}{2}\right)\boldsymbol{\xi}\cdot\hat{\nabla}\ln\hat{T}\right.$$
$$\left. + \frac{1}{(2\hat{T})^{1/2}}\left(1+\frac{8}{5}\nu g_0\right)2\overline{\boldsymbol{\xi}\boldsymbol{\xi}}:\hat{\nabla}\hat{\mathbf{v}}\right] \qquad (7.112)$$

which governs the behavior of Φ_K. Here the overline denotes the symmetric traceless part of a tensor, i.e., $\overline{\mathbf{a}} \equiv \frac{1}{2}(\mathbf{a}+\mathbf{a}^{\mathrm{T}}) - \frac{1}{3}\,\mathrm{tr}(\mathbf{a})\,\mathbf{I}$, and $\mathcal{L}$ is the linearized Boltzmann operator, defined as

$$\mathcal{L}(y) \equiv \frac{1}{\pi^{3/2}}\int_{\mathbf{k}\cdot\boldsymbol{\zeta}>0}[y(\boldsymbol{\xi})+y(\boldsymbol{\xi}_1)-y(\boldsymbol{\xi}')-y(\boldsymbol{\xi}_1')]\,e^{-\xi_1^2}\,\mathbf{k}\cdot\boldsymbol{\zeta}\,\mathrm{d}\mathbf{k}\,\mathrm{d}\boldsymbol{\xi}_1 \qquad (7.113)$$

To solve (7.112), we note that the left-hand side is linear in Φ_K, and the right-hand side is linear in the gradients of $\hat{\mathbf{v}}$ and $\hat{T}$. From the theory of linear equations, the solution of (7.112) is the sum of the most general solution of the homogeneous equation $\mathcal{L}(\Phi_K) = 0$, and any particular solution of (7.112). It is clear that

$$\Phi_K + \Phi_{K1} - \Phi_K' - \Phi_{K1}' = 0$$

satisfies the homogeneous equation; it can be shown that it is also a necessary condition (see Chapman and Cowling (1964, p. 87)). Consequently, the general solution of the homogeneous equation is a linear combination of the collisional invariants, namely, any constant, $\boldsymbol{\xi}$ and ξ^2 (the last because $\epsilon = 0$ at this order of the perturbation expansion). Since $\hat{\nabla}\hat{\mathbf{v}}$ and $\hat{\nabla}\ln\hat{T}$ are independently variable, the particular solution must depend linearly on them. Accordingly, we write

$$\Phi_K = -\frac{\nu_s}{\nu g_0}\left\{\left(1+\frac{12}{5}\nu g_0\right)\mathbf{A}\cdot\hat{\nabla}\ln\hat{T} + \frac{1}{(2\hat{T})^{1/2}}\left(1+\frac{8}{5}\nu g_0\right)\mathbf{B}:\hat{\nabla}\hat{\mathbf{v}}\right.$$
$$\left. + \alpha_1 + \boldsymbol{\alpha}_2\cdot\boldsymbol{\xi} + \alpha_3\xi^2\right\} \qquad (7.114)$$

Since this must hold for arbitrary values of $\hat{\nabla}\hat{\mathbf{v}}$ and $\hat{\nabla}\ln\hat{T}$, we substitute (7.114) in (7.112) and equate the coefficients of the gradients on the two sides. We then find that **A** and **B** satisfy the equations

$$\mathcal{L}(\mathbf{A}) = (\xi^2 - \frac{5}{2})\boldsymbol{\xi} \tag{7.115}$$

$$\mathcal{L}(\mathbf{B}) = 2\,\overline{\boldsymbol{\xi}\boldsymbol{\xi}} \tag{7.116}$$

It is clear that **A** and **B** are functions only of the variable $\boldsymbol{\xi}$. There being no vector or tensor parameter in the problem, it follows that the vector **A** must have the form

$$\mathbf{A} = \mathcal{A}(\xi)\,\boldsymbol{\xi} \tag{7.117}$$

Since the right-hand side of (7.116) is symmetric and traceless, it follows from the linearity of the operator $\mathcal{L}$ that the tensor **B** must also be so. Therefore, **B** must have the form

$$\mathbf{B} = \mathcal{B}(\xi)\,\overline{\boldsymbol{\xi}\boldsymbol{\xi}} \tag{7.118}$$

The solvability condition for (7.115) and (7.116) is that their right-hand sides be orthogonal to the linearly independent solutions of the homogeneous equation (Chapman and Cowling, 1964, pp. 110–111), which are 1, $\boldsymbol{\xi}$, and ξ^2. It is easily verified that this condition is satisfied, i.e.,

$$\int \hat{f}^0(\xi^2 - \frac{5}{2})\,\boldsymbol{\xi}\,\psi\,\mathrm{d}\boldsymbol{\xi} = 0, \qquad \int \hat{f}^0\,\overline{\boldsymbol{\xi}\boldsymbol{\xi}}\,\psi\,\mathrm{d}\boldsymbol{\xi} = 0, \tag{7.119}$$

for $\psi = 1$, $\boldsymbol{\xi}$ and ξ^2.

The constants α_1, $\boldsymbol{\alpha}_2$, and α_3 in (7.114) are determined by enforcing the moment constraint (7.86), which at O(K) can be written as

$$\int \Phi_K\, e^{-\xi^2}\,\psi\,\mathrm{d}\boldsymbol{\xi} = 0 \tag{7.120}$$

for $\psi = 1$, $\hat{\mathbf{c}}$, and $\hat{C}^2$. Equivalently, we may impose (7.120) for $\psi = 1$, $\boldsymbol{\xi}$, and ξ^2. Setting terms with odd powers of $\boldsymbol{\xi}$ to zero (as they vanish upon integration), and using (H.19) and (H.20), the moment constraint for $\psi = 1$, $\boldsymbol{\xi}$, and ξ^2, respectively, reduce to

$$\int \left(\alpha_1 + \alpha_3\xi^2\right) e^{-\xi^2}\mathrm{d}\boldsymbol{\xi} = 0 \tag{7.121}$$

$$\int \left\{\left(1 + \frac{12}{5}\nu g_0\right) \mathcal{A}(\xi)\hat{\nabla}\ln\hat{T} + \boldsymbol{\alpha}_2\right\} \xi^2\, e^{-\xi^2}\mathrm{d}\boldsymbol{\xi} = 0 \tag{7.122}$$

$$\int \left(\alpha_1\xi^2 + \alpha_3\xi^4\right) e^{-\xi^2}\mathrm{d}\boldsymbol{\xi} = 0 \tag{7.123}$$

Equations (7.121) and (7.123) imply $\alpha_1 = \alpha_3 = 0$, and (7.122) implies that $\boldsymbol{\alpha}_2$ is a multiple of $(1 + 12/5\nu g_0)\hat{\nabla}\ln\hat{T}$. Therefore, the term $\boldsymbol{\alpha}_2{\cdot}\boldsymbol{\xi}$ in (7.114) can be absorbed into the term proportional to $\hat{\nabla}\ln\hat{T}$, if we satisfy

$$\int \mathcal{A}(\xi)\,\xi^2 e^{-\xi^2}\mathrm{d}\boldsymbol{\xi} = 0 \tag{7.124}$$

This reduces the form of Φ_K to

$$\Phi_K = -\frac{\nu_s}{\nu g_0}\left\{\left(1 + \frac{12}{5}\nu g_0\right)\mathcal{A}(\xi)\,\boldsymbol{\xi}{\cdot}\hat{\nabla}\ln\hat{T} + \frac{1}{(2\hat{T})^{1/2}}\left(1 + \frac{8}{5}\nu g_0\right)\mathcal{B}(\xi)\,\overline{\boldsymbol{\xi}\boldsymbol{\xi}}{:}\hat{\mathbf{D}}\right\} \tag{7.125}$$

Note that we have replaced $\hat{\nabla}\hat{\mathbf{v}}$ with the dimensionless rate of deformation tensor $\hat{\mathbf{D}} \equiv \overline{\hat{\nabla}\hat{\mathbf{v}}}$, as it is contracted with the symmetric traceless tensor $\overline{\boldsymbol{\xi}\boldsymbol{\xi}}$.

7.3.10. Solution for Φ_K

The usual method of solving integral equations of the form (7.115) and (7.116) is by expanding the unknown function in a complete set of known basis functions†; here we write

$$\mathcal{A}(\xi) = \sum_{n=0}^{\infty} a_n \, \phi_n(\xi), \quad \mathcal{B}(\xi) = \sum_{n=0}^{\infty} b_n \, \phi_n(\xi), \tag{7.126}$$

where $\phi_n(\xi)$ are the basis functions. The series are truncated at $n = N$, and substituted in (7.115) and (7.116). The coefficients a_n, b_n are obtained either by *collocation*, whereby the equations are enforced at a discrete set of points ξ_n, or by *the method of moments*, in which of the moments of the equations with respect to the basis functions ϕ_n are enforced. A brief discussion of these methods follows.

The classical method of solution (Burnett, 1935; Chapman and Cowling, 1964) is to expand $\mathcal{A}$ and $\mathcal{B}$ in a series of Sonine polynomials in ξ^2, i.e., $\phi_n(\xi) = L_n^{(p)}(\xi^2)$, where p is a constant. More commonly known as the generalized Laguerre polynomials (see Abramowitz and Stegun (1965, pp. 773–775)), $L_n^{(p)}(x)$ are orthogonal polynomials, with the associated weight function $x^n e^{-x}$. The first few polynomials in the series are

$$L_0^{(p)}(x) = 1, \quad L_1^{(p)}(x) = p + 1 - x,$$
$$L_2^{(p)}(x) = \frac{(p+1)(p+2)}{2} - (p+1)x + \frac{x^2}{2} \tag{7.127}$$

The choice of Sonine polynomials is motivated by the fact that the integrand in $\mathcal{L}(y)$ can be written as a kernel $\mathcal{K}(\mathbf{k}, \boldsymbol{\xi}, \boldsymbol{\xi}_1)$ multiplied by a function of the form $\xi_1^2 e^{-\xi_1^2}$, the latter having the same form as the weight function of $L_1^{(p)}(\xi_1^2)$. The orthogonality relation for these polynomials is

$$\int_0^{\infty} x^p e^{-x} L_m^{(p)}(x) \, L_n^{(p)}(x) \, \mathrm{d}x = \begin{cases} 0 & \text{if } m \neq n \\ \dfrac{\Gamma(p+n+1)}{n!} & \text{if } m = n \end{cases} \tag{7.128}$$

where $\Gamma(x)$ is the Gamma function (Abramowitz and Stegun, 1965, p. 255). The series are truncated at $n = N$ and the coefficients a_n, b_n obtained by the method of moments. The choices of $p = 3/2$ for $\mathcal{A}(\xi)$ and $p = 5/2$ for $\mathcal{B}(\xi)$ effectively exploit the orthogonality property in determining the coefficients. A description of the method of moments can be found in the book of Reif (1985, pp. 536–545). In essence, (7.115) and (7.116) are not satisfied at every value of $\boldsymbol{\xi}$, but in an average sense; only the moments of the equations with respect to the basis functions are satisfied. For a given set of basis functions, this method gives the optimal set of coefficients for linear equations, i.e., one that minimizes the mean square error between the approximate and exact solutions (see Reif (1985, pp. 537–539)). Considering the series expansion for $\mathcal{A}(\xi)$, we note that the moment constraint (7.124), together with (7.128), implies $a_0 = 0$. To determine the other coefficients in the series, we determine the moments of (7.115): its mth moment is formed by taking its

† A set of functions is said be be complete if any continuous or piecewise-continuous function can be approximated to any desired degree of accuracy by a linear combination of the functions in the set (Arfken and Weber, 2001, p. 604).

inner (dot) product with $\boldsymbol{\xi}\, e^{-\xi^2} L_m^{(3/2)}(\xi^2)$ and integrating over $\boldsymbol{\xi}$, to get

$$\sum_{n=0}^{N} a_n\,[\boldsymbol{\xi} L_n^{(3/2)}(\xi^2),\ \boldsymbol{\xi} L_m^{(3/2)}(\xi^2)] = \int e^{-\xi^2}\xi^2\,(\xi^2 - \frac{5}{2})\,L_m^{(3/2)}(\xi^2)\,d\boldsymbol{\xi} \tag{7.129}$$

Here we have used the notation

$$[\mathbf{x}, \mathbf{y}] = \int \mathcal{L}(\mathbf{x}) * \mathbf{y}\, e^{-\xi^2} d\boldsymbol{\xi} \tag{7.130}$$

where $\mathbf{x} * \mathbf{y}$ indicates maximal contraction† between $\mathbf{x}$ and $\mathbf{y}$. The moment equation for $m = 1, N$ provides N algebraic equations for the coefficients a_1–a_N. Note that the right-hand side of (7.129) vanishes, except for $m = 1$, as a consequence of the orthogonality relation (7.128). Similarly, the inner (double dot) product of (7.116) with $e^{-\xi^2}\,\overline{\boldsymbol{\xi}\boldsymbol{\xi}}\,L_m^{(5/2)}(\xi)$ gives

$$\sum_{n=0}^{N} b_n\,[\overline{\boldsymbol{\xi}\boldsymbol{\xi}} L_n^{(5/2)}(\xi),\ \overline{\boldsymbol{\xi}\boldsymbol{\xi}} L_m^{(5/2)}(\xi)] = \frac{4}{3}\int e^{-\xi^2}\xi^4\,L_m^{(5/2)}(\xi)\,d\boldsymbol{\xi} \tag{7.131}$$

which, for $m = 0, N$, provides the $N+1$ algebraic equations for the coefficients b_0–b_N. The right-hand side of (7.131) vanishes, except for $m = 0$, as a consequence of (7.128).

The simplest approximations to $\mathcal{A}$ and $\mathcal{B}$ are obtained when only the first (nonvanishing) terms in their series are retained, i.e.,

$$\mathcal{A} = a_1(\frac{5}{2} - \xi^2), \quad \mathcal{B} = b_0 \tag{7.132}$$

When the coefficients are determined from (7.129) and (7.131) (with $m = 1$ and 0, respectively), we get

$$a_1 = -\frac{15\sqrt{2}}{32\sqrt{\pi}}, \quad b_0 = \frac{5\sqrt{2}}{8\sqrt{\pi}} \tag{7.133}$$

Chapman and Cowling (1964, p. 169) refined the solution by retaining the first four terms in each series. Interestingly, they found the viscosity determined from the one-term approximation to differ by less than 2% from that value obtained from the four-term solution! Similarly, the thermal conductivity determined from the two approximations differ by less than 3%. Thus even the simplest approximation for $\mathcal{A}$ and $\mathcal{B}$ provides reasonably accurate estimates of the transport properties.

Recently, Sela and Goldhirsch (1998) expanded $\mathcal{A}(\xi)$ and $\mathcal{B}(\xi)$ in a different set of basis functions, and used a collocation procedure to determine the coefficients. Writing

$$\mathbf{A} = \left[(\xi^2 - \frac{5}{2})\mathcal{A}'(\xi) + \alpha\right]\boldsymbol{\xi}, \quad \mathbf{B} = \mathcal{B}(\xi)\,\overline{\boldsymbol{\xi}\boldsymbol{\xi}} \tag{7.134}$$

where α is a constant (it does not vanish because the basis functions are not orthogonal to $(\xi^2 - \frac{5}{2})$), they showed that as $\xi \to \infty$, the asymptotic variation of $\mathcal{A}'(\xi)$ and $\mathcal{B}(\xi)$ is ξ^{-1}. They expanded $\mathcal{A}'(\xi)$ and $\mathcal{B}(\xi)$ in a set of basis functions $\phi_n(\xi) = e^{-\xi^2}\mathcal{I}_n(\xi^2)$, where $\mathcal{I}_n(x)$ are the modified Bessel functions of the first kind, which possess the required asymptotic behavior. In reality, this constraint on the basis functions is unnecessary, because the contribution to the transport properties from large ξ is negligibly small; for instance, the contribution to the thermal conductivity from $\xi > 5$ is less than 10^{-4}%. Sela and Goldhirsch determined the coefficients numerically using a collocation method, the details of which are

† In indicial notation, $*$ indicates the contraction of as many indices as possible. If $\mathbf{x}$ and $\mathbf{y}$ are both vectors, then it is the dot product; if they are both tensors, it is the double dot product (see Appendix A).

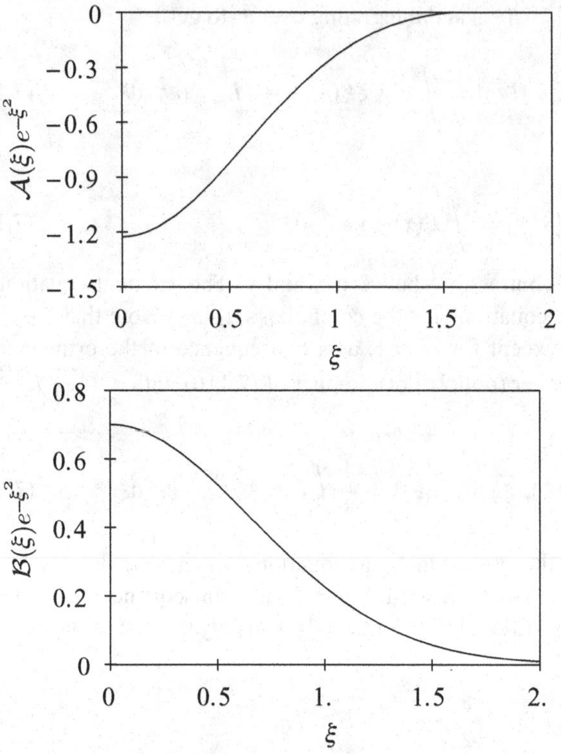

Figure 7.9. The functions $\mathcal{A}(\xi)$ and $\mathcal{B}(\xi)$ in Φ_K. (Adapted from Sela and Goldhirsch, 1998.)

somewhat involved; the reader is referred to the original work for the solution procedure. The functions $A(\xi)$ and $B(\xi)$ determined by retaining ten terms in each expansion are shown in Fig. 7.9. The viscosity obtained from this solution for $\mathcal{B}(\xi)$ differs by less than 2% from the one-term Sonine polynomial expansion described above.

Another class of methods determines the perturbation to the distribution function not by solving the Boltzmann equation, but by enforcing the Maxwell transport equation for certain higher velocity moments, i.e., for choices of ψ in (7.52) other than m_p, $m_p\mathbf{c}$, and $\frac{1}{2}m_p c^2$. Lun et al. (1984) attempted to extend for inelastic particles the procedure of Chapman (1916), and assumed a "trial function" of the form

$$\Phi = p_1\,\overline{\boldsymbol{\xi}\boldsymbol{\xi}}:\hat{\mathbf{D}} + p_2\,(\xi^2 - \frac{5}{2})\,\boldsymbol{\xi}\cdot\hat{\nabla}\ln\hat{T} + p_3\,(\xi^2 - \frac{5}{2})\,\boldsymbol{\xi}\cdot\hat{\nabla}\ln\nu \qquad (7.135)$$

where the coefficients p_1, p_2, p_3 are assumed to be independent of ξ. The values of the coefficients are determined by enforcing (7.52) for appropriate choices of ψ (see Problem 7.3). Note that, but for the last term, the trial function has the same form as Φ_K in (7.125). Lun et al. found the coefficients to be functions of the inelasticity ϵ, suggesting that O(K) and O($K\epsilon$) corrections were found. However, retaining some of the O($K\epsilon$) terms and dropping other terms of similar order, namely, the O(K^2) and O(ϵ^2) terms, results in an inconsistent expansion. Hence for consistency, ϵ must be set to zero, in which case p_3 vanishes and their result coincides with that of Chapman (1912, 1916). The latter is the one-term solution of Chapman and Cowling (1964), described earlier. Jenkins and Richman (1985) used the more formal procedure of Grad (1949), in which a trial function of the form

$$\Phi = \frac{1}{\hat{f}^0}\Big(-p_i\frac{\partial}{\partial\xi_i} + \frac{p_{ij}}{2!}\frac{\partial^2}{\partial\xi_i\partial\xi_j} - \frac{p_{ijk}}{3!}\frac{\partial^3}{\partial\xi_i\partial\xi_j\partial\xi_k} + \cdots\Big)\hat{f}^0 \qquad (7.136)$$

is assumed. The coefficients p_i, p_{ij} etc. may depend on $\mathbf{r}$ and t, but not on $\boldsymbol{\xi}$, and the corresponding functions represent multidimensional Hermite polynomials. The above is therefore an expansion of Φ in a series of Hermite polynomials. The coefficients are determined by the enforcing the Maxwell transport equation for higher velocity moments, as described above. The constraint (7.120) implies that $p_i = 0$ and $p_{ii} = 0$ (no sum on i). The solution for Φ_K obtained by retaining the first three terms is identical to that of Lun et al. (1984).

The one-term solution of Chapman and Cowling (1964), the solution of Lun et al. (1984) and the three-term solution of Jenkins and Richman (1985) yield identical results because, in essence, they all limit $\mathbf{A}(\xi)$ and $\mathbf{B}(\xi)$ to polynomial functions of third degree in $\boldsymbol{\xi}$, and they determine the coefficients by the method of moments.

With the solution for Φ_K at hand, we can now determine the constitutive relations for $\boldsymbol{\sigma}$ and $\mathbf{q}$ at O(K).

7.3.11. Constitutive Relations at O(*K*)

From (7.87), the streaming contribution to the dimensionless stress at O(K) is

$$\hat{\boldsymbol{\sigma}}^s_K = \frac{2\nu\hat{T}}{\pi^{3/2}} \int \boldsymbol{\xi}\boldsymbol{\xi}\, e^{-\xi^2}\, \Phi_K \, \mathrm{d}\boldsymbol{\xi} \tag{7.137}$$

Substituting for Φ_K from (7.125), we find that the term proportional to $\hat{\boldsymbol{\nabla}} \ln T$ contains odd powers of $\boldsymbol{\xi}$, and therefore vanishes on integration. This leaves the term proportional to $\hat{\mathbf{D}}$, which on integration using (H.21) and (H.23) gives

$$\hat{\boldsymbol{\sigma}}^s_K = -\frac{1}{g_0}\left(1 + \frac{8}{5}\nu g_0\right) \frac{6\nu_s}{\pi}\, 2\hat{\mu}^*\, \hat{\mathbf{D}} \tag{7.138}$$

where

$$\hat{\mu}^* \equiv \hat{T}^{1/2}\, \frac{2\sqrt{2\pi}}{45} \int_0^\infty \mathcal{B}(\xi)\, \xi^6\, e^{-\xi^2} \mathrm{d}\xi \tag{7.139}$$

is the dimensionless shear viscosity (i.e., the viscosity scaled by $\rho_p d_p v_s$) of a dilute gas of elastic spherical molecules (Chapman and Cowling, 1964, p. 122). Substituting the one-term Sonine polynomial series solution for $\mathcal{B}(\xi)$ from (7.132) and (7.133) and integrating over ξ using (H.23), we get

$$\hat{\mu}^* = \frac{5\sqrt{\pi}}{96}\, \hat{T}^{1/2} \tag{7.140}$$

As stated in §7.3.10, this differs by less than 2% from the values obtained from the four-term Sonine polynomial expansion of Chapman and Cowling (1964, p. 169) or the ten-term Bessel function expansion of Sela and Goldhirsch (1998) (the latter two differ by 0.01%).

The collisional part of the dimensionless stress comes from the second term in (7.87). Expanding g_0 and f^0 in a Taylor series and retaining only the O(K) term, we get

$$\begin{aligned}
\hat{\boldsymbol{\sigma}}^c_K = {} & \frac{6\nu^2 g_0 \hat{T}}{\pi^4} \int_{\mathbf{k}\cdot\boldsymbol{\zeta} > 0} \mathbf{k}\,(\boldsymbol{\xi}' - \boldsymbol{\xi})\, \mathbf{k}\cdot\boldsymbol{\zeta}\; e^{-\xi^2-\xi_1^2} (\Phi_K + \Phi_{K1})\, \mathrm{d}\mathbf{k}\, \mathrm{d}\boldsymbol{\xi}\, \mathrm{d}\boldsymbol{\xi}_1 \\
& + \frac{18\nu^2 \nu_s g_0 \hat{T}}{\pi^5} \int_{\mathbf{k}\cdot\boldsymbol{\zeta} > 0} \mathbf{k}\,(\boldsymbol{\xi}' - \boldsymbol{\xi})\, \mathbf{k}\cdot\boldsymbol{\zeta}\; e^{-\xi^2-\xi_1^2} \\
& \times \mathbf{k}\cdot\hat{\boldsymbol{\nabla}} \ln(\hat{f}^0/\hat{f}^0_1)\, \mathrm{d}\mathbf{k}\, \mathrm{d}\boldsymbol{\xi}\, \mathrm{d}\boldsymbol{\xi}_1
\end{aligned} \tag{7.141}$$

Substituting $\boldsymbol{\xi}' - \boldsymbol{\xi} = (\mathbf{k}\cdot\boldsymbol{\zeta})\mathbf{k}$ from (7.4) and for $\hat{\nabla} \hat{f}^0$ from (7.104), and integrating over $\mathbf{k}$ using (H.4), we find again that the only nonvanishing contributions in both integrals are the terms proportional to $\hat{\nabla}\hat{\mathbf{v}}$,

$$\hat{\boldsymbol{\sigma}}_K^c = -\frac{1}{g_0}\left(1 + \frac{8}{5}\nu g_0\right)\frac{6\sqrt{2}\nu\nu_s g_0 \hat{T}^{1/2}}{15\pi^3}\hat{\mathbf{D}}\mathbf{:}\int\Big[(2\zeta\boldsymbol{\zeta} + \zeta^2\mathbf{I})$$

$$\times e^{-\xi^2-\xi_1^2}(\mathcal{B}(\xi)\overline{\boldsymbol{\xi}\boldsymbol{\xi}} + \mathcal{B}(\xi_1)\overline{\boldsymbol{\xi}_1\boldsymbol{\xi}_1})\Big]\,\mathrm{d}\boldsymbol{\xi}\,\mathrm{d}\boldsymbol{\xi}_1$$

$$-\frac{3\nu^2\nu_s g_0 \hat{T}^{1/2}}{\sqrt{2}\pi^4}\int\Big[\hat{\nabla}\hat{\mathbf{v}}\mathbf{:}\boldsymbol{\zeta}\boldsymbol{\zeta}(\zeta\boldsymbol{\zeta} + \zeta^2\mathbf{I})/\zeta + \zeta\boldsymbol{\zeta}\hat{\nabla}\hat{\mathbf{v}}\cdot\boldsymbol{\zeta}$$

$$+\,\zeta\hat{\nabla}\hat{\mathbf{v}}\cdot\boldsymbol{\zeta}\boldsymbol{\zeta}\Big]\,e^{-\xi^2-\xi_1^2}\,\mathrm{d}\boldsymbol{\xi}\,\mathrm{d}\boldsymbol{\xi}_1 \tag{7.142}$$

To evaluate the first integral, we substitute $\boldsymbol{\zeta} = \boldsymbol{\xi}_1 - \boldsymbol{\xi}$ and discard terms involving odd powers of $\boldsymbol{\xi}$ or $\boldsymbol{\xi}_1$. Using (H.19), (H.21), and (H.22) to simplify the result, we get the first term on the right-hand side of (7.143). To evaluate the second integral in (7.142), we change the variables of integration to $\boldsymbol{\zeta}$ and $\mathbf{G} \equiv (\boldsymbol{\xi}_1 + \boldsymbol{\xi})/2$; as the determinant of the Jacobian for this transformation is unity (see §H.3), $\mathrm{d}\boldsymbol{\xi}\,\mathrm{d}\boldsymbol{\xi}_1 = \mathrm{d}\mathbf{G}\,\mathrm{d}\boldsymbol{\zeta}$. Discarding vanishing integrals and using (H.19), (H.21), and (H.22), we finally get

$$\hat{\boldsymbol{\sigma}}_K^c = -\frac{1}{g_0}\left(1 + \frac{8}{5}\nu g_0\right)\frac{8\nu g_0}{5} \times \frac{6\nu_s}{\pi}\,2\hat{\mu}^*\,\hat{\mathbf{D}}$$

$$-\,\hat{\mu}_b\frac{6\nu_s}{\pi}\left[\frac{6}{5}\hat{\mathbf{D}} + (\boldsymbol{\nabla}\cdot\hat{\mathbf{v}})\,\mathbf{I}\right] \tag{7.143}$$

where

$$\hat{\mu}_b \equiv \frac{8\nu^2 g_0 \hat{T}^{1/2}}{3\sqrt{\pi}} \tag{7.144}$$

is the bulk viscosity scaled by $\rho_p d_p v_s$. Adding the streaming and collisional contributions in (7.138) and (7.143), respectively, we get the total stress at O(K),

$$\hat{\boldsymbol{\sigma}}_K = -\frac{6\nu_s}{\pi}\left\{2\hat{\mu}\,\hat{\mathbf{D}} + \hat{\mu}_b(\boldsymbol{\nabla}\cdot\hat{\mathbf{v}})\,\mathbf{I}\right\} \tag{7.145}$$

with the dimensionless shear viscosity $\hat{\mu}$ given by

$$\hat{\mu} = \frac{1}{g_0}\left(1 + \frac{8}{5}\nu g_0\right)^2\hat{\mu}^* + \frac{3}{5}\hat{\mu}_b \tag{7.146}$$

We now turn to determining the pseudothermal energy flux at O(K). The streaming part of $\hat{\mathbf{q}}_K$ is (see (7.88))

$$\hat{\mathbf{q}}_K^s = \frac{\nu(2\hat{T})^{3/2}}{2\pi^{3/2}}\int\boldsymbol{\xi}\xi^2\,e^{-\xi^2}\,\Phi_K\,\mathrm{d}\boldsymbol{\xi} \tag{7.147}$$

Substituting for Φ_K from (7.125), we find that the term proportional to $\hat{\nabla}\hat{\mathbf{v}}$ contains odd powers of $\boldsymbol{\xi}$ and therefore vanishes on integration. Using (H.19) to integrate the remaining term, we get

$$\hat{\mathbf{q}}_K^s = -\frac{1}{g_0}\left(1 + \frac{12}{5}\nu g_0\right)\frac{6\nu_s}{\pi}\,\hat{\kappa}^*\,\hat{\nabla}\hat{T} \tag{7.148}$$

where

$$\hat{\kappa}^* \equiv \hat{T}^{1/2}\,\frac{2\sqrt{2\pi}}{9}\int_0^\infty \mathcal{A}(\xi)\,\xi^6\,e^{-\xi^2}\,\mathrm{d}\xi \tag{7.149}$$

is the dimensionless thermal conductivity (i.e., the thermal conductivity scaled by $\rho_p d_p v_s$) of a dilute gas of elastic spherical molecules. Substituting the one-term Sonine polynomial series solution for $\mathcal{A}(\xi)$ from (7.132) and (7.133) and integrating over ξ using (H.23), we get

$$\hat{\kappa}^* = \frac{25\sqrt{\pi}}{128}\hat{T}^{1/2} \tag{7.150}$$

This differs by less than 3% from the values obtained from the four-term Sonine polynomial expansion of Chapman and Cowling (1964, p. 169) or the ten-term Bessel function expansion of Sela and Goldhirsch (1998) (the latter two differ by 0.06%).

The collisional part of $\hat{\mathbf{q}}_K$ comes from the second term in (7.88). Following the same procedure as in the determination of $\hat{\boldsymbol{\sigma}}^c_K$, we find that the terms proportional to $\hat{\nabla}\hat{\mathbf{v}}$ and $\hat{\nabla}\nu$ contain odd powers of $\boldsymbol{\xi}$ or $\boldsymbol{\xi}_1$ and hence vanish upon integration. We then get

$$\hat{\mathbf{q}}^c_K = -\left\{\frac{1}{g_0}\left(1+\frac{12}{5}\nu g_0\right)\frac{12\nu g_0}{5}\hat{\kappa}^* + \frac{3}{2}\hat{\mu}_b\right\}\frac{6\nu_s}{\pi}\hat{\nabla}\hat{T} \tag{7.151}$$

Adding the streaming and collisional contributions from (7.148) and (7.151), respectively, we get the total energy flux at O(K)

$$\hat{\mathbf{q}}_K = -\frac{6\nu_s}{\pi}\hat{\kappa}\,\hat{\nabla}\hat{T} \tag{7.152}$$

with the dimensionless pseudothermal conductivity $\hat{\kappa}$ given by

$$\hat{\kappa} = \frac{1}{g_0}\left(1+\frac{12}{5}\nu g_0\right)^2\hat{\kappa}^* + \frac{3}{2}\hat{\mu}_b \tag{7.153}$$

The dimensionless dissipation rate $\hat{\Gamma}$ does not have an O(K) contribution, as indicated in (7.92).

7.3.12. Distribution Function and Constitutive Relations at O(ϵ)

We now determine Φ_ϵ and its contribution to the constitutive relations. That the left-hand side of the dimensionless Boltzmann equation (7.81) is multiplied by K does not mean that it has no O(ϵ) term. This is because $K\mathrm{D}\hat{T}/\mathrm{D}\hat{t}$ has one O(ϵ) term, namely, $-2\pi/(18\nu\nu_s)\,\epsilon\hat{\Gamma}_\epsilon$ (see (7.75) and (7.98)). Substituting for $\hat{\Gamma}_\epsilon$ from (7.98), the left-hand side of (7.81) at this order is

$$-2\sqrt{\pi}(\nu g_0/\nu_s)(\tfrac{2}{3}\xi^2 - 1)\hat{T}^{1/2}\hat{f}^0$$

To determine the right-hand side at O(ϵ), we expand in powers of ϵ the following:

(i) the factor $1/(1-\epsilon)$,
(ii) the product $\hat{f}^{0\prime\prime}\hat{f}_1^{0\prime\prime}$, using the relation $\xi''^2 + \xi_1''^2 = \xi^2 + \xi_1^2 + \epsilon/2\,(\mathbf{k}\cdot\boldsymbol{\zeta})^2$ (which comes from (7.38)), and
(iii) $\hat{f}$ itself, i.e., $\hat{f} = \hat{f}^0(1+\epsilon\Phi_\epsilon+\cdots)$.

These expansions are substituted in the right-hand side of (7.81), and the terms linear in ϵ are retained.

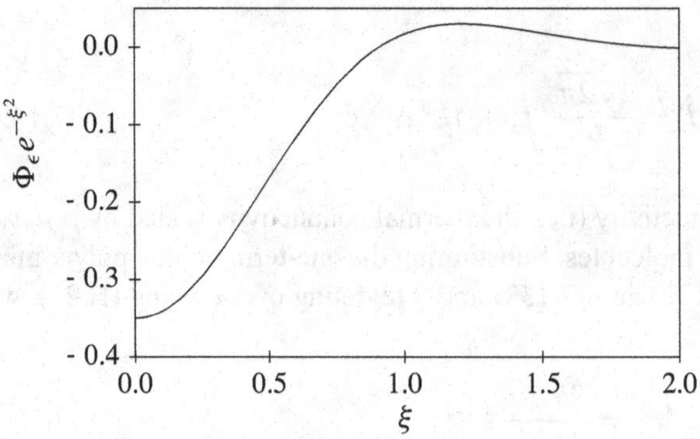

Figure 7.10. The function $\Phi_\epsilon(\xi)$. (Adapted from Sela and Goldhirsch, 1998.)

Equating the two sides and dividing by $\hat{f}^0$, we get (Sela and Goldhirsch, 1998)

$$\mathcal{L}(\Phi_\epsilon) = \sqrt{2\pi}\,(\frac{2}{3}\xi^2 - 1) + \frac{\sqrt{\pi}}{8}(3 - 2\xi^2)\,e^{-\xi^2} + \frac{\pi}{16\xi}(5 + 4\xi^2 - 4\xi^4)\,\text{erf}(\xi) \tag{7.154}$$

where $\mathcal{L}$ is the linearized Boltzmann operator defined in (7.113), and erf(x) is the error function (Abramowitz and Stegun, 1965, p. 297). Since the right-hand side of (7.154) is a function only of the scalar ξ, it follows from the linearity of the operator $\mathcal{L}$ that $\Phi_\epsilon = \Phi_\epsilon(\xi)$. Thus Φ_ϵ makes an isotropic contribution to Φ. This contribution was first recognized and computed by Sela and Goldhirsch (1998).

It can be verified that the right-hand side (7.154) is orthogonal to 1, $\boldsymbol{\xi}$, and ξ^2 (with $e^{-\xi^2}$ as the weighting function). Since these are the linearly independent solutions of the homogeneous equation $\mathcal{L}(\Phi_\epsilon) = 0$, the solvability condition is satisfied. As in (7.114), the solution of (7.154) is the sum of the general solution of the homogeneous equation and a particular solution $\mathcal{C}(\xi)$, i.e.,

$$\Phi_\epsilon = \alpha_1 + \boldsymbol{\alpha}_2\cdot\boldsymbol{\xi} + \alpha_3\xi^2 + \mathcal{C}(\xi) \tag{7.155}$$

Sela and Goldhirsch (1998) determined $\mathcal{C}(\xi)$ in the same manner as $\mathcal{A}'(\xi)$ and $\mathcal{B}(\xi)$ (see §7.3.10), but with a different set of basis functions. They showed that the asymptotic behavior of $\mathcal{C}(\xi)$ as $\xi \to \infty$ is $\mathcal{C}(\xi) \sim \xi^2 \log\xi$, and chose $\phi_n(\xi) = (1 + \log(1 + \xi^2))(1 + \xi^2)^{3/2}\,e^{-\xi^2}\mathcal{I}_n(\xi^2)$, where $\mathcal{I}_n$ are the modified Bessel functions of the first kind, as they have the same asymptotic behavior. The constants α_1, $\boldsymbol{\alpha}_2$ and α_3 are determined by the moment constraint (7.86), which gives $\boldsymbol{\alpha}_2 = 0$, and nonzero values for α_1 and α_3. The solution for Φ_ϵ is displayed in Fig. 7.10.

From (7.87), we see that the streaming part of $\hat{\boldsymbol{\sigma}}_\epsilon$ vanishes due to the orthogonality of Φ_ϵ to ξ^2 (recall that (7.86) is enforced at every order). The collisional part can have contributions from (a) the O(ϵ) part of $(\hat{\mathbf{c}}' - \hat{\mathbf{c}})$ and (b) Φ_ϵ. The latter vanishes due to the orthogonality of Φ_ϵ to 1 and ξ^2 (see Problem 7.4). Noting from (7.6) that the O(ϵ) part of $(\hat{\mathbf{c}}' - \hat{\mathbf{c}})$ is $-\epsilon(\mathbf{k}\cdot\hat{\mathbf{g}})\mathbf{k}/4$, and evaluating the second integral in (7.87) using (H.4), (H.19), and (H.23), we get

$$\hat{\boldsymbol{\sigma}}_\epsilon = -\nu^2 g_0 \hat{T}\,\mathbf{I} \tag{7.156}$$

From the form of Φ_ϵ and its orthogonality to ξ^2, we can deduce that (see Problem 7.4)

$$\hat{\mathbf{q}}_\epsilon = 0 \tag{7.157}$$

The O(ϵ) contribution to Γ was determined in §7.3.8.

7.3.13. Constitutive Relations to First Order in K and ϵ

We may now determine the constitutive relations up to linear order in the small parameters, by combining the results of §7.3.8, §7.3.11, and §7.3.12.

The total stress is the sum of the O(1) and O(K) and O(ϵ) contributions, given in (7.95), and (7.145), and (7.156) respectively. Reverting to the dimensional forms of the variables, the stress is

$$\sigma = \left[p - \mu_b \nabla \cdot \mathbf{v}\right] \mathbf{I} - 2\mu\, \mathbf{D} \tag{7.158}$$

where $\mathbf{D}$ is the rate of deformation tensor, the pressure p is given by

$$p = \rho_p \left(1 + 4\eta_1 \nu g_0\right) \nu T \tag{7.159}$$

and the bulk and shear viscosities by

$$\mu_b = \frac{8}{3\sqrt{\pi}} \nu^2 g_0\, \rho_p d_p\, T^{1/2} \tag{7.160}$$

$$\mu = \left(1 + \frac{8}{5}\nu g_0\right)^2 \frac{\mu^*}{g_0} + \frac{3}{5}\mu_b \tag{7.161}$$

Here μ^* is the shear viscosity of a dilute gas of elastic particles. The first approximation for μ^*, obtained by retaining only the first term in the Sonine polynomial expansion (and shown to be fairly accurate in §7.3.11), gives

$$\mu^* = \frac{5\sqrt{\pi}}{96} \rho_p d_p\, T^{1/2} \tag{7.162}$$

The flux of fluctuational kinetic energy is the O(K) contribution from (7.152), which in dimensional form is

$$\mathbf{q} = -\kappa\, \nabla T \tag{7.163}$$

with the pseudothermal conductivity κ given by

$$\kappa = \left(1 + \frac{12}{5}\nu g_0\right)^2 \frac{\kappa^*}{g_0} + \frac{3}{2}\mu_b \tag{7.164}$$

where κ^* is the thermal conductivity of a dilute gas of elastic particles. The first approximation for κ^* is

$$\kappa^* = \frac{25\sqrt{\pi}}{128} \rho_p d_p\, T^{1/2} \tag{7.165}$$

Finally, the rate of energy dissipation has a contribution only at O(ϵ), given in (7.98), which in dimensional form is

$$\Gamma = \frac{12}{\sqrt{\pi}} (1 - e_p^2) \frac{\rho_p\, T^{3/2}}{d_p} \nu^2 g_0 \tag{7.166}$$

Note that the coefficient of restitution e_p appears only in the relations for the pressure and the dissipation rate Γ. Thus to first order in K and ϵ, the only effects of inelasticity are in the pressure and the sink of pseudothermal energy. Otherwise, the equations of motion for a granular gas are identical to that of a dense gas of elastic particles. The modification of the pressure is small when $\epsilon \ll 1$, and deleting it does not significantly alter the form of the equations. Hence η_1 may be replaced by unity in (7.159). However, the sink of pseudothermal energy, though also an O(ϵ) term, is of paramount importance. Due to this term, there is no steady state without external forcing, no matter how small the inelasticity.

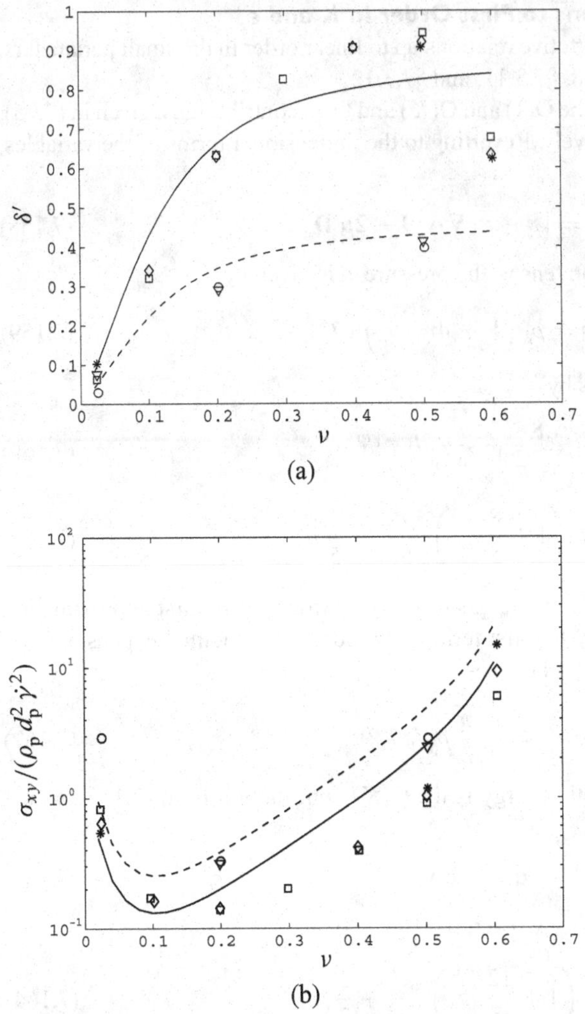

Figure 7.11. Comparison of the data from particle dynamics simulations (symbols) of Walton and Braun (1986) for uniform plane shear of smooth inelastic spheres with the predictions of the kinetic theory (lines). (a) The parameter δ', defined in (7.169), as a function of the solids fraction. (b) The dimensionless shear stress as a function of the solids fraction. The dashed line, circles and triangles are for $e_p = 0.95$, and the solid line, squares, diamonds and asterisks are for $e_p = 0.8$. For each value of e_p, the different symbols correspond to different values of the shear rate $\dot{\gamma}$.

It is this term that makes the behavior of a granular gas fundamentally different from that of molecular gases.

Experimental verification of the predictions of kinetic theory is lacking, due to the difficulties in noninvasive measurement of granular flows (see §8.3.1). Additionally, gravitational compaction usually results in nonuniform deformation of the material, making it difficult to measure properties at a particular value of the solids fraction or shear rate. However, the results of particle dynamics simulations, which can be conducted in the absence of gravity and in which all the relevant quantities can be easily inferred, may be readily compared with the theoretical predictions. Figure 7.11 compares the data from the simulations of Walton and Braun (1986) for uniform plane shear with the results of this section. Before considering the figure, we first derive the predictions of the theory for uniform plane shear. The analysis of uniform plane shear using the high density heuristic theory of Haff (1983) was discussed in §7.2.1, and it is a straightforward extension to use the constitutive relations derived from kinetic theory, given above. It is readily seen that the spatially uniform temperature is

$$T = \left(\frac{\eta}{\gamma}\right) d_p^2\,\dot{\gamma}^2 \tag{7.167}$$

where $\dot{\gamma}$ is the shear rate, and we have introduced the dimensionless functions of the solids fraction

$$\eta(\nu) \equiv \mu/(\rho_p d_p \sqrt{T}), \quad \gamma(\nu) \equiv \Gamma d_p/(\rho_p\, T^{3/2}) \tag{7.168}$$

Equation (7.167) may be expressed in terms of the parameter δ that was introduced in (7.49) as a measure of the departure from equilibrium. Here we modify it slightly by replacing s with d_p and introduce the factor $\sqrt{3}$ in the denominator, and write

$$\delta' = \frac{\dot{\gamma}\, d_p}{\sqrt{3T}} = \left(\frac{\gamma}{3\eta}\right)^{1/2} \tag{7.169}$$

This parameter is shown in Fig. 7.11a as a function of ν. The dimensionless shear stress σ_{yx} is

$$\frac{\sigma_{yx}}{\rho_p d_p^2\,\dot{\gamma}^2} = \left(\frac{\eta^3}{\gamma}\right)^{1/2} \tag{7.170}$$

which is shown in Fig. 7.11b.

We see that δ' is small only when $\nu \ll 1$, and increases with ν. Thus there is a significant departure from equilibrium at moderate to large ν. Despite this fact, there is good qualitative agreement between the predictions of the theory and the simulations. Secondly, there is reasonable quantitative agreement at $e_p = 0.95$, but the data deviate significantly from the theory at $e_p = 0.8$. Finally, the agreement is generally better at low ν than at high ν. The last two points are not surprising. We have derived only the leading order in a perturbation expansion in the inelasticity, and therefore expect the theory to be valid only for small inelasticity. Secondly, the assumption of molecular chaos, with the Enskog correction, is not expected to be accurate at high ν, where one expects strong correlations in the velocities of neighboring particles due to repeated collisions.

7.4. ANISOTROPY OF THE MICROSTRUCTURE

The issue of microstructural anisotropy was mentioned in §7.3.5, but not discussed at length. In the molecular chaos assumption with the Enskog correction (7.41), the pair distribution function $g(\mathbf{r}+\mathbf{k}\,d_p, \mathbf{r})$ characterizes the correlation in the positions of two particles in contact. If there is spatial homogeneity, it depends only on the separation vector $\mathbf{k}\,d_p$ and the fluid density; further, if the fluid is isotropic, it does not depend on $\mathbf{k}$. In §7.3.5 and thereafter, the isotropic equilibrium pair distribution function at contact g_0, which depends only on ν, was used. However, particle dynamics simulations suggest that the microstructure in sheared granular materials is not isotropic, and that the anisotropy increases with particle inelasticity (Hopkins and Louge, 1991). Unfortunately, determining the microstructure analytically is not a simple task in systems removed from equilibrium. Jenkins and Savage (1983) incorporated anisotropy in a simple and phenomenological way, following the earlier study of Savage and Jeffrey (1981). A slightly expanded version of their analysis is presented below.

In the presence of spatial variation of the hydrodynamic variables, $g(\mathbf{r}+\mathbf{k}\,d_p, \mathbf{r})$ may depend not just on $\mathbf{k}$, but also on $\nabla\mathbf{v}$, ∇T, and all the higher gradients of $\mathbf{v}$ and T. The pair distribution, being a scalar, must be a scalar function of $\mathbf{k}$ and the gradients. As we have thus far retained only the terms linear in the first gradients of the hydrodynamic variables, we shall do the same here for consistency. The only frame-indifferent scalars we can construct from $\mathbf{k}$ and the gradients, which are linear in the latter, are $\mathbf{kk}:\nabla\mathbf{v}$ and $\mathbf{k}\cdot\nabla T$. Hence the pair distribution may be expressed as

$$g(\mathbf{r}+\mathbf{k}\,d_p, \mathbf{r}) = g_0[\nu(r+\mathbf{k}\,d_p/2)]\left(1+\alpha\,\frac{d_p\,\mathbf{kk}:\nabla\mathbf{v}}{\sqrt{T}}+\beta\,\frac{d_p\,\mathbf{k}\cdot\nabla T}{T}\right) \tag{7.171}$$

where α and β are dimensionless functions of ν that characterize the degree of anisotropy. Using (7.171) to determine the constitutive relations, we get the following expressions for the bulk viscosity, shear viscosity, and pseudothermal conductivity (see Problem 7.5):

$$\mu_b = \left(1+\frac{\sqrt{\pi}\,\alpha}{2}\right)\frac{8}{3\sqrt{\pi}}\,\nu^2 g_0\,\rho_p d_p\,T^{1/2} \tag{7.172}$$

$$\mu = \left(1+\frac{8}{5}\nu g_0\right)^2\frac{\mu^*}{g_0}+\frac{3}{5}\mu_b \tag{7.173}$$

$$\kappa = \left(1+\frac{12}{5}\nu g_0\right)^2\frac{\kappa^*}{g_0}+\frac{3}{2}\left(\frac{1+21\beta/4}{1+\sqrt{\pi}\,\alpha/2}\right)\mu_b \tag{7.174}$$

These expressions may be compared with (7.160), (7.161), and (7.164), which were derived assuming an isotropic pair distribution.

The parameters α and β are not known *a priori*, but may be determined from particle dynamics simulations or experiments. Measurements of α have been made for molecular fluids (see, e.g., Clark and Ackerson (1980); Lutsko (1996)), but not for systems of inelastic particles. Further investigations on how the the inelasticity of particle collisions affects the pair distribution is necessary in order to fully understand its influence on the bulk properties of granular gases.

7.5. EXTENSION TO GRANULAR MIXTURES

The discussion so far has been limited to granular gases of uniformly sized spheres. Size differences are common in practical situations, so an extension of the kinetic theory to mixtures is useful. The extension of the Chapman–Enskog procedure for a dilute binary mixture gas of smooth elastic spheres was first made by Chapman (1917) (see also Chapman and Cowling (1964, Chapter 8)). However, the analysis for dense gases is more complicated, as explained below.

Recall that the Enskog-corrected assumption of molecular chaos (7.41) requires the pair distribution function at contact. In our consideration of uniformly sized spheres, we have used the equilibrium pair distribution evaluated at the point of contact (see (7.60)), which is the midpoint of the line joining the centers of the colliding particles. For particles of unequal size, this assumption leads to results that are in conflict with the description of diffusion in irreversible thermodynamics (Barajas et al., 1973), in that Onsager's symmetry relations for the transport coefficients are violated. Van Beijeren and Ernst (1973) proposed a formulation, called the *revised Enskog theory*, in which the pair distribution function is treated as a nonlocal *functional* of the density fields of the individual components. This theory yields fluxes whose forms are in conformity with Onsager's symmetry relations; it has also been verified in other ways (see Lopez do Haro et al. (1983)). This theory was employed by Lopez de Haro et al. to derive hydrodynamic equations for mixtures of smooth,

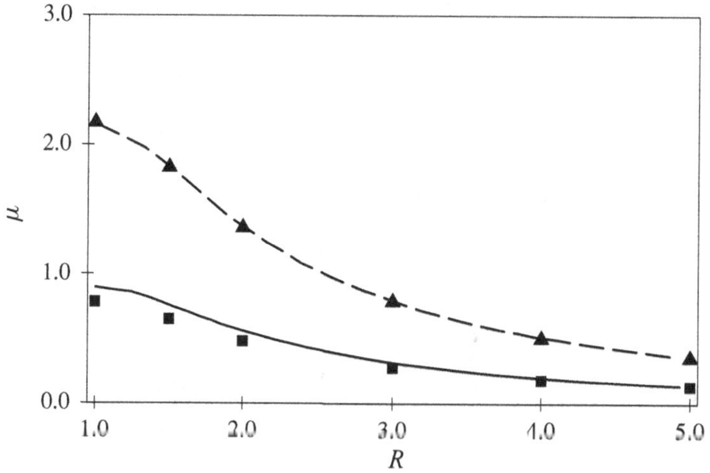

Figure 7.12. The dimensionless viscosity for a bidisperse granular gas as a function of the diameter ratio R of larger to smaller disks. The symbols are from the particle dynamics simulations of Alam and Luding (2003) (triangles, $e_\mathrm{p} = 0.99$; squares, $e_\mathrm{p} = 0.9$) and the lines are the predictions of the kinetic theory of Jenkins and Mancini (1989), corrected by Arnarson and Willits (1998). The total area fraction of the particles is 0.3, and the two species have equal area fraction and density. The viscosity is scaled by $\frac{1}{2}\rho_\mathrm{p} d_\mathrm{p}^2 \dot{\gamma} (1 + R^2)$, where d_p is the diameter of the smaller particles and $\dot{\gamma}$ is the shear rate. (Adapted from Alam and Luding, 2003.)

elastic spheres. It was subsequently extended to a granular gas of smooth, nearly elastic spheres by Jenkins and Mancini (1989); some errors in their expressions for the transport coefficients were corrected by Arnarson and Willits (1998).

The analysis for mixtures is not discussed here; the interested reader is referred to the studies mentioned above for the details. However, a comparison of the results of the theory with observations is illuminating. Experimental data for a uniformly sheared mixture are not available, so we compare with the results of the particle dynamics simulations of Alam and Luding (2003). The theory and simulations are for a binary mixture of inelastic *disks* of diameters d_p and Rd_p. Figure 7.12 shows the variation of the dimensionless viscosity with the size ratio R for $e_\mathrm{p} = 0.99$ and 0.9. The mass densities of the two species are assumed to be equal, and so are their coefficients of restitution. There is very good agreement between theory and simulations for $e_\mathrm{p} = 0.99$, and reasonably good agreement for $e_\mathrm{p} = 0.9$. The agreement gets poorer as e_p is further decreased (not shown).

However, the agreement is somewhat fortuitous, because one of the key assumptions of the theory is violated in the simulations. As we know from §7.3.7, the Chapman–Enskog method is a perturbation expansion about equilibrium. At equilibrium, there is equipartition of energy between the two species, and the temperature is defined such that this result carries over to the perturbed state. This is not observed in the simulations: denoting the mean kinetic energy of the small and large species by E^s and E^1, respectively, we see in Fig. 7.13 that the ratio E^1/E^s in the simulations increases, implying increasing deviation from equipartition, as the size ratio increases. While the curve for $e_\mathrm{p} = 0.9$ in Fig. 7.13 shows that $E^1/E^s > 4$ at $R = 5$, the viscosity for the same parameter set is in close agreement with the theoretical prediction (Fig. 7.12)! Despite the differences, Figs. 7.12 and 7.13 show good qualitative agreement between the theory and simulations for $e_\mathrm{p} \geq 0.9$, and even quantitative agreement for very small inelasticity. Alam and Luding (2003) observe that the deviation from equipartition is relatively insensitive to the solids fraction.

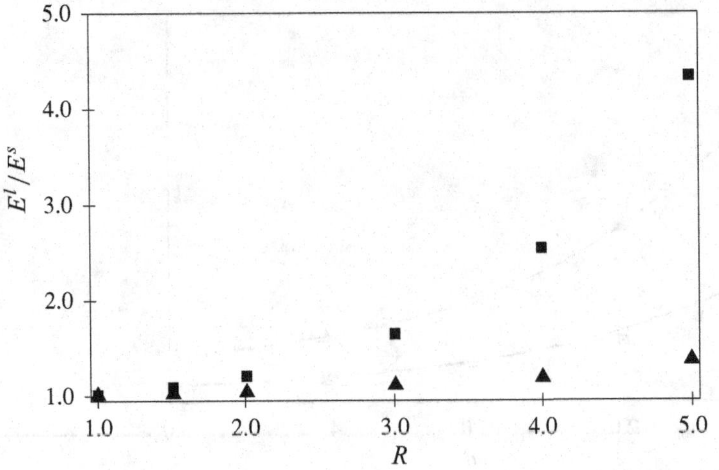

Figure 7.13. Ratio of the kinetic energy of the larger to smaller disks in uniform shear of a bidisperse granular material as a function of the ratio of particle diameters. The triangles are for $e_p = 0.99$, and the squares for $e_p = 0.9$ (assumed constant for all particles), other parameters as in Fig. 7.12. (Adapted from Alam and Luding, 2003.)

7.6. SUMMARY AND DISCUSSION

In this chapter we have considered the flow of granular materials in the regime of rapid flow. The physical picture of grain motion in this regime is similar to that of molecules in a dense gas, as grains are continuously in a state of fluctuational motion and interact only during brief collisions. However, there is a crucial difference between a granular material and a gas, which is that grains are inelastic and rough. Hence a part of the kinetic energy of the colliding grains is dissipated into internal energy of the grains. As a result, energy must be supplied continuously to sustain fluctuational motion, and therefore rapid granular flow is fundamentally far from equilibrium. For smooth particles, the extent of dissipation is characterized by the coefficient of restitution e_p.

A hydrodynamic description is normally applicable to quantities that are conserved during a collision, but it may be extended to quantities that are nearly conserved. For a granular gas, the kinetic energy is nearly conserved if the inelasticity is small, i.e., $\epsilon \equiv (1 - e_p^2) \ll 1$; therefore a hydrodynamic description is applicable under this constraint. We have used the heuristic theory of Haff (1983), limited to high density and smooth particles, to derive insight into the distinctive features of rapid granular flows and determine the limits of applicability of a hydrodynamic description. Following this, the more rigorous formalism of the kinetic theory of dense gases was used to derive the equations of motion, and the constitutive equations for the fluxes and sources that occur in the equations of motion. The Chapman–Enskog procedure for dense gases was extended by treating ϵ as additional small parameter. The constitutive relations were determined up to first order in ϵ and the inverse Knudsen number K.

At this order, the viscosity and pseudothermal conductivity are identical to the expressions for a dense molecular gas, and there is an $O(\epsilon)$ correction to the pressure. The most important respect in which the equations of motion differ from that of a dense molecular gas is in the presence of an energy sink in the balance of fluctuational kinetic energy. All the transport properties are functions of the grain temperature, which is not a thermodynamic quantity, but is determined from the balance of fluctuational energy. This is one of the key differences between granular and molecular gases.

The hydrodynamic description of fluids rests on the separation between the macroscopic and microscopic length and time scales. This separation of scales exists for fluids in most circumstances that we normally encounter, but it is problematic for dissipative granular materials. For the latter, a hydrodynamic description is valid if $K \ll 1$ and $\epsilon \equiv (1 - e_p^2) \ll 1$, where K is the inverse of the Knudsen number. However, one or both of these constraints are often violated in practical situations. The range of validity of the theory may be increased by retaining more terms in the Chapman–Enskog expansion, but its convergence is an open question (Goldhirsch, 2003). Nevertheless, predictions of the kinetic theory are in reasonably good agreement with the results of particle dynamics simulations (Figs. 7.11 and 7.12), even when the inelasticity is not small. Indeed, it is often the case that perturbation expansions are found to be accurate well beyond their expected range of validity. At sufficiently large inelasticity Glasser and Goldhirsch (2001) found evidence of scale dependence in a uniformly sheared granular material: properties such as the viscosity depend quite strongly on the length and time scales over which averages are determined. They speculate that nonlocal and history-dependent constitutive relations may be necessary to model granular materials. Studies of time-dependent and spatially nonuniform flows are necessary to resolve this issue.

A tentative attempt was made to incorporate the anisotropy in the microstructure in the constitutive theory. Adopting a simple phenomenological model for the anisotropy of the pair distribution function, its effect on the constitutive relations was determined. However, the effect of the inelasticity on the anisotropy is not known, and further investigation in this direction is necessary.

PROBLEMS

7.1. Determine the values of the coefficients a_1–a_4 occurring in the high density kinetic theory in §7.2 by comparing the constitutive relations with those of the full kinetic theory in §7.3.13. It may help to follow the steps given below.

(a) First, relate the interparticle distance s to the solids fraction. Do this by assuming a cell model for the particle distribution, in which the volume occupied by each particle is $k(d_p + s)^3$, where k is a constant. Then, relate the fluctuation velocity w to the grain temperature T, using their respective definitions.

(b) Using the above relations, pose the constitutive relations for the high density theory in terms of ν and T.

(c) Using the form for g_0 given by Ogawa et al. (1980),

$$g_0 = \frac{1}{1 - (\nu/\nu_{drp})^{1/3}}$$

determine the constitutive relations of the kinetic theory in the limit $\nu \to \nu_{drp}$.

(d) Compare the constitutive relations for the two theories and determine the coefficients a_1–a_4.

(e) Try to repeat the procedure using the expression for g_0 given by Ma and Ahmadi (1986). Why does it not work?

7.2. Equation (7.26) gives the thermal relaxation time for a sudden change in the shear rate in a granular gas, but it is clearly not valid in the limit $e_p \to 1$. Determine the thermal relaxation time for a fluid composed of elastic particles, enclosed between two plane parallel walls at a distance H from each other. One wall is stationary and the other moves at constant velocity v_0; the shear rate is changed suddenly by changing the velocity of the moving wall to v_1. The walls are maintained at constant temperature T_w. For simplicity, assume that the fluid is incompressible, and that its properties (such as viscosity and thermal conductivity) are independent of the temperature.

(a) Write the energy balance for the fluid.
(b) At the instant the shear rate changes, what is the rate of change of temperature?
(c) At steady state, what is the temperature profile?
(d) At intermediate times, all the terms in the energy balance must be of comparable magnitude (why?). By scaling distance by H, time by τ, and temperature by T_w, and comparing the magnitudes of the terms, determine τ in terms of the material constants and H. How does τ scale with H?

7.3. Determine Φ_K by enforcing the Maxwell transport equation (7.52) for certain higher moments, rather than by direct solution of the Boltzmann equation. Start with the trial function,

$$\Phi_K = p_1\,\overline{\boldsymbol{\xi}\boldsymbol{\xi}}\!:\!\hat{\mathbf{D}} + p_2\,(\xi^2 - \frac{5}{2})\,\boldsymbol{\xi}\cdot\hat{\nabla}\ln\hat{T} + p_3\,(\xi^2 - \frac{5}{2})\,\boldsymbol{\xi}\cdot\hat{\nabla}\ln\nu$$

To determine the value of p_1, substitute $\psi = c_x c_y$ in (7.52) and consider a situation where $v_x \equiv v_x(y)$, $v_y = v_z = 0$, $\nabla v = \dot{\gamma}\,\mathbf{e}_y\mathbf{e}_x$, and $\nabla T = \nabla\nu = 0$. Here $\dot{\gamma} \equiv du_x/dy$ is the constant shear rate, and $\mathbf{e}_x$ and $\mathbf{e}_y$ are the unit vectors in the x and y directions, respectively. Note that all the hydrodynamic variables are independent of time.

(a) First determine the left-hand side (LHS) of (7.52). How does it depend on $\dot{\gamma}$?
(b) Write $\boldsymbol{\theta}(c_x c_y)$ in terms of $\boldsymbol{\theta}(C_x C_y)$. To the same order in $\dot{\gamma}$ as the LHS, what hydrodynamic variables does $\boldsymbol{\theta}(C_x C_y)$ depend on? Hence what is $\frac{\mathrm{d}}{\mathrm{d}y}\boldsymbol{\theta}(c_x c_y)$?
(c) Now evaluate $\chi(c_x c_y)$, to the same order in $\dot{\gamma}$ as the LHS, with the help of the results in Appendix H.
(d) Putting all the above results together, determine p_1. Compare your result with the one-term Sonine polynomial expansion in §7.3.10.
(e) Similarly, determine p_2 and p_3 are by setting $\psi = c^2 c_x$ and considering T and ν, respectively, to have gradients in the x-direction while $\mathbf{v} = 0$.

7.4. From the form of Φ_ϵ, i.e., that it is a function only of the scalar ξ, and its orthogonality to ξ^2, show that it does not contribute to the stress and pseudothermal energy flux at $O(\epsilon)$.

7.5. Assuming that the pair distribution at contact is given by (7.171) show that the viscosity and thermal conductivity are given by (7.173) and (7.174). Use the simplest approximation for Φ_K given by (7.132) and (7.133).

8

Analysis of Rapid Flow in Simple Geometries

In this chapter, we apply the hydrodynamic equations derived in Chapter 7 to some simple flow problems, and compare the results with experimental data, where available. In each of the problems, we shall first employ the heuristic description of Haff (1983) (see §7.2), followed by the kinetic theory model that was described in Chapter 7. The simplifications made in the heuristic high-density theory allow relatively easy solution of the equations of motion, and give physical insight into the behavior. The results of the kinetic theory, when compared with those of the heuristic high-density theory, gives us an understanding of the effects of the compressibility of the granular medium. We also see how the results of the high-density theory appear as a certain limit of the kinetic theory.

All the problems that we consider, indeed all problems of practical interest, require the specification of boundary conditions at solid walls. In some cases, such as flow in inclined chutes, we also require boundary conditions at a free surface, i.e., the interface between the granular medium and the atmosphere. For the flow of fluids, the no-slip boundary condition is usually imposed at solid walls, thereby specifying the velocity of the fluid at the walls. The temperature of the fluid at the walls is set by specifying the temperature of the walls, or by specifying the flux of energy at the walls. In addition, the pressure at some of the boundaries must be specified. Boundary conditions for granular materials are significantly different: they usually display considerable slip at solid boundaries, and the grain temperature clearly has little to do with the thermodynamic temperature of the walls. The boundary conditions must be derived by considering the details of particle–wall collisions, which we do in the following section.

As stated in the Chapter 7, we shall neglect the influence of the interstitial fluid on the flowing granular material.

8.1. BOUNDARY CONDITIONS AT SOLID WALLS

Consider the layer of particles enclosed by the dashed line in Fig. 8.1, adjacent to the rigid wall. We define the unit normal $\mathbf{n}$ to the wall pointing away from the granular medium, and a unit tangent $\mathbf{t}$. We choose a reference frame in which the component of the wall velocity $\mathbf{v}_w$ along $\mathbf{n}$ is zero. The mean velocity of the particles in the layer can differ from the wall velocity, and we define the slip velocity $\mathbf{v}_{slip} \equiv \mathbf{v} - \mathbf{v}_w$. Particles in the layer collide with the wall, exchanging momentum and energy with it. We denote by $\mathcal{S}_w$ the stress transmitted to the wall by the particles in the layer; unless the wall is microscopically smooth and flat, the tangential component of the stress, $\mathcal{S}_w \cdot \mathbf{t}$, will in general be nonzero. In addition, particle–wall collisions are, in general, inelastic and result in a loss of kinetic energy at the rate of Γ_w per unit area of the wall. We shall determine the boundary conditions by considering the balances of momentum and energy for the layer, in the limiting case of the

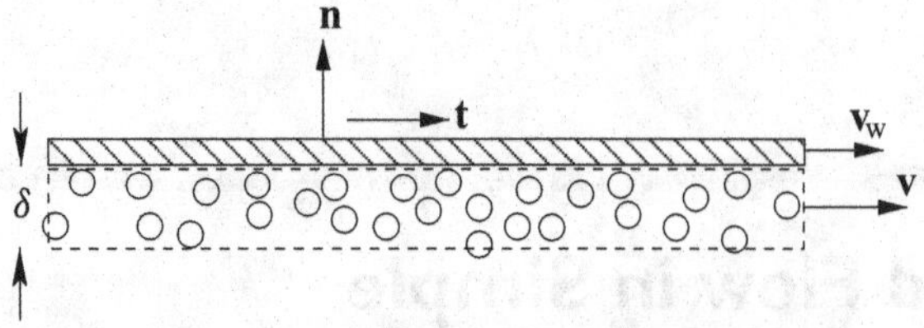

Figure 8.1. A thin layer of particles of thickness δ adjacent to a solid wall. Here $\mathbf{n}$ is the unit outward normal to the wall, $\mathbf{t}$ is any unit tangent to the wall, $\mathbf{v}_w$ is the velocity of the wall and $\mathbf{v}$ is the velocity of the material adjacent to the wall. The balances of energy and momentum for the layer give the boundary conditions.

layer thickness becoming infinitesimally small (Johnson and Jackson, 1987). It is assumed that averages are made over a large enough area of the wall and over a long enough duration that $\mathcal{S}_w$ and Γ_w are smooth functions of time and position on the wall.

As $\delta \to 0$, all body forces vanish, and the balance of tangential momentum for the layer requires $\mathcal{S}_w \cdot \mathbf{t}$ to equal the shear stress exerted on the lower boundary by the material below, i.e.,

$$\mathbf{n} \cdot \boldsymbol{\sigma} \cdot \mathbf{t} = \mathcal{S}_w \cdot \mathbf{t} \tag{8.1}$$

Similarly, the balance of fluctuational, or pseudothermal, energy in the layer is determined by a balance of the flux across its boundaries, the work done at the boundaries, and the energy sink at the wall. The flux of fluctuational energy across the lower boundary of the layer is $\mathbf{q} \cdot \mathbf{n}$, where $\mathbf{q}$ is the energy flux in the bulk, given by (7.163). Per unit area of the wall, the material below the layer does work on it at the rate $\mathbf{n} \cdot \boldsymbol{\sigma} \cdot \mathbf{v}$, and the wall does work on the layer at the rate $-\mathcal{S}_w \cdot \mathbf{v}_w$. We then have

$$\mathbf{q} \cdot \mathbf{n} + \mathbf{n} \cdot \boldsymbol{\sigma} \cdot \mathbf{v} = \Gamma_w + \mathcal{S}_w \cdot \mathbf{v}_w \tag{8.2}$$

The right-hand side is the net flux of fluctuational kinetic energy to the wall due to particle wall collisions, which we shall call $\mathbf{q}_w$ for future reference. With the use of (8.1), (8.2) may be written as

$$\mathbf{q} \cdot \mathbf{n} = \Gamma_w - \mathcal{S}_w \cdot \mathbf{v}_{slip} \tag{8.3}$$

The second term on the right-hand side represents the rate of generation of fluctuational energy at the wall due to slip. It was omitted in the boundary condition derived by Hui et al. (1984), but the analysis of Jenkins and Richman (1986) revealed its presence. It is an important term, for when e_w is close to unity, it causes the wall to appear as a source of fluctuational energy, and qualitatively alters the flow field (see §8.2).

Equations (8.1) and (8.3) provide the required boundary conditions at the wall, for which relations are required for $\mathcal{S}_w$ and Γ_w in terms of the hydrodynamic variables. These must be determined by averaging over a large number of particle-wall collisions. Hui et al. (1984) approached this in a simple and approximate manner, in the spirit of the high-density theory of Haff (1983). Jenkins and Richman (1986) presented a more formal derivation, using a specific form for the structure of the boundary, and assuming the velocity distribution function to be the local Maxwellian. We describe the heuristic approach below, and follow it with the more formal kinetic theory approach.

8.1.1. Heuristic Theory

The boundary conditions are written in terms of the variables used in the theory of Haff (1983), namely, the velocity $\mathbf{v}$, the mean fluctuation velocity w and the mean interparticle spacing s (see §7.2).

Following the arguments of §7.2, the flux of momentum to the wall is the product of the momentum impulse transferred to the wall per collision, the frequency of particle–wall collisions, and the number of particles per unit area of the wall. The tangential momentum impulse per collision J_t is the change in tangential momentum of particles during collisions with the wall. If the particles and the wall are smooth, their momentum in the tangential direction remains unchanged during collisions, and hence there is no momentum impulse. There is transfer of tangential momentum during a collision only if the particles, or the wall, or both, are rough or uneven. Assuming that every colliding particle transmits a fraction φ of its tangential momentum to the wall, the momentum impulse per collision is $J_t = \varphi\, m_{\mathrm{p}}\, \mathbf{v}_{\mathrm{slip}} \cdot \mathbf{t}$. The constant φ is the *specularity coefficient*, characterizing the roughness or unevenness of the wall. Collisions are said to be specular ($\varphi = 0$) if the wall is smooth and the colliding particles are deflected without any change in their tangential momentum, and diffuse ($\varphi = 1$) if particles are deflected in directions uncorrelated to their approach so that, on average, all their tangential momentum is absorbed by the wall. The frequency of collisions between the particles and the wall is $\sim w/s$, and there are $n(d_{\mathrm{p}}+s) \approx n d_{\mathrm{p}}$ particles per unit area in the layer, where n is the number density of grains. The momentum flux to the wall in the direction of $\mathbf{t}$ then is

$$\boldsymbol{\mathcal{S}}_{\mathrm{w}} \cdot \mathbf{t} = a_5 \varphi\, m_{\mathrm{p}} \mathbf{v}_{\mathrm{slip}} \cdot \mathbf{t}\, \frac{w}{s}\, n d_{\mathrm{p}} = a_5 \varphi\, \rho\, w\, \frac{d_{\mathrm{p}}}{s}\, \mathbf{v}_{\mathrm{slip}} \cdot \mathbf{t} \tag{8.4}$$

where a_5 is an O(1) constant. Substituting the above relation, and the expression for shear stress from (7.13) and (7.16) into the boundary condition (8.1), we get after cancelation of common factors

$$-a_2\, d_{\mathrm{p}}\, \nabla_n v_t = a_5 \varphi\, v_{\mathrm{slip}_t} \tag{8.5}$$

The subscripts n and t refer to components in the directions of the normal and the tangent, respectively, e.g., $\nabla_n = \mathbf{n} \cdot \nabla$. The above, in essence, is Maxwell's derivation of the slip boundary condition for a rarefied gas (Maxwell, 1879).

We can now see how the no-slip boundary condition arises: assuming that the velocity varies over a macroscopic length scale H, (8.5) can be written as

$$v_{\mathrm{slip}_t} = -\frac{a_2}{a_5 \varphi} \frac{d_{\mathrm{p}}}{H}\, \hat{\nabla}_n v_t$$

where $\hat{\nabla}_n \equiv H \nabla_n$ is the dimensionless gradient operator. Thus $v_{\mathrm{slip}_t} \approx 0$ when $H/d_{\mathrm{p}} \gg 1$.

For granular flows, the large separation between macroscopic and microscopic length scales is not always present. Even when there is such a separation, the dissipation of energy by inelastic collisions can cause the velocity to vary rapidly within a thin layer near the boundary, as shown in §8.2.1. We therefore use the more general boundary condition (8.5) that allows slip at the solid wall.

To determine Γ_{w}, we start with the fact that the energy lost in a single particle–wall collision is $\sim(1 - e_{\mathrm{w}}^2) m_{\mathrm{p}} w^2$, where e_{w} is the coefficient of restitution for particle–wall collisions. The product of this with the frequency of collisions w/s and the number of particles per unit area adjacent to the wall, $n(d_{\mathrm{p}} + s) \approx n d_{\mathrm{p}}$, gives the energy dissipation rate per unit area of the wall

$$\Gamma_{\mathrm{w}} = a_6 (1 - e_{\mathrm{w}}^2) \rho\, w^3 \frac{d_{\mathrm{p}}}{s} \tag{8.6}$$

Substituting this relation for Γ_{w} and the expression for $\mathbf{q}$ from (7.14) and (7.18) in (8.3), and canceling common factors, the energy boundary condition assumes the form

$$-a_3\, d_{\mathrm{p}} w \nabla_n w = a_6 (1 - e_{\mathrm{w}}^2)\, w^2 - a_5 \varphi\, v_{\mathrm{slip}}^2 \tag{8.7}$$

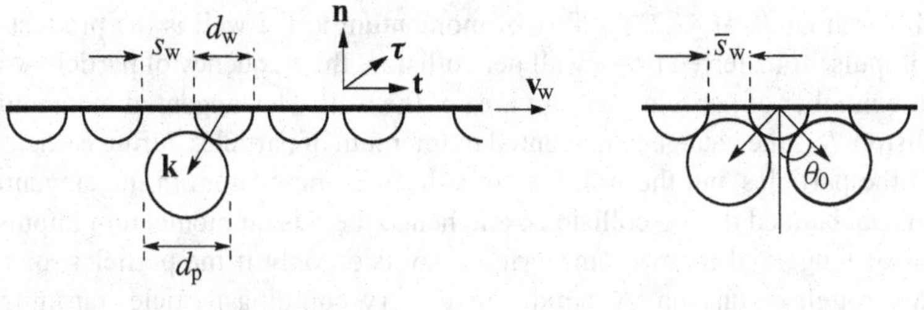

(a) (b)

Figure 8.2. (a) Model of the boundary used for deriving boundary conditions (adapted from Jenkins and Richman, 1986). Hemispheres of diameter d_w are stuck to the wall at an irregular spacing s_w, with mean $\bar{s}_w$. Neighboring hemispheres are close enough that grains never collide against the bare wall. The vectors $(\mathbf{n}, \mathbf{t}, \boldsymbol{\tau})$ form an orthonormal triad, **n** being the unit outward normal to the wall, and $\mathbf{t}$, $\boldsymbol{\tau}$ lying in the plane of the wall. The unit vector **k** points from the center of a wall hemisphere to the center of a grain colliding on it. (b) The fluxes of particle properties to the wall are computed by considering a "typical" wall hemisphere which is a distance $\bar{s}_w$ from its neighbors in all in-plane directions. The angle θ_0 is the maximum angle that **k** can subtend with the inward normal.

Equations (8.5) and (8.7) are the stress and energy boundary conditions, respectively, to be used in conjunction with the heuristic high-density theory of §7.2.

Johnson and Jackson (1987) too followed a heuristic approach, but without restriction to high density. They obtained the boundary conditions

$$-\mu \nabla_n v_t = \rho_p T^{\frac{1}{2}} \frac{\varphi \pi \sqrt{3}\, \nu g_0}{6\, \nu_{drp}} v_{slip_t} \tag{8.8}$$

$$-\kappa \nabla_n T = \rho_p (1 - e_w^2) \frac{\pi \sqrt{3} \nu g_0}{4 \nu_{drp}} T^{\frac{3}{2}} - \mathcal{S}_w \cdot \mathbf{v}_{slip} \tag{8.9}$$

where, μ, κ, and T are the viscosity, pseudothermal conductivity, and the grain temperature of the granular medium adjacent to the boundary, and g_0 the radial distribution at contact, defined in §7.3.5. These boundary conditions are identical in form to (8.5) and (8.7), but have functions of the solids fraction ν in place of the constants a_5 and a_6.

8.1.2. Kinetic Theory

A formal derivation of the boundary conditions using kinetic theory was first given by Jenkins and Richman (1986), assuming specific forms for the wall roughness and the distribution of particle velocities. They considered the boundary to be a randomly distributed array of smooth, inelastic hemispheres of diameter d_w attached to a flat wall, as shown in Fig. 8.2, and the particles to be smooth inelastic spheres. The spacing between the hemispheres is such that particles cannot strike the bare wall. Collisions between particles and wall hemispheres are described by a coefficient of restitution e_w, i.e., (7.3) with e_p replaced by e_w. It is clear that a particle colliding with a wall hemisphere will transmit tangential momentum if it collides at any point other than the tip of the hemisphere. Jenkins and Richman (1986) took the velocity distribution of the particles adjacent to the wall to be the local Maxwellian. We shall follow their treatment, but extend it by allowing the particle velocity distribution to depart from the Maxwellian.

Consider a grain approaching a wall hemisphere with velocity **c**. As the wall moves at a constant velocity $\mathbf{v}_w$, unchanged by collisions with particles, the collision model (7.3) gives

$$\mathbf{k} \cdot \mathbf{c}' = -e_w \, \mathbf{k} \cdot \mathbf{c} + (1 + e_w) \, \mathbf{k} \cdot \mathbf{v}_w \tag{8.10}$$

where $\mathbf{c}'$ is the postcollision velocity of the grain. The quantity of interest is ψ_w, the rate at which a property ψ is transmitted to the wall by particle-wall collisions. The procedure for deriving a relation for $\dot{\psi}_{\mathrm{w}}$ is similar to that in §7.3.5 for $\dot{\psi}_{\mathrm{coll}}$. We first consider the collision of particles on an infinitesimal surface, defined by the solid angle $\mathbf{dk}$, of a hemisphere whose center has position vector $\mathbf{r}$. During a short time interval dt before collision, the center of a particle approaching the hemisphere with velocity $\mathbf{c}$ must lie within a cylinder of volume $|\mathbf{k}\cdot\mathbf{g}|\,\overline{d}^2\,\mathbf{dk}\,dt$, where $\mathbf{g} \equiv \mathbf{c} - \mathbf{v}_{\mathrm{w}}$ and $\overline{d} \equiv \frac{1}{2}(d_{\mathrm{p}} + d_{\mathrm{w}})$. Hence the number of such collisions in the time interval dt is $g_{\mathrm{w}}\, f(t, \mathbf{r} + d_{\mathrm{p}}\mathbf{k}, \mathbf{c})|\mathbf{k}\cdot\mathbf{g}|\,\overline{d}^2\,\mathbf{dk}\,dt$, provided $\mathbf{k}\cdot\mathbf{g} < 0$. Here the function $f(t, \mathbf{r} + d_{\mathrm{p}}\mathbf{k}, \mathbf{c})$ is the singlet velocity distribution function, defined in §7.3.1. The factor g_{w} quantifies the enhancement in the number density of particles in contact with the wall from that in the bulk. It arises from the impenetrability of the wall and the finite size of the particles; its value must approach unity as the solids fraction ν decreases. It is analogos to the radial distribution function at contact g_0 for particles in the bulk, and for the same reasons as for g_0 (see §7.3.5), we assume g_{w} to be a function of ν only. The functional form of $g_{\mathrm{w}}(\nu)$ will be discussed later. If there are n_{w} hemispheres per unit area of the wall, the flux of any particle property ψ to the wall then is

$$\dot{\psi}_{\mathrm{w}} = n_{\mathrm{w}} g_{\mathrm{w}} \int_{\mathbf{k}\cdot\mathbf{g}<0} (\psi - \psi') f(\mathbf{r} + d_{\mathrm{p}}\mathbf{k}, \mathbf{c})|\mathbf{k}\cdot\mathbf{g}|\,\overline{d}^2\,\mathbf{dk}\,\mathbf{dc} \tag{8.11}$$

where ψ' is the postcollision value of ψ.

To compute $\dot{\psi}_{\mathrm{w}}$ we require the particle velocity distribution f at the wall. If particle–wall collisions are specular and the gas is at equilibrium, it can be shown that f at the wall is the local Maxwellian (Chapman and Cowling, 1964, p. 76–77). For any other case, determining the distribution function is a difficult task, and has been accomplished only for very simple wall models (Cercignani, 1988, chapter III). Lacking a more accurate description, we make the assumption that the distribution function at the wall equals that in the bulk. For the latter, we use the solution obtained from the Chapman–Enskog expansion in Chapter 7

$$f = f^0(1 + K\Phi_K + \epsilon\Phi_\epsilon + \cdots) \tag{8.12}$$

For consistency with our derivation of the constitutive relations for the bulk in §7.3.13, we shall determine $\dot{\psi}_{\mathrm{w}}$ to first order in the small parameters K and ϵ. To this end, we substitute in (8.11) the Taylor series expansion of $f(\mathbf{r} + d_{\mathrm{p}}\mathbf{k}, \mathbf{c})$ about $\mathbf{r}$ and discard terms of $\mathrm{O}(K^2)$ and higher. Then, substituting the expansion for f from (8.12), we get to first order in K and ϵ

$$\dot{\psi}_{\mathrm{w}} = -n_{\mathrm{w}}\overline{d}^2 g_{\mathrm{w}} \int_{\mathbf{k}\cdot\mathbf{g}<0} (\psi - \psi') f^0\Big(1 + K\Phi_K + \epsilon\Phi_\epsilon + \frac{6\nu_s K}{\pi}\mathbf{k}\cdot\hat{\nabla}\ln f^0\Big)(\mathbf{k}\cdot\mathbf{g})\,\mathbf{dk}\,\mathbf{dc} \tag{8.13}$$

We now determine the stress transmitted to the wall, $\mathcal{S}_{\mathrm{w}}$, by substituting $\psi = m_{\mathrm{p}}\mathbf{c}$ in (8.13). Noting that $\mathbf{c} - \mathbf{c}' = \mathbf{k}\cdot(\mathbf{c} - \mathbf{c}')\mathbf{k}$, as the particles and wall hemispheres are assumed to be smooth, (8.10) gives

$$m_{\mathrm{p}}(\mathbf{c} - \mathbf{c}') = m_{\mathrm{p}}(1 + e_{\mathrm{w}})\,\mathbf{k}\cdot(\mathbf{c} - \mathbf{v}_{\mathrm{w}})\,\mathbf{k} \tag{8.14}$$

Introducing the scaled relative velocities $\boldsymbol{\zeta} \equiv (\mathbf{c} - \mathbf{v}_{\mathrm{w}})/(2T)^{\frac{1}{2}}$, $\boldsymbol{\xi} = (\mathbf{c} - \mathbf{v})/(2T)^{\frac{1}{2}}$, we get

$$\mathcal{S}_{\mathrm{w}} = \frac{-2}{\pi^{\frac{3}{2}}} n_{\mathrm{w}}\overline{d}^2 g_{\mathrm{w}}\, n m_{\mathrm{p}}(1 + e_{\mathrm{w}})T \int_{\mathbf{k}\cdot\boldsymbol{\zeta}<0} (\mathbf{k}\cdot\boldsymbol{\zeta})^2\mathbf{k}\, e^{-\xi^2}$$

$$\left[1 + K\Phi_K + \epsilon\Phi_\epsilon + \frac{6\nu_s K}{\pi}\mathbf{k}\cdot\hat{\nabla}\ln f^0\right]\mathbf{dk}\,\mathbf{d}\boldsymbol{\xi} \tag{8.15}$$

Next, we note that $\boldsymbol{\xi} = \boldsymbol{\zeta} - \tilde{\mathbf{v}}_{\mathrm{slip}}$, where $\tilde{\mathbf{v}}_{\mathrm{slip}} \equiv (\mathbf{v} - \mathbf{v}_{\mathrm{w}})/(2T)^{\frac{1}{2}}$ is the scaled slip velocity. The estimate of $\tilde{\mathbf{v}}_{\mathrm{slip}}$ given by the heuristic theory is $\mathrm{O}(K)$ (see (8.5)); we shall verify this estimate later. Consequently, to retain only terms up to first order in K, we write $e^{-\xi^2}$ as

$e^{-\zeta^2}(1 + 2\boldsymbol{\zeta}\cdot\tilde{\mathbf{v}}_{\text{slip}})$ in one part of the integral, and replace $(\mathbf{k}\cdot\boldsymbol{\zeta})$ by $(\mathbf{k}\cdot\boldsymbol{\xi})$ in the other. The expression for $\boldsymbol{\mathcal{S}}_{\text{w}}$ then becomes

$$\boldsymbol{\mathcal{S}}_{\text{w}} = \frac{-2}{\pi^{\frac{3}{2}}} n_{\text{w}}\bar{d}^2 g_{\text{w}}\, \rho_{\text{p}} \nu\, (1 + e_{\text{w}}) T \left\{ \int\limits_{\mathbf{k}\cdot\boldsymbol{\zeta} < 0} (\mathbf{k}\cdot\boldsymbol{\zeta})^2 \mathbf{k}\, e^{-\zeta^2}(1 + 2\boldsymbol{\zeta}\cdot\tilde{\mathbf{v}}_{\text{slip}}) \mathrm{d}\mathbf{k}\, \mathrm{d}\boldsymbol{\zeta} \right.$$
$$\left. + \int\limits_{\mathbf{k}\cdot\boldsymbol{\xi} < 0} (\mathbf{k}\cdot\boldsymbol{\xi})^2 \mathbf{k}\, e^{-\xi^2} \left[K\,\Phi_K + \epsilon\,\Phi_\epsilon + \frac{6 \nu_s K}{\pi} \mathbf{k}\cdot\hat{\nabla} \ln f^0 \right] \mathrm{d}\mathbf{k}\, \mathrm{d}\boldsymbol{\xi} \right\} \tag{8.16}$$

From the geometry of the wall (see Fig. 8.2) we may write $n_{\text{w}}\bar{d}^2 = 1/(\pi \sin^2\theta_0)$. Also, at this level of approximation e_{w} must be replaced by unity in the above expression; this will become clear shortly, when we consider the flux of energy at the wall. The contribution of Φ_ϵ to the integral vanishes due to its orthogonality with ξ^2 (see §7.3.12). Substituting for Φ_K from (7.125), for f^0 from (7.78), and evaluating the integrals with the aid of (H.6)–(H.8), we get

$$\boldsymbol{\mathcal{S}}_{\text{w}} = \rho_{\text{p}} \nu g_{\text{w}} T\, \mathbf{n} + \frac{2^{\frac{3}{2}}}{3\pi^{\frac{1}{2}}} \rho_{\text{p}} \nu g_{\text{w}}\, T^{\frac{1}{2}}\, \mathbf{J}\cdot\mathbf{v}_{\text{slip}}$$
$$- \rho_{\text{p}} d_{\text{p}} g_{\text{w}} \left[\frac{2}{3\pi^{\frac{1}{2}} g_0} \left(1 + \frac{12}{5}\nu g_0\right) \int \xi^5 e^{-\xi^2} \mathcal{A}(\xi)\, \mathrm{d}\xi + \frac{\nu}{3} \right] \mathbf{J}\cdot\nabla T$$
$$- \frac{1}{3} \rho_{\text{p}} d_{\text{p}} g_{\text{w}}\, T\, \mathbf{J}\cdot\nabla \nu - \rho_{\text{p}} d_{\text{p}} g_{\text{w}}\, T^{\frac{1}{2}}$$
$$\times \nabla\mathbf{v} : \left[\frac{2^{\frac{5}{2}}}{15\pi^{\frac{1}{2}} g_0} \left(1 + \frac{8}{5}\nu g_0\right) \int \xi^6 e^{-\xi^2} \mathcal{B}(\xi)\, \mathrm{d}\xi\, \mathbf{K} + \frac{2^{\frac{3}{2}} \nu}{\pi^{\frac{1}{2}}} \mathbf{L} \right] \tag{8.17}$$

where $\mathcal{A}(\xi)$ and $\mathcal{B}(\xi)$ are the functions occurring in Φ_K, and

$$\begin{aligned} \mathbf{J} &= 2\,\mathrm{cosec}^2\theta_0 (1 - \cos^3\theta_0)\, \mathbf{nn} + \left[2\,\mathrm{cosec}^2\theta_0 (1 - \cos\theta_0) \right. \\ &\quad \left. - \cos\theta_0\right] (\mathbf{tt} + \boldsymbol{\tau\tau}) \\ \mathbf{K} &= (\tfrac{1}{3} + \cos^2\theta_0)\, \mathbf{nnn} + \tfrac{1}{2} \sin^2\theta_0 (\mathbf{ntt} + \mathbf{n}\boldsymbol{\tau\tau} + \mathbf{tnt} + \boldsymbol{\tau}\mathbf{n}\boldsymbol{\tau}) \\ &\quad - (\tfrac{1}{6} + \tfrac{1}{2}\cos^2\theta_0)(\mathbf{ttn} + \boldsymbol{\tau\tau}\mathbf{n}) \\ \mathbf{L} &= \tfrac{1}{2}(1 + \cos^2\theta_0)\, \mathbf{nnn} + \tfrac{1}{4} \sin^2\theta_0 (\mathbf{ntt} + \mathbf{n}\boldsymbol{\tau\tau} + \mathbf{tnt} + \mathbf{ttn} \\ &\quad + \boldsymbol{\tau}\mathbf{n}\boldsymbol{\tau} + \boldsymbol{\tau\tau}\mathbf{n}) \end{aligned} \tag{8.18}$$

Here $\mathbf{n}$, $\mathbf{t}$, and $\boldsymbol{\tau}$ are the orthogonal triad of unit vectors (see Fig. 8.2). To evaluate the integrals in (8.17), we retain only the first terms in the Sonine polynomial expansions of $\mathcal{A}$ and $\mathcal{B}$, as they yielded fairly accurate estimates of the transport coefficients in §7.3.11. The first and second integrals then evaluate to $15/(32\sqrt{2\pi})$ and $75/(64\sqrt{2})$, respectively.

The shear stress exerted on the wall by the granular material along the $\mathbf{t}$ direction then is

$$\boldsymbol{\mathcal{S}}_{\text{w}}\cdot\mathbf{t} = \frac{2^{\frac{3}{2}}\beta}{3\pi^{\frac{1}{2}}} \rho_{\text{p}} \nu g_{\text{w}}\, T^{\frac{1}{2}}\, v_{\text{slip}_t} - \frac{\beta}{3} \rho_{\text{p}} d_{\text{p}} g_{\text{w}}\, T\, \nabla_t \nu$$
$$- \beta \rho_{\text{p}} d_{\text{p}} g_{\text{w}} \left[\frac{5\sqrt{2}}{32\pi g_0} \left(1 + \frac{12}{5}\nu g_0\right) + \frac{\nu}{3} \right] \nabla_t T$$
$$- \rho_{\text{p}} d_{\text{p}} g_{\text{w}}\, T^{\frac{1}{2}} \sin^2\theta_0 \left[\frac{5}{32\pi^{\frac{1}{2}} g_0} \left(1 + \frac{8}{5}\nu g_0\right) + \frac{\nu}{(2\pi)^{\frac{1}{2}}} \right] \nabla_n v_t \tag{8.19}$$

where $\beta \equiv 2\mathrm{cosec}^2\theta_0(1 - \cos\theta_0) - \cos\theta_0$. In deriving the above, we have used the impenetrability condition $v_n = 0$ at the wall. Note that gradients of T or ν tangential to the wall contribute to the shear stress on the wall. These contributions do not appear if we take the distribution function at the wall to be the local Maxwellian, as in the study of Jenkins and Richman (1986). Their physical origin can be understood in the following manner. A temperature gradient in the **t** direction causes a fore-aft asymmetry in the momentum of particles striking a given wall hemisphere: for a positive gradient, particles will, on average, strike the right-hand side of a hemisphere in Fig. 8.2 harder and more frequently than the left-hand side, resulting in a net transfer of momentum to the wall in the negative **t** direction. Similarly, a gradient of ν along **t** causes an asymmetry in the frequency of collisions, resulting in the transmission of tangential momentum to the wall.

Next, we determine the flux of kinetic energy to the wall due to particle-wall collisions $\mathbf{q}_\mathrm{w}$, by substituting $\psi = 1/2 m_\mathrm{p} c^2$ in (8.13). The collision model (7.3) gives

$$\frac{1}{2}m_\mathrm{p}c^2 - \frac{1}{2}m_\mathrm{p}c'^2 = \frac{1}{2}m_\mathrm{p}(1 - e_\mathrm{w}^2)(\mathbf{k}\cdot\mathbf{g})^2 + m_\mathrm{p}(1 + e_\mathrm{w})(\mathbf{k}\cdot\mathbf{g})(\mathbf{k}\cdot\mathbf{v}_\mathrm{w}) \tag{8.20}$$

Recognizing the second term on the right-hand side to be $m_\mathrm{p}(\mathbf{c} - \mathbf{c}')\cdot\mathbf{v}_\mathrm{w}$ (see (8.14)), we see that its contribution to $\mathbf{q}_\mathrm{w}$ is the rate of working per unit area by the wall stress, $\boldsymbol{\mathcal{S}}_\mathrm{w}\cdot\mathbf{v}_\mathrm{w}$. The first term on the right-hand side is the energy dissipated in the collision, and hence the flux of energy is

$$\mathbf{q}_\mathrm{w} = \Gamma_\mathrm{w} + \boldsymbol{\mathcal{S}}_\mathrm{w}\cdot\mathbf{v}_\mathrm{w} \tag{8.21}$$

with

$$\Gamma_\mathrm{w} = \frac{1}{2\pi^{\frac{3}{2}}} n m_\mathrm{p}(1 - e_\mathrm{w}^2) n_\mathrm{w} \bar{d}^2 g_\mathrm{w}\,(2T)^{\frac{3}{2}} \int\limits_{\mathbf{k}\cdot\boldsymbol{\zeta} < 0} (\mathbf{k}\cdot\boldsymbol{\zeta})^3\, e^{-\xi^2}$$

$$\times \left[1 + K\Phi_K + \epsilon\Phi_\epsilon + \frac{6\nu_s K}{\pi}\mathbf{k}\cdot\hat{\nabla}\ln f^0\right] \mathrm{d}\mathbf{k}\,\mathrm{d}\boldsymbol{\xi} \tag{8.22}$$

Now all the terms in (8.2) other than Γ_w are O(K), hence Γ_w cannot be an O(1) quantity. We must therefore require that $(1 - e_\mathrm{w}^2) \ll 1$. For this reason, e_w was replaced with unity in (8.17). Consequently, we may determine the leading term in Γ_w by replacing $\boldsymbol{\zeta}$ with $\boldsymbol{\xi}$, and by setting K and ϵ to zero. Evaluating the integral with the aid of (H.9) and (H.23), we obtain

$$\Gamma_\mathrm{w} = \rho_\mathrm{p}\,\gamma_\mathrm{w} T^{\frac{3}{2}} \tag{8.23}$$

where

$$\gamma_\mathrm{w} \equiv \frac{2^{\frac{1}{2}}(1 - \cos\theta_0)}{\pi^{\frac{1}{2}}\sin^2\theta_0}(1 - e_\mathrm{w}^2)\,\nu g_\mathrm{w} \tag{8.24}$$

is the dimensionless wall dissipation function.

With (8.19) for the wall shear stress and (8.23) for the dissipation rate at the wall, boundary conditions (8.1) and (8.3) are complete if the functional form of $g_\mathrm{w}(\nu)$ is known. We get this by enforcing continuity of normal stress at the boundary, $\mathbf{n}\cdot\boldsymbol{\sigma}\cdot\mathbf{n} = \boldsymbol{\mathcal{S}}_\mathrm{w}\cdot\mathbf{n}$. At O(1) this gives

$$\nu(1 + 4\nu g_0)T = g_\mathrm{w}\nu\, T \tag{8.25}$$

and thereby $g_\mathrm{w} \approx (1 + 4\nu g_0)$. This is not an exact relation because we expect g_w to depend on the wall properties also (Jenkins and Richman, 1986); the relation is a result of our assumption that the velocity distribution at the wall is a small perturbation about the local

Maxwellian, as in the bulk. Nevertheless, we shall use this expression for g_w in the absence of more detailed knowledge of the structure near a wall.

Let us consider the common situation of no variation of the hydrodynamic variables in the tangential direction, such as in fully developed flow between parallel walls. The continuity of tangential stress at the boundary (8.1) reduces to

$$\mu \nabla_n v_t = \rho_p T^{\frac{1}{2}} \eta_w v_{slip_t} \tag{8.26}$$

where η_w is the dimensionless "wall viscosity" function

$$\eta_w = \frac{2^{\frac{3}{2}}\beta}{3\sqrt{\pi}} \nu g_w \left[1 - \frac{5 g_w \sin^2\theta_0}{32\sqrt{\pi}\, g_0 \eta}\left(1 + \frac{8}{5}\nu g_0\right) - \frac{\nu g_w \sin^2\theta_0}{\sqrt{2\pi}\,\eta}\right]^{-1} \tag{8.27}$$

By scaling the velocity with $(2T)^{1/2}$ and distance with the macroscopic length scale H in (8.26), we find that $\tilde{\mathbf{v}}_{slip}$ is $O(K)$, as assumed earlier in this section. The pseudothermal energy boundary condition (8.3), upon substituting for $\mathcal{S}_w$ from (8.17) and for Γ_w from (8.23), yields

$$-\kappa \nabla_n T = \rho_p \gamma_w T^{\frac{3}{2}} - \rho_p T^{\frac{1}{2}} \eta_w v_{slip}^2 \tag{8.28}$$

with the pseudothermal conductivity κ given by (7.164).

Equations (8.26) and (8.28) are the required boundary conditions. They have the same form as (8.8) and (8.9), respectively, in the heuristic theory. Comparing (8.8) and (8.26), we see that φ depends not just on the wall roughness (through θ_0), but also on the solids fraction. Similarly, on comparing (8.9) and (8.28), we see that Γ_w depends on θ_0 in the latter but not in the former.

While the kinetic theory analysis has given us boundary conditions in terms of the details of the wall roughness and inelasticity, it is often not possible to characterize walls in terms of the simple model adopted here. In such situations, it is useful to simplify the boundary conditions in the following manner: we introduce the specularity coefficient by rewriting (8.27) as

$$\eta_w = \varphi \nu g_w \tag{8.29}$$

where φ contains a part of the dependence of η_w on ν and its entire dependence on θ_0. Figure 8.3 shows that there is only a gentle variation of φ with ν, suggesting that it may be treated as a constant.† Similarly, the factor $\sqrt{2}(1 - \cos\theta_0)/(\sqrt{\pi}\sin^2\theta_0)$ in (8.24) (which varies between 0.4 and 0.46) may by absorbed into $(1 - e_w^2)$. Thus φ and e_w are treated as parameters that *together* characterize the roughness and inelasticity of the wall. This yields

$$\mathcal{S}_w\!\cdot\!\mathbf{t} = \varphi\rho_p \nu g_w T^{\frac{1}{2}} v_{slip_t}, \quad \Gamma_w = (1 - e_w^2)\nu g_w \rho_p T^{\frac{3}{2}} \tag{8.30}$$

which gives the boundary conditions

$$-\mu \nabla_n v_t = \varphi\rho_p \nu g_w T^{\frac{1}{2}} v_{slip} \tag{8.31}$$

$$-\kappa \nabla_n T = \rho_p (1 - e_w^2) \nu g_w T^{\frac{3}{2}} - \varphi\rho_p \nu g_w T^{\frac{1}{2}} v_{slip}^2 \tag{8.32}$$

Note that with this definition, φ remains an O(1) positive constant, but it is not restricted to be less than unity.

Thus we see that the rigorous kinetic theory analysis yields boundary conditions of the same form as those derived from the heuristic analysis. While it may seem that a lot of effort has been spent in recovering the boundary conditions that were formulated using the much

† We only consider θ_0 in the range $0 < \theta_0 < \frac{\pi}{4}$, as it is physically unrealistic to have wall hemispheres larger than the particles.

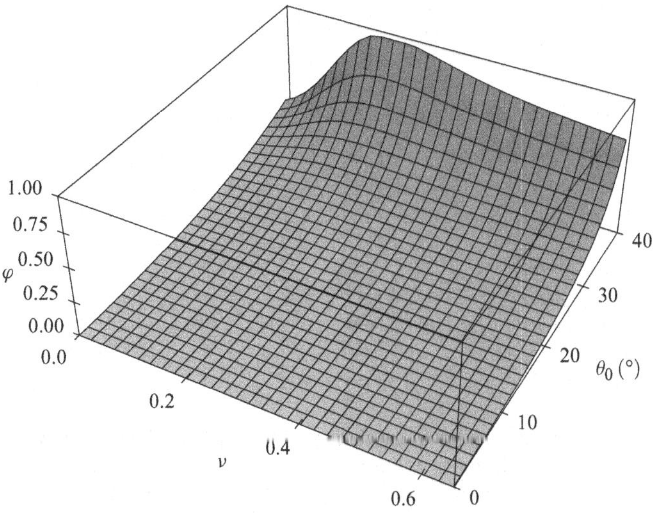

Figure 8.3. The specularity coefficient φ, as defined by (8.29) and (8.27), as a function of ν and θ_0. The functional form of $g_0(\nu)$ given by Ogawa et al. (1980) (see (7.62)) has been used, and the maximum packing solids fraction is taken as 0.65.

simpler heuristic analysis, it is not without utility: the theory provides a general framework that can be used for more sophisticated models for the wall roughness, or for extending the analysis to higher orders in K and ϵ. Moreover, the analysis shows that the tangential momentum boundary condition (8.19) has, in general, contributions from the tangential gradients of the flow variables; it reduces to the simple form given by the heuristic theory only when such gradients are absent. Thus we get results of greater generality by following the rigorous approach. Lastly, this approach acquaints us with the assumptions and approximations made in the analysis, thereby providing an idea of where improvements are possible.

8.2. PLANE COUETTE FLOW

In §7.2.1, we had considered the problem of spatially uniform shear of an unbounded granular medium. One of our findings was that the stress varies as the square of the shear rate, in accordance with the result of the simple dimensional analysis of §7.1. However, it was mentioned that this observation is not necessarily valid in shear flows bounded by rigid walls. In practical situations all flows are either driven or retarded by boundaries. Hence it is of interest to study the shear flow of a granular medium bounded by rigid walls.

Here we consider fully developed (i.e., no variation of any property in the flow direction), steady flow of a granular medium sheared between two rigid parallel walls (see Fig. 8.4). The upper wall translates at a constant speed u_w in the x-direction relative to the lower wall. It is assumed that the walls are of infinite extent in the z direction, and there is no variation of any quantity in this direction. Hence the velocity field has the form $v_x \equiv v_x(y)$, $v_y = v_z = 0$. Flows having a velocity field of this form are referred to as viscometric flows, as they are used for determining the viscosity and other rheological properties of fluids. In the presence of a gravitational field, the normal stress σ_{yy} increases with distance from the upper wall, leading to compaction of the material in the lower region. For this analysis, we ignore the influence of gravity. The normal stress on the walls is adjusted so that the distance between the walls is maintained at a constant value H.

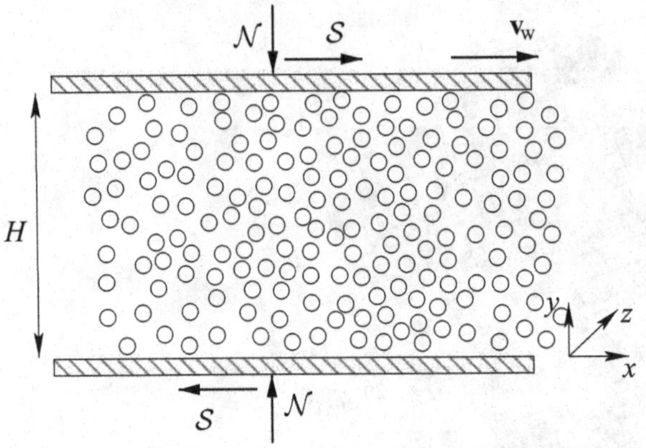

Figure 8.4. Plane Couette flow of a granular medium in the absence of gravity. The medium is sheared by moving the upper wall at a velocity v_W while the lower wall is held stationary. A normal stress $\mathcal{N}$ is applied on the two walls to keep them a fixed distance H apart, and shear stress $\mathcal{S}$ is applied to keep the velocities of the walls constant.

For this problem, the x and y components of the momentum balance (7.71) reduce to

$$\frac{d\sigma_{yx}}{dy} = 0 \tag{8.33}$$

$$\frac{d\sigma_{yy}}{dy} = 0 \tag{8.34}$$

respectively; the z component is identically satisfied. The balance of fluctuational kinetic energy (7.75) reduces to

$$-\frac{dq_y}{dy} - \sigma_{yx}\frac{dv_x}{dy} - \Gamma = 0 \tag{8.35}$$

Thus the shear and normal stresses are constant across the Couette gap, and the fluctuational kinetic energy is balanced between the flux, shear work, and dissipation due to inelastic collisions. To proceed further, constitutive relations for the stress, flux of pseudothermal energy, and dissipation rate are needed. For this we shall use the relations from the high-density theory, given in §7.2, and those from kinetic theory in §7.3.13. Regardless of which set of relations is used, the total order of the system of differential equations (8.33) – (8.35) is five, as the shear stress and the energy flux are themselves proportional to gradients of the velocity and grain temperature, respectively. Hence five boundary conditions are required to solve the problem. Equations (8.1) and (8.3) applied at the two walls provide four boundary conditions. The last condition comes from the specification of either the normal stress at the walls, or the average density in the Couette gap.

8.2.1. Predictions of the High-Density Theory

We first apply the high-density theory of Haff (1983), along with the boundary conditions given in §8.1.1, as it illustrates some important features of the solutions and throws light on the physical mechanisms at play. The analysis described below was first provided by Jenkins and Richman (1986), and in an expanded form by R. Jackson (1994, private communication).

Using the constitutive relations from §7.2, the y and x momentum balances take the form

$$\sigma_{yy} = a_1 \rho\, w^2 \frac{d_p}{s} = \mathcal{N} \tag{8.36}$$

$$\sigma_{yx} = -a_2 \rho\, d_p^2 \frac{w}{s}\frac{dv_x}{dy} = -\mathcal{S} \tag{8.37}$$

where the normal stress $\mathcal{N}$ and the shear stress $\mathcal{S}$ are (positive) constants. The fluctuational energy balance assumes the form

$$\frac{\mathrm{d}}{\mathrm{d}y}\left[a_3 d_\mathrm{p}^2 \frac{w}{s}\frac{\mathrm{d}}{\mathrm{d}y}(\frac{1}{2}\rho\, w^2)\right] - \sigma_{xy}\frac{\mathrm{d}v_x}{\mathrm{d}y} - a_4(1-e_\mathrm{p}^2)\rho\,\frac{w^3}{s} = 0 \tag{8.38}$$

Combining (8.36) and (8.37), we get

$$\frac{\mathrm{d}v_x}{\mathrm{d}y} = \frac{a_1\, r}{a_2}\,\frac{w}{d_\mathrm{p}} \tag{8.39}$$

where we have introduced the stress ratio $r \equiv \mathcal{S}/\mathcal{N}$. Substituting for w^2/s from (8.36) and for $\mathrm{d}v_x/\mathrm{d}y$ from (8.39) into (8.38), we get a differential equation for the fluctuation velocity w

$$\frac{\mathrm{d}^2 w}{\mathrm{d}y^2} - \frac{[L - Mr^2]}{d_\mathrm{p}^2}\,w = 0 \tag{8.40}$$

where $L \equiv a_4(1-e_\mathrm{p}^2)/a_3$ and $M \equiv a_1^2/(a_2\, a_3)$. For the constants a_1–a_6, we use the values

$$a_1 = \tfrac{4}{3}\,\nu_\mathrm{drp}, \quad a_2 = \frac{(3\pi)^{1/2}}{4}\left(\frac{8}{45}+\frac{32}{15\pi}\right)\nu_\mathrm{drp}, \quad a_3 = \left(\frac{\pi}{3}\right)^{1/2}\left(\frac{3}{4}+\frac{8}{3\pi}\right)\nu_\mathrm{drp}$$

$$a_4 = \frac{4}{(3\pi)^{1/2}}\,\nu_\mathrm{drp}, \quad a_5 = a_6 = \frac{\pi}{6\,\nu_\mathrm{drp}} \tag{8.41}$$

The values of a_1–a_4 are obtained by matching the constitutive relations of the high-density theory with those of the kinetic theory in the limit $\nu \to \nu_\mathrm{drp}$ (Problem 7.1), and those of a_5 and a_6 are are obtained by matching the boundary conditions of the high-density theory with the heuristic boundary conditions of Johnson and Jackson (1987) (Problem 8.1).

Equations (8.39) and (8.40) must be solved with boundary conditions (8.5) and (8.7) at the two walls. Though only three boundary conditions are needed to integrate (8.39) and (8.40), the fourth is required to determine the unknown constant r.

Combining the slip boundary condition (8.5) with (8.39), we get

$$v_\mathrm{slip} = v_x - v_\mathrm{w} = -\frac{a_1 r}{a_5 \varphi}\,w \quad \text{at } y = H \tag{8.42}$$

$$v_\mathrm{slip} = v_x = \frac{a_1 r}{a_5 \varphi}\,w \quad \text{at } y = 0 \tag{8.43}$$

Substituting the expressions for v_slip from the above into the energy boundary condition (8.7), we get

$$\frac{\mathrm{d}w}{\mathrm{d}y} = -\frac{\alpha}{d_\mathrm{p}}\,w \quad \text{at } y = H \tag{8.44}$$

$$\frac{\mathrm{d}w}{\mathrm{d}y} = \frac{\alpha}{d_\mathrm{p}}\,w \quad \text{at } y = 0 \tag{8.45}$$

where $\alpha \equiv (\ell - m\,r^2)$, with $\ell \equiv (1-e_\mathrm{w}^2)\,a_6/a_3$ and $m \equiv a_1^2/(a_3\, a_5\varphi)$. The above involve only w and can therefore be used with (8.40) to determine $w(y)$.

As gravity is absent, we expect w to be symmetric, and v_x antisymmetric, about the mid plane $y = H/2$. The form of the solution of (8.40) depends on the sign of $L - Mr^2$, and we consider the two cases separately. There is no solution (except the inadmissible trivial solution $w = 0$, see the following) when $L - Mr^2 = 0$.

Trigonometric solution
Let us first consider the case $L - Mr^2 < 0$. Defining $\Omega^2 \equiv Mr^2 - L$, and $\xi \equiv \Omega y/d_p$, we get the trigonometric solution for w

$$w = \mathcal{A}\cos\xi + \mathcal{B}\sin\xi \tag{8.46}$$

The boundary conditions (8.44) and (8.45) may be written as

$$\begin{bmatrix} \Omega\tan(\Omega H^*)-\alpha & -\Omega-\alpha\tan(\Omega H^*) \\ \alpha & -\Omega \end{bmatrix} \begin{bmatrix} \mathcal{A} \\ \mathcal{B} \end{bmatrix} = 0 \tag{8.47}$$

where $H^* \equiv H/d_p$ is the dimensionless Couette gap. The trivial solution $\mathcal{A} = \mathcal{B} = 0$ is not permissible, as it does not satisfy (8.39) with boundary conditions (8.42) and (8.43). Therefore the determinant of the coefficient matrix must vanish for a solution to exist. This yields an equation for the eigenvalue Ω, whose roots are

$$\Omega = \alpha\frac{\cos(\Omega H^*)\pm 1}{\sin(\Omega H^*)} \tag{8.48}$$

The root corresponding to the positive sign yields

$$\Omega = \alpha\cot[\Omega H^*/2] \tag{8.49}$$

As Ω and α are functions of r, the above is an implicit equation for r; it can be solved if the Couette gap and the material and wall parameters are specified. Together with the second of (8.47), this gives

$$w = \frac{\mathcal{A}\cos(\xi - \xi_m)}{\cos\xi_m} \tag{8.50}$$

where $\xi_m \equiv \Omega H^*/2$ is the value of ξ at the mid plane. The root corresponding to the negative sign in (8.48) is unphysical as the solution for w is negative in one-half of the Couette gap. We see from (8.50) that $\Omega H^* < \pi$ for w to remain nonnegative everywhere. Equation (8.49) then implies that a solution exists only if $\alpha > 0$. Thus

$$\frac{\ell}{m} > r^2 > \frac{L}{M} \tag{8.51}$$

is a necessary and sufficient for the trigonometric solution (8.50).

Equation (8.49) may be expressed as a relation between the dimensionless Couette gap and the stress ratio,

$$H^* = \frac{2}{(Mr^2 - L)^{1/2}}\tan^{-1}\left[\frac{\ell - mr^2}{(Mr^2 - L)^{1/2}}\right] \tag{8.52}$$

A plot of this relation is shown in Fig. 8.5a for a particular parameter set. The Couette gap vanishes as r approaches $(\ell/m)^{1/2}$, and diverges as r approaches $(L/M)^{1/2}$.

Substituting the solution for w from (8.50) into (8.39) and integrating, we get the velocity profile

$$v_x = \mathcal{A}\frac{a_1 r}{a_2\Omega}\frac{\sin(\xi - \xi_m)}{\cos\xi_m} + b \tag{8.53}$$

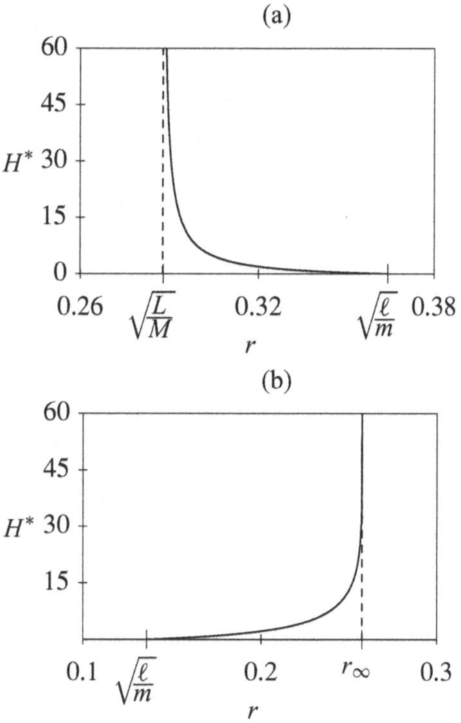

Figure 8.5. The dependence of the Couette gap H^* on the stress ratio $r \equiv \mathcal{S}/\mathcal{N}$ for the (a) trigonometric, and (b) hyperbolic solutions. The dashed lines are the asymptotes for infinitely large Couette gap, occurring as $r \to (L/M)^{1/2}$ in (a) and $r \to r_\infty$ (defined by (8.62)) in (b). Parameter values are $e_\mathrm{p} = 0.91$, $e_\mathrm{w} = 0.8$, $\varphi = 0.85$ in (a) and $e_\mathrm{p} = 0.91$, $e_\mathrm{w} = 0.91$, $\varphi = 0.25$ in (b).

where b is an integration constant. The constants $\mathcal{A}$ and b are determined by applying boundary conditions (8.42) and (8.43), and we finally get

$$v_x = \frac{\upsilon_\mathrm{w}}{2}\left\{1 + \left(1 + \frac{a_2\Omega^2}{\alpha\, a_5\varphi}\right)^{-1} \frac{\sin(\xi - \xi_\mathrm{m})}{\sin \xi_\mathrm{m}}\right\} \tag{8.54}$$

$$w = \frac{\upsilon_\mathrm{w}}{2}\frac{a_2\Omega}{a_1 r}\left(1 + \frac{a_2\Omega^2}{\alpha\, a_5\varphi}\right)^{-1} \frac{\cos(\xi - \xi_\mathrm{m})}{\sin \xi_\mathrm{m}} \tag{8.55}$$

The solution for $H^* = 50$ is shown in Fig. 8.6 (solid lines); we see that the shear rate and fluctuation velocity are maximum at the center ($y = H/2$), and decrease as the walls are approached.

The value of the dimensionless Couette gap H^* determines the extent of slip at the walls and the form of the profiles of v_x and w. The limit $H^* \to \infty$ is obtained when $\Omega \to 0$, or $\Omega H^* \to \pi$ (see (8.49)). In this limit, the slip at the walls becomes vanishingly small, and the solution reduces to

$$v_x \sim \frac{\upsilon_\mathrm{w}}{2}\,[1 - \cos(\pi y/H)] \tag{8.56}$$

$$w \sim \frac{\upsilon_\mathrm{w}}{H^*}\frac{a_2\pi}{2a_1(L/M)^{1/2}}\sin(\pi y/H) \tag{8.57}$$

As H^* decreases, the magnitude of slip increases, the velocity profile approaches a straight line, and the variation of the fluctuation velocity decreases. Note that the dimensionless shear rate $\frac{1}{\dot\gamma}\frac{\mathrm{d}v_x}{\mathrm{d}y}$, where $\dot\gamma \equiv \upsilon_\mathrm{w}/H$ is the nominal shear rate, is finite everywhere in the gap even as $H^* \to \infty$. We shall see that this is not so in the hyperbolic solution.

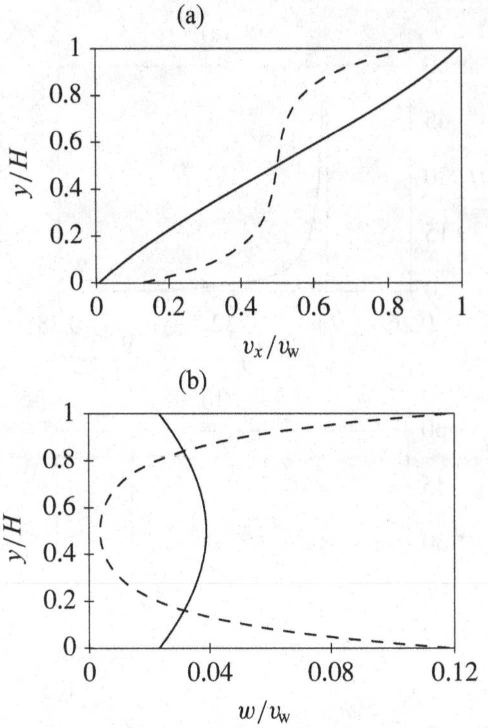

Figure 8.6. Profiles of (a) the velocity, and (b) the velocity fluctuation for the trigonometric (solid lines) and hyperbolic (dashed lines) solutions. Parameter values are $H^* = 50$, and the rest as in Fig. 8.5. For these parameter values, $\Omega = 0.037$, $\alpha = 0.05$ and $r = 0.289$ for the trigonometric case, and $\Omega = 0.167$, $\alpha = -0.167$ and $r = 0.257$ for the hyperbolic case.

Hyperbolic solution

Next, let us consider the case $L - Mr^2 > 0$. Defining $\Omega^2 \equiv L - Mr^2$ and $\xi \equiv \Omega y/d_p$, we get the solution for w in terms of hyperbolic functions

$$w = \mathcal{A}\cosh\xi + \mathcal{B}\sinh\xi \tag{8.58}$$

Repeating the procedure used for the trigonometric solution, we get the following relation for the eigenvalue Ω

$$\Omega = -\alpha\coth[\Omega H^*/2] \tag{8.59}$$

As $\Omega > 0$, the above implies that $\alpha < 0$, and hence

$$\frac{L}{M} > r^2 > \frac{\ell}{m} \tag{8.60}$$

Equation (8.59) may be recast as

$$H^* = \frac{2}{(L - Mr^2)^{1/2}}\tanh^{-1}\left[\frac{m r^2 - \ell}{(L - Mr^2)^{1/2}}\right] \tag{8.61}$$

which is shown in Fig. 8.5 for a particular parameter set. The Couette gap vanishes as r approaches $(\ell/m)^{1/2}$, and diverges as r approaches the limiting value r_∞, which is the solution of

$$mr^2 - \ell = (L - Mr^2)^{1/2} \tag{8.62}$$

Integrating (8.39) with boundary conditions (8.42) and (8.43) at the two walls, we finally get

$$v_x = \frac{v_w}{2}\left\{1 + \left(1 - \frac{a_2\Omega^2}{\alpha\, a_5\varphi}\right)^{-1}\frac{\sinh(\xi - \xi_m)}{\sinh\xi_m}\right\} \tag{8.63}$$

$$w = \frac{v_w}{2}\frac{a_2\Omega}{a_1 r}\left(1 - \frac{a_2\Omega^2}{\alpha\, a_5\varphi}\right)^{-1}\frac{\cosh(\xi - \xi_m)}{\sinh\xi_m} \tag{8.64}$$

This solution is depicted by the dashed lines in Fig. 8.6; we see that the shear rate and fluctuation velocity are highest near the walls, and decrease as the center is approached.

As in the trigonometric solution, H^* determines the extent of slip and the nature of the solution. The limit of infinitely large Couette gap is obtained when $r \to r_\infty$, or $\alpha/\Omega \to -1$ (see (8.59)). In this limit, the solution in the vicinity of the upper wall is approximated by

$$v_x \sim \frac{\upsilon_w}{2}\left[1 + \left(1 + \frac{a_2\Omega}{a_5\varphi}\right)^{-1} e^{-\Omega(H-y)/d_p}\right] \tag{8.65}$$

$$w \sim \frac{\upsilon_w}{2}\frac{a_2\Omega}{a_1 r}\left(1 + \frac{a_2\Omega}{a_5\varphi}\right)^{-1} e^{-\Omega(H-y)/d_p} \tag{8.66}$$

Near the lower wall it has the same form, but with the exponent replaced by $-\Omega y/d_p$. Note that the dimensionless shear rate $\frac{1}{\dot\gamma}\frac{dv_x}{dy}$ at the walls is $\sim H^*$; i.e., it is infinitely large. However, the shear rate and velocity fluctuation decay exponentially with distance from the walls, the decay length being $d_p/(L - Mr_\infty^2)^{1/2}$. As H^* decreases, the magnitude of slip increases, the material shears more uniformly, and the variation of the fluctuation velocity decreases.

8.2.2. Some Features of the High-Density Solutions

In the trigonometric solution, we see from (8.55) that the fluctuation velocity is highest at the center ($y = H/2$) and lowest at the walls, implying that fluctuation kinetic energy is conducted *to* the walls from the bulk. Thus the walls are a net sink of energy. On the other hand, in the hyperbolic solution (8.64) reveals that the fluctuation velocity is lowest at the center and highest at the walls, implying that fluctuation kinetic energy is conducted *from* the walls to the bulk. In this case, the walls are a net source of energy.

Inequalities (8.51) and (8.60) show that the relative magnitudes of L/M and ℓ/m decide which solution holds. The trigonometric solution holds if $L/M < \ell/m$, or

$$\frac{a_2 a_4(1 - e_p^2)}{a_6 a_5 \varphi(1 - e_w^2)} < 1 \tag{8.67}$$

and the hyperbolic solution holds if $L/M > \ell/m$, or

$$\frac{a_2 a_4(1 - e_p^2)}{a_6 a_5 \varphi(1 - e_w^2)} > 1 \tag{8.68}$$

Thus the magnitude of the composite parameter on the left-hand side of the above inequalities, which is a combination of e_p, e_w, and φ, decides whether the walls are sources or sinks of fluctuational energy.

Considering the limit $H^* \to \infty$, in the hyperbolic solution shearing is confined to a thin layer near the walls, within which fluctuational energy is produced and conducted to the interior. The thickness of this layer is $\sim d_p/(L - Mr_\infty^2)^{1/2}$; however it must be much larger than the d_p for the hydrodynamic description to be valid. This is the case only when $L/M - \ell/m \ll 1$ (see Problem 8.2). In the trigonometric solution, the shear rate is finite (though not uniform) throughout in the Couette gap even in the limit $H^* \to \infty$.

In both solutions, r lies between ℓ/m and L/M. Let us now consider the dependence of the normal and shear stresses on the nominal shear rate $\dot\gamma \equiv \upsilon_w/H$. From (8.36), we have

$$\bar{s}\mathcal{N} = a_1 d_p \rho \frac{1}{H}\int_0^H w^2 dy \tag{8.69}$$

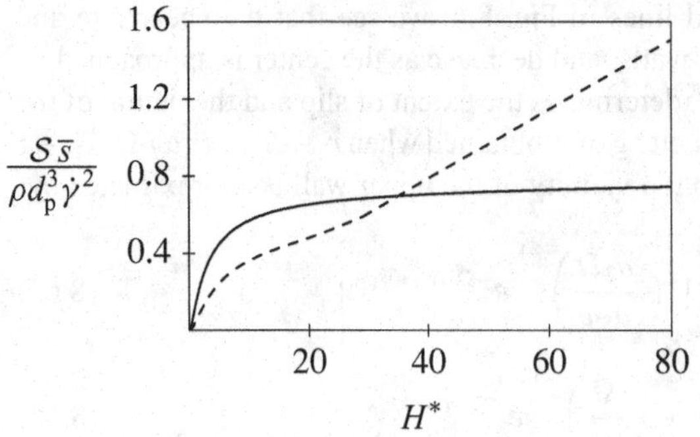

Figure 8.7. The variation of the dimensionless shear stress with the Couette gap. The solid and dashed lines represent the trigonometric and hyperbolic solutions, respectively. Parameter values are as in Fig. 8.5.

where $\bar{s}$ is the interparticle spacing averaged over the Couette gap. Substituting the solutions for w from (8.55) and (8.64) into (8.69), we get

$$\mathcal{N} = a_1 \rho\, \dot{\gamma}^2 d_p^2\, (d_p/\bar{s}) \left\{ \frac{1}{2} \frac{a_2 \Omega(H^*)}{a_1 r} \left(1 + \frac{a_2 \Omega^2}{|\alpha|\, a_5 \varphi} \right)^{-1} \right\}^2 \left[\pm \frac{1}{2} + \frac{1}{2} \frac{\Omega^2}{\alpha^2} + \frac{1}{|\alpha| H^*} \right] \tag{8.70}$$

with the positive sign in the first term within the square brackets for the trigonometric solution and the negative sign for the hyperbolic solution. We note that $\mathcal{N}$ varies as $\rho\, \dot{\gamma}^2 d_p^2$ in accordance with expectation and the analysis of §7.2. However, it has a rather complicated dependence on the Couette gap H, as r, α, and Ω are functions of H^*. The shear stress $\mathcal{S}$ is $\mathcal{N}r$, and hence it too varies as $\rho\, \dot{\gamma}^2 d_p^2$ times a function of H^*. The dependence of the shear stress on H^* is shown in Fig. 8.7. This behavior is unlike normal fluids, for which the stress depends on the Couette gap only through the shear rate; for instance, the viscosity of water does not depend on the thickness of the sheared layer. The stresses in rapid granular flow show a more complicated dependence on the Couette gap because the velocity profile is not linear, and its exact form depends on H^*.

In the limit $H^* \to \infty$, the term enclosed by the square brackets in (8.70) approaches the constant value $1/2$ for the trigonometric solution, while it decays as $(H^*)^{-1}$ for the hyperbolic solution; the term enclosed by the curly brackets approaches the constant value of $a_2\pi/(2a_1(L/M)^{1/2})$ for the trigonometric solution, but grows as $(H^*)^2$ for the hyperbolic solution. The stress ratio r approaches a constant asymptotic value as $H^* \to \infty$ in both solutions. Therefore as $H^* \to \infty$, $\mathcal{N}$, and $\mathcal{S}$ become independent of H^* in the trigonometric solution, but grow as H^* in the hyperbolic solution. This behavior is evident in Fig. 8.7. The physical explanation for this behavior is that the material shears everywhere in the trigonometric solution, but only in thin layers near the walls in the hyperbolic solution.

These features show that a granular material, as described by the high-density heuristic theory, resembles a normal fluid when the walls are net sinks of fluctuational energy, but behaves very differently when the walls are net sources of energy.

8.2.3. Predictions of the Kinetic Theory

We now discuss application of the kinetic theory, derived in Chapter 7, to plane Couette flow. We shall use the constitutive relations for smooth, inelastic particles, given in §7.3.13. The pressure, viscosity, pseudothermal conductivity, and the dissipation rate now have a more complicated dependence on the volume fraction ν, and we therefore expect the results to differ from those of the high-density theory described in the previous section.

We first define the following dimensionless variables

$$y^* = y/H, \quad v^* = v_x/v_w, \quad T^* = H^{*2}\, T/v_w^{\,2} \tag{8.71}$$

Table 8.1. Dimensionless Functions of ν Occurring in the Constitutive Relations Used in §8.2.3 and §8.3.5*

$$\mathcal{P}(\nu) = \nu(1 + 4\nu g_0)$$

$$\eta(\nu) = \frac{5\sqrt{\pi}}{96\, g_0}\left(1 + \tfrac{8}{5}\nu g_0\right)^2 + \frac{8}{5\sqrt{\pi}}\nu^2 g_0$$

$$\mathcal{K}(\nu) = \frac{25\sqrt{\pi}}{128\, g_0}\left(1 + \tfrac{12}{5}\nu g_0\right)^2 + \frac{4}{\sqrt{\pi}}\nu^2 g_0$$

$$\gamma(\nu) = \frac{12(1 - e_{\mathrm{p}}^2)\,\nu^2 g_0}{\sqrt{\pi}}$$

$$\eta_{\mathrm{w}}(\nu) = \varphi\,\frac{\pi\sqrt{3}\,\nu g_0}{6\nu_{\mathrm{drp}}}$$

$$\gamma_{\mathrm{w}}(\nu) = (1 - e_{\mathrm{w}}^2)\frac{\pi\sqrt{3}\,\nu g_0}{4\nu_{\mathrm{drp}}}$$

* The functions $\mathcal{P}$, η, $\mathcal{K}$, and γ are taken from the results of the kinetic theory, given in §7.3.13. The functions η_{w} and $\Gamma_{\mathrm{w}}(\nu)$ are derived from the heuristic boundary conditions of Johnson et al. (1990), given in (8.8) and (8.9).

where $H^* \equiv H/d_{\mathrm{p}}$ is the dimensionless Couette gap. The scaling for the grain temperature is arrived at by balancing the rate of energy dissipation with the stress work; for instance, this is the scaling of the fluctuation velocity w in (8.57).

In terms of these variables, the equations of motion (8.33)–(8.35) assume the form

$$\frac{\mathrm{d}}{\mathrm{d}y^*}\left(\eta(\nu)\sqrt{T^*}\,\frac{\mathrm{d}v^*}{\mathrm{d}y^*}\right) = 0 \tag{8.72}$$

$$\frac{\mathrm{d}}{\mathrm{d}y^*}\left(\mathcal{P}(\nu)T^*\right) = 0 \tag{8.73}$$

$$\frac{1}{H^{*2}}\frac{\mathrm{d}}{\mathrm{d}y^*}\left(\mathcal{K}(\nu)\sqrt{T^*}\,\frac{\mathrm{d}T^*}{\mathrm{d}y^*}\right) + \eta(\nu)\sqrt{T^*}\left(\frac{\mathrm{d}v^*}{\mathrm{d}y^*}\right)^2 - \gamma(\nu)\,T^{*3/2} = 0 \tag{8.74}$$

where $\mathcal{P}$, η, $\mathcal{K}$, and γ are dimensionless functions of ν that determine the pressure, viscosity, pseudothermal conductivity, and the dissipation rate, respectively. For smooth particles, their forms can be determined from the constitutive relations in §7.3.13, and are given in Table 8.1.

The boundary conditions (8.31) and (8.32) expressed in terms of the dimensionless variables are

$$v^* = \frac{\varepsilon}{H^*}\frac{\eta(\nu)}{\eta_{\mathrm{w}}(\nu)}\frac{\mathrm{d}v^*}{\mathrm{d}y^*} \tag{8.75}$$

$$\frac{1}{H^{*3}}\frac{\mathrm{d}T^*}{\mathrm{d}y^*} = -\frac{\eta_{\mathrm{w}}(\nu)}{\mathcal{K}(\nu)}v^{*2} + \frac{\varepsilon}{H^{*2}}\frac{\gamma_{\mathrm{w}}(\nu)}{\mathcal{K}(\nu)}T^* \tag{8.76}$$

at $y^* = 0$, and

$$v^* = 1 - \frac{\varepsilon}{H^*}\frac{\eta(\nu)}{\eta_{\mathrm{w}}(\nu)}\frac{\mathrm{d}v^*}{\mathrm{d}y^*} \tag{8.77}$$

$$\frac{1}{H^{*3}}\frac{\mathrm{d}T^*}{\mathrm{d}y^*} = \frac{\eta_{\mathrm{w}}(\nu)}{\mathcal{K}(\nu)}(v^* - 1)^2 - \frac{\varepsilon}{H^{*2}}\frac{\gamma_{\mathrm{w}}(\nu)}{\mathcal{K}(\nu)}T^* \tag{8.78}$$

at $y^* = 1$. Here η_{w} and γ_{w} are the dimensionless wall viscosity and dissipation functions, respectively, introduced in §8.1.2. Note, however, that η_{w} and γ_{w} given in Table 8.1 differ from the forms given in §8.1.2. They are derived from the heuristic boundary conditions of Johnson et al. (1990), for conformity with previous studies whose results we shall discuss in this section and in §8.3.5. We expect that the qualitative features of the solutions will remain unchanged even if the relations for η_w and γ_w given in §8.1.2 are used.

The parameter ε has been introduced in (8.75)–(8.78), after Nott et al. (1999), to elucidate the effect of the boundaries.† When $\varepsilon = 0$, the boundary conditions reduce to the no-slip and no-energy flux conditions, i.e., the walls are passive, and play a very limited role in the dynamics of the granular medium. By traversing the range of ε from zero to unity, we go from adiabatic walls and no slip to walls that allow slip and are sources or sinks of energy. The combination of no-slip and adiabatic walls cannot be realized from the boundary conditions for any physically realistic values of the parameters. However, it is useful to consider this case, as it is the condition used for "normal" fluids at insulating walls, and is therefore a limiting condition with which the results for the general case can be compared. The introduction of ε is simply a mathematical artifice to allow a smooth transition from the no-slip, adiabatic condition to the more general boundary condition that allows transport of momentum and energy to or from the walls.

In addition to (8.75)–(8.78), the confining pressure at the upper wall, or the mean solids fraction across the Couette gap, must be specified

$$\mathcal{P}(\nu(1))\, T^*(1) = \mathcal{N}^* \quad \text{or} \quad \int_0^1 \nu\, dy^* = \overline{\nu} \tag{8.79}$$

Equations (8.72)–(8.74) with boundary conditions (8.75)–(8.79) must, in general, be solved numerically. There is one case for which a simple analytical solution exists, namely, $\varepsilon = 0$, or adiabatic walls with no slip. For this case, it is easily verified that

$$\nu(y^*) = \overline{\nu}\ \text{(constant)}, \quad v^* = y^*, \quad T^* = \eta(\overline{\nu})/\gamma(\overline{\nu}) \tag{8.80}$$

is a solution of the governing equations, regardless of the channel width.‡ This is referred to as the "uniform" solution (Nott et al., 1999), as the density, temperature, and shear rate are constant everywhere. However, this is not the only solution: there are other, nonuniform, solutions bifurcating from the uniform solution at discrete intervals of H^* (Nott et al., 1999). The stability of the uniform solution and the nature of solutions that bifurcate from it are discussed in §8.4.

As ε increases from zero, i.e., as the walls become increasingly nonadiabatic, the variation in ν, T^*, and dv^*/dy^* between the walls increases. Solutions for $\varepsilon = 1$ are shown in Fig. 8.8 for two parameter sets. It is instructive to compare them with solutions of the high-density theory for the same parameter sets, shown in Fig. 8.6. Though there is substantial variation of ν in Fig. 8.8, some features of the solutions are in common with the solutions of the high-density theory. The walls act as sinks of pseudothermal energy when $e_w = 0.8$, $\varphi = 0.85$, and as sources when $e_w = 0.91$, $\varphi = 0.25$, as in the trigonometric and hyperbolic solutions, respectively, in the high-density theory. The shear rate is reduced in regions where the temperature is low.

Figure 8.9 shows that the density stratification becomes much more pronounced for large H^*. The formation of dense "plugs," regions where ν is close to ν_{drp} and the shear rate is very small, is apparent. The location of the plugs depends on the parameter values:

† Nott et al. (1999) introduced ε in the source term of the energy boundary condition also, i.e., the first term on the right-hand side of (8.76) and (8.78), but this is not necessary to recover the limit of adiabatic walls with no slip.

‡ There is another solution for $\varepsilon = 0$, namely, $T^* = 0$, and ν and v^* are any functions of y^* that satisfy $\int_0^1 \nu\, dy = \overline{\nu}$, $v^*(0) = 0$, $v^*(1) = 1$. However, this solution is unphysical: it represents a state where every particle moves along a streamline, with no fluctuational motion, and all components of the stress are zero.

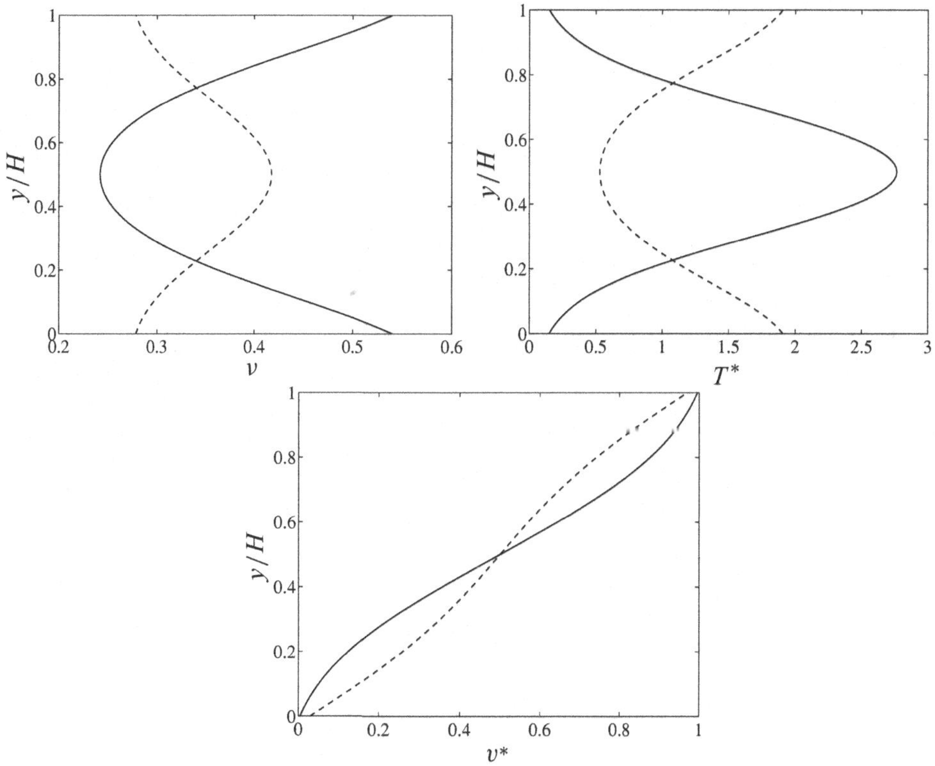

Figure 8.8. Predictions of the kinetic theory for plane Couette flow. The solid lines are for $e_w = 0.8$, $\varphi = 0.85$, and the dashed lines for $e_w = 0.91$, $\varphi = 0.25$. Other parameters are $e_p = 0.91$, $\overline{\nu} = 0.35$, $H^* = 50$. Note that the walls are sinks of fluctuational energy for the first set of parameters, and sources for the second, as in the high-density theory.

there is a plug at each wall for the case $e_w = 0.8$, $\varphi = 0.85$ (solid lines), but there is a single plug at the center when $e_w = 0.91$, $\varphi = 0.25$ (dashed lines). Physically, the reason for plug formation is that the diffusion of pseudothermal energy from regions of high to low shear rate decreases with increasing H^*; this causes the grain temperature to fall in the region of low shear rate, and hence the density to rise. Some of the assumptions of the kinetic theory, such as molecular chaos and binary collisions, break down at high densities, certainly near the close packing limit, and hence these predictions should be treated with caution. Note that the walls are no longer sources of energy for the case $e_w = 0.91$, $\varphi = 0.25$ (dashed lines), unlike in Fig. 8.8: the temperature gradient changes sign in thin layers near the walls, which makes them appear as sinks of energy. Thus the nature of the walls also depends on the Couette gap, unlike in the high-density solution.

8.3. FLOW IN INCLINED CHUTES

In the context of granular flows, chutes have come to mean open channels with a rigid base, inclined at an angle θ to the horizontal (see Fig. 8.10). Granular flow in chutes has been extensively investigated by experiments (Ishida and Shirai, 1979; Savage, 1979; Ishida et al., 1980; Johnson et al., 1990; Drake, 1991; Nott and Jackson, 1992; Pouliquen, 1999; Ancey, 2001), particle dynamics simulations (Walton, 1993a; Drake and Walton, 1995; Hanes and Walton, 2000; Silbert et al., 2001) and the application of continuum models (Savage, 1979; Johnson et al., 1990; Nott and Jackson, 1992; Louge, 2003). Its popularity can be attributed to several reasons. It is of industrial importance, as particulate materials are often transported

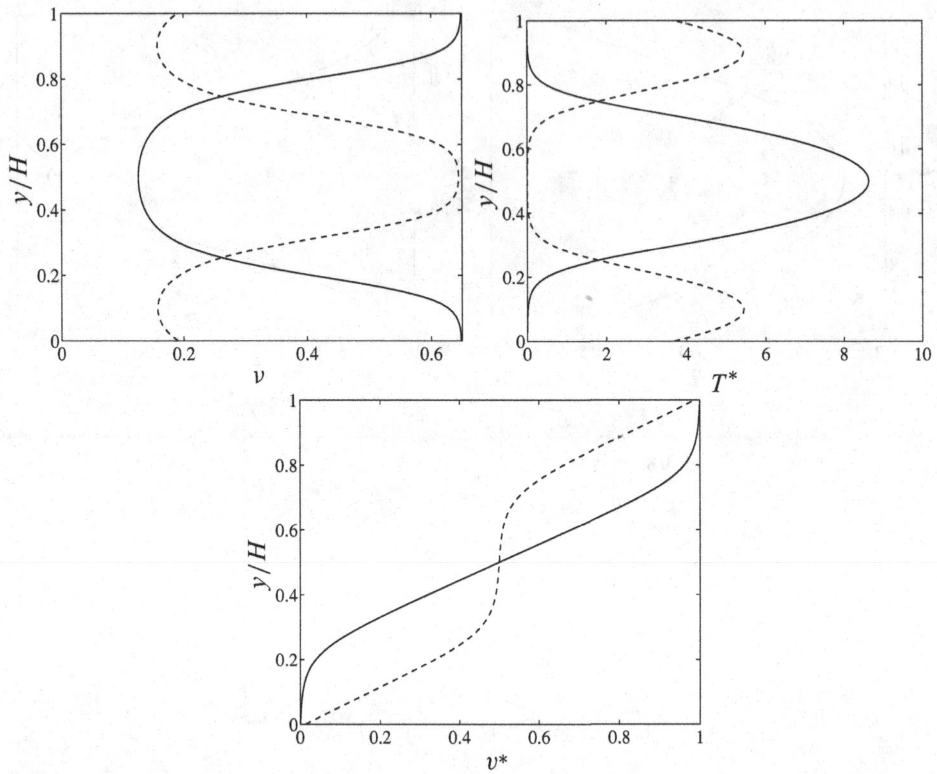

Figure 8.9. Predictions of the kinetic theory for plane Couette flow for $H^* = 120$. The solid lines are for $e_W = 0.8$, $\varphi = 0.85$, and the dashed lines for $e_W = 0.91$, $\varphi = 0.25$, other parameters remaining as in Fig. 8.8. Note the formation of the dense "plugs" in both cases, within which the shear rate is substantially reduced.

by the aid of gravity in open inclined channels in industrial processes. It is also of geological importance, as rock and snow avalanches, debris flows, and some aspects of the transport of sand in dunes may be thought of as flow through open channels, though erosion of the base and entrainment by the ambient air (in dunes) are added complications. Experiments of chute flow are relatively easy to conduct, as certain macroscopic properties of the flow, such as the flow rate, the loading (the mass of material per unit area of the chute base), and the flow depth can be readily measured. Though the difficulties associated with noninvasive measurement of granular flows remain, measurement of the velocity can be easily made from one of the boundaries, namely, the upper interface. From a theoretical viewpoint, it is a viscometric flow (see §8.2) if the influence of the side walls can be neglected, and

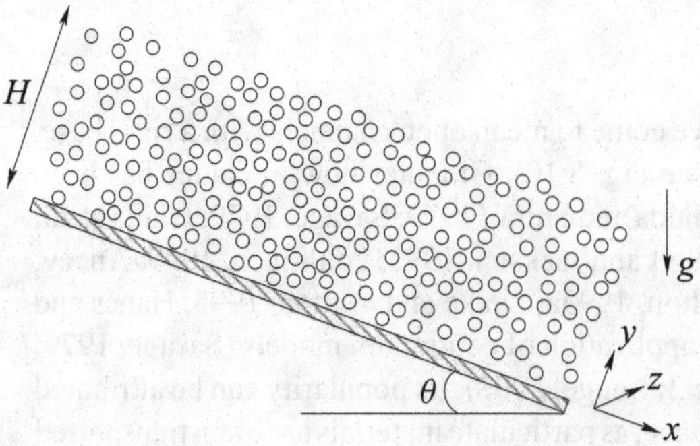

Figure 8.10. Schematic diagram of flow in an inclined chute. The base is inclined at an angle θ with the horizontal, and H is the thickness of the flowing layer.

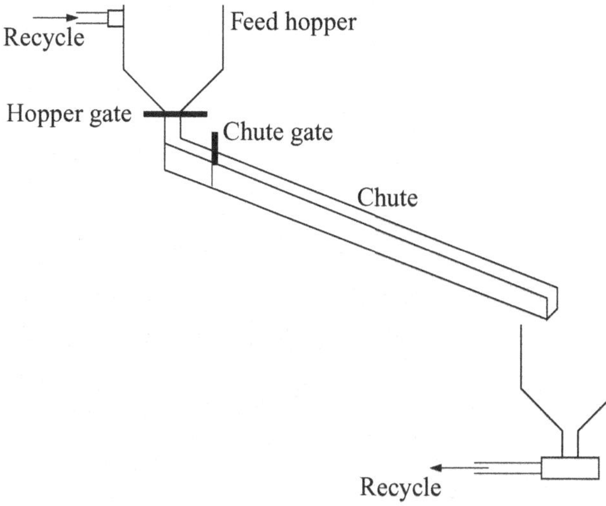

Figure 8.11. A typical experimental chute apparatus.

therefore easy to analyze. Despite its simplicity, chute flow demonstrates a rich variety of phenomena, such as multiplicity of steady states (Johnson et al., 1990), "granular jumps" (Savage, 1979; Brennen et al., 1983), surface undulations (Johnson, 1987b; Johnson et al., 1990; Forterre and Pouliquen, 2003), and other instabilities (Forterre and Pouliquen, 2001), many of which are yet to be fully understood.

8.3.1. Some Experimental Observations of Chute Flow

In an experimental chute flow apparatus, the material at the upstream end is fed by a storage device such as a hopper or bin (Fig. 8.11). The mass flow rate into the chute is controlled by adjusting the flow rate from the hopper, which in turn is accomplished by controlling the size of its exit slot. An additional degree of control is the state in which material is introduced into the chute. i.e., its density and state of agitation. A loose and agitated state is obtained by dropping the material from a height into the chute (Johnson et al., 1990; Azanza et al., 1999), while a dense and unagitated state is obtained by locating the hopper gate close to the head of the chute. These are referred to as the loose and dense entry conditions, respectively, in the discussion below.

The flow rate is measured by collecting the material issuing from the downstream end of the chute for a fixed-time interval. The flow depth H, or thickness of the flowing layer (see Fig. 8.11), has traditionally been measured by lowering a plate into the flowing layer until particles begin to strike it; recent studies have measured it by illuminating the layer with a laser sheet at grazing angle (Pouliquen, 1999) and by the use of an ultrasonic proximity sensor (Ancey, 2001). It is usually reported in dimensionless form as $H^* \equiv H/d_\text{p}$. When the flow is loose and rapid, the interface between the granular medium and the atmosphere is not very sharp; some "saltating" particles reach a fairly high elevation, making it difficult to measure the flow depth. In these conditions, the loading m_ℓ, which is the mass of material per unit area of the chute base, is more suited for characterizing the flow. The loading has been measured by trapping particles during flow in a small area of the chute (Johnson et al., 1990), and by the use of capacitance probes (Louge and Keast, 2001). The velocity field has been measured by various techniques: Shirai et al. (1977) designed an optical fibre probe, which has been used in many subsequent studies (Ishida and Shirai, 1979; Ishida et al., 1980; Johnson et al., 1990; Nott and Jackson, 1992), Azanza et al. (1999) used time lapse digital photography, and Ancey (2001) used high-speed video photography. Recent advances in high-speed digital videography and the reduction in the cost of the equipment has made it

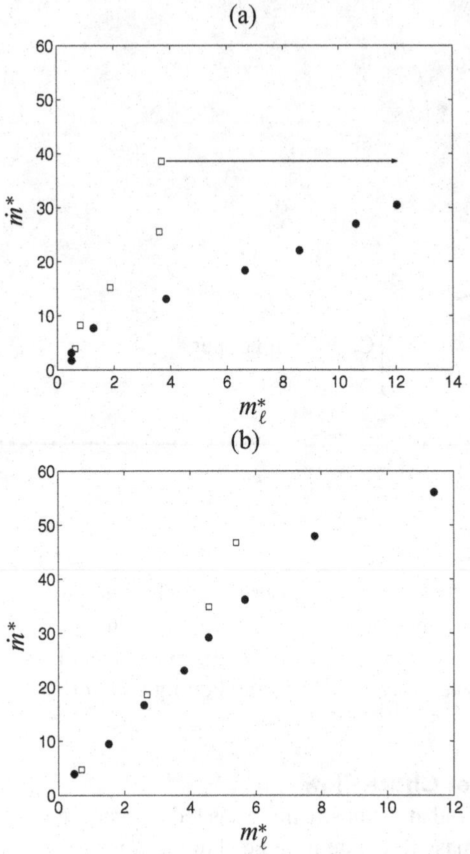

Figure 8.12. The variation of the dimensionless mass flow rate with the dimensionless loading for glass beads of mean diameter 1 mm flowing in (a) a smooth chute inclined at 15.5° from the horizontal, and (b) a rough chute inclined at 20°. The chute base for the smooth chute was a machined aluminium plate, and for the rough chute it was the same plate coated with sandpaper. The open squares and filled circles represent flows generated by loose and dense entry conditions, respectively. The horizontal arrow at the upper limit of the loose branch in panel (a) indicates a switch from the loose to the dense branch via a granular jump. Adapted from Johnson et al. (1990).

the choice of current techniques. The study of Nakagawa et al. (1993) has introduced the technique of magnetic resonance imaging for the noninvasive measurement of the density and velocity fields, though the response time of currently available instruments may restrict the range of measurable velocities.

In this section we discuss only steady, fully developed flow, i.e., time-independent flow in which the variables do not change in the flow direction. It is generally found that such flows exist only within a range of the inclination θ: the material ceases to flow below the lower limit of θ, and keeps accelerating down the length of the chute above the upper limit. However, whether the upper limit is a fundamental one or whether the observations were constrained by the use of finite length of the chute is open to question. The lower and upper limits of θ depend on the nature of the particles, the properties of the base, and the condition in which the material enters the chute, as described below.

Available data suggest that the roughness of the chute base and the condition of the material as it enters the chute plays an important role in determining the nature of the flow (Johnson et al., 1990). To study these features, we consider the flow curves, which are plots of the flow rate per unit width of the chute, $\dot{m}$, versus the loading, m_ℓ, for steady, fully developed flow. Figure 8.12a shows the data of Johnson et al. (1990) in terms of the dimensionless flow rate and loading

$$\dot{m}^* \equiv \frac{\dot{m}}{\rho_\mathrm{p}(g d_\mathrm{p}^3)^{1/2}}, \quad m_\ell^* \equiv \frac{m_\ell}{\rho_\mathrm{p} d_\mathrm{p}} \tag{8.81}$$

for a chute with a relatively smooth base. The data show two branches in the flow curve representing flows generated by loose and dense entry conditions. The loose and dense branches coincide at small $\dot{m}$, separate beyond some value of $\dot{m}$, after which the slope of

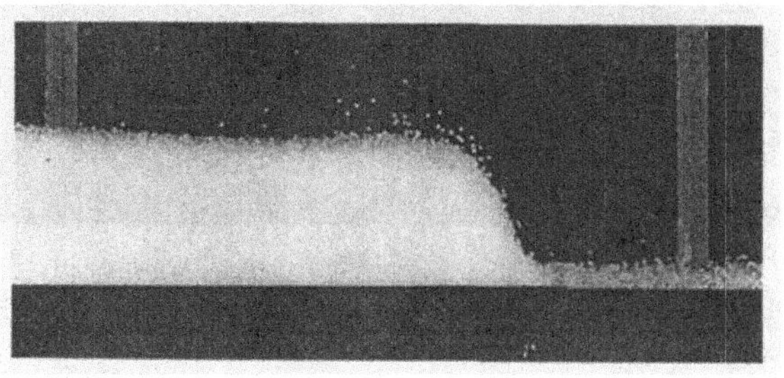

Figure 8.13. A granular jump observed in the flow of polystyrene beads (mean diameter 1.2 mm) in a chute of inclination 35°. Flow is from right to left. (Reproduced from Savage (1979), with permission from Cambridge University Press.)

the former is significantly higher. The loose branch terminates at an upper limit of $\dot{m}$; if $\dot{m}$ is increased beyond this limit, the flow switches to the dense branch via a granular jump. A granular jump, akin to hydraulic jumps observed in the flow of water and other liquids in open channels, is a region of sharp change in the layer thickness, density, and velocity; an example is shown in Fig. 8.13. With further increase in $\dot{m}$, the jump migrates all the way to the head of the chute, so that the entry condition is effectively dense. No upper limit has been observed for the dense branch. Johnson et al. (1990) found the flow in the loose branch to be fast and thin, with substantial slip at the base, while in the dense branch it was slow and thick, with a much smaller slip at the base (Fig. 8.14). They found the flow depth to decrease on increasing $\dot{m}$ in the loose branch, while it showed a monotonic rise in the dense branch.

At low angles of inclination, Johnson et al. found the dense branch to disappear completely for a smooth base. At even lower angles, even the loose branch disappears and there is no flow altogether, as stated earlier. As θ increases, the upper limit of the loose branch increases, the dense branch is displaced upwards, and the slopes of both branches increase slightly.

The behavior is quite different when the base is rough. Johnson et al. roughened the base by coating the aluminium plate with sandpaper, and found that, as with the smooth base, the loose branch is absent for small angles of inclination. At higher inclinations, the loose and dense branches are much closer to each other than in smooth chutes, as shown in Fig. 8.12b. Indeed, Nott and Jackson (1992) found no loose branch at all when they used a rough poly(propylene) sheet as the base plate; no matter what the inclination or flow rate, a loose entry condition always engendered a granular jump that migrated to the head of the chute. When the loose branch was present, Johnson et al. found the flow depth to increase with $\dot{m}$, unlike in the smooth chute.

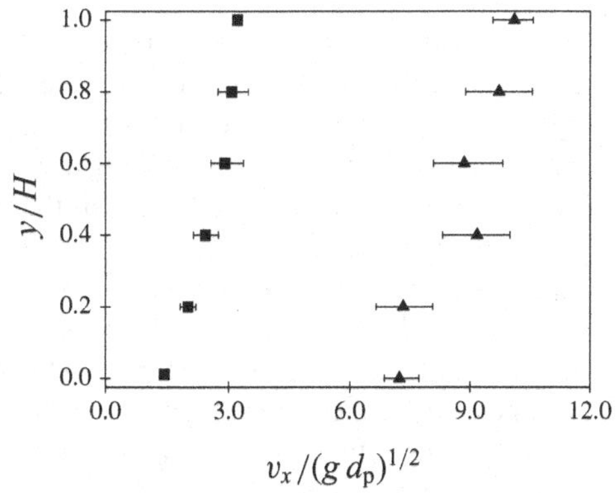

Figure 8.14. The velocity profile for flows with dense (squares) and loose (triangles) entry conditions, determined by Johnson et al. (1990) using a fiber optic probe. The horizontal line at each data point is the estimated measurement error. The triangles are for $\dot{m}^* = 46$, and the squares for $\dot{m}^* = 7.3$. Both sets of data are for glass beads of diameter 1 mm flowing in a chute of inclination $\theta = 17°$ with flow depth $H^* = 20$.

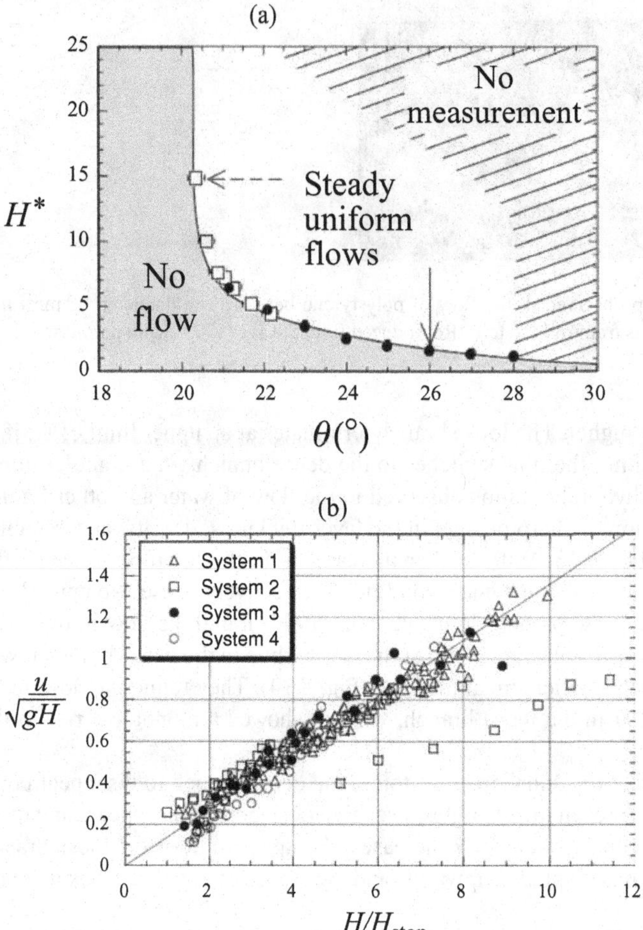

Figure 8.15. (a) A "phase diagram" in the H^*-θ plane, showing regions of steady, fully developed flow and no flow. Data are for dense assemblies of 0.5 mm diameter glass beads flowing down an incline roughened by sticking a layer of the same beads. The solid circles are the boundary between flow and no flow for experiments at constant θ and decreasing H^* (solid arrow), and the open squares are the same boundary for experiments at constant H^* and decreasing θ (dashed arrow). (b) The Froude number as a function of the scaled flow depth. Systems 1–4 refer to different combinations of particle size and wall roughness; the data in this plot are for a range of θ. The set of data that deviates from the straight line (open squares) are for relatively loose flows having a layer of saltating particles at the interface, which may have affected the measurement of H. (Reused with permission from Pouliquen, 1999. Copyright 1999, American Institute of Physics.)

For dense flows, the interface is well defined and the flow depth H can be measured quite accurately. As there is a one-to-one relation between the loading and the flow depth, the flow curves may be given as plots of $\dot{m}$ versus H. Using the dense entry condition and a base that was roughened by sticking a layer of particles to it, Pouliquen (1999) found that for each θ there is a minimum value of the flow depth, H_{stop}, below which there is no flow. He found that H_{stop} increases as θ is reduced, and it diverges as θ approaches a lower limit θ_{stop}. There is no flow for any finite depth when $\theta < \theta_{\text{stop}}$ (Fig. 8.15a). Silbert et al. (2001) also observed this behavior in DEM simulations of chute flow. Interestingly, Pouliquen found H_{stop} to be of significance even in the flowing state: if u is the average velocity across the chute depth, the Froude number $Fr \equiv u/\sqrt{gH}$ as a function of H/H_{stop} collapses onto a single straight line for all the granular materials, wall roughness (Pouliquen studied four

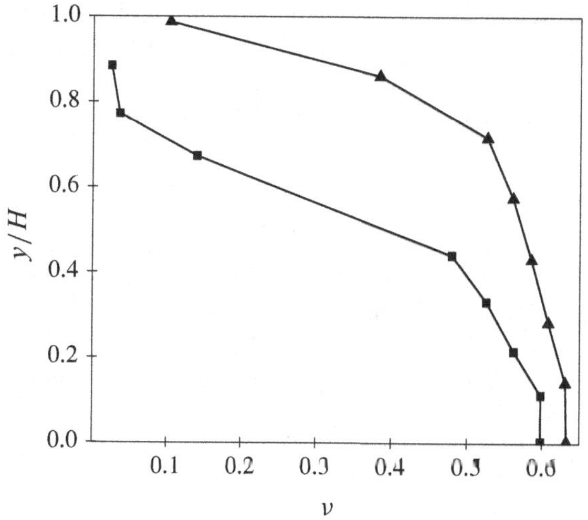

Figure 8.16. Profile of the solids fraction ν for the flow of glass beads of 1 mm diameter flowing in an inclined chute, determined by Ancey (2001) using gamma ray attenuation. The data points are connected by lines to guide the eye. The triangles are for $\theta = 37°$, $H^* = 8.1$, and the squares for $\theta = 33°$, $H^* = 14$; in both cases, $\dot{m}^* = 67$. The dense entry condition was used to feed the material.

combinations of glass bead size and wall roughness) and chute inclinations investigated, as shown in Fig. 8.15b. Thus H_{stop} appears to be a length scale that captures some important features of the flow in the dense regime.

The density field is more difficult to measure, and data are therefore rare. Ancey (2001) determined the density profile by measuring the attenuation of gamma rays passing through the flowing layer. Figure 8.16 shows the solids fraction profile for two angles of inclination, with the flow rate kept fixed. As Ancey only used the dense entry condition and kept the chute base rough, the density at the base was high. In both experiments, a rapid decay of ν beyond some distance from the chute base is evident.

8.3.2. Analysis of Steady, Fully Developed Flow

As in plane Couette flow, the mass balance is identically satisfied. The momentum balances are

$$\frac{\mathrm{d}\sigma_{yx}}{\mathrm{d}y} = \rho_{\mathrm{p}} \nu g \sin\theta \tag{8.82}$$

$$\frac{\mathrm{d}\sigma_{yy}}{\mathrm{d}y} = -\rho_{\mathrm{p}} \nu g \cos\theta \tag{8.83}$$

The balance of fluctuational kinetic energy remains unchanged from (8.35),

$$-\frac{\mathrm{d}q_y}{\mathrm{d}y} - \sigma_{yx}\frac{\mathrm{d}v_x}{\mathrm{d}y} - \Gamma = 0 \tag{8.84}$$

The boundary conditions at the base $y = 0$ are (8.5) and (8.7) for the high-density theory, and (8.31) and (8.32) for the full kinetic theory. The conditions at the upper interface $y = H$ are that the material is stress free and the energy flux vanishes,

$$\sigma_{yy} = \sigma_{yx} = 0; \quad \frac{\mathrm{d}T}{\mathrm{d}y} = 0. \tag{8.85}$$

We shall see that the condition of vanishing normal stress presents some difficulties, and describe an alternative condition in §8.3.5.

From the momentum balances, we see that the magnitudes of the shear and normal stresses increase with distance from the free surface, but their ratio remains constant, i.e., $|\sigma_{yx}|/\sigma_{yy} = \tan\theta$.

As in §8.2, we first apply the high-density theory to the problem, and follow it with the kinetic theory.

8.3.3. High-Density Theory

The high-density analysis for chute flow presented here closely follows that of Anderson and Jackson (1992). It is similar to the analysis for plane Couette flow, given in §8.2.1, but the presence of the gravitational body force alters the form of the solution. Note that the right-hand sides of (8.82) and (8.83) are treated as constants in the high-density limit.

Proceeding as in §8.2.1, we arrive at the differential equation governing the fluctuation velocity

$$\frac{d^2 w}{dy^2} - \frac{1}{(H-y)}\frac{dw}{dy} - \frac{(L - M\tan^2\theta)}{d_p^2} w = 0 \tag{8.86}$$

and the boundary condition

$$\frac{dw}{dy} = \frac{\alpha}{d_p} w \tag{8.87}$$

at the base ($y = 0$). Here L, M, and α are the dimensionless constants defined in §8.2.1, with r replaced by $\tan\theta$. At the upper free interface ($y = H$), we have the zero energy flux condition

$$\frac{dw}{dy} = 0 \tag{8.88}$$

The solution is given in terms of Bessel functions or modified Bessel functions of zeroth order, depending on the sign of $L - M\tan^2\theta$. As in the case of plane Couette flow, there is no solution for $L - M\tan^2\theta = 0$. The other cases are discussed below.

Case 1: $L - M\tan^2\theta < 0$

This case corresponds to the "trigonometric" solution in plane Couette flow, discussed in §8.2.1. Defining $\Omega^2 = (M\tan^2\theta - L)$ and $\xi = \Omega(H-y)/d_p$, (8.86) becomes

$$\frac{d^2 w}{d\xi^2} + \frac{1}{\xi}\frac{dw}{d\xi} + w = 0 \tag{8.89}$$

The general solution is in terms of the Bessel functions of order zero, $w = \mathcal{A} J_0(\xi) + \mathcal{B} Y_0(\xi)$, but we set $\mathcal{B} = 0$ so that w remains finite as $\xi \to 0$. Applying boundary condition (8.87), we find

$$-J_0'(\xi)/J_0(\xi) = \alpha/\Omega \quad \text{at } \xi = \Omega H^* \tag{8.90}$$

where the prime indicates differentiation with respect to ξ, and $H^* \equiv H/d_p$ is the dimensionless flow depth. As w must be positive, ΩH^* must be less than the first zero of J_0, for which the left-hand side of (8.90) is always positive. This implies that $\alpha > 0$ for a solution to exist, and hence

$$\frac{\ell}{m} > \tan^2\theta > \frac{L}{M} \tag{8.91}$$

This is precisely the condition for the trigonometric solution in plane Couette flow. Equation (8.90) gives the relation between the flow depth H^* and θ, shown graphically in Fig. 8.17a.

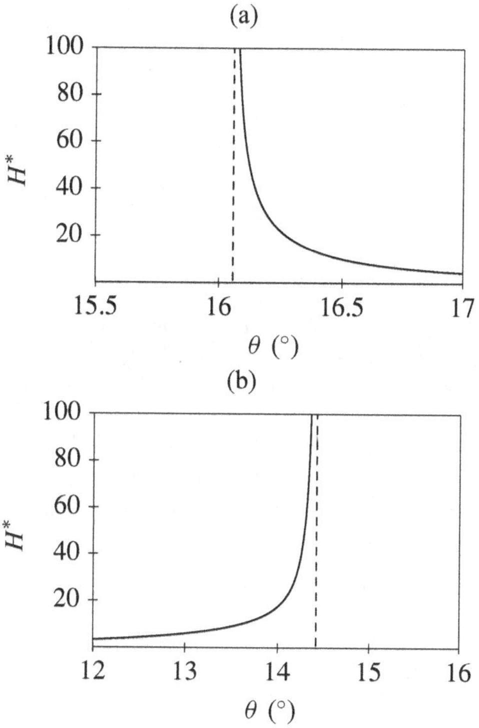

Figure 8.17. The dependence of the flow depth on the angle of inclination in the high density solution for flow in inclined chutes. Panel (a) is for case 1 and panel (b) for case 2. The dashed lines are the asymptotes for infinitely large H^*, occurring as $\theta \to \tan^{-1}\left[(L/M)^{1/2}\right]$ in (a) and $\theta \to \theta_\infty$ (see (8.96)) in (b). Parameters values are $e_p = 0.91$, $e_w = 0.8$, $\varphi = 0.85$ in (a), and $e_p = 91$, $e_w = 0.91$, $\varphi = 0.25$ in (b).

Substituting the solution for w into the momentum balances (8.82) and (8.83), and using the slip boundary condition (8.5) at the base, we get the velocity profile

$$v_x(\xi) = \mathcal{A}\,\frac{a_1 \tan\theta}{a_2}\left\{-\frac{1}{\Omega}\int_0^{\xi} J_0(\xi')\,\mathrm{d}\xi' + \frac{a_2}{\varphi} J_0(\Omega H^*)\right\} \tag{8.92}$$

where the constant $\mathcal{A}$ is determined by the flow rate $\dot{m}$.

Case 2: $L - M\tan^2\theta > 0$

This case corresponds to the "hyperbolic" solution in plane Couette flow, §8.2.1. Defining $\Omega^2 = (L - M\tan^2\theta)$ and $\xi = \Omega(H - y)/d_p$, (8.86) becomes

$$\frac{\mathrm{d}^2 w}{\mathrm{d}\xi^2} + \frac{1}{\xi}\frac{\mathrm{d}w}{\mathrm{d}\xi} - w = 0 \tag{8.93}$$

The general solution is in terms of the modified Bessel functions of order zero, $w = \mathcal{A}\, I_0(\xi) + \mathcal{B}\, K_0(\xi)$, but we set $\mathcal{B} = 0$ so that w remains finite as $\xi \to 0$. Applying boundary condition (8.45), we find

$$\frac{-I_0'(\xi)}{I_0(\xi)} = \frac{\alpha}{\Omega} \quad \text{at } \xi = \Omega H^* \tag{8.94}$$

where the prime indicates differentiation with respect to ξ. As the left hand side is bounded between 0 and -1, α/Ω must lie in the range $0 > \alpha/\Omega > -1$, and hence

$$\frac{\ell}{m} < \tan^2\theta < \tan^2\theta_\infty \,(< \frac{L}{M}) \tag{8.95}$$

where θ_∞ is the solution of the equation

$$m\tan^2\theta - \ell = (L - M\tan^2\theta)^{1/2} \tag{8.96}$$

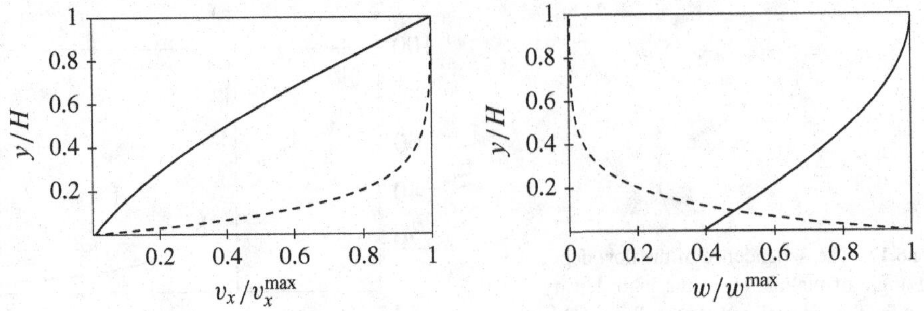

Figure 8.18. Profiles of the velocity and the velocity fluctuation for Cases 1 (solid lines) and 2 (dashed lines) of the high-density solution for flow in inclined chutes. Both the variables have been scaled by their respective maxima. Parameter values are $H^* = 50$, rest as in Fig. 8.17.

Recall that this is precisely the condition for the hyperbolic solution in plane Coutte flow, as $\tan\theta$ is the ratio of shear and normal stresses in chute flow. Equation (8.94) gives the relation between H^* and θ, shown graphically in Fig. 8.17b.

Proceeding as in case 1, we get the velocity field

$$v_x(\xi) = \mathcal{A}\,\frac{a_1 \tan\theta}{a_2}\left\{-\frac{1}{\Omega}\int_0^{\xi} I_0(\xi')\,\mathrm{d}\xi' + \frac{a_2}{\varphi} I_0(\Omega H^*)\right\} \tag{8.97}$$

with the constant $\mathcal{A}$ determined by specifying the flow rate $\dot{m}$.

8.3.4. Some Features of the High-Density Solutions

As in the solution for plane Couette flow, the relative magnitudes of L/M and ℓ/m decide which of the two solutions holds. Case 1 applies when

$$a_2 a_4(1 - e_{\mathrm{p}}^2) < a_6 a_5 \varphi(1 - e_{\mathrm{w}}^2) \tag{8.98}$$

For a given set of particle and wall inelasticities, this sets a lower bound for the specularity φ. Thus Case 1 applies for chutes with a relatively rough base. Similarly, Case 2 applies when $L/M > \ell/m$, or equivalently

$$a_2 a_4(1 - e_{\mathrm{p}}^2) > a_6 a_5 \varphi(1 - e_{\mathrm{w}}^2) \tag{8.99}$$

Hence it holds for chutes with a relatively smooth base. We shall therefore refer to the two cases as the "rough chute" and the "smooth chute" conditions.

Profiles of v_x and w are shown in Fig. 8.18. We see that the material shears almost uniformly through the flowing layer in Case 1, whereas it shears only near the base in Case 2. The fluctuation velocity profiles for the two cases are also markedly different; in Case 1 w *increases* from a finite value at the base and attains a maximum at the interface, while in Case 2, it *decreases* rapidly from a maximum at the base, and becomes negligibly small near the interface. These features are reminiscent of the trigonometric and hyperbolic solutions, respectively, in plane Couette flow.

Several features of the high-density solutions do not conform to intuitive expectation and, more importantly, to experimental observations. One such feature is that, for a given set of particle and wall properties, there is a unique flow depth for each angle of inclination. The data of Johnson et al. (1990), shown in Fig. 8.12, and others (Nott and Jackson, 1992; Pouliquen, 1999; Ancey, 2001) show that a range of flow depth is possible, with the flow rate $\dot{m}$ determining the flow depth. Another odd feature is in Case 2, where H^* *increases* as θ increases, whereas experiments show the flow getting thinner (and faster) with increasing θ. The solution in Case 1 does not suffer from this flaw. In both cases, steady, fully developed flow exists only for a limited range of the angle of inclination. We shall see that the kinetic theory provides solutions even outside this range.

In Case 1, w is maximum and $\mathrm{d}v_x/\mathrm{d}y$ is finite at the interface (Fig. 8.18). This does not appear to satisfy the stress-free boundary condition at the interface, but it indeed does: from the y momentum balance, we see that the interparticle spacing s varies as

$$\frac{s}{d_\mathrm{p}} = \frac{a_1 w^2}{g(H-y)\cos\theta} \tag{8.100}$$

Thus s becomes unbounded at the interface; it can be verified that this is also true for Case 2. As all components of the stress vary as $1/s$, they vanish at the interface. However, $s \to \infty$ violates the basic assumption of the high-density theory that $s/d_\mathrm{p} \ll 1$. It is shown in the following section that the stress-free boundary condition raises a difficulty in the kinetic theory also, and an alternative boundary condition that avoids the difficulty is presented there (see also Problem 8.3).

The mass flow rate per unit width of the chute is

$$\dot{m} = \int_0^H \rho\, v_x\, \mathrm{d}y \tag{8.101}$$

Though the density is assumed to be constant in the high-density theory, it is clear from the variation of s that it is not: indeed, the density vanishes at the interface. Ignoring the inconsistency for the moment, the density may be written in terms of s as

$$\rho = \frac{\rho_\mathrm{p}\nu_\mathrm{drp}}{(1+s/d_\mathrm{p})^3} \tag{8.102}$$

and hence the flow rate is

$$\dot{m} = \rho_\mathrm{p}\nu_\mathrm{drp}\int_0^H \frac{v_x}{(1+s/d_\mathrm{p})^3}\,\mathrm{d}y. \tag{8.103}$$

As both v_x and w vary as $\mathcal{A}$, we see that $\dot{m} \sim \mathcal{A}$ when $\mathcal{A}$ is small, and $\dot{m} \sim \mathcal{A}^{-5}$ when $\mathcal{A}$ is large. As $\dot{m}$ must be a continuous function of $\mathcal{A}$, it must vary between zero and a finite maximum. This feature too is at odds with experimental data, as there is no evidence of an upper bound on the mass flow rate for a given angle of inclination. (However, the length of the chute required to achieve fully developed flow increases with the flow rate (Johnson et al., 1990; Nott and Jackson, 1992)).

Notice that if ρ is assumed to be constant, there is no upper bound in $\dot{m}$. However, if we adopt the modified interface boundary condition proposed in §8.3.5.1 and enforce the condition of constant density $s/d_\mathrm{p} \ll 1$, we do find a constraint on the magnitude of $\dot{m}$ (see Problem 8.3).

Thus it is clear that several aspects of the high-density solution for chute flow are unsatisfactory. We shall see below that the kinetic theory corrects some of the flaws.

8.3.5. Predictions of the Kinetic Theory

The velocity scale in this problem is set not by a boundary, unlike in plane Couette flow, but by the mass flow rate $\dot{m}$. However, for the numerical solution of the governing equations, it is more convenient to specify the flow depth and determine $\dot{m}$ from the solution. We therefore choose $(gd_\mathrm{p})^{1/2}$ as the velocity scale. The scaled variables then are

$$y^* = y/H, \quad v^* = v_x/(gd_\mathrm{p})^{1/2}, \quad T^* = H^{*2}\,T/(gd_\mathrm{p}), \tag{8.104}$$

and equations (8.82)–(8.84) take the dimensionless forms

$$\frac{\mathrm{d}}{\mathrm{d}y^*}\left(\eta(\nu)\sqrt{T^*}\,\frac{\mathrm{d}v^*}{\mathrm{d}y^*}\right) = -H^{*3}\,\nu\sin\theta \tag{8.105}$$

$$\frac{\mathrm{d}}{\mathrm{d}y^*}\left(\mathcal{P}(\nu)T^*\right) = -H^{*3}\,\nu\cos\theta \tag{8.106}$$

$$H^{*-2}\frac{\mathrm{d}}{\mathrm{d}y^*}\left(\mathcal{K}(\nu)\sqrt{T^*}\,\frac{\mathrm{d}T^*}{\mathrm{d}y^*}\right) + \eta(\nu)\sqrt{T^*}\left(\frac{\mathrm{d}v^*}{\mathrm{d}y^*}\right)^2 - \gamma(\nu)\,T^{*3/2} = 0 \tag{8.107}$$

with the functions $\mathcal{P}$, η, $\mathcal{K}$, and γ defined in Table 8.1. The boundary conditions at the base remain as in (8.75) and (8.76).

The boundary conditions at the interface (8.85) require some consideration. The condition of vanishing normal stress implies that $\nu(1) = 0$ or $T^*(1) = 0$ (or both). If the former is assumed, (8.106) with the second of (8.85) leads to the result that all derivatives of ν vanish at $y^* = 1$. If H^* is finite, this yields the trivial solution $\nu(y^*) = 0$. A similar argument, using the energy balance (8.107), leads to the result that T^* vanishes everywhere if it is made to vanish at the free surface (see Problem 8.4). It appears that the condition of vanishing normal stress cannot be applied at the interface!

The problem arises because the thickness of the flowing layer is taken to be finite: in this case, if ν and all its derivatives vanish at the interface, then ν will vanish in the entire layer. This need not be the case if the layer were of infinite thickness; for example, if ν were to vary as $1/y$ (for $y \gg 1$), ν and all its derivatives would vanish as $y \to \infty$, but they would be nonzero when y is finite. This was recognized by Johnson et al. (1990), who proposed that the stress-free interface be applied at an infinite distance from the base. We must then have $\nu \to 0$, $\mathrm{d}v_x/\mathrm{d}y \to 0$ and $\mathrm{d}T/\mathrm{d}y \to 0$ as $y \to \infty$. As the flow depth is now infinite, another length must be found to scale the variables with; a suitable choice is discussed below. Denoting the asymptotic values of the velocity and grain temperature by v^*_∞ and T^*_∞, respectively, we write $T^* = T^*_\infty + \tilde{T}^*$ and $v^* = v^*_\infty + \tilde{v}^*$, such that ν, $\tilde{v}^*$ and $\tilde{T}^*$ are small for large y^*. We may then linearize the equations of motion in these variables to obtain the following asymptotic solution for $y^* \to \infty$ (Problem 8.5)

$$\nu = \nu_{\mathrm{m}}\exp\left[-\frac{H^{*3}\cos\theta}{T^*_\infty}\,y^*\right] \tag{8.108}$$

$$v^* = v^*_\infty - \frac{\sin\theta\, T^{*3/2}_\infty}{\cos^2\theta\,\eta_0 H^{*3}}\,\nu \tag{8.109}$$

$$\tilde{T}^* = T^*_\infty + \frac{T^{*3}_\infty}{4H^{*4}\cos^2\theta\,\kappa_0}(\gamma_0 - \tan^2\theta/\eta_0)\,\nu^2 \tag{8.110}$$

where η_0, κ_0 and γ_0 are the limits of $\eta(\nu)$, $\kappa(\nu)$ and $\gamma(\nu)/\nu^2$, respectively, as $\nu \to 0$. This asymptotic solution was first given by Johnson et al. (1990). The values of ν_{m}, T^*_∞ and v^*_∞, are determined by matching the asymptotic solution with the numerical solution for the region close to the chute base. For proper asymptotic matching, the solutions must overlap over a range of y^*; to do this numerically, one *patches* the two solutions at $y^* = y^*_{\mathrm{p}}$, and verifies that the composite solution is insensitive to the variation of y^*_{p} over a range. It is convenient to choose H as the distance from the base where the solutions are patched, i.e., $y^*_{\mathrm{p}} = 1$. Asymptotic matching is then achieved by ensuring that the solution is insensitive to varying H^* over a sufficiently large range. To close the problem, the loading

$$m_\ell \equiv \rho_{\mathrm{p}}\int_0^\infty \nu\,\mathrm{d}y \tag{8.111}$$

is specified.

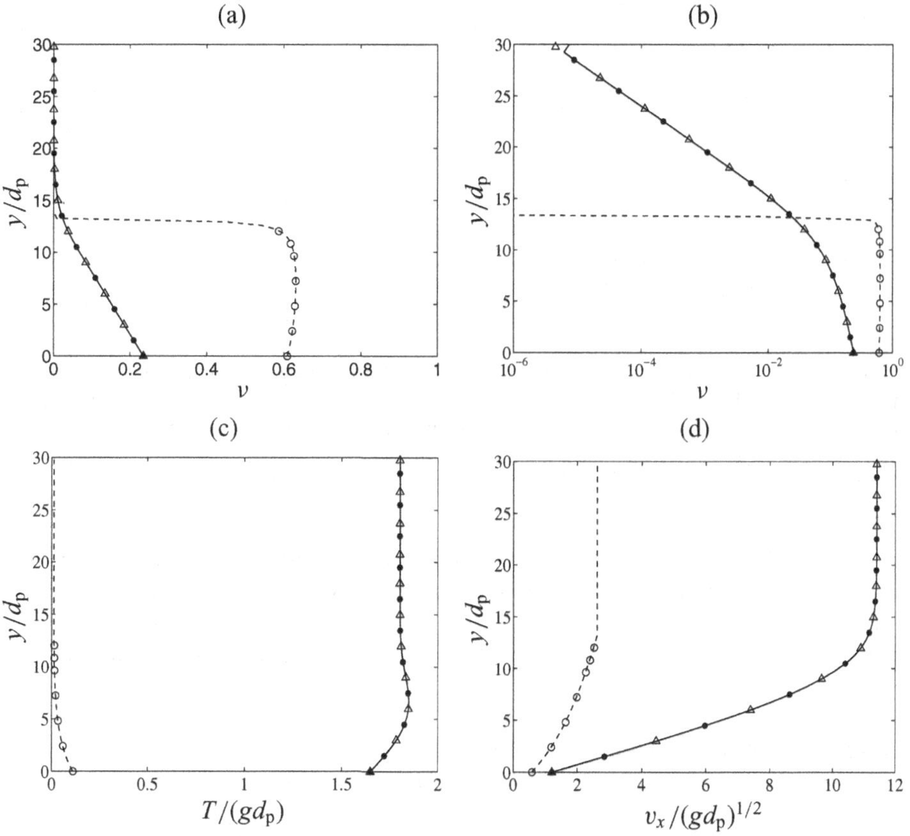

Figure 8.19. Comparison of the solutions obtained by the numerical-asymptotic method (lines and open triangles) and the modified interface boundary conditions (open and filled circles). Panel (b) shows the same data as in (a), but in log-linear axes. The solid lines, open triangles, and filled circles are for $\theta = 15°$, $m_\ell^* = 1.713$, and the dashed lines and open circles are for $\theta = 14°$, $m_\ell^* = 8.093$. For the numerical-asymptotic solution, matching was done at $H^* = 30$ for the solid line, at $H^* = 20$ for the open triangles, and at $H^* = 13$ for the dashed line. The solutions with the modified interface boundary conditions were obtained with $H^* = 30$ for the filled circles, and $H^* = 12$ for open circles. The values of other parameters are $e_p = 0.91$, $e_w = 0.91$, $\varphi = 0.25$.

Figure 8.19 shows the solutions for two values of the dimensionless loading m_ℓ^* (defined in (8.81)), represented by the solid and dashed lines. The higher value of m_ℓ^* results in a dense flow, and the lower value in a loose flow. The exponential decay of the solids fraction at large enough distance from the chute base is apparent in both cases (Fig. 8.19b). In the dense solution, the solids fraction is quite high near the base, and decays very sharply above a certain distance. Thus the solution appears to capture the sharp interface that is observed experimentally in dense flows. The open triangles and the solid line represent solutions for the loose flow obtained by matching the numerical and asymptotic solutions at very different values of H^*; their close agreement shows that an asymptotic match has been achieved. While matching can, in principle, be done over a range of H^* for the dense solution also, numerical difficulties restrict this range: matching must, on the one hand, be done when $\nu \ll 1$, in order that the asymptotic solution be valid, but on the other hand ν must remain above machine precision for the numerical solution be accurate. Figure 8.19b shows that the latter constraint restricts the range of H^* over which matching can be carried out for the dense flow, as ν decays very rapidly above a certain distance from the base. Panels (c) and (d) in Fig. 8.19 show that the temperature and velocity are much lower

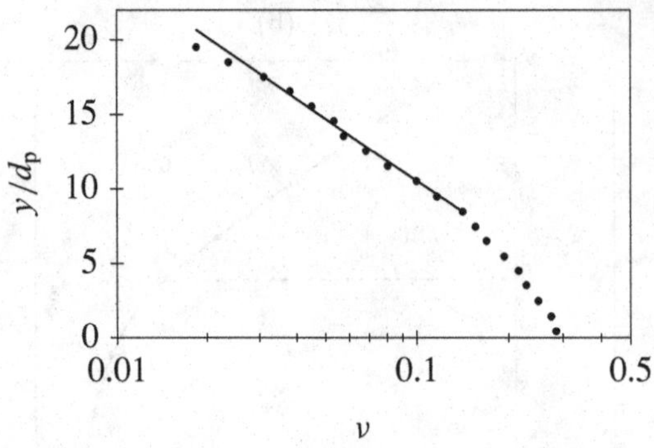

Figure 8.20. Data from Azanza et al. (1999) of the solids fraction profile for steel spheres of diameter 3 mm flowing in a chute. The flow rate was 1100 spheres/s and the chute inclination was 23°. The gap between the transparent front and back walls of the chute was small enough to hold only a single layer of spheres between them. The straight line is fitted through the points that appear to lie on the exponential part of the profile.

in the dense flow than in the loose flow. Both variables approach constant values as y becomes large, according to the asymptotic solution in (8.109) and (8.110).

Figure 8.20 gives experimental evidence of the exponential decay of ν when it is small. The data were reported by Azanza et al. (1999) for the flow of a monolayer of steel shots in a chute.

8.3.5.1 An alternative boundary condition. Enforcing the free-surface conditions at infinity makes the problem mathematically well posed, but it is unsatisfactory on physical grounds: experiments show a sharp interface for dense flows (Johnson et al., 1990; Nott and Jackson, 1992; Pouliquen, 1999; Ancey, 2001), and despite the presence of a diffuse "saltating" layer at the top for loose flows, the flow depth is finite. The difficulty can be overcome if we recognize that the continuum approximation breaks down at the free surface: the continuum equations (8.105)–(8.107) do not permit a finite density in a plane of vanishing normal stress. Johnson et al. (1990) proposed a set of modified interface boundary conditions that bridges the continuum with the discrete nature of particles at the interface. They treated the uppermost layer of particles separately from the continuum, and determined the stress in the medium immediately below it by balancing the forces acting of the layer. Referring to Fig. 8.21, a balance of the forces on the layer in the x direction gives

$$\sigma_{yx} = -\rho_p \nu g \, d_p \sin\theta \tag{8.112}$$

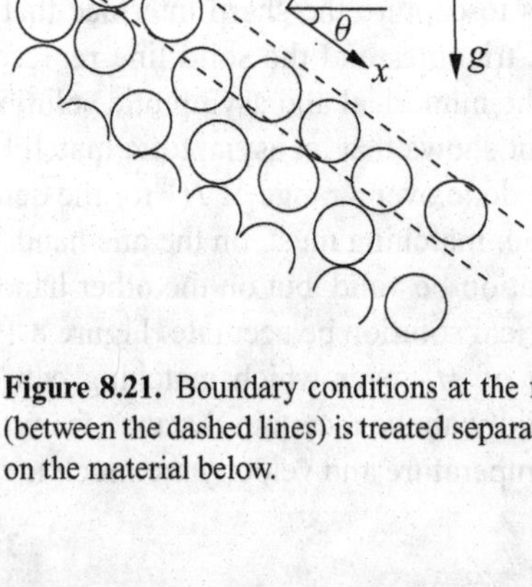

Figure 8.21. Boundary conditions at the free interface. The uppermost layer of particles at the interface (between the dashed lines) is treated separately from the continuum, and its weight results in a stress exerted on the material below.

and in the y direction it gives

$$\sigma_{yy} = \rho_p \nu g \, d_p \cos\theta \tag{8.113}$$

These conditions, in addition to the zero energy flux condition $\mathrm{d}T/\mathrm{d}y = 0$, hold at $y = H$. They are self-consistent, in that ν at the interface is not specified, but determined as part of the solution. Substituting the constitutive relation for the stress, and scaling all variables, we get the nondimensional form of the modified interface boundary conditions at $y^* = 1$,

$$\mathcal{P}(\nu)T^* = H^{*2} \, \nu \cos\theta \tag{8.114}$$

$$\eta(\nu)\sqrt{T^*}\,\frac{\mathrm{d}v^*}{\mathrm{d}y^*} = H^{*2} \, \nu \sin\theta \tag{8.115}$$

$$\frac{\mathrm{d}T^*}{\mathrm{d}y^*} = 0 \tag{8.116}$$

8.3.5.2 Numerical solutions. With the modified interface boundary conditions, the solution can be determined if H^* is prescribed; the loading and the flow rate can be computed from the solution. As the equations are nonlinear, the solution must in general be obtained numerically. As an illustrative example, we consider the solutions for the parameter set used in Fig. 8.19. The dilute and dense solutions in this figure, obtained by the numerical-asymptotic method, are for $m_\ell^* = 1.713$ and 8.093, respectively. It is found that the same loading is obtained using the modified interface boundary conditions with $H^* = 30$ and $H^* = 12$, respectively. The filled and open circles in Fig. 8.19 are the numerical solutions obtained using the modified interface boundary conditions. For both values of the loading, the solutions are in excellent agreement with the numerical-asymptotic solution. Thus the modified interface boundary conditions capture with accuracy all the features of the flow, without requiring a more elaborate procedure of asymptotic matching. Moreover, they are physically more realistic, as the flow depth is finite. We shall therefore use these boundary conditions in the remainder of this section.

The nature of the solution depends on the three material parameters e_p, e_w, and φ, and the flow parameter H^*. For the purpose of our discussion, we retain the classification of the parameter sets defined in the high-density theory (see §8.3.3), i.e., a given set of parameters is said to represent a smooth chute if $\ell/m < L/M$, and a rough chute if $\ell/m > L/M$, where ℓ, m, L, and M are defined in §8.3.3. We shall see below that the other constraints required by the high-density theory for the existence of solutions are not necessary in the kinetic theory. For a given angle of inclination θ, we shall refer to the dimensionless flow depth predicted by the high-density theory as H_{hd}^*.

We first consider the combination of parameters $e_p = 0.91$, $e_w = 0.91$, and $\varphi = 0.25$, which by our definition above constitutes a smooth chute. In the high-density theory (Case 2 in §8.3.3), solutions exist only in the range of inclination $7.76° < \theta < 14.41°$ for this parameter set. Figure 8.22a shows the variation of the dimensionless mass flow rate $\dot{m}^*$ (defined in (8.81)) with the dimensionless flow depth at an inclination of $\theta = 14°$. The flow rate increases with H^*, reaches a maximum, and then decreases to vanish altogether at $H^* = H_{\mathrm{hd}}^* - 1 \approx 16.6$. The subtraction of unity from the H_{hd}^* accounts for the interface layer of thickness d_p that provides the modified interface boundary conditions. We see from Figs. 8.22b–d that as H^* approaches $H_{\mathrm{hd}}^* - 1$, the solids fraction approaches ν_{drp} throughout the flowing layer, and the shape of the temperature and velocity profiles approach that of the high-density solution, shown in Fig. 8.18. Thus the kinetic theory asymptotes to the high-density solution, shown by the dotted line in Fig. 8.22a. This, however, is a drawback of the theory, as it predicts solutions only for a limited range of H^*; as the data in §8.3.1 show, experiments do not indicate an upper limit in the flow depth or the flow rate.

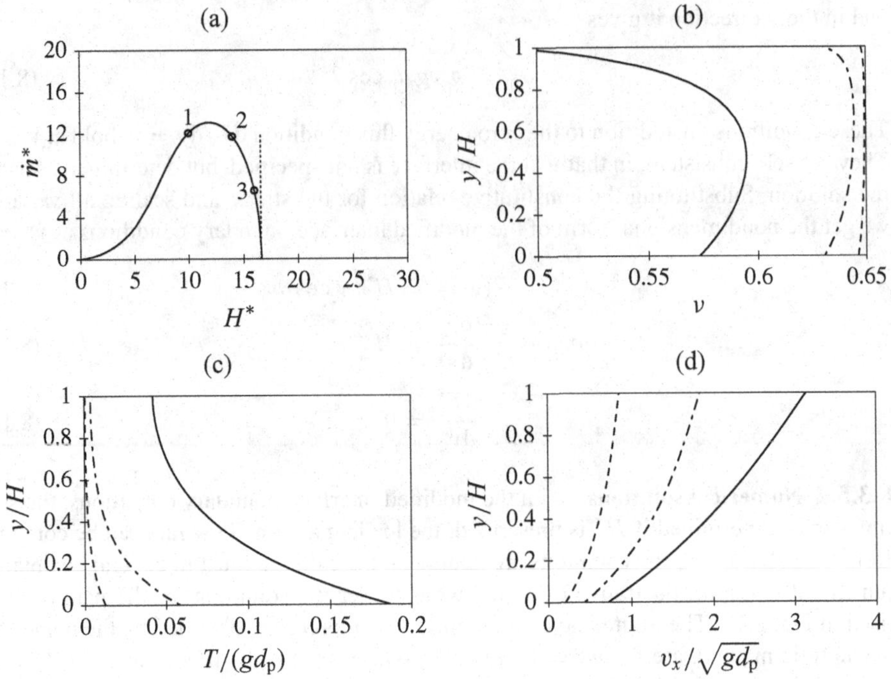

Figure 8.22. (a) The mass flow rate as a function of the flow depth for a "smooth" chute ($e_p = 0.91$, $e_w = 0.91$, $\varphi = 0.25$) inclined at an angle of 14° from the horizontal. The dotted line is the high-density solution. (b)–(d) Profiles of the flow variables at the points labeled 1(solid lines), 2 (dash-dot lines), and 3 (dashed lines) in panel (a).

Next, we consider the set of parameters $e_p = 0.91$, $e_w = 0.8$, $\varphi = 0.85$, which by our definition constitutes a rough chute. The high-density theory (Case 1 in §8.3.3) yields solutions only for $16.06 < \theta < 19.98°$. Figure 8.23 shows the results for $\theta = 18.25°$. The dotted line in panel (a) shows the flow depth, $H^*_{hd} - 1 \approx 18.1$, of the high-density solution. Unlike the smooth chute, the kinetic theory solution does not asymptote to the high-density solution as $H^* \to H^*_{hd} - 1$. An interesting feature in the flow curve is that it exhibits a fold; the point at which the curve turns around (marked "2" on the plot) is called a *turning point* (Strogatz, 1994, p. 47). The equations are singular at a turning point, as a small change in the value of the parameter (here H*) in one direction does not yield a unique solution, and in the other direction yields no solutions at all. Thus both H^* and $\dot{m}$ have a limited range, and hence the flow curve for a rough chute is also not in conformity with experimental observation. The profiles of the variables at the points labeled 1, 2, and 3 in Fig. 8.23a are shown in panels (b)–(d); the solids fraction is much lower than in the smooth chute, and decreases monotonically with distance from the base. Beyond some distance from the base, ν decays exponentially and T^* and $v*$ reach constant values, which are all features of the asymptotic solution for low-density (see (8.108) and (8.110)). The density near the interface remains low no matter how large the flow depth, which is why the solution never approaches the high-density solution.

Finally, we consider angles outside the range defined by (8.95) for smooth chutes, and by (8.91) for rough chutes. Unlike the high-density theory, the kinetic theory does not necessarily preclude solutions for such inclinations. For the smooth chute parameter set used above, the dashed line in Fig. 8.24 shows the solutions for $\theta = 15°$. Recall that the maximum inclination allowed by the high-density theory for this parameter set is 14.41°. For this case, the flow rate increases monotonically with H*, and appears to grow linearly for large H*. Profiles of the flow variables for $H^* = 30$, shown by the solid lines in Fig. 8.19

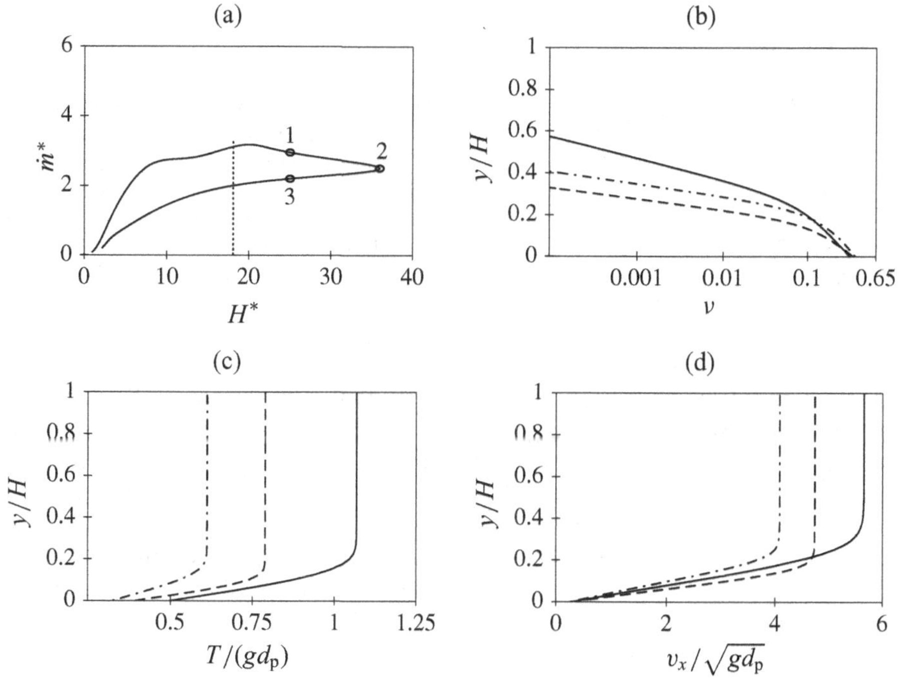

Figure 8.23. (a) The mass flow rate as a function of the flow depth for a rough chute ($e_p = 0.91$, $e_w = 0.8$, $\varphi = 0.25$) at an inclination of 18.25° (solid line). The dotted line is the high-density solution. (b)–(d) Profiles of the solids fraction, grain temperature, and velocity, respectively, for the points labeled 1 (solid lines), 2 (dashed lines), and 3 (dash-dot lines) in panel (a). The point labeled 2 in panel (a) is the limit point.

reveal that the material is loose and fast flowing, and the density decays rapidly away from the base. Figure 8.24 also shows the results for $\theta = 21°$ for the rough chute (solid line), which is above the maximum angle of 19.98° allowed by the high-density theory. Here too, as in 8.23, the flow curve exhibits a turning point, restricting the range of $\dot{m}$ and H^*.

Though we have by no means attempted a full parametric exploration of the theory, the four cases we have examined illustrate its main features. We find that the kinetic theory is a significant improvement over the high-density theory, as it provides solutions for a range of the flow depth for each angle of inclination. However, it is still flawed. With the exception of the smooth chute at large inclinations, it predicts an upper bound on H*, which is contrary to experimental observation. Even when there is no upper bound on H*, such as in the dashed line in Fig. 8.24, the nature of the flow is quite different from what is observed. Johnson et al. (1990) found the material to be dense and slowly shearing at large H*, but the theory predicts a relatively loose, rapidly shearing flow with a very dilute saltating layer (solid lines in Fig. 8.19).

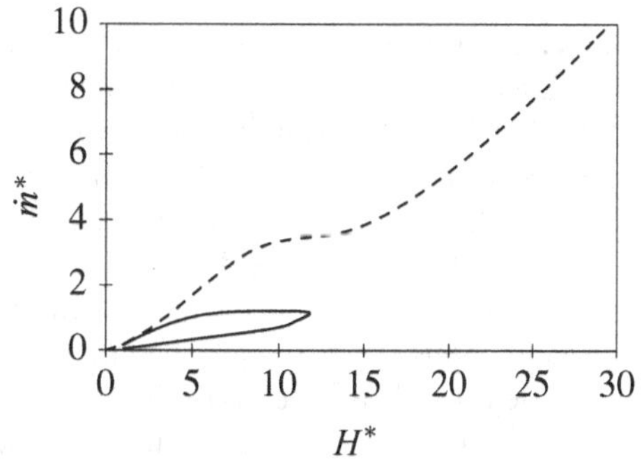

Figure 8.24. (a) The mass flow rate as a function of the flow depth for a rough chute (parameters as in Fig. 8.23) at an inclination of 21° (solid line), and a smooth chute (parameters as in Fig. 8.22) at an inclination of 15° (dashed line).

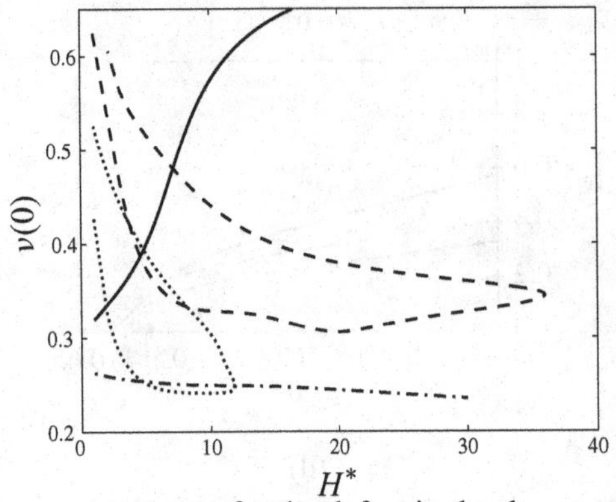

Figure 8.25. The variation of the solids fraction at the chute base with the flow depth for the four sets of parameter values considered in figures 8.22–8.24. The solid and dash-dot lines are for the smooth chute with inclinations of 14° and 15°, respectively, and the dashed and dotted lines are for the rough chute with inclinations of 18.25° and 21°, respectively.

A reason for the defect in the theory suggests itself if we examine the variation of the density at the base as a function of H^* for each of the four parameter sets considered so far. The results, displayed in Fig. 8.25, show that in some cases, the solids fraction at the base exceeds that of loose random packing ($\nu \approx 0.56$, see §1.32) for a range of H^*. At such high densities, some key assumptions of the theory, such as binary collisions and molecular chaos, are likely to break down. Thus a modification to the theory is perhaps necessary if such situations are to be described. A few examples of modified theories that have been proposed in the literature are discussed in Chapter 10.

8.4. STABILITY OF RAPID SHEAR FLOWS

In the preceding sections we have considered steady, fully developed flow in simple geometries. However, experimental observations have shown that steady, fully developed flow cannot be realized under some circumstances, and sometimes a variety of interesting dynamics is observed. In flow down inclined chutes, for example, undulations of the interface have been observed (Johnson, 1987b; Johnson et al., 1990; Forterre and Pouliquen, 2003). In the same geometry, Forterre and Pouliquen (2001) observed periodically spaced elongated vortices for dense flows, which are shown in Fig. 8.26. Nonuniform microstructures and density stratification have been observed in particle dynamics simulations of sheared granular flows (Hopkins and Louge, 1991; Tan, 1995), which are believed to be a result of instabilities of the steady state. Thus there appears to be a rich variety of instabilities in granular flows, and the application of linear stability analysis to these problems is a worthwhile exercise. In this section, we consider as illustrative examples the stability of unbounded plane shear flow, followed by the stability of plane Couette flow, both in the absence of gravity. The stability of other problems, such as flow in vertical channels (Wang et al., 1997), inclined chutes (Forterre and Pouliquen, 2002) and plane Couette flow in the presence of gravity (Alam et al., 2005), have also been analyzed, and we refer the reader to the original works for details. A brief introduction to linear stability theory is given in Appendix I.

8.4.1. Stability of Unbounded Plane Shear Flow

The first investigations of the linear stability of rapidly shearing granular materials (Savage, 1992; Babić, 1993; Schmid and Kytömaa, 1994) considered the base state of uniform unbounded shear, i.e., constant shear rate in the absence of boundaries. Assuming that the base state flow is such that the velocity is in the x-direction, and velocity gradient in the y-direction, the velocity field has the form $\mathbf{v}^0 = (\dot{\gamma}\, y,\ 0,\ 0)$, where $\dot{\gamma}$ is the shear rate. As the state is assumed to be spatially uniform, the solids fraction and temperature are also constants. Thus the base state has translational symmetry in all directions with respect to the density, shear rate and grain temperature fields. However, the perturbation fields do not possess the same symmetry, because the coefficients of some advective terms in

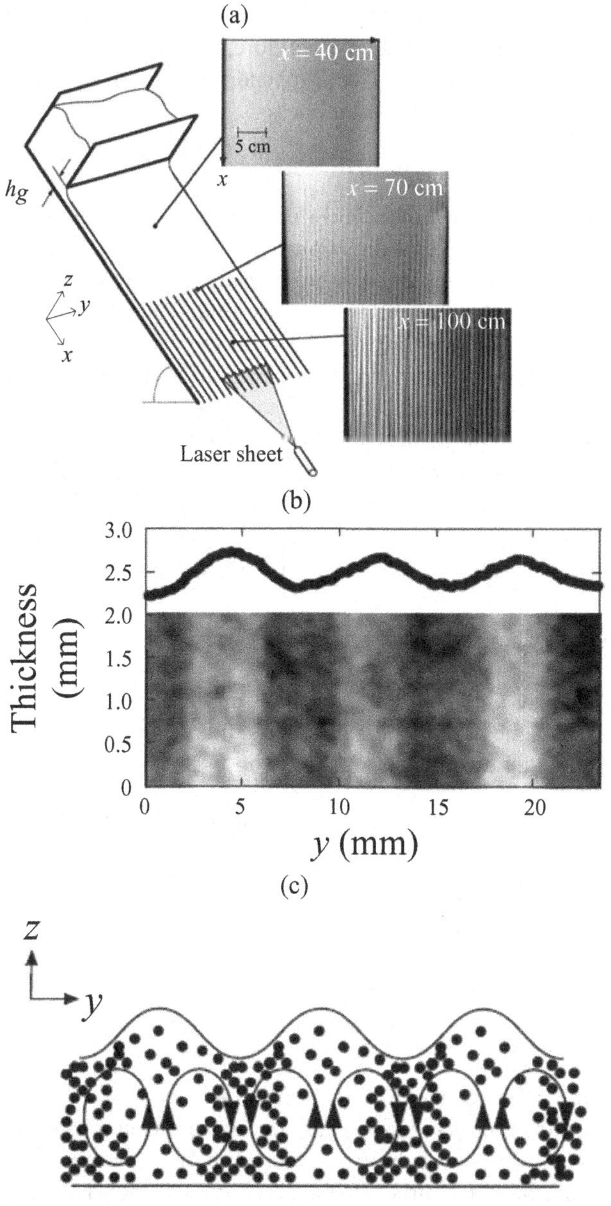

Figure 8.26. Experimental observations of an instability in chute flow. The grains (sand or glass beads) are fed in a dense state from a reservoir at the upper end of the chute. Panel (a) shows a schematic of the assembly. When the material is illuminated from the side and viewed from above, striations aligned with the flow (x) direction are observed in the downstream end of the chute, shown in the insets of panel (a). The solid line in (b) is the profile of the layer thickness, measured by bouncing a laser sheet off the free surface at grazing angle, as shown in (a). Superimposed on this plot is the image obtained by illuminating the chute from below, through the transparent glass base; the light and dark regions correspond to regions of low and high solids fraction. Panel (c) shows a schematic of the secondary flow in the y–z plane. (Reprinted figures with permission from Forterre and Pouliquen, 2001. Copyright 2001, American Institute of Physics.)

the linearized equations depend explicitly on y. For example, in the linearized continuity equation

$$\frac{\partial \nu'}{\partial t} + u_x^0 \frac{\partial \nu'}{\partial x} + \nu^0 \frac{\partial v_x'}{\partial x} + \nu^0 \frac{\partial v_y'}{\partial y} + + \nu^0 \frac{\partial v_z'}{\partial z} = 0 \tag{8.117}$$

the coefficient of $\partial v'/\partial x$ is $\dot{\gamma}\, y$. Here the superscript "0" is used to denote a base state variable, and the prime is used to denote a perturbation. Therefore, the linearized equations do not admit solutions in the form of normal modes (see Appendix I) in the y-direction. It turns out that the solution is a slight variant of normal modes,

$$\mathbf{X}' = \hat{\mathbf{X}}(t)e^{ik_x x + ik_y(t)y + ik_z z} \quad \text{with } k_y(t) = k_{y_0} - k_x \dot{\gamma}\, t \tag{8.118}$$

where $\mathbf{X} \equiv (\nu, v_x, v_y, v_z, T)$ is the vector of all the field variables, and k_x, k_{y0}, and k_z are the (constant) wavenumbers for the three spatial directions (see Appendix I). This form of the perturbation works because the advective terms mentioned above are canceled by a part of the time derivative $\partial \mathbf{X}'/\partial t$, and hence the coefficients are independent of y (see Problem 8.6). Note that the magnitude of k_y increases as $\dot{\gamma}\, t$; as the wavenumber vector $\mathbf{k} \equiv (k_x, k_y, k_z)$ is normal to the perturbation wavefronts (surfaces of constant phase), the above form of the perturbation gives plane waves that are rotated by the mean shear, and as $t \to \infty$ the wavefronts are aligned in the x-direction. Perturbations of the form (8.118) are called "Kelvin modes", because they were first introduced by Thomson (1887, later titled "Lord Kelvin"), in a study of the stability of a sheared Newtonian fluid. They were adopted for the study of instability in granular shear flow by Savage (1992). Substitution of the Kelvin mode in the linearized equations yields a set of linear equations of the form

$$\frac{\mathrm{d}\hat{\mathbf{X}}}{\mathrm{d}t} = \mathbf{A}\hat{\mathbf{X}}, \quad \text{where } \mathbf{A} = \mathbf{A}_0 + \mathbf{A}_1 k_x t + \mathbf{A}_2 k_x^2 t^2 \tag{8.119}$$

is the coefficient matrix, with $\mathbf{A}_0$, $\mathbf{A}_1$, and $\mathbf{A}_2$ being constant matrices. The solution of (8.119) is not of the form $\hat{\mathbf{X}}(0)e^{st}$, as the coefficient matrix $\mathbf{A}$ is a function of time. The full solution must be obtained numerically for a given initial condition $\hat{\mathbf{X}}(0) = \hat{\mathbf{X}}_0$. However, if $k_x > 0$, Alam and Nott (1997) showed that the long-time asymptotic behavior can be obtained analytically. For all initial conditions, the asymptotic behavior is (see Problem 8.7)

$$\hat{\nu} \sim e^{-Ct}, \quad (\hat{v}_x, \hat{v}_y, \hat{v}_z, \hat{T}) \sim \frac{1}{t}e^{-Ct} \quad \text{as } t \to \infty \tag{8.120}$$

where

$$C \equiv \frac{\nu^0\, \mathcal{P}_\nu(\nu^0)\sqrt{T^0}}{d_{\mathrm{p}}\left(\eta_{\mathrm{b}}(\nu^0) + 4\eta(\nu^0)/3\right)} > 0$$

and $\mathcal{P}_\nu \equiv \mathrm{d}\mathcal{P}/\mathrm{d}\nu$. Thus all perturbations with $k_x > 0$ decay as $t \to \infty$.

This, however, does mean that the flow is stable, because the above asymptotic behavior is not valid for the singular limit $k_x = 0$. In this limit, the Kelvin modes reduce to normal modes, and $\mathbf{A} = \mathbf{A}_0$ becomes a constant matrix. Hence stability is determined by the eigenvalues s of $\mathbf{A}_0$. As $\mathbf{X}$ is a five-dimensional vector, there are five eigenvalues for each set of wavenumbers. Stability is determined by only two parameters, namely, the solids fraction ν^0 and the particle coefficient of restitution e_{p}.

Wang et al. (1996) found that, with the exception of a narrow range of intermediate solids fraction, the flow is unstable for $e_{\mathrm{p}} < 1$. The nature of the instability is quite different in the lower and higher range of ν^0. For the higher range of ν^0 in which the flow is unstable, the fastest growing normal mode is one with $k_z = 0$, $k_y > 0$. Modes with $k_x = 0$, $k_z = 0$ correspond to a sinusoidal variation of the variables in the y-direction, with no variation in the x- and z-directions. They can be thought of as alternating layers in the x–z plane with relatively high and low values of the flow variables, stacked in the y-direction, and hence they have been called "layering modes" (Wang et al., 1996). For the lower range of ν^0 in which the flow is unstable, the fastest growing normal mode is one with $k_y = 0$, $k_z > 0$. This is also a layering mode, but the layers are in x–y plane, stacked in the z-direction.

The range of ν^0 for which the flow is stable increases with e_p, and encompasses the entire range of solids fraction when $e_p = 1$. Thus the instability is caused by the inelasticity of grain collisions. However, the compressibility of the material is an essential factor for instability: if the solids fraction of the material is treated as a constant, there is no instability (Alam and Nott, 1997). These features suggest the following physical mechanism of the instability: a perturbation that causes a spatial variation in, say, the density leads to an imbalance between the production and dissipation of fluctuational energy, such that more energy is dissipated than produced in the dense regions, and vice versa in the loose regions. This lowers the temperature in the dense regions, which in turn further increases the density there; the reverse happens in the loose regions.

8.4.2. Stability of Plane Couette Flow

In §8.2.3, we had discussed the application of the kinetic theory to the problem of steady, fully developed plane Couette flow (see Fig. 8.4) in the absence of gravity. We now consider the stability of these solutions.

For this problem, the stability analysis proceeds in the customary manner of seeking normal mode solutions for the perturbations, i.e., of the form given in (I.3), which reduces the linearized equations (I.1) to ordinary differential equations in y. They may be solved by a variety of methods. For ease of exposition, let us consider one method, which is to discretize the independent variable, $y = (y_1, y_2, \ldots, y_n)$, and approximate derivatives with respect to y by finite differences. For example, $\mathrm{d}\hat{T}/\mathrm{d}y$ at y_i is replaced by the first-order forward difference $(\hat{T}_{i+1} - \hat{T}_i)/(y_{i+1} - y_i)$, where the notation $\hat{T}_i \equiv \hat{T}(y_i)$ is used. This reduces (I.1) to a system of homogeneous linear algebraic equations

$$s\hat{\mathbf{X}}^d = \mathbf{L}^d(k_x, k_z)\cdot\hat{\mathbf{X}}^d \tag{8.121}$$

which is an eigenvalue problem. Here $\mathbf{L}^d$ is the matrix resulting from discretizing of the linear operator, s is the eigenvalue that must be determined to assess stability of the normal mode (see Appendix I), and the eigenvector $\hat{\mathbf{X}}^d$ is the vector of variables at the n discrete points. The details of the discretization procedure, the manner in which the boundary conditions are incorporated, and some intricacies regarding discretization for compressible flows, are not given here but may be found in Alam and Nott (1998). For each normal mode, the number of eigenvalues equals the number of discretized equations. The accuracy of the numerical procedure is ensured by having a large enough number of discretization points n, so that the eigenvalues do not change significantly on further increasing n.

The stability of bounded Couette flow (Fig. 8.4) was first analyzed by Wang et al. (1996). However, their numerical solution of the base state, and hence the stability characteristics, was found to be incorrect by Alam and Nott (1998). Both the studies considered only perturbations in the x–y plane, i.e., they set $k_z = 0$. The parameters that determine stability are the material property e_p, the wall properties e_w and φ, the dimensionless Couette gap H^* and the mean solids fraction $\overline{\nu}$. A typical stability map in the $\overline{\nu}$–H^* plane, with other parameters kept constant, is shown in Fig. 8.27a. The curves are contours of constant growth rate $s_r \equiv Re(s)$ of the fastest growing (i.e., most unstable) normal modes. The four types of curves represent four distinct types of normal modes: layering modes, which were introduced in the context of unbounded shear, are perturbations with $k_x = 0$; stationary waves, which are perturbations whose spatial variation remains stationary because $s_i \equiv Im(s)$ vanishes; traveling waves, which are perturbations with $s_i \neq 0$, and therefore translate in the x-direction with phase velocity $-s_i/k_x$; and long waves, which are perturbations with large wavelength in the flow direction, $2\pi/k_x \gg H$.

The curves marked "0" for each type are their *neutral stability* contours, i.e., contours of $s_r = 0$, representing states for which perturbations neither grow nor decay with time.

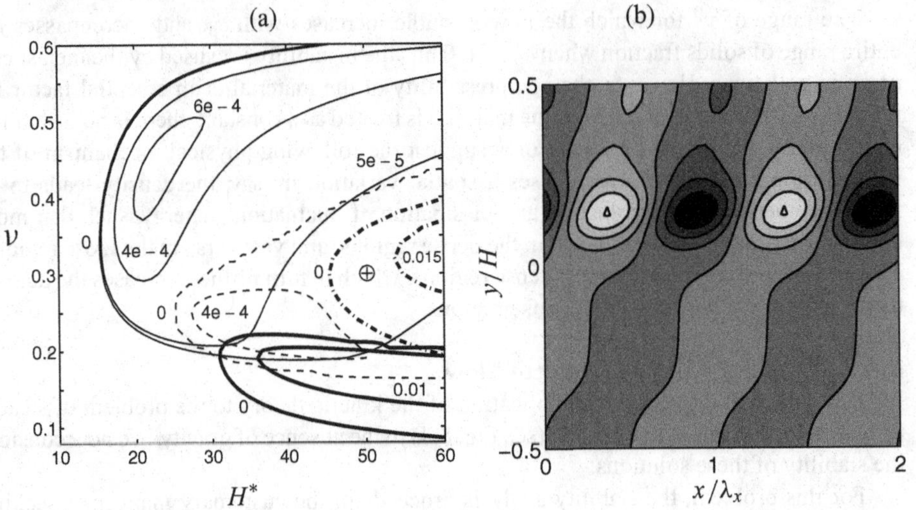

Figure 8.27. (a) The stability map for plane Couette flow (Fig. 8.4) for the parameter set $e_p = 0.8$, $e_W = 0.97$, $\varphi = 0.6$, for which the walls act as sources of fluctuational energy. The thin solid lines are contours of constant growth rate for layering modes ($k_x = 0$), the dotted lines for long-wave modes, the thick solid lines for strong stationary waves, and the thick dashed lines for travelling waves. (b) A gray scale map of the solids fraction perturbation for the fastest growing mode at the point marked ⊕ in (a). The x coordinate is normalised by the wavelength $\lambda_x \equiv 2\pi/k_x$. In the gray scale, black represents the maximum density and white the minimum. The pattern translates in the x-direction with phase velocity 0.03 v_W, where v_W is the velocity of the upper wall. (Reproduced from Alam and Nott, 1998, with permission from Cambridge University Press.)

The regions to the left of the neutral stability contours are stable. We see that the flow is stable if either H^* or $\overline{\nu}$ is small enough, but is unstable otherwise. Stability for small H^* or $\overline{\nu}$ is due to the stabilizing influence of the diffusion of fluctuational energy. If we consider the dimensionless energy balance (see (8.74)), we see that the term representing energy diffusion is multiplied by $1/H^{*2}$; its magnitude increases as H^* decreases. Similarly, when we decrease $\overline{\nu}$ the rate of energy dissipation due to inelastic collisions decreases in magnitude, as it varies as ν^2, and a balance is achieved between the stress work and the gradient of diffusion flux. Energy diffusion acts to even out temperature differences, thereby reducing stratification and stabilizing the flow.

The solids fraction variation $\nu'(x, y) = \mathrm{Re}\left(\hat{\nu}(y)\, e^{ik_x x}\right)$ of the fastest growing perturbation at the point marked ⊕ in the stability map is given in Fig. 8.27b. This instability is a traveling wave with wavenumber $k_x = 2.56/H$ and phase velocity (see Appendix I) 0.03 v_w. The localized buildup of particle density is evident. Indeed, inhomogeneous density distribution is one of the primary manifestations of all the types of instabilities (layering, stationary, or traveling waves). The stability map differs in the details for other parameter values, but the broad features are similar to those in Fig. 8.27a.

As in unbounded shear, the inelasticity of grain collisions and the compressibility of the medium are the causes of instability. These features make the instability of a rapidly shearing granular medium quite different from that of an incompressible Newtonian fluid.

In summary, plane Couette flow of a granular material in the rapid flow regime is linearly unstable, except for a small range of the mean solids fraction $\overline{\nu}$ and the dimensionless Couette gap H/d_p. One of the prominent features of the instability is clustering or stratification of particles. Experimental verification of the theoretical predictions of the base state and the instabilities is lacking, as the density and velocity fields are difficult to measure noninvasively. In chute flow, where measurement of the velocity at the free surface is possible, Forterre and Pouliquen (2002) found qualitative agreement between their

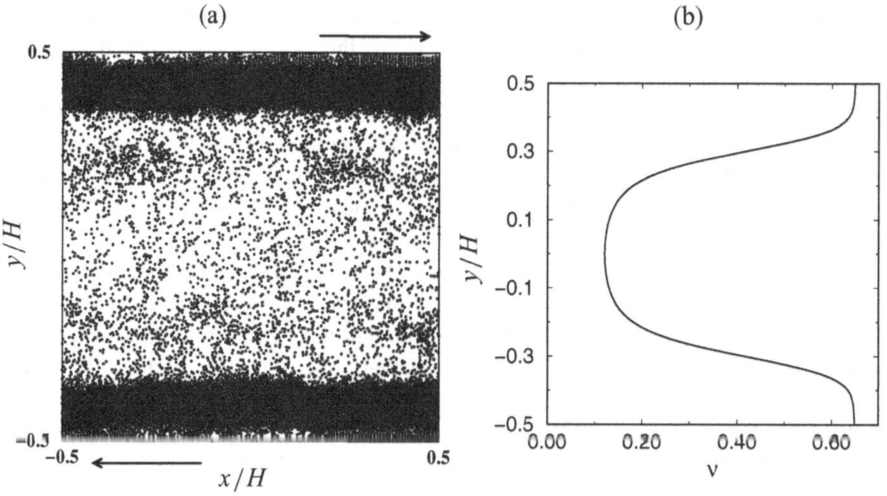

Figure 8.28. (a) Snapshot from a dynamic simulation of plane Couette flow. Particles are represented as dots, and the arrows indicate the direction of motion of the walls. The simulation was performed with 15,000 particles using the parameter values $e_p = 0.8, \overline{\nu} = 0.3, H^* = 198$ and adiabatic walls. (Reproduced from Alam et al., 2005, with permission from Cambridge University Press). (b) The solid fraction profile of the final steady state resulting from a layering mode instability. Parameter values are $e_p = 0.8, \overline{\nu} = 0.35$, $H^* = 71.43$, adiabatic walls. (Adapted from Nott et al., 1999.)

experimental observations of longitudinal vortices and the results of their linear stability analysis. Determination of the density and velocity fields is, of course, possible in dynamic simulations, and several studies have reported clustering (Hopkins and Louge, 1991) and density inhomogeneity in simple shear. Figure 8.28a is an example of density stratification observed in simulations – the particle concentration is high near the walls, and low at the center. The pattern appears to have no variation in the flow direction, suggesting a similarity with the layering modes observed in the stability analysis. The density variation in this snapshot is very similar to that of the final steady state of a layering instability determined by Nott et al. (1999), shown alongside in Fig. 8.28b.

8.5. SUMMARY

In §8.2-8.4, we have applied the equations of motion for rapid granular flows to two simple problems, namely, shear in a plane Couette device and flow in an inclined chute. Boundary conditions at a rigid wall are required to solve these problems, and they were formulated in §8.1 by considering balances of momentum and energy in a thin layer near the wall. A heuristic analysis, in the spirit of the theory of Haff (1983), was first employed to derive the boundary conditions for the high-density limit. In this analysis, the coefficient of particle-wall collisions e_w and the specularity coefficient φ are parameters that characterize the wall. This was followed by the rigorous kinetic theory approach of Jenkins and Richman (1986), where a simple model of the wall roughness was used to derive the boundary conditions for arbitrary density.

When the high-density theory is applied to plane Couette flow in the absence of gravity, it is found that the solution is one of two distinct types, depending on the value of the composite parameter $(1 - e_p^2)/[\varphi(1 - e_w^2)]$; when it is less than a critical value, the walls are a net sink of pseudothermal energy, and the shear rate is finite everywhere. On the other hand, when $(1 - e_p^2)/[\varphi(1 - e_w^2)]$ is greater than the critical value, the walls are a net source of pseudothermal energy, and shearing is confined to thin layers adjacent to the walls. In both cases, the shear and normal stresses vary as the square of the nominal shear rate, but they have a rather complicated dependence on the Couette gap H. Application of the kinetic theory to this problem also yields the two types of solutions, but there are substantial variation of the density within the sheared layer. The density stratification increases with H, and when H becomes large, there are regions where the density approaches that of maximum packing and the shear rate is negligibly small. At such densities, the validity of some of the assumptions of kinetic theory, such as binary instantaneous collisions, and molecular chaos, is suspect, and hence these predictions must be treated with caution.

There is a considerable body of experimental investigation on inclined chute flow. Most of the experimental studies have been conducted on chutes with a roughened base, but there is limited data on chutes with a smooth base. The nature of the flow depends on whether the material is fed to the chute in a loose agitated condition, or as a dense mass. The former results in thin, fast moving layers of low density, whereas the latter gives dense, thick layers that move relatively slowly. The flow rate $\dot{m}$ in the chute is an increasing function of the holdup under some conditions, interesting features such as hysteresis loops and granular jumps are observed. Measurements of the velocity indicate significant slip at the base, especially in loose, agitated flows. The limited data on the density field indicate an increase in the density from the free surface to the base.

As in plane Couette flow, the high-density theory predicts two types of solutions for steady, fully developed flow in inclined chutes, depending on the value of $(1 - e_{\mathrm{p}}^2)/[\varphi(1 - e_{\mathrm{w}}^2)]$. However, its main predictions contradict experimental observations: the theory predicts a unique flow depth H for each angle of inclination, but experiments show that a wide range of H is possible. Further, for each angle of inclination, the theory predicts only a limited range of the flow rate, whereas no evidence of an upper bound on the flow rate has been found in experiments. The kinetic theory fares a little better, but its predictions too depart qualitatively from experimental observations: for a given angle of inclination, it generally allows flows only for a limited range of H (and therefore $\dot{m}$). There are some exceptions, such the dashed line in Fig. 8.24, in which the flow depth does not seem to be restricted, but the flow in this case differs very much from what is observed: the theory predicts a loose, rapidly shearing flow, but observations indicate a dense, thick, slowly shearing flow.

PROBLEMS

8.1. In the same manner as in problem 7.1, determine the values of the coefficients a_5 and a_6 occurring in the boundary conditions derived from the heuristic high-density theory in §8.1.1, by comparing the boundary conditions with:

(a) the simplified form of the boundary conditions derived from kinetic theory, given by (8.31) and (8.32);

(b) the heuristic boundary conditions of Johnson and Jackson (1987), given by (8.8) and (8.9).

8.2. In the solution for plane Couette flow using the high-density heuristic theory (§8.2.1), shearing occurs only in thin layers adjacent to the walls when $H/d_{\mathrm{p}} \to \infty$ in the hyperbolic solution. However, the thickness of the layer must be much larger than the d_{p} if the hydrodynamic description is to be valid. Show that this is the case only when $L/M - \ell/m \ll 1$.

8.3. Using the modified interface boundary conditions (8.112) and (8.113) in place of the condition of vanishing stress at the free surface, determine the solution of the high-density heuristic theory for flow in an inclined chute. How does it differ from the solutions given in §8.3.2? Show that the assumption $s/d_p \ll 1$ can now be satisfied, provided a constraint is placed on the flow rate.

8.4. It was shown in §8.3.5 that the assumption of vanishing solids fraction at the free surface in chute flow, which was made to satisfy the stress-free boundary condition, leads to the conclusion that ν vanishes everywhere in the flowing layer, if the layer is of finite thickness. Instead, if we assume that $T(1) = 0$ and $\nu(1)$ is finite, show using the balance of fluctuation energy that T vanishes everywhere.

8.5. Show that the asymptotic solution for large y^* for chute flow is given by (8.108)–(8.110). As per the discussion in §8.3.5, we expect $\nu \to 0$ as $y^* \to \infty$. In addition, the velocity and temperature approach constant values, say v^*_∞ and T^*_∞, respectively. Hence we may express them as $v^* = \tilde{v}^* + v^*_\infty$ and $T^* = \tilde{T}^* + T^*_\infty$, where $\tilde{v}^* \ll 1$, $\tilde{T}^* \ll 1$. Now obtain the asymptotic solution by linearizing the governing equations (8.105)–(8.107) in the variables ν, $\tilde{v}^*$ and $\tilde{T}^*$, and solving them.

8.6. Show that the appropriate form of the perturbation fields in unbounded shear are the Kelvin modes (8.118). In other words, show that the time dependence of the wavenumber k_y leads to a system of linear differential equations whose coefficients do not depend explicitly on y, thereby resulting in a set of linear ordinary differential equations in t.

8.7. Determine the long-time ($t \to \infty$) asymptotic behaviour of the perturbation variables in uniform, unbounded shear. Proceed in the following manner.

(a) Substitute the Kelvin modes (8.118) in the linearized equations of motion and obtain the equations governing $\hat{\nu}(t)$, $\hat{v}_x(t)$, $\hat{v}_y(t)$, $\hat{v}_z(t)$, and $\hat{T}(t)$. Note that the right-hand sides of all the equations, except the continuity equation, have terms that vary as t^2. In particular, show that the linearized x momentum balance has the form

$$\frac{d\hat{v}_x}{dt} = a_1\hat{\nu} + a_2\hat{v}_x + a_3\hat{v}_y + a_4\hat{v}_z + a_5\hat{T}$$

(b) As $t \to \infty$, show that $a_2 \sim t^2$, while $a_1, a_3, a_4, a_5 \sim t$. It appears reasonable to assume that $a_2\hat{v}_x$ is asymptotically much larger than the other terms on the right-hand side. Hence the asymptotic behavior of v_x may be determined by assuming that the *dominant balance* (see Bender and Orszag (1984, p. 83)) is between $d\hat{v}_x/dt$ and $a_2\hat{v}_x$. Using the same argument for the other components of the velocity and the temperature, determine their asymptotic solutions. What is their functional form?

(c) Substitute the above in the linearized continuity equation, and determine the asymptotic solution for $\hat{\nu}$. Show that $\hat{\nu}$ does not decay with time, but at leading order remains constant. Is this consistent with the assumed dominant balance?

(d) You will find that the correct dominant balance in the momentum and energy balances is between all the terms on the right-hand side of each equation. Thus in the x momentum balance, it is

$$-a_2\hat{v}_x \sim a_1\hat{\nu} + a_3\hat{v}_y + a_4\hat{v}_z + a_5\hat{T}$$

and similar asymptotic expressions ensue for the other components of the momentum balance and the energy balance. Solve them to get expressions for $\hat{v}_x$, $\hat{v}_y$, $\hat{v}_x$, and $\hat{T}$ in terms of $\hat{\nu}$. Substitute these in the continuity equation and determine the asymptotic solution for $\hat{\nu}$, and thereby for the other variables. Show that the solution has the form given by (8.120). Check the validity of the assumed dominant balance, i.e., check whether the terms that were dropped are asymptotically smaller than the ones that were retained.

9

Theory for Rapid Flow of Rough, Inelastic Particles

In Chapter 7, we considered the kinetic theory of a granular gas composed of smooth inelastic spheres. However, the particles in most granular materials one normally encounters are rough, and it is therefore desirable to extend the theory to rough spheres. It is clear that particle roughness will result in the transmission of a tangential impulse during collision. We shall see that the tangential impulse is partly determined by the angular velocities of the colliding pair. As a consequence, the hydrodynamic balance of the angular momentum field, which was implicitly satisfied for smooth particles, must be enforced here.

The dimensional analysis of §7.1 does not depend on whether the particles are smooth or rough. Hence the scaling of the stress with the shear rate, particle size, and density for a granular gas composed of rough particles will be the same as that for smooth particles. However, we shall see that the form and symmetry of the stress tensor is different and that additional hydrodynamic variables are required to describe the flow of a rough granular gas.

Pidduck (1922) was the first to develop a kinetic theory for a gas of rough spherical molecules. He considered "perfectly rough" spheres, which conserve the total kinetic energy of a particle pair during collision. He determined the equilibrium distribution function and applied the Chapman–Enskog analysis to determine the first correction from equilibrium for a *dilute* gas; his analysis is presented (with a few corrections) in Chapman and Cowling (1964, chap. 11). McCoy et al. (1966) extended the analysis to dense gases, and Lun (1991) further extended it to a dense gas of nearly elastic, nearly perfectly rough spheres. Although Lun's analysis is claimed to hold for particles of arbitrary roughness, it is in fact correct only for nearly perfectly rough particles. Recently, Goldhirsch et al. (2005a) considered the contrasting case of "nearly smooth" spheres and applied the Chapman–Enskog method to a *dilute* gas of these particles.

In this chapter, we consider the hydrodynamics of nearly perfectly rough and nearly smooth particles. We shall demonstrate that these are the only two limits for which a hydrodynamic description is valid. In the spirit of the analysis for smooth spheres, a simple collision model that captures the important qualitative effects of particle roughness is chosen in §9.1. The hydrodynamic equations of motion are derived in §9.2. Relations for the fluxes and sources that arise in the equations of motion are derived. We identify the equilibrium distribution in each case, carry out the Chapman–Enskog expansion, and determine the constitutive relations to first order in the inverse Knudsen number K, the inelasticity ϵ, and the roughness parameter ε.

Most of the chapter is devoted to the derivation of the hydrodynamic equations and of the constitutive relations for the fluxes and sinks of certain quantities that occur in the hydrodynamic equations. The reader who is not interested in the derivation, but only in the final result, may proceed directly to the relevant section. The equations of motion are summarized at the end of §9.2. The constitutive relations are given in §9.4.1 and §9.4.2 for nearly perfectly rough and nearly smooth particles, respectively.

9.1. COLLISION MODELS FOR ROUGH PARTICLES

The simplest hard-sphere collision model that incorporates particle roughness was proposed by Jenkins and Richman (1985), as an extension of the collision model for smooth spheres (7.3). In this model, collisions are characterized by two parameters: the coefficient of restitution e_p, introduced in §7.1.1, and a roughness coefficient β. Consider the collision of two spheres 1 and 2 (Fig. 9.1), each of diameter d_p, mass m_p, and moment of inertia I. Before collision, the spheres have translational velocities $\mathbf{c}_1$ and $\mathbf{c}$, and angular velocities about their own axes,† or "spin," $\boldsymbol{\omega}_1$ and $\boldsymbol{\omega}$. We define $\mathbf{h}$ as the relative velocity *of the point of contact* of particle 1 with respect to that of particle 2,

$$\mathbf{h} = \mathbf{g} - \frac{d_p}{2}\mathbf{k} \times (\boldsymbol{\omega}_1 + \boldsymbol{\omega}) \tag{9.1}$$

where $\mathbf{g}$ is the relative translational velocity $\mathbf{c}_1 - \mathbf{c}$. The collision model relates the normal and tangential components of $\mathbf{h}$ after collision to the respective components before collision, i.e.,

$$\mathbf{k}\cdot\mathbf{h}' = -e_p\,(\mathbf{k}\cdot\mathbf{h}) \tag{9.2}$$

$$(\mathbf{k}\times\mathbf{h}')\times\mathbf{k} = -\beta\,(\mathbf{k}\times\mathbf{h})\times\mathbf{k} \tag{9.3}$$

the primed quantities referring to postcollisional velocities. The coefficient of restitution is restricted to the range $0 \le e_p \le 1$ (as for smooth particles), and for reasons that will become apparent shortly, the roughness coefficient is restricted to the range $-1 \le \beta \le 1$. The limit of smooth particles, in which a collision leaves the tangential component of the relative velocity at the point of contact unchanged, is achieved by setting $\beta = -1$. Increasing β from -1 corresponds to an increasing degree of surface friction and a decreasing tangential relative velocity after impact. When $\beta = 0$, the tangential relative velocity is reduced to zero, and when $\beta > 0$, the direction of the tangential relative velocity is reversed after collision. The latter case can be observed in a child's toy called "super ball" or "crazy ball": when a spinning ball strikes a surface, the direction of its spin is reversed after collision and the ball executes a zigzag motion after successive bounces. When $\beta = 1$, the tangential relative velocity is reversed with its magnitude remaining unchanged; this limit is referred to in the literature as "perfectly rough" particles. It was proposed by Bryan (1894, cited in Chapman and Cowling, 1964, p. 199) as a model for gases that can interconvert translational kinetic energy and other modes of internal energy. A physical picture of collisions in this limit was given by Pidduck (1922):

> Imagine two spheres to collide and grip each other, so as to bring the points of contact to relative rest. A small elastic deformation is produced, which we suppose to be released immediately afterwards, the force during release being equal to that at the corresponding stage of compression. Thus the relative velocity of the points of contact is reversed by collision.

In general, collisions are said to be of the "sticking" type when $0 < \beta \le 1$ and of the "sliding" type when $-1 \le \beta \le 0$ (Jenkins, 1992; Walton, 1993b; Herbst et al., 2000).

† Here the axes of a particle refer to the axes of a coordinate frame whose origin coincides with the particle center and whose axes are parallel to those of the laboratory reference frame.

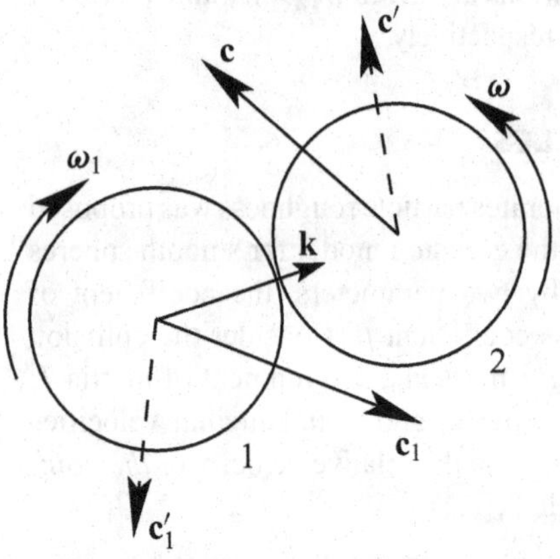

Figure 9.1. A collision between two rough, inelastic spheres. The translational velocities of the spheres before collision are $\mathbf{c}_1$ and $\mathbf{c}$, respectively, and their spin (i.e., their angular velocities about their own respective axes) are $\boldsymbol{\omega}_1$ and $\boldsymbol{\omega}$, respectively. The unit vector along the line joining the center of particle 1 to the center of particle 2 is $\mathbf{k}$. The translational velocities after collision are indicated by the dashed arrows.

Within the range of values for e_p and β mentioned above, the postcollision translational velocities and spin are given by

$$\mathbf{c}' = \mathbf{c} + \mathbf{J}; \qquad \mathbf{c}_1' = \mathbf{c}_1 - \mathbf{J} \tag{9.4}$$

$$\boldsymbol{\omega}' = \boldsymbol{\omega} - \frac{m_p d_p}{2I}\mathbf{k}\times\mathbf{J}; \qquad \boldsymbol{\omega}_1' = \boldsymbol{\omega}_1 - \frac{m_p d_p}{2I}\mathbf{k}\times\mathbf{J} \tag{9.5}$$

where $\mathbf{J}$ is the impulse per unit mass on particle 2. Note that the spin angular momentum $I\boldsymbol{\omega}$ of the colliding pair is not conserved during collision, but the total angular momentum relative to the lab frame $I\boldsymbol{\omega} + m_p \mathbf{r}\times\mathbf{c}$, where r is the position vector of the particle center, is conserved. From (9.2)–(9.5), we find the impulse to be

$$\mathbf{J} = \eta_1(\mathbf{k}\cdot\mathbf{h})\,\mathbf{k} + \eta_2(\mathbf{k}\times\mathbf{h})\times\mathbf{k} \tag{9.6}$$

where $\eta_1 \equiv \frac{1}{2}(1+e_p)$, $\eta_2 \equiv \frac{1}{2}(1+\beta)\hat{I}/(1+\hat{I})$, and $\hat{I} \equiv 4I/(m_p d_p^2)$ is the dimensionless moment of inertia of the particles. Substitution of the above in (9.4) and (9.5) gives the postcollision velocities. Note that the normal and tangential components of the impulse, which are the first and second terms, respectively, on the right-hand side of (9.6), are independent; they depend on the normal and tangential components of $\mathbf{h}$, respectively.

From (9.4)–(9.6), we find that the change in translational kinetic energy of the pair during collision is

$$\frac{\Delta E}{m_p} = -\frac{1}{4}(1-e_p^2)(\mathbf{k}\cdot\mathbf{g})^2 + \eta_2^2\,|\mathbf{k}\times\mathbf{h}|^2 - \eta_2\,(\mathbf{k}\times\mathbf{h})\cdot(\mathbf{k}\times\mathbf{g}) \tag{9.7}$$

and the change in rotational kinetic energy $\Delta E_r \equiv \Delta(\frac{1}{2}I\omega^2 + \frac{1}{2}I\omega_1^2)$ is

$$\frac{\Delta E_r}{m_p} = \frac{1}{\hat{I}}\eta_2^2\,|\mathbf{k}\times\mathbf{h}|^2 - \frac{d_p}{2}\eta_2\,(\mathbf{k}\times\mathbf{h})\cdot(\boldsymbol{\omega}+\boldsymbol{\omega}_1) \tag{9.8}$$

The change in the total kinetic energy $\Delta E_{tot} \equiv \Delta E + \Delta E_r$ is

$$\frac{\Delta E_{tot}}{m_p} = -\frac{1}{4}(1-e_p^2)(\mathbf{k}\cdot\mathbf{g})^2 - \frac{1}{4}\frac{\hat{I}}{(1+\hat{I})}(1-\beta^2)|\mathbf{k}\times\mathbf{h}|^2 \tag{9.9}$$

Note that ΔE and ΔE_r are zero when the particles are elastic and smooth ($e_p = 1$, $\beta = -1$), but not when they are elastic and perfectly rough ($e_p = 1$, $\beta = 1$). In both cases, $\Delta E_{tot} = 0$ or the total kinetic energy is conserved, implying that in the latter case there is transfer of energy between translational and rotational modes. There is dissipation of the total kinetic energy for any other combination of e_p and β.

Another collision model was proposed by Walton (1993b), in which the effects of Coulomb friction and tangential bounce-back are integrated. In the collision model defined by (9.2) and (9.3), a given value of β results in either sticking or sliding collisions, regardless of the particle velocities. Walton (1993b) argued that a realistic model should allow both types of collisions. He proposed that for a given combination of precollision velocities, the type of collision is decided by the angle of incidence θ of the surface velocity, which is the angle between $\mathbf{k}$ and $\mathbf{h}$. Walton assumed that sliding collisions occur for θ greater than a critical angle θ_0 and sticking collisions for θ below θ_0. In other words, the roughness coefficient β is not a constant but depends on θ. Additionally, he proposed that the tangential impulse in a sliding collision arises from Coulomb friction and is therefore proportional to the normal impulse. As a result, the total impulse for a sliding collision is

$$\mathbf{J} = \eta_1(\mathbf{k}\cdot\mathbf{h})\,\mathbf{k} + \mu_{\mathrm{f}}\eta_1\,(\mathbf{k}\cdot\mathbf{h})\frac{(\mathbf{k}\times\mathbf{h})\times\mathbf{k}}{|\mathbf{k}\times\mathbf{h}|} \tag{9.10}$$

where μ_{f} is the friction coefficient. The first term is the normal impulse (compare (9.6)) and the second term the tangential impulse. The impulse for sticking collisions continues to be that given in (9.6) with $\beta = \beta_0$, a positive constant. The value of θ_0 is determined by assuming that $\mathbf{J}$ is continuous at $\theta = \theta_0$. Equating the tangential impulse in (9.6) and (9.10) (the normal impulse is already the same), we get

$$\tan\theta_0 = \frac{\mu_{\mathrm{f}}\eta_1}{\eta_2} \tag{9.11}$$

where we have recognized that $\cot\theta = (\mathbf{k}\cdot\mathbf{h})/|\mathbf{k}\times\mathbf{h}|$. The impulse for both sliding and sticking collisions can be represented by (9.6) if we express β as a function of θ,

$$\beta(\theta) = \min\left[\beta_0,\ -1 + \frac{(1+\hat{I})}{\hat{I}}2\mu_{\mathrm{f}}\eta_1\cot\theta\right] \tag{9.12}$$

The physical interpretation of this model is the following: when the ratio of tangential to normal impulse is below μ_{f}, friction is not mobilized and there is sticking contact. When the ratio equals μ_{f}, friction is mobilized and the particles begin to slide with respect to each other at the point of contact. The sliding lasts only for an infinitesimal period of time, as the particles are assumed to be spheres that are in contact at a single point. Sliding ensures that the ratio does not exceed μ_{f}.

Recently, Kumaran (2006) proposed a "partially rough" collisional model, by the ad hoc combination of the smooth and perfectly rough cases of the model of Jenkins and Richman (1985). The roughness coefficient β depends on the angle of incidence θ, but it can assume only one of the two discrete values -1 or 1. The particles are smooth ($\beta = -1$) if $\theta > \theta_0$ and perfectly rough ($\beta = 1$) if $\theta < \theta_0$. The critical angle θ_0 is an arbitrary constant, which Kumaran took to be $\pi/4$.

The models discussed above are simplifications; the experiments of Goldsmith (1960) and Maw et al. (1976, 1981), and the analysis of Johnson (1987a, pp. 361–369) show that the mechanics of collision is far more complex. To make the models more realistic, the parameters e_{p}, β, and μ_{f} would have to be functions of the magnitude of the total relative velocity $|\mathbf{h}|$ and the angle of incidence θ, and depend on the size of the particles and their elastic and plastic properties (Stronge, 2000, chap. 6). However, the accuracy gained would be at the cost of increasing the complexity of the analysis. We are primarily interested in understanding the qualitative effects of particle roughness, so the simple models described above may suffice; this has been the rationale of all the studies that have used these models. In the rest of this chapter, we adopt the collision model of Jenkins and Richman (1985) (represented by (9.2) and (9.3)), with the parameters e_{p} and β assumed to be constants.

Kinetic theory analyses that have adopted the model of Walton (1993b) may be found in Jenkins and Zhang (2002) and Yoon and Jenkins (2005).

With a collision model at hand, we are now in a position to determine the hydrodynamic equations of motion by following the procedure of §7.3.6–§7.3.13.

9.2. EQUATIONS OF MOTION FOR A GRANULAR GAS OF ROUGH, INELASTIC SPHERES

As discussed in §7.3.6, we may write hydrodynamic equations for particle properties that are either conserved or nearly conserved during a collision. Recall that for nearly elastic, smooth particles, the mass and linear momentum are conserved and the translational kinetic energy is nearly conserved. Any function of the spin is also conserved, because $\boldsymbol{\omega}$ of each particle is unaltered by collisions. This implies that *all* the infinite moments of the spin distribution $\vartheta(\boldsymbol{\omega})$ are conserved. Hence $\vartheta(\boldsymbol{\omega})$ itself is a hydrodynamic variable, though in a trivial sense: it is transported only by advection, and hence $\mathrm{D}\vartheta/\mathrm{D}t = 0$. Moreover, we were not concerned with the evolution of $\vartheta(\boldsymbol{\omega})$ or its moments in §7.3.6 because they do not appear in the balances of mass, linear momentum, and energy.

For rough particles, in addition to the balances of mass and linear momentum, the balance of angular momentum must also be enforced. Unlike the case of smooth particles, the balance of angular momentum has significance: we shall see that the mean spin appears in the hydrodynamic balance of linear momentum. For nearly elastic particles, the kinetic energy is not conserved, but it is nearly conserved in the following two cases: "nearly smooth" particles, for which $(1+\beta)$ is a small parameter, and "nearly perfectly rough" particles, for which $(1-\beta)$ is a small parameter. In both cases, $\varepsilon \equiv (1-|\beta|)$ is a small parameter. For nearly smooth, nearly elastic particles, the translational kinetic energy and rotational kinetic energy (see (9.7) and (9.8)) are individually nearly conserved. Indeed, in this case any function of $\boldsymbol{\omega}$ is nearly conserved, and hence $\vartheta(\boldsymbol{\omega})$ itself is a hydrodynamic variable (Goldhirsch et al., 2005b). However, we shall see that if the expansion of the constitutive relations is stopped at linear order in the small parameters K, ϵ, and ε, only the hydrodynamic balances for the mean spin and the mean rotational kinetic energy are needed; these are uncoupled from the balances for higher moments of $\vartheta(\boldsymbol{\omega})$. For nearly elastic, nearly perfectly rough particles, the translational and rotational kinetic energy are not individually conserved, but the total kinetic energy is nearly conserved (see (9.9)).

Consequently, the hydrodynamic fields for the two cases discussed above are the density $\rho \equiv n m_\mathrm{p}$, the mean velocity $\mathbf{v} \equiv \langle \mathbf{c} \rangle$, the mean spin $\overline{\boldsymbol{\omega}} \equiv \langle \boldsymbol{\omega} \rangle$, the translational temperature $T \equiv \frac{1}{3}\langle C^2 \rangle$, and the rotational temperature $T_r \equiv I\langle \Omega^2 \rangle/(3m_\mathrm{p})$. Here $\mathbf{C} \equiv \mathbf{c} - \mathbf{v}$ is the peculiar velocity and $\boldsymbol{\Omega} \equiv \boldsymbol{\omega} - \overline{\boldsymbol{\omega}}$ is the peculiar spin. For nearly elastic, nearly rough particles, T and T_r are individually not hydrodynamic variables, but the "total" temperature $T_{\mathrm{tot}} \equiv T + T_r$ is.

The average value $\langle \psi \rangle$ of the particle property ψ is defined as

$$n(t,\mathbf{r})\,\langle \psi \rangle(t,\mathbf{r}) = \int \psi f(t,\mathbf{r},\mathbf{c},\boldsymbol{\omega})\,\mathrm{d}\mathbf{c}\,\mathrm{d}\boldsymbol{\omega} \tag{9.13}$$

where $f(t,\mathbf{r},\mathbf{c},\boldsymbol{\omega}) \equiv f^{(1)}$ is the singlet velocity distribution function, whose definition is similar to that in §7.3.1: the number of particles in the volume $\mathrm{d}\mathbf{r}$, velocity interval $\mathrm{d}\mathbf{c}$, and spin interval $\mathrm{d}\boldsymbol{\omega}$, centerd at $\mathbf{r}$, $\mathbf{c}$, and $\boldsymbol{\omega}$, respectively, is $f(t,\mathbf{r},\mathbf{c},\boldsymbol{\omega})\,\mathrm{d}\mathbf{r}\,\mathrm{d}\mathbf{c}\,\mathrm{d}\boldsymbol{\omega}$.

To derive the hydrodynamic balances for these variables, we first extend the Boltzmann equation for rough particles. The state of each particle is now determined by its position, velocity, *and* spin; the phase space comprises the three dimensions each of position, velocity, and spin. We do not include particle orientation in the phase space, because our interest is in spheres, for which orientation is of no consequence. The Boltzmann equation then is a simple extension of (7.42),

$$\frac{\partial f}{\partial t} + \mathbf{c}\cdot\nabla f + \mathbf{b}\cdot\nabla_c f + \frac{m_p}{I}\boldsymbol{\tau}\cdot\nabla_\omega f = d_p^2 \int_{\mathbf{k}\cdot\mathbf{g}>0} \left[\frac{|\mathcal{J}|}{e_p} f_1''(\mathbf{r}+d_p\mathbf{k}) f''(\mathbf{r}) \right.$$
$$\left. \times\, g(\mathbf{r}+d_p\mathbf{k},\mathbf{r}) - f_1(\mathbf{r}-d_p\mathbf{k}) f(\mathbf{r})\, g(\mathbf{r}-d_p\mathbf{k},\mathbf{r}) \right] \mathbf{k}\cdot\mathbf{g}\, d\mathbf{k}\, d\mathbf{c}_1\, d\boldsymbol{\omega}_1 \qquad (9.14)$$

where $\boldsymbol{\tau}$ is the external torque per unit mass on the particles and $\mathcal{J}$ is the determinant of the Jacobian for the transformation $(\mathbf{c}'', \boldsymbol{\omega}'', \mathbf{c}_1'', \boldsymbol{\omega}_1'') \longrightarrow (\mathbf{c}, \boldsymbol{\omega}, \mathbf{c}_1, \boldsymbol{\omega}_1)$. The third term on the left-hand side of (9.14), which was not present in (7.42), represents advection in spin space. It arises because, in the absence of collisions, the spin of a particle changes from $\boldsymbol{\omega}$ to $\boldsymbol{\omega} + (m_p\boldsymbol{\tau}/I)\,dt$ in a time interval dt. In the collision integral on the right-hand side, f_1'' and f'' refer to the probability distributions of the precollision velocities of the inverse collision (see §7.3.2), which are given by

$$\mathbf{c}'' = \mathbf{c} + \frac{\eta_1}{e_p}(\mathbf{k}\cdot\mathbf{h})\,\mathbf{k} + \frac{\eta_2}{\beta}(\mathbf{k}\times\mathbf{h})\times\mathbf{k} \qquad (9.15)$$

$$\mathbf{c}_1'' = \mathbf{c}_1 - \frac{\eta_1}{e_p}(\mathbf{k}\cdot\mathbf{h})\,\mathbf{k} - \frac{\eta_2}{\beta}(\mathbf{k}\times\mathbf{h})\times\mathbf{k} \qquad (9.16)$$

$$\boldsymbol{\omega}'' = \boldsymbol{\omega} - \frac{m_p d_p}{2I\beta}\eta_2\mathbf{k}\times\mathbf{h} \qquad (9.17)$$

$$\boldsymbol{\omega}_1'' = \boldsymbol{\omega}_1 - \frac{m_p d_p}{2I\beta}\eta_2\mathbf{k}\times\mathbf{h} \qquad (9.18)$$

Using (9.15)–(9.18), it is straightforward (though algebraically tedious) to show that $|\mathcal{J}| = 1/(e_p\beta^2)$.

From the Boltzmann equation (9.14), we can derive the Maxwell transport equation for rough particles. We note that total angular momentum relative to the laboratory reference frame $I\boldsymbol{\omega} + m_p\mathbf{r}\times\mathbf{c}$, which is conserved in a collision, is explicitly a function of the spatial position. Therefore, we must derive the transport relation for a particle property ψ that may depend explicitly on the spatial position. We multiply (9.14) by $\psi(t, \mathbf{r}, \mathbf{c}, \boldsymbol{\omega})$ and integrate over all $\mathbf{c}$ and $\boldsymbol{\omega}$. Transforming the integrals in the same manner as in §7.3.5, we get the transport equation

$$\frac{\partial}{\partial t}(n\langle\psi\rangle) + \nabla\cdot(n\langle\mathbf{c}\psi\rangle) - n\langle\frac{\partial\psi}{\partial t}\rangle - n\langle\mathbf{c}\cdot\nabla\psi\rangle = n\langle\mathbf{b}\cdot\nabla_c\psi\rangle + \frac{m_p n}{I}\langle\boldsymbol{\tau}\cdot\nabla_\omega\psi\rangle + \dot{\psi}_{\text{coll}} \qquad (9.19)$$

The expression for the collisional rate of change $\dot{\psi}_{\text{coll}}$ is derived in the same manner as in §7.3.5. It may be written as $-\nabla\cdot\boldsymbol{\theta} + \chi$, with the flux $\boldsymbol{\theta}$ and source χ given by

$$\boldsymbol{\theta} = \frac{d_p^3}{2}\int_{\mathbf{k}\cdot\mathbf{g}>0} \mathbf{k}\,(\psi' - \psi)\,\mathbf{k}\cdot\mathbf{g}\left[1 - \frac{1}{2!}(d_p\mathbf{k}\cdot\nabla) + \cdots\right]$$
$$\times \left[g_0(\nu(\mathbf{r}+d_p\mathbf{k}/2)) f(t, \mathbf{r}+d_p\mathbf{k}, \mathbf{c}, \boldsymbol{\omega}) f(t, \mathbf{r}, \mathbf{c}_1, \boldsymbol{\omega}_1)\right] d\mathbf{k}\, d\mathbf{c}\, d\mathbf{c}_1\, d\boldsymbol{\omega}\, d\boldsymbol{\omega}_1 \qquad (9.20)$$

and

$$\chi = \frac{d_p^2}{2}\int_{\mathbf{k}\cdot\mathbf{g}>0} (\psi' + \psi_1' - \psi - \psi_1)\,\mathbf{k}\cdot\mathbf{g}\, g_0(\nu(\mathbf{r}+d_p\mathbf{k}/2))$$
$$\times f(t, \mathbf{r}+d_p\mathbf{k}, \mathbf{c}, \boldsymbol{\omega}) f(t, \mathbf{r}, \mathbf{c}_1, \boldsymbol{\omega}_1)\, d\mathbf{k}\, d\mathbf{c}\, d\mathbf{c}_1\, d\boldsymbol{\omega}\, d\boldsymbol{\omega}_1 \qquad (9.21)$$

The mass, linear momentum, and translational kinetic energy balances are determined by substituting $\psi = m_p$, $m_p\mathbf{c}$, and $\frac{1}{2}m_p c^2$ in (9.19), yielding

$$\frac{D\rho}{Dt} = -\rho\nabla\cdot\mathbf{v} \tag{9.22}$$

$$\rho\frac{D\mathbf{v}}{Dt} = -\nabla\cdot\sigma + \rho\,\mathbf{b} \tag{9.23}$$

$$\frac{3}{2}\rho\frac{DT}{Dt} = -\nabla\cdot\mathbf{q} - \sigma^{T}:\nabla\mathbf{v} - \Gamma \tag{9.24}$$

Although they remain unchanged from the corresponding balances for smooth particles (see §7.3.6), the constitutive relations for σ, $\mathbf{q}$, and Γ may differ from the corresponding relations for smooth particles.

To derive the angular momentum balance, we substitute $\psi = I\omega + m_p\mathbf{r}\times\mathbf{c}$ in (9.19), to get

$$\frac{I}{m_p}\left[\frac{\partial(\rho\,\bar{\omega})}{\partial t} + \nabla\cdot(\rho\langle\mathbf{c}\,\omega\rangle)\right] + \frac{\partial(\rho\mathbf{w})}{\partial t} + \nabla\cdot(\rho\,\mathbf{c}\,\mathbf{r}\times\mathbf{c}) = \rho\,\mathbf{r}\times\mathbf{b}$$

$$+\,\rho\,\tau - \nabla\cdot\theta(I\omega + m_p\mathbf{r}\times\mathbf{c}) \tag{9.25}$$

where $\mathbf{w} \equiv \mathbf{r}\times\mathbf{v}$. We then subtract from (9.25) the vector (cross) product of $\mathbf{r}$ with the linear momentum balance (7.69) to obtain the balance for the intrinsic angular momentum, or angular momentum due to spin,

$$\frac{I}{m_p}\rho\frac{D\bar{\omega}}{Dt} = -\nabla\cdot\mathbf{M} + \epsilon:\sigma + \rho\,\tau \tag{9.26}$$

where ϵ is the alternating tensor, defined in §A.6.1, and

$$\mathbf{M} \equiv \frac{I}{m_p}\rho\langle\mathbf{C}\Omega\rangle + \theta(I\Omega) \tag{9.27}$$

is the flux of intrinsic angular momentum, usually referred to as the *couple stress*. On a plane with unit normal $\mathbf{n}$, the torque per unit area acting on the material into which $\mathbf{n}$ points is $\mathbf{n}\cdot\mathbf{M}$. Note that the stress asymmetry $\epsilon:\sigma$ is a source of angular momentum; from our discussion at the beginning of this section, $\bar{\omega}$ is a hydrodynamic variable only if the magnitude of $\epsilon:\sigma$ is very small. We shall see that this is indeed the case for nearly perfectly rough and nearly smooth particles. The body couple, if present, has the same effect as the body force in the linear momentum balance: at equilibrium, there is a "hydrostatic" balance with the gradient of tr($\mathbf{M}$). The body couple term may be dropped if $\mathbf{M}$ is defined as the deviation from the hydrostatic part.

In a classical continuum, body couples and surface couples are absent and angular momentum is conserved by securing symmetry of the Cauchy stress σ; the angular momentum balance is therefore not presented in most books on fluid mechanics. The exchange of tangential momentum during collisions in general leads to an asymmetry of the Cauchy stress and a nonzero couple stress. These features are present in continuum mechanical models of *micropolar* or *Cosserat* continua (Jaunzemis, 1967; Eringen and Kafadar, 1976), and the angular momentum balance given there has precisely the form of (9.26).

The balance of rotational fluctuational kinetic energy is determined by substituting $\psi = \frac{1}{2}I\omega^2$ in (9.19) and subtracting from it the scalar (dot) product of $\bar{\omega}$ with (9.26). We then obtain

$$\frac{3}{2}\rho\frac{DT_r}{Dt} = -\nabla\cdot\mathbf{q}_r - \mathbf{M}^{T}:\nabla\bar{\omega} - (\epsilon:\sigma)\cdot\bar{\omega} - \Gamma_r \tag{9.28}$$

where

$$\mathbf{q}_r \equiv \frac{1}{2}(I/m_{\rm p})\rho\langle\mathbf{C}\Omega^2\rangle + \boldsymbol{\theta}(\frac{1}{2}I\Omega^2) \tag{9.29}$$

and

$$\Gamma_r \equiv -\chi(\frac{1}{2}I\omega^2) \tag{9.30}$$

are the flux and volumetric dissipation rate, respectively, of the rotational fluctuational kinetic energy.

For nearly elastic, nearly perfectly rough particles, it is the total kinetic energy that is nearly conserved. The hydrodynamic balance for this quantity is the sum of (9.24) and (9.28),

$$\frac{3}{2}\rho\frac{\mathrm{D}T_{\rm tot}}{\mathrm{D}t} = -\nabla\cdot(\mathbf{q}+\mathbf{q}_r) - \mathbf{M}^{\rm T}:\nabla\overline{\boldsymbol{\omega}} - \boldsymbol{\sigma}^{\rm T}:\nabla\mathbf{v} - (\boldsymbol{\epsilon}:\boldsymbol{\sigma})\cdot\overline{\boldsymbol{\omega}} - (\Gamma+\Gamma_r) \tag{9.31}$$

A few words on the frame indifference of (9.28) and (9.31) are in order. We assume that the couple stress $\mathbf{M}$, like the Cauchy stress $\boldsymbol{\sigma}$, is a frame indifferent tensor. The energy fluxes $\mathbf{q}$ and $\mathbf{q}_r$ are expected to vary as the gradient of a temperature. Hence the first two terms on the right-hand side of each equation are frame indifferent. The third term on the right-hand side of (9.28) is clearly not frame indifferent, as $\overline{\boldsymbol{\omega}}$ changes with the angular velocity of the reference frame. Hence for the equation to be frame indifferent, the sum of the third and fourth terms on the right-hand side must be frame indifferent. Similarly, the third and fourth terms on the right-hand side of (9.31) are not frame indifferent, as $\nabla\mathbf{v}$ is not frame indifferent. However, the sum of the two terms may be written as $-\boldsymbol{\sigma}:\mathbf{D} - (\boldsymbol{\epsilon}:\boldsymbol{\sigma})\cdot(\overline{\boldsymbol{\omega}} - \frac{1}{2}\nabla\times\mathbf{v})$ and is therefore frame indifferent. Thus an equivalent form of (9.31) is

$$\frac{3}{2}\rho\frac{\mathrm{D}T_{\rm tot}}{\mathrm{D}t} = -\nabla\cdot(\mathbf{q}+\mathbf{q}_r) - \boldsymbol{\sigma}:\mathbf{D} - \mathbf{M}^{\rm T}:\nabla\overline{\boldsymbol{\omega}} - (\boldsymbol{\epsilon}:\boldsymbol{\sigma})\cdot(\overline{\boldsymbol{\omega}} - \frac{1}{2}\nabla\times\mathbf{v}) - (\Gamma+\Gamma_r) \tag{9.32}$$

For frame indifference of this equation, we require $\Gamma + \Gamma_r$ to be frame indifferent. We shall return to the issue of frame indifference of the equations of motion in §9.4.2.

To summarize, the equations of motion for a granular gas of nearly elastic, nearly smooth particles are

$$\frac{\mathrm{D}\rho}{\mathrm{D}t} = -\rho\nabla\cdot\mathbf{v} \tag{9.33}$$

$$\rho\frac{\mathrm{D}\mathbf{v}}{\mathrm{D}t} = -\nabla\cdot\sigma + \rho\,\mathbf{b} \tag{9.34}$$

$$\frac{I}{m_{\rm p}}\rho\frac{\mathrm{D}\overline{\boldsymbol{\omega}}}{\mathrm{D}t} = -\nabla\cdot\mathbf{M} + \boldsymbol{\epsilon}:\boldsymbol{\sigma} + \rho\,\boldsymbol{\tau} \tag{9.35}$$

$$\frac{3}{2}\rho\frac{\mathrm{D}T}{\mathrm{D}t} = -\nabla\cdot\mathbf{q} - \boldsymbol{\sigma}^{\rm T}:\nabla\mathbf{v} - \Gamma \tag{9.36}$$

$$\frac{3}{2}\rho\frac{\mathrm{D}T_r}{\mathrm{D}t} = -\nabla\cdot\mathbf{q}_r - \mathbf{M}^{\rm T}:\nabla\overline{\boldsymbol{\omega}} - \Gamma_r - (\boldsymbol{\epsilon}:\boldsymbol{\sigma})\cdot\overline{\boldsymbol{\omega}} \tag{9.37}$$

The equations for a granular gas of nearly elastic, nearly perfectly rough particles are the same, but with (9.36) and (9.37) replaced by (9.32).

To close the equations, constitutive relations for the fluxes $\boldsymbol{\sigma}$, $\mathbf{M}$, $\mathbf{q}$, and $\mathbf{q}_r$, and the energy sinks Γ and Γ_r are required. These relations must be derived by computing the appropriate integrals, for which the distribution function $f(t, \mathbf{r}, \mathbf{c}, \boldsymbol{\omega})$ is necessary. The determination of f for rough particles is the subject of the next section.

9.3. THE VELOCITY DISTRIBUTION FUNCTION

To cast the Boltzmann equation (9.14) in dimensionless form, we use the scaled variables defined in (7.79), but with the scaling of f modified as

$$\hat{f} = f(m_\mathrm{p}/I)^{3/2} v_s^6 / n_s \tag{9.38}$$

and introduce the additional scaled variables

$$\hat{\omega} = \omega(I/m_\mathrm{p})^{1/2}/v_s, \quad \hat{\boldsymbol{\Omega}} = \boldsymbol{\Omega}(I/m_\mathrm{p})^{1/2}/v_s, \quad \hat{\boldsymbol{\tau}} = (\boldsymbol{\tau} H/v_s^2)(m_\mathrm{p}/I)^{1/2} \tag{9.39}$$

In terms of these scaled variables, the Boltzmann equation for rough particles (9.14) assumes the dimensionless form

$$\begin{aligned} \hat{f} K\left(\frac{\hat{\mathrm{D}}}{\mathrm{D}\hat{t}} + \hat{\mathbf{C}}\cdot\hat{\nabla} + \hat{\mathbf{b}}\cdot\hat{\nabla}_c + \hat{\boldsymbol{\tau}}\cdot\hat{\nabla}_\omega\right) \ln \hat{f} &= \int_{\mathbf{k}\cdot\hat{\mathbf{g}}>0} \left[\frac{1}{(1-\epsilon)\beta^2} \hat{f}_1''(\hat{\mathbf{r}} + \frac{6\nu_s}{\pi} K\mathbf{k}) \right. \\ &\times \hat{f}''(\hat{\mathbf{r}}) g_0(\nu(\hat{\mathbf{r}} + \frac{3\nu_s}{\pi}K\mathbf{k})) - \hat{f}(\hat{\mathbf{r}})\hat{f}_1(\hat{\mathbf{r}} - \frac{6\nu_s}{\pi}K\mathbf{k}) g_0(\nu(\hat{\mathbf{r}} - \frac{3\nu_s}{\pi}K\mathbf{k}))\Big] \\ &\times \mathbf{k}\cdot\hat{\mathbf{g}}\, \mathrm{d}\mathbf{k}\, \mathrm{d}\hat{\mathbf{c}}_1\, \mathrm{d}\hat{\omega}_1 \end{aligned} \tag{9.40}$$

We follow the same procedure as for smooth particles (see §7.3.7) and seek a solution for $\hat{f}$ as a small perturbation from the equilibrium distribution function $\hat{f}^0$, i.e.,

$$\hat{f} = \hat{f}^0(1+\Phi), \quad \Phi \ll 1 \tag{9.41}$$

For smooth particles, Φ was expanded as a power series in the inverse Knudsen number K and the inelasticity parameter $\epsilon \equiv (1 - e_\mathrm{p}^2)$. For rough particles, it was shown in §9.2 that two cases qualify as small departures from equilibrium: nearly smooth particles, for which $(1+\beta) \ll 1$, and nearly perfectly rough particles, for which $(1-\beta) \ll 1$. For both cases, $\varepsilon \equiv (1 - |\beta|)$ is the small parameter characterizing roughness, and we may therefore write

$$\Phi = K\Phi_K + \epsilon\,\Phi_\epsilon + \varepsilon\,\Phi_\varepsilon + \cdots \tag{9.42}$$

To determine the nonequilibrium velocity distribution function, we follow the extended Chapman–Enskog procedure described in §7.3.7. Since the equilibrium distribution function, and therefore the perturbation from equilibrium, for nearly perfectly rough and nearly smooth particles are in general different, we consider the two cases separately.

As in the case of smooth particles, we wish to determine the constitutive relations up to linear order in K, ϵ, and ε. We had concluded in §7.3.12 that Φ_ϵ does not contribute to the relations at linear order; this conclusion remains valid when the particles are rough, and similar arguments show that Φ_ε too makes no contribution. It therefore suffices to determine only Φ_K.

9.3.1. Nearly Elastic, Nearly Perfectly Rough Particles

It is first necessary to determine the equilibrium distribution function. At equilibrium, $e_\mathrm{p} = 1$, $\beta = 1$, and spatial gradients of all properties must be absent. There could, however, be a spatial gradient of the velocity that corresponds to rigid body rotation, as this is compatible with our definition of an equilibrium state being one in which no external forces act, and the properties do not change with time.

The collisional invariants are the mass, linear momentum, total angular momentum $(I\boldsymbol{\omega} + m_\mathrm{p}\mathbf{r}\times\mathbf{c})$, and the total kinetic energy $(\frac{1}{2}m_\mathrm{p}c^2 + \frac{1}{2}I\omega^2)$. By following the procedure of §7.3.3, it is straightforward to show that the singlet distribution function at equilibrium is (Chapman and Cowling, 1964, pp. 204–206)

$$f^0 = \frac{n\,(I/m_p)^{3/2}}{(2\pi T)^3} \exp\left[\frac{-C^2 - (I/m_p)\Omega^2}{2T}\right] \tag{9.43}$$

and the mean spin is equal to half the vorticity, or the macroscopic rotation rate of the gas, i.e.,

$$\overline{\omega} = \frac{1}{2}\nabla \mathbf{x}\,\mathbf{v} \tag{9.44}$$

Note that $T_r = T$ and that the kinetic energy is partitioned equally between all the translational and rotational modes.

The scaled form of the equilibrium singlet distribution function is

$$\hat{f}^0 = \frac{\nu/\nu_s}{(2\pi\hat{T})^3}\exp(-\xi^2 - s^2) \tag{9.45}$$

where s is the magnitude of the rescaled peculiar spin $\mathbf{s} \equiv \hat{\mathbf{\Omega}}/(2\hat{T})^{1/2}$ and ξ is the magnitude of the rescaled peculiar translational velocity defined in (7.111).

As mentioned above, for the present purpose we require only the O(K) correction to $\hat{f}^0$. At this order ϵ and ε are set to zero; hence Φ_K is the first correction to $\hat{f}^0$ for elastic, perfectly rough spheres. This was first determined for a *dilute gas* by Pidduck (1922), and his analysis is reproduced in the book of Chapman and Cowling (1964). It was extended for a dense gas by McCoy et al. (1966) and Theodosopulu and Dahler (1974), who assumed a "trial function" for Φ_K and determined the unknown coefficients by enforcing the transport equation for certain higher-velocity moments. Lun (1991) extended the analysis for rough, inelastic particles and also allowed the translational and rotational temperatures to differ; however, when terms other than of linear order in the small parameters are deleted, which is required for consistency with the assumptions of the theory, his results coincide with those of Theodosopulu and Dahler (1974) (but for some errors in the latter).

The linearized Boltzmann equation is obtained by proceeding in the same manner as in §7.3.9; the algebra is more tedious, but in the end we get (P. R. Nott, unpublished manuscript; see also McCoy et al., 1966)

$$\begin{aligned}
\mathcal{L}_r(\Phi_K) = &-\frac{\nu_s}{\nu g_0}\left\{\left[\left(1+\frac{8}{5}\nu g_0\frac{3+5\hat{I}}{1+\hat{I}}\right)\xi^2 + \left(1+\frac{16\nu g_0}{5(1+\hat{I})}\right)s^2\right.\right.\\
&\left. - 4(1-5\nu g_0)\right]\boldsymbol{\xi} + \frac{6\nu g_0\sqrt{\hat{I}}}{(1+\hat{I})}\left[(\xi^2+1/2)\frac{e^{-\xi^2}}{\sqrt{\pi}\,\xi^2}\right.\\
&\left.\left. + (\xi^4+\xi^2-1/4)\frac{\mathrm{erf}(\xi)}{\xi^3}\right]\boldsymbol{\xi}\mathbf{x}\mathbf{s} - \frac{16\nu g_0}{5(1+\hat{I})}\boldsymbol{\xi}\mathbf{x}\mathbf{s}\mathbf{x}\mathbf{s}\right\}\cdot\hat{\nabla}\ln\hat{T}\\
&-\frac{\nu_s}{\nu g_0}\frac{1}{(2\hat{T})^{1/2}}\left\{\left(1+4\nu g_0\frac{2+5\hat{I}}{5(1+\hat{I})}\right)2\overline{\boldsymbol{\xi}\boldsymbol{\xi}}\right.\\
&+\frac{1}{3}(1+4\nu g_0)(\xi^2-s^2)\mathbf{I} - \frac{3\nu g_0\sqrt{\hat{I}}}{(1+\hat{I})}\left[(\xi^2-\frac{3}{2})\frac{e^{-\xi^2}}{\sqrt{\pi}\,\xi^4}\right.\\
&\left.\left.+(\xi^4-\xi^2+\frac{3}{4})\frac{\mathrm{erf}(\xi)}{\xi^5}\right]\boldsymbol{\xi}(\boldsymbol{\xi}\mathbf{x}\mathbf{s})\right\}:\hat{\nabla}\hat{\mathbf{v}}
\end{aligned} \tag{9.46}$$

Here erf(x) is the error function (Abramowitz and Stegun, 1965, p. 297), and the operator $\mathcal{L}_r$ is a slightly altered form of $\mathcal{L}$ (compare (7.113)),

$$\mathcal{L}_r(y) \equiv \frac{1}{\pi^3} \int_{\mathbf{k}\cdot\boldsymbol{\zeta} > 0} [y(\boldsymbol{\xi}) + y(\boldsymbol{\xi}_1) - y(\boldsymbol{\xi}') - y(\boldsymbol{\xi}_1')]\, e^{-\xi_1^2 - s_1^2}\, \mathbf{k}\cdot\boldsymbol{\zeta}\, d\mathbf{k}\, d\boldsymbol{\xi}_1 d\mathbf{s}_1 \tag{9.47}$$

Proceeding in the same manner as in §7.3.9, we find that Φ_K has the form

$$\Phi_K = -\frac{\nu_s}{\nu g_0}\left[\mathbf{A}\cdot\hat{\nabla}\ln\hat{T} + \frac{1}{(2\hat{T})^{1/2}}\mathbf{B}:\hat{\nabla}\hat{\mathbf{v}} + \alpha_1^K + \boldsymbol{\alpha}_2^K\cdot\boldsymbol{\xi} + \boldsymbol{\alpha}_3^K\cdot\mathbf{s} + \alpha_4^K(\xi^2 + s^2)\right] \tag{9.48}$$

with the vector $\mathbf{A}$ and the tensor $\mathbf{B}$ satisfying the equations

$$\mathcal{L}_r(\mathbf{A}) = \left[\left(1 + \frac{8}{5}\nu g_0 \frac{3 + 5\hat{I}}{1 + \hat{I}}\right)\xi^2 + \left(1 + \frac{16\nu g_0}{5(1 + \hat{I})}\right)s^2 - 4(1 - 5\nu g_0)\right]\boldsymbol{\xi} + \frac{6\nu g_0\sqrt{\hat{I}}}{(1 + \hat{I})}\left[(\xi^2 + 1/2)\frac{e^{-\xi^2}}{\sqrt{\pi}\,\xi^2} + (\xi^4 + \xi^2 - 1/4)\frac{\mathrm{erf}(\xi)}{\xi^3}\right]\boldsymbol{\xi}\times\mathbf{s} - \frac{16\nu g_0}{5(1 + \hat{I})}(\boldsymbol{\xi}\times\mathbf{s})\times\mathbf{s} \tag{9.49}$$

$$\mathcal{L}_r(\mathbf{B}) = \frac{1}{3}(1 + 4\nu g_0)(\xi^2 - s^2)\mathbf{I} + \left(1 + 4\nu g_0\frac{2 + 5\hat{I}}{5(1 + \hat{I})}\right)2\boldsymbol{\xi}\boldsymbol{\xi} - \frac{3\nu g_0\sqrt{\hat{I}}}{(1 + \hat{I})}\left[(\xi^2 - \frac{3}{2})\frac{e^{-\xi^2}}{\sqrt{\pi}\,\xi^4} + (\xi^4 - \xi^2 + \frac{3}{4})\frac{\mathrm{erf}(\xi)}{\xi^5}\right]\boldsymbol{\xi}(\boldsymbol{\xi}\times\mathbf{s}) \tag{9.50}$$

In the limit $\nu \to 0$, we recover the equations obtained by Chapman and Cowling (1964, pp. 210, 212) for a dilute gas. Note that the right-hand side of (9.50) is a symmetric tensor for a dilute gas ($\nu \to 0$, $g_0 \to 1$), but not in general.

Now, $\mathbf{A}$ and $\mathbf{B}$ are functions only of $\boldsymbol{\xi}$ and $\mathbf{s}$, and do not depend on any vector or tensor parameter. However, $\mathbf{s}$ is a pseudovector, while $\mathbf{A}$ is a polar vector and $\mathbf{B}$ a polar tensor (see Appendix J for the definition of polar and pseudo scalars, vectors, and tensors). The polar nature of $\mathbf{A}$ and $\mathbf{B}$ arises from the fact that Φ_K is a polar scalar. Therefore, $\mathbf{s}$ can appear only in a cross product with a polar vector or tensor, or as the dyad $\mathbf{ss}$. From considerations of tensor symmetry, it can be shown that the most general forms of $\mathbf{A}$ and $\mathbf{B}$ are (Chapman and Cowling, 1964, pp. 209–210; see also Problem 9.1)

$$\mathbf{A} = \mathcal{A}_1\,\boldsymbol{\xi} + \mathcal{A}_2\,\boldsymbol{\xi}\times\mathbf{s} + \mathcal{A}_3\,(\boldsymbol{\xi}\times\mathbf{s})\times\mathbf{s} \tag{9.51}$$

$$\mathbf{B} = \mathcal{B}_1\,\mathbf{I} + \mathcal{B}_2\,\boldsymbol{\epsilon}\cdot\mathbf{s} + \mathcal{B}_3\,\boldsymbol{\xi}\boldsymbol{\xi} + \mathcal{B}_4\,\mathbf{s}\mathbf{s} + \mathcal{B}_5\,\boldsymbol{\xi}(\boldsymbol{\xi}\times\mathbf{s}) + \mathcal{B}_6\,\boldsymbol{\xi}\,[(\boldsymbol{\xi}\times\mathbf{s})\times\mathbf{s}] + \mathcal{B}_7\,(\boldsymbol{\xi}\times\mathbf{s})(\boldsymbol{\xi}\times\mathbf{s}) + \mathcal{B}_8\,(\boldsymbol{\xi}\times\mathbf{s})\,[(\boldsymbol{\xi}\times\mathbf{s})\times\mathbf{s}] + \mathcal{B}_9\,[(\boldsymbol{\xi}\times\mathbf{s})\times\mathbf{s}]\,[(\boldsymbol{\xi}\times\mathbf{s})\times\mathbf{s}] \tag{9.52}$$

where $\mathcal{A}_1$–$\mathcal{A}_3$ and $\mathcal{B}_1$–$\mathcal{B}_9$ are functions only of ξ^2, s^2, and $(\boldsymbol{\xi}\cdot\mathbf{s})^2$, the last of these being the square of the "helicity." Chapman and Cowling (1964, p. 210) had omitted the terms involving $\boldsymbol{\epsilon}\cdot\mathbf{s}$ and $\mathbf{s}\,\mathbf{s}$ in (9.52). Omission of the former is justified because of the symmetry of $\mathbf{B}$ for a dilute gas; we shall retain it, to allow the possibility of an asymmetric stress for a dense granular gas. However, omission of the latter term appears to be incorrect.

Analogous to (7.120), we impose the moment constraint

$$\int e^{-\xi^2-s^2}\,\Phi_K\,\psi\,\mathrm{d}\hat{\mathbf{c}}\,\mathrm{d}\hat{\boldsymbol{\omega}} = 0 \tag{9.53}$$

for $\psi = 1$, $\mathbf{c}$, $\boldsymbol{\omega}$, and $C^2+\Omega^2$, as the density, mean translational and rotational velocities, and the temperature are defined as moments of f^0 (see (9.43)). These conditions are used to determine the coefficients α_1^K–α_4^K in (9.48). It is straightforward to show that (i) α_1^K and α_4^K are proportional to $\boldsymbol{\nabla}\cdot\hat{\mathbf{v}}$, and hence the corresponding terms in (9.48) can be absorbed into the term involving $\mathcal{B}_1$; (ii) $\boldsymbol{\alpha}_2^K$ is proportional to $\hat{\boldsymbol{\nabla}}\hat{T}$, and hence the corresponding term can be absorbed into the term involving $\mathcal{A}_1$; (iii) $\boldsymbol{\alpha}_3^K$ is proportional to $\hat{\boldsymbol{\nabla}}\times\hat{\mathbf{v}}$, and hence the corresponding term can be absorbed into the term involving $\mathcal{B}_2$. Consequently, (9.48) reduces to

$$\Phi_K = -\frac{\nu_s}{\nu g_0}\left[\mathbf{A}\cdot\hat{\boldsymbol{\nabla}}\ln\hat{T} + \frac{1}{(2\hat{T})^{1/2}}\mathbf{B}:\hat{\boldsymbol{\nabla}}\hat{\mathbf{v}}\right] \tag{9.54}$$

and the moment constraints reduce to the following constraints on the forms of $\mathbf{A}$ and $\mathbf{B}$ (see Problem 9.1):

$$\begin{aligned}&\int e^{-\xi^2-s^2}\boldsymbol{\xi}\,\mathbf{A}\,\mathrm{d}\boldsymbol{\xi}\,\mathrm{d}\mathbf{s} = 0, \quad \int e^{-\xi^2-s^2}\mathbf{B}\,\mathrm{d}\boldsymbol{\xi}\,\mathrm{d}\mathbf{s} = 0,\\ &\int e^{-\xi^2-s^2}\mathbf{s}\,\mathbf{B}\,\mathrm{d}\boldsymbol{\xi}\,\mathrm{d}\mathbf{s} = 0, \quad \int e^{-\xi^2-s^2}\mathbf{B}\,(\xi^2+s^2)\,\mathrm{d}\boldsymbol{\xi}\,\mathrm{d}\mathbf{s} = 0.\end{aligned} \tag{9.55}$$

The solution for the coefficients $\mathcal{A}_i$ and $\mathcal{B}_i$ in (9.51) and (9.52) may be obtained by expanding them in a complete set of functions of ξ^2, s^2, and $(\boldsymbol{\xi}\cdot\mathbf{s})^2$ and determining the coefficients by the method of moments, as in §7.3.10. Lun (1991) chose a simpler method to obtain an approximate solution, in the spirit of the earlier work of Lun et al. (1984) for a granular gas of smooth particles. He assumed the following simple forms for $\mathbf{A}$ and $\mathbf{B}$:

$$\mathbf{A} = \left[a_1(\xi^2-5/2) + a_1'(s^2-3/2)\right]\boldsymbol{\xi} \tag{9.56}$$

$$\mathbf{B} = b_0\,\overline{\boldsymbol{\xi}\boldsymbol{\xi}} \tag{9.57}$$

where a_1, a_1', and b_0 are constants. This form identically satisfies the constraints (9.55). The constants a_1, a_1', and b_0 are determined by satisfying the Maxwell transport equation for certain higher moments. For instance, a_1 is determined by setting $\psi = \frac{1}{2}m_p C^2\mathbf{c}$ in (9.19) and setting $\boldsymbol{\nabla}\mathbf{v} = \boldsymbol{\nabla}\nu = 0$ (see Problem 7.3). However, Lun (1991) used a form for $\hat{f}^0$ in which the rotational temperature T_r could differ from the translational temperature T. While deviation from equipartition has been observed in particle dynamics simulations (see, e.g., McNamara and Luding, 1998), the formal Chapman–Enskog procedure does not allow it, as it is a small perturbation about equilibrium, where equipartition holds. Lun also allowed ϵ and ε to have nonzero values, which is again not permitted at O(K). For a result that is consistent with the assumptions of the theory, we must therefore equate T and T_r and set $\epsilon = \varepsilon = 0$ in Lun's result. With these corrections, the solution is

$$\begin{aligned} a_1 &= \frac{3(1+\hat{I})}{4\sqrt{2\pi}}\,\frac{15(1+\hat{I})^2(1+2\hat{I}) + 4\nu g_0(9+43\hat{I}+48\hat{I}^2+30\hat{I}^3)}{(12+75\hat{I}+101\hat{I}^2+102\hat{I}^3)}\\ a_1' &= \frac{3(1+\hat{I})}{\sqrt{2\pi}}\,\frac{(3+19\hat{I})(1+\hat{I}) + \nu g_0(8+49\hat{I}+25\hat{I}^2)}{(12+75\hat{I}+101\hat{I}^2+102\hat{I}^3)}\\ b_0 &= \frac{3(1+\hat{I})}{2\sqrt{2\pi}}\,\frac{5(1+\hat{I})+4\nu g_0(2+5\hat{I})}{6+13\hat{I}}\end{aligned} \tag{9.58}$$

When the particles have no rotary inertia, i.e., $\hat{I} = 0$, the relations for a_1 and b_0 coincide with the solution obtained for a granular gas of smooth particles, given in (7.133). This corresponds to the situation where the entire mass of each particle is concentrated at its center, and therefore tangential momentum is not exchanged during collision.

9.3.2. Nearly Elastic, Nearly Smooth Particles

This case was considered by Goldhirsch et al. (2005b) for dilute gases and extended to dense gases by P. R. Nott (unpublished manuscript). The equilibrium distribution function here is that of elastic, smooth particles. When a pair of smooth particles collide, their spin remains unchanged and the change in their translational velocities is unrelated to their spin. Thus as collisions drive the granular gas toward equilibrium, the spin of each particle, and therefore the distribution of spin, remains unchanged. Hence the equilibrium distribution function is

$$f^0 = \frac{n}{(2\pi T)^{3/2}} \exp\left[\frac{-C^2}{2T}\right] \vartheta(\boldsymbol{\omega}) \tag{9.59}$$

Here $\vartheta(\boldsymbol{\omega})$ is the spin distribution, fixed by the initial condition of the granular gas. It may be any frame-indifferent scalar function of $\boldsymbol{\omega}$ that satisfies the constraint

$$\int \vartheta(\boldsymbol{\omega})\, \mathrm{d}\boldsymbol{\omega} = 1 \tag{9.60}$$

so that $n = \int f^0 \,\mathrm{d}\mathbf{c}\, \mathrm{d}\boldsymbol{\omega}$. The dimensionless form of the equilibrium distribution function is

$$\hat{f}^0 = \frac{\nu/\nu_s}{(2\pi \hat{T})^{3/2}} e^{-\xi^2}\, \hat{\vartheta} \tag{9.61}$$

where $\hat{\vartheta} \equiv (m_\mathrm{p}/I)^{3/2} v_s^3 \vartheta$ is the dimensionless spin distribution.

We now determine the distribution function for nearly smooth particles at O(K) by following the procedure of §7.3.9. It is readily seen that the left-hand side (LHS) of the Boltzmann equation (9.40) at O(K) is

$$\begin{aligned} \mathrm{LHS}_K = \hat{f}^0 &\left[\frac{1}{\nu}\frac{\hat{\mathrm{D}}\nu}{\mathrm{D}\hat{t}} - \left(\frac{3}{2\hat{T}} - \frac{\hat{C}^2}{2\hat{T}^2}\right)\frac{\hat{\mathrm{D}}\hat{T}}{\mathrm{D}\hat{t}} + \frac{1}{\hat{T}}\hat{\mathbf{C}}\hat{\mathbf{C}}\mathbf{:}\hat{\nabla}\hat{\mathbf{v}}\right] \\ &+ \hat{f}^0\hat{\mathbf{C}}\cdot\left[\frac{1}{\nu}\hat{\nabla}\nu - \left(\frac{3}{2\hat{T}} - \frac{\hat{C}^2}{2\hat{T}^2}\right)\hat{\nabla}\hat{T} + \frac{1}{\hat{T}}\left(\frac{\hat{\mathrm{D}}\hat{\mathbf{v}}}{\mathrm{D}\hat{t}} - \hat{\mathbf{b}}\right)\right] \\ &+ \hat{f}^0\left[\frac{\hat{D}\ln\hat{\vartheta}}{\hat{D}\hat{t}} + \hat{\mathbf{C}}\cdot\hat{\nabla}\ln\hat{\vartheta}\right] \end{aligned} \tag{9.62}$$

with all the material derivatives evaluated at O(1). The only difference between (9.62) and the corresponding expression for smooth particles (7.99) is the term in the last line of the former. As discussed in §9.2, the material derivative of $\hat{\vartheta}$ must vanish at O(1). This can also be shown by constructing the transport equation for ϑ (see Problem 9.2). The material derivatives of ν, $\hat{\mathbf{v}}$, and $\hat{T}$ are given by (7.110).

In the right-hand side (RHS) of (9.40), retention of only the O(K) terms yields the same three terms as in (7.101), but with the integrals evaluated also over $\hat{\omega}_1$,

$$\mathrm{RHS}_K = \frac{6\nu_s}{\pi} g_0 \int_{\mathbf{k}\cdot\hat{\mathbf{g}}>0} \hat{f}^0 \hat{f}_1^0 \left(\Phi_K'' + \Phi_{K1}'' - \Phi_K - \Phi_{K1}\right) \mathbf{k}\cdot\hat{\mathbf{g}}\, \mathrm{d}\mathbf{k}\, \mathrm{d}\hat{\mathbf{c}}_1\, \mathrm{d}\hat{\omega}_1$$
$$+ \frac{6\nu_s}{\pi} g_0 \int_{\mathbf{k}\cdot\hat{\mathbf{g}}>0} \hat{f}^0 \hat{f}_1^0\, \mathbf{k}\cdot\hat{\nabla}\ln(\hat{f}_1^0 \hat{f}_1^{0''})\, \mathbf{k}\cdot\hat{\mathbf{g}}\, \mathrm{d}\mathbf{k}\, \mathrm{d}\hat{\mathbf{c}}_1\, \mathrm{d}\hat{\omega}_1$$
$$+ \frac{6\nu_s}{\pi} \int_{\mathbf{k}\cdot\hat{\mathbf{g}}>0} \hat{f}^0 \hat{f}_1^0\, \mathbf{k}\cdot\hat{\nabla} g_0\, \mathbf{k}\cdot\hat{\mathbf{g}}\, \mathrm{d}\mathbf{k}\, \mathrm{d}\hat{\mathbf{c}}_1\, \mathrm{d}\hat{\omega}_1 \tag{9.63}$$

Integrating over $\mathbf{k}$ and $\hat{\mathbf{c}}_1$ (see the text following (7.101)), the third integral in (9.63) becomes

$$-4\nu \hat{f}^0 \hat{\mathbf{C}}\cdot\hat{\nabla} g_0 \tag{9.64}$$

On substituting for $\hat{f}^0$ from (9.61), the second term in (9.63) reduces to

$$\frac{6\nu_s}{\pi} g_0 \int_{\mathbf{k}\cdot\hat{\mathbf{g}}>0} \mathbf{k}\cdot \left[\frac{2}{\nu}\hat{\nabla}\nu - \frac{3}{\hat{T}}\hat{\nabla}\hat{T} + \frac{1}{2\hat{T}^2}\hat{\nabla}\hat{T}(\hat{C}_1^2 + \hat{C}_1'^2)\right.$$
$$\left. + \frac{1}{\hat{T}}\hat{\nabla}\hat{\mathbf{v}}\cdot(\hat{\mathbf{C}}_1 + \hat{\mathbf{C}}_1')\right] \hat{f}^0 \hat{f}_1^0\, \mathbf{k}\cdot\hat{\mathbf{g}}\, \mathrm{d}\mathbf{k}\, \mathrm{d}\hat{\mathbf{C}}_1\, \mathrm{d}\hat{\mathbf{\Omega}}_1$$
$$+ \frac{6\nu_s}{\pi} g_0 \int_{\mathbf{k}\cdot\hat{\mathbf{g}}>0} \mathbf{k}\cdot\hat{\nabla}(\hat{\vartheta}_1 + \hat{\vartheta}_1') \hat{f}^0 \hat{f}_1^0\, \mathbf{k}\cdot\hat{\mathbf{g}}\, \mathrm{d}\mathbf{k}\, \mathrm{d}\hat{\mathbf{c}}_1\, \mathrm{d}\hat{\omega}_1 \tag{9.65}$$

The second term in the above expression is the additional contribution due to particle roughness (compare (7.105)). However, $\hat{\vartheta}_1' = \hat{\vartheta}_1$, and it follows from (9.60) that

$$\int \hat{\nabla}\hat{\vartheta}(\hat{\omega}_1)\, \mathrm{d}\hat{\omega}_1 = \hat{\nabla}\int \hat{\vartheta}(\hat{\omega}_1)\, \mathrm{d}\hat{\omega}_1 = 0 \tag{9.66}$$

As a result, the additional contribution vanishes, and the second term in (9.63) coincides with the corresponding expression for smooth particles, given by (7.107).

Assembling the expressions for LHS_K and RHS_K, above, and collecting the terms as in §7.3.9, we arrive at the Boltzmann equation for nearly smooth particles at O(K),

$$\int_{\mathbf{k}\cdot\boldsymbol{\zeta}>0} \hat{f}_1^0 \hat{\vartheta}_1 \left(\Phi_K + \Phi_{K1} - \Phi_K' - \Phi_{K1}'\right) \mathbf{k}\cdot\boldsymbol{\zeta}\, \mathrm{d}\mathbf{k}\, \mathrm{d}\boldsymbol{\xi}_1\, \mathrm{d}\mathbf{s}_1 = -\frac{1}{g_0}\left[\left(1 + \frac{12}{5}\nu g_0\right)\right.$$
$$\left. \times (\xi^2 - 5/2)\boldsymbol{\xi}\cdot\hat{\nabla}\ln\hat{T} + \frac{1}{(2\hat{T})^{1/2}}\left(1 + \frac{8}{5}\nu g_0\right) 2\overline{\boldsymbol{\xi}\boldsymbol{\xi}}:\hat{\mathbf{D}} + \boldsymbol{\xi}\cdot\hat{\nabla}\ln\hat{\vartheta}\right] \tag{9.67}$$

which is the equation governing Φ_K. The frame indifference of ϑ implies that its argument must be $\boldsymbol{\omega} - \overline{\boldsymbol{\omega}}$; we may therefore, without loss of generality, express it as a function of $\mathbf{s}$.

The solution of (9.67) is the sum of the general solution of the homogeneous equation and a particular solution that is linear in $\hat{\mathbf{D}}$, $\hat{\nabla}\ln\hat{T}$, and $\hat{\nabla}\ln\hat{\vartheta}(\mathbf{s})$. The collisional invariants at this order are $\boldsymbol{\xi}$, ξ^2, and any function of $\mathbf{s}$. Therefore, Φ_K has the form

$$\Phi_K = -\frac{\nu_s}{\nu g_0}\left\{\left(1+\frac{12}{5}\nu g_0\right)\mathbf{A}\cdot\hat{\nabla}\ln\hat{T} + \frac{1}{(2\hat{T})^{1/2}}\left(1+\frac{8}{5}\nu g_0\right)\mathbf{B}:\hat{\mathbf{D}} + \mathbf{R}\cdot\hat{\nabla}\ln\hat{\vartheta}(\mathbf{s}) + \alpha_1^K(\mathbf{s}) + \boldsymbol{\alpha}_2^K\cdot\boldsymbol{\xi} + \alpha_3^K\,\xi^2\right\} \tag{9.68}$$

The coefficients **A**, **B**, and **R** are functions only of $\boldsymbol{\xi}$ and $\mathbf{s}$. Hence **A** and **B** have the general forms given in (9.51) and (9.52), respectively, and **R** has the general form

$$\mathbf{R} = \mathcal{R}_1\,\boldsymbol{\xi} + \mathcal{R}_2\,\boldsymbol{\xi}\times\mathbf{s} + \mathcal{R}_3\,(\boldsymbol{\xi}\times\mathbf{s})\times\mathbf{s} \tag{9.69}$$

However, since the coefficients of $\hat{\nabla}\ln\hat{T}$, $\hat{\nabla}\hat{\mathbf{v}}$, and $\hat{\nabla}\ln\hat{\vartheta}$ on right-hand side of (9.67) are independent of $\mathbf{s}$, the terms involving $\mathbf{s}$ in **A**, **B**, and **R** must vanish upon integration in the left-hand side of (9.67). Hence **A**, **B**, and **R** may depend on $\mathbf{s}$ only through collisional invariants, i.e., functions having $\mathbf{s}$ alone as the argument. Such terms are not present in the general forms of **A** and **R**, but the term $\mathcal{B}_4\,\mathbf{s}\,\mathbf{s}$ in (9.52) is of this form. However, such a term is already included in the (9.68) (see the term $\alpha_1^K(\mathbf{s})$) as part of the solution of the homogeneous equation. Therefore, we may treat **A**, **B**, and **R** as functions of $\boldsymbol{\xi}$ alone and write

$$\mathbf{A} = \mathcal{A}\,\boldsymbol{\xi},\quad \mathbf{B} = \mathcal{B}\,\overline{\boldsymbol{\xi}\boldsymbol{\xi}},\quad \text{and}\quad \mathbf{R} = \mathcal{R}\,\boldsymbol{\xi} \tag{9.70}$$

The scalar coefficients $\mathcal{A}$, $\mathcal{B}$, and $\mathcal{R}$ are functions only of ξ^2, and satisfy the equations

$$\mathcal{L}(\mathcal{A}\,\boldsymbol{\xi}) = (\xi^2 - \frac{5}{2})\boldsymbol{\xi} \tag{9.71}$$

$$\mathcal{L}(\mathcal{B}\,\overline{\boldsymbol{\xi}\boldsymbol{\xi}}) = 2\,\overline{\boldsymbol{\xi}\boldsymbol{\xi}} \tag{9.72}$$

$$\mathcal{L}^*(\mathcal{R}\,\boldsymbol{\xi}) = \boldsymbol{\xi} \tag{9.73}$$

where $\mathcal{L}$ is the operator defined in (7.113), and $\mathcal{L}^*$ is defined as

$$\mathcal{L}^*(y) \equiv \frac{1}{\pi^{3/2}}\int_{\mathbf{k}\cdot\boldsymbol{\zeta}>0}\left[y(\boldsymbol{\xi}) - y(\boldsymbol{\xi}')\right]\,e^{-\xi_1^2}\,\mathbf{k}\cdot\boldsymbol{\zeta}\;\mathrm{d}\mathbf{k}\,\mathrm{d}\boldsymbol{\xi}_1 \tag{9.74}$$

The operator on the left-hand side of (9.73) is $\mathcal{L}^*$ instead of $\mathcal{L}$ as a consequence of (9.66).

The function $\alpha_1^K(\mathbf{s})$ and constants $\boldsymbol{\alpha}_2^K$ and α_3^K are determined by satisfying the moment constraint (9.53). The density, mean velocity, translational temperature, and the distribution of spin are all defined as velocity moments of f^0; hence the corresponding moments must vanish at higher orders. This is enforced by satisfying (9.53) for $\psi = \hat{\mathbf{c}}, \hat{C}^2$, and imposing the condition

$$\int f^0\,\Phi_K\,\mathrm{d}\mathbf{c} = 0 \tag{9.75}$$

Note that the above integral is only over translational velocities. These conditions imply, as in the case of smooth particles (see (7.121) and (7.123) and the text following them), that $\alpha_1^K = \alpha_3^K = 0$ and that the term $\boldsymbol{\alpha}_2^K\cdot\hat{\boldsymbol{\xi}}$ in (9.68) can be absorbed into the term proportional to $\hat{\nabla}\ln\hat{T}$, provided the constraint

$$\int \mathcal{A}(\xi)\,\xi^2 e^{-\xi^2}\,\mathrm{d}\boldsymbol{\xi} = 0 \tag{9.76}$$

is satisfied.

The solutions for $\mathcal{A}$, $\mathcal{B}$, and $\mathcal{R}$ may be obtained by expanding them in a complete set of functions, as in §7.3.10. However, our experience with smooth particles in §7.3.10 suggests that we may assume simple forms, without sacrificing much on the accuracy of the transport properties. In this spirit, we assume

$$\mathcal{A} = a_1 \left(\xi^2 - \frac{5}{2}\right), \quad \mathcal{B} = b_0, \quad \mathcal{R} = r_0 \tag{9.77}$$

where a_1, b_0, and r_0 are constants. The above are simply the first (nonvanishing) terms in the respective Sonine polynomial expansions. The values of a_1 and b_0 have already been determined in §7.3.10. The value of r_0 is determined by taking the inner product of (9.73) with $\boldsymbol{\xi}$ and integrating the result over all $\boldsymbol{\xi}$, yielding

$$r_0 = \frac{3\sqrt{2}}{16\sqrt{\pi}} \tag{9.78}$$

9.4. CONSTITUTIVE RELATIONS UP TO FIRST ORDER IN K, ϵ, AND ε

We shall now use the distribution function determined in the previous section to derive the constitutive relations to first order in the three small parameters. Since the distribution functions for the cases of nearly perfectly rough particles and nearly smooth particles are different, we consider the two cases separately.

9.4.1. Nearly Elastic, Nearly Perfectly Rough Particles

We note from (9.44) that $\overline{\omega}$ at equilibrium varies as the inverse of the macroscopic length scale; we expect the same to hold away from equilibrium. Therefore terms involving gradients of $\overline{\omega}$, or $\overline{\omega}$ multiplied by ϵ or ε, are not included in the constitutive relations at this order.

The constitutive relations for the flux and the source of the hydrodynamic variables are determined in the same manner as in §7.3.11. Although the distribution function here differs from that of smooth granular gases, and the integrals must be carried out over the spin in addition to the translation velocity, the procedure is in essence the same. We therefore omit the details of the calculations here and give only the results.

Determination of the fluxes $\boldsymbol{\sigma}$, $\mathbf{M}$, and $\mathbf{q}$ requires Φ_K, for which we use the result from §9.3.1. The O(1) contribution to the stress is unchanged from that of a smooth granular gas (7.95). At first order in the small parameters, the only nonvanishing contribution to the stress are $\hat{\boldsymbol{\sigma}}_{\mathrm{K}}$ and $\hat{\boldsymbol{\sigma}}_{\epsilon}$. The total stress tensor up to first order in the small parameters is

$$\boldsymbol{\sigma} = [p - \mu_b \nabla\cdot\mathbf{v}]\,\mathbf{I} - 2\mu\,\mathbf{D} \tag{9.79}$$

where $\mathbf{D} \equiv -\mathbf{C}$ (see (2.49)) is the rate of deformation tensor defined in the tensile sense. Here the pressure p is given by

$$p = \rho_{\mathrm{p}} \left(1 + 4\eta_1 \nu g_0\right) \nu T \tag{9.80}$$

and the bulk and shear viscosities by

$$\mu_b = \frac{8}{3\sqrt{\pi}} \nu^2 g_0\, \rho_{\mathrm{p}} d_{\mathrm{p}}\, T^{1/2} \tag{9.81}$$

$$\mu = \frac{6}{25} \frac{\left[5(1+\hat{I}) + 4\nu g_0 (2 + 5\hat{I})\right]^2}{(6 + 13\hat{I})} \frac{\mu^*}{g_0} + \frac{3(4 + 7\hat{I})}{20(1+\hat{I})} \mu_b \tag{9.82}$$

respectively, where μ^* is the shear viscosity of a dilute gas of smooth particles, defined in (7.162). Note that the stress is symmetric at this order. Lun (1991) found a contribution to the stress of the form $\varepsilon\, \boldsymbol{\epsilon}\cdot(\overline{\boldsymbol{\omega}} - \frac{1}{2}\nabla \times \mathbf{v})$, which is antisymmetric. However, this is an O$(K\varepsilon)$ contribution and is therefore not included here.

The couple stress tensor vanishes identically at this order of the expansion, i.e.,

$$\mathbf{M} = 0 \tag{9.83}$$

The flux of fluctuational energy takes the form

$$\mathbf{q} = -\kappa \nabla T \tag{9.84}$$

with the pseudothermal conductivity given as

$$\begin{aligned}\kappa = {} & \frac{\kappa^*}{25 g_0} \frac{1}{(12 + 75\hat{I} + 101\hat{I}^2 + 102\hat{I}^3)} \Big[4(111 + 552\nu g_0 + 688\nu^2 g_0^2) \\ & + 4\hat{I}(675 + 3448\nu g_0 + 4352\nu^2 g_0^2) + 4\hat{I}^2(1167 + 5416\nu g_0 + 6544\nu^2 g_0^2) \\ & + 4\hat{I}^3(753 + 3720\nu g_0 + 5280\nu^2 g_0^2) + 600\hat{I}^4(1 + 4\nu g_0)^2\Big] \\ & + \frac{3}{2}\frac{(2\hat{I} + 1 + \sqrt{\pi})}{(\hat{I} + 1)} \mu_b \end{aligned} \tag{9.85}$$

where κ^* is the thermal conductivity of a dilute gas of smooth particles, defined in (7.165).

From §7.3.12 we recall that determination of the dissipation rate of translational kinetic energy Γ requires only the unperturbed (i.e., equilibrium) distribution function; it is easily seen that this is also the case for the rate of dissipation of rotational kinetic energy. Carrying out the integrals, we get the expression for the dissipation rate of total fluctuational kinetic energy,

$$\Gamma_{\text{tot}} = \frac{12}{\sqrt{\pi}} \nu^2 g_0 \frac{\rho_p T^{3/2}}{d_p} \left[(1 - e_p^2) + (1 - \beta^2)\right] \tag{9.86}$$

Equations (9.79)–(9.86) are the constitutive relations for a granular gas of nearly elastic, nearly perfectly rough particles. In the limit $\hat{I} \to 0$, the viscosity reverts that of a granular gas of smooth particles (compare (7.161)), but the thermal conductivity remains different. The reason for the latter is that the rotational kinetic energy does not vanish when the rotary inertia vanishes, because equipartition of energy is assumed to hold. In other words, $\langle \Omega^2 \rangle \to \infty$ as $\hat{I} \to 0$, such that $\frac{1}{2} I \langle \Omega^2 \rangle$ remains finite. Consequently, there is transport of rotational kinetic energy even as $\hat{I} \to 0$.

9.4.2. Nearly Elastic, Nearly Smooth Particles

For this case, the mean spin is not constrained to equal half the vorticity at equilibrium; indeed, it can assume any value. Therefore, it need not vary as the inverse of the macroscopic length scale. Hence we must retain terms involving $\overline{\omega}$ and its gradients in the constitutive relations.

We use the distribution function from §9.3.2 to determine the fluxes $\boldsymbol{\sigma}$, $\mathbf{M}$, $\mathbf{q}$, and $\mathbf{q}_r$. The O(1) and O(K) contributions to the stress remain unchanged from that of a smooth granular gas (7.95). In addition, there is an O(ε) contribution in the collisional part of the stress, which in dimensionless form is

$$\hat{\boldsymbol{\sigma}}_\varepsilon = \frac{3}{\pi} \nu_s^2 g_0 \int_{\mathbf{k}\cdot\hat{\mathbf{g}} > 0} (\hat{\mathbf{C}}' - \hat{\mathbf{C}})_\varepsilon \, \mathbf{k}\cdot\hat{\mathbf{g}} \, \mathbf{k} \hat{f}^0 \hat{f}_1^0 \, d\mathbf{k} \, d\hat{\mathbf{C}} \, d\hat{\mathbf{C}}_1 \, d\hat{\boldsymbol{\Omega}} \, d\hat{\boldsymbol{\Omega}}_1 \tag{9.87}$$

where the $(\hat{\mathbf{C}}' - \hat{\mathbf{C}})_\varepsilon$ is the O(ε) part of $\hat{\mathbf{C}}' - \hat{\mathbf{C}}$. From (9.4) and (9.6), we find that

$$(\hat{\mathbf{C}}' - \hat{\mathbf{C}})_\varepsilon = \frac{\hat{I}}{2(1 + \hat{I})} \left[\hat{\mathbf{g}} - (\mathbf{k}\cdot\hat{\mathbf{g}})\mathbf{k} + \hat{I}^{-1/2} \mathbf{k} \times (\hat{\boldsymbol{\Omega}}_1 + \hat{\boldsymbol{\Omega}})\right] \tag{9.88}$$

On substituting the above in (9.87) and evaluating the integral, we get $\hat{\sigma}_\varepsilon = -2\hat{\zeta}\,\boldsymbol{\epsilon}\cdot\hat{\overline{\omega}}$, where $\hat{\zeta}$ is the dimensionless *spin viscosity*, the expression for which will be given shortly. We note that while the other contributions to the stress are frame indifferent, $\hat{\sigma}_\varepsilon$ is not: if the angular velocity of the reference frame is altered, so is the shear stress. This is unacceptable on physical grounds. This problem is resolved if we retain the $O(K\varepsilon)$ term $\hat{\zeta}\,\boldsymbol{\epsilon}\cdot(\hat{\nabla}\mathbf{x}\hat{\mathbf{v}})$; the sum of this term and $\hat{\sigma}_\varepsilon$ is proportional to the *difference* $(\overline{\omega} - \frac{1}{2}\nabla\,\mathbf{x}\,\mathbf{v})$, which is indeed frame indifferent. Thus the inclusion of a higher-order term is necessary to maintain frame indifference of the stress. Reverting to dimensional form, the stress tensor up to first order in the small parameters (with the exception of the term discussed above) is

$$\boldsymbol{\sigma} = [p - \mu_b \nabla\cdot\mathbf{v}]\,\mathbf{I} - 2\mu\,\mathbf{D} - 2\zeta\boldsymbol{\epsilon}\cdot(\overline{\omega} - \frac{1}{2}\nabla\,\mathbf{x}\,\mathbf{v}) \tag{9.89}$$

It is clear that the stress tensor is not symmetric when $\overline{\omega}$ differs from half the vorticity. The expressions for p, μ_b, and μ are identical to those for smooth particles, given in (7.159), (7.160) and (7.161), respectively. The spin viscosity ζ is given by

$$\zeta = \frac{(1+\beta)}{\sqrt{\pi}}\,\frac{\hat{I}}{(1+\hat{I})}\,\nu^2 g_0\,\rho_p\,d_p\,T^{1/2} \tag{9.90}$$

The couple stress has a contribution only from the streaming part (the leading contribution to the collisional part is $O(K\varepsilon)$), which is

$$\mathbf{M} = -\mu_r\,\nabla\overline{\omega} \tag{9.91}$$

where

$$\mu_r = \frac{\sqrt{\pi}}{128\,g_0}\hat{I}\rho_p\,d_p^3\,T^{1/2} \tag{9.92}$$

characterizes the diffusion of intrinsic angular momentum.

The fluxes of translational and rotational fluctuational energy are

$$\mathbf{q} = -\kappa\nabla T \quad \text{and} \quad \mathbf{q}_r = -\kappa_r\nabla T_r \tag{9.93}$$

respectively. It is easily seen that κ is the same as that for smooth particles, given by (7.164). The rotational kinetic energy flux has a contribution only from the streaming part, and the conductivity is

$$\kappa_r = \frac{12}{25 g_0}\,\kappa^* \tag{9.94}$$

Finally, the rates of dissipation of translational and rotational kinetic energy are

$$\Gamma = \frac{12}{\sqrt{\pi}}\,\nu^2 g_0 \frac{\rho_p\,T^{3/2}}{d_p}\left[(1-e_p^2) + \frac{\hat{I}}{1+\hat{I}}(1-\beta^2)\right] \tag{9.95}$$

$$\Gamma_r = \frac{12}{\sqrt{\pi}}\,\nu^2 g_0 \frac{\rho_p\,T^{1/2}\,T_r}{d_p}\,\frac{1}{1+\hat{I}}(1-\beta^2) \tag{9.96}$$

$$+\frac{1}{\sqrt{\pi}}\,\nu^2 g_0\,\rho_p\,d_p\,T^{1/2}\frac{\hat{I}}{1+\hat{I}}(1-\beta^2)\,\overline{\omega}^2$$

Note that the second term in Γ_r is not frame indifferent, as it varies as $\overline{\omega}^2$. As discussed in §9.2, for the rotational energy balance to be frame indifferent, the sum of Γ_r and $(\boldsymbol{\epsilon} : \boldsymbol{\sigma})\cdot\overline{\omega}$ must be frame indifferent. It is clear that this is not the case at leading order in the small parameters. Thus it appears that, as in the case of $\boldsymbol{\sigma}$, higher-order terms must be retained to recover frame indifference of these terms.

The above discussion brings out an important difference between the kinetic theories for smooth and rough particles. In the former, frame indifference of the constitutive relations is satisfied at every order of the perturbation expansion that has been determined (we have considered only the $O(1)$ and $O(K)$ terms in Chapter 7, but some higher order contributions have been determined recently (Sela and Goldhirsch, 1998; Kumaran, 2004). For rough particles, however, to achieve frame indifference of the constitutive relations at first order in K, ϵ, and ε, it is necessary to include certain higher order terms. Though the reason for this is not entirely clear, it appears likely that the additional terms may not be truly of higher order everywhere in the flow field. For instance, we have seen in §8.2.1 and §8.2.3 that the velocity varies sharply near the walls in plane Couette flow; hence the dimensionless vorticity in this region is larger than $O(K)$. In regions far from the walls, where it is $O(K)$, it is likely that $\hat{\bar{\boldsymbol{\omega}}}$ too is of similar magnitude, and hence the term $\hat{\zeta}\,\boldsymbol{\epsilon}\cdot(\hat{\bar{\boldsymbol{\omega}}} - \frac{1}{2}\hat{\boldsymbol{\nabla}}\mathbf{x}\hat{\mathbf{v}})$ is effectively of higher order (see Problem 9.3). This was indeed the finding of Campbell (1993) in particle dynamics simulations of plane Couette flow. Nevertheless, further investigation is necessary to resolve this issue.

9.5. SUMMARY

In this chapter we have considered the hydrodynamics of a granular material composed of rough, inelastic spheres. The impulse transmitted during a collision between two rough particles depends on their angular velocities about their own axes, or spin, apart from their translational velocities. In the collision model that we have used, the roughness coefficient β quantifies the roughness of the particles. While β may, in general, depend on the velocity and incident angle of the impact, we have assumed it to be constant, in the spirit of the analysis of Chapter 7.

A hydrodynamic description of rough gases is applicable when the kinetic energy is nearly conserved: this is the case if the inelasticity is small ($\epsilon \equiv (1 - e_{\rm p}^2) \ll 1$) and if the particles are either nearly smooth ($1 + \beta \ll 1$) or nearly perfectly rough ($1 - \beta \ll 1$). The Chapman–Enskog procedure for dense gases was extended by treating ϵ and $\varepsilon \equiv 1 - |\beta|$, in addition to the inverse Knudsen number K, as small parameters. The constitutive relations were determined up to first order in these parameters.

For nearly perfectly rough particles, the Chapman–Enskog expansion is carried out about the equilibrium state for perfectly rough, elastic particles, for which the translation velocities and spin follow the Maxwell–Boltzmann distribution. There is equipartition of energy between translational and rotational modes, and at leading order in the expansion, the mean spin equals half the vorticity. The constitutive relations have the same form as that of a granular gas of smooth inelastic particles, but with altered viscosity, thermal conductivity, and dissipation rate.

For particles that are nearly smooth, the Chapman–Enskog expansion is about the equilibrium state for smooth, elastic particles, for which the translation velocities of the particles follow the Maxwell–Boltzmann distribution, but there is no restriction on distribution of spin. Hence energy need not be equally partitioned between the translational and rotational modes, and the mean spin is not necessarily equal to the vorticity. In this case, the granular gas has features of a *micropolar* or *Cosserat* continuum: (i) the stress tensor is in general not symmetric; (ii) the angular momentum balance is not identically satisfied, but must be enforced; and (iii) there is in general a *couple stress*, or angular momentum flux. The viscosity and thermal conductivity are identical to those of a granular gas of smooth particles, as the tangential impulse transmitted during collision does not contribute to the flux of momentum or energy to first order in the small parameters. An important difference with the analysis for smooth particles is that certain higher-order terms in the expansion have to be retained to keep the momentum and energy balances frame indifferent.

PROBLEMS

9.1. Show that considerations of tensor symmetry lead to the forms for **A** and **B** given in (9.51) and (9.52), respectively. Show that the moment constraint (9.53) reduces to the forms given in (9.55).

9.2. Construct a hydrodynamic balance for ϑ for nearly smooth, nearly elastic particles. First integrate the Boltzmann equation (9.40) over **c**. Use the continuity equation to simplify the result and write the material derivative of ϑ in terms of the divergence of $\rho\,\vartheta\,\delta\mathbf{v}$ and an integral, where $\delta\mathbf{v} \equiv \mathbf{v} - \mathbf{v}_{\omega}$ is the difference between the mean velocity and the mean velocity of particles with spin $\boldsymbol{\omega}$. What do $\delta\mathbf{v}$ and the integral evaluate to at equilibrium (i.e., when $K = \epsilon = \varepsilon = 0$)? Using these results, show that $D\vartheta/Dt = 0$ at $O(K)$.

9.3. Consider steady, uniform, unbounded shear flow of a granular gas of nearly smooth particles. The flow is in the x direction and the velocity gradient is in the y direction. There is no variation of any variable in the x direction, and except for u_x (which varies linearly with y), there is no variation of any variable in the y direction. Determine the value of the mean spin $\overline{\boldsymbol{\omega}}$ and compare it with half the vorticity.

10

Hybrid Theories

We have seen in §8.2.3 and 8.3.5 that the kinetic theory for rapid granular flows predicts regions of high density, even when the mean density is low. The stratification of density is caused by a boundary that either absorbs or produces pseudothermal energy. Even in the absence of boundaries, it has been shown in §8.4 that spontaneous instabilities may arise that lead to localized enhanced dissipation by inelastic collisions, and thereby the formation of dense clusters. In the problem of plane Couette flow in the absence of gravity, the solids fraction ν in the plug-like region approaches that of maximum packing when the Couette gap becomes large (see Fig. 8.28). In situations of practical interest, gravitational compaction often causes ν to exceed the value corresponding to loose random packing; for example, Fig. 8.16 shows that ν near the base of an inclined chute can approach ν_{drp}, the solids fraction corresponding to dense random packing. At these densities, some key assumptions of kinetic theory, such as binary collisions and molecular chaos, break down. It is reasonable to expect that a part of the grain assembly is in sustained contact. In such situations, the accuracy of the kinetic theory–based model for granular flow is in doubt. Indeed, the data for the stress in simple shear in Fig. 7.2 show significant deviation at high solids fraction from the scaling $\boldsymbol{\sigma} \sim \dot{\gamma}^2$ predicted by the rapid flow theories. Similarly, the data for chute flow in Fig. 8.12 differ qualitatively from the theoretical predictions in Figs. 8.22 and 8.23.

On the other hand, in flows that are, on average, slowly deforming and dense, there are often regions of rapid deformation and substantial dilation. An example is the exit region in a hopper or bunker (see Fig. 3.11a), where the density falls below that corresponding to loose random packing, and it is therefore unlikely that sustained contacts alone are responsible for transmitting momentum between grains.3

Thus in many situations of practical interest, the flow is in an intermediate regime, between slow quasistatic flow on the one hand and rapid streaming–collisional flow on the other. Grain contacts can be thought of neither as permanent nor as instantaneous; the duration of contact depends on many factors, such as their elastic properties, shape, and roughness. Most theoretical studies have focused on the limits of slow and rapid flows, mainly because there is insufficient understanding of the mechanics and statistics of grain contact in the intermediate regime. However, a few studies have taken simple, but crude, approaches by describing the mechanics in the intermediate regime as a combination of the effects of short-term collisions and sustained contact between grains. The next section is devoted to the discussion of one hybrid model. Other hybrid models that have been proposed in the literature (Savage, 1998; Aranson and Tsimring, 2002; Jop et al., 2006) are discussed briefly in the following section.

10.1. THE FRICTIONAL-KINETIC MODEL

The first attempts at evolving a hybrid model were by Kaza (1982, pp. 317–320) and Savage (1983), who proposed that the net stress may be written as the sum of a rate-independent, quasi-static part, and a rate-dependent part due to short-term collisions between grains. This idea was adopted and refined by Johnson and Jackson (1987), who called it the "frictional–collisional" model; we find the name "frictional–kinetic" used by some authors (Jyotsna and Rao, 1997; Mohan et al., 1997) to be more appropriate as the rapid flow component of the stress includes streaming as well as collisional contributions. Johnson and Jackson (1987) also proposed hybrid boundary conditions for the momentum and energy fluxes, and applied the model to plane shearing between parallel walls. Subsequently, Johnson et al. (1990) and Nott and Jackson (1992) applied this model to flow in inclined chutes.

The first assumption of the model is that the net stress in the granular medium is simply the sum of the "frictional" stress $\boldsymbol{\sigma}^{\mathrm{f}}$ and the "kinetic" stress $\boldsymbol{\sigma}^{\mathrm{k}}$, *each determined as though it acts alone*. The former is the stress generated by sustained grain contacts, discussed in Chapter 2. The latter is the stress arising from streaming of grains and short-term collisions, discussed in Chapter 7. We therefore have

$$\boldsymbol{\sigma} = \boldsymbol{\sigma}^{\mathrm{f}} + \boldsymbol{\sigma}^{\mathrm{k}} \tag{10.1}$$

The term "frictional" was used by Johnson and Jackson (1987) and others to describe the stress arising from sustained contact between grains, though friction only provides tangential forces at the points of grain contact – normal forces can be transmitted even in the absence of friction. However, we retain this terminology for conformity with previous studies. It is clear that (10.1) is a crude approximation; the assumption that the frictional and kinetic components of the stress are determined as though each acts independently implies that grains are simultaneously in sustained contact and in rapid fluctuational motion, which is unphysical. Nevertheless, we shall find that the model is a good starting point. Its predictions are qualitatively in agreement with some experimental observations, and the model is a significant improvement over the frictional and kinetic theories.

The second assumption of the model, proposed by Johnson and Jackson (1987), relates to the balance of pseudothermal energy: it is that the work done by the frictional stress dissipates directly as true heat (i.e., the incoherent thermal motion of the molecules constituting the grains). In other words, the frictional stress does not participate in the pseudothermal energy balance and the rate of production of pseudothermal energy is $-(\boldsymbol{\sigma}^{\mathrm{k}})^{\mathrm{T}}\!:\!\nabla\mathbf{v}$. With this, the pseudothermal energy balance (7.75) becomes

$$\frac{3}{2}\rho\frac{DT}{Dt} = -(\boldsymbol{\sigma}^{\mathrm{k}})^{\mathrm{T}}\!:\!\nabla\mathbf{v} - \nabla\cdot\mathbf{q} - \Gamma \tag{10.2}$$

with the constitutive relations for $\mathbf{q}$ and Γ given in §7.3.11.

The third assumption of the model is an extension of the first two to the boundaries. The shear stress generated by particle–boundary interactions is taken to be the sum of the frictional and collisional contributions, with the former given by the friction boundary condition (3.34) and the latter by (8.19). Referring to Fig. 8.1, the boundary condition (8.1) is replaced by

$$\mathbf{n}\cdot(\boldsymbol{\sigma}^{\mathrm{f}} + \boldsymbol{\sigma}^{\mathrm{k}})\cdot\mathbf{t} = \boldsymbol{\mathcal{S}}_{\mathrm{w}}\cdot\mathbf{t} + \mathbf{n}\cdot\boldsymbol{\sigma}^{\mathrm{f}}\cdot\mathbf{n}\tan\delta\ \operatorname{sgn}[(\mathbf{v}-\mathbf{v}_{\mathrm{w}})\cdot\mathbf{t}] \tag{10.3}$$

where δ is the angle of wall friction, introduced in §3.4, and $\operatorname{sgn}(x)$ is the signum function,

$$\operatorname{sgn}(x)\begin{cases} = +1, & \text{if } x > 0 \\ = -1, & \text{if } x < 0 \end{cases} \tag{10.4}$$

The signum function is necessary to ensure that the frictional stress exerted on the wall opposes the movement of the wall relative to the adjacent granular material.

As in the bulk, it is assumed that the frictional shear stress dissipates directly into heat and therefore does not enter the pseudothermal energy balance at the wall; the energy boundary condition therefore remains unchanged from (8.3),

$$\mathbf{q}\cdot\mathbf{n} = \Gamma_w - \boldsymbol{\mathcal{S}}_w\cdot\mathbf{v}_{slip} \tag{10.5}$$

10.2. APPLICATION TO FLOW IN CHUTES

The frictional–kinetic model has been applied to plane Couette flow (Johnson and Jackson, 1987), flow in inclined chutes (Johnson et al., 1990; Nott and Jackson, 1992; Louge and Keast, 2001), vertical channels (Mohan et al., 1997), hoppers (Jyotsna and Rao, 1997), and bins (Srivastava and Sundaresan, 2003). We consider its application to chute flow as an illustrative example. For the frictional stress, we use the critical state theory in conjunction with the associated flow rule (see §2.14.3). For the kinetic stress we use the form given in §7.3.11 for smooth particles. Thus we assume that frictional forces arise only from long-term contacts between particles; the tangential impulse generated during short-term collisions is assumed to be absent. This is merely a simplifying assumption, not based on a microstructural model of the collisions; however, the analysis may be readily extended to account for the tangential impulse by using the constitutive relations for rough granular gases, given in Chapter 9.

As in §8.3, we consider steady, fully developed flow (see Fig. 8.10), for which the velocity field has the form $v_x = v_x(y)$, $v_y = v_z = 0$. This is a plane flow, as defined in Chapter 2; hence to obtain relations for the frictional stress we may use the equations for plane flow given in §2.14. As the diagonal components of the rate of deformation tensor C_{ij} vanish, the forms of the flow rule and coaxiality condition given by (2.105) and (2.104), respectively, are not useful. Instead, we note that $c_* \equiv C_{ii}/2\ (i = x, y) = 0$. From the flow rule (2.73), we therefore have

$$\dot{\lambda}\,\frac{\partial f}{\partial \sigma} = 2c_* = 0 \tag{10.6}$$

where $f(\sigma, \tau, \nu) = 0$ is the yield condition, σ is the mean stress, and τ the deviator stress, as defined in §2.7. We expect the material to shear or be in a state of incipient shear everywhere and therefore assume that $\dot{\lambda} \neq 0$. Then, (10.6) implies that $\partial f/\partial\sigma = 0$. Thus the material is at critical state, and hence

$$\sigma = \sigma_c(\nu); \qquad \tau = \sigma_c(\nu)\sin\phi \tag{10.7}$$

where $\sigma_c(\nu)$ is the mean frictional stress at critical state and ϕ is the angle of internal friction. We expect the frictional stress to contribute only when ν exceeds ν_{lrp}, the solids fraction at loose random packing, and we expect it to diverge as $\nu \to \nu_{drp}$ (see Fig. 3.12). A few forms of $\sigma_c(\nu)$ that represent this behavior are given in §3.8. Johnson et al. (1990) used the form

$$\sigma_c(\nu) = \begin{cases} a\,\frac{(\nu-\nu_{lrp})^b}{(\nu_{drp}-\nu)^c}, & \text{if } \nu > \nu_{lrp} \\ 0, & \text{if } \nu \le \nu_{lrp} \end{cases} \tag{10.8}$$

where a, b, and c are constants, $\nu_{drp} = 0.65$, and $\nu_{lrp} = 0.5$. Although this form is not obtained by fitting experimental data, we use it here for a qualitative study of the hybrid model. Johnson et al. (1990) used the values $a = 0.05\,\mathrm{Nm}^{-2}$, $b = 2$, and $c = 5$ for glass beads, so that the stress at $\nu = 0.58$ supports a head equal to a few centimeters of the material; we retain these values here.

For the coaxiality condition, we note from (2.79) that $c_* = 0$ implies that $s\cos 2\gamma = 0$. Together with (2.78), this results in

$$\gamma = \mathrm{sgn}(\mathrm{d}v_x/\mathrm{d}y)\,\pi/4 \tag{10.9}$$

For the sake of convenience, we henceforth denote $\mathrm{sgn}(\mathrm{d}v_x/\mathrm{d}y)$ by ϱ. From (10.7), (10.9) and (2.102), we get components of the frictional stress tensor as,

$$\begin{aligned} \sigma^{\mathrm{f}}_{xx} &= \sigma^{\mathrm{f}}_{yy} = \sigma_c(\nu) \\ \sigma^{\mathrm{f}}_{yx} &= -\varrho\,\tau = -\varrho\,\sigma_c(\nu)\sin\phi \end{aligned} \tag{10.10}$$

Thus the frictional stress reduces to a simple form for chute flow; indeed, this is the form for any viscometric flow.

For this flow, the constitutive relations for the kinetic stress, energy flux, and dissipation rate reduce to

$$\sigma^{\mathrm{k}}_{xx} = \sigma^{\mathrm{k}}_{yy} = \rho_{\mathrm{p}}\mathcal{P}(\nu)\,T, \quad \sigma^{\mathrm{k}}_{xy} = \sigma^{\mathrm{k}}_{yx} = \rho_{\mathrm{p}}d_{\mathrm{p}}\eta(\nu)\,T^{1/2}\frac{\mathrm{d}v_x}{\mathrm{d}y}, \tag{10.11}$$

$$q_y = \rho_{\mathrm{p}}d_{\mathrm{p}}\mathcal{K}(\nu)\,T^{1/2}\frac{\mathrm{d}T}{\mathrm{d}y}, \quad \Gamma = \frac{\rho_{\mathrm{p}}}{d_{\mathrm{p}}}\gamma(\nu)\,T^{3/2} \tag{10.12}$$

where the functions $\mathcal{P}(\nu)$, $\eta(\nu)$, and $\gamma(\nu)$ are defined in Table 8.1.

On scaling the variables as

$$y^* = y/H, \quad v^* = v_x/(gd_{\mathrm{p}})^{1/2}, \quad T^* = H^{*2}\,T/(gd_{\mathrm{p}}) \tag{10.13}$$

the equations of motion assume the following dimensionless forms

$$\frac{\mathrm{d}}{\mathrm{d}y^*}\left(\mathcal{P}(\nu)T^* + H^{*2}\sigma_c^*(\nu)\right) = -H^{*3}\,\nu\cos\theta \tag{10.14}$$

$$\frac{\mathrm{d}}{\mathrm{d}y^*}\left(\eta(\nu)T^{*1/2}\frac{\mathrm{d}v^*}{\mathrm{d}y^*} + H^{*2}\varrho\,\sigma_c^*(\nu)\sin\phi\right) = -H^{*3}\,\nu\sin\theta \tag{10.15}$$

$$\frac{1}{H^{*2}}\frac{\mathrm{d}}{\mathrm{d}y^*}\left(\mathcal{K}(\nu)T^{*1/2}\frac{\mathrm{d}T^*}{\mathrm{d}y^*}\right) + \eta(\nu)T^{*1/2}\left(\frac{\mathrm{d}v^*}{\mathrm{d}y^*}\right)^2 - \gamma(\nu)\,T^{*3/2} = 0 \tag{10.16}$$

where $\sigma_c^* \equiv \sigma_c/\rho_{\mathrm{p}}d_{\mathrm{p}}g$ is the dimensional mean frictional stress at critical state and $H^* \equiv H/d_{\mathrm{p}}$ is the dimensionless flow depth.

Boundary conditions (10.3) and (10.5) at the base ($y^* = 0$) take the form

$$H^*\eta_{\mathrm{w}}(\nu)\,T^{*1/2}\,v^* = H^{*2}\sigma_c^*(\nu)(\varrho\,\sin\phi - \mathrm{sgn}(v^*)\tan\delta) \tag{10.17}$$

$$+\,\eta(\nu)\,T^{*1/2}\,\frac{\mathrm{d}v^*}{\mathrm{d}y^*} \tag{10.18}$$

$$\frac{1}{H^{*2}}\mathcal{K}(\nu)\frac{\mathrm{d}T^*}{\mathrm{d}y^*} = -H^*\eta_{\mathrm{w}}(\nu)v^{*2} + \frac{\varepsilon}{H^*}\gamma_{\mathrm{w}}(\nu)\,T^* \tag{10.19}$$

At the free surface ($y^* = 1$), boundary conditions (8.114) and (8.115), are extended to include the frictional stress,

$$H^{*2}\sigma_c^*(\nu) + \mathcal{P}(\nu)T^* = \nu\cos\theta \tag{10.20}$$

$$H^{*2}\varrho\,\sigma_c^*(\nu)\sin\phi + \eta(\nu)T^{*1/2}\,\frac{\mathrm{d}v^*}{\mathrm{d}y^*} = \nu\sin\theta \tag{10.21}$$

and the boundary condition for pseudothermal energy is that of no flux,

$$\frac{\mathrm{d}T^*}{\mathrm{d}y^*} = 0 \tag{10.22}$$

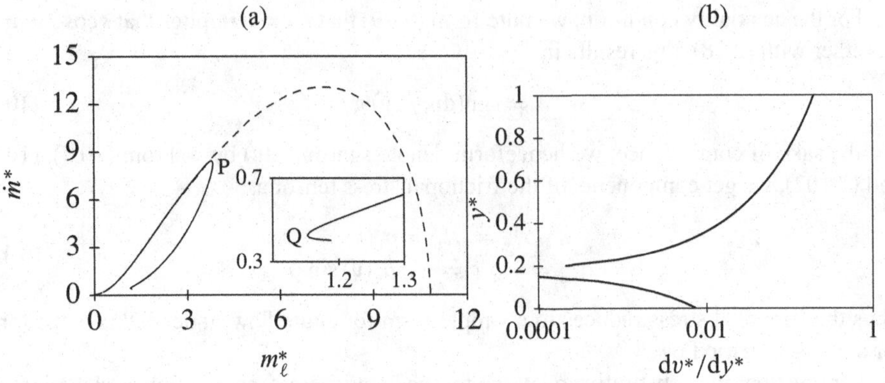

Figure 10.1. (a) The mass flow rate as a function of the loading, predicted by the frictional–kinetic model (solid line) for a chute inclined at 14° from the horizontal. The prediction of the kinetic theory (dashed line) is given for comparison. The inset gives a magnified view of the lower turning point. Parameters are $e_p = 0.91$, $e_w = 0.91$, $\varphi = 0.25$, $a/(\rho_p d_p g) = 1.76 \times 10^{-3}$, $\phi = 28.5°$, $\delta = 12.3°$. (b) Profile of the velocity gradient at the point where the flow curve terminates. Note that dv^*/dy^* vanishes near $y^* = 0.15$.

The solution of equations (10.14)–(10.16), with boundary conditions (10.18)–(10.22), must be determined numerically. Here we have used a Galerkin finite element method, and solved the resulting algebraic equations iteratively by the Newton–Raphson method. The flow curve for the parameter values corresponding to a smooth chute (see §8.3.5.2) is shown in Fig. 10.1a, along with the prediction of the kinetic theory. Starting from a low initial value of m_ℓ and increasing it gradually, we proceed along the upper branch. Initially, this branch coincides with the kinetic theory solution (dashed line), as ν does not exceed ν_{lrp} anywhere. As we traverse the upper branch, the mean density keeps rising, and at a particular point on the flow curve, ν reaches ν_{lrp} at some location in the flowing layer. A short distance further, the solution exhibits a turning point at the point marked P and begins to deviate from the kinetic theory solution. Both m^*_ℓ and $\dot{m}^*$ decrease along the curve, until another turning point occurs at the point marked Q (see inset of Fig. 10.1a). However, a short distance from the second limit point, the solution ceases to exist. At the point of termination, the shear rate vanishes at some distance above the base (Fig. 10.1b). As there is a jump in the value of sgn(x) at $x = 0$, its derivative becomes unbounded. Hence the Jacobian matrix of the discretized equations become ill defined, and a solution cannot be found. Deleting the factor sgn(dv^*/dy^*) from the frictional stress does not solve the problem, rather it worsens it by yielding a solution in which shear rate is negative in a part of the layer. This is an unphysical solution, because the kinetic shear stress in that region acts to aid gravity rather than oppose it, and the kinetic and frictional shear stresses act in opposite directions.

To get around this problem, Johnson et al. (1990) and Nott (1991) enforced a "lockup" condition, which assumes that where the shear rate vanishes, the frictional shear stress adjusts itself so as to ensure a balance of the x momentum. In other words, the frictional stress is not $\varrho\, \sigma^*_c(\nu) \sin\phi$, but assumes a value that balances the forces in the x direction. Thus in regions where $dv^*/dy^* = 0$, (10.15) is discarded from the system of equations, and (10.14) and (10.16) are solved to determine ν and T^*. However, there is a flaw in this argument: although it implicitly relies on the assumption that friction is not fully mobilized in regions of vanishing shear rate, yet the frictional normal stress is taken to be $\sigma^*_c(\nu)$, the mean stress at critical state. This is an internal contradiction in the model, but correcting it is not simple. In regions where the rate of deformation vanishes, the stress is determined by the elastic deformation. Although models that integrate plastic and elastic deformation are available, some of which are discussed in Chapter 5, they are rather complicated. Nevertheless, the idea that friction is not fully mobilized in regions when the shear rate

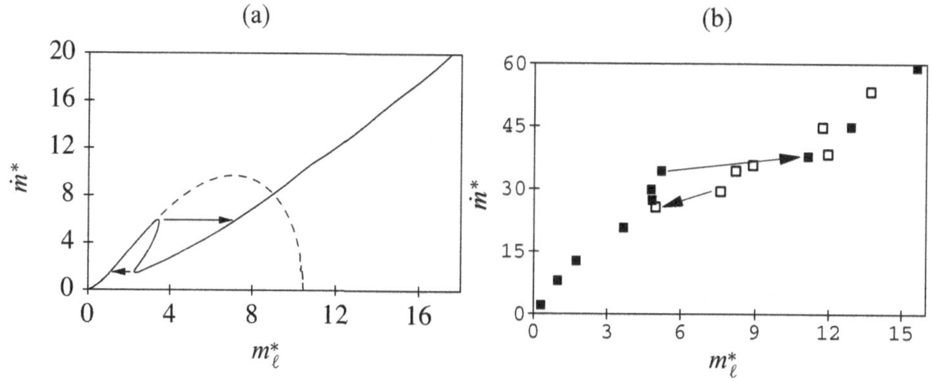

Figure 10.2. (a) The mass flow rate as a function of the loading predicted by the frictional–kinetic model with the "lockup" condition. The results are for a chute inclined at 15° from the horizontal. Parameters as in Fig. 10.1. The prediction of the kinetic theory (dashed line) is given for comparision. (b) Data of Johnson et al. (1990) for 1 mm glass beads flowing in a chute with a smooth aluminum base inclined at 17° from the horizontal. The solid squares are the data for increasing $\dot{m}^*$ and the open squares are for decreasing $\dot{m}^*$.

vanishes is sound, and formulating an approach that avoids the above inconsistency would be useful.

Its internal inconsistency notwithstanding, we discuss the predictions of the frictional–kinetic model with the lockup condition, but with the caveat that it may, at best, be treated as an improvised and ad hoc closure. Figure 10.2a shows the numerical results of Nott (1991) for a chute inclined at 15°. The results of the kinetic theory (dashed line) are shown for comparison. Nott (1991) enhanced the dimensionless viscosity $\eta(\nu)$ (see Table 8.1) by a factor of 4/3 as an approximate way of capturing the anisotropic distribution of collisions on each grain (see §7.4). Without the lockup condition, the flow curve terminates just a little past the lower turning point, as shown in Fig. 10.1a; with the lockup condition, solutions can be found for large m_ℓ or H^*, as in the experimental data of Johnson et al. (1990) and Ancey (2001). At large m_ℓ, the solids fraction is high and the material is locked everywhere except in a thin layer adjacent to the base, a feature that was observed in the experiments of Johnson et al. (1990) and Nott and Jackson (1992). For the range of m_ℓ^* between the two turning points, we observe that there are three solutions. Although their stability has not been analyzed, it is likely that the intermediate branch is unstable. In any case, one should observe a hysteretic variation of $\dot{m}^*$ when m_ℓ^* is increased and decreased, or of m_ℓ^* when $\dot{m}^*$ is increased and decreased. The latter case is indicated by the arrows in the Fig. 10.2a. This feature too was observed by Johnson et al. (1990); their data for $\theta = 17°$ are shown in Fig. 10.2b, the arrows indicating the hysteretic jumps. In their data for decreasing flow flow rate (open squares), they observed surface waves on the flowing layer just before it jumped to a lower m_ℓ^*, suggesting the existence of an instability, as mentioned in §8.4. In the theoretical results as well as experiments, the hysteresis loop exists only for a small range of the chute inclination.

Nott and Jackson (1992) also applied the frictional–kinetic model with the lockup condition to flow in an aerated chute and compared it with experimental data. In their experiments, aeration was effected by pumping air through a porous base plate at a controlled flow rate. In the theory, the effect of aeration was modeled by incorporating a drag force on particles in the direction perpendicular to the base, i.e., in the y direction. The aeration rate is quantified by the dimensionless air velocity $v* \equiv v_{\rm air}/v_{\rm mf}$, where $v_{\rm mf}$ is the *minimum fluidization* velocity. At minimum fluidization, the weight of the granular material is balanced by air drag, and hence $v*$ indicates the extent to which air flow counteracts gravity. Figure 10.3 shows the close agreement between their data and model predictions.

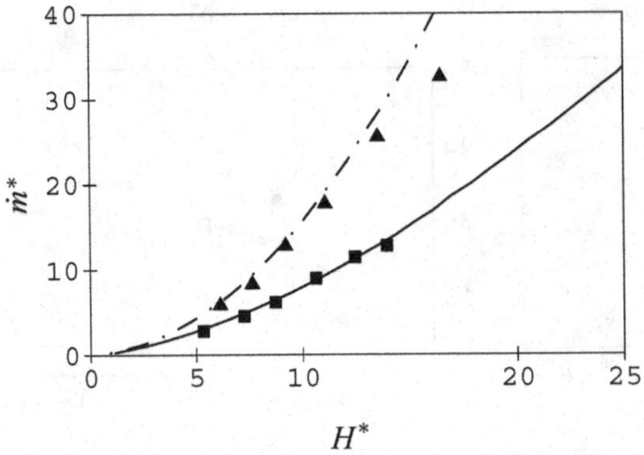

Figure 10.3. The mass flow rate as a function of the flow depth predicted by the frictional–kinetic model with the "lockup" condition, for flow in an aerated chute inclined at 17° to the horizontal. The symbols are the experimental data of Nott and Jackson (1992) for glass beads of diameter $d_{\mathrm{p}} = 1$ mm, and the lines are the predictions of the frictional kinetic model. The squares and solid line are for an aeration rate $v* = 0.42$, and the triangles and dash-dot line are for $v* = 0.74$. Parameter values are $e_{\mathrm{p}} = 0.8$, $e_{\mathrm{w}} = 0.5$, $\phi' = 0.25$, $\phi = 28.5°$, $\delta = 14°$. (Adapted from Nott and Jackson, 1992.)

The parameter φ could not be measured independently, and the value used in the model was chosen to achieve a good fit with the data. Hence one should not read too much into the quantitative agreement, but it is reasonable to say that the model correctly captures the trend of the data.

Thus the frictional–kinetic model with the simple ansatz of the lockup condition appears to reflect not just the gross trends in the experimental data, but even some of the details. However, the theory suffers from the drawback of being extremely sensitive to the values of the parameters e_{w} and φ; Nott (1991) found that under some conditions a very small change in one of these parameters causes a large change in the flow rate. This sensitivity is perhaps an indication that some physical mechanism has been omitted in the theory. The results so far indicate that the idea of combining the frictional and kinetic theories holds promise, but further investigation is necessary to rectify the flaws of the existing model.

10.3. OTHER HYBRID MODELS

Several studies have employed minor variants of the frictional–kinetic model described in the previous section to study specific problems (see, e.g., Louge and Keast, 2001; Srivastava and Sundaresan, 2003). There are a few other hybrid models which, though somewhat speculative and without firm basis, have features that have the potential of making improvements to the frictional–kinetic model.

Savage (1998) proposed a hybrid model that combines the frictional and kinetic theories, but in a manner that differs from that of Johnson and Jackson (1987). The model starts with a yield condition and the associated flow rule, but Savage argued that the rate of deformation tensor **D** at any location fluctuates about the mean $\langle \mathbf{D} \rangle$. He assumed that the fluctuations of **D** follow a Gaussian distribution with standard deviation ϵ, and determined the average stress tensor $\langle \boldsymbol{\sigma} \rangle$ by averaging over the entire range of **D**. With the assumption that $|\langle D_{ij} \rangle| \ll \epsilon$, he obtained a Newtonian constitutive relation for $\langle \boldsymbol{\sigma} \rangle$, with the bulk and shear viscosities given by

$$\mu_{\mathrm{b}} = \frac{\sigma_c(\nu)\,(B - A)}{\epsilon}, \quad \mu = \frac{\sigma_c(\nu)\,A}{\epsilon} \tag{10.23}$$

where σ_c is the mean stress at a critical state, and A and B are material constants that depend only on the angle of internal friction ϕ. To bring in the kinetic theory into the model, Savage replaced σ_c with $\sigma_c + p^k$, where p^k is the kinetic pressure

$$p^k = \rho_p \mathcal{P}(\nu) T \qquad (10.24)$$

$\mathcal{P}(\nu)$ being the "pressure function" defined in Table 8.1. Thus he too has added the frictional and kinetic contributions, but only for the pressure. As a result, the relations for the bulk and shear viscosities become

$$\mu_b = \frac{(\sigma_c(\nu) + p^k)(B - A)}{\epsilon}, \quad \mu = \frac{(\sigma_c(\nu) + p^k) A}{\epsilon} \qquad (10.25)$$

To achieve closure, Savage associated ϵ with the grain temperature T by assuming the relation

$$\epsilon = \beta T^{1/2}/d_p \qquad (10.26)$$

The constant β is determined by requiring the viscosity to approach μ^k, the viscosity derived from kinetic theory, in the limit $p^k \gg \sigma_c$, giving

$$\beta = \frac{\mathcal{P}(\nu) A}{\eta(\nu)} \qquad (10.27)$$

where $\eta(\nu)$ is the viscosity function defined in Table 8.1. Thus only the isotropic parts of the frictional and kinetic stresses are combined in a simple additive manner; the deviatoric parts of the two stresses are combined in a more complicated way.

This model has some desirable features. The idea of accounting for fluctuations in the deformation rate has merit, as sizable fluctuations of the stress have been observed in experiments on slow flows (Tüzün and Nedderman, 1985b; Howell et al., 1999). The dependence of the viscosity on the grain temperature – it decreases as T increases when T is low, as in liquids, but increases with T when T is high, as in gases – is a useful feature in describing situations where slowly and rapidly shearing regions coexist. However, some key assumptions of the model appear to be without adequate basis. It is not clear how macroscopic fluctuations in the deformation rate can be directly related to microscopic fluctuations of the particle velocity: The former is a cooperative phenomenon between many grains, but the latter is a property of single grains. Secondly, the physical origin of fluctuations in the velocity gradient that are much larger than the mean is not clear, and without experimental evidence. Finally, though the constitutive relations are derived using the assumption $|\langle \mathbf{D} \rangle|/\epsilon \ll 1$, the balance of fluctuational energy gives a value of O(1) for this ratio. However, as Savage (1998) himself noted, this model is a preliminary step toward formulating a theory for the intermediate flow regime. The model shows promise, and its refinement seems to be a worthwhile pursuit.

Two other hybrid models have recently been proposed (Aranson and Tsimring, 2002; Jop et al., 2006), but are not sufficiently developed. The model of Aranson and Tsimring (2002) is based on an equation for an order parameter that distinguishes between "solid-like" and "fluid-like" regions. However, the basis for such an equation is not clear. Jop et al. (2006) rely on an empirical rheological model that is based on the experimental data of GDR MiDi (2004); the theory appears to reflect some experimental observations, but lacks a theoretical basis. Both models have ingredients that may be useful in the development of a more comprehensive model for the intermediate flow regime.

10.4. SUMMARY

The occurrence of regions of high density, even when the overall mean solids fraction is below that of loose random packing, indicates that long-term contacts play an important role in the rheology of granular materials. Motivated by this consideration, and the poor agreement between the results of the kinetic theory and experimental observations in chute flow, hybrid frictional–kinetic models have been formulated. In the model of Johnson and Jackson (1987), the kinetic stress due to instantaneous inelastic collisions is combined with the frictional stress that arises from long-term frictional contact in a simple additive manner. When applied to chute flow, it is found that at a particular value of H^*, the shear rate vanishes at a point in the flowing layer, making the derivative of the frictional stress ill defined. Thus solutions exist only for a finite range of H^*, as in the kinetic theory. An ad hoc closure proposed by Johnson et al. (1990), though formally incorrect, allows the continuation of the solution to large H^*; with this assumption, the qualitative features of the solution, and even some of the details, are in agreement with experiment. This is an indication that the model has some desirable features, and an attempt to refine it and eliminate the existing flaws would be a worthwhile pursuit.

Other hybrid models were briefly mentioned, in which the frictional and collisional effects were combined in a slightly different way. Some of these models possess features that may improve the frictional–kinetic model, and therefore warrant further investigation.

APPENDIX A

Operations with Vectors and Tensors

A.1. VECTORS

Consider two points A and B in the physical, three-dimensional space. The *displacement* of a particle moving from A to B is represented by the straight line joining A and B, with an arrow indicating the direction of movement (Fig. A.1). The length of the line AB represents the *magnitude* of the displacement. Thus the displacement has both magnitude and direction and is denoted by a boldface letter, say, $\mathbf{a}$. The magnitude of $\mathbf{a}$ is denoted by $|\mathbf{a}|$. The sum $\mathbf{c}$ of two displacements $\mathbf{a}$ and $\mathbf{b}$ is denoted by $\mathbf{c} = \mathbf{a} + \mathbf{b}$ and is evaluated using the parallelogram rule (Fig. A.1). In other words, the sum is represented by the diagonal of the parallelogram $ABCD$. *Vectors* or *first-order tensors* are quantities that behave like displacements (Resnick and Halliday, 1966, p. 16). Thus vectors have both magnitude (or length) and direction, and the sum of two vectors is a vector obtained using the parallelogram rule. Quantities such as velocity, acceleration, and force are examples of vectors. Multiplication of a vector $\mathbf{a}$ by a real number or *scalar* β produces a vector $\beta\mathbf{a}$, which has a magnitude $|\beta|\,|\mathbf{a}|$, and a direction which is parallel to $\mathbf{a}$ if $\beta > 0$ and antiparallel to $\mathbf{a}$ if $\beta < 0$.

In an n-dimensional space, a set of vectors $\mathbf{b}_i$, $i = 1, n$, forms a *basis* if any vector $\mathbf{a}$ in this space can be expressed as a linear combination of these vectors, i.e.,

$$\mathbf{a} = a_1\mathbf{b}_1 + a_2\mathbf{b}_2 + \cdots + a_n\mathbf{b}_n \tag{A.1}$$

where a_i, $i = 1, n$, are real numbers. If (A.1) holds, the $\{\mathbf{b}_i\}$ are called *basis vectors*.

Consider a rectangular Cartesian coordinate system with coordinates (x_1, x_2, x_3) (Fig. A.1). The vectors $\mathbf{e}_1$, $\mathbf{e}_2$, and $\mathbf{e}_3$, which are directed along the coordinate axes, constitute a set of basis vectors for a three-dimensional space as (A.1) holds. It is convenient to choose these vectors as *unit vectors*, i.e., vectors of unit length. Any vector $\mathbf{a}$ can be expressed in terms of these basis vectors as

$$\mathbf{a} = a_1\mathbf{e}_1 + a_2\mathbf{e}_2 + a_3\mathbf{e}_3 = \sum_{k=1}^{3} a_k\mathbf{e}_k \tag{A.2}$$

The quantities a_i, $i = 1, 3$, are called the *components* of $\mathbf{a}$ relative to a rectangular Cartesian coordinate system with basis vectors $\mathbf{e}_1$, $\mathbf{e}_2$, and $\mathbf{e}_3$. Henceforth, such a coordinate system will be called a Cartesian coordinate system.

In a space of n dimensions, the components of a vector can be represented by a *column vector* or *column matrix*, which is an array containing 1 column and n rows. For example,

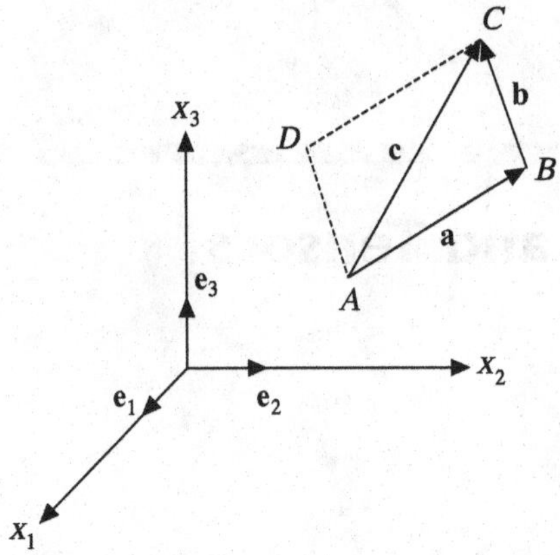

Figure A.1. A rectangular Cartesian coordinate system with basis vectors $\mathbf{e}_1$, $\mathbf{e}_2$, and $\mathbf{e}_3$. The vector **a** represents the displacement of a particle from the point A to the point B. The vector **c** is the sum of the vectors **a** and **b**.

if $n = 3$, the column vector corresponding to the vector **a** is given by

$$[\mathbf{a}] \equiv [a_i] = \begin{pmatrix} a_1 \\ a_2 \\ a_3 \end{pmatrix} \tag{A.3}$$

where a_i, $i = 1, 3$, are the components of **a** relative to a coordinate system.

A.2. THE SUMMATION CONVENTION

Equation (A.2) can be written more compactly in *index notation* by using the *summation convention*, which states that indices which are repeated on any side of an equation imply summation over all possible values of the index, provided the index is a letter. The following material has been adapted from Hunter (1983, pp. 64–69).

Considering a Cartesian coordinate system, (A.2) can be written in index notation as

$$\mathbf{a} = a_i \mathbf{e}_i \tag{A.4}$$

The index i varies from 1 to n, where n is the dimension of the space of interest. In the preceding example, $n = 3$. Some rules regarding the use of the summation convention are discussed below.

Repeated indices are called *dummy* indices, whereas those that are not repeated are called *free* indices. For example, in the equation

$$a_i = b_{ij}\, c_j \tag{A.5}$$

i is a free index and j is a dummy index. Hence this equation can also be written as $a_i = b_{ik}\, c_k$. In general, an equation can contain both dummy and free indices. Each term in an equation should contain the *same* free indices, and dummy indices should occur in pairs only. Equation (A.5) satisfies these requirements, but the equation $a_i = b_{jj}\, c_j$ does not, and hence cannot be evaluated. If the indices are numbers, or if the subscript is enclosed by a bracket, repeated indices do *not* imply summation. Thus $\lambda_{(k)}\, n_{(k)}$ represents a single term, and not a sum of terms.

A.3. THE SCALAR PRODUCT OF TWO VECTORS

It is convenient to introduce a quantity δ_{ij} called the *Kronecker delta* such that

$$\begin{aligned} \delta_{ij} &= 1, \quad i = j \\ &= 0, \quad i \neq j \end{aligned} \tag{A.6}$$

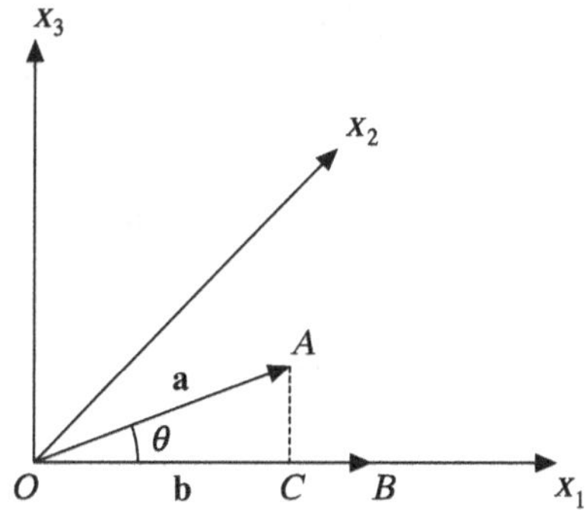

Figure A.2. The scalar product of two vectors **a** (OA) and **b** (OB) is given by the product of the lengths of OC and OB. Here the lengths of OB and OC are b_1 and a_1, respectively, and the length of OA is $|\mathbf{a}|$.

Thus $\delta_{11} = 1$ and $\delta_{12} = 0$, but $\delta_{ii} = n$, where n is the dimension of the space.

The *scalar product* or *dot product* of two vectors **a** and **b** is denoted by $\mathbf{a} \cdot \mathbf{b}$ and is defined by

$$\mathbf{a} \cdot \mathbf{b} \equiv |\mathbf{a}|\, |\mathbf{b}| \cos\theta \tag{A.7}$$

where $|\mathbf{a}|$ denotes the magnitude of **a** and θ is the angle between the vectors. As the basis vectors of a Cartesian coordinate system are of unit length and are mutually orthogonal, we have

$$\mathbf{e}_i \cdot \mathbf{e}_j = \delta_{ij} \tag{A.8}$$

Using (A.8), the scalar product of two vectors $\mathbf{a} = a_i \mathbf{e}_i$ and $\mathbf{b} = b_j \mathbf{e}_j$ is given by

$$\begin{aligned} \mathbf{a} \cdot \mathbf{b} &= (a_i\, \mathbf{e}_i) \cdot (b_j\, \mathbf{e}_j) \\ &= a_i\, b_j\, (\mathbf{e}_i \cdot \mathbf{e}_j) = a_i\, b_j \delta_{ij} \\ &= a_i\, b_i = a_j\, b_j \end{aligned} \tag{A.9}$$

The equivalence of (A.7) and (A.9) can be seen by considering a Cartesian coordinate system whose x_1 axis is directed along the vector **b** (Fig. A.2). Then the only nonzero component of **b** is b_1, and hence $\mathbf{a} \cdot \mathbf{b} = a_1\, b_1$. Referring to Fig. A.2, $OA = |\mathbf{a}|$, $OB = b_1 = |\mathbf{b}|$, and $OC = a_1 = |\mathbf{a}| \cos\theta$. Hence $\mathbf{a} \cdot \mathbf{b} = a_1\, b_1 = |\mathbf{a}| \cos\theta\, |\mathbf{b}| = |\mathbf{a}|\, |\mathbf{b}| \cos\theta$.

The magnitude or length of a vector **a** can be expressed in terms of its components by

$$|\mathbf{a}| = \sqrt{\mathbf{a} \cdot \mathbf{a}} = \sqrt{a_i\, a_i} \tag{A.10}$$

A.4. SECOND-ORDER TENSORS

A *second-order tensor* **A** is a linear mapping that maps vectors into vectors (Slattery, 1999, p. 644). To every vector **u**, the tensor **A** assigns another vector $\mathbf{v} \equiv \mathbf{A} \cdot \mathbf{u}$. As **A** is a linear mapping, it satisfies

$$\mathbf{A} \cdot (\mathbf{u} + \mathbf{w}) = \mathbf{A} \cdot \mathbf{u} + \mathbf{A} \cdot \mathbf{w}; \quad \mathbf{A} \cdot (\beta\, \mathbf{u}) = \beta\, (\mathbf{A} \cdot \mathbf{u})$$

where **u** and **w** are any two vectors, and β is a scalar.

According to Brillouin (1964, pp. 2–3), the origin of the word tensor is as follows:

> When we study the stresses or tensions in the interior of a deformed body, we discover a collection of six numbers, inseparable from one another, which behave like the six components of a certain new quantity. Physicists hesitated for a long time to give a name to this entity, but the study of the physics of crystals revealed a great many analogous examples. The great crystal physicist, Voigt, was the first to recognize the kinship of these various quantities, insist on their common character, and baptize them tensors.

Let $\{\mathbf{e}_i\}$ denote the basis vectors of a Cartesian coordinate system. As $\mathbf{A} \cdot \mathbf{e}_j$ is a vector, it can be expressed as a linear combination of the basis vectors. Hence

$$\mathbf{A} \cdot \mathbf{e}_j = A_{ij}\, \mathbf{e}_i \tag{A.11}$$

where A_{ij} is a scalar. For any vector $\mathbf{u}$,

$$\mathbf{A} \cdot \mathbf{u} = \mathbf{A} \cdot (u_j\, \mathbf{e}_j) = u_j\, (\mathbf{A} \cdot \mathbf{e}_j) = u_j\, A_{ij}\, \mathbf{e}_i = A_{ij}\, u_j\, \mathbf{e}_i \tag{A.12}$$

where (A.11) has been used. Similarly,

$$\mathbf{u} \cdot \mathbf{A} = (u_i\, \mathbf{e}_i) \cdot \mathbf{A} = u_i\, A_{ik}\, \mathbf{e}_k \tag{A.13}$$

To simplify the notation, we shall henceforth denote $\mathbf{A} \cdot \mathbf{u}$ by $\mathbf{Au}$.

It is convenient to express the tensor $\mathbf{A}$ in terms of the tensor product of two vectors, which is as defined follows. Given two vectors $\mathbf{u}$ and $\mathbf{w}$, their *tensor product* or *dyadic product* $\mathbf{uw}$ is defined as a linear mapping of vectors into vectors such that (Slattery, 1999, p. 644)

$$(\mathbf{uw}) \cdot \mathbf{v} \equiv \mathbf{u}\, (\mathbf{w} \cdot \mathbf{v}) \tag{A.14}$$

In particular,

$$(\mathbf{e}_i \mathbf{e}_j) \cdot \mathbf{v} = \mathbf{e}_i\, (\mathbf{e}_j \cdot \mathbf{v}) = \mathbf{e}_i\, [\mathbf{e}_j \cdot (v_k\, \mathbf{e}_k)] = v_j\, \mathbf{e}_i$$

The tensor $\mathbf{A}$ can be defined in terms of the tensor products of the basis vectors $\{\mathbf{e}_i\}$ as

$$\mathbf{A} \equiv A_{ij}\, \mathbf{e}_i \mathbf{e}_j \tag{A.15}$$

Hence $\mathbf{Au} \equiv \mathbf{A} \cdot \mathbf{u} = A_{ij} (\mathbf{e}_i \mathbf{e}_j\ \cdot \mathbf{u}) = A_{ij}\, u_j\, \mathbf{e}_i$, which agrees with (A.12).

As the $\{\mathbf{e}_i\}$ form the basis vectors of a Cartesian coordinate system, the quantities $\{A_{ij}\}$ are said to be the *Cartesian components* of the tensor $\mathbf{A}$. Thus $\mathbf{A}$ has n^2 components in an n-dimensional space. In the applications of interest to us, $n = 2$ or 3.

The *matrix representation* of a second-order tensor $\mathbf{A}$ is given by the square matrix or array formed from the components of the tensor, such that A_{ij} denotes the element in row i and column j of the matrix. Thus for $n = 3$, the matrix representation of $\mathbf{A}$ is given by

$$[\mathbf{A}] \equiv [A_{ij}] = \begin{bmatrix} A_{11} & A_{12} & A_{13} \\ A_{21} & A_{22} & A_{23} \\ A_{31} & A_{32} & A_{33} \end{bmatrix} \tag{A.16}$$

A.4.1. The Unit Tensor

The *unit tensor* of second order is defined by

$$\mathbf{I} \equiv \delta_{ij}\, \mathbf{e}_i\, \mathbf{e}_j \tag{A.17}$$

where δ_{ij} is the Kronecker delta (see (A.6)).

A.4.2. The Trace of a Second-Order Tensor

The *trace* of a second-order tensor $\mathbf{A}$ is the sum of its diagonal components. It is denoted by $\mathrm{tr}(\mathbf{A})$ and is defined by

$$\mathrm{tr}(\mathbf{A}) \equiv A_{ii} \tag{A.18}$$

where the summation convention applies. Thus in a space of n dimensions,

$$\begin{aligned} \mathrm{tr}(\mathbf{A}) &= A_{11} + A_{22} + A_{33}, \quad n = 3 \\ &= A_{11} + A_{22}, \quad n = 2 \end{aligned}$$

A.5. CARTESIAN TENSOR NOTATION

Quantities involving vectors and tensors, such as balance equations and constitutive equations, are said to be expressed in *Cartesian tensor notation* if they are written in terms of the components of vectors and tensors relative to a Cartesian coordinate system. For example, the dot product of two vectors **a** and **b** can be expressed in Cartesian tensor notation as

$$\mathbf{a} \cdot \mathbf{b} = a_i \, b_i \tag{A.19}$$

where $\{a_i\}$ and $\{b_i\}$ denote the components of **a** and **b**, respectively, relative to a Cartesian coordinate system. Cartesian tensor notation will be used henceforth, except in §A.8.

A.6. THIRD AND HIGHER ORDER TENSORS

These tensors are defined by a generalization of (A.15). For example, a third-order tensor **B** is defined as

$$B \equiv B_{ijk} \, \mathbf{e}_i \mathbf{e}_j \mathbf{e}_k \tag{A.20}$$

and the product of **B** with a vector **u** is a second-order tensor C given by

$$\mathbf{C} = \mathbf{B} \cdot \mathbf{u} = (B_{ijk} \, \mathbf{e}_i \mathbf{e}_j \mathbf{e}_k) \cdot (u_l \, \mathbf{e}_l) = B_{ijk} \, u_k \mathbf{e}_i \mathbf{e}_j \tag{A.21}$$

A.6.1. The Alternating Tensor

In a three-dimensional space, the *alternating* tensor $\boldsymbol{\epsilon}$ is defined by

$$\boldsymbol{\epsilon} \equiv \epsilon_{ijk} \, \mathbf{e}_i \mathbf{e}_j \mathbf{e}_k \tag{A.22}$$

where ϵ_{ikj} is called the *permutation symbol.* It is defined by

$$\begin{aligned} \epsilon_{ijk} &= 0, \text{ if any two indices are equal} \\ &= 1, \text{ if } ijk \text{ is an even permutation of 123} \\ &= -1, \text{ if } ijk \text{ is an odd permutation of 123} \end{aligned} \tag{A.23}$$

Thus $\epsilon_{231} = 1$ and $\epsilon_{321} = -1$.

A.7. OPERATIONS WITH VECTORS AND TENSORS

Here we discuss some operations that are used in this book.

A.7.1. The Vector Product of Two Vectors

In a three-dimensional space, the *vector product* or *cross product* of two vectors **a** and **b** is a vector defined by

$$\mathbf{a} \times \mathbf{b} \equiv (|\mathbf{a}| \, |\mathbf{b}| \, \sin\theta) \, \mathbf{c} \tag{A.24}$$

where $|\mathbf{a}|$ denotes the magnitude of **a** and θ is the included angle between **a** and **b**, chosen such that $0 \le \theta \le \pi$. The vector **c** is a unit vector perpendicular to the plane formed by **a** and **b**, directed such that **a**, **b**, and **c** form a right-handed system. In other words, if the fingers of the right hand are directed along an arc connecting **a** to **b**, chosen such that the included angle between the vectors is θ, **c** points in the direction of the thumb. This is called the *right-hand rule*; it reflects an arbitrary but common choice.

Yet another definition of the cross product is given in terms of the determinant of a matrix $[\mathbf{C}] = [C_{ij}]$, which is formed from the unit vectors and the components of **a** and **b** by setting $C_{1j} = \mathbf{e}_j,\ C_{2j} = a_j,\ C_{3j} = b_j,\ j = 1, 3$ (Goodbody, 1982, p. 39):

$$\mathbf{a} \times \mathbf{b} \equiv \det([C]) = \det \begin{bmatrix} \mathbf{e}_1 & \mathbf{e}_2 & \mathbf{e}_3 \\ a_1 & a_2 & a_3 \\ b_1 & b_2 & b_3 \end{bmatrix} \tag{A.25}$$

where "det" denotes the determinant, which is evaluated using (A.41).

The cross product can also be expressed in terms of the components of the alternating tensor as

$$\mathbf{a} \times \mathbf{b} \equiv \epsilon_{ijk}\, a_j\, b_k\, \mathbf{e}_i \tag{A.26}$$

where ϵ_{ijk} is defined by (A.23). The direction of $\mathbf{a} \times \mathbf{b}$, given by (A.26), agrees with that given by the right-hand rule only for a right-handed coordinate system. (For such a coordinate system, if the fingers of the right hand are directed along an arc connecting the basis vector $\mathbf{e}_1$ to $\mathbf{e}_2$, the thumb points in the direction of $\mathbf{e}_3$.) If a left-handed coordinate system is used, either a left-hand rule must be used to define the direction of the vector product or (A.26) must be replaced by $\mathbf{a} \times \mathbf{b} \equiv -\epsilon_{ijk}\, a_j\, b_k\, \mathbf{e}_i$. Unless otherwise stated, we shall assume that a right-handed coordinate system is used.

Using the definitions of the dot and vector products, the permutation symbol ϵ_{ijk} can be expressed as (Goodbody, 1982, p. 114)

$$\epsilon_{ijk} = \mathbf{e}_i \cdot (\mathbf{e}_j \times \mathbf{e}_k) \tag{A.27}$$

A.7.2. The Product of Two Second-Order Tensors

The product of two second-order tensors **A** and **B** is a second-order tensor **C** defined by

$$\begin{aligned} \mathbf{C} \equiv \mathbf{AB} \equiv \mathbf{A} \cdot \mathbf{B} &= (A_{ij}\, \mathbf{e}_i \mathbf{e}_j) \cdot (B_{kl}\, \mathbf{e}_k \mathbf{e}_l) = (A_{ij}\, B_{kl}\, \mathbf{e}_i)(\mathbf{e}_j \cdot \mathbf{e}_k)\, \mathbf{e}_l \\ &= A_{ij}\, B_{jl}\, \mathbf{e}_i\, \mathbf{e}_l \end{aligned} \tag{A.28}$$

Hence the Cartesian components of **C** are $C_{il} = A_{ij}\, B_{jl}$.

In view of the matrix representation of tensors (see §A.4), (A.28) can also be written in matrix form as

$$[\mathbf{C}] = [C_{il}] = [\mathbf{A}][\mathbf{B}] = [A_{ij}\, B_{jl}] \tag{A.29}$$

Thus the product of two matrices [**A**] and [**B**] is a matrix $[\mathbf{C}] = [C_{il}]$, and the element in row i and column l of [**C**] is $A_{ij}\, B_{jl}$. Here [**A**] and [**B**] need not be square matrices, but the number of columns of [**A**] must be equal to the number of rows of [**B**]. It should be noted that matrix multiplication is not commutative in general, i.e., $[\mathbf{A}][\mathbf{B}] \neq [\mathbf{B}][\mathbf{A}]$ in general.

A.7.3. The Scalar Product of Two Second-Order Tensors

The scalar or double dot product of two second-order tensors **A** and **B** is a scalar denoted by **A** : **B** and is defined by

$$\begin{aligned} \mathbf{A} : \mathbf{B} &= (A_{ij} \mathbf{e}_i \mathbf{e}_j) : (B_{kl} \mathbf{e}_k \mathbf{e}_l) \\ &\equiv (A_{ij} B_{kl})(\mathbf{e}_j \cdot \mathbf{e}_k)(\mathbf{e}_i \cdot \mathbf{e}_l) = A_{ij}\, B_{ji} \end{aligned} \tag{A.30}$$

The cross product of two vectors **a** and **b** has been defined by (A.24) and (A.25). It can be expressed in terms of the double dot product of two tensors as (Slattery, 1999, p. 674)

$$\mathbf{a} \times \mathbf{b} \equiv \boldsymbol{\epsilon} : (\mathbf{ba}) \tag{A.31}$$

where $\boldsymbol{\epsilon}$ is the *alternating* tensor defined by (A.22), and a right-handed coordinate system is used. In index notation, (A.31) takes the form

$$\begin{aligned}\mathbf{a} \times \mathbf{b} &= (\epsilon_{ijk}\, \mathbf{e}_i \mathbf{e}_j \mathbf{e}_k) : (b_l\, a_m\, \mathbf{e}_l \mathbf{e}_m) \\ &= (\epsilon_{ijk}\, b_l\, a_m\, \mathbf{e}_i)(\mathbf{e}_j \mathbf{e}_k) : (\mathbf{e}_l \mathbf{e}_m) \\ &= \epsilon_{ijk}\, a_j\, b_k\, \mathbf{e}_i \\ &= (a_2\, b_3 - a_3\, b_2)\, \mathbf{e}_1 + (a_3\, b_1 - a_1\, b_3)\, \mathbf{e}_2 \\ &\quad + (a_1\, b_2 - a_2\, b_1)\, \mathbf{e}_3 \end{aligned} \tag{A.32}$$

The equivalence of (A.24) and (A.32) can be seen by considering a coordinate system with $\mathbf{e}_1$ directed along $\mathbf{a}$, and $\mathbf{e}_2$ chosen to lie in the plane formed by $\mathbf{a}$ and $\mathbf{b}$.

A.7.4. The Transpose of a Tensor

The transpose of a tensor $\mathbf{A}$ is denoted by $\mathbf{A}^{\mathrm{T}}$, and its elements are defined by

$$A^{\mathrm{T}}{}_{ij} = A_{ji} \tag{A.33}$$

If $\mathbf{A}^{\mathrm{T}} = \mathbf{A}$, the tensor is said to be *symmetric*.

The transpose of a matrix $[\mathbf{A}] = [A_{ij}]$ is a matrix $[\mathbf{A}^{\mathrm{T}}] = [A^{\mathrm{T}}{}_{ij}] = [A_{ji}]$.

A.7.5. The Inverse of a Second-Order Tensor

Given a tensor $\mathbf{A}$, the tensor $\mathbf{A}^{-1}$ is said to be the *inverse* of $\mathbf{A}$ if the product of these two tensors is the identity tensor, i.e.,

$$\mathbf{A}^{-1}\mathbf{A} \equiv \mathbf{A}^{-1} \cdot \mathbf{A} = \mathbf{A} \cdot \mathbf{A}^{-1} \equiv \mathbf{A}\mathbf{A}^{-1} = \mathbf{I} \tag{A.34}$$

Equation (A.34) can be written in index notation as

$$A_{ij}^{-1}\, A_{jk} = A_{ij}\, A_{jk}^{-1} = \delta_{ik} \tag{A.35}$$

where δ_{ik} denotes the Kronecker delta.

The inverse of a square matrix $[\mathbf{A}] = [A_{ij}]$ is a square matrix $[\mathbf{A}^{-1}] = [A_{ij}^{-1}]$, which satisfies

$$[\mathbf{A}^{-1}][\mathbf{A}] = [\mathbf{A}][\mathbf{A}^{-1}] = [\mathbf{I}] \tag{A.36}$$

where $[\mathbf{I}] = [\delta_{ij}]$ is the unit matrix. The elements of $[\mathbf{A}^{-1}]$ satisfy (A.35).

A.7.6. The Determinant of a Second-Order Tensor

In a space of three dimensions, the *determinant* of a second-order tensor $\mathbf{A}$ is defined by (Slattery, 1999, p. 678)

$$\begin{aligned}\det(\mathbf{A}) &\equiv \epsilon_{ijk}\, A_{i1}\, A_{j2}\, A_{k3} \\ &= A_{11}(A_{22}\, A_{33} - A_{32}\, A_{23}) - A_{21}(A_{12}\, A_{33} - A_{32}\, A_{13}) \\ &\quad + A_{31}(A_{12}\, A_{23} - A_{22}\, A_{13}) \end{aligned} \tag{A.37}$$

where ϵ_{ijk} is the permutation symbol (see (A.23)) and A_{ij} is a Cartesian component of $\mathbf{A}$. It can be seen by inspection that

$$\det(\mathbf{A}) = \epsilon_{ijk}\, A_{1i}\, A_{2j}\, A_{3k} \tag{A.38}$$

Hence

$$\det(\mathbf{A}) = \det(\mathbf{A}^{\mathrm{T}}) \tag{A.39}$$

Similarly, in a space of two dimensions,

$$\det(\mathbf{A}) \equiv \epsilon_{ij3}\, A_{1i}\, A_{2j} = A_{11}\, A_{22} - A_{12}A_{21} \tag{A.40}$$

The determinant of a square matrix $[\mathbf{A}] = [A_{ij}]$ is given by the same expression as the determinant of a second-order tensor **A**. Thus

$$\det([\mathbf{A}]) = \epsilon_{ijk}\, A_{1i}\, A_{2j}\, A_{3k} \tag{A.41}$$

in three dimensions, and

$$\det([\mathbf{A}]) \equiv \epsilon_{ij3}\, A_{1i}\, A_{2j} = A_{11}\, A_{22} - A_{12}A_{21} \tag{A.42}$$

in two dimensions.

The determinant of a matrix is usually denoted by a pair of vertical lines enclosing the elements of the matrix. In a three-dimensional space, we have

$$\begin{aligned}\det([\mathbf{A}]) = \begin{vmatrix} A_{11} & A_{12} & A_{13} \\ A_{21} & A_{22} & A_{23} \\ A_{31} & A_{32} & A_{33} \end{vmatrix} &= A_{11} \begin{vmatrix} A_{12} & A_{13} \\ A_{22} & A_{23} \end{vmatrix} \\ &\quad - A_{12} \begin{vmatrix} A_{21} & A_{23} \\ A_{31} & A_{33} \end{vmatrix} + A_{13} \begin{vmatrix} A_{21} & A_{22} \\ A_{31} & A_{32} \end{vmatrix} \\ &= A_{11}\,(A_{22}\, A_{33} - A_{23}\, A_{32}) - A_{12}\,(A_{21}\, A_{33} - A_{23}\, A_{31}) \\ &\quad + A_{13}\,(A_{21}\, A_{32} - A_{22}\, A_{31}) \end{aligned} \tag{A.43}$$

A.7.7. Orthogonal Second-Order Tensors

An *orthogonal* tensor **Q** is one whose transpose is equal to its inverse, i.e., $\mathbf{Q}^{\mathrm{T}} = \mathbf{Q}^{-1}$. Hence (A.34) implies that

$$\mathbf{Q}^{\mathrm{T}} \cdot \mathbf{Q} = \mathbf{Q} \cdot \mathbf{Q}^{\mathrm{T}} = \mathbf{I}$$

To simplify the notation, this equation will be rewritten as

$$\mathbf{Q}^{\mathrm{T}}\mathbf{Q} = \mathbf{Q}\mathbf{Q}^{\mathrm{T}} = \mathbf{I} \tag{A.44}$$

Considering a Cartesian coordinate system with basis vectors $\mathbf{e}_i$, $i = 1, 3$, (A.44) can be written in index notation as

$$Q^{\mathrm{T}}{}_{ij}\, Q_{jk} = Q_{ij}\, Q^{\mathrm{T}}{}_{jk} = \delta_{ik} \tag{A.45}$$

As $Q^{\mathrm{T}}{}_{ij} = Q_{ij}$, (A.45) reduces to

$$Q_{ji}\, Q_{jk} = Q_{ij}\, Q_{kj} = \delta_{ik} \tag{A.46}$$

Equation (A.44) implies that

$$\det(\mathbf{Q}^{\mathrm{T}}\, \mathbf{Q}) = \det(\mathbf{I}) = 1$$

As $\det(\mathbf{Q}^{\mathrm{T}}\, \mathbf{Q}) = \det(\mathbf{Q}^{\mathrm{T}}) \det(\mathbf{Q})$ (Problem A.1), and $\det(\mathbf{Q}^{\mathrm{T}}) = \det(\mathbf{Q})$ (see (A.39)), it follows from (A.44) that

$$\det(\mathbf{Q}) = \pm 1 \tag{A.47}$$

Analogous to an orthogonal tensor, an *orthogonal* matrix $[\mathbf{Q}] = [Q_{ij}]$ satisfies

$$[\mathbf{Q}][\mathbf{Q}^{\mathrm{T}}] = [\mathbf{Q}^{\mathrm{T}}][\mathbf{Q}] = [\mathbf{I}] \tag{A.48}$$

where $[\mathbf{I}] = [\delta_{ij}]$ is the unit matrix. The elements of $[\mathbf{Q}]$ satisfy (A.46), and (A.47) implies that

$$\det([\mathbf{Q}]) = \pm 1 \tag{A.49}$$

A.7.8. The Gradient Operator

The *gradient* operator is a vector differential operator denoted by ∇. With respect to a Cartesian coordinate system, it is defined by

$$\nabla \equiv \mathbf{e}_i \frac{\partial}{\partial x_i} \tag{A.50}$$

where the $\{\mathbf{e}_i\}$ are the basis vectors. It should be emphasized that gradient operator has a different form in a curvilinear coordinate system. In general, if such coordinate systems are used, the expressions for quantities involving vector differential operators differ from those obtained using Cartesian tensor notation, as indicated in §A.8.

The Laplacian operator is defined by

$$\nabla^2 \equiv \nabla \cdot \nabla \tag{A.51}$$

In Cartesian tensor notation, (A.51) takes the form

$$\nabla^2 = \frac{\partial}{\partial x_i}\frac{\partial}{\partial x_i} = \frac{\partial^2}{\partial x_1^2} + \frac{\partial^2}{\partial x_2^2} + \frac{\partial^2}{\partial x_3^2} \tag{A.52}$$

in a three-dimensional space.

A.7.9. The Gradient of Scalars and Vectors

The gradient of a scalar c is a vector denoted by ∇c and is given by

$$\nabla c = (\mathbf{e}_i \frac{\partial}{\partial x_i})(c) = \frac{\partial c}{\partial x_i}\,\mathbf{e}_i \tag{A.53}$$

Thus the 2-component of ∇c is $\frac{\partial c}{\partial x_2}$.

The gradient of a vector $\mathbf{v}$ is a second-order tensor denoted by $\nabla\mathbf{v}$. For a vector which depends on the position $\mathbf{x}$ and the time t, it is defined by (Slattery, 1999, p. 657)

$$(\nabla\mathbf{v}) \cdot \mathbf{a} \equiv \lim_{s \to 0} \frac{\mathbf{v}(\mathbf{x} + s\,\mathbf{a}, t) - \mathbf{v}(\mathbf{x}, t)}{s} \tag{A.54}$$

where $\mathbf{a}$ is an arbitrary vector joining two position vectors and s is a scalar. Setting $\mathbf{a} = \mathbf{e}_1$, (A.54) reduces to

$$(\nabla\mathbf{v}) \cdot \mathbf{e}_1 = \lim_{s \to 0} \frac{\mathbf{v}(\mathbf{x} + s\,\mathbf{e}_1, t) - \mathbf{v}(\mathbf{x}, t)}{s} \tag{A.55}$$

By definition, the right-hand side of (A.55) is the partial derivative of $\mathbf{v}$ with respect to x_1, and $\mathbf{v} = v_k\,\mathbf{e}_k$. Hence (A.55) reduces to

$$(\nabla\mathbf{v}) \cdot \mathbf{e}_1 = \frac{\partial \mathbf{v}}{\partial x_1} = \frac{\partial v_k}{\partial x_1}\,\mathbf{e}_k \tag{A.56}$$

Using (A.56), $\nabla\mathbf{v}$ can be expressed in terms of its Cartesian components to obtain

$$\nabla\mathbf{v} = \frac{\partial v_k}{\partial x_i}\,\mathbf{e}_k\,\mathbf{e}_i \tag{A.57}$$

Note that the alternative expression

$$\nabla\mathbf{v} = \frac{\partial v_k}{\partial x_i}\,\mathbf{e}_i\,\mathbf{e}_k \tag{A.58}$$

is also admissible; it corresponds to the definition

$$\mathbf{a} \cdot (\nabla\mathbf{v}) \equiv \lim_{s \to 0} \frac{\mathbf{v}(\mathbf{x} + s\,\mathbf{a}, t) - \mathbf{v}(\mathbf{x}, t)}{s} \tag{A.59}$$

The expression (A.58) will be used in this book.

A.7.10. The Divergence of a Second-Order Tensor

The *divergence* of a tensor **A** is a vector denoted by $\nabla \cdot \mathbf{A}$ and is given by

$$\nabla \cdot \mathbf{A} = (\mathbf{e}_i \frac{\partial}{\partial x_i}) \cdot (A_{jk} \mathbf{e}_j \mathbf{e}_k) = \frac{\partial A_{ik}}{\partial x_i} \mathbf{e}_k \qquad (A.60)$$

A.7.11. The Curl of a Vector

In a space of three dimensions, the *curl* of a vector **v** is defined by

$$\text{curl } \mathbf{v} \equiv \nabla \times \mathbf{v} \equiv - \boldsymbol{\epsilon} : (\nabla \mathbf{v}) \qquad (A.61)$$

where $\nabla \mathbf{v}$ is defined by (A.58). Using (A.58), (A.61) can be expressed in Cartesian tensor notation as

$$\text{curl } \mathbf{v} = \nabla \times \mathbf{v} = \epsilon_{ijk} \frac{\partial v_k}{\partial x_j} \mathbf{e}_i \qquad (A.62)$$

for a right-handed coordinate system. If **v** is the velocity vector, curl **v** is called the *vorticity* vector.

A.8. EQUATIONS IN ORTHOGONAL CURVILINEAR COORDINATE SYSTEMS

The balance laws and expressions for some tensors will be presented for two examples of such systems, namely, cylindrical coordinates and spherical coordinates. In these cases, the directions of the basis vectors vary with position. Hence the expressions for the balance laws are more complicated than in the case of Cartesian coordinates. Details regarding the derivation of the relevant expressions are given in books (see, e.g., Happel and Brenner, 1965, pp. 474–508; Spiegel, 1974, pp. 135–207; Goodbody, 1982, pp. 132–142; Bird et al., 2002, pp. 825–831).

In the equations given later, $\rho = \rho_p \, \nu$ is the density of the granular material, ρ_p is the particle density, ν is the solids fraction, $\{\sigma_{ij}\}$, $\{v_i\}$, and $\{b_i\}$ are the components of the stress tensor, defined in the compressive sense, the velocity vector, and the body force per unit mass, respectively. The equations in §A.8.1 and A.8.2 have been adapted from Goodbody (1982, pp. 137–141) and Bird et al. (2002, pp. 846–847).

A.8.1. Cylindrical Coordinates

The coordinates (r, θ, z) are chosen as indicated in Fig. A.3.

Mass balance

$$\frac{\partial \rho}{\partial t} + \frac{1}{r} \frac{\partial (\rho \, r \, v_r)}{\partial r} + \frac{1}{r} \frac{\partial (\rho \, v_\theta)}{\partial \theta} + \frac{\partial (\rho \, v_z)}{\partial z} = 0 \qquad (A.63)$$

Momentum balance (r component)

$$\rho \left(\frac{\partial v_r}{\partial t} + v_r \frac{\partial v_r}{\partial r} + \frac{v_\theta}{r} \frac{\partial v_r}{\partial \theta} + v_z \frac{\partial v_r}{\partial z} - \frac{v_\theta^2}{r} \right)$$
$$+ \frac{1}{r} \frac{\partial (r \, \sigma_{rr})}{\partial r} + \frac{1}{r} \frac{\partial \sigma_{\theta r}}{\partial \theta} + \frac{\partial \sigma_{zr}}{\partial z} - \frac{\sigma_{\theta\theta}}{r} - \rho \, b_r = 0 \qquad (A.64)$$

Momentum balance (θ component)

$$\rho \left(\frac{\partial v_\theta}{\partial t} + v_r \frac{\partial v_\theta}{\partial r} + \frac{v_\theta}{r} \frac{\partial v_\theta}{\partial \theta} + v_z \frac{\partial v_\theta}{\partial z} + \frac{v_r \, v_\theta}{r} \right)$$
$$+ \frac{1}{r^2} \frac{\partial (r^2 \, \sigma_{r\theta})}{\partial r} + \frac{1}{r} \frac{\partial \sigma_{\theta\theta}}{\partial \theta} + \frac{\partial \sigma_{z\theta}}{\partial z} + \frac{\sigma_{\theta r} - \sigma_{r\theta}}{r} - \rho \, b_\theta = 0 \qquad (A.65)$$

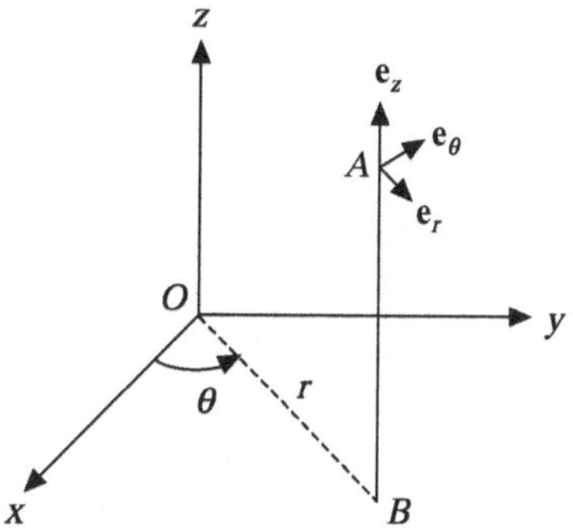

Figure A.3. The cylindrical coordinate system. The point A has coordinates (r, θ, z).

Momentum balance (z component)

$$\rho\left(\frac{\partial v_z}{\partial t}+v_r\,\frac{\partial v_z}{\partial r}+\frac{v_\theta}{r}\,\frac{\partial v_z}{\partial \theta}+v_z\,\frac{\partial v_z}{\partial z}\right)$$
$$+\frac{1}{r}\,\frac{\partial(r\,\sigma_{rz})}{\partial r}+\frac{1}{r}\,\frac{\partial\sigma_{\theta z}}{\partial\theta}+\frac{\partial\sigma_{zz}}{\partial z}-\rho\,b_z=0 \tag{A.66}$$

The preceding momentum balances are valid even if the stress tensor is not symmetric.

The gradient operator

$$\nabla=\mathbf{e}_r\,\frac{\partial}{\partial r}+\frac{\mathbf{e}_\theta}{r}\,\frac{\partial}{\partial\theta}+\mathbf{e}_z\,\frac{\partial}{\partial z} \tag{A.67}$$

Here $\mathbf{e}_r$, $\mathbf{e}_\theta$, and $\mathbf{e}_z$ are orthogonal unit vectors directed along the coordinate directions (see Fig. A.3).

Components of the rate of deformation tensor $\mathbf{C}=-(\nabla\mathbf{v}+(\nabla\mathbf{v})^{\mathrm{T}})/2$

$$C_{rr}=-\frac{\partial v_r}{\partial r};\quad C_{\theta\theta}=-\left(\frac{1}{r}\frac{\partial v_\theta}{\partial\theta}+\frac{v_r}{r}\right);\quad C_{zz}=-\frac{\partial v_z}{\partial z}$$
$$C_{r\theta}=-\frac{1}{2}\left(r\,\frac{\partial}{\partial r}\left(\frac{v_\theta}{r}\right)+\frac{1}{r}\frac{\partial v_r}{\partial\theta}\right);\quad C_{rz}=-\frac{1}{2}\left(\frac{\partial v_z}{\partial r}+\frac{\partial v_r}{\partial z}\right)$$
$$C_{\theta z}=-\frac{1}{2}\left(\frac{1}{r}\frac{\partial v_z}{\partial\theta}+\frac{\partial v_\theta}{\partial z}\right) \tag{A.68}$$

Note that $\mathbf{C}$ is defined in the compressive sense.

Components of the vorticity tensor $\mathbf{W}=(\nabla\mathbf{v}-(\nabla\mathbf{v})^{\mathrm{T}})/2$

$$W_{r\theta}=\frac{1}{2}\left(\frac{1}{r}\,\frac{\partial(r\,v_\theta)}{\partial r}-\frac{1}{r}\frac{\partial v_r}{\partial\theta}\right);\quad W_{rz}=\frac{1}{2}\left(\frac{\partial v_z}{\partial r}-\frac{\partial v_r}{\partial z}\right)$$
$$W_{\theta z}=\frac{1}{2}\left(\frac{1}{r}\frac{\partial v_z}{\partial\theta}-\frac{\partial v_\theta}{\partial z}\right) \tag{A.69}$$

Components of the Eulerian infintesimal strain tensor $\boldsymbol{\epsilon}_*=-(\nabla\mathbf{u}+(\nabla\mathbf{u})^{\mathrm{T}})/2$

$$\epsilon_{*rr}=-\frac{\partial u_r}{\partial r};\ \epsilon_{*\theta\theta}=-\left(\frac{1}{r}\frac{\partial u_\theta}{\partial\theta}+\frac{u_r}{r}\right);\quad \epsilon_{*zz}=-\frac{\partial u_z}{\partial z}$$
$$\epsilon_{*r\theta}=-\frac{1}{2}\left(r\,\frac{\partial}{\partial r}\left(\frac{u_\theta}{r}\right)+\frac{1}{r}\frac{\partial u_r}{\partial\theta}\right)$$
$$\epsilon_{*rz}=-\frac{1}{2}\left(\frac{\partial u_z}{\partial r}+\frac{\partial u_r}{\partial z}\right);\quad \epsilon_{*\theta z}=-\frac{1}{2}\left(\frac{1}{r}\frac{\partial u_z}{\partial\theta}+\frac{\partial u_\theta}{\partial z}\right) \tag{A.70}$$

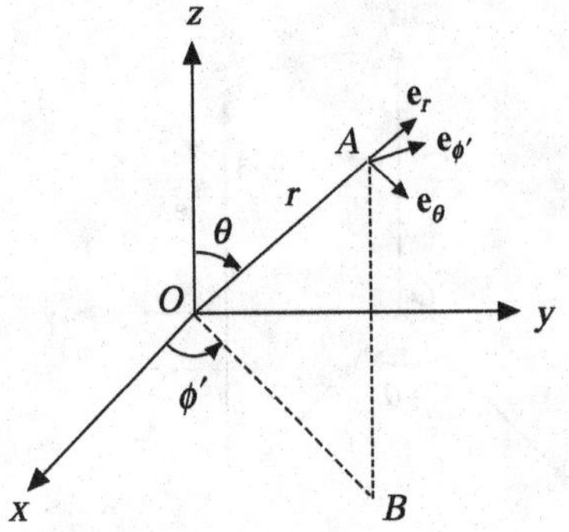

Figure A.4. The spherical coordinate system. The point A has the coordinates (r, θ, ϕ').

Here **u** represents the displacement (see Fig. 1.27). Note that ϵ_* is defined in the compressive sense.

A.8.2. Spherical Coordinates

The coordinates (r, θ, ϕ') are chosen as indicated in Fig. A.4. We use ϕ' instead of the standard symbol ϕ, as the latter has been used to denote the angle of internal friction.

Mass balance

$$\frac{\partial \rho}{\partial t} + \frac{1}{r^2}\frac{\partial(\rho\, r^2\, v_r)}{\partial r} + \frac{1}{r\,\sin\theta}\frac{\partial(\rho\, v_\theta\,\sin\theta)}{\partial\theta} + \frac{1}{r\,\sin\theta}\frac{\partial(\rho\, v_{\phi'})}{\partial\phi'} = 0 \tag{A.71}$$

Momentum balance (r component)

$$\begin{aligned}&\rho\left(\frac{\partial v_r}{\partial t} + v_r\frac{\partial v_r}{\partial r} + \frac{v_\theta}{r}\frac{\partial v_r}{\partial\theta} + \frac{v_{\phi'}}{r\,\sin\theta}\frac{\partial v_r}{\partial\phi'} - \frac{(v_\theta^2 + v_{\phi'}^2)}{r}\right)\\ &+\frac{1}{r^2}\frac{\partial(r^2\,\sigma_{rr})}{\partial r} + \frac{1}{r\,\sin\theta}\frac{\partial(\sigma_{\theta r}\,\sin\theta)}{\partial\theta} + \frac{1}{r\,\sin\theta}\frac{\partial\sigma_{\phi' r}}{\partial\phi'} - \frac{(\sigma_{\theta\theta}+\sigma_{\phi'\phi'})}{r}\\ &-\rho\, b_r = 0\end{aligned} \tag{A.72}$$

Momentum balance (θ component)

$$\begin{aligned}&\rho\left(\frac{\partial v_\theta}{\partial t} + v_r\frac{\partial v_\theta}{\partial r} + \frac{v_\theta}{r}\frac{\partial v_\theta}{\partial\theta} + \frac{v_{\phi'}}{r\,\sin\theta}\frac{\partial v_\theta}{\partial\phi'} + \frac{v_r\, v_\theta - v_{\phi'}^2\,\cot\theta}{r}\right)\\ &+\frac{1}{r^3}\frac{\partial(r^3\,\sigma_{r\theta})}{\partial r} + \frac{1}{r\,\sin\theta}\frac{\partial(\sigma_{\theta\theta}\,\sin\theta)}{\partial\theta}\\ &+\frac{1}{r\,\sin\theta}\frac{\partial\sigma_{\phi'\theta}}{\partial\phi'} + \frac{(\sigma_{\theta r}-\sigma_{r\theta}) - \sigma_{\phi'\phi'}\,\cot\theta}{r} - \rho\, b_\theta = 0\end{aligned} \tag{A.73}$$

Momentum balance (φ′ component)

$$\begin{aligned}&\rho\left(\frac{\partial v_{\phi'}}{\partial t} + v_r\frac{\partial v_{\phi'}}{\partial r} + \frac{v_\theta}{r}\frac{\partial v_{\phi'}}{\partial\theta} + \frac{v_{\phi'}}{r\,\sin\theta}\frac{\partial v_{\phi'}}{\partial\phi'} + \frac{v_{\phi'}\,(v_r + v_\theta\,\cot\theta)}{r}\right)\\ &+\frac{1}{r^3}\frac{\partial(r^3\,\sigma_{r\phi'})}{\partial r} + \frac{1}{r\,\sin\theta}\frac{\partial(\sigma_{\theta\phi'}\,\sin\theta)}{\partial\theta} + \frac{1}{r\,\sin\theta}\frac{\partial\sigma_{\phi'\phi'}}{\partial\phi'}\\ &+\frac{(\sigma_{\phi' r}-\sigma_{r\phi'}) + \sigma_{\phi'\theta}\,\cot\theta}{r} - \rho\, b_{\phi'} = 0\end{aligned} \tag{A.74}$$

The preceding momentum balances are valid even if the stress tensor is not symmetric.

The gradient operator

$$\nabla = \mathbf{e}_r \frac{\partial}{\partial r} + \frac{\mathbf{e}_\theta}{r}\frac{\partial}{\partial \theta} + \frac{\mathbf{e}_{\phi'}}{r \sin\theta}\frac{\partial}{\partial \phi'} \tag{A.75}$$

Here $\mathbf{e}_r$, $\mathbf{e}_\theta$, and $\mathbf{e}_{\phi'}$ are orthogonal unit vectors directed along the coordinate directions (see Fig. A.4).

Components of the rate of deformation tensor $\mathbf{C} = -(\nabla\mathbf{v} + (\nabla\mathbf{v})^{\mathrm{T}})/2$

$$\begin{aligned}
C_{rr} &= -\frac{\partial v_r}{\partial r}; \; C_{\theta\theta} = -\left(\frac{1}{r}\frac{\partial v_\theta}{\partial \theta} + \frac{v_r}{r}\right)\\
C_{\phi'\phi'} &= -\left(\frac{1}{r\sin\theta}\frac{\partial v_{\phi'}}{\partial \phi'} + \frac{v_r + v_\theta \cot\theta}{r}\right)\\
C_{r\theta} &= -\frac{1}{2}\left(r\frac{\partial}{\partial r}\left(\frac{v_\theta}{r}\right) + \frac{1}{r}\frac{\partial v_r}{\partial \theta}\right)\\
C_{r\phi'} &= -\frac{1}{2}\left(r\frac{\partial}{\partial r}\left(\frac{v_{\phi'}}{r}\right) + \frac{1}{r\sin\theta}\frac{\partial v_r}{\partial \phi'}\right)\\
C_{\theta\phi'} &= -\frac{1}{2}\left(\frac{\sin\theta}{r}\frac{\partial}{\partial \theta}\left(\frac{v_{\phi'}}{\sin\theta}\right) + \frac{1}{r\sin\theta}\frac{\partial v_\theta}{\partial \phi'}\right)
\end{aligned} \tag{A.76}$$

Note that $\mathbf{C}$ is defined in the compressive sense.

Components of the vorticity tensor $\mathbf{W} = (\nabla\mathbf{v} - (\nabla\mathbf{v})^{\mathrm{T}})/2$

$$\begin{aligned}
W_{r\theta} &= \frac{1}{2}\left(\frac{1}{r}\frac{\partial(r\, v_\theta)}{\partial r} - \frac{1}{r}\frac{\partial v_r}{\partial \theta}\right)\\
W_{r\phi'} &= \frac{1}{2}\left(\frac{1}{r}\frac{\partial(r\, v_{\phi'})}{\partial r} - \frac{1}{r\sin\theta}\frac{\partial v_r}{\partial \phi'}\right)\\
W_{\theta\phi'} &= \frac{1}{2}\left(\frac{1}{r\sin\theta}\frac{\partial(v_{\phi'}\sin\theta)}{\partial \theta} - \frac{1}{r\sin\theta}\frac{\partial v_\theta}{\partial \phi'}\right)
\end{aligned} \tag{A.77}$$

Components of the Eulerian infintesimal strain tensor $\boldsymbol{\epsilon}_* = -(\nabla\mathbf{u} + (\nabla\mathbf{u})^{\mathrm{T}})/2$

$$\begin{aligned}
\epsilon_{*rr} &= -\frac{\partial u_r}{\partial r}; \; \epsilon_{*\theta\theta} = -\left(\frac{1}{r}\frac{\partial u_\theta}{\partial \theta} + \frac{u_r}{r}\right)\\
\epsilon_{*\phi'\phi'} &= -\left(\frac{1}{r\sin\theta}\frac{\partial u_{\phi'}}{\partial \phi'} + \frac{u_r + u_\theta \cot\theta}{r}\right)\\
\epsilon_{*r\theta} &= -\frac{1}{2}\left(r\frac{\partial}{\partial r}\left(\frac{u_\theta}{r}\right) + \frac{1}{r}\frac{\partial u_r}{\partial \theta}\right)\\
\epsilon_{*r\phi'} &= -\frac{1}{2}\left(r\frac{\partial}{\partial r}\left(\frac{u_{\phi'}}{r}\right) + \frac{1}{r\sin\theta}\frac{\partial u_r}{\partial \phi'}\right)\\
\epsilon_{*\theta\phi'} &= -\frac{1}{2}\left(\frac{\sin\theta}{r}\frac{\partial}{\partial \theta}\left(\frac{u_{\phi'}}{\sin\theta}\right) + \frac{1}{r\sin\theta}\frac{\partial u_\theta}{\partial \phi'}\right)
\end{aligned} \tag{A.78}$$

Here $\mathbf{u}$ represents the displacement (see Fig. 1.27). Note that $\boldsymbol{\epsilon}_*$ is defined in the compressive sense.

PROBLEMS

A.1. Determinant of the product of two second-order tensors

(a) Consider a second-order tensor $\mathbf{A} = A_{ij}\,\mathbf{e}_i\mathbf{e}_j$ and the quantity $C_{mnp} \equiv \epsilon_{ijk}\,A_{im}\,A_{jn}\,A_{kp}$, where ϵ_{ijk} is the permutation symbol defined by (A.23). Show that C_{mnp} changes sign whenever any two of the dummy indices are interchanged. Hence show that the determinant of $\mathbf{A}$, defined by (A.37), can be expressed as

$$C_{mnp} = \epsilon_{ijk}\,A_{im}\,A_{jn}\,A_{kp} = \det(\mathbf{A})\,\epsilon_{mnp} \qquad \text{(A.79)}$$

Similarly, in a two-dimensional space, show that

$$D_{mn3} = \epsilon_{ij3}\,A_{im}\,A_{jn} = \det(\mathbf{A})\,\epsilon_{mn3} \qquad \text{(A.80)}$$

(b) Consider the product of the determinants of two second-order tensors $\mathbf{A}$ and $\mathbf{B}$, given by

$$\det(\mathbf{A})\det(\mathbf{B}) = \det(\mathbf{A})\,\epsilon_{ijk}\,B_{i1}\,B_{j2}\,B_{k3}$$

Using (A.38) and (A.79), show that

$$\det(\mathbf{A})\det(\mathbf{B}) = \det(\mathbf{AB}) \qquad \text{(A.81)}$$

Similarly, for two square matrices $[\mathbf{A}]$ and $[\mathbf{B}]$,

$$\det([\mathbf{A}])\det([\mathbf{B}]) = \det([\mathbf{A}][\mathbf{B}]) \qquad \text{(A.82)}$$

This problem has been adapted from Slattery (1999, pp. 628, 678).

A.2. Identities involving the vector triple product

For any three vectors $\mathbf{a}$, $\mathbf{b}$, and $\mathbf{c}$ in a three-dimensional space, show that the *vector triple products* $\mathbf{a} \times (\mathbf{b} \times \mathbf{c})$ and $(\mathbf{a} \times \mathbf{b}) \times \mathbf{c}$ can be expressed as

$$\mathbf{a} \times (\mathbf{b} \times \mathbf{c}) = \mathbf{b}\,(\mathbf{a} \cdot \mathbf{c}) - \mathbf{c}\,(\mathbf{a} \cdot \mathbf{b}) \qquad \text{(A.83)}$$

$$(\mathbf{a} \times \mathbf{b}) \times \mathbf{c} = \mathbf{b}\,(\mathbf{a} \cdot \mathbf{c}) - \mathbf{a}\,(\mathbf{b} \cdot \mathbf{c}) \qquad \text{(A.84)}$$

Hint: Prove the result for the 1-component of (A.83) by evaluating the right-hand and left-hand sides of the dot product of $\mathbf{e}_1$ with (A.83). Results for the other components follow by permutation of the indices 1, 2, and 3.

This problem has been adapted from Spiegel (1974, pp. 28–29).

A.3. An identity involving the premutation symbol

Using (A.27), we have

$$\epsilon_{ijk}\,\epsilon_{ilm} = [\mathbf{e}_i \cdot (\mathbf{e}_j \times \mathbf{e}_k)]\,[\mathbf{e}_i \cdot (\mathbf{e}_l \times \mathbf{e}_m)] \qquad \text{(A.85)}$$

(a) Noting that $\mathbf{e}_i$ is a basis vector for a Cartesian coordinate system, show by inspection that

$$\begin{aligned}[\mathbf{e}_i \cdot (\mathbf{e}_j \times \mathbf{e}_k)]\,[\mathbf{e}_i \cdot (\mathbf{e}_l \times \mathbf{e}_m)] &= (\mathbf{e}_j \times \mathbf{e}_k) \cdot (\mathbf{e}_l \times \mathbf{e}_m) \\ &= [(\mathbf{e}_j \times \mathbf{e}_k) \times \mathbf{e}_l] \cdot \mathbf{e}_m \end{aligned} \qquad \text{(A.86)}$$

(b) Using (A.84) and (A.86), show that

$$\epsilon_{ijk}\,\epsilon_{ilm} = \delta_{jl}\,\delta_{km} - \delta_{jm}\,\delta_{kl} \qquad \text{(A.87)}$$

This problem has been adapted from Goodbody (1982, p. 116).

A.4. The inverse of a second-order tensor

(a) In a three-dimensional space, use (A.87) to show that

$$\epsilon_{mnp}\,\epsilon_{rnp} = 2\,\delta_{rm} \qquad \text{(A.88)}$$

where δ_{rm} is the Kronecker delta. Similarly, in a two-dimensional space, show that

$$\epsilon_{mn3}\,\epsilon_{rn3} = \delta_{rm} \qquad \text{(A.89)}$$

In (A.89), the indices m, n, and r can only have the values 1 and 2.

(b) Using (A.79), and considering a three-dimensional space, show that

$$B_{ri}\,A_{im} = \det(\mathbf{A})\,\delta_{rm} \tag{A.90}$$

where B_{ri} is the *cofactor* of a_{ir}, defined by

$$B_{ri} \equiv (1/2)\,\epsilon_{rnp}\,\epsilon_{ijk}\,A_{jn}\,A_{kp} \tag{A.91}$$

(c) Considering a Cartesian coordinate system with basis vectors $\mathbf{e}_i$, $i = 1, 3$, (A.90) can be regarded as the component form of the tensor equation

$$\mathbf{BA} = \det(\mathbf{A})\,\mathbf{I}$$

Hence show that

$$\mathbf{A}^{-1} = \mathbf{B}/\det(\mathbf{A})$$

provided $\det(\mathbf{A}) \neq 0$. Conversely, if $\det(\mathbf{A}) = 0$, $\mathbf{A}^{-1}$ does not exist.

(d) In a two-dimensional space, use (A.80) to show that the result obtained in part (c) is valid, provided the cofactor B_{ri} of A_{ir} is defined by

$$B_{ri} = \epsilon_{rn3}\,\epsilon_{ij3}\,a_{jn}$$

This problem has been adapted (with certain modifications for the two-dimensional case) from Slattery (1999, p. 628).

APPENDIX B

The Stress Tensor

The linear momentum balance can be used to express the stress vector $\mathbf{t}_{(\mathbf{n})}$ in terms of a second-order tensor $\boldsymbol{\sigma}$, as shown below. The material has been adapted from Slattery (1999, pp. 32–33).

Consider a material volume *OABC* (Fig. B.1), which is chosen for convenience to be a tetrahedron. As discussed in Chapter 1, the orientations of the normals to the surfaces of the tetrahedron are chosen so as to ensure that compressive normal stresses are assigned positive values. Thus the stress vector $\mathbf{t}_{(\mathbf{n})}$ is the force per unit area exerted on *ABC* by the material above it. Similar remarks apply to the other stress vectors such as $\mathbf{t}_{(\mathbf{e}_2)}$.

As the size of the tetrahedron shrinks to zero, contact forces dominate both the body forces and the acceleration terms. Hence the momentum balance reduces to a balance of forces, which is given by

$$\mathbf{t}_{(\mathbf{n})}\, A + \mathbf{t}_{(\mathbf{e}_i)}\, A_i = 0 \tag{B.1}$$

where the summation convention (see Appendix A) has been used, $\mathbf{n}$ is the unit inward normal to the face *ABC*, and $\mathbf{e}_1$, $\mathbf{e}_2$, and $\mathbf{e}_3$ are normalized basis vectors. Here A is the area of *ABC* (Fig. B.1) and A_i is the area of that face of the tetrahedron whose normal is parallel to the x_i axis. Equation (B.1) can be rewritten as

$$\mathbf{t}_{(\mathbf{n})} = -\mathbf{t}_{(\mathbf{e}_i)}\, (A_i/A) \tag{B.2}$$

As A_i is the projected area of the plane *ABC* on the coordinate plane with unit normal $\mathbf{e}_i$, we have (see Problem 1.6)

$$A_i/A = -n_i = -\mathbf{n}\cdot\mathbf{e}_i$$

The minus sign in the preceding equation arises because $\mathbf{n}$ has been chosen as the inward normal (see Fig. B.1). Thus (B.2) can be rewritten as

$$\mathbf{t}_{(\mathbf{n})} = \mathbf{t}_{(\mathbf{e}_i)}\, n_i \tag{B.3}$$

Vectors such as $\mathbf{t}_{(\mathbf{e}_i)}$ can be expressed as a linear combination of the basis vectors to obtain

$$\mathbf{t}_{(\mathbf{e}_i)} = \sigma_{ij}\mathbf{e}_j \tag{B.4}$$

where σ_{ij} is the component of the vector $\mathbf{t}_{(\mathbf{e}_i)}$ along the x_j axis. Using (B.4), (B.3) can be rewritten as

$$\mathbf{t}_{(\mathbf{n})} = \sigma_{ij}\, \mathbf{e}_j\, n_i = n_i\, \sigma_{ij}\, \mathbf{e}_j$$

Introducing a second-order tensor

$$\boldsymbol{\sigma} \equiv \sigma_{ij}\, \mathbf{e}_i\, \mathbf{e}_j \tag{B.5}$$

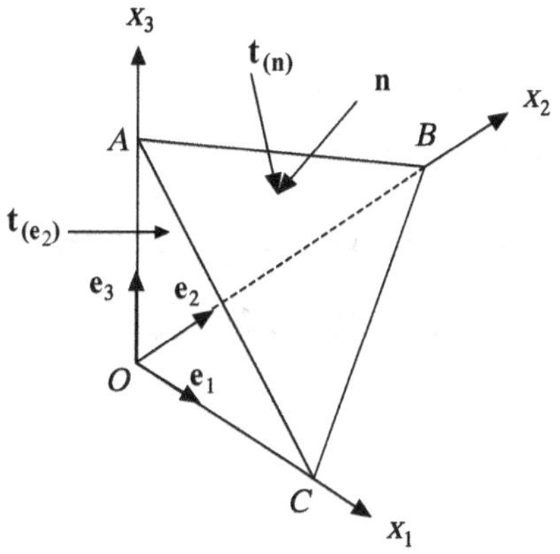

Figure B.1. Balance of forces for a material volume of tetrahedral shape. Here **n** is the unit inward normal to ABC.

(see §A.4 for a brief discussion of tensors), the preceding equation reduces to

$$\mathbf{t}_{(\mathbf{n})} = \mathbf{n} \cdot \boldsymbol{\sigma} \tag{B.6}$$

This is the desired relation between the stress vector and the stress tensor.

APPENDIX C

Hyperbolic Partial Differential Equations of First Order

The frictional constitutive equations discussed in Chapters 2 and 3 often lead to hyperbolic partial differential equations. Some features of hyperbolic equations have been used by several workers to construct stress and velocity fields in hoppers and bunkers. These aspects are discussed briefly later for the case of two independent variables.

Denoting the independent variables by x and y, consider a system of n equations in n dependent variables $u_1, u_2, \ldots, u_n$. The system can be written in matrix form as

$$\mathbf{A}\frac{\partial \mathbf{u}}{\partial x} + \mathbf{B}\frac{\partial \mathbf{u}}{\partial y} + \mathbf{c} = 0 \tag{C.1}$$

where $\mathbf{u}$ is a column matrix representing the vector of dependent variables, $\mathbf{A}$ and $\mathbf{B}$ are $n \times n$ matrices, and $\mathbf{c}$ is a column matrix with n components. (In Appendix A, matrices are denoted by boldface letters enclosed by square brackets; the brackets are omitted here to simplify the notation.) The quantities $\mathbf{A}$, $\mathbf{B}$, and $\mathbf{c}$ are assumed to depend only on x, y, and $\mathbf{u}$, and not on the derivatives of $\mathbf{u}$. Thus (C.1) constitutes a *quasilinear* system, as it is linear in the derivatives of $\mathbf{u}$.

Given initial data along a curve Γ in the x–y plane, the solution can be found at a point Q close to Γ (Fig. C.1), provided the *normal* derivatives $\mathbf{n} \cdot \nabla u_i$, $i = 1, n$, are known. Here $\mathbf{n}$ is the unit normal to Γ at the point P and ∇u_1 is a vector with components $(\partial u_1/\partial x, \partial u_1/\partial y)$. For some systems of equations, there are curves across which the normal derivatives cannot be solved for uniquely. Such curves are called *characteristics*. Along a real characteristic, the system (C.1) reduces to an ordinary differential equation or *compatibility* condition. Thus information propagates along characteristics, and this feature provides a natural scheme for the construction of the solution.

To determine the characteristics, it is convenient to change the independent variables (x, y) to new independent variables (x', y') (Fig. C.1), as suggested by Hill (1950, pp. 347–348). Equation (C.1) can be rewritten as

$$\mathbf{A}\left(\cos\psi\frac{\partial \mathbf{u}}{\partial x'} + \sin\psi\frac{\partial \mathbf{u}}{\partial y'}\right) + \mathbf{B}\left(-\sin\psi\frac{\partial \mathbf{u}}{\partial x'} + \cos\psi\frac{\partial \mathbf{u}}{\partial y'}\right) + \mathbf{c} = 0$$

or

$$\tilde{\mathbf{A}}\frac{\partial \mathbf{u}}{\partial x'} + \tilde{\mathbf{B}}\frac{\partial \mathbf{u}}{\partial y'} + \mathbf{c} = 0 \tag{C.2}$$

where

$$\tilde{\mathbf{A}} = \mathbf{A}\cos\psi - \mathbf{B}\sin\psi; \quad \tilde{\mathbf{B}} = \mathbf{A}\sin\psi + \mathbf{B}\cos\psi \tag{C.3}$$

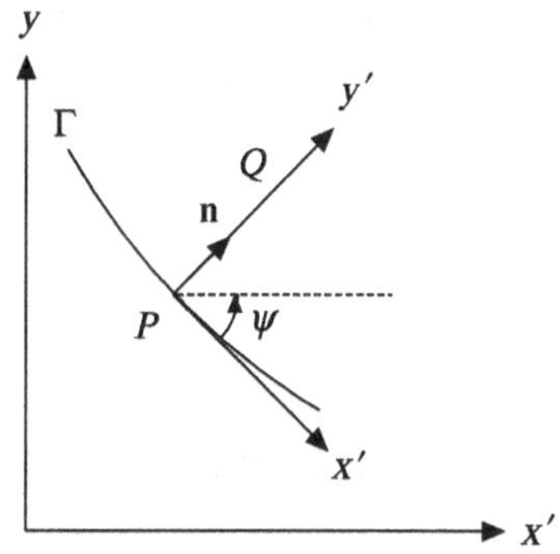

Figure C.1. The locally Cartesian coordinate system (x', y') at the point P. The x' axis is tangential to the curve Γ at P, and $\mathbf{n}$ is the unit normal to Γ.

Equation (C.2) cannot be solved uniquely for the normal derivatives $\partial \mathbf{u}/\partial y'$ if

$$\det[\tilde{\mathbf{B}}] = 0 \tag{C.4}$$

where "det" denotes the determinant (see §A.7.6). The classification of the system (C.1) depends on the roots $\psi_i, i = 1, n$, of the *characteristic equation* (C.4). The system is said to be *hyperbolic* (Courant and Hilbert, 1962, p. 425; Prasad and Ravindran, 1985, p. 153) if (i) there are n real roots, and (ii) there are n *linearly independent* left eigenvectors $\mathbf{l}_j$, such that

$$\mathbf{l}^T_{(j)}\, \tilde{\mathbf{B}}(\psi_{(j)}) = 0, \quad j = 1, n \tag{C.5}$$

where the brackets around the subscript j indicate that the summation convention does not apply. Condition (ii) is automatically satisfied if there are n real and *distinct* roots.

In the hyperbolic case, (C.2) can be rewritten in a simpler form, as follows. Setting $\psi = \psi_j$, premultiplying (C.2) by $\mathbf{l}^T_j$, and using (C.5), we obtain

$$\mathbf{l}^T_{(j)}\, \tilde{\mathbf{A}}_{(j)} \frac{\partial \mathbf{u}}{\partial x'} + \mathbf{l}^T_j\, \mathbf{c} = 0 \tag{C.6}$$

where

$$\tilde{\mathbf{A}}_j \equiv \tilde{\mathbf{A}}(\psi_j) = \mathbf{A} \cos\psi_j - \mathbf{B} \sin\psi_j \tag{C.7}$$

As

$$\frac{\partial \mathbf{u}}{\partial x'} = \cos\psi_j \frac{\partial \mathbf{u}}{\partial x} - \sin\psi_j \frac{\partial \mathbf{u}}{\partial y}$$

(C.6) reduces to

$$\mathbf{l}^T_{(j)}\, \tilde{\mathbf{A}}_{(j)} \left(\cos\psi_j \frac{\partial \mathbf{u}}{\partial x} - \sin\psi_j \frac{\partial \mathbf{u}}{\partial y}\right) + \mathbf{l}^T_j\, \mathbf{c} = 0 \tag{C.8}$$

The jth-characteristic is defined by

$$\frac{\mathrm{d}x}{\mathrm{d}s_j} = \cos\psi_j, \quad \frac{\mathrm{d}y}{\mathrm{d}s_j} = -\sin\psi_j, \quad j = 1, n \tag{C.9}$$

where s_j is the arc length measured along this characteristic. Hence (C.8) can be rewritten as

$$\mathbf{l}^T_{(j)}\, \tilde{\mathbf{A}}_{(j)} \frac{\mathrm{d}\mathbf{u}}{\mathrm{d}s_j} + \mathbf{l}^T_j\, \mathbf{c} = 0, \quad j = 1, n \tag{C.10}$$

and (C.1) has been reduced to an equivalent system of n ordinary differential equations or compatibility conditions along the characteristics. The procedure is illustrated in the following example.

Appendix C

Consider steady, plane flow parallel to the x–y plane. Assuming that the y axis is directed vertically upward, gravity is the only body force, and the inertial terms are negligible, the momentum balances (2.100) and (2.101) reduce to

$$\frac{\partial \sigma_{xx}}{\partial x} + \frac{\partial \sigma_{yx}}{\partial y} = 0$$

$$\frac{\partial \sigma_{xy}}{\partial x} + \frac{\partial \sigma_{yy}}{\partial y} + \rho_p \, \nu \, g = 0 \tag{C.11}$$

Assuming that the solids fraction ν is a known function of (x, y), and considering a cohesionless material satisfying the Mohr–Coulomb yield condition (3.10), the characteristics and the compatiblity conditions can be determined as follows.

The yield condition is given by

$$\tau = \sigma \, \sin\phi \,, \quad \phi = \text{constant} \tag{C.12}$$

and the stress components are given by

$$\sigma_{xx} = \sigma + \tau \, \cos 2\gamma; \quad \sigma_{xy} = \sigma_{yx} = -\tau \, \sin 2\gamma$$

$$\sigma_{yy} = \sigma - \tau \, \cos 2\gamma;$$

$$\sigma \equiv (\sigma_1 + \sigma_2)/2; \quad \tau \equiv (\sigma_1 - \sigma_2)/2 \tag{C.13}$$

Let $\mathbf{u}^T = (\sigma, \gamma)$. The matrices **A**, **B**, and the vector **c** are given by

$$\mathbf{A} = \begin{bmatrix} 1 + \sin\phi \, \cos 2\gamma & -2\sigma \, \sin\phi \, \sin 2\gamma \\ -\sin\phi \, \sin 2\gamma & -2\sigma \, \sin\phi \, \cos 2\gamma \end{bmatrix}$$

$$\mathbf{B} = \begin{bmatrix} -\sin\phi \, \sin 2\gamma & -2\sigma \, \sin\phi \, \cos 2\gamma \\ 1 - \sin\phi \, \cos 2\gamma & 2\sigma \, \sin\phi \, \sin 2\gamma \end{bmatrix}$$

$$\mathbf{c} = \begin{bmatrix} 0 \\ \rho_p \, \nu \, g \end{bmatrix} \tag{C.14}$$

where ρ_p is the particle density and g is the acceleration due to gravity. Hence

$$\tilde{\mathbf{A}} = \begin{bmatrix} \cos\psi + \sin\phi \, \cos(2\gamma - \psi) & -2\sigma \, \sin\phi \, \sin(2\gamma - \psi) \\ -\sin\psi - \sin\phi \, \sin(2\gamma - \psi) & -2\sigma \, \sin\phi \, \cos(2\gamma - \psi) \end{bmatrix}$$

$$\tilde{\mathbf{B}} = \begin{bmatrix} \sin\psi - \sin\phi \, \sin(2\gamma - \psi) & -2\sigma \, \sin\phi \, \cos(2\gamma - \psi) \\ \cos\psi - \sin\phi \, \cos(2\gamma - \psi) & 2\sigma \, \sin\phi \, \sin(2\gamma - \psi) \end{bmatrix} \tag{C.15}$$

The characteristics are given by $\det[\tilde{\mathbf{B}}] = 0$, or

$$2\sigma \, \sin\phi \, (\cos(2\gamma - 2\psi) - \sin\phi) = 0 \tag{C.16}$$

Equation (C.16) has three roots: (i) $\sigma = 0$, and (ii) $\cos(2\gamma - 2\psi) = \sin\phi = \cos((\pi/2) - \phi)$, or $\psi = \gamma \pm \Omega$, where $\Omega \equiv (\pi/4) - (\phi/2)$. Let us consider each of these roots in turn.

(i) $\sigma = 0$

For a cohesionless material, the shear and normal stresses vanish along a surface on which $\sigma = 0$. Thus the characteristic condition is satisfied along any stress free surface of arbitrary orientation. It can be shown that γ cannot be arbitrarily prescribed along such a surface, but must satisfy a certain condition (Problem C.1).

(ii) $\sigma \neq 0$, $\psi = \gamma \pm \Omega$

There are two stress characteristics, defined by

$$\psi_1 = \gamma + \Omega; \quad \psi_2 = \gamma - \Omega \tag{C.17}$$

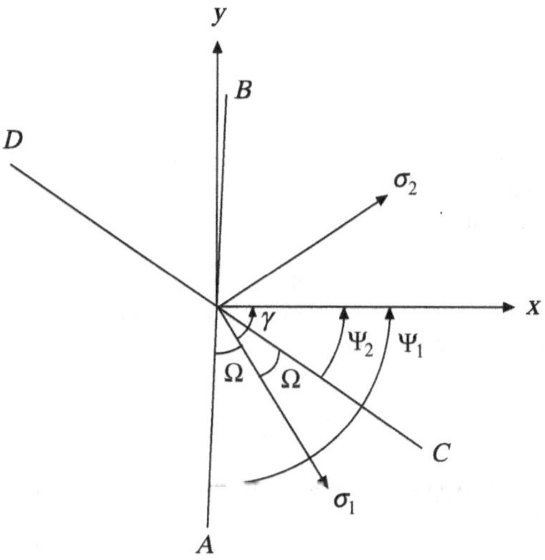

Figure C.2. Orientation of the stress characteristics AB and CD. Here AB and CD correspond to the roots $\psi = \psi_1 = \gamma + \Omega$ and $\psi = \psi_2 = \gamma - \Omega$, respectively, where $\Omega \equiv (\pi/4) - (\phi/2)$ and ϕ is the angle of internal friction.

These are symmetrically disposed relative to the major principal stress axis (Fig. C.2). As there are two real and distinct roots ψ_j, the system is hyperbolic. Using (C.9), the characteristics are given by

$$\frac{\mathrm{d}x}{\mathrm{d}s_j} = \cos(\gamma \pm \Omega); \quad \frac{\mathrm{d}y}{\mathrm{d}s_j} = -\sin(\gamma \pm \Omega), \quad j = 1, 2 \tag{C.18}$$

where the upper sign corresponds to $j = 1$ and the lower sign to $j = 2$.

The compatibility conditions can be determined as follows. Using (C.5) and (C.15), the left eigenvectors $\mathbf{l}_j^T$ are given by

$$-l_{1j}\cos(2\gamma - \psi_j) + l_{2j}\sin(2\gamma - \psi_j) = 0 \tag{C.19}$$

(As $\det[\tilde{\mathbf{B}}(\psi_j)] = 0$, only one of the two equations (C.5) is linearly independent. Equation (C.19) is based on the second column of $\tilde{\mathbf{B}}$.) Thus l_{1j} and l_{2j} can be chosen as

$$l_{1j} = \sin(2\gamma - \psi_j); \quad l_{2j} = \cos(2\gamma - \psi_j), \quad j = 1, 2 \tag{C.20}$$

Hence the expressions for $(\mathbf{l}_{(j)}^T \tilde{\mathbf{A}}_{(j)})$ and $\mathbf{l}_j^T \mathbf{c}$ are given by

$$\begin{aligned}
(\mathbf{l}_j^T\, \tilde{\mathbf{A}}_{(j)})_1 &= \sin(2\gamma - \psi_j)(\cos\psi_j + \sin\phi\, \cos(2\gamma - \psi_j)) \\
&\quad - \cos(2\gamma - \psi_j)(\sin\psi_j + \sin\phi\, \sin(2\gamma - \psi_j)) \\
&= \sin(2\gamma - 2\psi_j) = \mp\cos\phi \\
(\mathbf{l}_j^T\, \tilde{\mathbf{A}}_{(j)})_2 &= -2\sigma\, \sin\phi \\
\mathbf{l}_j^T\, \mathbf{c} &= \rho_\mathrm{p}\, \nu\, g\, \cos(2\gamma - \psi_j) \\
&= \rho_\mathrm{p}\, \nu\, g\, \cos(\gamma \mp \Omega), \quad j = 1, 2
\end{aligned} \tag{C.21}$$

Using the preceding results, the compatibility conditions (C.10) reduce to

$$\mp\cos\phi\, \frac{\mathrm{d}\sigma}{\mathrm{d}s_j} - 2\sigma\, \sin\phi\, \frac{\mathrm{d}\gamma}{\mathrm{d}s_j} + \rho_\mathrm{p}\, \nu\, g\, \cos(\gamma \mp \Omega) = 0, \quad j = 1, 2$$

or

$$\cos\phi\, \frac{\mathrm{d}\sigma}{\mathrm{d}s_j} \pm 2\sigma\, \sin\phi\, \frac{\mathrm{d}\gamma}{\mathrm{d}s_j} \mp \rho_\mathrm{p}\, \nu\, g\, \cos(\gamma \mp \Omega) = 0, \quad j = 1, 2 \tag{C.22}$$

In (C.22), upper sign corresponds to $j = 1$ and the lower sign to $j = 2$.

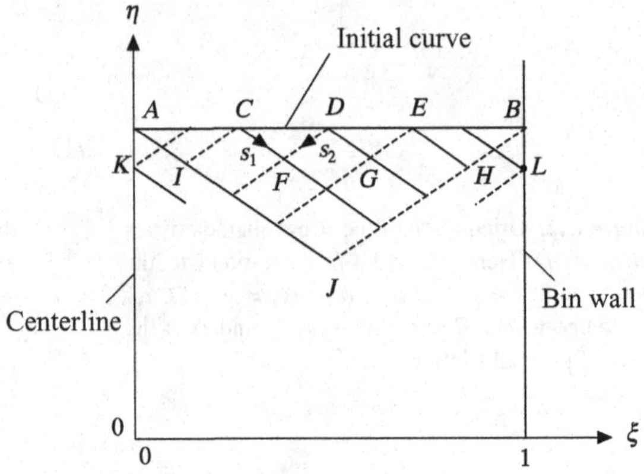

Figure C.3. Characteristics corresponding to $j = 1$ (the solid lines such as *AI* and *CF*) and $j = 2$ (the broken lines such as *BH* and *EG*). The characteristics are shown as straight lines, but are curved in general. Here s_1 and s_2 denote arc lengths along the 1- and 2-families of characteristics, respectively, $\xi = x/W$, $\eta = y/W$, W is the half-width of the bin section, and the coordinate system is chosen as shown in Fig. 4.17.

C.1. SOLUTION BY THE METHOD OF CHARACTERISTICS

Here we discuss how the characteristic curves (4.10) and the compatibility conditions (4.12) can be integrated numerically to obtain the stress field in a bunker.

Consider a horizontal line AB in the ξ–η plane, along which the values of $\overline{\sigma}$ and γ are specified (Fig. C.3). For the moment, it is assumed that AB is not tangential to a characteristic curve at any point. Let γ_C denote the value of γ at the point C on AB. If $\gamma_C + \Omega = \gamma_C + (\pi/4) - (\phi/2) < \pi/2$, the characteristic corresponding to $j = 1$ has a negative slope at C (see (4.10)). Similarly, if $\gamma_C < \Omega$, the characteristic corresponding to $j = 2$ has a positive slope at C.

Confining attention to symmetric solutions, the stress field at points which are below AB can be constructed as follows. The 1-characteristic passing through C (Fig. C.3) can be approximated by a straight line with a slope equal to the slope at C. Hence (4.10) implies that

$$\xi - \xi_C = \cos(\gamma_C + \Omega)\, s_1; \quad \eta - \eta_C = -\sin(\gamma_C + \Omega)\, s_1 \tag{C.23}$$

Similarly, the 2-characteristic passing through D (Fig. C.3) can be approximated by

$$\xi - \xi_D = \cos(\gamma_D - \Omega)\, s_2; \quad \eta - \eta_D = -\sin(\gamma_D - \Omega)\, s_2 \tag{C.24}$$

Equations (C.23) and (C.24) can then be solved to determine the arc lengths $s_1 = CF$ and $s_2 = DF$ and the coordinates $\xi = \xi_F$ and $\eta = \eta_F$ of the point of intersection F of the two characteristics passing through C and D.

Having located F, the compatibility conditions (4.12) can be integrated along the characteristics CF and DF by evaluating all the coefficients in differential equations at C. Thus we have

$$\begin{aligned} \cos\phi\,(\overline{\sigma}_F - \overline{\sigma}_C) + 2\overline{\sigma}_C \sin\phi\,(\gamma_F - \gamma_C) - \cos(\gamma_C - \Omega)\, s_1 = 0 \\ \cos\phi\,(\overline{\sigma}_F - \overline{\sigma}_D) - 2\overline{\sigma}_D \sin\phi\,(\gamma_F - \gamma_D) + \cos(\gamma_D + \Omega)\, s_2 = 0 \end{aligned} \tag{C.25}$$

Equations (C.25) can be solved for the mean stress $\overline{\sigma}_F$ and the angle γ_F. This completes the determination of the stress field at F.

The procedure used earlier for solving the hyperbolic partial differential equations (4.8) is called the *method of characteristics* (Lister, 1965; Zucrow and Hoffman, 1976, pp. 588–591; Nedderman, 1992, pp. 190–199). It is discussed in greater detail in these references, and also in Sokolovskii (1965, pp. 22–24, 31–35).

A few remarks regarding the solution of (4.10) and (4.12) are in order. Equations (C.23)–(C.25) represent an Euler difference scheme, which is of first-order accuracy. In other words, the truncation error involved in approximating the solution to the first of (4.10) by (C.23) is proportional to ${s_1}^2$. Alternatively, the truncation error in approximating derivatives such as $d\xi/ds_1$ is proportional to s_1. As discussed below, schemes of second-order accuracy can also be constructed.

For example, consider the modified Euler predictor–corrector scheme (Zucrow and Hoffman, 1976, pp. 598–599). Let

$$\overline{\gamma}_{CF} \equiv (\gamma_C + \gamma_F)/2; \quad \overline{\sigma}_{CF} \equiv (\overline{\sigma}_C + \overline{\sigma}_F)/2 \tag{C.26}$$

denote the arithmetic mean of the values of γ at C and F (Fig. C.3). Then (4.10) can be approximated by

$$\xi_F - \xi_C = \cos(\overline{\gamma}_{CF} + \Omega)\, s_1; \quad \eta_F - \eta_C = -\sin(\overline{\gamma}_{CF} + \Omega)\, s_1 \tag{C.27}$$

along $j = 1$. Similarly, the compatibility conditions along CF and DF can be approximated by

$$\begin{aligned} &\cos\phi\,(\overline{\sigma}_F - \overline{\sigma}_C) + 2\overline{\sigma}_{CF}\,\sin\phi\,(\gamma_F - \gamma_C) - \cos\phi\,(\overline{\gamma}_{CF} - \Omega)\, s_1 = 0 \\ &\cos\phi\,(\overline{\sigma}_F - \overline{\sigma}_D) - 2\overline{\sigma}_{DF}\,\sin\phi\,(\gamma_F - \gamma_D) + \cos\phi\,(\overline{\gamma}_{DF} + \Omega)\, s_2 = 0 \end{aligned} \tag{C.28}$$

Equations (C.27) and (C.28) are nonlinear in $\overline{\sigma}_F$ and γ_F and can be solved iteratively as follows:

$$\begin{aligned} \xi_F^{n+1} - \xi_C &= \cos\left(\frac{\gamma_F^n + \gamma_C}{2} + \Omega\right) s_1^{n+1} \\ \eta_F^{n+1} - \eta_C &= -\sin\left(\frac{\gamma_F^n + \gamma_C}{2} + \Omega\right) s_1^{n+1} \end{aligned} \tag{C.29}$$

$$\begin{aligned} &\cos\phi\,(\overline{\sigma}_F^{n+1} - \overline{\sigma}_C) + (\overline{\sigma}_C + \overline{\sigma}_F^n)\sin\phi\,(\gamma_F^{n+1} - \gamma_C) \\ &\qquad - \cos\phi\left(\frac{\gamma_F^n + \gamma_C}{2} - \Omega\right) s_1^{n+1} = 0 \\ &\cos\phi\,(\overline{\sigma}_F^{n+1} - \overline{\sigma}_D) - (\overline{\sigma}_D + \overline{\sigma}_F^n)\sin\phi\,(\gamma_F^{n+1} - \gamma_D) \\ &\qquad + \cos\phi\left(\frac{\gamma_F^n + \gamma_D}{2} + \Omega\right) s_2^{n+1} = 0, \quad n = 1, 2, \ldots \end{aligned} \tag{C.30}$$

Here γ_F^n is the value of γ_F after n iterations. For $n = 1$, we set γ_F^1 equal to the value of γ_F obtained by solving the linear equations (C.23)–(C.25). This represents the *predictor* step, and (C.29) and (C.30) represent the *corrector* step.

PROBLEMS

C.1. The variation of γ along a traction-free curve

Consider a cohesionless material satisfying the Mohr–Coulomb yield condition (2.38).

(a) If the mean stress σ vanishes along a curve Γ in the x–y plane, show that the variation of γ along this curve is given by

$$\sin(2\gamma - \psi) = (\sin\psi)/\sin\phi_* \tag{C.31}$$

where the angles ψ and γ are defined as shown in Figs. C.1 and C.2, respectively, and ϕ_* is the angle of internal friction.

(b) Show that the steepest possible slope (traction-free curve) is one which is inclined at an angle ϕ_* to the horizontal.

C.2. Stress characteristics for a material with curved yield loci

Consider the example discussed in the text, with the Mohr–Coulomb yield condition replaced by a more general yield condition of the form $\tau = \tau(\sigma, \nu)$, where σ and τ denote the mean and deviator stresses, respectively. These quantities are defined by (2.33). Show that (C.11) represents a hyperbolic system if $|\partial\tau/\partial\sigma| < 1$ and $\tau \neq 0$. Hint: Instead of repeating the derivation given earlier, replace "$\sin\phi$" in the first column of $\tilde{\mathbf{B}}$ (see (C.15)) by $\partial\tau/\partial\sigma$ and replace "$\sigma\sin\phi$" in the second column of $\tilde{\mathbf{B}}$ by τ. In the hyperbolic case, show that the stress characteristics are inclined at angles of $\pm[(\pi/4) - (\omega/2)]$ relative to the major principal stress axis, where $\sin\omega \equiv \partial\tau/\partial\sigma$.

C.3. Characteristics of the velocity equations for plane flow

Consider the coaxiality condition (2.104) and the flow rule (2.105), subject to the additional assumption that the angle of dilation ν_d is a constant. If the orientation γ of the major principal stress axis is a known function of position, these equations constitute a linear system.

(a) Show that this system is hyperbolic, with characteristics given by

$$\psi_{1,2} = \gamma \pm \Omega_v \tag{C.32}$$

where $\Omega_v \equiv (\pi/4) - (\nu_d/2)$ and ψ is the angle between the characteristic and the x axis (Fig. C.1).

(b) Show that the compatibility conditions are given by

$$-(\cos 2\gamma + \sin\nu_d)\,dv_x + (\sin 2\gamma + \cos\nu_d)\,dv_y = 0, \quad \text{along} \quad \psi = \psi_1$$
$$-(\cos 2\gamma + \sin\nu_d)\,dv_x + (\sin 2\gamma - \cos\nu_d)\,dv_y = 0, \quad \text{along} \quad \psi = \psi_2 \tag{C.33}$$

(c) Let v_1 and v_2 denote the orthogonal projections of the velocity vector onto the 1- and 2-characteristics, respectively (see Fig. C.4). Show that (C.33) can be rewritten as

$$dv_1 + (v_2 \sec\nu_d - v_1 \tan\nu_d)\,d\gamma = 0, \quad \text{along} \quad \psi = \psi_1$$
$$dv_2 + (v_2 \tan\nu_d - v_1 \sec\nu_d)\,d\gamma = 0, \quad \text{along} \quad \psi = \psi_2 \tag{C.34}$$

These equations were derived by Shield (1953).

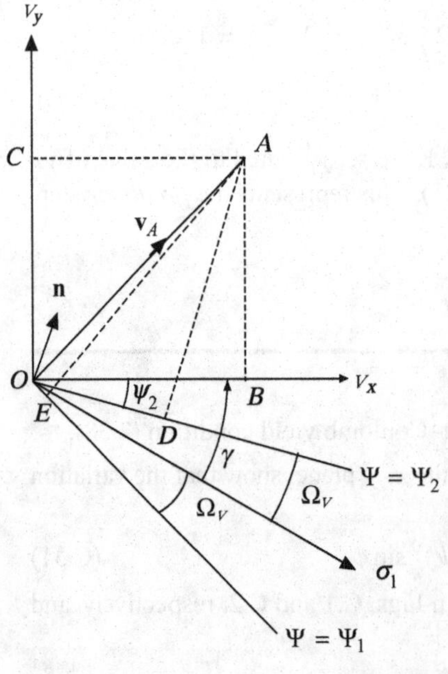

Figure C.4. Projections of the velocity vector $\mathbf{v}_A$ onto the coordinate axes and onto the tangents OE and OD to the 1 - and 2-characteristics passing through O. Here $OB = v_x$, $OC = v_y$, $OD = v_2$, $OE = v_1$, and $\mathbf{n}$ is the unit normal to the 2-characteristic OD.

(d) Consider the flow rule (2.105). If the x axis is locally aligned with a characteristic direction $\psi = \psi_j$, show that

$$\frac{\partial v_x}{\partial x} = 0, \quad j = 1, 2 \tag{C.35}$$

In other words, the rate of stretching vanishes along a velocity characteristic. Hence the velocity characteristics are called *zero extension* lines. Shield (1953) used (C.35) to deduce (C.34). This provides a simpler route than the one indicated in part (c).

APPENDIX D

Jump Balances

As discussed in Chapters 3 and 4, it may be necessary in certain cases to consider discontinuous solutions, wherein the variables such as velocity and density exhibit jumps in their values on crossing a discontinuity surface Γ. Even though the variables are discontinuous across Γ, the jumps cannot be of arbitrary magnitude, but must conform to the *jump balances*, which are given later. If the accumulation of mass and momentum on Γ are neglected, these balances require the mass flux and the sum of the momentum flux and the stress vector to be *continuous* across Γ. The jump mass balance is derived below. The derivation has been adapted from Slattery (1999, pp. 23–25).

D.1. THE JUMP MASS BALANCE

Consider a material volume $V_m(t)$, which has a bounding surface $S_m(t)$ (Fig. D.1), where t denotes the time. A material volume is one whose bounding surface moves in such a manner that the volume contains the same set of material particles at all times. Let $V_m(t)$ enclose a part of the discontinuity surface Γ, as shown in Fig. D.1. The latter divides $V_m(t)$ into two volumes $V_1(t)$ and $V_2(t)$, such that $V_m(t) = V_1(t) + V_2(t)$. Similarly, the bounding surface $S_m(t)$ is divided into two surfaces $S_1(t)$ and $S_2(t)$, such that $S_m(t) = S_1(t) + S_2(t)$.

Integrating the differential mass balance (1.25) over $V_1(t)$ and using the divergence theorem (1.81), we obtain

$$\int_{V_1(t)} \frac{\partial \rho}{\partial t}\, \mathrm{d}V + \int_{S_1(t)} \mathbf{n}_* \cdot (\rho\, \mathbf{v})\, \mathrm{d}S - \int_{S_d(t)} \mathbf{n}_d \cdot (\rho_1\, \mathbf{v}_1)\, \mathrm{d}S = 0 \tag{D.1}$$

where $\mathbf{n}_*$ is the unit outward normal at any point of S_1 or S_2, $S_d(t)$ is that part of the discontinuity surface which is enclosed by $V_m(t)$, and $\mathbf{n}_d$ is the unit normal at any point of S_d (Fig. D.1). Here $\mathbf{n}_d$ has been chosen to point from region 2 to region 1 and ρ_1 is the limiting value of ρ as Γ is approached from region 1.

The first term on the left-hand side of (D.1) may be rewritten using the generalized transport theorem (1.82):

$$\frac{\mathrm{d}}{\mathrm{d}t} \int_{V(t)} \psi\, \mathrm{d}V = \int_{V(t)} \frac{\partial \psi}{\partial t}\, \mathrm{d}V + \int_{S(t)} (\mathbf{n}_* \cdot \mathbf{w})\, \psi\, \mathrm{d}S \tag{D.2}$$

Here $V(t)$ is an arbitrary time-dependent volume whose bounding surface $S(t)$ moves with a velocity $\mathbf{w}$ and $\mathbf{n}_*$ is the unit outward normal to $S(t)$. To apply (D.2) to $V_1(t)$, we note that the boundary of $V_1(t)$ consists of a surface $S_1(t)$ which moves with velocity $\mathbf{v}$ and a surface

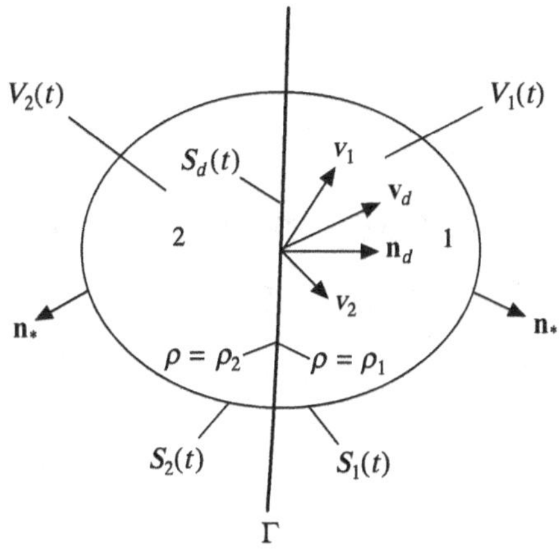

Figure D.1. A material volume $V_m(t) = V_1(t) + V_2(t)$ which encloses a part of the discontinuity surface Γ. Here $\mathbf{n}_*$ is the unit outward normal to the material surface $S_m(t) = S_1(t) + S_2(t)$, $\mathbf{n}_d$ is the unit normal to Γ, and $\mathbf{v}_d$ is the velocity of Γ relative to the laboratory frame.

$S_d(t)$ which moves with velocity $\mathbf{v}_d$. Setting $\psi = \rho$ in (D.2), we obtain

$$\frac{\mathrm{d}}{\mathrm{d}t} \int_{V_1(t)} \rho \, \mathrm{d}V = \int_{V_1(t)} \frac{\partial \rho}{\partial t} \, \mathrm{d}V + \int_{S_1(t)} (\mathbf{n}_* \cdot \mathbf{v}) \, \rho \, \mathrm{d}S - \int_{S_d(t)} (\mathbf{n}_d \cdot \mathbf{v}_d) \, \rho_1 \, \mathrm{d}S \tag{D.3}$$

Using (D.1), (D.3) reduces to

$$\frac{\mathrm{d}}{\mathrm{d}t} \int_{V_1(t)} \rho \, \mathrm{d}V - \int_{S_d(t)} \mathbf{n}_d \cdot [\rho_1 (\mathbf{v}_1 - \mathbf{v}_d)] \, \mathrm{d}S = 0 \tag{D.4}$$

Similarly, integration of (1.25) over $V_2(t)$ gives

$$\frac{\mathrm{d}}{\mathrm{d}t} \int_{V_2(t)} \rho \, \mathrm{d}V + \int_{S_d(t)} \mathbf{n}_d \cdot [\rho_2 (\mathbf{v}_2 - \mathbf{v}_d)] \, \mathrm{d}S = 0 \tag{D.5}$$

Adding (D.4) and (D.5), and noting that $V_1 + V_2 = V_m$, we obtain

$$\frac{\mathrm{d}}{\mathrm{d}t} \int_{V_m(t)} \rho \, \mathrm{d}V + \int_{S_d(t)} [\, \mathbf{n}_d \cdot (\rho_2 (\mathbf{v}_2 - \mathbf{v}_d)) - \mathbf{n}_d \cdot (\rho_1 (\mathbf{v}_1 - \mathbf{v}_d)) \,] \, \mathrm{d}S = 0 \tag{D.6}$$

Using the integral mass balance (1.24) and assuming that the integrand in the second integral in (D.6) is continuous, (D.6) reduces to the *jump mass balance*

$$\mathbf{n}_d \cdot [\rho_2 (\mathbf{v}_2 - \mathbf{v}_d)] - \mathbf{n}_d \cdot [\rho_1 (\mathbf{v}_1 - \mathbf{v}_d)] = 0 \tag{D.7}$$

at every point of Γ. As $\rho = \rho_p \, \nu$, where ρ_p is the particle density (assumed constant) and ν is the solids fraction, (D.7) can also be written as

$$\mathbf{n}_d \cdot [\nu_2 (\mathbf{v}_2 - \mathbf{v}_d)] - \mathbf{n}_d \cdot [\nu_1 (\mathbf{v}_1 - \mathbf{v}_d)] = 0 \tag{D.8}$$

Equation (D.7) implies that the mass flux is *continuous* across the discontinuity surface.

PROBLEMS

D.1. The jump momentum balance

Starting with the differential linear momentum balance (1.32), show that the jump linear momentum balance is given by (Slattery, 1999, p. 34)

$$\mathbf{n}_d \cdot [\rho_2 (\mathbf{v}_2 - \mathbf{v}_d) \, \mathbf{v}_2 + \boldsymbol{\sigma}_2] - \mathbf{n}_d \cdot [\rho_1 (\mathbf{v}_1 - \mathbf{v}_d) \, \mathbf{v}_1 + \boldsymbol{\sigma}_1] = 0 \tag{D.9}$$

where $\boldsymbol{\sigma}_1$ is the stress tensor obtained in the limit as the discontinuity surface is approached from region 1 (Fig. D.1). Note that the stress tensor is defined in the compressive sense and the integral form of the linear momentum balance is given by (1.29).

APPENDIX E

Discontinuous Solutions of Hyperbolic Equations

Hyperbolic partial differential equations admit both continuous and discontinuous solutions. The jumps in the values of the dependent variables across a discontinuity curve C cannot be arbitrary, but must satisfy certain *jump* conditions. These conditions will now be derived for a system of quasilinear equations of first order in two independent variables x and y. (Quasilinear equations are those which are linear in the derivatives.) The corresponding conditions for linear equations will then be derived as a special case. The derivation has been adapted from Prasad and Ravindran (1985, pp. 202–207).

The discussion is confined to systems of equations which can be written in the *divergence* or *conservation law* form

$$\frac{\partial \mathbf{E}}{\partial x} + \frac{\partial \mathbf{F}}{\partial y} + \mathbf{G} = 0 \tag{E.1}$$

where $\mathbf{E}$, $\mathbf{F}$, and $\mathbf{G}$ are vector functions of x, y, and the vector of dependent variables $\mathbf{u}$.

The derivation proceeds as follows. Using a test function (to be defined shortly), (E.1) is rewritten in an alternative *weak* form. A *genuine* solution $\mathbf{u}(x, y)$ has continuous partial derivatives, and it satisfies both (E.1) and the weak form corresponding to (E.1). As the latter does not involve the derivatives of $\mathbf{u}$, it can also be satisfied by a *weak* solution that may be discontinuous across a curve C in the x–y plane. Using the concepts of a weak solution and a test function, (E.1) can be manipulated to obtain the desired jump conditions.

E.1. WEAK SOLUTION

Consider a closed domain D in the x–y plane and another closed domain D_t which is contained entirely within D (Fig. E.1). A *test function* $V(x, y)$ is any function which has continuous derivatives in D_t and which vanishes outside D_t.

Let $\mathbf{u}(x, y)$ be a genuine solution of (E.1). Multiplying (E.1) by $V(x, y)$ and integrating over D, we obtain

$$\int\!\!\int_D V\left(\frac{\partial \mathbf{E}}{\partial x} + \frac{\partial \mathbf{F}}{\partial y} + \mathbf{G}\right) \mathrm{d}x\,\mathrm{d}y = \int\!\!\int_D \left(\frac{\partial (V\mathbf{E})}{\partial x} + \frac{\partial (V\mathbf{F})}{\partial y}\right) \mathrm{d}x\,\mathrm{d}y$$
$$+ \int\!\!\int_D \left(-\mathbf{E}\frac{\partial V}{\partial x} - \mathbf{F}\frac{\partial V}{\partial y} + \mathbf{G}V\right) \mathrm{d}x\,\mathrm{d}y = 0$$

or, using Green's theorem (see, e.g., Kaplan, 1952, pp. 239–241)

$$\int_B (n_{x*}V\mathbf{E} + n_{y*}V\mathbf{F})\,\mathrm{d}s + \int\!\!\int_D \left(-\mathbf{E}\frac{\partial V}{\partial x} - \mathbf{F}\frac{\partial V}{\partial y} + \mathbf{G}V\right) \mathrm{d}x\,\mathrm{d}y = 0 \tag{E.2}$$

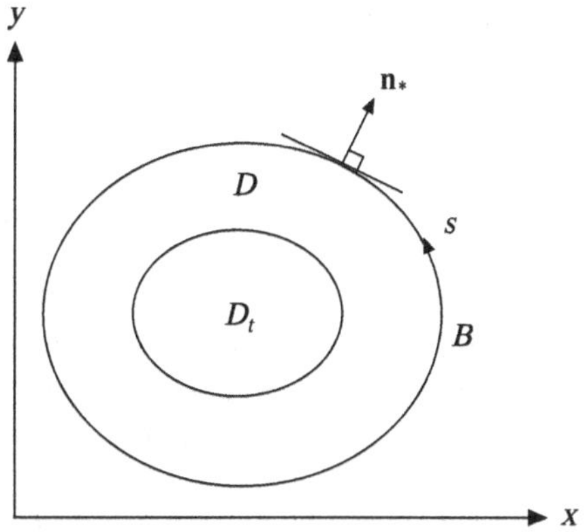

Figure E.1. A domain D containing a subdomain D_t. A test function has continuous derivatives in D_t and vanishes outside D_t. Here $\mathbf{n}_*$ is the unit outward normal to D and B is the bounding curve of D.

Here n_{x*} and n_{y*} are the components of the unit outward normal to the boundary B of the domain D, s is the arc length measured along B, and the line integral is evaluated in the anticlockwise direction (Fig. E.1). (As noted by Kaplan (1952, p. 242), Green's theorem can be regarded as a two-dimensional version of the divergence theorem (1.81), with the volume integral in (1.81) replaced by a surface integral and the area integral by a line integral.) As V vanishes outside the domain D_t (Fig. E.1), (E.2) reduces to

$$\int\int_D \left(-\mathbf{E} \frac{\partial V}{\partial x} - \mathbf{F} \frac{\partial V}{\partial y} + \mathbf{G} V \right) \, dx \, dy = 0 \tag{E.3}$$

A genuine solution satisfies both (E.1) and (E.3). However, derivatives of $\mathbf{u}(x, y)$ do not occur in (E.3), and hence it can be satisfied by a larger class of functions than those satisfying (E.1). A function $\mathbf{u}(x, y)$ that satisfies (E.3) for all admissible test functions $V(x, y)$ is called a weak solution.

E.2. JUMP CONDITIONS

Consider a weak solution $\mathbf{u}(x, y)$ that coincides with a genuine solution in the subdomains D_1 and D_2 (Fig. E.2) and is discontinuous across the curve C. Let $V(x, y)$ be a test function that vanishes outside a domain D_t, the latter being contained entirely within the union of

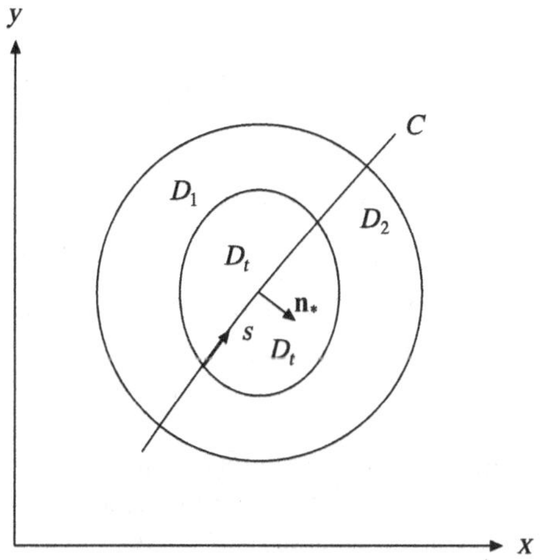

Figure E.2. A discontinuity curve C separating the domains D_1 and D_2. Here $\mathbf{n}_*$ is the unit normal to C, s is the arc length measured along C, and the test function $V(x, y)$ vanishes outside the subdomain D_t.

D_1 and D_2. Multiplying (E.1) by V and integrating over D_1, we obtain

$$\int\int_{D_1} V \left(\frac{\partial \mathbf{E}}{\partial x} + \frac{\partial \mathbf{F}}{\partial y} + \mathbf{G} \right) \mathrm{d}x\, \mathrm{d}y = \int_C V\,(n_{x*}\,\mathbf{E}_1 + n_{y*}\,\mathbf{F}_1)\,\mathrm{d}s$$

$$+ \int\int_{D_1} \left(-\mathbf{E}\,\frac{\partial V}{\partial x} - \mathbf{F}\,\frac{\partial V}{\partial y} + \mathbf{G}\,V \right) \mathrm{d}x\, \mathrm{d}y = 0 \qquad \text{(E.4)}$$

where s is the arc length measured along C, the subscript 1 denotes the limiting value attained as C is approached from the left, and n_{x*} and n_{y*} are the components of the unit normal to C (Fig. E.2). Similarly, multiplication of (E.1) with V and integration over D_2 gives

$$-\int_C V\,(n_{x*}\,\mathbf{E}_2 + n_{y*}\,\mathbf{F}_2)\,\mathrm{d}s + \int\int_{D_2} \left(-\mathbf{E}\,\frac{\partial V}{\partial x} - \mathbf{F}\,\frac{\partial V}{\partial y} + \mathbf{G}\,V \right) \mathrm{d}x\, \mathrm{d}y = 0 \qquad \text{(E.5)}$$

Adding (E.4) and (E.5), and noting that a weak solution satisfies (E.3), we obtain

$$\int_C V\,[n_{x*}\,(\mathbf{E}_1 - \mathbf{E}_2) + n_{y*}\,(\mathbf{F}_1 - \mathbf{F}_2)]\,\mathrm{d}s = 0 \qquad \text{(E.6)}$$

Assuming continuity of the integrand, (E.6) reduces to the *jump conditions*

$$n_{x*}\,(\mathbf{E}_1 - \mathbf{E}_2) + n_{y*}\,(\mathbf{F}_1 - \mathbf{F}_2) = 0 \qquad \text{(E.7)}$$

which hold at every point of C.

The derivation of jump conditions does not require the system to be hyperbolic. The latter requirement is necessary for the discontinuous solution to be *admissible* (Lax, 1957; Gelfand, 1963; Prasad and Ravindran, 1985, pp. 207–209). An admissible discontinuous solution or *shock* is one that can be obtained as the limit of continuous solutions of perturbed systems. For a system involving n dependent variables, this requirement can be translated to the condition that $n + 1$ characteristics should be incident on a shock from either side of it (Gelfand, 1963).

E.3. JUMP CONDITIONS FOR LINEAR EQUATIONS

Referring to (E.1), consider the special case where $\mathbf{E}$, $\mathbf{F}$, and $\mathbf{G}$ are linear in the dependent variables $\mathbf{u}$. In particular, let $\mathbf{E} = \mathbf{A}(x, y)\,\mathbf{u} + \mathbf{H}(x, y)$ and $\mathbf{F} = \mathbf{B}(x, y)\,\mathbf{u} + \mathbf{I}(x, y)$, where $\mathbf{A}$ and $\mathbf{B}$ are matrices and $\mathbf{H}$ and $\mathbf{I}$ are vectors. Then (E.1) can be rewritten as

$$\mathbf{A}\,\frac{\partial \mathbf{u}}{\partial x} + \mathbf{B}\,\frac{\partial \mathbf{u}}{\partial y} + \mathbf{J} = 0 \qquad \text{(E.8)}$$

where $\mathbf{J}(x, y, \mathbf{u})$ is linear in $\mathbf{u}$.

The jump conditions (E.7) reduce to

$$(n_{x*}\,\mathbf{A} + n_{y*}\,\mathbf{B})\,(\mathbf{u}_1 - \mathbf{u}_2) = 0 \qquad \text{(E.9)}$$

or, if $\mathbf{u}_1 - \mathbf{u}_2 \neq 0$ at any point on C,

$$\det([n_{x*}\,\mathbf{A} + n_{y*}\,\mathbf{B}]) = 0 \qquad \text{(E.10)}$$

where "det" denotes the determinant (see (A.41)). If the curve C is identified with the curve Γ in Fig. C.1, (E.10) reduces to

$$\det([\sin\psi\,\mathbf{A} + \cos\psi\,\mathbf{B}]) = 0 \qquad \text{(E.11)}$$

As shown in Appendix C, this is the condition satisfied by a characteristic curve (see (C.4) and (C.3)). Hence for linear equations, the dependent variables can be discontinuous only across the characteristic curves.

PROBLEMS

E.1. The inclination of the velocity jump across a velocity characteristic
Consider the coaxiality condition (2.104) and the flow rule (2.105), subject to the additional assumptions that the angle of dilation ν_d is a constant and γ is a known continuous function of position. As shown in Problem C.3, these equations are hyperbolic, and the characteristics are inclined at angles of $\pm\Omega_v = \pm[(\pi/4) - (\nu_d/2)]$ relative to the major principal stress axis (Fig. E.3). Suppose that the discontinuity curve coincides with the 2-characteristic CD. Using (E.9), show that the vector representing the velocity jump is directed along the normal to the 1-characteristic, i.e., along the line OE in Fig. E.3. Hence show that this vector is inclined at an angle ν_d relative to the discontinuity curve.

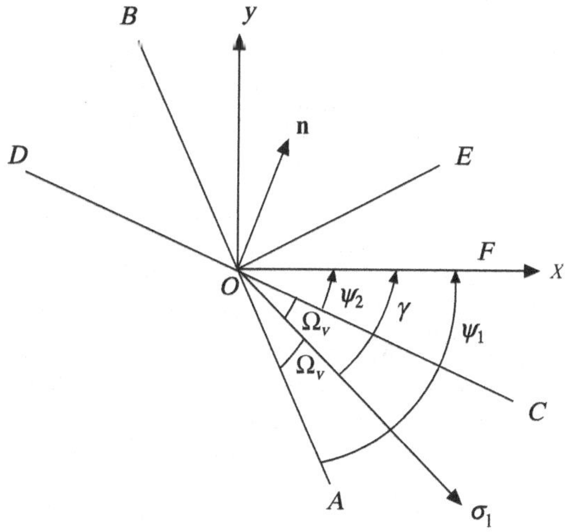

Figure E.3. Orientation of the velocity characteristics AB and CD. Here AB corresponds to the root $\psi = \psi_1 = \gamma + \Omega_v$ and CD corresponds to the root $\psi = \psi_2 = \gamma - \Omega_v$, where $\Omega_v \equiv (\pi/4) - (\nu_d/2)$, ν_d is the angle of dilation, $O\sigma_1$ represents the major principal stress axis, **n** is the unit normal to the 2-characteristic CD, and the line OE is normal to the 1-characteristic AB.

APPENDIX F

Proof of the Coaxiality Condition

The definition of an isotropic material is used below to deduce the coaxiality condition. The proof has been adapted from Hunter (1983, pp. 136–137).

Consider a constitutive equation of the form

$$\boldsymbol{\sigma} = \mathbf{A}(\mathbf{C}, \nu) \tag{F.1}$$

where $\boldsymbol{\sigma}$ is the stress tensor, $\mathbf{C}$ is the rate of deformation tensor, defined in the compressive sense, ν is the solids fraction, and $\mathbf{A}$ is a tensor-valued function of its arguments. For isotropic materials, the constitutive equation must be unaffected by translation, rotation, and reflection of the coordinate axes.

Consider two coordinate systems that have coordinates (x_1', x_2', x_3') and (x_1, x_2, x_3), respectively (Fig. F.1). Let $\mathbf{e}_i'$, $i = 1, 3$, and $\mathbf{e}_k$, $k = 1, 3$, denote the normalized basis vectors for the primed and unprimed coordinate systems, respectively. For any point P, we have

$$\mathbf{x}' = \boldsymbol{\beta} + \mathbf{x} \tag{F.2}$$

where $\mathbf{x}' = x_i'\, \mathbf{e}_i'$ and $\mathbf{x} = x_k\, \mathbf{e}_k$ are the position vectors corresponding to P in the primed and unprimed coordinate systems, respectively, and $\boldsymbol{\beta}$ is the position vector corresponding to the point O in the primed coordinate system (Fig. F.1). Equation (F.2) can be written in index notation (see §A.2) as

$$x_i'\, \mathbf{e}_i' = \beta_i\, \mathbf{e}_i' + x_k\, \mathbf{e}_k \tag{F.3}$$

where the summation convention has been used.

The vector $\mathbf{e}_i'$ can be expressed as a linear combination of the basis vectors $\mathbf{e}_j$, $j = 1, 3$, to obtain

$$\mathbf{e}_i' = Q_{ij}\, \mathbf{e}_j \tag{F.4}$$

where

$$Q_{ij} = \mathbf{e}_i' \cdot \mathbf{e}_j \tag{F.5}$$

Thus Q_{ij} is the cosine of the angle between the vectors $\mathbf{e}_i'$ and $\mathbf{e}_j$. It can be shown that (Problem F.1)

$$\mathbf{e}_j = Q_{ij}\, \mathbf{e}_i' \tag{F.6}$$

$$Q_{ij}\, Q_{ik} = \delta_{jk}; \;\; Q_{ji}\, Q_{ki} = \delta_{jk} \tag{F.7}$$

where δ_{jk} denotes the Kronecker delta. Equation (F.7) implies that the matrix $[\mathbf{Q}] = [Q_{ij}]$ is an orthogonal matrix (see §A.7.7). Note that (F.7) is valid even if the $\{Q_{ij}\}$ are functions of time, as in Appendix G.

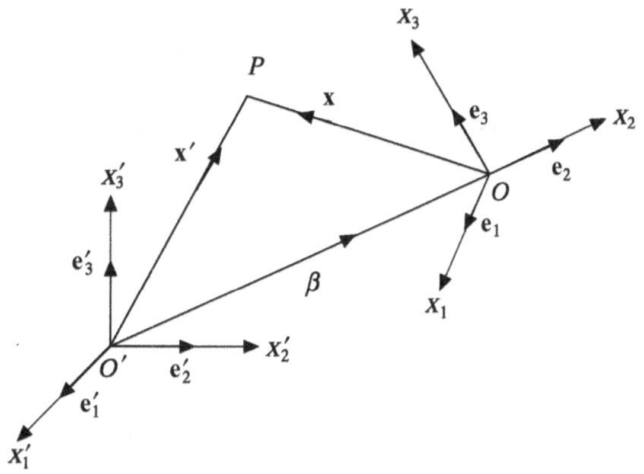

Figure F.1. A time-independent coordinate transformation. The point P has coordinates (x_1, x_2, x_3) in the unprimed coordinate system and (x_1', x_2', x_3') in the primed coordinate system. The coordinate systems do not move relative to each other. (Adapted from Fig. 4.1 of Hunter, 1983, p. 70.)

Using (F.6), (F.3) can be rewritten as

$$x_i' \, \mathbf{e}_i' = \beta_i \, \mathbf{e}_i' + x_k \, Q_{ik} \, \mathbf{e}_i'; \; = \beta_i \, \mathbf{e}_i' + Q_{ik} \, x_k \, \mathbf{e}_i' \tag{F.8}$$

or

$$x_i' = \beta_i + Q_{ik} \, x_k \tag{F.9}$$

The quantities $\{\beta_i\}$ and $\{Q_{ik}\}$ are taken to be *independent* of time, and hence (F.9) defines a time-independent *coordinate transformation*. Consider the special case $Q_{ik} = \delta_{ik}$, where δ_{ik} represents the Kronecker delta. Then (F.9) reduces to $x_i' = \beta_i + x_i$, and hence the first term on the right-hand side of (F.9) represents a *translation* of the coordinate axes. As discussed in Problem F.2, the second term represents the effect of either a rotation or a combination of a rotation and a reflection of the axes. Hence the constitutive equation (F.1) for an isotropic material must be invariant under the coordinate transformation (F.9).

As (F.1) is a tensor equation, it is valid in all coordinate systems that are related to each other by time-independent coordinate transformations of the form (F.9). Hence it can be written in terms of components as

$$\sigma_{ij} = A_{ij}(\{C_{rs}\}, \nu) \tag{F.10}$$

relative to the unprimed coordinate system, and as

$$\sigma_{ij}' = A_{ij}'(\{C_{rs}'\}, \nu) \tag{F.11}$$

relative to the primed coordinate system. For an *isotropic* material

$$A_{ij}'(\{C_{rs}'\}, \nu) = A_{ij}(\{C_{rs}'\}, \nu)$$

and (F.11) takes the form

$$\sigma_{ij}' = A_{ij}(\{C_{rs}'\}, \nu) \tag{F.12}$$

To proceed further, consider the transformation of the components of second-order tensors under a change of coordinates defined by (F.9). As $\boldsymbol{\sigma} = \sigma_{ij}' \mathbf{e}_i' \mathbf{e}_j' = \sigma_{kl} \mathbf{e}_k \mathbf{e}_l$, (F.6) implies that

$$\sigma_{ij}' \mathbf{e}_i' \mathbf{e}_j' = \sigma_{kl} \, Q_{ik} \, Q_{jl} \, \mathbf{e}_i' \mathbf{e}_j'$$

or

$$\sigma_{ij}' = \sigma_{kl} \, Q_{ik} \, Q_{jl} = Q_{ik} \, Q_{jl} \, \sigma_{kl} \tag{F.13}$$

Similarly,

$$C_{rs}' = Q_{rp} \, Q_{st} \, C_{pt} \tag{F.14}$$

We now discuss the derivation of the coaxiality condition for the three-dimensional case. Let the coordinate axes of the unprimed coordinate system be directed along the eigenvectors or principal axes of **C**. Then, as discussed in §2.7, the off-diagonal components of **C** vanish relative to the unprimed coordinate system, i.e., $C_{rs} = 0$ for $r \neq s$. Choosing $Q_{11} = 1$, $Q_{22} = -1$, $Q_{33} = -1$, and $Q_{rs} = 0$ for $r \neq s$, (F.14) implies that $C'_{12} = Q_{1p}\, Q_{2t}\, C_{pt} = Q_{11}\, Q_{22}\, C_{12} = 0$. Using similar results, it follows that $C'_{rs} = C_{rs}$. Hence (F.12) reduces to

$$\sigma'_{ij} = A_{ij}(\{C_{rs}\}, \nu)$$

or using (F.10),

$$\sigma'_{ij} = \sigma_{ij} \tag{F.15}$$

Substituting for σ'_{ij} from (F.13) into (F.15), we obtain

$$Q_{ik}\, Q_{jl}\, \sigma_{kl} = \sigma_{ij} \tag{F.16}$$

Using the preceding choice for the $\{Q_{ij}\}$, (F.16) implies that $Q_{11}\, Q_{22}\, \sigma_{12} = \sigma_{12}$ or $-\sigma_{12} = \sigma_{12}$. Hence $\sigma_{12} = 0$. Similarly, $\sigma_{13} = 0$. Further, by choosing $Q_{11} == -1$, $Q_{22} = -1$, $Q_{33} = 1$, and $Q_{rs} = 0$ for $r \neq s$, it follows that $\sigma_{23} = 0$. Hence the off-diagonal components of $\boldsymbol{\sigma}$ vanish in a coordinate system whose axes are aligned with the principal axes of **C**. This implies that the principal axes of $\boldsymbol{\sigma}$ and **C** are aligned with each other.

A similar proof can also be used in the two-dimensional case to establish coaxiality.

PROBLEMS

F.1. Relations to be satisfied by the $\{Q_{ij}\}$

Consider two rectangular Cartesian coordinate systems with normalized basis vectors $\mathbf{e}'_i$, $i = 1, 3$, and $\mathbf{e}_j$, $j = 1, 3$, respectively.

(a) Expressing $\mathbf{e}_j$ in terms of the basis vectors of the primed coordinate system, and using (F.5), show that

$$\mathbf{e}_j = Q_{ij}\, \mathbf{e}'_i$$

(b) Using the relations

$$\mathbf{e}'_i \cdot \mathbf{e}'_j = \delta_{ij}; \quad \mathbf{e}_k \cdot \mathbf{e}_l = \delta_{kl}$$

where δ_{ij} denotes the Kronecker delta, show that

$$Q_{ik}\, Q_{jk} = \delta_{ij}; \quad Q_{ij}\, Q_{ik} = \delta_{jk} \tag{F.17}$$

In matrix notation, (F.17) can be rewritten as

$$[Q][Q]^{\mathrm{T}} = [\mathbf{I}] = [Q]^{\mathrm{T}}[Q] \tag{F.18}$$

where $[\mathbf{I}]$ is the unit matrix.

F.2. Examples of coordinate transformations

Consider a special case of the coordinate transformation (F.9), with $\beta_i = 0$, $i = 1, 3$, so that the origins of the two coordinate systems coincide (Fig. F.2a). Consider a position vector OP, which is of length L and is inclined at an angle α to the x_2 axis (Fig. F.2a).

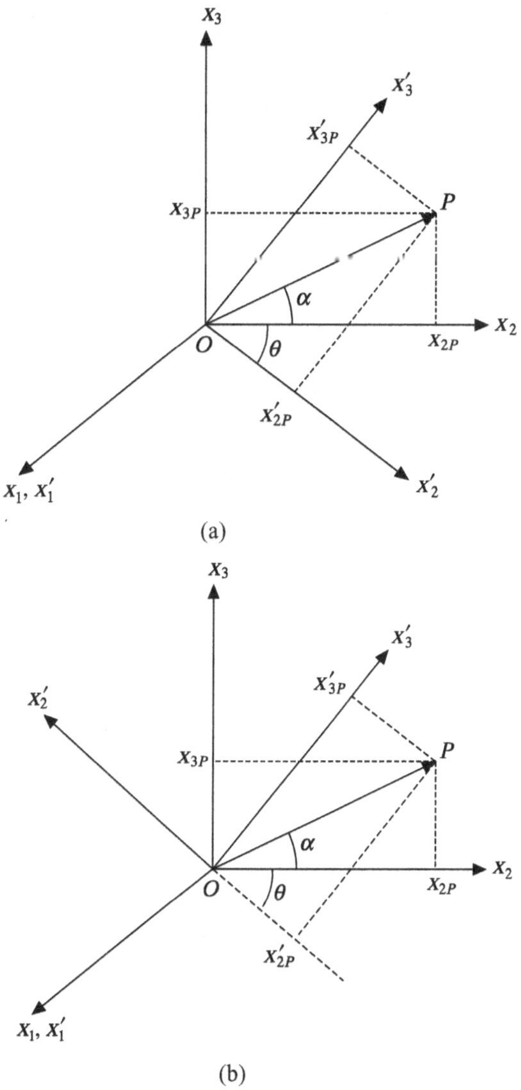

Figure F.2. Transformation of coordinates under (a) a rotation of the x_2' and x_3' axes about the x_1 axis by an angle θ, and (b) a combination of a rotation of the x_2' and x_3' axes about the x_1 axis by an angle θ, and a reflection of the x_2' axis about the x_1'–x_3' plane. The point P has coordinates (x_{1P}, x_{2P}, x_{3P}) in the unprimed coordinate system and $(x_{1P}' = x_{1P}, x_{2P}', x_{3P}')$ in the primed coordinate system. The origins of the two coordinate systems coincide, and the x_1' axis is aligned with the x_1 axis. The length of the position vector OP is L.

(a) Let the matrix $[\mathbf{Q}]$ be defined by

$$[\mathbf{Q}] \equiv [Q_{ij}] = \begin{pmatrix} 1 & 0 & 0 \\ 0 & \cos\theta & -\sin\theta \\ 0 & \sin\theta & \cos\theta \end{pmatrix} \tag{F.19}$$

Show that $\det([\mathbf{Q}]) = 1$. With the preceding choice for the $\{Q_{ij}\}$, show that the coordinate transformation $x_i' = Q_{ij}\,x_j$ corresponds to a *rotation* of the x_2' and x_3' axes about the x_1 axis by an angle θ in the clockwise direction, as indicated in Fig. F.2a.

(a) Let

$$[\mathbf{Q}] \equiv [Q_{ij}] = \begin{pmatrix} 1 & 0 & 0 \\ 0 & -\cos\theta & \sin\theta \\ 0 & \sin\theta & \cos\theta \end{pmatrix} \tag{F.20}$$

Show that det([**Q**]) = −1. With the preceding choice for the $\{Q_{ij}\}$, show that the coordinate transformation $x'_i = Q_{ij}\,x_j$ corresponds to a combination of a *rotation* of the x'_2 and x'_3 axes by an angle θ in the clockwise direction and a *reflection* of the x'_2 axis about the x'_1–x'_3 plane, as indicated in Fig. F.2b.

APPENDIX G

Material Frame Indifference

As noted in §1.5.1, a reference frame consists of a spatial coordinate system and clock. The principle of material frame indifference requires constitutive equations to be invariant or unaffected by an arbitrary *time-dependent* translation, rotation, and reflection of the coordinate axes and by an arbitrary translation of the time axis. In contrast, constitutive equations expressed in tensor form are invariant under *time-independent* transformations of the coordinate axes, but they are not necessarily invariant under time-dependent transformations.

The basis for frame indifference is the intuitive idea that the response of a material should be independent of the motion of the observer (Oldroyd, 1950; Noll, 1955, p. 45; Noll, 1959). The principle cannot be proved, but simple examples have been given (Truesdell and Noll, 1965; Hunter, 1983, p. 123) where its validity appears plausible. An example taken from Hunter (1983, p. 123) is discussed below.

A pilot bails out of an aircraft and opens his parachute. Suppose that the force **P** exerted by the parachute on the pilot is measured by the extension of a spring attached to the harness of the parachute. Consider two reference frames, one fixed to the pilot and the other to an observer standing on the ground. Suppose that the spring is visible in both the frames. When relativistic effects are ignored, it is usually implicitly *assumed* that at any time t, the distance between two points in space has the same value regardless of the motion of the reference frame (see, e.g., Resnick, 1968, p. 5). This assumption implies that the pilot and the observer on the ground will record the same extension for the spring. Hence the contact force **P**, which depends on the extension of the spring, also has the same value in both the frames. In such a case, the force is said to be frame indifferent.

To proceed further, the concept of a change of frame will be quantified, followed by a definition of the frame-indifferent quantities. The following material has been adapted, with some modifications, from Hunter (1983, pp. 122–126).

G.1. CHANGE OF FRAME

Consider two reference frames, denoted as the unprimed and primed frames, which move relative to each other and have embedded in them Cartesian coordinate systems with coordinates (x_1, x_2, x_3) and (x_1', x_2', x_3'), respectively (Fig. G.1). Let t be the time measured by a clock in the unprimed frame. The time measured by a clock in the primed frame is given by

$$t' = t - b \tag{G.1}$$

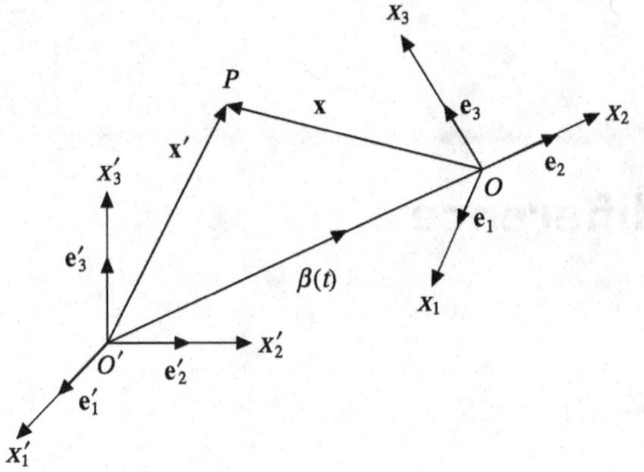

Figure G.1. A time-dependent coordinate transformation. The coordinate systems are embedded in two reference frames that move relative to each other. The point P has coordinates (x_1, x_2, x_3) in the unprimed coordinate system and (x_1', x_2', x_3') in the primed coordinate system.

where b is an arbitrary constant. For any point P, we have

$$\mathbf{x}' = \boldsymbol{\beta}(t) + \mathbf{x} \tag{G.2}$$

where $\mathbf{x}' = x_i' \, \mathbf{e}_i'$ and $\mathbf{x} = x_k \, \mathbf{e}_k$ are the position vectors corresponding to P in the primed and unprimed coordinate systems, respectively, and $\boldsymbol{\beta}(t)$ is the instantaneous position vector corresponding to the point O in the primed coordinate system (Fig. G.1). Here $\mathbf{e}_i'$, $i = 1, 3$, and $\mathbf{e}_k$, $k = 1, 3$, are the normalized basis vectors for the primed and unprimed coordinate systems, respectively. Equation (G.2) can be written in index notation (see §A.2) as

$$x_i' \, \mathbf{e}_i' = \beta_i(t) \, \mathbf{e}_i' + x_k \, \mathbf{e}_k \tag{G.3}$$

where the summation convention has been used.

As discussed in Appendix F, the vector $\mathbf{e}_i'$ can be expressed as a linear combination of the basis vectors $\mathbf{e}_i$, $i = 1, 3$, to obtain

$$\mathbf{e}_i' = Q_{ij} \, \mathbf{e}_j \tag{G.4}$$

where

$$Q_{ij} = \mathbf{e}_i' \cdot \mathbf{e}_j \tag{G.5}$$

Similarly, it can be shown (Problem F.1) that

$$\mathbf{e}_j = Q_{ij} \, \mathbf{e}_i' \tag{G.6}$$

It should be noted that the $\{Q_{ij}\}$ are functions of the time t in general, as the two frames are in relative motion.

Using (G.6), (G.3) can be rewritten as

$$x_i' \, \mathbf{e}_i' = \beta_i(t) \, \mathbf{e}_i' + x_k \, Q_{ik}(t) \, \mathbf{e}_i' = \beta_i(t) \, \mathbf{e}_i' + Q_{ik}(t) \, x_k \, \mathbf{e}_i' \tag{G.7}$$

or, in terms of components, as

$$x_i' = \beta_i(t) + Q_{ik}(t) \, x_k \tag{G.8}$$

The first term on the right-hand side of (G.8) represents a time-dependent translation of the coordinate axes, and, as discussed in Problem F.2, the second term represents either a time-dependent rotation of the axes, or a combination of a reflection and a time-dependent rotation of the axes. Equations (G.1) and (G.8) define a *change of frame* (Noll, 1959).

G.2. FRAME INDIFFERENT SCALARS, VECTORS, AND TENSORS

G.2.1. Scalars

A scalar b is said to be *frame indifferent* if its value is unaffected by a change of frame, i.e., if

$$b'(\mathbf{x}', t') = b(\mathbf{x}, t) \tag{G.9}$$

where $\mathbf{x}'$and $\mathbf{x}$ denote the position vectors corresponding to any point P in the primed and unprimed reference frames (Fig. G.1), and t' is related to t by (G.1). The density is an example of a frame indifferent quantity (Problem G.6). On the other hand, as noted by Astarita and Marucci (1974, p. 29), the time t is not frame indifferent, whereas the time interval between two events is (see (G.1)).

G.2.2. Vectors

Consider two position vectors $\mathbf{x} = x_i\, \mathbf{e}_i$ and $\mathbf{y} = y_i\, \mathbf{e}_i$ in the unprimed frame. Under a change of frame, (G.8) implies that the corresponding position vectors $\mathbf{x}' = x_i'\, \mathbf{e}_i'$ and $\mathbf{y}' = y_i'\, \mathbf{e}_i'$ have components

$$x_i' = \beta_i + Q_{ij}\, x_j; \quad y_i' = \beta_i + Q_{ij}\, y_j$$

As

$$y_i' - x_i' = Q_{ij}\,(y_j - x_j) \tag{G.10}$$

the difference $y_j - x_j$ between the components of the two position vectors is unaffected by a translation of the coordinate axes. Using (G.10) and (F.7), the square of the length of the vector $\mathbf{y}' - \mathbf{x}' = (y_i' - x_i')\, \mathbf{e}_i'$ is given by

$$\begin{aligned}(\mathbf{y}' - \mathbf{x}') \cdot (\mathbf{y}' - \mathbf{x}') &= (y_i' - x_i')\,(y_i' - x_i') = Q_{ij}\, Q_{ik}\,(y_j - x_j)\,(y_k - x_k) \\ &= \delta_{jk}\,(y_j - x_j)\,(y_k - x_k) = (y_j - x_j)\,(y_j - x_j) \\ &= (\mathbf{y} - \mathbf{x}) \cdot (\mathbf{y} - \mathbf{x})\end{aligned}$$

Thus a change of frame preserves the length of the vector representing the difference between two position vectors and, as shown by (G.10), only changes its orientation. Further, this change of orientation is caused only because the two coordinate systems are not aligned with each other and is identical to that caused by a time-independent coordinate transformation. The latter result can be inferred by comparing (G.10) with the corresponding equation obtained from (F.9). In other words, the components of a vector defined as the difference between two position vectors transform in the same manner under time-independent and time-dependent coordinate transformations.

Using (G.8) and (G.6), it follows that

$$\begin{aligned}\mathbf{y}' - \mathbf{x}' &= (y_i' - x_i')\mathbf{e}_i' = Q_{ij}\,(y_j - x_j)\,\mathbf{e}_i' \\ &= (y_j - x_j)\,\mathbf{e}_j = \mathbf{y} - \mathbf{x}\end{aligned}$$

Generalizing this result, a vector $\mathbf{w}$ is said to be *frame indifferent* if

$$\mathbf{w}' = \mathbf{w} \tag{G.11}$$

Hence as shown by (G.10), its components transform as

$$w_i' = Q_{ij}(t)\, w_j \tag{G.12}$$

Equation (J.7) can be written more compactly in matrix notation as

$$[\mathbf{w}'] = [\mathbf{Q}][\mathbf{w}] \tag{G.13}$$

where $[\mathbf{Q}] \equiv [Q_{ij}]$ is a matrix formed from the $\{Q_{ij}\}$. Comparison of (G.12) and (G.8) shows that position vectors in physical space are *not* frame indifferent.

Let us examine how the components of the velocity vector transform under a change of frame. Consider the primed reference frame and the primed coordinate system embedded in it (Fig. G.1). As discussed in Chapter 1, the motion of a material point P relative to this frame is described by $\mathbf{x}' = \mathbf{x}'(\mathbf{X}, t')$, where $\mathbf{X}$ is the position vector corresponding to P in the reference configuration and $\mathbf{x}'$ is the position vector corresponding to P at time t'. The velocity of P relative to the primed frame is defined by

$$\mathbf{v}' = \left(\frac{\partial \mathbf{x}'}{\partial t'}\right)_{\mathbf{X}} \tag{G.14}$$

Similarly, if $\mathbf{x}$ denotes the position vector corresponding to P in the unprimed frame at time t, the velocity of P relative to this frame is defined by $\mathbf{v} = (\partial \mathbf{x}/\partial t)_{\mathbf{X}}$.

Expressing (G.14) in component form, and using (G.1) and (G.8), we obtain

$$v_k' = \left(\frac{\partial x_k'}{\partial t'}\right)_{\mathbf{X}} = \dot{\beta}_k(t) + Q_{kl}(t)\, v_l + \dot{Q}_{kl}(t)\, x_l \tag{G.15}$$

where $v_l = \partial x_l/\partial t$ is the lth component of the velocity of P relative to the unprimed frame, and

$$\dot{\beta}_k(t) \equiv \frac{\mathrm{d}\beta_k}{\mathrm{d}t}; \quad \dot{Q}_{kl}(t) \equiv \frac{\mathrm{d}Q_{kl}}{\mathrm{d}t} \tag{G.16}$$

As $v_k' \neq Q_{kl}\, v_l$, the velocity vector is not a frame indifferent vector. Similarly, it can be shown that the acceleration is not a frame indifferent vector (Problem G.2).

G.2.3. Second-Order Tensors

By analogy with (G.9) and (G.11), a second-order tensor $\mathbf{A}$ is said to be *frame indifferent* if

$$\mathbf{A}' = \mathbf{A} \tag{G.17}$$

Expressing (G.17) in terms of components, and using (G.6), we obtain

$$A_{ik}'\, \mathbf{e}_i' \mathbf{e}_k' = A_{ml}\, \mathbf{e}_m \mathbf{e}_l = A_{ml}\, Q_{im}\, Q_{kl}\, \mathbf{e}_i' \mathbf{e}_k'$$

or

$$A_{ik}' = Q_{im}\, A_{ml}\, Q_{kl} \tag{G.18}$$

Equation (G.18) defines the transformation rule for the components of a frame indifferent second-order tensor. It can be written more compactly in matrix notation as

$$[\mathbf{A}'] = [\mathbf{Q}][\mathbf{A}][\mathbf{Q}^{\mathrm{T}}] \tag{G.19}$$

where $[\mathbf{A}] = [A_{ik}]$ denotes the matrix representation of $\mathbf{A}$ (see §A.4) and $[\mathbf{Q}] \equiv [Q_{ml}]$.

It is *assumed* on intuitive grounds that the stress tensor $\boldsymbol{\sigma}$ is a frame indifferent tensor. Thus

$$\sigma_{ik}' = Q_{im}\, \sigma_{ml}\, Q_{kl} \tag{G.20}$$

or equivalently,

$$[\boldsymbol{\sigma}'] = [\mathbf{Q}][\boldsymbol{\sigma}][\mathbf{Q}^{\mathrm{T}}] \tag{G.21}$$

On the other hand, the velocity gradient tensor $\nabla \mathbf{v}$ is not a frame indifferent tensor (Problem G.1).

G.3. THE PRINCIPLE OF MATERIAL FRAME INDIFFERENCE

The principle states that constitutive relations must be unaffected by a change of frame (Noll, 1955; Noll, 1959; Truesdell and Noll, 1965, p. 44). For example, suppose that the stress tensor $\boldsymbol{\sigma}$ is specified by the equation

$$\sigma_{ij} = F_{ij}(\{A_{kl}\}, \{w_k\}, s) \tag{G.22}$$

or in matrix form by

$$[\boldsymbol{\sigma}] = [F]([\mathbf{A}], [\mathbf{w}], s) \tag{G.23}$$

where $\{A_{kl}\}$ and $\{w_k\}$ represent the components of a second-order tensor $\mathbf{A}$ and a vector $\mathbf{w}$, respectively, s is a scalar, and $[\mathbf{w}]$ is a column vector (see §A.1). If (G.22) is an admissible constitutive equation, it must satisfy

$$\sigma'_{ij} = F_{ij}(\{A'_{kl}\}, \{w'_k\}, s') \tag{G.24}$$

or in matrix form

$$[\boldsymbol{\sigma}'] = [F]([A'], [w'], s')$$

where the prime denotes components in the primed frame. An application of this principle is discussed in Chapter 5.

Admissible constitutive equations can also be formulated using alternative approach due to Oldroyd (1950), which is based on a coordinate system which is convected with the material. Oldroyd notes that only those kinematic quantities which depend on the distance between pairs of neighboring points in the convected coordinate system can occur in constitutive relations. More generally, any constitutive equation containing only tensor components referred to the convected coördinate system automatically satisfies the principle of material frame indifference (Astarita and Marucci, 1974, p. 98). A detailed discussion of this approach is given in Oldroyd (1950) and Astarita and Marucci (1974, pp. 98–103).

G.4. AN ALTERNATIVE INTERPRETATION OF A CHANGE OF FRAME

So far, we have viewed a change of frame as a time-dependent transformation of the coordinates and a translation of the time axis. Alternatively, as noted by Schowalter (1978, p. 178) and Haupt (2000, p. 164), we can regard a change of frame as a time-dependent transformation of vectors and tensors in a *single* coordinate system and a translation of the time axis. Using a superscript * to denote such a change of frame, (G.7) can be rewritten as

$$\mathbf{x}^* = \boldsymbol{\beta}(t) + \mathbf{Q}(t) \cdot \mathbf{x} \tag{G.25}$$

where $\mathbf{x}^* = x_i^* \, \mathbf{e}_i$, $\mathbf{x} = x_i \, \mathbf{e}_i$, and $\mathbf{Q}(t)$ is an orthogonal tensor (see §A.7.7) defined by $\mathbf{Q}(t) = Q_{ij}(t) \, \mathbf{e}_i \mathbf{e}_j$, which satisfies

$$\mathbf{Q}^{\mathrm{T}}\mathbf{Q} = \mathbf{Q}\mathbf{Q}^{\mathrm{T}} = \mathbf{I} \tag{G.26}$$

Here $\mathbf{I}$ is the unit tensor. Similarly, (G.1) can be rewritten as

$$t^* = t - a \tag{G.27}$$

where a is an arbitrary scalar.

As discussed in §G.1, the first term on the right-hand side of (G.25) represents a translation and the second term represents either a rotation or a reflection, or a combination of the two. Here however, there is no transformation of the coordinate axes, but only of the vectors and tensors (Fig. G.2).

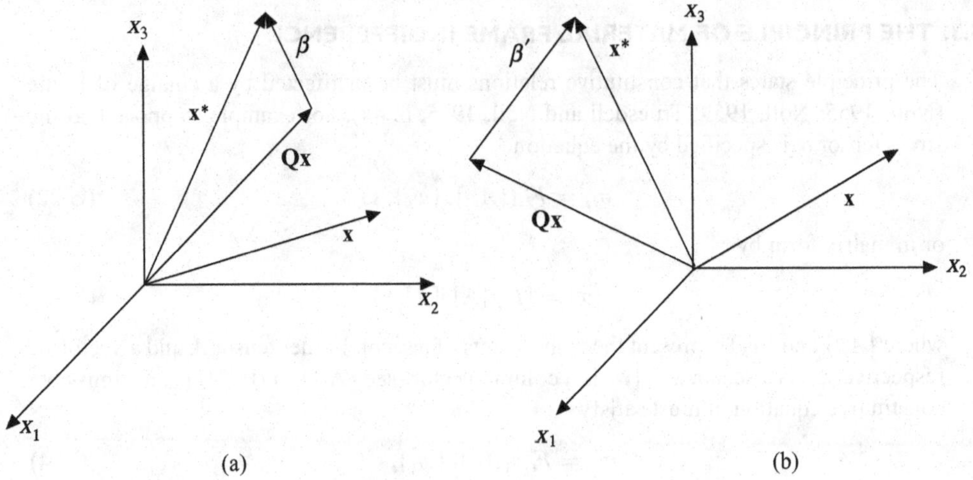

Figure G.2. Changes of frame consisting of (a) a rotation followed by translation and (b) a reflection about the x_1–x_3 plane followed by translation.

Adopting this viewpoint, a scalar b, a vector $\mathbf{w}$, and a second-order tensor $\mathbf{A}$ transform as

$$b^*(\mathbf{x}^*, t^*) = b(\mathbf{x}, t); \quad \mathbf{w}^* = \mathbf{Q} \cdot \mathbf{w} \equiv \mathbf{Qw}; \quad \mathbf{A}^* = \mathbf{Q} \cdot \mathbf{A} \cdot \mathbf{Q}^{\mathrm{T}} \equiv \mathbf{QAQ}^{\mathrm{T}} \tag{G.28}$$

under a change of frame, if they are frame indifferent. These equations are analogs of (G.9), (G.13), and (G.19). Similarly, the analogs of (G.21), (G.31), (G.35), and (G.37) are

$$\begin{aligned} \boldsymbol{\sigma}^* &= \mathbf{Q}\boldsymbol{\sigma}\mathbf{Q}^{\mathrm{T}}; \quad \nabla^*\mathbf{v}^* = \mathbf{Q}(\nabla\mathbf{v})\mathbf{Q}^{\mathrm{T}} + \mathbf{Q}\dot{\mathbf{Q}}^{\mathrm{T}}; \\ \mathbf{W}^* &= \mathbf{QWQ}^{\mathrm{T}} - \dot{\mathbf{Q}}\mathbf{Q}^{\mathrm{T}}; \\ \frac{\mathrm{D}\boldsymbol{\sigma}^*}{\mathrm{D}t^*} &= \mathbf{Q}\,\frac{\mathrm{D}\boldsymbol{\sigma}}{\mathrm{D}t}\,\mathbf{Q}^{\mathrm{T}} + \dot{\mathbf{Q}}\,\boldsymbol{\sigma}\,\mathbf{Q}^{\mathrm{T}} + \mathbf{Q}\,\boldsymbol{\sigma}\,\dot{\mathbf{Q}}^{\mathrm{T}} \end{aligned} \tag{G.29}$$

If $\boldsymbol{\sigma} = \mathbf{F}(c, \mathbf{u}, \mathbf{B})$ is a frame indifferent constitutive equation for the stress tensor $\boldsymbol{\sigma}$, where c, $\mathbf{u}$, and $\mathbf{B}$ represent a scalar, vector, and a second-order tensor, respectively, it must satisfy $\boldsymbol{\sigma}^* = \mathbf{F}(c^*, \mathbf{u}^*, \mathbf{B}^*)$.

PROBLEMS

G.1. Transformation of some tensors under a change of frame

(a) Show that (G.8) can be rewritten as

$$x_j = Q_{ij}\,(x_i' - \beta_i) \tag{G.30}$$

(b) As noted in §A.7.9, the velocity gradient tensor $\nabla\mathbf{v}$ is defined by $\nabla\mathbf{v} = B_{ik}\mathbf{e}_i\,\mathbf{e}_k$, where

$$B_{ik} \equiv \frac{\partial v_k}{\partial x_i}$$

Using (G.15), show that

$$B_{ik}' = \frac{\partial v_k'}{\partial x_i'} = Q_{im}\,B_{ml}\,Q_{kl} + Q_{il}\,\dot{Q}_{kl} \tag{G.31}$$

Hence $\nabla\mathbf{v}$ is not a frame indifferent tensor.

(c) The rate of deformation tensor is defined in the compressive sense by $\mathbf{C} = C_{ik}\,\mathbf{e}_i\mathbf{e}_k$, where

$$C_{ik} \equiv -\frac{1}{2}\left(\frac{\partial v_k}{\partial x_i} + \frac{\partial v_i}{\partial x_k}\right) \tag{G.32}$$

Using (G.31) and (F.7), show that $\mathbf{C}$ is a frame indifferent tensor, i.e.,

$$C'_{ik} = Q_{im}\,C_{ml}\,Q_{kl} \tag{G.33}$$

(d) The vorticity tensor is defined by $\mathbf{W} = W_{ik}\,\mathbf{e}_i\mathbf{e}_k$, where

$$W_{ik} \equiv \frac{1}{2}\left(\frac{\partial v_k}{\partial x_i} - \frac{\partial v_i}{\partial x_k}\right) \tag{G.34}$$

Show that

$$W'_{ik} = Q_{im}\,W_{ml}\,Q_{kl} - \dot{Q}_{il}\,Q_{kl} \tag{G.35}$$

Hence $\mathbf{W}$ is not a frame indifferent tensor.

(e) Show that

$$\begin{aligned}\frac{\mathrm{D}}{\mathrm{D}t'} &\equiv \frac{\partial}{\partial t'} + v'_i\,\frac{\partial}{\partial x'_i}\\ &= \frac{\partial}{\partial t} + v_k\,\frac{\partial}{\partial x_k} \equiv \frac{\mathrm{D}}{\mathrm{D}t}\end{aligned} \tag{G.36}$$

(f) Show that

$$\frac{\mathrm{D}\sigma'_{ik}}{\mathrm{D}t'} = Q_{im}\,\frac{\mathrm{D}\sigma_{ml}}{\mathrm{D}t}\,Q_{kl} + \dot{Q}_{im}\,\sigma_{ml}\,Q_{kl} + Q_{im}\,\sigma_{ml}\,\dot{Q}_{kl} \tag{G.37}$$

where $\{\sigma_{ml}\}$ are the components of the stress tensor and $\mathrm{D}/\mathrm{D}t$ denotes the material derivative defined by (G.36). Hence the material derivative of the stress tensor is not a frame indifferent tensor.

(g) The Jaumann derivative $\overset{\circ}{\boldsymbol{\sigma}}$ of the stress tensor is defined by

$$\overset{\circ}{\sigma}_{ik} \equiv \frac{\mathrm{D}\sigma_{ik}}{\mathrm{D}t} + W_{im}\,\sigma_{mk} - \sigma_{im}\,W_{mk} \tag{G.38}$$

where $\{W_{im}\}$ are the components of the vorticity tensor, defined by (G.34). Using (G.20), (G.33), and (G.37), show that $\overset{\circ}{\boldsymbol{\sigma}}$ is a frame indifferent tensor, i.e.,

$$\overset{\circ}{\sigma}{}'_{ik} = Q_{im}\,\overset{\circ}{\sigma}_{ml}\,Q_{kl} \tag{G.39}$$

It may be helpful to use matrix notation to simplify the algebra, in which case (G.39) takes the form

$$[\overset{\circ}{\boldsymbol{\sigma}}{}'] = [\mathbf{Q}][\overset{\circ}{\boldsymbol{\sigma}}][\mathbf{Q}^{\mathrm{T}}] \tag{G.40}$$

Here the square brackets denote the matrix representation of a tensor (see §A.4). For convenience, the square brackets can be omitted while solving the problem.

G.2. Transformation of the acceleration under a change of frame

(a) Using (G.15), show that the acceleration

$$a'_k \equiv \left(\frac{\partial v'_k}{\partial t'}\right)_{\mathbf{X}} \equiv \frac{\mathrm{D}v'_k}{\mathrm{D}t'} = \ddot{\beta}_k + Q_{kl}\,a_l + 2\,\dot{Q}_{kl}\,v_l + \ddot{Q}_{kl}\,x_l \tag{G.41}$$

where

$$\ddot{\beta}_k \equiv \frac{\mathrm{d}^2\beta_k}{\mathrm{d}t^2}$$

(b) Using (F.17), show that (G.41) can be solved for a_i to obtain

$$a_i = Q^{\mathrm{T}}_{ik}\,a'_k - Q^{\mathrm{T}}_{ik}\,\ddot{\beta}_k - 2\,A_{il}\,v_l - Q^{\mathrm{T}}_{ik}\,\ddot{Q}_{kl}\,x_l \tag{G.42}$$

where

$$A_{il} \equiv Q^{\mathrm{T}}_{ik}\,\dot{Q}_{kl} \tag{G.43}$$

Show that (G.42) can be rewritten as

$$a_i = Q^{\mathrm{T}}_{ik}\, a'_k - Q^{\mathrm{T}}_{ik}\, \ddot{\beta}_k - 2\, A_{il}\, v_l - \dot{A}_{il}\, x_l - A_{in}\, A_{nl}\, x_l \tag{G.44}$$

where

$$\dot{A}_{il} \equiv \frac{\mathrm{d}A_{il}}{\mathrm{d}t}$$

Equation (G.44) can be written in matrix form as

$$[\mathbf{a}] = [\mathbf{Q}^{\mathrm{T}}][\mathbf{a}'] - [\mathbf{Q}^{\mathrm{T}}][\ddot{\boldsymbol{\beta}}] - 2\,[\mathbf{A}][\mathbf{v}] - [\dot{\mathbf{A}}][\mathbf{x}] - [\mathbf{A}^2][\mathbf{x}]$$

As $a_i \neq Q^{\mathrm{T}}_{ik}\, a'_k$, the acceleration is not a frame indifferent vector.

(c) Introduce a vector $\boldsymbol{\omega} = \omega_k \mathbf{e}_k$, such that

$$A_{il} \equiv Q^{\mathrm{T}}_{ik}\, \dot{Q}_{kl} \equiv -\epsilon_{ilk}\, \omega_k \tag{G.45}$$

It can be shown (Problem G.3) that $\boldsymbol{\omega}$ represents the *angular velocity* of the unprimed frame relative to the primed frame. Show that

$$A_{il}\, v_l\, \mathbf{e}_i = \boldsymbol{\omega} \times \mathbf{v}; \quad A_{in}\, A_{nl}\, x_l\, \mathbf{e}_i = \boldsymbol{\omega} \times (\boldsymbol{\omega} \times \mathbf{x})$$

where

$$\mathbf{v} = v_l\, \mathbf{e}_l; \quad \mathbf{x} = x_l\, \mathbf{e}_l$$

and $\mathbf{e}_i$ is a basis vector for the unprimed coordinate system (see Fig. G.1).

(d) Multiplying (G.44) by $\mathbf{e}_i$, show that

$$\mathbf{a} = a_i\, \mathbf{e}_i = \mathbf{a}' - \ddot{\boldsymbol{\beta}} - 2\,(\boldsymbol{\omega} \times \mathbf{v}) - \dot{\boldsymbol{\omega}} \times \mathbf{x} - \boldsymbol{\omega} \times (\boldsymbol{\omega} \times \mathbf{x}) \tag{G.46}$$

where

$$\mathbf{a}' = a_j\, \mathbf{e}'_j; \quad \ddot{\boldsymbol{\beta}} = \ddot{\beta}_j\, \mathbf{e}'_j; \quad \boldsymbol{\omega} \equiv \omega_k\, \mathbf{e}_k; \quad \dot{\boldsymbol{\omega}} \equiv \frac{\mathrm{d}\boldsymbol{\omega}}{\mathrm{d}t}$$

(e) Let $\mathbf{e}_\omega$ denote a unit vector directed along $\boldsymbol{\omega}$. Show that

$$\boldsymbol{\omega} \times (\boldsymbol{\omega} \times \mathbf{x}) = -|\boldsymbol{\omega}|^2\, \mathbf{R}$$

where

$$\mathbf{R} \equiv \mathbf{x} - (\mathbf{e}_\omega \cdot \mathbf{x})\, \mathbf{e}_\omega$$

is the component of $\mathbf{x}$ that is normal to $\boldsymbol{\omega}$ (Fig. G.3). Hence (G.46) can be rewritten as

$$\mathbf{a} = a_i\, \mathbf{e}_i = \mathbf{a}' - \ddot{\boldsymbol{\beta}} - 2\,(\boldsymbol{\omega} \times \mathbf{v}) - \dot{\boldsymbol{\omega}} \times \mathbf{x} + |\boldsymbol{\omega}|^2\, \mathbf{R} \tag{G.47}$$

or equivalently, as

$$\mathbf{a}' = \mathbf{a} + \ddot{\boldsymbol{\beta}} + 2\,(\boldsymbol{\omega} \times \mathbf{v}) + \dot{\boldsymbol{\omega}} \times \mathbf{x} - |\boldsymbol{\omega}|^2\, \mathbf{R} \tag{G.48}$$

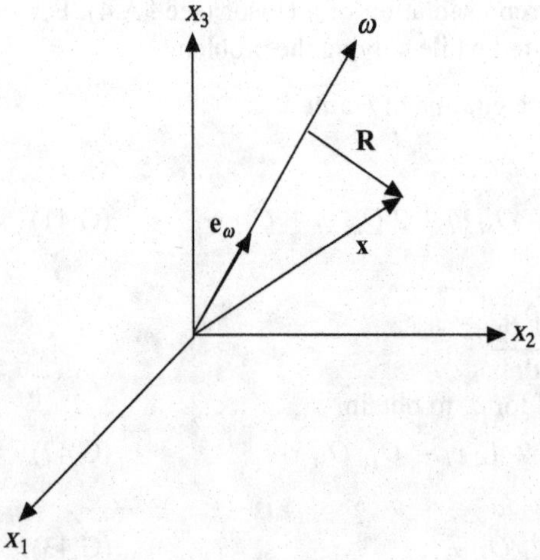

Figure G.3. The vector **R** represents the component of **x** that is normal to the direction of the angular velocity vector $\boldsymbol{\omega}$.

Equation (G.48) shows that the acceleration of a material point relative to the primed frame is the sum of its acceleration relative to the unprimed frame and terms representing the translation and rotation of the unprimed frame relative to the primed frame. The terms $2\,(\boldsymbol{\omega} \times \mathbf{v})$, $\dot{\boldsymbol{\omega}} \times \mathbf{x}$, and $-|\boldsymbol{\omega}|^2\,\mathbf{R}$ represent the *Coriolis acceleration*, the effect of the angular acceleration $\dot{\boldsymbol{\omega}}$, and the *centripetal acceleration*, respectively.

Parts of this problem have been adapted from Truesdell and Toupin (1960, p. 45), Noll (1955), and Hunter (1983, p. 126).

G.3. The angular velocity of the unprimed frame relative to the primed frame
Consider a special case for which the origins of the two coordinate systems shown in Fig. G.1 coincide and the x_1' axis coincides with the x_1 axis at all times. At time t, let the x_2 axis be inclined at an angle θ relative to the x_2' axis (Fig. G.4). Using (G.45), show that

$$A_{32} = \omega_1 = \dot{\theta} \equiv \frac{\mathrm{d}\theta}{\mathrm{d}t}$$

where $\dot{\theta}$ represents the angular velocity of the unprimed frame relative to the primed frame. It may be helpful to use matrix notation.

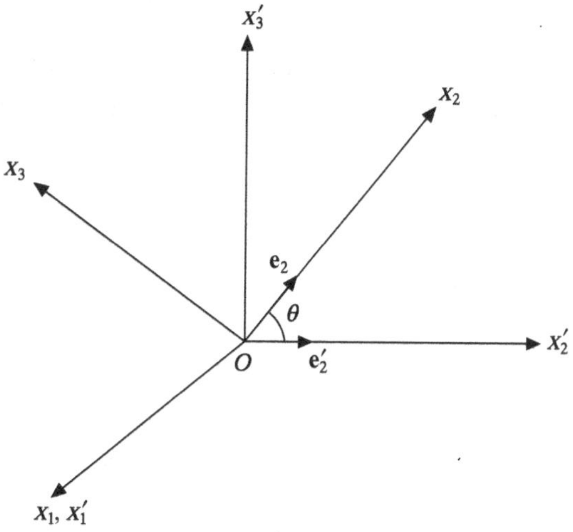

Figure G.4. The origins of the unprimed and primed coordinate systems coincide at all times, and the unprimed coordinate system rotates in an anticloclwise direction about the x_1' axis with an angular velocity $\dot{\theta} \equiv \mathrm{d}\theta/\mathrm{d}t$.

G.4. Transformation of the body force under a change of frame
Consider two reference frames in relative motion (Fig. G.1). In the primed frame, the linear momentum balance (1.30) is given by

$$\rho'\,\mathbf{a}' = -\nabla' \cdot \boldsymbol{\sigma}' + \rho'\,\mathbf{b}' \tag{G.49}$$

where

$$\mathbf{a}' \equiv \frac{\mathrm{D}\mathbf{v}'}{\mathrm{D}t'}; \quad \nabla' \equiv \mathbf{e}_j'\,\frac{\partial}{\partial x_j'}$$

(a) Assuming that the stress tensor is frame indifferent, show that

$$\frac{\partial \sigma_{kj}'}{\partial x_k'} = Q_{jn}\,\frac{\partial \sigma_{mn}}{\partial x_m} \tag{G.50}$$

Multiplying (G.50) by $\mathbf{e}_j'$ and using (G.6), we obtain

$$\nabla' \cdot \boldsymbol{\sigma}' = \nabla \cdot \boldsymbol{\sigma} \tag{G.51}$$

Hence the stress gradient is a frame indifferent vector.

(b) Multiplying (G.46) by ρ', and noting that $\rho' = \rho$ (see Problem G.6), show that

$$\rho\,\mathbf{a} = -\nabla\cdot\boldsymbol{\sigma} + \rho\,[\mathbf{b}' - \ddot{\boldsymbol{\beta}} - 2\,(\boldsymbol{\omega}\times\mathbf{v}) - \dot{\boldsymbol{\omega}}\times\mathbf{x} - \boldsymbol{\omega}\times(\boldsymbol{\omega}\times\mathbf{x})] \tag{G.52}$$

where

$$\mathbf{a} \equiv \frac{\mathrm{D}\mathbf{v}}{\mathrm{D}t}$$

(c) If the linear momentum balance is to be frame indifferent, show that the body force per unit mass in the unprimed frame is given by

$$\mathbf{b} = \mathbf{b}' - \ddot{\boldsymbol{\beta}} - 2\,(\boldsymbol{\omega}\times\mathbf{v}) - \dot{\boldsymbol{\omega}}\times\mathbf{x} - \boldsymbol{\omega}\times(\boldsymbol{\omega}\times\mathbf{x}) \tag{G.53}$$

Hence **b** transforms in the same manner as the acceleration under a change of frame and is not a frame indifferent vector. The last three terms on the right-hand side of (G.53) represent the Coriolis force per unit mass, the effect of the angular acceleration, and the centrifugal force per unit mass, respectively.

G.5. Transformation of the components of the rate of deformation and vorticity tensors under a particular change of frame

(a) Referring to Fig. 2.49, let the x–y coordinate system be embedded in a "fixed" reference frame that is stationary with respect to the laboratory frame and the ξ–η coordinate system be embedded in a reference frame that translates and rotates relative to the fixed frame. The origin of the ξ–η coordinate system moves with the velocity (relative to the fixed frame) of the material point at that location. Let $\mathbf{e}_1$ and $\mathbf{e}_2$ denote unit vectors directed along the x and y axes, respectively, and $\mathbf{e}_1'$ and $\mathbf{e}_2'$ denote unit vectors directed along the ξ and η axes, respectively. If $[\mathbf{Q}_*]$ is a matrix with elements $Q_{ij} = \mathbf{e}_i'\cdot\mathbf{e}_j$, show that

$$[\mathbf{Q}_*] = \begin{bmatrix} \cos\psi & \sin\psi \\ -\sin\psi & \cos\psi \end{bmatrix} \tag{G.54}$$

where ψ is the inclination of the ξ axis relative to the x axis. Show that $[\mathbf{Q}]_*$ is an orthogonal matrix.

(b) As discussed in §G.4, (G.25) can be regarded as representing *either* the transformation relating two position vectors in a given frame or the transformation relating the coordinates of a point in two frames which move relative to each other. Adopting the latter viewpoint, (G.33) and (G.35) can be regarded as relations between the components of the tensors **C** and **W** in the two frames.

Let C_{ij} denote a component of the rate of deformation tensor, defined by (G.32). Using (G.33), (G.35), and (G.54), show that

$$\begin{aligned}
C_{11}' &= C_{\xi\xi} = \cos^2\psi\, C_{xx} + 2\cos\psi\,\sin\psi\, C_{xy} + \sin^2\psi\, C_{yy} \\
C_{22}' &= C_{\eta\eta} = \sin^2\psi\, C_{xx} - 2\cos\psi\,\sin\psi\, C_{xy} + \cos^2\psi\, C_{yy} \\
C_{12}' &= C_{\xi\eta} = \sin\psi\,\cos\psi\,(C_{yy} - C_{xx}) + \cos 2\psi\, C_{xy} \\
W_{12}' &= W_{\xi\eta} = W_{xy} - \dot{\psi}
\end{aligned} \tag{G.55}$$

where $\dot{\psi} \equiv \mathrm{D}\psi/\mathrm{D}t$ is the material derivative of ψ.

G.6. Proof that the density is a frame-indifferent scalar

Let **x** and **x**$'$ denote the position vectors corresponding to a material point in two reference frames (Fig. G.1) at any time t measured in the unprimed frame. Using (1.93), (G.8), and (A.82), show that $\rho'(\mathbf{x}', t') = \rho(\mathbf{x}, t)$. Hence ρ is a frame indifferent scalar.

APPENDIX H

The Evaluation of Some Integrals

We discuss here the methods adopted for evaluating the integrals that occur in Chapters 7–9.

H.1. INTEGRATION OVER k

Integrals of the form

$$\int_{\mathbf{k}\cdot\mathbf{g}>0} (\mathbf{k}\cdot\mathbf{g})^n \mathbf{k}\cdots\mathbf{k}\,\mathrm{d}\mathbf{k}$$

occur in several places in Chapters 7 and 9, where $\mathrm{d}\mathbf{k}$ is the solid angle and the integration is over the surface of a sphere of unit radius.

It is convenient to use a coordinate system defined by the orthonormal triad ($\mathbf{n}$, $\mathbf{t}$, $\boldsymbol{\tau}$) such that $\mathbf{n}$ is the direction of the vector $\mathbf{g}$, i.e., $\mathbf{n} = \mathbf{g}/g$ (Fig. H.1). The unit vector $\mathbf{k}$ may be expressed in terms of the polar angles θ and ϕ as

$$\mathbf{k} = \cos\theta\,\mathbf{n} + \sin\theta\cos\phi\,\mathbf{t} + \sin\theta\sin\phi\,\boldsymbol{\tau} \tag{H.1}$$

where θ is the angle between $\mathbf{k}$ and $\mathbf{n}$ in the plane containing the two vectors and ϕ is the angle between the projection of $\mathbf{k}$ on the t–τ plane and the t axis. The constraint $\mathbf{k}\cdot\mathbf{g} > 0$ in the integral implies that $0 \leq \theta < \pi/2$. The limits of ϕ are 0 and 2π, and the solid angle $\mathrm{d}\mathbf{k}$ is $\sin\theta\,\mathrm{d}\theta\,\mathrm{d}\phi$. It follows that terms in the integrand with odd powers of $\sin\phi$ or $\cos\phi$ vanish upon integration. Thus for example,

$$\begin{aligned}\int_{\mathbf{k}\cdot\mathbf{g}>0} \mathbf{k}(\mathbf{k}\cdot\mathbf{g})\mathrm{d}\mathbf{k} &= \int_{\theta=0}^{\pi/2}\int_{\phi=0}^{2\pi} \mathbf{n}\,\cos\theta\,g\cos\theta\sin\theta\,\mathrm{d}\theta\,\mathrm{d}\phi \\ &= \frac{2\pi}{3} g\,\mathbf{n} = \frac{2\pi}{3}\mathbf{g}\end{aligned} \tag{H.2}$$

In a similar manner, the following results can be readily derived:

$$\int_{\mathbf{k}\cdot\mathbf{g}>0} (\mathbf{k}\cdot\mathbf{g})^3 \mathrm{d}\mathbf{k} = \frac{\pi}{2} g^3 \tag{H.3}$$

$$\int_{\mathbf{k}\cdot\mathbf{g}>0} \mathbf{k}\mathbf{k}(\mathbf{k}\cdot\mathbf{g})^2 \mathrm{d}\mathbf{k} = \frac{2\pi}{15}(2\mathbf{g}\mathbf{g} + g^2\mathbf{I}) \tag{H.4}$$

$$\int_{\mathbf{k}\cdot\mathbf{g}>0} \mathbf{k}\mathbf{k}(\mathbf{k}\cdot\mathbf{g})^2(\mathbf{k}\cdot\mathbf{h})\mathrm{d}\mathbf{k} = \frac{\pi}{12}\left[\frac{\mathbf{h}\cdot\mathbf{g}}{g}(\mathbf{g}\mathbf{g} + g^2\mathbf{I}) + g(\mathbf{g}\mathbf{h} + \mathbf{h}\mathbf{g})\right] \tag{H.5}$$

for any vector $\mathbf{h}$.

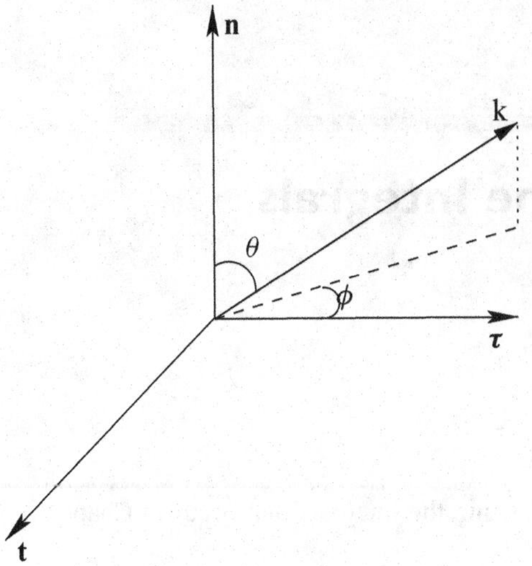

Figure H.1. The orthogonal triad and the polar angles used for the integration over **k**.

H.2. INTEGRATION OVER k FOR BOUNDARY CONDITIONS

In the derivation of boundary conditions in §8.1.2, the polar angle θ that defines $\mathbf{k}$ is restricted to the range $\pi - \theta_0 < \theta < \pi$ (see Fig. H.2). Therefore, a slightly different procedure is adopted to evaluate integrals of the form

$$\int_{\mathbf{k}\cdot\mathbf{g}<0} (\mathbf{k}\cdot\mathbf{g})^m F(g)\,\mathbf{g}\cdots\mathbf{g}\,\mathbf{k}\cdots\mathbf{k}\,\mathrm{d}\mathbf{k}\mathrm{d}\mathbf{g}$$

Expanding $\mathbf{g}$ in an orthonormal basis $(\mathbf{k}, \mathbf{i}, \mathbf{j})$, i.e., $\mathbf{g} = (\cos\theta'\,\mathbf{k} + \sin\theta'\cos\phi'\,\mathbf{i} + \sin\theta'\sin\phi'\,\mathbf{j})\,g$, the above integral is reduced to the form

$$\int \mathbf{k}\cdots\mathbf{k}\mathbf{i}\cdots\mathbf{i}\mathbf{j}\cdots\mathbf{j}\,\mathrm{d}\mathbf{k} \int \mathcal{F}(g,\theta',\phi')\,\mathrm{d}\mathbf{g}$$

The constraint $\mathbf{k}\cdot\mathbf{g} < 0$ implies that $\pi/2 < \theta' < \pi$. The first integral is evaluated by writing $\mathbf{k}$ in terms of polar angles, as in (H.1), but now we identify $\mathbf{n}$ as the normal to the boundary (see Fig. 8.2a) and $\mathbf{t}$, $\boldsymbol{\tau}$ as tangents to the boundary. Thus for example,

$$\int_{\mathbf{k}\cdot\mathbf{g}<0} \mathbf{k}(\mathbf{k}\cdot\mathbf{g})^2 e^{-g^2}\,\mathrm{d}\mathbf{k}\,\mathrm{d}\mathbf{g} = \int \mathbf{k}\,\mathrm{d}\mathbf{k} \int g^2\cos^2\theta' e^{-g^2}\,\mathrm{d}\mathbf{g}$$

$$= \mathbf{n}\int_0^{2\pi}\int_{\pi-\theta_0}^{\pi} \cos\theta\,\sin\theta\,\mathrm{d}\theta\,\mathrm{d}\phi \int_0^{\infty}\int_0^{2\pi}\int_{\pi/2}^{\pi} g^4 e^{-g^2}\cos^2\theta'\,\sin\theta'\,\mathrm{d}\theta'\,\mathrm{d}\phi'\,\mathrm{d}g$$

$$= -\frac{\pi^{5/2}}{4}\sin^2\theta_0\,\mathbf{n} \tag{H.6}$$

In a similar manner, the following results can be derived:

$$\int_{\mathbf{k}\cdot\mathbf{g}<0} \mathbf{g}\mathbf{k}(\mathbf{k}\cdot\mathbf{g})^2 F(g)\,\mathrm{d}\mathbf{k}\,\mathrm{d}\mathbf{g} = -\frac{\pi}{2}\int g^5 F(g)\,\mathrm{d}g \times \left[\frac{2\pi}{3}(1-\cos^3\theta_0)\mathbf{n}\mathbf{n}\right.$$

$$\left. + \frac{\pi}{3}(2+\cos^3\theta_0 - 3\cos\theta_0)(\mathbf{t}\mathbf{t}+\boldsymbol{\tau}\boldsymbol{\tau})\right], \tag{H.7}$$

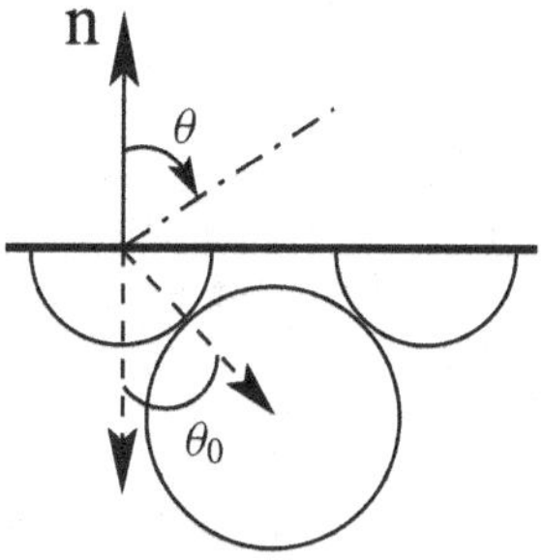

Figure H.2. The range of the polar angle θ in the integrals over **k** that arise in the derivation of the boundary conditions at a solid wall. The two dashed arrows circumscribe the range of θ. Here **n** is the unit outward normal on the boundary and θ_0 is the maximum angle a striking sphere can subtend with the inward normal (see Fig. 8.2).

$$\int_{\mathbf{k}\cdot\mathbf{g}<0} \mathbf{k}\,\overline{\mathbf{gg}}(\mathbf{k}\cdot\mathbf{g})^2 F(g)\,\mathrm{d}\mathbf{k}\,\mathrm{d}\mathbf{g} = \int g^6 F(g)\,\mathrm{d}g \times \frac{2\pi^2 \sin^2\theta_0}{15}$$
$$\times \left[(2+\cos^2\theta_0)\mathbf{nnn} + \frac{1}{2}(3-\cos^2\theta_0)(\boldsymbol{\tau\tau}\mathbf{n} + \mathbf{ttn}) + \frac{1}{2}\sin^2\theta_0(\mathbf{n}\boldsymbol{\tau\tau} + \mathbf{ntt} + \boldsymbol{\tau}\mathbf{n}\boldsymbol{\tau} + \mathbf{tnt})\right] \tag{H.8}$$

$$\int_{\mathbf{k}\cdot\mathbf{g}<0} (\mathbf{k}\cdot\mathbf{g})^3 F(g)\,\mathrm{d}\mathbf{k}\,\mathrm{d}\mathbf{g} = -\pi^2(1-\cos\theta_0)\int g^5 F(g)\,\mathrm{d}g \tag{H.9}$$

H.3. CHANGE OF VARIABLES

It is often convenient or necessary to change the variables of integration, as, for example, in §7.3.11. Consider the integral

$$\int f(\mathbf{x})\,\mathrm{d}\mathbf{x} \tag{H.10}$$

over the n dimensional vector of variables $\mathbf{x} = (x_1, x_2, \ldots x_n)$. Here $\mathrm{d}\mathbf{x} \equiv \prod_{i=1}^{n} \mathrm{d}x_i$ is a differential volume in **x** space. Given the relation $\mathbf{x} = \mathbf{x}(\mathbf{y})$, we wish to change the variables of integration to **y**. Using the above relation, we may replace $f(\mathbf{x})$ by $F(\mathbf{y})$, and hence the problem reduces to expressing the volume $\mathrm{d}\mathbf{x}$ in terms of $\mathrm{d}\mathbf{y}$.

A change in one component of **x** may be expressed in terms of changes in the components of **y** as

$$\mathrm{d}x_i = \frac{\partial x_i}{\partial y_j}\,\mathrm{d}y_j \tag{H.11}$$

where repeated indices imply summation, as discussed in Appendix A. This may be written in tensor notation as

$$\mathrm{d}_v\mathbf{x} = \mathbf{J}\cdot\mathrm{d}_v\mathbf{y} \tag{H.12}$$

where we have used the symbol $\mathrm{d}_v\mathbf{x}$ for the vector $(\mathrm{d}x_1, \mathrm{d}x_2, \ldots \mathrm{d}x_n)$, to distinguish it from the volume $\mathrm{d}\mathbf{x}$, and **J** is the Jacobian tensor whose components are

$$J_{ij} = \frac{\partial x_i}{\partial y_j} \tag{H.13}$$

For the transformation to exist, **J** must be non-singular.

The Jacobian relates the volumes dx and dy as

$$\mathbf{dx} = |\det(\mathbf{J})|\,\mathbf{dy} \tag{H.14}$$

The proof of this result may be found in many books on linear algebra or advanced calculus (see, e.g., Axler, 1997, pp. 236–243).

As a result, the integral (H.10) may be transformed to

$$\int f(\mathbf{x})\,\mathbf{dx} = \int F(\mathbf{y})\,|\mathcal{J}|\,\mathbf{dy} \tag{H.15}$$

As an example, consider the transformation $(\mathbf{C}, \mathbf{C}_1) \longrightarrow (\mathbf{G}, \mathbf{g})$ that is used in §7.3.8 and §7.3.11. The two sets of variables are related by

$$\mathbf{C} = \mathbf{G} - \frac{\mathbf{g}}{2}, \quad \mathbf{C}_1 = \mathbf{G} + \frac{\mathbf{g}}{2} \tag{H.16}$$

The transformation tensor in matrix form is

$$[\mathbf{J}] = \begin{bmatrix} 1 & 0 & 0 & -\frac{1}{2} & 0 & 0 \\ 0 & 1 & 0 & 0 & -\frac{1}{2} & 0 \\ 0 & 0 & 1 & 0 & 0 & -\frac{1}{2} \\ 1 & 0 & 0 & \frac{1}{2} & 0 & 0 \\ 0 & 1 & 0 & 0 & \frac{1}{2} & 0 \\ 0 & 0 & 1 & 0 & 0 & \frac{1}{2} \end{bmatrix} \tag{H.17}$$

It is a straightforward exercise to show that the determinant of this matrix is unity.

H.4. VOLUME INTEGRALS

Integrals of the form

$$\int \mathbf{F}(\mathbf{C})\,\mathbf{dC}$$

are often required in kinetic theory, where the integration is over all possible values of $\mathbf{C}$, i.e., $-\infty < C_i < \infty$, C_i $(i = 1, 2, 3)$ being the Cartesian components of $\mathbf{C}$. Recall from §7.3.1 that the differential volume element $\mathbf{dC}$ is the product $dC_1\,dC_2\,dC_3$. If $\mathbf{F}(\mathbf{C})$ is an odd function of $\mathbf{C}$, this integral vanishes, as the contribution to the integral of $C_i < 0$ cancels that of $C_i > 0$. Thus

$$\int f(C)\mathbf{C}\,\mathbf{dC} = 0, \quad \int f(C)\mathbf{CCC}\,\mathbf{dC} = 0 \tag{H.18}$$

If $\mathbf{F}(\mathbf{C})$ is an even function of $\mathbf{C}$, the integral takes a symmetric tensorial form. For instance, consider

$$\int f(C)\mathbf{CC}\,\mathbf{dC}$$

The nondiagonal components of the dyadic $\mathbf{CC}$ are odd functions of the components C_i and therefore vanish upon integration. The diagonal components are even functions, the integral of component jj being

$$\int f(C)C_j^2\,\mathbf{dC}$$

By symmetry, the above is equal to the integrals of the other two diagonal components. Hence we may write the integral of the diagonal components as

$$\int f(C)C_j^2\,\mathbf{dC} = \frac{1}{3}\int f(C)C^2\mathbf{dC}$$

and therefore

$$\int f(C)\mathbf{CC}\,\mathrm{d}C = \frac{1}{3}\mathbf{I}\int f(C)C^2\mathrm{d}\mathbf{C} \tag{H.19}$$

It follows from (H.19) that

$$\int f(C)\overline{\mathbf{CC}}\,\mathrm{d}\mathbf{C} = 0 \tag{H.20}$$

In a similar manner, it is straightforward to show that

$$\begin{aligned}\int f(C)\mathbf{CC}(\overline{\mathbf{CC}}{:}\mathbf{w})\,\mathrm{d}\mathbf{C} &= \int f(C)\overline{\mathbf{CC}}(\mathbf{CC}{:}\mathbf{w})\,\mathrm{d}\mathbf{C} \\ &= \frac{2}{15}\mathbf{w}\int f(C)C^4\,\mathrm{d}\mathbf{C}\end{aligned} \tag{H.21}$$

where $\mathbf{w}$ is any second-order tensor. This relation is used in the determination of certain collisional fluxes.

Integrals such as (H.19) and (H.21) over the three dimensions of velocity space may be evaluated by transforming to polar coordinates (C, θ, ϕ); for example,

$$\int f(C)\,\mathrm{d}\mathbf{C} = \int\limits_{C=0}^{\infty}\int\limits_{\theta=0}^{\pi}\int\limits_{\phi=0}^{2\pi} f(C)C^2\sin\theta\,\mathrm{d}\theta\,\mathrm{d}\phi\,\mathrm{d}C = 4\pi\int\limits_{0}^{\infty} f(C)C^2\mathrm{d}C \tag{H.22}$$

H.5. GAUSSIAN INTEGRALS

All the important integrals that occur in kinetic theory reduce to the form (H.22), with the function $f(C)$ having the form $f(C) = C^n e^{-C^2}$. These are called Gaussian integrals and are evaluated using the following result:

$$\int\limits_{0}^{\infty} e^{-C^2}C^n\mathrm{d}C = \frac{1}{2}\Gamma(\frac{n+1}{2}) \quad \text{for } n > -1 \tag{H.23}$$

where $\Gamma(n)$ is the gamma function. In particular, if n is an integer,

$$\int\limits_{0}^{\infty} e^{-C^2}C^n\mathrm{d}C = \begin{cases} \frac{\sqrt{\pi}}{2}\cdot\frac{1}{2}\cdot\frac{3}{2}\cdot\frac{5}{2}\cdots\frac{n-1}{2}, & \text{if } n \text{ is even} \\ \frac{1}{2}\left(\frac{n-1}{2}\right)!, & \text{if } n \text{ is odd}\end{cases}$$

APPENDIX I

A Brief Introduction to Linear Stability Theory

Linear stability analysis of fluid flows is discussed in considerable detail in several texts (Chandrasekhar, 1981; Drazin, 2002; Drazin and Reid, 2004), hence only a brief introduction to the subject is given here. First, we identify the state of the system, known as the "base state" or the "basic flow," whose stability we wish to investigate. Usually the base state is a steady-state solution of the governing equations; here we restrict attention to such base states. We denote by $\mathbf{X}$ the vector of variables describing the states of the system and $\mathbf{X}^0$ the base state. For rapid granular flow, the variables are

$$\mathbf{X} = (\nu, v_x, v_y, v_z, T)$$

which in general are functions of time and the spatial coordinates.

As an illustrative example, consider the problem of plane Couette flow (see Fig. 8.4). The base state of steady, fully developed flow has the form $\mathbf{X}^0 = (\nu^0(y), v_x^0(y), 0, 0, T(y))$. To investigate its stability, perturbations are imposed on $\mathbf{X}$, i.e.,

$$\mathbf{X} = \mathbf{X}^0(y) + \mathbf{X}'(t, x, y, z,)$$

where the prime denotes perturbations, and the time evolution of the perturbation is studied. The base state is said to be stable if all perturbations that are initially small remain small for all time and is unstable if even one perturbation does not remain small. A precise definition of what is meant by a "small perturbation" can be found in Drazin (2002, p. 29) or Drazin and Reid (2004, pp. 8–9). For the purpose of this discussion, we say that the perturbations are small enough that in a series expansion of any function of $\mathbf{X}$ about $\mathbf{X}^0$, only the terms linear in $\mathbf{X}'$ are retained and terms of higher order are discarded. By this process, we obtain the linearized form of the governing equations and boundary conditions,

$$\frac{\partial \mathbf{X}'}{\partial t} = \mathcal{L}(\mathbf{X}'), \quad \mathcal{B}(\mathbf{X}') = 0 \tag{I.1}$$

respectively. The operators $\mathcal{L}$ and $\mathcal{B}$ are, in general, partial differential equations in the spatial coordinates, with coefficients that may depend on the base state variables $\mathbf{X}^0$ and all the material and flow parameters $\mathbf{p} \equiv (p_1, p_2, \ldots)$. For instance, the linearized continuity equation is

$$\frac{\partial \nu'}{\partial t} = -\mathbf{v}^0 \cdot \nabla \nu' - \mathbf{v}' \cdot \nabla \nu^0 - \nu^0 \nabla \cdot \mathbf{v}' - \nu' \nabla \cdot \mathbf{v}^0 \tag{I.2}$$

The symmetries of the base state determine the form of the solution for $\mathbf{X}'$. For instance, the base state and the linearized equations in plane Couette flow have translational symmetry in the x and z directions, i.e., they are invariant to translation in the x and z directions. The general solution is therefore a periodic function of x and z. Hence we may obtain the solution in terms of *normal modes*

$$\mathbf{X}' = \hat{\mathbf{X}}(y)e^{ik_x x + ik_z z + st} \tag{I.3}$$

where $i \equiv \sqrt{-1}$. Above, the wavenumbers k_x and k_z are real and nonnegative, and s and $\hat{\mathbf{X}}$ are, in general, complex. The wavenumber k_x is related to the wavelength λ_x for variation in the x direction by $k_x = 2\pi/\lambda_x$, and a similar relation holds for k_z. If the variables $\mathbf{X}$ are real, only the real part of $\mathbf{X}'$ is relevant. Substitution of the above in (I.1) yields an eigenvalue problem, with eigenvalues s and eigenfunctions $\hat{\mathbf{X}}(y)$.

The growth or decay of a normal mode is determined by the real part of the eigenvalues, $s_r \equiv \mathrm{Re}(s)$. It decays if $s_r < 0$ for all the eigenvalues, and the mode is said to be stable. It grows without bound† if $s_r > 0$ for even one eigenvalue, and the mode is said to be unstable. If the maximum of all the s_r for a mode is zero, it is said to be neutrally stable. The imaginary part of an eigenvalue $s_i \equiv Im(s)$ determines the phase velocity of the corresponding eigenfunction, or the velocity with which the perturbation translates in space. The phase velocity in the x direction is $-s_i/k_x$ and it is similarly defined in the z direction.

A general perturbation can be expressed as a linear combination of all the normal modes, but the linearity of the equation governing $\mathbf{X}'$ allows each normal mode to be analyzed separately. Hence the flow is linearly stable if all the normal modes are stable and unstable if even one normal mode is unstable.

† In reality, perturbations do not grow without bound because the nonlinear terms in the governing equations become important when their amplitude becomes finite, and a new steady or time-dependent state is reached.

APPENDIX J

Pseudo Scalars, Vectors, and Tensors

Consider a *time-independent* transformation of the unprimed reference frame to a primed reference frame. In the notation of Appendix F, this implies that $\boldsymbol{\beta}(t) = 0$ and $\mathbf{Q}$ is a constant. In particular, we are interested in a transformation arising from a *reflection* of one or more of the coordinate axes. We take the unprimed reference frame to have a right-handed coordinate system (see §A.7.1).

If $\mathbf{w}$ is a vector, the change to the primed reference frame causes its components to transform as (compare (F.9) with $\boldsymbol{\beta}(t) = 0$)

$$w_i' = Q_{ij}\, w_j \tag{J.1}$$

For example, for a reflection about the x_2–x_3 plane (i.e., an inversion of the x_1 axis),

$$[\mathbf{Q}] = \begin{bmatrix} -1 & 0 & 0 \\ 0 & 1 & 0 \\ 0 & 1 & 0 \end{bmatrix} \tag{J.2}$$

which gives

$$w_1' = -w_1, \quad w_2' = w_2, \quad w_3' = w_3 \tag{J.3}$$

For each additional axis reflected, the sign of the corresponding component changes. (Note that each reflection results in the handedness of the primed coordinate system switching between right-handed and left-handed.) Thus the vector $\mathbf{w}$ remains unchanged in physical space (Fig. J.1), but the sign of its components corresponding to the axes that are inverted have changed. All vectors that behave in this manner are called *polar* vectors. Physical examples of polar vectors are the position vector, velocity, force, and electric field. Similarly, scalars and tensors that transform as

$$b' = b, \quad A_{ij\cdots pq}' = Q_{ik}\, Q_{jl} \cdots Q_{pr}\, Q_{qs}\, A_{kl\cdots rs} \tag{J.4}$$

are called polar scalars and tensors, respectively.

There is a class of vectors that do not transform as (J.1) under reflection of an axis but are of considerable importance, as they occur frequently in physical problems. Consider the vector product (cross product) of two polar vectors $\mathbf{v}$ and $\mathbf{w}$. On reflection of the x_1 axis, the two polar vectors transform according to (J.3). However, as the coordinate system of the primed reference frame is left-handed, the components of the vector $\boldsymbol{\Omega} = \mathbf{v} \times \mathbf{w}$ are (see the text following (A.26))

$$\Omega_1' = v_2' w_3' - v_3' w_2' = v_2 w_3 - v_3 w_2 = \Omega_1, \quad \text{and similarly } \Omega_2' = -\Omega_2,\ \Omega_3' = -\Omega_3 \tag{J.5}$$

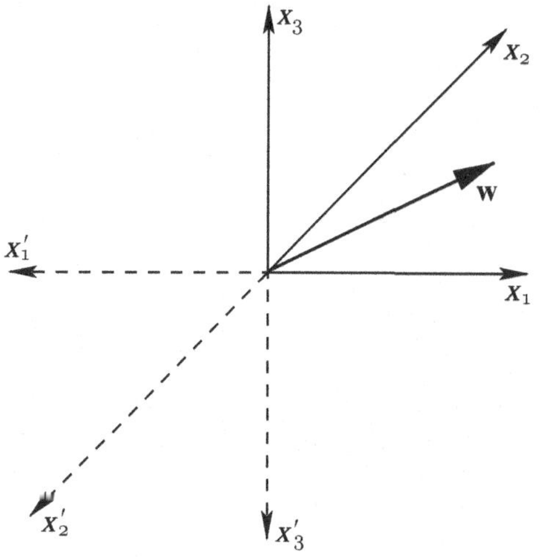

Figure J.1. The transformation of a polar vector **w** under reflection of one or more of the coordinate axes. The vector **w** remains unchanged in physical space, but for each reflected axis its corresponding component changes sign.

This transformation is quite different from that of a polar vector. If we now reflect the x_2 axis also, the coordinate system switches back to a right-handed one and the components of $\mathbf{\Omega}'$ become

$$\Omega_1' = -\Omega_1, \quad \Omega_2' = -\Omega_2, \quad \Omega_3' = \Omega_3 \tag{J.6}$$

Thus $\mathbf{\Omega}$ now transforms like a polar vector. It is clear that the transformation of $\mathbf{\Omega}$ depends on the handedness of the primed coordinate system compared to the unprimed one. Vectors that transform in this manner are called pseudovectors or axial vectors.† Physical examples of pseudovectors are the angular velocity, torque, and magnetic field. If the transformation of the reference frame arises from a pure rotation, the handedness of the coordinate system does not change, and hence pseudovectors transform like polar vectors.

Let us take a spinning top as an example and see how the above transformation arises. In the unprimed reference frame, the axis of the top is aligned with the x_1 axis, as shown in Fig. J.2. Since the unprimed frame is right handed, the direction of the angular velocity vector $\mathbf{\Omega}$ must be determined using the right-hand rule: if the fingers of the right hand are curled in the direction of rotation of the top, the thumb points in the direction of $\mathbf{\Omega}$. Thus $\mathbf{\Omega}$ points in the positive x_1 direction in the unprimed frame. If the x_1 axis of the reference frame is reflected to obtain the primed reference frame, the coordinate system of the latter becomes left handed. Hence the direction of the angular velocity must be determined using the left-hand rule, which differs from the right-hand rule only in that the left hand is used. We then find that the angular velocity vector $\mathbf{\Omega}'$ in the primed frame points in the positive x_1' direction. If the x_1 and x_2 axis are both reflected, the primed coordinate system is right handed and $\mathbf{\Omega}'$ points in the negative x_1' direction.

It is clear from the above discussion that a pseudovector $\mathbf{\Omega}$ transforms as

$$\Omega_i' = \det(\mathbf{Q})\, Q_{ij}\, \Omega_j \tag{J.7}$$

Similarly, we define a pseudoscalar β and a pseudotensor $\mathbf{\Lambda}$ as quantities that transform as

$$\beta' = \det(\mathbf{Q})\, \beta, \quad \Lambda_{ij\cdots pq}' = \det(\mathbf{Q})\, Q_{ik}\, Q_{jl} \cdots Q_{pr}\, Q_{qs}\, \Lambda_{kl\cdots rs} \tag{J.8}$$

In this appendix, roman letters are used for polar quantities and greek letters for pseudo (axial) quantities, so as to distinguish between the two. Note that det($\mathbf{Q}$) is -1 for a reflection of an odd number of coordinate axes, and it is $+1$ for a reflection of an even number of axes.

† The term axial vector is often used because these vectors are usually used to describe rotation about an axis (Arfken and Weber, 2001, p. 142).

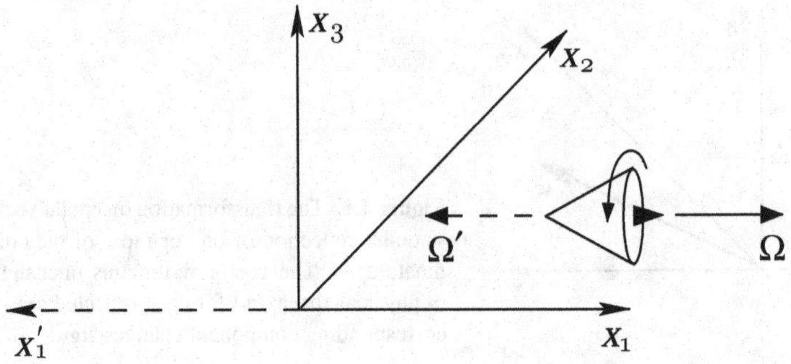

Figure J.2. The transformation of the angular velocity vector Ω of a spinning top, a pseudovector, under reflection of a coordinate axes. The angular velocity transforms to Ω' when the x_1 axis is reflected. Note that the direction of spin remains unchanged in physical space, but the change in handedness of the primed coordinate system results in the sign of that component of Ω remaining unchanged.

An example of a pseudoscalar is the scalar triple product $(\mathbf{v}\times\mathbf{w})\cdot\mathbf{z}$. The dyad (see §A.4) $\mathbf{v}\,\boldsymbol{\Omega}$ is a second-order pseudotensor. The alternating tensor $\boldsymbol{\epsilon}$ is a third-order pseudotensor.

As polar vectors and pseudovectors transform differently under a change of frame, they cannot be added. The same holds for scalars and tensors. For example, a relation of the form

$$\mathbf{w} = \mathbf{v} + b\,\boldsymbol{\Omega} \tag{J.9}$$

where b is a polar scalar, is not permitted. However, the relation

$$\mathbf{w} = \mathbf{v} + b\,\mathbf{u}\times\boldsymbol{\Omega} \tag{J.10}$$

is permitted if $\mathbf{u}$ is a polar vector, as the second term on the right hand is a polar vector. Similarly, the dyad $\boldsymbol{\Omega}\,\boldsymbol{\omega}$ is a polar tensor.

APPENDIX K

Answers to Selected Problems

1.1 2.7 km

1.2 (d) Neck: 29.3; base: 31.9

(e) 2.28×10^{-5} N; 99% of $F_{l,\max}$

1.3 (a) $Bo = \sin(\beta_r - \beta_{ro})/(n \cos\psi \sin\beta_{ro})$

1.4 6.3×10^{-4} C m^{-2}

1.9 (a) 30°; (d) 24.0°; 23.8° by numerical integration

1.11 (b) $N = [(k+2)\, m\, g]/2$; $H = [(2i - k)\, m\, g \tan\theta]/2$

1.12 (a) $\mathrm{d}^2 x/\mathrm{d}t^2 + 2\, h\, \mathrm{d}x/\mathrm{d}t + \omega^2 x = 0$

(b) $h^2 < \omega^2 : x = e^{-h\,t}\,(B_1 \cos(\beta\, t) + B_2 \sin(\beta\, t))$, where $\beta \equiv \sqrt{\omega^2 - h^2}$.
Alternative form: $x = B_3\, e^{-h\,t} \cos(\beta\, t + B_4)$, where B_1–B_4 are constants.
$h^2 > \omega^2 : x \equiv x_1 = B_5\, e^{(-h+\beta)\,t} + B_6\, e^{-(h+\beta)\,t}$, where B_5 and B_6 are constants and $\beta \equiv \sqrt{h^2 - \omega^2}$.

(d) $x(0) > 0,\ (\mathrm{d}x/\mathrm{d}t)(0) > 0$

(e) $x \equiv x_2 = e^{-h\,t}\,(B_7 + B_8\, t)$, where B_7 and B_8 are constants.

(f) $x_2/x_1 \propto t\, e^{-\sqrt{h^2-\omega^2}\,t}$ for large t. Hence $x_2/x_1 \to 0$ as $t \to \infty$.

1.13 (a) 43–45 N for solids fractions in the range 0.56–0.64

(b) $-m\, g + \int_{CD} (\rho\, v_y^2 + \sigma_{yy})\, \mathrm{d}S - \mathrm{d}/\mathrm{d}t \int_{V(t)} \rho\, v_y\, \mathrm{d}V$

1.16 (b) Air: 1.9×10^{-3} for $w = 0.01$ m s^{-1}; 2.6 for $w = 6$ m s^{-1}
Water: 3×10^{-3} for $w = 5 \times 10^{-4}$ m s^{-1}; 1.4 for $w = 0.1$ m s^{-1}

2.1 The pressure exerted by the foot causes the sand to shear and dilate. Hence water flows in from the surface to fill the extra void space created. When the foot is lifted, elastic effects lead to a partial recovery of the original configuration. This causes some of the water to be squeezed out.

2.2 The applied force H and the frictional force F_t generate a couple. This is opposed by the couple generated by the weight W and the normal force F_n. If F_n passes through the center of mass, the opposing couple vanishes.

2.4 (a) (i) $N = (\sigma_{xx} + \sigma_{zz})/2 + [(\sigma_{xx} - \sigma_{zz})/2] \cos 2\theta_w - \sigma_{xz} \sin 2\theta_w$
$T = [(\sigma_{zz} - \sigma_{xx})/2] \sin 2\theta_w - \sigma_{xz} \cos 2\theta_w$
(ii) $N = \sigma_{\theta\theta},\ T = -\sigma_{\theta r}$

(c) $2\,\sigma_{1,2} = \sigma_{rr} + \sigma_{\theta\theta} \pm \sqrt{(\sigma_{rr} + \sigma_{\theta\theta})^2 + 4\,({\sigma_{r\theta}}^2 - \sigma_{rr}\,\sigma_{\theta\theta})}; \sigma_3 = \sigma_{\phi'\phi'}$

(d) The material is not cohesionless, as one of the principal stresses is negative.

(e) $F_b = \rho\, g\, V;\ V - \pi\, R^3/3\ \left[(2\,(1 - \cos\theta_w)/\sin^3\theta_w)(1 + (H/R)\tan\theta_w)^3 - \cot\theta_w\right]$
$F_v = <\sigma_{zz}>\ \pi\, R^2 + [<\sigma_{\theta\theta}>\ \sin\theta_w - <\sigma_{\theta r}>\ \cos\theta_w]\, S_w$
$S_w = \pi\, [(R + H \tan\theta_w)^2 - R^2]/\sin\theta_w$

(f) 6.2 kPa

2.5 (a) $\sin\theta_c/(\cos\theta_c)^n = a\,(m\,g)^{n-1}$; (b) 0, as the sphere rolls for $\theta > 0$

2.6 (a) $(\rho\,g\,h)/(2\,c) \leq (\cos\phi_*\,\sin\theta)/[\sin(\alpha-\phi_*)\,\sin(\theta-\alpha)]$
$h_* = (4\,c\,\cos\phi_*)/[\rho\,g\,(1-\sin\phi_*)]$; $c = 0.45$ kPa

2.9 (a) There are two values for a, given by $a_1 = 11.22$ and $a_2 = 3.233$.
$|T| = 3.50$ kPa, $N = 4.17$ kPa, for $a = a_1$; $|T| = 1.54$ kPa, $N = 1.83$ kPa, for $a = a_2$
(b) $a = a_1 : n_{11} = 0.9746,\ n_{21} = 0.2242,\ n_{12} = 0.2245,\ n_{22} = -0.9743$
$a = a_2 : n_{11} = 0.7274,\ n_{21} = 0.6863,\ n_{12} = 0.6862,\ n_{22} = -0.7274$
(c) 13° for $a = a_1$; 43° for $a = a_2$

2.10 (a) $\gamma_w = (\delta+\omega)/2 \equiv \gamma_{wp}$ (passive); $\gamma_w = (\pi+\delta-\omega)/2 \equiv \gamma_{wa}$ (active)
Here $\sin\omega \equiv \sin\delta/\sin\phi_*$.
(b) $\gamma_{wp} = 32°$; $\gamma_{wa} = 78°$

2.11 (a) $\tau = 229$ kPa; $\sigma = 254$ kPa; $\sin\phi = 0.9$

2.14 (a) Cohesionless material; (b) 0.48

2.16 (b) $\pi/4$

2.17 (a) $f_u = \dfrac{2\,(\sin\omega-\sin\phi_*)\,\sigma_{1C}}{(1-\sin\phi_*)\,(1+\sin\omega)}$
(b) $D = (2\,f_u\,\sin 2\beta)/(\rho\,g)$
(c) (i) The arch will collapse; (ii) the arch will be stable.

2.18 (a) Body force $\rho\,g\,V$, centrifugal force $\rho\omega^2\,r\,V$, normal force F_n, and shear force F_t
(b) $F_n = \rho\,V(g\,\cos\beta + \omega^2\,r\,\sin\beta)$
$F_t = \rho\,V(g\,\sin\beta - \omega^2\,r\,\cos\beta)$
(c) $\tan\beta = \mathrm{d}z_f/\mathrm{d}r = (\alpha\,\mu\,g + \omega^2\,r)/(g - \alpha\,\mu\,\omega^2\,r)$
(d) $\mathrm{d}\eta/\mathrm{d}\xi = (\alpha\,\mu + G\,\xi)/(1 - \alpha\,\mu\,G\,\xi)$; (e) flat ($\eta = 0$)
(f) $\xi_c = \mu/G$;
$$\eta = \eta_0 + \frac{\xi-\xi_c}{\mu} - \frac{1+\mu^2}{\mu^2\,G}\ln\left(\frac{1+\mu\,G\,\xi}{1+\mu\,G\,\xi_c}\right)$$
(g) The solution implies a reduction in density relative to the initial state.

3.1 (a) 8.56 kPa; (b) no

3.2 (a) Passive state; (b) $r/r_0 = (\sec[(\gamma_w\,\theta/\theta_w)+\Omega]/\sec\Omega)^{-\theta_w/\gamma_w}$ for $\gamma_w + \Omega < \pi/2$; $\Omega \equiv (\pi/4) - (\phi/2)$; $r(\theta_w)/r_0 = 0.78$

3.3 (a) (i) 0.024 kg s^{-1}; (ii) 0.014 kg s^{-1}; (b) 0.0098 kg s^{-1}

3.6 (a) 35°; (c) 1.29, for $\nu = 0.63$; 1.11, for $\nu = 0.56$

3.8 (a) $v = v_0/\sqrt{1+(2\,\xi/F_*)}$; $F_* \equiv v_0^2/(g\,W)$
(b) 0.57 m s^{-1}; (c) 0.32 (assuming that $v_0(0) \approx \overline{v}$)

4.1 (b) $(1/M') + (V_0 - (1/M'))e^{-M'\eta}$; $M' \equiv K\,F\,\tan\delta$
(f) $\eta_* = 63$ (active); 7.4 (passive); $\overline{\sigma}_{xx} = 7.1$, $\overline{\sigma}_{xy} = 1.0$

4.2 (c) 10.5 (active); 2.3 (passive)
(d) The load cell cannot be used for case (i), as the stress is less than the resolution of the load cell.

5.9 $p_c = p_0/2$

5.10 (c) 149 kPa (at $\xi = 0$); 818 kPa (at $\xi = 1$)

5.11 (g) $\phi_* = 35°$; $c = 105$ kPa

5.12 (a) r component: $\mathrm{d}\sigma_{rr}/\mathrm{d}r + 2\,(\sigma_{rr} - \sigma_{\theta\theta})/r = 0$

θ component: $((\sigma_{\theta\theta} - \sigma_{\phi\phi})\cot\theta)/r = 0$

(c) $\tilde{\sigma}_{rr} = \dfrac{(1/\xi^3) - 1}{(1/\xi_0^3) - 1}$; $\tilde{\sigma}_{\theta\theta} = -\dfrac{(1 + 1/(2\,\xi^3))}{(1/\xi_0^3) - 1}$

(d) σ_{rr} : compressive; $\sigma_{\theta\theta}$: tensile

(e) $(2/3)\,\sigma_0\,(1 - \xi_0^3)$

5.14 (a) Incompressible flow

(b) $\sigma_{xx} = \sigma_{yy} = \sigma_{zz} = p$; $\sigma_{xz} = 0 = \sigma_{yz}$; $\sigma_{xy} = \sigma_{yx} = p\,\sin\phi_*$

(d) $\pi\,R^2\,\rho\,g\,\sin\phi_*\,L^2$

(e) The r component of the momentum balance implies that $\partial p/\partial r \geq 0$, whereas the θ component implies that $\partial p/\partial r \leq 0$.

(f) Unacceptable, as $\sigma_{r\theta} < 0$ at the outer wall whereas it should be ≥ 0 for rotation in the anticlockwise direction.

6.1 (a) $(3\,\dot{Q}\,t_b - \pi\,R_b{}^3\,\tan\beta_r)/(3\,\pi\,R_b{}^2)$, (b) $r/R_b = 0.721$; $z/R_b = 1.86$

References

Abramowitz, M. and Stegun, I. (1965). *Handbook of Mathematical Functions*. (Dover, New York).

Adamson, A. W. and Gast, A. P. (1997). *Physical Chemistry of Surfaces*, 6 edn. (John Wiley, New York).

Airey, D. W., Budhu, M., and Wood, D. M. (1985). Some aspects of the behaviour of soils in simple shear. In *Developments in Soil Mechanics and Foundation Engineering – 2: Stress-Strain Modelling of Soils*, P. Banerjee and R. Butterfield, eds. (Elsevier, Amsterdam), 185–213.

Al-Din, N. and Gunn, D. J. (1984). The flow of non-cohesive solids through orifices. *Chem. Eng. Sci.* **39**, 121–127.

Alam, M., Arakeri, V. H., Nott, P. R., Goddard, J. D., and Herrmann, H. J. (2005). Instability-induced ordering, universal unfolding and the role of gravity in granular Couette flow. *J. Fluid Mech.* **523**, 277–306.

Alam, M. and Luding, S. (2003). Rheology of bidisperse granular mixtures via event-driven simulations. *J. Fluid Mech.* **476**, 69–103.

Alam, M. and Nott, P. R. (1997). The influence of friction on the stability of unbounded granular shear flow. *J. Fluid Mech.* **343**, 267–301.

Alam, M. and Nott, P. R. (1998). Stability of plane Couette flow of a granular material. *J. Fluid Mech.* **377**, 99–136.

Albert, R., Albert, I., Hornbaker, D., Schiffer, P., and Barabasi, A. (1997). Maximum angle of stability in wet and dry spherical granular media. *Phys. Rev. E* **6**, R6271–R6274.

Albert, R., Pfeifer, M. A., Barabasi, A. L., and Schiffer, P. (1999). Slow drag in a granular medium. *Phys. Rev. Lett.* **82**, 205–208.

Amontons, G. (1699). De la resistance causée dans les machines (Concerning the resistance caused in machines). *Mem. Acad. Roy.* **A**, 257–282.

Anand, L. and Gu, C. (2000). Granular materials: constitutive equations and strain localization. *J. Mech. Phys. Solids* **48**, 1701–1733.

Ancey, C. (2001). Dry granular flows down an inclined channel: experimental investigations on the frictional-collisional regime. *Phys. Rev. E* **65**, 011304–1–19.

Anderson, K. G. and Jackson, R. (1992). A comparison of the solutions of some proposed equations of motion of granular materials for fully developed flow down inclined planes. *J. Fluid Mech.* **241**, 145–168.

Anderson, T. B. and Jackson, R. (1967). A fluid mechanical description of fluidized beds. Equations of motion. *Ind. Eng. Chem. Fundam.* **6**, 527–539.

Andronov, A. and Chaikin, C. E. (1949). *Theory of Oscillations*. (Princeton University Press, Princeton).

Aranson, I. S. and Tsimring, L. S. (2002). Continuum theory of partially fluidized granular flows. *Phys. Rev. E* **65**, 061303–1–20.

Arfken, G. B. and Weber, H. J. (2001). *Mathematical Methods for Physicists*, 5 edn. (Academic Press, New Delhi).

Arnarson, B. O. and Willits, J. T. (1998). Thermal diffusion in binary mixtures of smooth, nearly elastic spheres with and without gravity. *Phys. Fluids* **10**, 1324–1328.

References

Astarita, G. and Marucci, G. (1974). *Principles of Non-Newtonian Fluid Mechanics.* (McGraw-Hill, Maidenhead).

Atkinson, J. H. and Bransby, P. L. (1982). *The Mechanics of Soils*, ELBS edn. (McGraw-Hill, Maidenhead).

Axler, S. (1997). *Linear Algebra Done Right* (Springer-Verlag, New York).

Azanza, E., Chevoir, F., and Moucheront, P. (1999). Experimental study of collisional granular flows down an inclined plane. *J. Fluid Mech.* **400**, 199–227.

Babić, M. (1993). On the stability of rapid granular flows. *J. Fluid Mech.* **254**, 127–150.

Bagchi, P. and Balachander, S. (2003). Inertial and viscous forces on a rigid sphere in straining flows at moderate Reynolds numbers. *J. Fluid Mech.* **481**, 105–148.

Bagnold, R. A. (1954). Experiments on a gravity-free dispersion of large solid spheres in a Newtonian fluid under shear. *Proc. Roy. Soc. Lond.* **A225**, 49–63.

Balasubramaniam, A. S. (1969). Some factors influencing the stress-strain behavior of clay. Ph.D. thesis, University of Cambridge.

Barajas, L., Garcia-Colin, L. S., and Piña, E. (1973). On the Enskog–Thorne theory for a binary mixture of dissimilar rigid spheres. *J. Stat. Phys.* **7**, 161.

Bardet, J. P. (1997). *Experimental Soil Mechanics*. (Prentice Hall, Upper Saddle River).

Batchelor, G. K. (1967). *An Introduction to Fluid Dynamics*. (Cambridge University Press, Cambridge).

Bates, L. (2006). The need for industrial education in bulk technology. *Bulk Solids Handl.* **26**, 464–473.

Bauer, E. (1996). Calibration of a comprehensive hypoplastic model for granular materials. *Soils Found.* **36**, 13–26.

Bauer, E., Huang, W., and Wu, W. (2004). Shearbanding in anisotropic hypoplasticity. *Int. J. Solids Strut.* **36**, 5903–5919.

Bauer, E. and Wu, W. (1995). A hypoplastic constitutive model for cohesive powders. *Powder Technol.* **85**, 1–9.

Baxter, G. W. and Behringer, R. P. (1990). Cellular automata models of granular flow. *Phys. Rev. A* **42**, 1017–1020.

Beetstra, R., van der Hoef, M. A., and Kuipers, J. A. M. (2007). Drag force of intermediate Reynolds number flows past mono and bidisperse arrays of spheres. *AIChE J.* **53**, 489–501.

Bender, C. M. and Orszag, S. A. (1984). *Advanced Mathematical Methods for Scientists and Engineers*. (McGraw-Hill, Singapore).

Benink, E. (1989). Flow and stress analysis of cohesionless bulk materials in silos related to codes. Ph.D. thesis, University of Twente.

Berryman, J. G. (1983). Definition of dense random packing. In *Advances in the Mechanics and the Flow of Granular Materials*, M. Shahinpoor, ed., Vol. I. (Trans Tech Publications, Clausthal-Zellerfeld), 1–18.

Beverloo, W. A., Leniger, H. A., and van de Velde, J. (1961). The flow of granular solids through orifices. *Chem. Eng. Sci.* **15**, 260–269.

Bingham, E. C. and Wikoff, R. W. (1931). The flow of dry sand through capillary tubes. *J. Rheol.* **2**, 395–400.

Bird, R. B., Armstrong, R. C., and Hassager, O. (1977). *Dynamics of Polymeric Liquids*, Vol. 1. (John Wiley, New York).

Bird, R. B., Stewart, W. E., and Lightfoot, E. N. (2002). *Transport Phenomena*. (John Wiley, New York).

Bishop, A. W. (1966). The strength of soils as engineering materials. *Géotechnique.* **16**, 91–128.

Bishop, A. W., Webb, D. L., and Skinner, A. E. (1965). Triaxial tests on soils at elevated cell pressure. In *Proc. 6th Int. Conf. Soil Mech. Found. Eng.* Vol. I. Montreal, 170–174.

Bishop, J. F. W. and Hill, R. (1951). A theory of the plastic distortion of a polycrystalline aggregate under combined stresses. *Phil. Mag. Ser. 7* **42**, 414–427.

Blair-Fish, P. M. and Bransby, P. L. (1973). Flow patterns and wall stresses in a mass flow bunker. *Trans. ASME B J. Eng. Ind.* **95**, 17–26.

Blight, G. E. (1986). Pressures exerted by materials stored in silos: part I, coarse materials. *Géotech.* **36**, 33–46.

References

Bocquet, L., Charlaix, E., Ciliberto, S., and Crassous, J. (1998). Moisture-induced ageing in granular media and the kinetics of capillary condensation. *Nature* **396**, 735–737.

Bocquet, L., Losert, W., Schalk, D., Lubensky, T. C., and Gollub, J. P. (2002). Granular shear flow dynamics and forces: experiments and continuum theory. *Phys. Rev. E* **65**, 11307–1–11307–19.

Bolton, M. D. (1986). The strength and dilatancy of sands. *Géotech.* **36**, 65–78.

Boltzmann, L. (1872). Further studies on the thermal equilibrium among gas molecules. *Wien. Ber.* **66**, 275.

Boltzmann, L. (1995). *Lectures on Gas Theory*, Reprint edn. (Dover, New York).

Bosley, J., Schofield, C., and Shook, C. A. (1969). An experimental study of granule discharge from model hoppers. *Trans. Inst. Chem. Eng.* **47**, T147– T153.

Bouchaud, J.-P., Cates, M. E., and Claudin, P. (1995). Stress distribution in granular media and nonlinear wave equation. *J. Phys. I France* **5**, 639–656.

Bowden, F. P. and Leben, L. (1939). The nature of sliding and the analysis of friction. *Proc. R. Soc. London* **A169**, 371–391.

Bowden, F. P., Moore, A. J. W., and Tabor, D. (1943). The ploughing and adhesion of metals. *J. Appl. Phys.* **14**, 80–91.

Bowden, F. P. and Tabor, D. (1939). The area of contact between stationary and moving surfaces. *Proc. R. Soc. London* **A169**, 391–413.

Bransby, P. L. and Blair-Fish, P. M. (1974). Wall stresses in mass-flow bunkers. *Chem. Eng. Sci.* **29**, 1061–1074.

Bransby, P. L. and Blair-Fish, P. M. (1975). Deformations near rupture surfaces in flowing sand. *Géotech.* **25**, 384–389.

Bransby, P. L., Blair-Fish, P. M., and James, R. G. (1973). An investigation of the flow of granular materials. *Powder Technol.* **8**, 197–206.

Brennen, C. and Pearce, J. C. (1978). Granular material flow in two-dimensional hoppers. *Trans. ASME E J. Appl. Mech.* **45**, 43–50.

Brennen, C. E., Sieck, K., and Paslaski, J. (1983). Hydraulic jumps in granular material flow. *Powder Technol.* **35**, 31–37.

Brilliantov, N. V., Spahn, F., Hertzsch, J. M., and Poschel, T. (1996). Model for collisions in granular gases. *Phys. Rev. E* **53**, 5382–5392.

Brillouin, L. (1964). *Tensors in Mechanics and Elasticity* (Academic Press, New York).

Briscoe, B. (1982). Friction, fact and fiction. *Chem. Ind.* **14**, 467–474.

Brockbank, R., Huntley, J. M., and Ball, R. C. (1997). Contact force distribution beneath a three-dimensional granular pile. *J. Phys. II France* **7**, 1521–1532.

Brown, R. L. and Richards, J. C. (1959). Exploratory study of the flows of granules through apertures. *Trans. Instn. Chem. Engrs.* **37**, 108–119.

Brown, R. L. and Richards, J. C. (1970). *Principles of Powder Mechanics*. (Pergamon, Oxford).

Bryan. (1894). *Brit. Assoc. Rep.* 83.

Budny, T. J. (1979). Stick-slip friction as a method of powder flow characterization. *Powder Technol.* **23**, 197–201.

Burnett, D. (1935). The distribution of velocities in a slightly non-uniform gas. *Proc. Lond. Math. Soc.* **39**, 385–430.

Campbell, C. S. (1982). Shear flows of granular materials. Ph.D. thesis, California Institute of Technology.

Campbell, C. S. (1993). Boundary interaction for two-dimensional granular flows. Part 1. Flat boundaries, asymmetric stresses and couple stresses. *J. Fluid Mech.* **247**, 111–136.

Campbell, C. S. (2006). Granular material flows–an overview. *Powder Technol.* **162**, 208–229.

Campbell, C. S. and Brennen, C. E. (1985). Chute flows of granular materials: some computer simulations. *Trans. ASME E J. Appl. Mech.* **52**, 172–178.

Carnahan, N. F. and Starling, K. E. (1969). Equation of state of nonattracting rigid spheres. *J. Chem. Phys.* **51**, 635–636.

Carslaw, H. S. and Jaeger, J. C. (1959). *Conduction of Heat in Solids*, 2 edn. (Oxford University Press, Oxford).

Cates, M. E., Wittmer, J. P., Bouchaud, J. P., and Claudin, P. (1998). Development of stresses in cohesionless poured sand. *Phil. Trans. R. Soc. Lond. A* **356**, 2535–2560.

Cattaneo, C. (1938). Sul contatto di due corpi elastici: distribuzione locale degli sforzi (On the contact of two elastic bodies: distribution of local forces). *Rendiconti dell' Accademia nazionale dei Lincei* **27**, 342–348, 434–436, 474–478.

Cercignani, C. (1988). *The Boltzmann Equation and Its Applications*. (Springer-Verlag, New York).

Chaikin, P. M. and Lubensky, T. C. (1997). *Principles of Condensed Matter Physics*. (Cambridge University Press, Cambridge).

Chakrabarthy, J. (1987). *Theory of Plasticity*. (McGraw-Hill, Singapore).

Chambon, R., Desrues, J., Hammad, W., and Charlier, R. (1994). Cloe, a new rate-type constitutive model for geomaterials: theoretical basis and implementation. *Int. J. Num. Anal. Meth. Geomech.* **18**, 253–278.

Chandrasekhar, S. (1981). *Hydrodynamic and Hydromagnetic Stability*. (Dover, New York).

Chapman, S. (1912). The kinetic theory of a gas constituted of spherically symmetrical molecules. *Phil. Trans. Roy. Soc. A* **211**, 433–483.

Chapman, S. (1916). On the law of distribution of velocities, and on the theory of viscosity and thermal conduction in a non-uniform simple monatomic gas. *Phil. Trans. Roy. Soc. A* **216**, 279–348.

Chapman, S. (1917). On the kinetic theory of a gas; part II, a composite monatomic gas, diffusion, viscosity and thermal conduction. *Phil. Trans. Roy. Soc. A* **217**, 115–197.

Chapman, S. and Cowling, T. G. (1964). *The Mathematical Theory of Nonuniform Gases*, 2 edn. (Cambridge University Press, Cambridge).

Chen, Y. M., Rangachari, S., and Jackson, R. (1984). Theoretical and experimental investigation of fluid and particle flow in a vertical standpipe. *Ind. Eng. Chem. Fundam.* **23**, 354–370.

Chung, T. J. (1988). *Continuum Mechanics*. (Prentice-Hall, New York).

Clague, K. (1973). The effects of stresses in bunkers. Ph.D. thesis, University of Nottingham.

Clark, N. A. and Ackerson, B. J. (1980). Observation of the coupling of concentration fluctuations to steady-state shear flow. *Phys. Rev. Lett.* **44**, 1005–1008.

Cleary, P. (1998). Discrete element modelling of industrial granular flow applications. *TASK Q.* **2**, 385–415.

Cleaver, J. A. S. and Nedderman, R. M. (1993a). Measurement of velocity profiles in conical hoppers. *Chem. Eng. Sci.* **48**, 3703–3712.

Cleaver, J. A. S. and Nedderman, R. M. (1993b). Theoretical prediction of stress and velocity profiles in conical hoppers. *Chem. Eng. Sci.* **48**, 3693–3702.

Clift, R., Grace, J. R., and Weber, M. E. (1978). *Bubbles, Drops and Particles*. (Academic Press, New York).

Cole, E. R. L. (1967). The behaviour of soils in the simple shear apparatus. Ph.D. thesis, University of Cambridge.

Collins, I. F. (1990). Plane strain characteristics theory for soils and granular materials with density dependent yield criteria. *J. Mech. Phys. Solids* **38**, 1–25.

Condon, E. U. (1967). Kinematics and dynamics. In *Handbook of Physics*, E. U. Condon and H. Odishaw, eds. (McGraw-Hill, New York).

Connelly, L. M. (1979). Wall-pressure and material-velocity measurements for the flow of granular materials under plane strain conditions. In *Mechanics Applied to the Transport of Bulk Materials*, S. C. Cowin, ed. (ASME, New York), 35–59.

Coulomb, C. A. (1776). Essai sur une application des règles de maximis et minimis à quelques problèmes de statique, relatifs à ĺ architecture (Essay on the rules of maxima and minima applied to some problems of equilibrium related to architecture). *Mém. Math. Acad. Roy. Sci.* **7**, 343–382.

Coulomb, C. A. (1785). Théorie des machines simples, en ayant égard au frottement de leur parties, et á la roideur de cordages (Theory of simple machines, by taking care of the rubbing of their parts and the straightening of cords). *Mém. Acad. Roy. Sci.* **10**, 161–332.

Coumoulos, D. G. (1968). A radiographic study of soils. Ph.D. thesis, University of Cambridge.

Courant, R. and Hilbert, D. (1962). *Methods of Mathematical Physics*. Vol. 2. (Interscience, New York).

Cowin, S. C. (1977). The theory of static loads in bins. *Trans. ASME E J. Appl. Mech.* **44**, 409–412.

References

Crewdson, B. J., Ormond, A. L., and Nedderman, R. M. (1977). Air-impeded discharge of fine particles from a hopper. *Powder Technol.* **16**, 197–207.

Cundall, P. A. (1971). A computer model for simulating progressive, large-scale movements in blocky rock systems. In *Proc. Symp. Int. Soc. Rock Mech. 2*. Vol. 8. Nancy.

Cundall, P. A. (1974). A computer model for rock-mass behavior using interactive graphics for the input and output of geometrical data. MRD-2-74, Missouri River Division of US Army Corps of Engineers.

Cundall, P. A. and Strack, O. D. L. (1979). A discrete numerical model for granular assemblies. *Géotech.* **29**, 47–65.

Darton, R. C. (1976). The structure and dispersion of jets of solid particles falling from a hopper. *Powder Technol.* **13**, 241–250.

Das Gupta, S., Khakhar, D. V., and Bhatia, S. K. (1991). Axial segregation of particles in a horizontal rotating cylinder. *Chem. Eng. Sci.* **46**, 1513–1517.

Davidson, J. F. and Nedderman, R. M. (1973). The hour-glass theory of hopper flow. *Trans. Inst. Chem. Eng.* **51**, 29–35.

Davis, R. O. and Mullenger, G. (1978). A rate-type constitutive model for soil with a critical state. *Int. J. Num. Anal. Meth. Geomech.* **2**, 255–282.

de Gennes, P. G. (1998). Reflections on the mechanics of granular matter. *Physica A* **261**, 267–293.

de Josselin de Jong, G. (1959). Statics and kinematics of the failable zone of a granular material. Ph.D. thesis, University of Delft.

de Josselin de Jong, G. (1971). The double sliding, free rotating model for granular assemblies. *Géotech.* **21**, 155–163.

de Josselin de Jong, G. (1976). Rowe's stress-dilatancy relation based on friction. *Géotech.* **26**, 527–534.

de Josselin de Jong, G. (1977). Mathematical elaboration of the double sliding, free rotating model. *Arch. Mech.* **29**, 561–591.

de Saint-Venant, A. J. C. B. (1871). Mémoire sur l'establissement des equations differentielles des mouvements intérieurs opérés dans les corps solides ductiles au dela' des limites où l'elasticitè pourrait les ramener à leur premier ètat (Concerning interior movements taking place in ductile solid bodies beyond the limits at which elasticity can restore them to their initial state). *J. Math. Pures Appl. Ser. II* **16**, 308–316.

Delanges, P. (1788). Statica e meccanica dé semifluidi. *Memorie di matematica e fisica della società Italiana* **IV**, 329–368.

Deming, W. E. and Mehring, A. L. (1929). The gravitational flow of fertilizers and other comminuted solids. *Ind. Eng. Chem.* **21**, 661–665.

Desai, C. S. (1972). Overview, trends, and projections: theory and applications of the finite element method in geotechnical engineering. In *Proc. Symp. on Application of the FEM in Geotechnical Engineering*. Waterways Experiment Station, Vicksburg.

Desai, C. S. and Salami, M. R. (1987). Constitutive model for rocks. *J. Geotech. Eng. ASCE* **113**, 407–423.

Desai, C. S. and Siriwardane, H. J. (1984). *Constitutive Laws for Engineering Materials*. (Prentice-Hall, Englewood Cliffs).

Desai, C. S. and Zhang, D. (1987). Viscoplastic model for geologic materials with generalized flow rule. *Int. J. Num. Anal. Meth. Geomech.* **11**, 603–620.

Desrues, J., Chambon, R., Mokni, M., and Mazerolle, F. (1996). Void ratio evolution inside shear bands in triaxial sand specimens by computed tomography. *Géotech.* **46**, 529–546.

Dhoriyani, M. L., Jonnalagadda, K. K., Kandikatla, R. K., and Rao, K. K. (2006). Silo music: sound emission during the flow of granular materials through tubes. *Powder Technol.* **167**, 55–71.

Didwania, A. K., Cantelaube, F., and Goddard, J. D. (2000). Static multiplicity of stress states in granular heaps. *Proc. R. Soc. Lond. A* **456**, 2569–2588.

Dieterich, J. H. (1979). Modeling of rock friction. I. Experimental results and constitutive equations. *J. Geophys. Res.* **84**, 2161–2175.

Dieterich, J. H. (1981). Constitutive properties of faults with simulated gouge. In *Mechanical Behavior of Crustal Rocks*, N. L. Carter, M. Friedman, J. M. Logan, and D. W. Stearns, eds., Vol. 24. (American Geophysical Union, Washington, DC).

DiMaggio, F. L. and Sandler, I. S. (1971). Material model for granular soils. *J. Eng. Mech. ASCE* **97**, 935–950.

Donald, M. B. and Roseman, B. (1962). Mixing and demixing of solid particles. I. Mechanisms in a horizontal drum mixer. *Br. Chem. Eng.* **7**, 749–753.

Dowson, D. (1979). *History of Tribology*. (Longman, New York).

Draad, A. A. and Nieuwstadt, F. T. M. (1998). The earth's rotation and laminar pipe flow. *J. Fluid Mech.* **361**, 297–308.

Drake, T. G. (1991). Granular flow: physical experiments and their implications for microstructural theories. *J. Fluid Mech.* **225**, 121–152.

Drake, T. G. and Walton, O. R. (1995). Comparison of experimental and simulated grain flows. *Trans. ASME E J. Appl. Mech.* **62**, 131–135.

Drazin, P. G. (2002). *Introduction to Hydrodynamic Stability*. (Cambridge University Press, Cambridge).

Drazin, P. G. and Reid, W. H. (2004). *Hydrodynamic Stability*, 2 edn. (Cambridge University Press, Cambridge).

Drescher, A. (1976). An experimental investigation of flow rules for granular materials using optically sensitive glass particles. *Géotech.* **26**, 591–601.

Drescher, A. (1991). *Analytical Methods in Bin Load Analysis*. (Elsevier, Amsterdam).

Drescher, A., Cousens, T. W., and Bransby, P. L. (1978). Kinematics of the mass flow of granular material through a plane hopper. *Géotech.* **28**, 27–42.

Drucker, D. C., Gibson, R. E., and Henkel, D. J. (1957). Soil mechanics and the work-hardening theories of plasticity. *Trans. ASCE* **122**, 338–346.

Drucker, D. C. and Prager, W. (1952). Soil mechanics and plastic analysis or limit design. *Q. Appl. Math.* **10**, 157–165.

Duran, J. (2000). *Sands, Powders, and Grains*. (Springer Verlag, New York).

Edwards, S. F. and Oakeshott, R. B. S. (1989). The transmission of stress in an aggregate. *Physica D* **38**, 88–92.

Eibl, J. (1997). Silo loads: numerical results. In *Containment Structures – Proceedings of the Henderson Colloquium*, B. Simpson, ed. (E. and F. N. Spon, London), 43–63.

Eibl, J. and Rombach, G. (1988). Numerical investigations on discharging silos. In *Numerical Methods in Geomechanics*, G. Swoboda, ed. (Balkema, Rotterdam).

Elaskar, S. A., Godoy, L. A., Gray, D. D., and Stiles, J. M. (2000). A viscoplastic approach to model the flow of granular solids. *Int. J. Solids Struct.* **37**, 2185–2214.

Ennis, B. J., Green, J., and Davies, R. (1994). Particle technology. The legacy of neglect in the U.S. *Chem. Eng. Prog.* **90**, 32–43.

Enskog, D. (1917). The kinetic theory of phenomena in fairly rare gases. Ph.D. thesis, Uppsala. English translation of title taken from Chapman & Cowling (1964).

Enskog, D. (1922). Kinetic theory of thermal conduction, viscosity, and self-diffusion in certain dense gases and liquids. *Kungliga Svenska Vetenskaps Akademiens Handlingar* **63**. English translation in S. Brush, Kinetic Theory, Vol. 3, Pergamon Press, London, 1972.

Enstad, G. (1975). On the theory of arching in mass flow hoppers. *Chem. Eng. Sci.* **30**, 1273–1283.

Eringen, A. C. and Kafadar, C. B. (1976). Part I. Polar field theories. In *Continuum Physics, Volume IV – Polar and nonlocal field theories*, A. C. Eringen, ed. (Academic Press, New York).

Evesque, P. (1997). Stress in static sandpiles: role of the deformation in the case of silos and oedometers. *J. Phys. I France* **7**, 1–12.

Fan, L. S. and Zhu, C. (1998). *Principles of Gas-Solid Flows*. (Cambridge University Press, Cambridge).

Feda, J. (1982). *Mechanics of Particulate Materials*. (Elsevier, Amsterdam).

Feilden, G. B. R. (1986). Mechanical paradoxes. *Proc. R. Inst. Gt. Britain* **58**, 95–112.

Feise, H. J. (1998). A review of anisotropy and steady state flow in powders. *Powder Technol.* **98**, 191–200.

Fickie, K. E., Mehrabi, R., and Jackson, R. (1989). Density variation in a granular material flowing from a wedge-shaped hopper. *AIChE J.* **35**, 853–855.

Fillunger, P. (1936). *Erdbaumechanik? (Soil Mechanics?)*. (Author's own publication, Wein).

Fisher, R. A. (1926). On the capillary forces in an ideal soil: correction of formulae given by W. B. Haines. *J. Agric. Sci.* **16**, 492–503.

Forterre, Y. and Pouliquen, O. (2001). Longitudinal vortices in granular flows. *Phys. Rev. Lett.* **86**, 5886–5889.

Forterre, Y. and Pouliquen, O. (2002). Stability analysis of rapid granular chute flows: formation of longitudinal vortices. *J. Fluid Mech.* **467**, 361–387.

Forterre, Y. and Pouliquen, O. (2003). Long-surface-wave instability in dense granular flows. *J. Fluid Mech.* **486**, 21–50.

Fowler, R. T. and Glastonbury, J. R. (1959). The flow of granular solids through orifices. *Chem. Eng. Sci.* **10**, 150–156.

Fung, Y. C. (1977). *A First Course in Continuum Mechanics.*, 2 edn. (Prentice-Hall, Englewood Cliffs).

Galeli, G. (2003). Dialogues concerning two sciences. In *On the Shoulders of Giants*, S. Hawking, ed. (Viva Books, New Delhi), 399–626.

GDR MiDi. (2004). On dense granular flows. *Eur. Phys. J. E* **14**, 341–365.

Gelfand, I. M. (1963). Some problems in the theory of quasilinear equations. *Am. Math. Soc. Trans. Ser. 2* **29**, 295–381.

Geng, J., Reydellet, G., Clement, E., and Behringer, R. P. (2003). Green's function measurements of force transmission in 2D granular materials. *Physica D* **182**, 274–303.

Gerald, C. F. and Wheatley, P. O. (1994). *Applied Numerical Analysis*, 5 edn. (Addison-Wesley, Reading).

Gerogiannopoulos, N. G. and Brown, E. T. (1978). The critical state concept applied to rock. *Int. J. Rock Mech. Min. Sci. (& Geomech. Abstr.)* **15**, 1–10.

Gidaspow, D. (1994). *Multiphase Flow and Fluidization.* (Academic Press, New York).

Glasser, B. J. and Goldhirsch, I. (2001). Scale dependence, correlations, and fluctuations of stresses in rapid granular flows. *Phys. Fluids* **13**, 407–420.

Goldhirsch, I. (2003). Rapid granular flows. *Annu. Rev. Fluid Mech.* **35**, 267–293.

Goldhirsch, I., Noskowicz, S. H., and Bar-Lev, O. (2005a). Hydrodynamics of nearly smooth granular gases. *J. Phys. Chem.* **109**, 21449–21470.

Goldhirsch, I., Noskowicz, S. H., and Bar-Lev, O. (2005b). Nearly smooth granular gases. *Phys. Rev. Lett.* **95**, 068002–1–4.

Goldhirsch, I. and Sela, N. (1996). Origin of normal stress differences in rapid granular flows. *Phys. Rev. E* **54**, 4458–4461.

Goldsmith, W. (1960). *Impact.* (Edward Arnold, London).

Goldstein, H. (1980). *Classical Mechanics*, 2 edn. (Narosa Publishing House, New Delhi).

Goodbody, A. M. (1982). *Cartesian Tensors: With Applications to Mechanics, Fluid Mechanics and Elasticity*. (Ellis Horwood, Chichester).

Goodman, M. A. and Cowin, S. C. (1971). Two problems in the gravity flow of granular materials. *J. Fluid Mech.* **45**, 321–339.

Grad, H. (1949). On the kinetic theory of rarefied gases. *Comm. Pure. Appl. Math.* **2**, 331–407.

Graham, D. P., Tait, A. R., and Wadmore, R. S. (1987). Measurement and prediction of flow patterns of granular solids in cylindrical vessels. *Powder Technol.* **50**, 65–76.

Graton, L. C. and Fraser, H. J. (1935). Systematic packing of spheres, with particular relation to porosity and permeability. *J. Geol.* **43**, 785–909.

Gray, W. G. and Hassanizadeh, M. (1998). Macroscale continuum mechanics for multiphase porous-media flow including phases, interfaces, common lines and common points. *Adv. Water Resour.* **21**, 261–281.

Green, G. E. and Bishop, A. W. (1969). A note on the drained strength of sand under generalized strain conditions. *Géotech.* **19**, 144–149.

Griffiths, D. J. (2002). *Introduction to Electrodynamics*. (Pearson Education (Singapore), New Delhi).

Gu, J. C., Rice, J. R., Ruina, A. L., and Tse, S. T. (1984). Slip motion and stability of a single degree of freedom elastic system with rate and state dependent friction. *J. Mech. Phys. Solids* **32**, 167–196.

Gu, Z. H., Arnold, P. C., and McLean, A. G. (1993). The influence of surcharge level on the flowrate of bulk solids from mass flow bins. *Powder Technol.* **74**, 141–151.

Gudehus, G. (1996). A comprehensive constitutive equation for granular materials. *Soils and Foundations* **36**, 1–12.
Haar, A. and von Karman, T. (1909). Zur theorie der spannungszustaende in plastischen und sandartigen medien (On the theory of tensile states in plastic and sand-like media). *Nachr. Ges. Wiss. Goettingen Math. Phys. Klasse* **2**, 204–218.
Haff, P. K. (1983). Grain flow as a fluid-mechanical phenomenon. *J. Fluid Mech.* **134**, 401–430.
Hagen, G. (1852). Druck und bewegung des trocken un sandes (Pressure and movement of dry sand). *Berliner Monatsberichte Akad. d. Wiss.* S35–S42.
Haigh, B. T. (1920). The strain energy function and the elastic limit. *Engineering* **109**, 158–160.
Halliday, D. and Resnick, R. (1969). *Physics Part II.* (Wiley Eastern, New Delhi).
Hancock, A. W. and Nedderman, R. M. (1974). Prediction of stresses on vertical bunker walls. *Trans. Inst. Chem. Eng.* **52**, 170–179.
Handley, M. F. and Perry, M. G. (1967/1968). Stresses in granular materials flowing in converging hopper sections. *Powder Technol.* **1**, 245–251.
Hanes, D. M. and Inman, D. L. (1985). Observations of rapidly flowing granular-fluid materials. *J. Fluid Mech.* **150**, 357–380.
Hanes, D. M. and Walton, O. R. (2000). Simulations and physical measurements of glass spheres flowing down a bumpy incline. *Powder Technol.* **109**, 133–144.
Hansen, B. (1958). Line ruptures regarded as narrow rupture zones. In *Proc. Conf. Earth Pressure Problems.* Vol. 1. Brussels, 39–48.
Happel, J. and Brenner, H. (1965). *Low Reynolds Number Hydrodynamics*. (Prentice-Hall, Englewood Cliffs).
Harris, D. (1995). A unified formulation of plasiticity models of granular and other materials. *Proc. R. Soc. Lond. A* **450**, 37–49.
Hassanizadeh, M. and Gray, W. G. (1979a). General conservation equations for multi-phase systems: 1. Averaging procedure. *Adv. Water Resour.* **2**, 131–144.
Hassanizadeh, M. and Gray, W. G. (1979b). General conservation equations for multi-phase systems: 2. Mass, momenta, energy, and entropy equations. *Adv. Water Resour.* **2**, 191–203.
Haupt, P. (2000). *Continuum Mechanics and Theory of Materials*. (Springer, Berlin).
Heertjes, P. M., Khoe, G. K., and Kuster, D. (1978). A condition diagram for some non-cohesive round glass particles. *Powder Technol.* **21**, 63–71.
Henkel, D. J. (1960). The relationships between effective stresses and water content in saturated clays. *Géotech.* **10**, 41–54.
Herbst, O., Huthmann, M., and Zippelius, A. (2000). Dynamics of inelastically colliding spheres with Coulomb friction: relaxation of translational and rotational energy. *Granul. Matter* **2**, 211–219.
Herbst, T. F. and Winterkorn, H. F. (1965). Shear phenomena in granular random packings. Princeton Soil Engineering Research No. 2. Princeton University, Princeton.
Herle, I. and Gudehus, G. (1999). Determination of parameters of a hypoplastic constitutive model from properties of a grain assembly. *Int. J. Num. Anal. Meth. Geomech.* **4**, 461–486.
Herrmann, H. J. and Luding, S. (1998). Modelling granular media on the computer. *Continuum Mech. Thermodyn.* **10**, 189–232.
Hertz, H. (1882). Über die berührung fester elastische körper and über die harte (On the contact of elastic solids). *J. Reine und Angewandte Mathematik* **92**, 156–171.
Heslot, F., Baumberger, T., Perrin, B., Caroli, B., and Caroli, C. (1994). Creep, stick-slip, and dry-friction dynamics: experiments and a heuristic model. *Phys. Rev. E* **49**, 4973–4988.
Heyman, J. (1972). *Coulomb's Memoir on Statics.* (Cambridge University Press, Cambridge).
Hill, J. M. and Selvadurai, A. P. S. (2005). Mathematics and mechanics of granular materials. *J. Eng. Math.* **52**, 1–9.
Hill, J. M. and Spencer, A. J. M. (1999). Some dynamic shear flow problems for granular materials. *Q. J. Mech. Appl. Math.* **52**, 253–267.
Hill, R. (1950). *The Mathematical Theory of Plasticity.* (Clarendon Press, Oxford).
Hirschfelder, J. O., Curtiss, C. F., and Bird, R. B. (1964). *Molecular Theory of Gases and Liquids.* (John Wiley, New York).
Hopkins, M. A. and Louge, M. Y. (1991). Inelastic microstructure in rapid granular flows of smooth disks. *Phys. Fluids* **A3**, 47–57.

References

Horne, R. M. and Nedderman, R. M. (1976). Analysis of stress distributions in two-dimensional bins by the method of characteristics. *Powder Technol.* **14**, 93–102.

Horne, R. M. and Nedderman, R. M. (1978). An analysis of switch stresses in two-dimensional bunkers. *Powder Technol.* **19**, 235–241.

Hosseini-Ashrafi, M. E. and Tüzün, U. (1993). A tomographic study of voidage profiles in axially symmetric granular flow. *Chem. Eng. Sci.* **48**, 53–67.

Howell, D., Behringer, R. P., and Veje, C. (1999). Stress fluctuations in a 2d granular Couette experiment: a continuous transition. *Phys. Rev. Lett.* **82**, 5241–5244.

Hsiau, S.-S. and Shieh, Y.-M. (2000). Effect of solid fraction on fluctuations and self-diffusion of sheared granular flows. *Chem. Eng. Sci.* **55**, 1969–1979.

Huang, K. (1985). *Statistical Mechanics*. (McGraw Hill, Singapore).

Huber, M. T. (1904). The specific shear strain work as a criterion of material strength. *Czasopismo Techniczne* **22**, 181. The title of the article is an English translation, given in http://www.scholar.google.com.

Huber-Burnand. (1829). Sur l'ecoulement et la pression du sable (Concerning flow and pressure of sand). *Geneve Bibliotheque Universelle, Science et Art* **40**.

Hui, K., Haff, P. K., Ungar, J. E., and Jackson, R. (1984). Boundary conditions for high-shear grain flows. *J. Fluid Mech.* **145**, 223–233.

Hunter, S. C. (1983). *Mechanics of Continuous Media*. (Ellis Horwood, Chichester).

Hutter, K. and Rajagopal, K. R. (1994). On flows of granular materials. *Continuum Mech. Thermodyn.* **6**, 81–139.

Hvorslev, M. J. (1937). Über die festigkeitseigenschaften gestorter bindiger böden (On the physical properties of disturbed cohesive soils). *Ingeniorvidenskabelige Skrifter* **A45**. For an English version, see Translation 69-5, U.S. Army Corps of Engineers Waterways Experiment Station, Vicksburg, Mississippi, June 1969.

Ishida, M., Hatano, H., and Shirai, T. (1980). The flow of solid particles in an aerated inclined channel. *Powder Technol.* **27**, 7–12.

Ishida, M. and Shirai, T. (1979). Velocity distributions in the flow of solid particles in an inclined open channel. *J. Chem. Eng. Japan* **12**, 46–50.

Israelachvili, J. N. (1992). *Intermolecular and Surface Forces*. (Academic Press, London).

Iwashita, K. and Oda, M. (1998). Rolling resistance at contacts in simulation of shear band development by DEM. *J. Eng. Mech. ASCE* **124**, 285–292.

Jackson, R. (1983). Some mathematical and physical aspects of continuum models for the motion of the granular materials. In *Theory of Dispersed Multiphase Flow*, R. E. Meyer, ed. (Academic Press, New York), 291–337.

Jackson, R. (1986). Some features of the flow of granular materials and aerated granular materials. *J. Rheol.* **30**, 907–930.

Jackson, R. (1997). Locally averaged equations of motion for a mixture of identical particles and a Newtonian fluid. *Chem. Eng. Sci.* **52**, 2457–2469.

Jackson, R. (2000). *The Dynamics of Fluidized Particles*. (Cambridge University Press, Cambridge).

Jaeger, H. M. and Nagel, S. R. (1992). Physics of the granular state. *Science* **255**, 1523–1531.

Jaeger, H. M., Nagel, S. R., and Behringer, R. P. (1996). Granular solids, liquids, and gases. *Rev. Mod. Phys.* **68**, 1259–1273.

Janssen, H. A. (1895). Versuche ueber Getreidedruck in Silozellen (Experiments on grain pressure in silos). *Z. Ver. Deut. Ing.* **39**, 1045–1049.

Jarrett, N. D., Brown, C. J., and Moore, D. B. (1995). Pressure measurements in a rectangular silo. *Géotech.* **45**, 95–104.

Jaumann, G. (1911). Geschlossenes system physikalisher und chemischer differentialgesetze (Closed system of physical and chemical differential laws). *Sitzungsberichte Akad. Wiss. Wien (IIa)* 120, 385–530.

Jaunzemis, W. (1967). *Continuum Mechanics*. (The Macmillan Company, New York).

Jeffries, M. G. (1993). Nor sand: a simple critical state model for sand. *Géotech.* **43**, 91—104.

Jenike, A. W. (1961). Gravity flow of bulk solids. Bulletin 108, University of Utah Engineering Experiment Station.

Jenike, A. W. (1964a). Steady gravity flow of frictional-cohesive solids in converging channels. *Trans. ASME E J. Appl. Mech.* **31**, 5–11.
Jenike, A. W. (1964b). Storage and flow of bulk solids. Bulletin 123, University of Utah Engineering Experiment Station.
Jenike, A. W. (1987). A theory of flow of particulate solids in converging and diverging channels based on a conical yield function. *Powder Technol.* **50**, 229–236.
Jenike, A. W. and Johanson, J. R. (1968). Bin loads. *J. Struct. Div. ASCE* **94**, 1011–1041.
Jenike, A. W. and Shield, R. T. (1959). On the plastic flow of Coulomb solids beyond original failure. *Trans ASME E J. Appl. Mech.* **26**, 599–602.
Jenkins, J. T. (1992). Boundary-conditions for rapid granular flow – flat, frictional walls. *J. Appl. Mech.* **59**, 120–127.
Jenkins, J. T. and Mancini, F. (1989). Kinetic theory for binary mixtures of smooth, nearly elastic spheres. *Phys. Fluids* A **1**, 2050–2057.
Jenkins, J. T. and Richman, M. W. (1985). Grad's 13-moment system for a dense gas of inelastic particles. *Arch. Rat. Mech. Anal.* **87**, 355–377.
Jenkins, J. T. and Richman, M. W. (1986). Boundary conditions for plane flows of smooth, nearly elastic, circular disks. *J. Fluid Mech.* **171**, 53–69.
Jenkins, J. T. and Savage, S. B. (1983). A theory for the rapid flow of identical, smooth, nearly elastic, spherical particles. *J. Fluid Mech.* **130**, 187–202.
Jenkins, J. T. and Zhang, C. (2002). Kinetic theory for identical, frictional, nearly elastic spheres. *Phys. Fluids* **14**, 1228–1235.
Johanson, J. R. (1964). Stress and velocity fields in the gravity flow of bulk solids. *Trans. ASME E J. Appl. Mech.* **31**, 499–506.
Johnson, K. L. (1987a). *Contact Mechanics*. (Cambridge University Press, Cambridge).
Johnson, P. C. (1987b). Frictional-collisional equations of motion for granular flows with applications to chutes and shear cells. Ph.D. thesis, Princeton University.
Johnson, P. C. and Jackson, R. (1987). Frictional-collisional constitutive relations for granular materials, with application to plane shearing. *J. Fluid Mech.* **176**, 67–93.
Johnson, P. C., Nott, P., and Jackson, R. (1990). Frictional-collisional equations of motion for particulate flows and their application to chutes. *J. Fluid Mech.* **210**, 501–535.
Jop, P., Forterre, Y., and Pouliquen, O. (2006). A constitutive law for dense granular flows. *Nature* **441**, 727–730.
Joseph, D. D. and Ocando, D. (2002). Slip velocity and lift. *J. Fluid Mech.* **454**, 263–286.
Jyotsna, R. and Rao, K. K. (1997). A frictional-kinetic model for the flow of granular materials through a wedge-shaped hopper. *J. Fluid Mech.* **346**, 239–270.
Kachanov, L. M. (1974). *Fundamentals of the Theory of Plasticity*. (Mir Publishers, Moscow).
Kamath, S. and Puri, V. M. (1997). Measurement of powder flow constitutive model parameters using a cubical triaxial tester. *Powder Technol.* **90**, 59–70.
Kanatani, K.-I. (1979). A micropolar continuum theory for the flow of granular materials. *Int. J. Eng. Sci.* **17**, 419–432.
Kaplan, W. (1952). *Advanced Calculus*. (Addison-Wesley, Reading).
Karlsson, T., Klisinski, M., and Runneson, K. (1998). Finite element simulation of granular material flow in plane silos with complicated geometry. *Powder Technol.* **99**, 29–39.
Kaza, K. R. (1982). The mechanics of flowing granular materials. Ph.D. thesis, University of Houston.
Kaza, K. R. and Jackson, R. (1982a). A problem in the flow of granular materials. In *Proc. 9th US Natl. Cong. Appl. Mech.* Ithaca, 389–394.
Kaza, K. R. and Jackson, R. (1982b). The rate of discharge of coarse granular material from a wedge-shaped mass flow hopper. *Powder Technol.* **33**, 223–237.
Kaza, K. R. and Jackson, R. (1984). Boundary conditions for a granular material flowing out of a hopper or bin. *Chem. Eng. Sci.* **39**, 915–916.
Keiter, T. and Rombach, G. (2001). Numerical aspects of FE simulations of granular flow in silos. *J. Eng. Mech. ASCE* **127**, 969–1074.
Khakhar, D. V., McCarthy, J. J., Shinbrot, T., and Ottino, J. M. (1997). Transverse flow and mixing of granular materials in a rotating cylinder. *Phys. Fluids* **9**, 31–43.
Khan, A. S. and Huang, S. (1995). *Continuum Theory of Plasticity*. (John Wiley, New York).

References

Ko, H. Y. and Scott, R. F. (1967). Deformation of sand in shear. *J. Soil Mech. Found. Div. ASCE* **93**, 283–310.

Koch, D. L. and Sangani, A. S. (1999). Particle pressure and marginal stability limits for a homogeneous monodisperse gas fluidized bed: kinetic theory and numerical simulation. *J. Fluid Mech.* **400**, 229–263.

Kolymbas, D. (1987). A novel constitutive law for soils. In *Proc. 2nd Int. Conf. on Constitutive Laws for Engineering Materials*. Vol. 1. Tucson, 319–326.

Kolymbas, D. (1991). An outline of hypoplasticity. *Arch. Appl. Mech.* **61**, 143–151.

Krenk, S. (2000). Characteristic state plasticity for granular materials. Part I: basic theory. *Int. J. Solids Struct.* **37**, 6343–6360.

Krim, J. (1996). Friction at the atomic scale. *Sci. Am.* **275**, 48–56.

Krim, J. (2002). Friction at the macroscopic and microscopic length scales. *Am. J. Phys.* **70**, 890–897.

Kumar, J., Rao, C. L., and Massoudi, M. (2003). Couette flow of granular materials. *Int. J. Nonlin. Mech.* **38**, 11–20.

Kumaran, V. (2004). Constitutive relations and linear stability of a sheared granular flow. *J. Fluid Mech.* **506**, 1–43.

Kumaran, V. (2006). The constitutive relation for the granular flow of rough particles, and its application to the flow down an inclined plane. *J. Fluid Mech.* **561**, 1–42.

Lacombe, F., Zapperi, S., and Herrmann, H. (2000). Dilatancy and friction in sheared granular media. *Eur. Phys. J. E* **2**, 181–189.

Lade, P. V. (1977). Elasto-plastic stress-strain theory for cohesionless soil with curved yield surfaces. *Int. J. Solids Struct.* **13**, 1019–1035.

Lade, P. V. (1988). Effects of voids and volume changes on the behaviour of frictional materials. *Int. J. Num. Anal. Meth. Geomech.* **12**, 351–370.

Lade, P. V. and De Boer, R. (1997). The concept of effective stress for soil, concrete and rock. *Géotech.* **47**, 61–78.

Lade, P. V. and Duncan, J. M. (1973). Cubical triaxial tests on cohesionless soil. *J. Soil Mech. Found. Div. ASCE* **99**, 793–811.

Lade, P. V. and Duncan, J. M. (1975). Elastoplastic stress-strain theory for cohesionless soil. *J. Geotech. Engng. Div. ASCE.* **101**, 1037–1053.

Lambe, T. W. and Whitman, R. V. (1969). *Soil Mechanics*. (John Wiley, New York).

Landau, L. D. and Lifshitz, E. M. (1986). *Theory of Elasticity*, 3 edn. (Butterworth-Heinemann, Oxford).

Langston, P. A., Tüzün, U., and Heyes, D. M. (1995). Discrete element simulation of granular flow in 2D and 3D hoppers: dependence of discharge rate and wall stress on particle interactions. *Chem. Eng. Sci.* **50**, 967–987.

Lax, P. D. (1957). Hyperbolic systems of conservation laws II. *Comm. Pure Appl. Math.* **10**, 537–566.

Le Pennec, T., J., M. K., Flekkøy, E. G., Messager, J. C., and Ammi, M. (1998). Silo hiccups: dynamic effects of dilatancy in granular flow. *Phys. Fluids* **10**, 3072–3079.

Lebedev, N. N. (1972). *Special Functions and Their Applications*. (Dover, New York).

Lee, J., Cowin, S. C., and Templeton, III, J. S. (1974). An experimental study of the kinematics of flow through hoppers. *Trans. Soc. Rheol.* **18**, 247–269.

Lévy, M. (1871). Mémoire sur les équationes géneralés des mouvements intérieurs des corps solides ductile au delá limites où l'élasticité pourrait les ramener à leur premier état (Report on general equations for the movement of ductile solid bodies beyond the limits at which elasticity can restore them to their initial state). *J. Math.* **16**, 369–372.

Lian, G., Thornton, C., and Adams, M. J. (1993). A theoretical study of the liquid bridge force between two rigid spherical bodies. *J. Coll. Int. Sci.* **161**, 138–147.

Lianis, G. and Ford, H. (1957). An experimental investigation of the yield criterion and the stress-strain law. *J. Mech. Phys. Solids* **5**, 215–222.

Liffman, K., Chan, D. Y. C., and Hughes, B. D. (1992). Force distribution in a two-dimensional sandpile. *Powder Technol.* **72**, 255–267.

Lister, M. (1965). The numerical solution of hyperbolic partial differential equations by the method of characteristics. In *Mathematical Methods for Digital Computers*, A. Ralston and H. S. Wilf, eds. (John Wiley, New York), 165–179.

Litwiniszyn, J. (1956). Application of the equation of stochastic processes to the mechanics of loose bodies. *Archiwum Mechaniki Stosowaneij* **8**, 393–411.

Liu, C.-H., Nagel, S. R., Schecter, D. A., Coppersmith, S. N., Majumdar, S., Narayan, O., and Witten, T. A. (1995). Force fluctuations in bead packs. *Science* **269**, 513–515.

Lopez de Haro, M., Cohen, E. G. D., and Kincaid, J. M. (1983). The Enskog theory for multicomponent mixtures. I. Linear transport theory. *J. Chem. Phys.* **78**, 2746–2759.

Louge, M. Y. (2003). Model for dense granular flows down bumpy inclines. *Phys. Rev. E* **67**, 061303–1–11.

Louge, M. Y. and Keast, S. C. (2001). On dense granular flows down flat frictional inclines. *Phys. Fluids* **13**, 1213–1232.

Lun, C. K. K. (1991). Kinetic theory for granular flow of dense, slightly inelastic, slightly rough spheres. *J. Fluid Mech.* **223**, 539–559.

Lun, C. K. K. and Savage, S. B. (1987). A simple kinetic theory for granular flow of rough, inelastic spherical particles. *Trans. ASME E J. Appl. Mech.* **54**, 47–53.

Lun, C. K. K., Savage, S. B., Jeffrey, D. J., and Chepurniy, N. (1984). Kinetic theories for granular flow: inelastic particles in Couette flow and slightly inelastic particles in a general flow field. *J. Fluid Mech.* **140**, 223–256.

Lutsko, J. F. (1996). Molecular chaos, pair correlations, and shear-induced ordering of hard spheres. *Phys. Rev. Lett.* **77**, 2225–2228.

Ma, D. and Ahmadi, G. (1986). An equation of state for dense, rigid sphere gases. *J. Chem. Phys.* **84**, 3449–3450.

MacCurdy, E. (1956). *The Notebooks of Leonardo da Vinci*. Vol. I. (Jonathan Cape, London).

Malvern, L. E. (1969). *Introduction to the Mechanics of a Continuous Medium*. (Prentice-Hall, Englewood Cliffs).

Mandel, J. (1947). Sur les lignes de glissement et le calcul des déplacements dans la déformation plastique (On the line of sliding and the calculation of displacements in plastic deformation). *C. R. Acad. Sci. Paris* **225**, 1272–1273.

Mangwandi, C., Cheong, Y. S., Adams, M. J., Hounslow, M. J., and Salmanan, A. D. (2007). The coefficient of restitution of different representative types of granules. *Chem. Eng. Sci.* **62**, 437–450.

Manjunath, K. S. (1987). Silo/Feeder Interfacing. (Chr. Michelsen Institute, Bergen).

Matsuoka, H., Hoshikawa, T., and Ueno, K. (1989). A general failure criterion and constitutive relation for granular materials to metals. In *Powders and Grains*, J. Biarez and R. Gourves, eds. (Balkema, Rotterdam), 339–346.

Maw, N., Barber, J. R., and Fawcett, J. N. (1976). The oblique impact of elastic spheres. *Wear* **38**, 101–114.

Maw, N., Barber, J. R., and Fawcett, J. N. (1981). The role of tangential elastic compliance in oblique impact. *Trans. ASME J. Lubr. Technol.* **103**, 74–80.

Maxey, M. R. and Riley, J. J. (1983). Equation of motion for a small rigid sphere in a nonuniform flow. *Phys. Fluids* **26**, 883–889.

Maxwell, J. C. (1867). On the dynamical theory of gases. *Phil. Trans. Roy. Soc.* **157**, 49–88.

Maxwell, J. C. (1879). On stresses in rarefied gases arising from inequalities of temperature. *Phil. Trans. Roy. Soc.* **170**, 231–256.

McCabe, R. P. (1974). Flow patterns in granular material in circular silos. *Géotech.* **24**, 45–62.

McCoy, B. J., Sandler, S. I., and Dahler, J. S. (1966). Transport properties of polyatomic fluids. IV. The kinetic theory of a dense gas of perfectly rough spheres. *J. Chem. Phys.* **45**, 3485–3512.

McNamara, S. and Luding, S. (1998). Energy nonequipartition in systems of inelastic, rough spheres. *Phys. Rev. E* **58**, 2247–2250.

McQuarrie, D. A. (2003). *Statistical Mechanics*. (Viva Books, New Delhi).

McTigue, D. F. (1978). A model for stresses in shear flow of a granular material. In *Proc. US-Japan Seminar on Continuum Mechanical and Statistical Approaches in the Mechanics of Granular Materials*. Gakujutsu Bunken Fukyukai, 266–271.

Mühlhaus, H. B. (1986). Shear band analysis in granular materials by Cosserat theory. *Ing. Arch.* **56**, 389–399. In German.

Mühlhaus, H. B. and Vardoulakis, I. (1987). The thickness of shear bands in granular materials. *Géotech.* **37**, 271–283.

References

Medina, A., Luna, E., Alvarado, R., and Trevino, C. (1995). Axisymmetrical rotation of a sand heap. *Phys. Rev. E* **51**, 4621–4625.

Mehrabadi, M. M. and Cowin, S. C. (1978). Intital planar deformation of dilatant granular materials. *J. Mech. Phys. Solids* **26**, 269–284.

Mendelson, A. (1983). *Plasticity: Theory and Application*. (Krieger, Huntington).

Meric, R. A. and Tabarrok, B. (1982). On the gravity flow of granular materials. *Int. J. Mech. Sci.* **24**, 469–478.

Michalowski, R. L. (1987). Flow of granular media through a plane parallel/converging bunker. *Chem. Eng. Sci.* **42**, 2587–2596.

Mindlin, R. D. (1949). Compliance of elastic bodies in contact. *Trans. ASME E J. Appl. Mech.* **16**, 259–268.

Mohan, L. S., Nott, P. R., and Rao, K. K. (1997). Fully developed flow of coarse granular materials through a vertical channel. *Chem. Eng. Sci.* **52**, 913–933.

Mohan, L. S., Nott, P. R., and Rao, K. K. (1999). A frictional Cosserat model for the flow of granular materials through a vertical channel. *Acta Mech.* **138**, 75–96.

Mohan, L. S., Rao, K. K., and Nott, P. R. (2002). A frictional Cosserat model for the slow shearing of granular materials. *J. Fluid Mech.* **138**, 75–96.

Mohr, O. (1882). über die parstellung des spannungszustaendes und des deformations-zustandes eines koerper elementes und über die anwendung derselben in der festigkeitslehre (On the presentation of tensile states and the deformation state of a body element and its application in solid mechanics). *Zivilingenieur* **28**, 112–155.

Molerus, O. (1978). Effect of interparticle cohesive forces on the flow behaviour of powders. *Powder Technol.* **20**, 161–175.

Mueth, D. M., Debregeas, G. F., Karczmar, G. S., Eng, P. J., Nagel, S. R., and Jaeger, H. M. (2000). Signatures of granular microstructure in dense shear flows. *Nature* **406**, 385–389.

Mullins, W. W. (1972). Stochastic theory of particle flow under gravity. *J. Appl. Phys.* **43**, 665–678.

Munch-Andersen, J. (1987). The boundary layer in rough silos. *Mech. Eng. Trans. (The Instn Engrs. Australia)* **ME12**, 167–170.

Munch-Andersen, J. and Askegaard, V. (1993). Silo model tests with sand – and grain. In *Proc. Int. Symp. Reliable Flow of Particulate Solids II*. Oslo, 269–282.

Nakagawa, M., Altobelli, S. A., Caprihan, A., and Fukushima, E. (1997). NMRI study: axial migration of radially segregated core of granular mixtures in a horizontal rotating cylinder. *Chem. Eng. Sci.* **52**, 4423–4428.

Nakagawa, M., Altobelli, S. A., Caprihan, A., Fukushima, E., and Jeong, E. K. (1993). Noninvasive measurements of granular flows by magnetic-resonance-imaging. *Exp. Fluids* **16**, 54–60.

Nase, S. T., Vargas, W. L., Abatan, A. A., and McCarthy, J. J. (2001). Discrete characterization tools for cohesive granular material. *Powder Technol.* **116**, 214–223.

Nasuno, S., Kudrolli, A., and Gollub, J. P. (1997). Friction in granular layers: hysteresis and precursors. *Phys. Rev. Lett.* **79**, 949–952.

Natarajan, V., Hunt, M., and Taylor, E. (1995). Local measurements of velocity fluctuations and diffusion coefficients for a granular material flow. *J. Fluid Mech.* **304**, 1–25.

Nedderman, R. M. (1988). The measurement of the velocity profile in a granular material discharging from a conical hopper. *Chem. Eng. Sci.* **43**, 1507–1516.

Nedderman, R. M. (1992). *Statics and Kinematics of Granular Materials*. (Cambridge University Press, Cambridge).

Nedderman, R. M. (1995). The use of the kinematic model to predict the development of the stagnant zone boundary in the batch discharge of a bunker. *Chem. Eng. Sci.* **50**, 959–965.

Nedderman, R. M. and Laohakul, C. (1980). The thickness of the shear zone of flowing granular materials. *Powder Technol.* **25**, 91–100.

Nedderman, R. M. and Tüzün, U. (1979). A kinematic model for the flow of granular materials. *Powder Technol.* **22**, 243–253.

Nedderman, R. M., Tüzün, U., Savage, S. B., and Houlsby, G. T. (1982). The flow of granular materials – I : discharge rates from hoppers. *Chem. Eng. Sci.* **37**, 1597–1609.

Negi, S. C., Rong, G., and Jofriet, J. C. (1992). Discrete particle simulation of gravity flow of particulate material. *Powder Handl. Process.* **4**, 375–380.

Nguyen, T. V., Brennen, C., and Sabersky, R. H. (1980). Funnel flow in hoppers. *Trans. ASME E J. Appl. Mech.* **47**, 729–735.

Nguyen, T. V., Brennen, C. E., and Sabersky, R. H. (1979). Gravity flow of granular materials in conical hoppers. *Trans. ASME E J. Appl. Mech.* **46**, 529–535.

Noll, W. (1955). On the continuity of the solid and fluid states. *Arch. Rat. Mech. Anal.* **4**, 3–81.

Noll, W. (1958/1959). A mathematical theory of the mechanical behaviour of continuous media. *Arch. Rat. Mech. Anal.* **2**, 197–226.

Nott, P. and Jackson, R. (1992). Frictional-collisional equations of motion for granular materials and their application to flow in aerated chutes. *J. Fluid Mech.* **241**, 125–144.

Nott, P. R. (1991). Analysis of granular flow in aerated and vibrated chuetes. Ph.D. thesis, Princeton University.

Nott, P. R., Alam, M., Agrawal, K., Jackson, R., and Sundaresan, S. (1999). The effect of boundaries on the plane Couette flow of granular materials: a bifurcation analysis. *J. Fluid Mech.* **397**, 203–229.

Ogawa, S. (1978). Multitemperature theory of granular materials. In *Proc. US-Japan Seminar on Continuum Mechanical and Statistical Approaches in the Mechanics of Granular Materials*, S. C. Cowin and M. Satake, eds. Gakujutsu Bunken Fukyu-kai, 208–217.

Ogawa, S., Umemura, A., and Oshima, N. (1980). On the equations of fully fluidized granular materials. *ZAMP* **31**, 483–493.

Oldroyd, J. G. (1950). On the formulation of rheological equations of state. *Proc. R. Soc. London A* **200**, 523–541.

Onoda, G. Y. and Liniger, E. G. (1990). Random loose packings of uniform spheres and the dilatancy onset. *Phys. Rev. Lett.* **64**, 2727–2730.

Ovarlez, G. and Clément, E. (2003). Slow dynamics and aging of a confined granular flow. *Phys. Rev. E* **68**, 031302–1–031302–17.

Ovarlez, G. and Clément, E. (2005). Elastic medium confined in a column vs. the Janssen experiment. *Eur. Phys. J. E* **16**, 421–438.

Ovarlez, G., Fond, C., and Clément, E. (2003). Overshoot effect in the Janssen granular column: a crucial test for granular mechanics. *Phys. Rev. E* **67**, 060302–1–060302–4.

Oyama, Y. (1939). Studies on mixing of solids. Mixing of binary system of two sizes by ball mill motion. *Sc. Pap. I. P. C. R.* **37**, 17–29. The 179th Report from Okochi Research Laboratory I.P.C.R.

Parry, R. H. G. (1960). Triaxial compression and extension tests on remoulded saturated clay. *Géotech.* **10**, 166–180.

Pennisi, S. and Trovato, M. (1987). On the irreducibility of Professor G. F. Smith's representation for isotropic functions. *Int. J. Eng. Sci.* **25**, 1059–1065.

Perry, M. G. and Handley, M. F. (1967). The dynamic arch in free flowing granular material discharging from a model hopper. *Trans. Inst. Chem. Eng.* **45**, T367–T371.

Perry, M. G. and Jangda, H. A. S. (1970). Pressures in flowing and static sand in model bunkers. *Powder Technol.* **4**, 89–96.

Phillips, C. E. S. (1910). Electrical and other properties of sand. *Proc. Roy. Inst. Gt. Britain* **19**, 742–752.

Pidduck, F. B. (1922). The kinetic theory of a special type of rigid molecule. *Proc. Roy. Soc. A* **101**, 101–112.

Pipes, L. A. (1963). *Matrix Methods for Engineering*. (Prentice-Hall, Englewood Cliffs).

Piskunov, N. (1969). *Differential and Integral Calculus*. (Mir Publishers, Moscow).

Pitman, E. B. (1986). Stress and velocity fields in two- and three-dimensional hoppers. *Powder Technol.* **47**, 219–231.

Pitman, E. B. (1988). The stability of granular flow in converging hoppers. *SIAM J. Appl. Math.* **38**, 1033–1053.

Pitman, E. B. and Schaeffer, D. G. (1987). Stability of time dependent compressible granular flow in two dimensions. *Comm. Pure Appl. Math.* **40**, 421–447.

Platonov, P. and Poltorak, V. (1969/1970). Investigation of shear of a granular material along a bordering surface. *Powder Technol.* **3**, 361–363.

Popken, L. and Cleary, P. W. (1999). Comparison of kinetic theory and discrete element schemes for modelling granular Couette flows. *J. Comp. Phys.* **155**, 1–25.

Pouliquen, O. (1999). Scaling laws in granular flows down rough inclined planes. *Phys. Fluids* **11**, 542–548.

Pouliquen, O., Nicholas, M., and Weidman, P. D. (1997). Crystallization of non-Brownian spheres under horizontal shaking. *Phys. Rev. Lett.* **79**, 3640–3643.

Prager, W. (1961). *Introduction to the Mechanics of Continua*. (Ginn and Co., Boston).

Prakash, J. R. (1989). Steady compressible plane flow of cohesionless granular materials in hoppers and bunkers. Ph.D. thesis, Indian Institute of Science.

Prakash, J. R. and Rao, K. K. (1988). Steady compressible flow of granular materials through a wedge-shaped hopper: the smooth-wall, radial gravity problem. *Chem. Eng. Sci.* **43**, 479–494.

Prakash, J. R. and Rao, K. K. (1991). Steady compressible flow of cohesionless granular materials through a wedge-shaped bunker. *J. Fluid Mech.* **225**, 21–80.

Prandtl, L. (1924). Spannungsverteilung in plastischen koerpern (Tension distribution in plastic bodies). In *Proc. First Int. Congr. Appl. Mech.* Delft, 43–54.

Prasad, P. and Ravindran, R. (1985). *Partial Differential Equations*. (Wiley Eastern, New Delhi).

Rajagopal, K. R. and Massoudi, M. (1990). A method for measuring the material moduli of granular materials: flow in an orthogonal rheometer. DOE/PETC/TR-90/3, Pittsburgh Energy Technology Center.

Rajchenbach, J. (2001). Stress transmission through textured granular packings. *Phys. Rev. E* **63**, 041301/1–5.

Rankine, W. J. M. (1857). On the stability of loose earth. *Phil. Trans. R. Soc.* **147**, 9–27.

Rao, V. L. and Venkateswarlu, D. (1974). Static and dynamic wall pressures in experimental mass flow hoppers. *Powder Technol.* **10**, 143–152.

Rathbone, T., Nedderman, R. M., and Davidson, J. F. (1987). Aeration, deaeration, and flooding of fine powders. *Chem. Eng. Sci.* **42**, 725–736.

Reif, F. (1985). *Fundamentals of Statistical and Thermal Physics*. (McGraw-Hill, Singapore).

Reinhardt, W. and Dubey, R. (1996). Application of objective rates in mechanical modelling of solids. *Trans. ASME E J. Appl. Mech.* **118**, 692–698.

Rendulic, L. (1936). Pore-index and pore-water pressure. *Bauingenieur* **17**, 559.

Resnick, R. (1968). *Introduction to Special Relativity*. (John Wiley, New York).

Resnick, R. and Halliday, D. (1966). *Physics, Part I*. (John Wiley, New York).

Reuss, E. (1930). Beruecksichtigung der elastischen formaenderungen in der plastizitaetstheorie (Consideration of elastic deformation in plasticity theory). *Z. Angew. Math. Mech.* **10**, 266–274.

Reynolds, O. (1885). On the dilatancy of media composed of rigid particles in contact. With experimental illustrations. *Phil. Mag.* **20**, 469–481.

Richards, P. C. (1977). Bunker design. Part 1: bunker outlet design and initial measurement of wall pressures. *Trans. ASME B J. Eng. Ind.* **99**, 809–813.

Richardson, J. F. and Zaki, W. N. (1954). Sedimentation and fluidization. Part I. *Trans. Inst. Chem. Eng. London* **32**, 35–53.

Rivlin, R. S. and Ericksen, J. L. (1955). Stress-deformation relations for isotropic materials. *J. Rat. Mech. Anal.* **4**, 323–425.

Roberts, A. W. (1998). Particle technology–reflections and horizons: an engineering perspective. *Trans. Inst. Chem. Eng.* **76**, 775–796.

Romano, M. (1974). A continuum theory for granular media with a critical state. *Arch. Mech.* **26**, 1011–1028.

Rombach, G. and Eibl, J. (1988). Consistent modelling of filling and discharging processes in silos. In *Proc. Silos – Research and Experience*. Karlsruhe, 1–15.

Roscoe, K. H. (1970). The influence of strains in soil mechanics. *Géotech.* **20**, 129–170.

Roscoe, K. H. and Burland, J. B. (1968). On the generalized stress-strain behavior of wet clay. In *Engineering Plasticity*, J. Heyman and R. A. Leckie, eds. (Cambridge University Press, Cambridge), 535–609.

Roscoe, K. H., Schofield, A. N., and Wroth, C. P. (1958). On the yielding of soils. *Géotech.* **8**, 22–53.

Rose, H. E. and Tanaka, T. (1959). Rate of discharge of granular materials from bins and hoppers. *The Engineer (Lond.)* **208**, 465–469.

Rotter, J. M., Holst, J. M. F. G., Ooi, J. Y., and Sanad, A. M. (1998). Silo pressure predictions using discrete-element and finite-element analyses. *Phil. Trans. R. Soc. Lond. A* **356**, 2685–2712.

Rowe, P. W. (1962). The stress-dilatancy relation for static equilibrium of an assembly of particles in contact. *Proc. R. Soc. Lond. A* **269**, 500–527.

Ruina, A. (1983). Slip instability and state variable friction laws. *J. Geophys. Res.* **88**, 10359–10370.

Savage, J. C. and Lockner, D. A. (1997). A test of the double-shearing model of flow for granular materials. *J. Geophys. Res.* **102**, 12287–12294.

Savage, S. B. (1965). The mass flow of granular material from coupled velocity-stress fields. *Brit. J. Appl. Phys.* **16**, 1885–1888.

Savage, S. B. (1967). Gravity flow of a cohesionless bulk solid in a converging conical channel. *Int. J. Mech. Sci.* **9**, 651–659.

Savage, S. B. (1979). Gravity flow of cohesionless granular materials in chutes and channels. *J. Fluid Mech.* **92**, 53–96.

Savage, S. B. (1983). Granular flow down rough inclines – review and extension. In *Mechanics of Granular Materials: New Models and Constitutive Relations*, J. T. Jenkins and M. Satake, eds. (Elsevier Science, Amsterdam), 261–282.

Savage, S. B. (1992). Instability of unbounded uniform granular shear flow. *J. Fluid Mech.* **241**, 109–123.

Savage, S. B. (1998). Analyses of slow high-concentration flows of granular materials. *J. Fluid Mech.* **377**, 1–26.

Savage, S. B. and Hutter, K. (1989). The motion of a finite mass of granular material down a rough incline. *J. Fluid Mech.* **199**, 177–215.

Savage, S. B. and Jeffrey, D. J. (1981). The stress tensor in a granular flow at high shear rates. *J. Fluid Mech.* **110**, 255–272.

Savage, S. B. and McKeown, S. (1983). Shear stress developed during rapid shear of dense concentrations of large spherical particles between concentric cylinders. *J. Fluid Mech.* **127**, 453–472.

Savage, S. B. and Sayed, M. (1984). Stresses developed by dry cohesionless granular materials sheared in an annular shear cell. *J. Fluid Mech.* **142**, 391–430.

Savage, S. B. and Yong, R. N. (1970). Stresses developed by cohesionless granular materials in bins. *Int. J. Mech. Sci.* **12**, 675–693.

Sawicki, A. and Swidzinski, W. (1995). Cyclic compaction of soils, grains, powders. *Powder Technol.* **85**, 97–104.

Sawicki, A. and Swidzinski, W. (1998). Elastic moduli of non-cohesive particulate materials. *Powder Technol.* **96**, 24–32.

Schaeffer, D. G. (1987). Instability in the evolution equations describing incompressible granular flow. *J. Differ. Equ.* **66**, 19–50.

Schaeffer, D. G. (1992). A mathematical model for localization in granular flow. *Proc. R. Soc. Lond. A* **436**, 217–250.

Schleicher, F. (1925). Die energiegrenze der elastizitaet (plastizitaetsbedingung) (The energy boundary of elasticity (plasticity condition)). *Zeit. fur ang. Math. Mech.* **5**, 478–479.

Schleicher, F. (1926). Der spannungfzustand an der fliessgrenze (plastizitaetsbedingung) (The tensile state at the flow limit (plasticity condition)). *Zeit. fur ang. Math. Mech.* **6**, 199–216.

Schmid, P. J. and Kytömaa, H. K. (1994). Transient and asymptotic stability of granular shear flow. *J. Fluid Mech.* **264**, 255–275.

Schofield, A. and Wroth, C. P. (1968). *Critical State Soil Mechanics*. (McGraw-Hill, London).

Scholz, C. H. (1998). Earthquakes and friction laws. *Nature* **391**, 37–41.

Schowalter, W. R. (1978). *Mechanics of Non-Newtonian Fluids*. (Pergamon Press, Oxford).

Schwedes, J. (2000). Testers for measuring flow properties of particulate solids. *Powder Handl. Process.* **12**, 337–354.

Scott, G. D. (1960). Packing of spheres. *Nature* **188**, 908–909.

Scott, G. D. and Kilgour, D. M. (1969). The density of random close packing of spheres. *Brit. J. Appl. Phys.* **2**, 863–866.

Scott, R. F. (1963). *Principles of Soil Mechanics*. (Addison-Wesley, Reading).

Sela, N. and Goldhirsch, I. (1998). Hydrodynamic equations for rapid flows of smooth inelastic spheres, to Burnett order. *J. Fluid Mech.* **361**, 41–74.

Sela, N., Goldhirsch, I., and Noskowicz, S. H. (1996). Kinetic theoretical study of a simply sheared two dimensional granular gas to Burnett order. *Phys. Fluids* **8**, 2337–2353.

Serrin, J. (1959). Mathematical principles of classical fluid mechanics. In *Handbuch der Physik*, S. Flugge, ed., Vol. VIII/1. (Springer-Verlag, Berlin).

Seville, J. P. K., Willett, C. D., and Knight, P. C. (2000). Interparticle forces in fluidisation: a review. *Powd. Technol.* **113**, 261–268.

Shames, I. H. (1962). *Mechanics of Fluids*, Revised first edn. (McGraw-Hill, Tokyo).

Shames, I. H. and Cozzarelli, F. A. (1992). *Elastic and Inelastic Stress Analysis*. (Prentice-Hall, Englewood Cliffs).

Shield, R. T. (1953). Mixed boundary value problems in soil mechanics. *Q. Appl. Mech.* **11**, 61–75.

Shirai, T., Ishida, M., Ito, Y., Inoue, N., and Kobayashi, S. (1977). Rotation of a horizontal disk within an aerated particle bed – behavior of solid particles and torque requirements. *J. Chem. Eng. Japan* **10**, 40–45.

Silbert, L. E., Ertas, D., Grest, G. S., Halsey, T. C., Levine, D., and Plimpton, S. J. (2001). Granular flow down an inclined plane: Bagnold scaling and rheology. *Phys. Rev. E* **64**, 051302.

Skempton, A. W. (1960). Effective stresses in soils, concrete, and rocks. In *Conference on Pore Pressure and Suction in Soils*, (Butterworths, London), 4–16.

Slattery, J. C. (1999). *Advanced Transport Phenomena*. (Cambridge University Press, Cambridge).

Smid, J. and Novosad, J. (1981). Pressure distribution under heaped bulk solids. In *Proc. 1981 Powtech Conference (Inst. Chem. Eng. Symp.)*. Vol. 63. Birmingham, D3V 1– 12.

Sokolovskii, V. V. (1965). *Statics of Granular Media*. (Pergamon, Oxford).

Spencer, A. J. M. (1964). A theory of the kinematics of ideal soils under plane strain conditions. *J. Mech. Phys. Solids* **12**, 337–351.

Spencer, A. J. M. (1982). Deformation of ideal granular materials. In *Mechanics of Solids*, H. G. Hopkins and M. J. Sewell, eds. (Pergamon, Oxford), 607–652.

Spencer, A. J. M. (1986). Instability of steady shear flow of granular materials. *Acta Mech.* **64**, 77–87.

Sperl, M. (2006). Experiments on corn pressure in silo cells – translation and comment of Janssen's paper from 1895. *Granul. Matter* **8**, 59–65.

Spiegel, M. R. (1974). *Vector Analysis*. (McGraw-Hill, New York).

Spink, C. D. and Nedderman, R. M. (1978). Gravity discharge rate of fine particles from hoppers. *Powder Technol.* **21**, 245–261.

Srivastava, A. and Sundaresan, S. (2003). Analysis of a frictional-kinetic model for gas-particle flow. *Powder Technol.* **129**, 72–85.

Stewart, R. L., Bridgwater, J., Zhou, Y. C., and Yu, A. B. (2001). Simulated and measured flow of granules in a bladed mixer – a detailed comparison. *Chem. Eng. Sci.* **56**, 5457–5471.

Stokes, G. G. (1851). On the effect of internal friction of fluids on the motion of pendulums. *Trans. Camb. Phil. Soc.* **9**, 8–106.

Strogatz, S. H. (1994). *Nonlinear Dynamics and Chaos*. (Perseus Books, Cambridge, Massachusetts).

Stronge, W. J. (2000). *Impact Mechanics*. (Cambridge University Press, Cambridge).

Stroud, M. A. (1971). The behaviour of sand at low stress levels in the simple shear apparatus. Ph.D. thesis, University of Cambridge.

Stutz, P. (1973). Comportement élasto-plastique des milieux granulaires (Elasto-plastic behaviour of granular media). In *Foundations of Plasticity*, A. Sawczuk, ed., Vol. 1. (Noordhoff, Leyden), 37–49.

Sullivan, W. N. (1972). Heat transfer to flowing granular media. Ph.D. thesis, California Institute of Technology.

Sundaresan, S. (2001). Some outstanding questions in handling of cohesionless particles. *Powder Technol.* **115**, 2–7.

Tabor, D. (1955). The mechanism of rolling friction. II. The elastic range. *Proc. Roy. Soc. London A* **229**, 198–220.

Tan, M.-L. (1995). Microstructures and macrostructures in rapid granular flows. Ph.D. thesis, Princeton University.

Tardos, G. I., Khan, M. I., and Schaeffer, D. G. (1998). Forces on a slowly rotating, rough cylinder in a Couette device containing a dry, frictional powder. *Phys. Fluids* **10**, 335–341.

Tardos, G. I., McNamara, S., and Talu, I. (2003). Slow and intermediate flow of a frictional bulk powder in the Couette geometry. *Powder Technol.* **131**, 23–39.

Taylor, D. W. (1948). *Fundamentals of Soil Mechanics*. (John Wiley, New York).

Tejchman, J. (1998a). FE-simulation of rapid silo flow with a polar elastoplastic constitutive model. *TASK Q.* **2**, 473–501.

Tejchman, J. (1998b). Numerical simulation of filling in silos with a polar hypoplastic constitutive model. *Powder Technol.* **96**, 227–239.

Tejchman, J. and Bauer, E. (1996). Numerical simulation of shear band formation with a polar hypoplastic constitutive model. *Comput. Geotech.* **19**, 221–244.

Tejchman, J. and Gudehus, G. (2001). Shearing of a narrow granular layer with polar quantities. *Int. J. Num. Anal. Meth. Geomech.* **25**, 1–28.

Tejchman, J. and Wu, W. (1993). Numerical study on patterning of shear bands in a Cosserat continuum. *Acta Mech.* **99**, 61–74.

Terzaghi, K. (1923). Die berchnung der durchlaessigkeitsziffer des tones aus dem verlauf der hydrodynamischen spannungsercheinungen (The calculation of the permeability number of clay from the course of the hydrodynamic tension effects). *Sitzungber. Akad. Wiss. Wien.* **132**, 125–138.

Terzaghi, K. (1936). The shearing resistance of saturated soil, and the angle between the planes of shear. In *Proc. First Int. Conf. Soil Mech. Found. Engng.* Vol. I. Harvard, 54–56.

Theodosopulu, M. and Dahler, J. S. (1974). The kinetic theory of polyatomic liquids. II. The rough sphere, rigid ellipsoid, and square-well ellipsoid models. *J. Chem. Phys.* **60**, 4048–4057.

Thomas, N. (2000). Reverse and intermediate segregation of large beads in dry granular media. *Phys. Rev. E* **62**, 961–974.

Thompson, P. A. and Grest, G. S. (1991). Granular flow: friction and dilatancy transition. *Phys. Rev. Lett.* **67**, 1751–1754.

Thomson, W. (1887). Stability of fluid motion: rectilinear motion of viscous fluid between two parallel plates. *Phil. Mag.* **24**, 188–196.

Thornton, C. (2000). Numerical simulations of deviatoric shear deformation of granular media. *Géotech.* **50**, 43–53.

Timoshenko, S. P. and Goodier, J. N. (1970). *Theory of Elasticity*, 3 edn. (McGraw-Hill, Singapore).

Torquato, S., Truskett, T. M., and Debenedetti, P. G. (2000). Is random close packing of spheres well defined? *Phys. Rev. Lett.* **84**, 2064–2067.

Tresca, H. (1864). Sur l'e'coulement des corps solides soumis a' de fortes pressions (Memoir on the flow of solid bodies under strong pressure). *Comptes Rendus Acad. Sci. Paris* **59**, 754.

Trollope, D. H. (1968). The mechanics of discontinua or clastic mechanics in rock problems. In *Rock Mechanics in Engineering Practice*, K. G. Stagg and O. C. Zienkiewicz, eds. (John Wiley, New York), 275–320.

Truesdell, C. (1955). Hypoelasticity. *J. Rat. Mech. Anal.* **4**, 83–133.

Truesdell, C. and Muncaster, R. G. (1980). *Fundamentals of Maxwell's Kinetic Theory of a Simple Monatomic Gas*. (Academic Press, New York).

Truesdell, C. and Noll, W. (1965). The nonlinear field theories of mechanics. In *Handbuch der Physik*, S. Flügge, ed., Vol. III/3. (Springer-Verlag, Berlin), 1–659.

Truesdell, C. and Toupin, R. A. (1960). The classical field theories of mechanics. In *Handbuck der Physik*, S. Flügge, ed., Vol. III/1. (Springer Verlag, Berlin), 226–647.

Tsuji, Y., Tanaka, T., and Ishida, T. (1992). Lagrangian numerical simulation of plug flow of cohesionless particles in a horizontal pipe. *Powder Technol.* **71**, 239–250.

Turton, R. and Levenspiel, O. (1986). A short note on the drag correlation for spheres. *Powder Technol.* **47**, 83–86.

Tüzün, U. (1979). Velocity distributions in funnel flow bins. Ph.D. thesis, University of Cambridge.

Tüzün, U., Adams, M., and Briscoe, B. J. (1988). An interface dilation model for the prediction of wall friction in a particulate bed. *Chem. Engng Sci.* **43**, 1083–1098.

Tüzün, U. and Nedderman, R. M. (1985a). Gravity flow of granular materials round obstacles – I. *Chem. Eng. Sci.* **40**, 325–336.

Tüzün, U. and Nedderman, R. M. (1985b). Gravity flow of granular materials round obstacles – II. *Chem. Eng. Sci.* **40**, 337–351.

Van Beijeren, H. and Ernst, M. H. (1973). The modified Enskog equation. *Physica* **68**, 437–456.

van Eekelen, H. A. M. (1980). Isotropic yield surfaces in three dimensions for use in soil mechanics. *Int. J. Num. Anal. Meth. Geomech.* **4**, 89–101.

Van Zanten, D. C. and Mooij, A. (1977). Bunker design part 2: wall pressures in mass flow. *Trans. ASME Ser. B J. Eng. Ind.* **99**, 814–818.

Vanel, L., Claudin, P., Bouchaud, J. P., Cates, M. E., Clément, E., and Wittmer, J. P. (2000). Stresses in silos: comparison between theoretical models and new experiments. *Phys. Rev. Lett.* **84**, 1439–1442.

Vanel, L., Howell, D., Clark, D., Behringer, R. P., and Clement, E. (1999). Memories in sand: experimental tests of construction history on stress distributions under sandpiles. *Phys. Rev. E* **60**, R5040–R5043.

Vardoulakis, I. and Aifantis, E. C. (1991). A gradient flow theory of plasticity for granular materials. *Acta Mech.* **87**, 197–217.

Vardoulakis, I. and Graf, B. (1985). Calibration of constitutive models for granular materials using data from biaxial experiments. *Géotech.* **35**, 299–317.

Vardoulakis, I., Graf, B., and Hettler, A. (1985). Shear band formation in a fine-grained sand. In *Proc. 5th Int. Conf. Numer. Meth. Geomech.* Vol. 1. Nagoya, 517–521.

Verghese, T. M. and Nedderman, R. M. (1993). A theoretical prediction of the discharge rate from conical hoppers in both the mass flow and core flow regimes. *Trans. Inst. Chem. Eng.* **71A**, 637–642.

von Mises, R. (1913). Mechanik der festen körper im plastisch deformablen zustand (Mechanics of solid bodies in the plastically deformable state). *Nachrichten Akad. Wiss. Göttingen Math. Phys. Klasse* **H4**, 582–592.

von Mises, R. (1926). Footnote to the paper by Schleicher (1926). *Zeit. Ang. Math. Mech.* **6**, 199.

von Mises, R. (1928). Mechanik der plastischen formaenderung von kristallen (Mechanics of plastic deformation in crystals). *Z. Angew. Math. Mech.* **8**, 161–185.

von Wolffersdorf, P. A. (1996). A hypoplastic relation for granular materials with a predefined limit state surface. *Mech. Cohes. Frict. Mater.* **1**, 251–271.

Wakabayashi, T. (1957). Photoelastic method for determination of stress in powdered mass. In *Proc. 7th Japan Natl. Congr. Appl. Mech.* Tokyo, 153–158.

Walker, D. M. (1966). An approximate theory for pressures and arching in hoppers. *Chem. Eng. Sci.* **21**, 975–997.

Walters, J. K. (1973). A theoretical analysis of stresses in silos with vertical walls. *Chem. Eng. Sci.* **28**, 13–21.

Walton, O. R. (1993a). Numerical simulation of inclined chute flows of monodisperse, inelastic, frictional spheres. *Mech. Mater.* **16**, 239–247.

Walton, O. R. (1993b). Numerical simulation of inelastic frictional particle-particle interactions. In *Particle Two-Phase Flow*, M. C. Roco, ed. (Butterworth-Heinemann, Boston).

Walton, O. R. and Braun, R. L. (1986). Stress calculations for assemblies of inelastic spheres in uniform shear. *Acta Mech.* **63**, 73–86.

Wang, C.-H., Jackson, R., and Sundaresan, S. (1996). Stability of bounded rapid shear flows of a granular material. *J. Fluid Mech.* **308**, 31–62.

Wang, C.-H., Jackson, R., and Sundaresan, S. (1997). Instabilities of fully developed rapid flow of a granular material in a channel. *J. Fluid Mech.* **342**, 179–197.

Wang, Y. and Hutter, K. (1999). Shearing flows in a Goodman–Cowin type granular material – theory and numerical results. *Part. Sci. Tech.* **17**, 978–124.

Watson, G. R. (1993). Flow patterns in flat bottomed silos. Ph.D. thesis, University of Edinburgh.

Westergaard, H. M. (1920). On the resistance of ductile metals to combined stresses. *J. Franklin Inst.* **189**, 627–640.

Whitaker, S. (1968). *Introduction to Fluid Mechanics*. (Prentice-Hall, Englewood Cliffs).

Whitaker, S. (1973). The transport equations for multiphase systems. *Chem. Eng. Sci.* **28**, 139–147.

Wieckowski, Z. (1994). Finite deformation analysis of motion of bulk material in silo. TULEA 1994:26, Lulea University of Technology.

Wieghardt, K. (1952). Ueber einige versuche an strömungen in sand (About several experiments on flows in sand). *Ing. Arch.* **20**, 109–115.

Wieghardt, K. (1975). Experiments in granular flow. *Ann. Rev. Fluid Mech.* **7**, 89–114.

Williams, J. C. (1977). The rate of discharge of coarse granular materials from conical mass flow hoppers. *Chem. Eng. Sci.* **32**, 247–255.

References

Williams, J. C. and Birks, A. H. (1965). The preparation of samples for shear cell testing. *Rheol. Acta* **4**, 170–180.

Wilms, H. and Schwedes, J. (1985). Analysis of the active field in hoppers. *Powder Technol.* **42**, 15–25.

Woods, L. C. (1975). *The Thermodynamics of Fluid Systems*. (Clarendon Press, Oxford).

Xiao, H., Bruhns, O. T., and Meyers, A. (1997). Logarithmic strain, logarithmic spin and logarithmic rate. *Acta Mech.* **124**, 89–105.

Yoon, D. K. and Jenkins, J. T. (2005). Kinetic theory for identical, frictional, nearly elastic disks. *Phys. Fluids* **17**, 083301–1–10.

Yu, M. (2002). Advances in strength theories for materials under complex stress state in the 20th century. *Appl. Mech. Rev.* **55**, 169–219.

Yuu, S., Abe, T., Saitoh, T., and Umekage, T. (1995). Three-dimensional numerical simulation of the motion of particles discharging from a rectangular hopper using distinct element method and comparison with experimental data (effects of time step and material properties). *Adv. Powder Technol.* **6**, 259–269.

Zagainov, L. (1967). Equations of plane steady-state motion of a granular medium (in Russian); translated as *Mech. Solids* **2** (1970) 130–134. *Inzh. Zh. Mekh. Tverdova Tela* **2**, 188–196.

Zeininger, G. and Brennen, C. (1985). Intersitial fluid effects in hopper flows of granular material. In *Cavitation and Multiphase Flow Forum – 1985*. (American Society of Mechanical Engineers, New York), 132–136.

Zhang, D. Z. and Rauenzahn, R. M. (1997). A viscoelastic model for dense granular flows. *J. Rheol.* **41**, 1275–1298.

Zhou, Y. C., Wright, B. D., Yang, R. Y., Xu, B. H., and Yu, A. B. (1999). Rolling friction in the dynamic simulation of sandpile formation. *Physica A* **269**, 536–553.

Zhou, Y. C., Xu, B. H., Yu, A. B., and Zulli, P. (2002). An experimental and numerical study of the angle of repose of coarse spheres. *Powder Technol.* **125**, 45–54.

Zucrow, M. J. and Hoffman, J. D. (1976). *Gas Dynamics*. Vol. 2. (John Wiley, New York).

Index

Index

Index

Index

Printed in the United States
by Baker & Taylor Publisher Services